INFRASTRUCTURE PLANNING AND FINANCE

Infrastructure Planning and Finance is designed for the local practitioner or student in America who wants to learn the basics of how to develop an infrastructure plan, a program, or an individual infrastructure project. A team of authors with experience in public works, planning, and city government explain the history and economic environment of infrastructure and capital planning, addressing common tools such as the comprehensive plan, sustainability plans, and local regulations. The book guides readers through the preparation and development of comprehensive plans and infrastructure projects, and through major funding mechanisms, from bonds, user fees, and impact fees to privatization and competition.

The seven parts of the book describe the individual infrastructure systems: their elements, current issues and a "how-to-do-it" section that covers the system and the comprehensive plan, development regulations and how it can be financed. Innovations such as decentralization, green and blue-green technologies are described, and local policy actions to achieve a more sustainable city are also addressed. Chapters include water, wastewater, solid waste, streets, transportation, airports, ports, community facilities, parks, schools, energy and telecommunications. Attention is given to how local policies can ensure a sustainable and environment-friendly infrastructure system, and how planning for this can be integrated across disciplines.

Infrastructure Planning and Finance is a non-technical guide to the engineering, planning, and financing of major infrastructure projects, providing both step-by-step guidance, and a broad overview of the technical, political, and economic challenges of creating lasting infrastructure in the twenty-first century.

Vicki Elmer is the Director of the Oregon Leadership in Sustainability graduate program at the University of Oregon where she teaches climate action planning and water and the urban environment. Prior to that she taught housing, infrastructure policy and finance at the University of California's Department of City and Regional Planning. Dr. Elmer was the Planning Director as well as the Public Works Director in the City of Berkeley before serving as City Manager in Eugene, Oregon. She was also the head of HUD's San Francisco research office for many years. Elmer taught science as a Peace Corps Volunteer in Nepal from 1965 to 1967. Her B.A. is from the University of Michigan (1964), M.S.: Columbia University (1970), PhD: University of California (1991) in planning.

Adam Leigland is currently the Director of Public Works for Santa Fe County, New Mexico, where he is responsible for the provision, operation, and maintenance of roads and trails, water and sewer service, solid waste management, public facilities, and parks and open space. Prior to that, he was the Deputy Director of Public Works for a NATO air base in northern Italy, a senior transportation planner at the San Francisco County (California) Transportation Authority, and a civil engineering officer in the U.S. Air Force with service all over the world. He holds a B.S. in civil engineering from the University of Notre Dame, and an M.S. in civil engineering and an M.C.P. in city planning, both from the University of California, Berkeley. He is a registered professional engineer in New Mexico, California, and Arizona, and a member of the American Institute of Certified Planners.

INFRASTRUCTURE PLANNING AND FINANCE: A SMART AND SUSTAINABLE GUIDE FOR LOCAL PRACTITIONERS

Vicki Elmer and Adam Leigland

LONDON AND NEW YORK

First edition published 2014
by Routledge

and by Routledge
711 Third Avenue, New York, NY 10017

Routledge is an imprint of the Taylor & Francis Group, an informa business

British Library Cataloguing in Publication Data
A catalogue record for this book is available from the British Library

Library of Congress Cataloging-in-Publication Data
Infrastructure planning and finance : a smart and sustainable guide / Vicki Elmer, Adam Leigland.
p. cm.
1. Infrastructure (Economics)–United States–Planning. 2. Infrastructure (Economics)–United States–Finance. 3. City planning–United States. 4. Municipal services–United States–Planning. 5. Municipal services–United States–Finance. I. Leigland, Adam. II. Title.
HC110.C3E46 2014 363.6068–dc23

ISBN13: 978-0-415-69318-9 (hbk)
ISBN13: 978-0-203-55239-1 (ebk)

Typeset in Avenir by
Cenveo Publisher Services

Printed and bound in the United States of America by Sheridan Books, Inc. (a Sheridan Group Company).

To Anya Marcella, 2/20/2003, Solana Rem,
8/22/2005 and Dax Shackelford, 12/13/2008.
We're planning infrastructure for your generation!
(V. Elmer)

To my wife, Liz, and my parents, Jim and Vickie,
who provided me the tools and to Eva and Isabel whose
future this book is for.
(A. Leigland)

Contents

List of Illustrations ix
Preface xvii
Notes on Contributors xxi
Acknowledgments xxiii
List of Abbreviations xxv

Section 1 Planning and Financing Infrastructure

Part I INTRODUCING INFRASTRUCTURE 3

1 Infrastructure and Today's Challenges 5
2 The Emergence of Infrastructure in the United States 19
3 Growth, Demand, and the Need for Infrastructure 37
4 Institutions of Infrastructure: The Providers 51

Part II PLANNING AND DEVELOPING INFRASTRUCTURE 69

5 Local Plans and Infrastructure 71
6 The Infrastructure Program and its Preparation 87
7 Smart and Sustainable Development Rules 103
8 Developing the Public Infrastructure Project 119
9 Capital Improvements Plan and the Capital Budget 137
10 Managing Infrastructure 153

Part III FINANCING INFRASTRUCTURE 171

11 The Financial Context for Infrastructure 173
12 Bonds and Borrowing 189
13 User Fees and Public Pricing 207
14 Exactions and Impact Fees 227
15 Competition and Privatization 247

Section 2 The Major Infrastructure Systems

Part IV ENVIRONMENTAL SYSTEMS 267

16 Water Supply
Jeff Loux, Adam Leigland, and Vicki Elmer 269
17 Wastewater and New Paradigms 295
18 Stormwater and Flooding 325
19 Solid Waste Management 351

Part V TRANSPORTATION 383

20 Streets and Streetscapes 385
21 Automobiles and Mass Transit 411
22 Airports 451
23 Ports and Waterfronts
Peter H. Brown and Peter V. Hall 481

Part VI COMMUNITY FACILITIES 505

24 Public and Quasi-Public Buildings 507
25 Public Schools as Public Infrastructure: Schools, Community, and Land Use Planning
Jeffrey Vincent 531
26 Parks, Recreation, and Open Space 553

Part VII ENERGY AND TELECOMMUNICATIONS 577

27 Energy and Power 579
28 Telecommunications 613

Section 3 Conclusion

Part VIII CONCLUSION 641

29 A New Paradigm for Infrastructure 643

Glossary 653
Notes 665
Illustration Credits 695
Index 707

List of Illustrations

Figures

1.1 Differences from 1850 to 2000 in key global warming indicators 12
1.2 Carbon dioxide emissions in 2009 by sector and fuel type in the United States 13
2.1 Federal capital outlays for transportation and water and wastewater treatment from 1956 to 2007 (est.) 33
3.1 Thirty largest metro areas, 2010 44
4.1 Organizational structure of a small or medium-sized city or county 58
4.2 Organizational structure of the planning department 59
4.3 Organizational structure of a typical public works department in a small or medium-sized city or county 60
4.4 Percentages of special districts by number and by expenditures in FY 2000–2001 for major infrastructure areas 62
4.5 Megaregions in the United States 67
5.1 Ideal relationship of comprehensive plan to implementation tools 74
5.2 Systems-based view of U.S. GHG emissions, 2006 81
6.1 Vertical, horizontal, and internal plan consistency at the local and state level 91
6.2 Infrastructure costs by assumptions about growth in a comprehensive planning process 100
8.1 Overview of infrastructure project development process 120
8.2 Organization chart of an idealized project development team 121
8.3 Flow chart of the environmental review process 123
9.1 Federal government guidelines for capital decisions 139
9.2 FY1998 sidewalks gap in Berkeley, CA 143
9.3 Relationship of key plans and budgets for infrastructure purposes 146
10.1 Public spending on the operation and maintenance of transportation and water infrastructure from 1956 to 2007 154
10.2a, b Comparing a planned approach to system maintenance with "run to deterioration" 156
10.3 Asset management system and relationship to budget 158
10.4 Street lighting grades from LA City report card 160
10.5 Street assessment of Edmonds, Washington, in 2012 161
11.1 Total public spending on transportation and water infrastructure in constant dollars and as a share of the GDP, 1956 to 2007 174
11.2 Breakdown of capital spending in 2004 by public–private institutions 175
11.3 Investment ratios, federal compared to state and local 177
11.4 State and local government tax revenue, 2009 180
11.5 2011 Dekalb County real estate tax statement 183
11.6 Fairfax County, CIP revenue sources, 2013 185
12.1 Dollar volume of annual municipal bond sales from 1986 through 2010 191
12.2 Holders of U.S. municipal securities from 1996 through 2010 193
12.3 A complicated sale and leaseback bond issuance in Philadelphia 196
13.1 Trends in consumer price index for public utility services, 1970–2007 211
13.2 Consumer expenditures on utilities by income quintile in 2007 (all consumers %) 212
14.1 Developing an impact fee program as part of the general plan update 235
14.2 Developing an impact fee in response to a specific project development application 235
15.1 Composition of local government service delivery, 1992–2007 251
15.2 Dynamics of local government service 252
15.3 Factors necessary for a successful privatization effort 258
16.1 Changing the water paradigm 271
16.2 Variations in the Earth's temperatures from 1000 to 2000 AD and predictions for 2100 272
16.3 The water cycle 276
16.4 Water table drawdown from pumping and time to recovery 277

16.5 New York City's water supply system showing the Catskill and Croton watersheds 279
16.6 A typical urban water distribution system 280
16.7 U.S. water withdrawals in 2005 282
16.8 U.S. GDP and water withdrawals, 1900–2008 283
16.9 Percentage of agencies using different fee structures in 1996 286
16.10 Water rate structures by residential, commercial, and industrial use 286
17.1 Sewer system conceptual schematic 300
17.2 Building sewer system connection detail 300
17.3 Sewer system lift station detail 301
17.4 Map of U.S. sanitary sewer overflows 304
17.5 Sewer overflow, Rhode Island 304
17.6 Estimated occurrence of sanitary sewer overflows by cause 305
17.7 Vactor pump truck keeps sewr lines pumped out 306
17.8 Private septic system schematic 308
17.9 Global sources of phosphorus over time and alternatives to avoid peak phosphorus 309
17.10 False Creek energy center diagram 313
17.11 Schematic for a distributed wastewater recycling system 315
17.12 The ideal CPNK cycle 316
17.13 Household wastewater types 316
17.14 Three types of urine-diverting toilets 317
17.15 Regional wastewater services plan, 2000–2030 320
18.1 Components of a watershed 328
18.2 Development effects on river flow 334
18.3 Change in number of Category 4 and 5 hurricanes by ocean basin for the 15-year periods, 1975–1989 and 1990–2004 336
18.4 Observed change in annual average precipitation, 1958–2008 336
18.5 Recent sea level rise, 1880–2010 337
18.6 Stormwater runoff by major land use category 339
18.7 Pre- and post-development hydraulics 340
18.8 Key elements of low impact development 341
18.9 Pollutant accumulation on impervious street and highway surfaces 343
19.1 Journey of solid waste from the curb to the county landfill 355
19.2 Cross-section of a landfill 358
19.3 Solid waste incinerator 359
19.4 Per capita fees in selected cities around the world compared to the average in the United States 364
19.5 Tonnage of solid waste disposal methods from 1956 to 1999 367
19.6 Desired policy movement in solid and hazardous waste 370
19.7 Truck delivers food waste to EBMUD biodigester 373
19.8 Fluorescent light bulbs 374
20.1 Cross-section of a Roman standard for construction of a side road 386
20.2 Construction drawings of a cross-section of a Roman military road 386
20.3 Sunlight reflecting off high-albedo pavement but being absorbed and heating up conventional pavement 390
20.4 Two alley light fixtures 393
20.5 Roadside zone schematic from the New ITE Standards 395
20.6 New street classification system (context-sensitive solutions) from the Institute of Transportation Engineers 396
20.7 Four urban context zones ranging from suburban to dense urban 398
20.8 Pavement deterioration curves 403
20.9 Network budget analysis 403
20.10 Placement of water main, sanitary sewer, storm drain, and other right-of-way requirements 406
21.1 Chicago commuter rail map 421
21.2 Levels of service for automobiles on freeways 429
21.3 Range of congestion pricing schemes 435
21.4 Electronic toll collection 435
21.5 General transportation environmental impacts 437
21.6 Transportation energy use by mode in 2006 438
22.1 Basic runway configurations 460
22.2 The four forces balanced in flight 462
22.3 Airfoils and flight 462
22.4 Airport layout plan for Minneapolis Airport 471
22.5 Safety compatibility zone example 3 473

22.6 Typical DNL (day–night average sound level) values 475
22.7 Map of airport zoning overlay district 477
24.1 Blaine, MN, town square park plan 509
24.2 Site plan of a corporation yard prior to renovation 511
24.3 Energy-efficient windows 521
24.4 Schematic of an interior plumbing system 522
24.5 Green Infrastructure schematic 523
24.6a Design competition winner for Berkeley Public Safety Building, 1996 527
24.6b Exterior of new Berkeley Police Dept building as built, 2005 527
26.1 Planning map of District 5 parks in Arlington, TX 560
26.2 Proceeds of local government bond measures for parks 563
26.3 Map of open space in San Francisco 564
26.4 Importance of various parks and recreation functions to local citizens, Northville, MI 567
26.5 Tree canopy coverage in the City of Santa Monica, CA 574
27.1 Sources and uses of energy in the United States in 2010 583
27.2 Changes in the concentration of carbon dioxide in the past ten thousand years and changes from 1970 to 2005 in the inset 584
27.3 Transporting electricity 585
27.4 How electricity is generated from fuel, wind, and water sources 586
27.5 Energy consumption history in the United States by source from 1980 and projections to 2035 in quadrillion BTU per year 587
27.6 U.S. net electric generation, 2006 587
27.7 Diagram of flow of incoming solar energy to the Earth and its disposition 592
27.8 North American Electrical Reliability Corporation Regions 596
27.9 Sources of Scopes 1, 2, and 3 emissions 606
28.1 Cell phone subscribers in the United States from 1995 to 2010 616
28.2 Cell phone system 616
28.3 The electromagnetic spectrum 617
28.4 Wi-Fi signals in downtown Salt Lake City in 2004 620

Photos

Arch Dam in Arizona 6
San Mateo-Hayward Bridge in California 6
Industrial pipes leading into refining plant 6
Construction on Portland, Oregon's Westside Big Pipe Sewer Project at Fremont Bridge 7
Mercer Island Crossing, Seattle, Washington 7
Maryland Bus Rapid Transit 7
Ashbridges Bay wastewater treatment plant in Toronto, Canada 9
Airport Terminal, Chicago 10
Denver downtown mall 10
City of Berkeley Marina K dock 10
Telecommunications mast 11
A group of satellite dishes point skyward 11
Oil pump at Bakken oil fields, North Dakota 11
I-35W Mississippi River Bridge collapse in Minneapolis, 2007 14
a, b, c Solid waste being collected by hand into horse-drawn collection vehicles 20
Horse-drawn road construction 21
Historic photo of refuse collection truck dumping into another truck that will transfer the refuse to a disposal site 24
Chicago Public Library (now Chicago Cultural Center), built 1897 25
Chicago's Union Station opened in 1925 replacing one built in 1881 25
Bay Shore, New York, Hose Company No. 1 Firehouse 25
Tunnel construction 26
Road construction 26
Grand Coulee Dam 28
Hoover Dam hydroelectric power plant 28
Los Angeles railroad system 29
Curb cuts for postwar subdivision 30
Street construction in subdivision around 1946 to 1948 31
More street construction in 1950s 31

Man with jackhammer in urban setting breaking up curbs 31
Chattanooga,TN, Park 105
Road construction in Eugene, OR 105
Cement truck for curbs, gutters and sidewalks as part of street construction 106
Construction of local bus stop, Eugene, OR 115
Berkeley Public Library, New Addition 121
Bridge in Chattanooga, TN, part of waterfront redevelopment area 124
Picnic table near wildlife center in Berkeley 124
Transit stop and Bus Rapid Transit, Eugene, OR 129
Bike lane, and bike box in Eugene, OR 129
Local building construction project, Eugene, OR 131
Renovation of the California Palace of Honor 133
Typical structural steel erection 133
Sobrante Ozonation Facility, El Sobrante, CA 275
Exchequer Dam in Merced, CA 275
Water line under construction 278
Artistic manhole covers 301
The Crescent City water pollution control facility, a complex of three wastewater treatment buildings 301
Sewage treatment plant 302
Orange County's groundwater replenishment system 311
Orange County staff check out the reverse osmosis connections for the GWRS 311
Close-up of a rack of microfiltration membranes 312
Cutaway of microfiltration membranes 312
It tastes like water because it is water 312
Membrane bioreactor 314
Lowering a drainage pipe at Sandlake Galloway Road Project 329
Drainage fills at Sandlake Galloway Road Project 329
A stormwater culvert under construction 329
Mayor Carrie Hoyt in 1944 standing with "recyclable" metal for the war effort in front of the Old City Hall, Berkeley, California 354
Automated garbage pick-up truck 357
Bulldozers moving trash around at a landfill 359
Bike boulevard 391
Street scene with people gathering 391
Urban avenue 392
Traffic control signs 392
A bus shelter, Eugene, Oregon 393
Street trees in Portland, OR 394
Tree trimming with a cherry picker truck 395
Building the subgrade 402
Laying the asphalt 402
Smoothing out the asphalt 402
Street sweeping 404
Repairing street signals 405
A city worker fills one of many potholes 405
Transit tunnel, Bay Area, CA 416
BART Extension to Milpitas, San Jose and Santa Clara, CA 416
The Geneva Yard in San Francisco houses light rail transit cars from different decades and cities 419
Peachtree Station Concourse in the MARTA Heavy Rail System in Atlanta is the deepest subterranean station in the system 419
Public art is often found in transit stations since it affects the individual's perception of wait times 419
Inbound and outbound commuter rail cars on the METRA system in Chicago 419
Airport control tower in Cherry Pointe, NC 461
Inside the airport 461
Airport approach lights 473
Oakland's Container Port, looking west towards San Francisco 483
Rotting finger piers at the urban waterfront in San Francisco, with the port of Oaklands container cranes visible in the distance 485
Hotels, marinas, parks, a convention center, and a ballpark combine to create San Diego's trourism driven diversified waterfront 495
Seattle Public Library, Ballard Branch 508
Denver Art Museum Expansion doubled the size of the existing facility when it opened in 2006. 508
Merced Multicultural Arts Center, Merced, CA 509
Adeline Maintenance Facility interior 510
Interior of the Alameda County EOC, Oakland, CA 510
Modesto Police Headquarters, Modesto, CA 511
Police Station in Southwick, MA 511

Historic Portland Hotel Gate at Pioneer Courthouse Square in Portland, Oregon 512
Cherokee County, NC, Courthouse in 1927 512
Santa Barbara County Courthouse 513
Rendering of exterior of San Francisco Jail No. 3, San Bruno, CA 513
Cutaway of design for San Francisco Jail No. 3 513
The Birthing Center of the Cleveland Regional Medical Center in Shelby, NC 514
Community Health Center Palm Springs, FL 514
Exterior of McCormick Place, Chicago 515
Austin Convention Center 515
Portland, Oregon, Convention Center 516
Coors Field Baseball Stadium, Denver, CO 516
Gillette Stadium 516
Cowboy Stadium under construction 516
Denver Lower Downtown Historic District. An historic streetlamp is visible in the foreground and the historic renovated Union Station in the background 517
Denver Lower Downtown Historic District 517
LEED certification mark 519
Diagram of heat loss in conventional building 520
Green roof diagram 520
Minneapolis Central Library green roof 520
Interior view of window wall system of the Community School of Music and Arts in Mountain View, CA 522
A crane lifting a heating, ventilation, and air conditioning unit 523
Wheelchair access to balcony in San Antonio Municipal Building 527
Golden Gate Park playground 554
Sign for César Chávez Park 557
Wild kites casting a shadow in César Chávez Park 557
Kite festival in César Chávez Park for kids 557
Welcome sign at Adventure Playground 558
Outside Adventure Playground 558
Inside Adventure Playground 558
Multiple transmission lines 590
Utility pole with transformer, power and telephone lines 590
Sly Creek power plant, Easter, CA 598

Tables

1.1 American Society of Civil Engineers Infrastructure Report Cards, 1988–2009 15
3.1 Changes in proportions of households in three age groups, 2000–2025 40
3.2 U.S. growth by U.S. Census region, 2000–2025 41
4.1 Responsibilities of general purpose governments and other infrastructure service providers 55
4.2 Number of government agencies in the United States in 1952 and 2002 56
4.3 Local government infrastructure and capital expenditures in 2000–2001 by type of local government 56
5.1 Significant state planning statute reforms 77
5.2 Identification of expected climate change impacts on key infrastructure systems in King County, WA 83
8.1 Key questions when preparing a facility design review 127
9.1 Fiscal year 2004 debt affordability projections for Anne Arundel County, MD 142
9.2 Estimated annual expenditure needed to eliminate street maintenance backlog over 10 years in the City of Los Angeles, 2003 144
9.3 Five-year capital improvement budget summary for Santa Barbara County, CA, through fiscal year ending June 30, 2014 145
10.1 Street light report card for the City of Los Angeles in 2003 160
10.2 Infrastructure cost measures 161
10.3 Expected service life of various infrastructure facilities 162
10.4 Infrastructure reliability measures 164
10.5 Infrastructure output measures 166
11.1 Investment expenditures on infrastructure from 1995 to 2010 in chained (2005$) millions for federal and state and local governments 175
11.2 Capital outlays by state and local governments in 2009 178
11.3 Sources of revenue for state and local governments in 2009 180

11.4 State and local government revenues by source 1927 to 1998 184
11.5 Local criteria for assessing infrastructure expenditure increases 187
11.6 Comparison of annual capital needs estimates at the local level 187
12.1 New security issues, state and local governments, selected years 192
12.2 U.S. municipal bond issuance from 1996 through 2010: general obligation and revenue 194
12.3 Cumulative default rates of municipal and corporate bonds by bond rating category, as of February 2010 200
12.4 U.S. municipal bond issuance in 2000, 2005 and 2010 by type of sale 204
13.1 Sources of revenue for state and local government from 1980 to 1996 emphasizing fees and user charges 208
14.1 States with enabling legislation for impact fees 231
14.2 Average impact fees by type and land use, 2011 233
14.3 Facilities eligible for impact fees by state in 2011 237
16.1 The world's water 276
17.1 Sewer-related land use policies 319
18.1 Causes of different types of pollution in surface water bodies 333
18.2 Runoff coefficients for types of land uses and soil in Knox County, Tennessee 340
18.3 Planning tools for water quality protection 343
18.4 Stormwater fees by land use category for a city of 3,000 acres 346
19.1 MSW recycling rates 355
19.2 Solid waste generated in the United States from 1960 to 2010 355
19.3 Material recovered at San Leandro Resource Recovery Park 360
19.4 Municipal waste stream composition for the USA from 1960 to 2010 361
19.5 Overview of a waste management program 370
19.6 Diversion targets for a county by sector in tons by ranking in the waste stream 371
19.7 Comparing actual performance of Washington, Oregon, and California for recycling, recovery, and diversion 371
19.8 Example of carbon emissions reductions from waste management practices 379
21.1 U.S. public transit ridership 414
21.2 Operating characteristics of public transit modes 420
21.3 Usage of public transit modes in 2002 420
21.4 Non-motorized transportation trip purpose 420
21.5 Household trip production 425
21.6 Trip attraction rates for various land uses 426
21.7 Purpose of trips 426
21.8 Trips by mode for all purposes in the United States 427
21.9 Journey to work trips by mode in the United States 427
21.10 Road usage by classification, 1997 427
21.11 Florida State Department of Transportation level of service definitions 429
21.12 Effect of housing density on gasoline consumption 439
21.13 Transportation air pollutants 440
21.14 Revenue potential of various funding options for transportation facilities in a metropolitan area 446
22.1 Number of U.S. airports in 2010 459
22.2 Ownership of public airports, 1992 464
22.3 Area of selected airports 469
22.4 Incompatible land use types and issues 472
23.1 U.S. port rankings, 2009 484
23.2 Port activities on the diversified waterfront 488
23.3 Potential sources of revenue on the diversified waterfront 496
23.4 Processes in cargo handling, infrastructure requirements, and key interest groups 500
24.1 Sports facility construction by decade from 1950 through 2005 517
24.2 LEED certification points by category 519
24.3 Local controls over public buildings 525
25.1 Examples of small schools compared to larger, conventional school sizes 545

26.1 Children's park access in seven major cities 561
26.2 Parks and open space classifications and their location and size criteria 568
26.3 Greenways and parkways classification and description 569
26.4 City of Santa Monica's open space and land use goals for sustainability 573
27.1 U.S. electric utility statistics, 2000 597
27.2 BTU conversion 598
27.3 Percentage use of energy by U.S. household, 2001 599
27.4 Residential energy consumption by age of house, 2005 599
27.5 Residential energy consumption by household income, 2005 599
27.6 Selected energy conservation land use policies at the local level 600
27.7 Cost savings and payback periods for selected energy-efficient building improvements 602
27.8 Linkages between energy and other general plan elements 604
28.1 Internet usage 619
28.2 Average broadband speeds (in Mbps) of OECD countries in 2007 620
28.3 Measures of data speed 621
28.4 Speed of service for different content packages by broadband technology type 621
28.5 Land use mechanisms to achieve local telecom objectives 631
28.6 Should the locality become involved in telecom provision? 632
28.7 Percentage of government entities offering public participation on their websites from 2000 to 2008 635

Preface

Purpose: Current System Not Adequate for Today's Challenges

The United States enjoys one of the highest standards of infrastructure provision in the world. Yet, today local infrastructure systems are at risk with serious consequences for our quality of life. Most infrastructure in the United States is planned and provided at the local level, but the existing system of planning and financing is inadequate for the challenges of a rapidly changing world and increasingly interconnected regions and nations.

During the past several decades, the benefits of the post-World War II system of infrastructure provision which relied upon technical experts in single purpose agencies or departments have been eclipsed by its problems. The proliferation of local infrastructure providers has led to piecemeal development and urban sprawl. The low priority of maintenance coupled with the need to replace most if not all of our large-scale infrastructure systems will result in system failures—costly to the environment and to local budgets. The lack of cross-functional coordination between specialized infrastructure providers and local planning agencies has led to the lack of key facilities in growth areas and missed opportunities for piggybacking improvements in already urbanized areas. Reliance on supply-side infrastructure provision rather than demand management and conservation approaches has resulted in unnecessary public expenditures. Resistance to green and low carbon emission building methods by entrenched bureaucracies perpetuates environmental stress. These problems play out against severe budget constraints, with results that will determine the future of localities for decades to come.

Smart and Sustainable Approach to Infrastructure

We wrote this book to help practitioners solve these problems. While the book covers the basics of these issues, it also highlights a new view of infrastructure—one that is interdisciplinary and that looks at infrastructure from a systems point of view. Just as smart growth has emphasized the conscious use of land, so smart and sustainable infrastructure emphasizes the conscious look at synergies between systems to develop infrastructure that respects the metabolism of the city.

Our approach to infrastructure planning and finance emphasizes learning the basics of infrastructure systems in order to use the capital budgeting process as an entry point for strategic infrastructure investments and life cycle budgeting. This approach promotes a "virtual" infrastructure jurisdiction to ensure that land use, economic, environmental and equity impacts are considered as part of the planning process. Demand management and environmental friendly tools are a part of smart and sustainable planning.

Finally, we recognize that the form and scale of infrastructure will change radically in the coming years. The centralized, vertically integrated model of single system infrastructure provision is being challenged by the advent of climate change and the need for new ways of addressing the built environment. At the same time, new technologies are opening up a world where decentralized and distributed infrastructure solutions with a smaller ecological footprint are now possible. We have tried to suggest various paths that might be taken in the individual chapters.

Plan of the Book

This book is divided into seven parts.

Part I: Introducing Infrastructure. Chapters in Part I examine the historical, social, economic, and institutional contexts in which infrastructure planning and financing occurs. The initial chapter in this section defines infrastructure and outlines current issues. Chapter 2 documents the history of infrastructure provision in the United States. Chapter 3 analyzes social and economic factors that influence infrastructure demand, while Chapter 4 addresses the way in which governments have been structured to provide infrastructure.

Part II: Planning and Developing Infrastructure. Part II begins with an overview of the various types of plans used at the local level where infrastructure might be addressed. Chapter 5 details how to prepare a long-range infrastructure plan. Subsequent chapters describe the variety of development rules that affect the provision of infrastructure, as well as outlining how to develop an infrastructure project from initial conception through construction. Chapter 9 on capital improvement planning and budgeting is followed by Chapter 10 on managing infrastructure.

Part III: Financing Infrastructure. This part of the book turns to the various financing mechanisms for infrastructure. Chapter 11 is an overview of local level budgets, sources of revenue and expenditures. Next are chapters on bonds, user fees and public pricing, exactions and impact fees, and finally competition and privatization. Each chapter provides a brief analysis of the state of the art at the local level, along with the issues and institutional context surrounding the use of the financial tool before moving to basic steps in the process.

Part IV: Environmental Systems. The chapters in Part IV address water, sewer, storms and flooding, and solid waste management. Chapter 16 describes water supply, treatment, and distribution systems, and discusses conservation and pricing along with land use planning issues. Chapters 17 and 18 on wastewater and stormwater also describes the basic infrastructure as well as innovations that local governments are taking to address a changing climate and the older systems. Chapter 19 on solid waste management addresses both land use and conservation issues of the "stepchild" of sanitary engineering.

Part V: Transportation. Chapter 20 looks at local-level infrastructure such as sidewalks, street lighting, traffic control devices, and street furniture and right-of-way management. Chapter 21 examines automobiles and transit and their planning. Chapter 22 on airports describes the elements in an airport and addresses the thorny topic of land use compatibility including noise and safety. The final chapter in Part V outlines the origin of public port authorities before turning to current financing approaches and planning process issues for this important type of infrastructure.

Part VI: Community Facilities. Chapter 24 covers public and quasi-public buildings, parks and open space, and other public facilities that are built once the major land use decisions have been made. Chapter 25 on public schools and land use planning follows, which is one of the few of its kind in print today. Finally, this section of the book contains Chapter 26 on parks, recreation, and open space and outlines elements of a park and innovative financing strategies and capital planning issues.

Part VII: Energy and Telecommunications. The two chapters in Part VII examine the deregulated "private" infrastructure systems of energy and telecommunications and the small but vigorous sector of publicly owned systems. Chapter 27 outlines the history of electrical power and natural gas and describes the generation, transmission, and distribution of each system, plus land use and policy considerations. Chapter 28 addresses the institutional context of telecommunications and deregulation, then turns to land use issues of siting equipment, undergrounding facilities, and right-of-way management issues.

Part VIII: Conclusion. The final chapter of the book develops a general approach to infrastructure planning. Reaching beyond the traditional single system perspective, this chapter presents a short-term strategy for the local practitioner with actions appropriate within the existing system, followed by a long-term strategy that involves restructuring local institutions and state laws, in order to practice a smart and sustainable approach to infrastructure.

Contributors

Peter Hendee Brown is an architect and planner with over 30 years of experience in private architectural practice, city government administration, and real estate development. He currently works as a consultant in the Twin Cities, advising public agencies and private developers on urban redevelopment projects. Brown has taught site planning and urban design and he currently teaches Private Sector Real Estate Development at the University of Minnesota's Hubert H. Humphrey School of Public Affairs. He earned his Bachelor of Architecture degree from Cornell University and his Master of Governmental Administration, MA in City and Regional Planning, and Ph.D. from the University of Pennsylvania. In 2009, the University of Pennsylvania Press published Dr. Brown's book *America's Waterfront Revival: Port Authorities and Urban Redevelopment*.

Vicki Elmer is the Director of the Oregon Leadership in Sustainability graduate program at the University of Oregon where she teaches climate action planning and water and the urban environment. Prior to that she taught housing, infrastructure policy and finance at the University of California's Department of City and Regional Planning. Dr. Elmer was the Planning Director as well as the Public Works Director in the City of Berkeley before serving as City Manager in Eugene, Oregon. She was also the head of HUD's San Francisco research office for many years. Elmer taught science as a Peace Corps Volunteer in Nepal from 1965 to 1967. Her B.A. is from the University of Michigan (1964), M.S.: Columbia University (1970), PhD: University of California (1991) in planning.

Peter Hall is Associate Professor of Urban Studies at Simon Fraser University in Vancouver, Canada, where he teaches economic development, transportation geography and research methods. He holds a doctorate in city and regional planning from the University of California at Berkeley, and he worked previously in local government in the port city of Durban. His research examines the connections between port cities, seaports and logistics, as well as local economic, employment and community development. He is the co-editor of *Integrating Seaports and Trade Corridors* (Ashgate, 2011) and *Cities, Regions and Flow* (Routledge, 2013). His research has been published in edited volumes and academic journals including *Environment and Planning A, Urban Studies, Regional Studies, Journal of Urban Technology, GeoJournal, Economic Development Quarterly, Maritime Policy and Management, The Canadian Geographer* and *Economic Geography*.

Adam Leigland is currently the Director of Public Works for Santa Fe County, New Mexico, where he is responsible for the provision, operation, and maintenance of roads and trails, water and sewer service, solid waste management, public facilities, and parks and open space. Prior to that, he was the Deputy Director of Public Works for a NATO air base in northern Italy, a senior transportation planner at the San Francisco County (California) Transportation Authority, and a civil engineering officer in the U.S. Air Force with service all over the world. He holds a B.S. in civil engineering from the University of Notre Dame, and an M.S. in civil engineering and an M.C.P. in city planning, both from the University of California, Berkeley. He is a registered professional engineer in New Mexico, California, and Arizona, and a member of the American Institute of Certified Planners.

Jeff Loux is Chair of the Science, Agriculture and Natural Resources Department at University of California Davis Extension, Director of the Land Use and Natural Resources Program and the Sustainability Studies Program. He is also an Adjunct faculty, teaching and conducting research in Sustainable Communities design at UC Davis in Human Ecology. He has 30 years of experience in the public, private and university sectors in urban and environmental planning, resource management and water policy. He directs a professional education program with 140 classes and conferences and 4,500 participants annually. Dr. Loux co-authored the book *Water and Land Use* (Solano Press, 2004) and *The Open Space and Land Conservation Handbook* (Solano Press, 2011), and has lectured in Europe, Australia, Thailand, and throughout the United States. He received his doctorate from University of California, Berkeley, in Environmental Planning in 1987, specializing in water policy in California.

Jeffrey Vincent is Deputy Director and cofounder of the Center for Cities & Schools (CC&S) at the University of California, Berkeley. CC&S is an action-oriented policy and technical assistance think tank, promoting high quality education as an essential component of urban and metropolitan vitality. Jeff's research interests lie at the intersection of land use planning, community development, and educational improvement, with a focus on how school facilities serve as educational and neighborhood assets. Jeff's work has been published in peer-reviewed journals, practitioner-oriented journals, books, and other outlets on a variety of issues. He is also a researcher with Building Educational Success Together (BEST), a national collaborative providing research and resources to improve public school facilities. Jeff has a PhD in City and Regional Planning from Berkeley.

Acknowledgments

Vicki Elmer writes: Professors emeriti Fred Collignon and David Dowall at University of California at Berkeley inspired this book by asking me to teach a class on infrastructure at the Department of City and Regional Planning in Spring 2000. Fred also reviewed many drafts throughout the evolution of this book. Others at UC Berkeley were equally generous with their time—Karen Christensen, Jan Whittington, Mike Teitz, Ruth Steiner (former DCRP graduate, now professor at University of Florida) also provided extensive and helpful comments. Judith Innes, Director of the Institute of Urban and Regional Development at UC Berkeley when I wrote many of the chapters as a Visiting Scholar there, provided many resources. Thanks also to Warren Jones and Natalie Macris of Solano Press for their enthusiasm about the book and their encouragement during the early years of the writing process.

The students in the infrastructure classes were a constant source of information, questions, and insights which have informed this volume. I am grateful for their enthusiasm and opinions. Michael Brown, CEO of Santa Barbara County, Vivian Kahn, former APA board member and colleague at the City of Berkeley, and Kurt Svendsen, Capital Budget Officer of Anne Arundel, Maryland County, also provided helpful comments for revisions. David Booher provided inspiration for the smart approach to infrastructure along with other helpful comments.

For review of the technical chapters, thanks are due to Uri Avin of PlaceMatters, Vit Troyen of Parsons, and Bob Guletz, a board member of APWA and also with Harris and Associates. Chris Mead, Director of Information Services at the City of Berkeley, and Wendy Cosin, Deputy Planning Director at the City of Berkeley, reviewed chapters in their fields of expertise. Patrick Keilch, former Deputy Public Works Director at the City of Berkeley, provided helpful comments for Chapter 19 on Solid Waste, as did Jeff Egeberg, Director of Engineering for the City of Berkeley on Chapter 20 on Streets. James Polk, Retired, US Air Force, and Marlin Beckwith reviewed portions of Chapter 22 on Airport, while Gil Kelly and his staff at the Portland Planning Department looked over the planning chapters and provided useful insights. Rene Cardinaux, former Public Works Director at the City of Berkeley commented on Chapter 8 on developing the infrastructure project.

The Lincoln Institute of Land Policy provided research support for the chapters on capital budgets, bonds, and borrowing. The American Public Works Association graciously allowed us to use many of their copyrighted illustrations (acknowledged in the text) as did the American Planning Association. Others too numerous to mention (but recognized in the source material) generously provided images for the various chapters at no cost or for a token contribution.

A heartfelt thanks goes to John Morris of Editide for rescuing this manuscript and for his patient and unending attention to the details of figures, photos, headings as well as keeping every aspect of this book organized. Kelly Whitford, who obtained permissions and oversaw the footnotes was a wizard beyond compare. Thanks also to Beth Goldstein who obtained many of the figures and photos.

Finally, to Tom Grant, my gratitude for his support and encouragement. Any errors, omissions, or faults are mine.

Adam Leigland writes: Robert Cervero and Samer Madanat, both of University of California, Berkeley, reviewed early drafts of my work. Ryan Leigland, Will Baumgardner, Doug Johnson, and Isaiah Stackhouse, all professionals in their respective fields, provided invaluable input to the chapters relevant to their expertise. Finally, I thank my wife, Liz Berdugo, who as an environmental engineer not only reviewed the wastewater and solid waste chapters, but also put up with me working on this.

List of Abbreviations

AASHTO American Association of State Highway and Transportation Officials
ADA Americans with Disabilities Act
ADAAG ADA Architectural Guidelines
AFO animal feeding operation
AIP Airport Improvement Program
ALP airport layout plan
APA American Planning Association
APFO Adequate Public Facilities ordinance
APWA American Public Works Association
ARPA Advanced Research Projects Agency
ARRA American Recovery and Reinvestment Act of 2009
ASCE American Society of Civil Engineers
AT&T American Telegraph and Telephone Company
BAB Build America Bonds
BID business improvement district
BLS Bureau of Labor Statistics
BOOT Build–Own–Operate–Transfer
BOT Build–Operate–Transfer
BPA Bonneville Power Administration
BTO Build–Transfer–Operate
CAFE corporate average fuel economy
CAFOs concentrated animal feeding operations
CAPs climate action plans
CATV Community Antenna Television
CBA community benefits agreement
CBO Congressional Budget Office
CCS carbon capture and storage
c&d construction & demolition
CEFPI Council of Educational Facility Planners International
CEQA California Environmental Quality Act
CERCLA Comprehensive Environmental Response, Compensation, and Liability Act
CERL Construction Engineering Research Lab
CHC community health center
CHPS Collaborative for High Performance Schools
CIP capital improvements plan
CMMS computerized maintenance management systems
COGs councils of governments
COOL Composting Organics Out of Landfills
COPs Certificates of Participation
CPI continuous process improvement
CPNK Carbon, Phosphorus, Nitrogen, Potassium cycles
CRS community rating system
CRT cathode ray tube
CSO combined sewer overflow
CSS combined sewer systems
CSS context-sensitive solutions
CTC Community Technology Center
CWA Clean Water Act
CWSRF Clean Water State Revolving Fund
dB decibels
DBO Design–Build–Operate
DESAR Decentralized Sanitation and Reuse
DNL Day–Night Average Sound Level
DOT Department of Transportation
DSW Definition of Solid Waste
EA environmental assessment
EBMUD East Bay Municipal Utility District
EDA Economic Development Administration
EIS environmental impact statement
EM electromagnetic spectrum
EOC emergency operations center
EPA Environmental Protection Agency
EPAct Energy Policy Act
EPR Extended Product Responsibility
ERO electric reliability organization
ESECA Energy Supply and Environmental Coordination Act
FAA Federal Aviation Administration
FAA Federal Aviation Agency
FEMA Federal Emergency Management Agency
FERC Federal Energy Regulatory Commission
FHA Federal Housing Administration
FHWA Federal Highway Administration
FONSI finding of no significant impact
FPC Federal Power Commission
FTA Federal Transit Authority
FTTP fiber optic to the premises
FY fiscal year
GAO General Accounting Office
GASB Government Accounting Standards Board
GFOA Government Financial Officers Association
GHG greenhouse gas
GIS Geographic information systems

GNP Gross National Product
GO general obligation
GPG general purpose government
GPS Global Positioning System
GWRS Groundwater Replenishment System
HBO Home Box Office
HEW Health, Education, and Welfare
HHW Household Hazardous Waste
HLZO Height Limitation Zoning Ordinance
HOT High Occupancy Toll
HTF Highway Trust Fund
HVAC heating, ventilation, and air conditioning
I/I inflow and infiltration
ICC Interstate Commerce Commission
ICLEI International Council for Local Environmental Initiatives (Local Governments for Sustainability)
ICMA International City/County Managers Association
IEA International Energy Agency
IPes Internet Protocol Enabled Services
IPR indirect potable reuse
IRP integrated resource planning
IRR Internal Rate of Return
IRWM Integrated Regional Water Management
ISTEA Intermodal Surface Transportation Efficiency Act
ITE Institute of Transportation Engineers
ITFA Internet Tax Freedom Act
ITS intelligent transportation systems
LANs Local Area Networks
LASH lighter-aboard-ship
LCA life-cycle analysis
LCD local distribution company
LFG landfill gas
LID low impact development
LOS level of service
LULU locally undesirable land use
MBR membrane bioreactor
MGD million gallons per day
MOC maintenance operations center
MPO Metropolitan Planning Organization
MRF materials recovery facility
MS4 municipal separate stormwater systems
MSW municipal solid waste
NACWA National Association of Clean Water Agencies
NATOA National Association of Telecommunications Officers and Advisors
NEPA National Environmental Policy Act
NERC North American Electric Reliability Council
NEU Neighborhood Energy Utility
NHTS National Household Travel Survey
NLA New Large Aircraft
NPDES National Pollutant Discharge Elimination System
NPIAS National Plan of Integrated Airport Systems
NPV Net Present Value
NRPA National Recreation and Parks Association
NTSC National Television System Committee
OS official statement
PAB private-activity bond
PC personal computer
PCBS polychlorinated biphenyls
PCI pavement condition index
PFI Private Finance Initiative
PFU plumbing fixture unit
PMA Power Marketing Administrations
PMS pavement management system
POS preliminary official statement
POTW publicly operated treatment works
PPIC Public Policy Institute of California
PRP proposed recommended practices
PSAP public safety access point
PUC Public Utilities Commission
PUD Planned Unit Development
PUHCA Public Utilities Holding Company Act
PURPA Public Utility Regulatory Policies Act
PV photovoltaic
PWA Public Works Administration
QF qualifying facility
RAM Recreation Activity Menu
RCA Radio Corporation of America
RCRA Resource Conservation and Recovery Act
RFPs requests for proposals
ROW right-of-way
SDC system development charge
SDWA Safe Drinking Water Act
SEC Securities and Exchange Commission

SIFMA	Securities Industry and Financial Markets Association
SO	special obligation
S&P	Standard & Poor's Corporation
SRF	State Revolving Loan Fund
SSO	sanitary sewer overflow
SUDS	sustainable development system
SWAT	smart water application technology
SWMP	stormwater management plan
SZEA	Standard State Zoning Enabling Act
TARP	Tunnel and Reservoir Plan
TAZ	transportation or traffic analysis zones
TDM	transportation demand management
TIFIA	Transportation Infrastructure Finance and Innovation Act
TMDL	total maximum daily load
TOD	Transit-Oriented Development
TVA	Tennessee Valley Authority
UAP	utility access port
UMTA	Urban Mass Transportation Act
UPWP	unfunded project work plan
U.S.	United States
USA	the United States
USGBC	United States Green Building Council
VMT	vehicle miles traveled
VOC	volatile organic compound
WAPA	Western Area Power Administration
WMS	work order management systems
WPA	Works Progress Administration
WTE	waste-to-energy
WTTP	wastewater treatment plant
WWI	World War I
WWII	World War II

Section ONE

Planning and Financing Infrastructure

PART ONE

Introducing Infrastructure

Part I describes the overall context for local level infrastructure with four chapters that address infrastructure issues, tell its history in the United States, describe the drivers of demand for infrastructure, and the institutions that provide infrastructure.

Chapter 1 (Infrastructure and Today's Challenges) defines infrastructure from several perspectives before describing the major infrastructure issues facing local officials today. These include climate change, capital investment needs, and problems with the institutional arrangements for infrastructure provision and maintenance.

Chapter 2 (The Emergence of Infrastructure in the United States) begins with infrastructure provision in colonial times and moves to the rise of major transportation and water and sewer systems as the country's economic base changed from predominantly agricultural to industrial. The impact of the Great Depression on infrastructure provision is outlined as well as the post-World War II explosive growth of infrastructure systems.

Chapter 3 (Growth, Demand and the Need for Infrastructure) focuses on the social, economic, and demographic forces driving the demand for infrastructure. It also addresses the relationship between infrastructure and growth.

Chapter 4 (Institutions of Infrastructure) concentrates on the providers of infrastructure. The segmented nature of infrastructure providers is documented, with strategies to help the practitioner create a "virtual" institution in order to maximize capital investment dollars in a more sustainable way.

chapter 1

Infrastructure and Today's Challenges

In this chapter...

Importance to Local Practitioners. 5
What Is Infrastructure?. 6
The Big Picture 6
Current Definitions of Infrastructure. 6
Economic Views of Infrastructure. 7
Developers' Definition of Infrastructure. 8
What Are the Major Infrastructure Systems? 9
Environmental Systems: Water and Waste 9
Streets and Transportation. 10
Community Facilities 10
Telecommunications, Energy, and Power 10
Infrastructure Challenges for the City of the Future. 11
Climate Change and the Environment 11
Capital Investment Needs 12
Local Fiscal Pressures: Boom and Bust. 16
Inadequate Institutional Structure 16
Conclusion: A New Paradigm for Infrastructure . . . 18

Why Is Infrastructure Important to Local Practitioners?

Infrastructure is at the heart of vibrant, innovative cities and metropolitan areas. It is responsible for our health and safety, our prosperity, and for our ability to socialize and live together in communities. During the next 20 years, the population of the United States is expected to increase by 48 million, and by almost 50 percent by 2050. Much of this increase will be accommodated in areas without streets, roads, schools, or sanitary sewers, at the edges of existing metropolitan areas. The remainder will occur in the heart of major cities and in the inner suburbs, adding burdens to existing infrastructure already at capacity or deteriorating due to lack of maintenance or replacement. Even in relatively young cities or counties with recent growth, infrastructure built only a few decades ago can already be obsolete or in need of repair. Climate change effects have enormous implications for infrastructure. Major local responses are critical to address mitigation and adaptation needs and also to ensure that our cities and counties are ready to compete in the global economy in a sustainable way.

However, the institutions that supply infrastructure are not always prepared to deal with planning and financing needs. They are fragmented, overly bureaucratic, and single purpose in scope, when increasingly the challenge is to provide multisector responses on a regional, metropolitan, or state basis, or even between states and nations. The urgent need to reduce energy use and carbon emissions may provide the necessary stimulus to bring together the diverse interests and actors in the infrastructure sector to address these problems. This chapter begins by describing infrastructure from different points of view, then provides an overview of the major systems. The chapter concludes with a discussion of the infrastructure challenges facing local officials today.

What Is Infrastructure?

The Big Picture

Arch Dam in Arizona

San Mateo-Hayward Bridge in California

Industrial pipes leading into refining plant

Infrastructure is the built foundation of our cities and towns. It is what we cannot buy or build on our own. Infrastructure includes the common places, buildings, roads, pipes, and systems that we must join together as a society to plan, finance, and care for. In the Middle Ages, cathedrals represented the quintessential infrastructure of the time. Cathedrals were used by all and considered necessary for the maintenance of life and culture. Too big for a single generation to build, many were constructed over the course of centuries. They were built to be used by everyone, and meant to last for many generations.

The infrastructure of our society today includes the large-scale capital structures we all share that are usually publicly financed and built. Infrastructure is capital-intensive and might not be built were it not for the intervention of government. Because infrastructure is so costly, building only "one" of it often makes sense, and usually it is owned and operated by a governmental agency.

Infrastructure can be the invisible cable underground connecting our homes to the Internet, or the colossal dam that generates megawatts of electricity and protects downstream areas from flooding. It can be the library downtown, the sidewalk and street in a neighborhood, the local park, or the invisible network of pipes and drains underground. Infrastructure includes museums, schools, hospitals, and municipal landfills. None of these can be bought by an individual, but are shared by the whole community; they require large-scale investment, and can last for generations. They also deteriorate over time and need to be replaced or improved.

Current Definitions of Infrastructure

One writer called infrastructure "the connective tissue that knits people, places, social institutions, and the natural environment into coherent urban relationships ... It is shorthand for the structural underpinnings of the public realm."[1] The infrastructure historian, Joel Tarr, defines infrastructure as "the 'sinews' of the city: its road, bridge, and transit networks; its water and sewer lines and waste disposal facilities; its power systems; its public buildings; and its parks and recreation areas."[2]

Prior to the 1980s, engineers usually called infrastructure "public works," a term which is still used today for sewers, water and waste systems among others. Infrastructure was also a military term used to describe base camps, ports, airstrips, or other physical structures. David Perry, in his short and excellent history of infrastructure financing, notes that Presidents Washington and Jefferson referred to infrastructure as "internal improvements."[3]

Claire Felbinger, an infrastructure expert, writes that the term

> [refers to] those physical structures and processes that support community development and enhance physical health and safety—roads, streets, bridges, water treatment and distribution systems, wastewater treatment facilities, irrigation systems, waterways, airports, and mass transit—as well as other utility services that are regulated by the public sector ... [T]hey are distinguished from private-sector infrastructure such as manufacturing facilities. Hence public infrastructure traditionally includes public, physical systems that ensure a healthy quality of life and enhance movement of people, goods, and services.[4]

Stephen Graham and Simon Marvin assert that the configuration of infrastructure services represents the "key physical and technological assets of modern cities." They define infrastructure as the networked systems, such as transport, streets, communications, energy, water, and waste systems that together "constitute the largest and most sophisticated technological artifacts ever devised by humans." Graham and Marvin argue that the form of the infrastructure technology is "bundled" with the economic and cultural base of the period. Consequently, "these vast lattices of technological and material connections have been necessary to sustain the ever-expanding demands of contemporary societies for increasing levels of exchange, movement and transaction across distance."[5] This view of infrastructure as "shaping the sublime terrain for production" is echoed by architects today who use the lens of infrastructure through which to view the city.[6]

Economic Views of Infrastructure

Economists such as W. W. Rostow used the word to designate the capital invested in major physical projects such as

Construction on Portland, Oregon's Westside Big Pipe Sewer Project at Fremont Bridge

Mercer Island Crossing, Seattle, Washington

Maryland Bus Rapid Transit

A Regional Planning Agency's Definition of Infrastructure

According to the San Diego Association of Governments (SANDAG), infrastructure must do the following:

- Be a public facility or regulated monopoly.
- Be a publicly shared system, network, or resource used by or benefiting a majority of the region on a regular and consistent basis.
- Provide for equal opportunity for all residents and businesses to benefit.
- Be run, regulated, or overseen by state or local elected officials or their appointed representatives.
- Ensure that the level of service available and the price of the service to be about the same for all comparable users.
- Play an integral part in maintaining the quality of everyday life for the average resident.

Using these criteria, SANDAG's Infrastructure Plan addressed the following seven areas, recognizing that, while others met these criteria, resources did not permit their inclusion at the time the plan was prepared:

- Transportation (including regional airport, maritime port, transit, highways, and international ports of entry)
- Water Supply and Delivery System
- Wastewater (sewage collection, treatment, and discharge system)
- Solid Waste Collection, Recycling, and Disposal
- Energy Supply and Delivery System
- Education (including K-12, community colleges, and universities)
- Parks and Open Space (including parks and recreation, shoreline preservation, and habitat preservation)

SANDAG, "Technical Appendix," in *Comprehensive Plan* (San Diego Association of Governments, 2004).

highways, utilities, and other government facilities.[7] The U.S. Bureau of Economic Statistics distinguishes between expenditures on consumption and those on fixed assets. Fixed asset expenditures, excluding residential buildings, is defined as infrastructure in federal statistical analyses. Perry notes that the current view of public infrastructure extends beyond the capital project itself to include the entire system of planning, building, financing, and maintenance of infrastructure.[8]

Economists also characterize some infrastructure systems as public goods or natural monopolies, which justify public investment and/or regulation if provided by a private entity. In this case, the general public sees the merit of providing the good to everyone, because its use by those who would not normally pay for the service benefits all. For example, clean water or sewage treatment results in better health and less disease for all individuals. The transportation network is also a public good, enabling industry to produce and deliver goods that are consumed by all, and enabling workers to travel to work to produce those goods.

Natural monopolies usually refer to networked facilities covering a certain territory, often so capital-intensive that developers are deterred from building a competing system. Natural monopolies are thought to have economies of scale—the larger the facility and the more persons served, the lower the unit cost for everyone.

Developers' Definition of Infrastructure

Developers frequently classify infrastructure by whether it is onsite—part of a particular development such as a subdivision—or offsite. These distinctions are often made for financial reasons. The costs of onsite infrastructure can be bundled in the price of the house or in the rent of a commercial structure. Financing offsite and community provided infrastructure such as improvements to a wastewater treatment plant or a school can be a contentious issue which subsequent chapters cover in detail.

Onsite infrastructure for an urban area project—even in a new subdivision for residential, commercial, or industrial uses—consists of the systems within the boundaries of the parcel(s). In an urban area where streets already exist and the project reuses existing sites, infrastructure includes the onsite connecting pipes for the water, sanitary, and storm sewer systems; gas, electrical, and telecommunication hookups; and the repair of sidewalks and curbs that might have been dug up or damaged during construction. For a new subdivision, onsite project infrastructure includes streets, roads, and sidewalks. It might also include community centers, parks, a school, or other facilities that only the eventual owners and occupants of the new development will use.[9]

Community (or offsite) infrastructure consists of facilities that must be built or expanded as a result of the development but which will be used by others beyond the development in question. It can include all of the major infrastructure systems.

What Are the Major Infrastructure Systems?

The following categories describe the infrastructure which local governments must provide or ensure is provided in a sustainable, efficient, and equitable manner:[10]

- environmental systems: water and waste[11]
- streets and transportation
- community facilities
- energy and telecommunications.

Environmental Systems: Water and Waste

Planning for water and the removal of waste has been the concern of city builders spanning all civilizations. Ancient China had intricate dikes and irrigation systems, while some of the waterworks of imperial Rome are standing today. The tunnels of Afghanistan, still extant, were part of a vast network of pipes bringing water to Persian capitals.

Today's local practitioner must think about how to transport water or how to connect existing water systems to the development, and about water purification and how to pay for it. Today's official must continue the tradition of the turn-of-the-twentieth-century reformers to ensure that wastewater infrastructure is adequate for disposal of sanitary wastes. Solid waste has seen a new consciousness in the past 30 years as land for disposal becomes scarce and concerns about toxics have arisen.

Ashbridges Bay wastewater treatment plant in Toronto, Canada

A more recent concern has been the pollution and ecological degradation caused by stormwater runoff. Local officials also must be concerned with the impact of rising coastal waters, droughts and other extreme events caused by climate change on water supply and treatment.

Streets and Transportation

Transportation systems such as highways, airports, canals, and railroad lines have shaped the form and perimeter of cities since they were first located along rivers and natural ports. Transportation is the ultimate infrastructure. Local officials do not have control over the large-scale subnational transportation projects, or sometimes even ones at the state level.

However, local governments do have responsibility for streets and local bridges. Local officials must plan, finance, and maintain the local street grid, and establish local standards for private developers when building new streets, sidewalks, bike paths, and even a mall. Local governments also control the rights-of-way for streets, sidewalks, and medians that are used by networked infrastructures. Many local and regional governments are turning to mass transit systems and "walkable" neighborhoods.

Airport Terminal, Chicago

Denver downtown mall

Community Facilities

Community facilities are public buildings, parks and recreational facilities, and quasi-public facilities that include civic centers, libraries, public safety buildings, jails, recreation centers, schools, pools, and tennis courts. Although some public buildings (schools, for example) may be under the jurisdiction of the local school district, an area library commission, or area-wide park district—or even county or state jails and courthouses—usually are locally planned and financed.

Community facilities also include parks, marinas, recreation areas, and open space. They may encompass quasi-public buildings like concert halls, convention centers, sports stadiums, and hospitals, often financed with local government support but owned and operated by the private or nonprofit sector.

City of Berkeley Marina K dock

Telecommunications, Energy, and Power

The provision of energy and telecommunications is crucial to the overall infrastructure system in today's urban developments. In their modern form, energy and telecommunications have been on the agendas of local politicians since the beginning of this century when cities gave franchises to local private operators. The past century saw energy and telecommunications utilities pass from small-scale but highly competitive entities to large-scale monopolies regulated by federal and state governments.

A decade ago, the major concern of local land use planners may have been only to ensure that a subdivision developer made appropriate arrangements with private energy and telecommunication utilities. Some county and city planners may have been involved with the siting of large-scale power plants in remote areas. Today many local officials must deal with franchises for wireless repeaters, satellite TV, and small-scale power generation facilities inside the jurisdiction's boundaries. Indeed, many cities are evaluating how best to provide fiber optic to the premises (FTTP) in order to promote economic development.

Telecommunications mast

A group of satellite dishes point skyward

Oil pump at Bakken oil fields, North Dakota

Infrastructure Challenges for the City of the Future

Four major challenges face urban officials in the United States regarding infrastructure in the coming decades: (1) climate change and the environment; (2) the need for large amounts of capital investment for new growth and replacement needs; (3) the local fiscal crisis and its impact upon infrastructure planning; and (4) the lack of a coherent institutional system or framework to address these challenges.

Climate Change and the Environment

Several decades ago researchers began to notice that the temperature of the Earth was slowly increasing, that the glaciers were retreating, sea levels were rising, and severe weather events were becoming more common. In 2007, a group of 650 highly renowned scholars concluded that the "warming of the climate system is unequivocal." They noted that from 1906 to 2005 the Earth's temperature had increased by 1.3 degrees Fahrenheit, with most of the increase occurring in the past fifty years. Their report also projected further increases in the global average temperature of 3 to 7 degrees F by the year 2100. This will result in continued sea ice and snow cover losses, and the rise of sea levels during the twenty-first century from 7 to 23 inches. Other impacts will be more

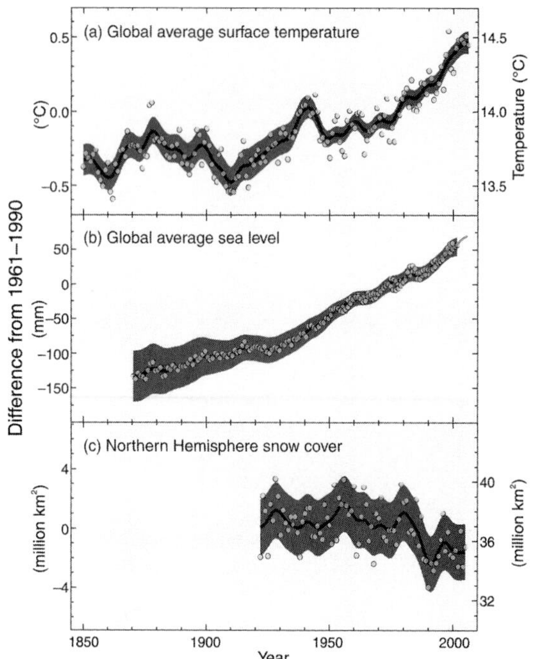

Figure 1.1
Differences from 1850 to 2000 in key global warming indicators

Infrastructure and Environmental Quality

"The development of infrastructure, be it a new highway or expansion of sewer lines, alters the environment through the infrastructure use of materials, energy, and land. This alteration can be seen immediately in runoff patterns, sedimentation in streams and rivers, and increased temperature due to the heat island effect of impervious surfaces .... [it] also has enduring long-term effects including climate shifts, hydrological changes, nutrient cycles, and a decline in biodiversity .... [This] is of great concern as these major natural shifts have a direct effect on human health."

Elisa Mayes, *Infrastructure and Sustainability* (2012), available at: www.unc.edu/~shashi/Infrastructure/general.html.

extreme weather events, including increased frequency of droughts and heat waves, along with more precipitation and flooding (Figure 1.1). An estimated 20–30 percent of known animal and plant species will be at risk of extinction.[12]

This body of scientists also concluded that the cause of global warming is increased greenhouse gases (GHGs) produced by human activities, particularly with respect to the use of fossil fuels. The major GHGs—including carbon dioxide, methane, and nitrous oxide—are called greenhouse gases because they trap some of the heat radiating outward from the Earth, like a greenhouse traps heated air. These gases have increased rapidly since 1750, when the Industrial Revolution began.

Reducing energy use and carbon emissions is seen as the major means in the USA to address the issue, since carbon emissions account for 85 percent of all GHGs in the country. For the most part they are produced by fossil fuel consumption that occurs during the process of generating electricity (used in residential, commercial, and industrial sites) and in transportation-related activities. Coal is relied on heavily to produce both electricity and petroleum for transportation (Figure 1.2).

Infrastructure systems lie at the heart of solutions to climate change. The design and structure of transportation systems, waste management, public facilities, and energy infrastructure directly affect energy consumption in the USA and how much GHG is emitted. Transportation infrastructure also plays an indirect role because of its importance in facilitating more compact and sustainable development. Water supply and wastewater infrastructure is the first major infrastructure system to be negatively and seriously affected by climate change. The need to adapt to the more extreme drought and storm cycles places huge burdens on older systems. Projected rises of up to a meter in the sea level call for a paradigm shift in infrastructure and city planning.[13]

Climate change concerns, however, have only brought to the forefront serious concerns by environmentalists over the past decades about the degradation of the environment due to rapid growth and old-style infrastructure.

Capital Investment Needs

Large amounts of capital investment must be made in infrastructure over the next decades to accommodate growth and

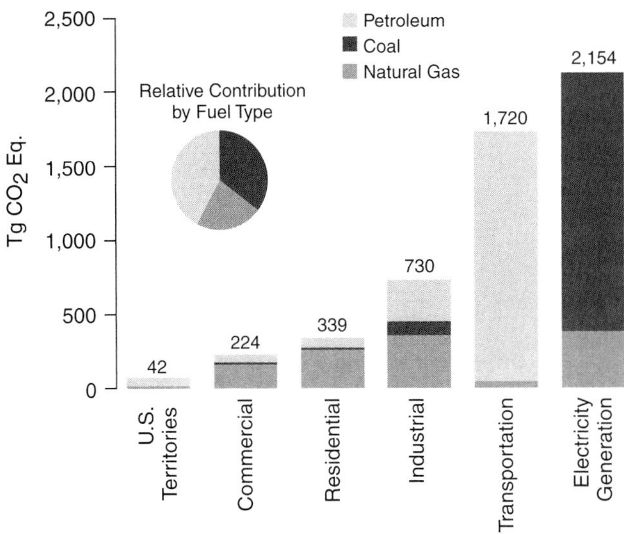

Figure 1.2
Carbon dioxide emissions in 2009 by sector and fuel type in the United States

to replace deteriorated infrastructure—a challenge for local practitioners under any circumstances. However, this is also an opportunity to explore fundamentally different approaches that promote compact growth and reduce GHG emissions. To paraphrase from the manifesto produced by a conference in 2008 on the implications of peak oil for urban design, "changing climate patterns and diminishing supplies of inexpensive oil require us to design our ... [infrastructure] ... in radically different ways."[14]

New and unevenly distributed growth. Despite the economic downturn at the end of the first decade of the twenty-first century, population growth will continue to drive the need for new infrastructure. Population growth in the South and the West will continue, but not as dramatically as during the 1980s and 1990s. Part of the growth will be caused by in-migration as a result of jobs generated by economic recovery, but the driver for most of the expected population increase through 2050 is already in place—the birth rate of those already here. In 2011, the U.S. Census projected a 50 percent increase in the U.S. population from 1990 levels by 2050.[15]

The explosive growth in some areas of the United States in the 1980s and 1990s required huge amounts of infrastructure. It also gave rise to sprawl, which in turn led to efforts by planners and the design community to develop new ways of providing transportation and other infrastructure. New financing mechanisms were developed, as well as design tools from transportation planners: Transit Oriented Development (TOD) and the urban designers—the New Urbanism—all emphasized more compact infrastructure systems. While these positive trends are expected to continue, local governments will struggle to ensure sufficient infrastructure to support needed infill and greenfield development.

Deteriorated infrastructure. The second driver for capital investment in infrastructure is the need to replace or upgrade older infrastructure. In many urbanized areas, infrastructure built 50 to 100 years ago has reached the end of its useful life or no longer meets current environmental standards. Water and sewer systems are reaching "the dawn of the replacement era"[16] where existing infrastructure will need replacement because of the confluence of people and pipe demographics. Many of the nation's roads, highways, and bridges are in poor condition. Local school buildings are overcrowded and many need rehabilitation, as will other civic buildings in older urban areas. National infrastructure systems such as the energy distribution grid and the air traffic control system have urgent replacement needs. Advances in science and technology also have rendered current infrastructure systems out of date. Even relatively new infrastructure may be deteriorated because of lack of maintenance. Many of those experts indicate writing about infrastructure that the largest dollar need for infrastructure comes from these categories, rather than from new development.

I-35W Mississippi River Bridge collapse in Minneapolis, 2007

The collapse of the bridge in Minneapolis and the explosion of the Con Ed steam pipe in New York City in 2007 are emblematic of a huge backlog of deteriorated infrastructure in the USA.

Although not a comprehensive analysis (excluding most public buildings except schools, for example), in 2009, the American Society of Civil Engineers issued a "report card" on some major categories of national infrastructure needs, and came up with a total of $2.2 trillion needed for the five years thereafter, or about 2.5 percent of the estimated U.S. Gross National Product (GNP) for these years. Generally, industrialized nations spend 2–3 percent of GNP for infrastructure, while developing nations may have a percentage as high as 5–7 percent. The American Society of Civil Engineers Report Card noted that the nation's infrastructure has shown little to no improvement since 1988, with areas such as transit and wastewater sliding backward.[17] See Table 1.1.

Table 1.1
American Society of Civil Engineers Infrastructure Report Cards, 1988–2009

Infrastructure Category		1988 Grade	2001 Grade	2009 Grade	Comments
Transportation	Roads	C+	D+	D–	One third of the nation's major roads are in poor or mediocre condition; congestion wastes $78 billion yearly and in 2007, 42,000 people were killed in car accidents.
	Bridges	C+	C	C	26 percent of bridges are structurally deficient or obsolete, but an improvement from 31 percent in 1996. It will cost $17 billion a year for 20 years to eliminate deficiencies.
	Transit	C–	C–	D	Transit ridership has increased 25 percent since 1995, but capital spending must increase 41 percent to maintain system in present condition.
	Rail	n/a	n/a	C–	More trains are needed ($200 billion) to meet growth needs.
	Aviation	n/a	D	D	There were 370 runways in service in 2007, but only 3 percent of runways rated poor. Air traffic control system needs upgrade.
	Inland Waterways	n/a	D+	D–	Key deep channels are inadequate for mega-container ships. Demand expected to double by 2020.
Water and Environment	Drinking Water	B–	D	D–	These systems face an annual shortfall of $11 billion to replace facilities nearing the end of their lives and to comply with federal water regulations.
	Wastewater	C	D	D–	Some systems are 100 years old, and there is a $12 billion annual shortfall in funding. One-third of US surface waters are polluted.
	Solid Waste	D–	C+	C+	Amount of solid waste sent to landfills declined 13 percent since 1990, and the amount recycled has doubled. Electronic waste is an issue.
	Hazardous Waste	D	D+	D	Most contamination practices stopped, but the backlog of superfund sites is growing. Brownfield redevelopment creates jobs.
Civic Facilities	Schools	n/a	D–	D	75 percent of nation's school buildings overcrowded or deteriorated. Since1998, total capital need increased from $112 billion to $127 billion.
	Parks & Rec	n/a	n/a	C–	Per capita parkland declined due to rapid population incursion.
	Other	n/a	n/a	n/a	Civic buildings, hospitals, library expenditure up during boom times.
Energy	Energy (Private and Public)		D+	D+	The U.S. energy transmission and generation infrastructure relies on old technology, but increased investment has occurred in the last five years.
	Dams		D	D	In 2007 there were 4,095 deficient dams and, of these, 1,826 are high hazard.

Source: American Society of Civil Engineers, 2009.

Local Fiscal Pressures: Boom and Bust

After facing severe growth pressures in the booming 1990s and 2000s, which strained the abilities of local governments to provide adequate infrastructure, these same cities are now faced with an economic bust. Most U.S. cities, counties, and states were hard hit by the collapse of the housing market in 2008 and the subsequent global recession. Local revenues dropped precipitously and new capital commitments declined. When faced with the need to lay off teachers and public safety officers, local governments tightened the belt on infrastructure expenditures. Some localities shut down fire stations, and one community even turned off portions of the local street lighting grid to save money.[18] Cities and counties with lower incomes were disproportionally affected—this on top of equity issues already facing the provision of infrastructure.[19]

Another factor in the fiscal crisis of the cities is the unfunded liability that many face as obligations to pension plans promised in the height of the boom in the 1990s are now coming due. The economic downturn is forcing many cities and states to attempt to cap these costs.

Nationally, the total amount of public spending on water and transportation infrastructure (major components of all infrastructure spending) declined from 2003 to 2007. Although the American Recovery and Reinvestment Act allocated significant amounts of funding for infrastructure in 2009, these funds will come to a virtual standstill by 2014.[20] In late 2011, revenues at the state level showed signs of recovery but employment conditions at the local level were not at pre-recession levels.[21]

Inadequate Institutional Structure

The institutional framework to address increased demand for local infrastructure is inadequate to address future growth issues, the need to replace and upgrade existing systems, and equally inadequate to transform conventional expenditures into ones that will promote carbon-neutral cities. The array of local agencies with infrastructure responsibilities has been described as "splintered,"[22] "fragmented,"[23] or "fractured."[24] Water, transportation, and solid waste infrastructure systems have regional boundaries, and should be planned and financed regionally or even at the state level. Many of these systems are the responsibility of special districts which may or may not be regional. Additionally, they are independent from local general purpose governments which have the responsibility to plan for private sector infrastructure and municipal infrastructure. Although a cross-functional regional planning agency may exist, in most localities it has no power to resolve disagreements between member governments.

Not only are institutions fragmented and isolated. Within the general purpose government, the two disciplines primarily responsible for addressing bread-and-butter infrastructure issues such as streets, sewers, and sidewalks are planning and civil engineering. However, neither discipline recognizes the need to understand the other—most evident in academia where civil engineers who design and build most local streets, roads, bridges, and are responsible for most local infrastructure facilities, lack training in land use planning. Similarly, planners are trained in land use, but are not familiar with the technical system requirements of sewers, water, and even streets. Recent improvements have been made in bridging this gap for transportation programs, but much remains to be done in the other infrastructure fields to bring the two disciplines together.

Within the local general purpose government, infrastructure has traditionally been the bailiwick of engineers, and when it comes time to replace or upgrade public buildings, of architects. Capital budgeting is usually a private affair between the finance and public works departments, or a *fait accompli* by the local water utility with little involvement of the local general purpose government. Planners have generally been concerned with infrastructure as a result of the permit process, which does coordinate with the public works department, but many cities and counties lack a long-range strategic plan that provides for infrastructure.

At the state level. The structure in which the local government operates is determined by the state. A survey done by the American Planning Association in 1999 revealed that most states operate with outdated planning ordinances, and many make local planning optional. This practice is starting to change in areas such as Florida, Texas, California, and Maryland, where rapid growth caused by the strong economy of the 1990s and the continuing deterioration of streets, sewers, and public buildings built over 40 years ago are beginning to converge. Other states, such as New Jersey, Washington, and Oregon, have a history of providing strong frameworks for local planning.[25]

Most states lack interdisciplinary long-range infrastructure plans and budgets for infrastructure investments that examine local land use impacts. Minnesota, New Jersey, and the State of Washington are notable exceptions. The lack of such plans hinders local officials in planning their own infrastructure, since at the state level, decisions about transportation networks and water issues impact local growth decisions. Historically, the lack of such a system in California affected billions of dollars in public assets.[26] In the early years of the decade beginning in 2010, climate change concerns have jump-started state-wide efforts to coordinate infrastructure investments to reduce energy use and carbon emissions in leading edge states.

Lack of a national framework. National infrastructure expenditures since World War II have been based on the premise that the sum of local, state, and special interests would add up to a coherent national interest. Exceptions such as the highway system and the national air traffic control system came from safety and defense needs. Although these needs resulted in the relatively speedy rollout of infrastructure to connect the country from coast to coast, many of the impacts of this system are not appropriate for today's concerns about energy independence, climate change, and preservation of green spaces. Two centuries ago the Gallatin Plan attempted to map out a national approach to infrastructure and was roundly defeated by state and local commercial interests.[27] Today infrastructure professionals are again asking the federal government to set up an overall framework for infrastructure that states and local governments can work within.[28] It remains to be seen whether the twenty-first century can forge a coalition to make more effective infrastructure choices.

Conclusion: A New Paradigm for Infrastructure

Local officials and practitioners are aware of these challenges within the current system. However, the task of dealing with day-to-day operations often trumps major long-term efforts. Many of the challenges to infrastructure described above are not new. The urgency, however, to address climate change may spark the coalition of interests necessary to transform the various infrastructure fields. This urgency will be aided by the fruits of information technologies and other innovations that will permit a move toward a more decentralized, holistic model that views the built environment as a whole system.[29] Many innovative best practices are now available to enable local practitioners to develop a sustainable approach to infrastructure.

In all these areas, the local practitioner will be challenged to ensure that every dollar spent on infrastructure results in progress toward a carbon-neutral state and a sustainable community. Just as the "smart growth" movement emphasized the conscious use of land, so the new paradigm for infrastructure demands a deliberate use of infrastructure investments to minimize GHG emissions and energy use, and to promote more sustainable cities and regions. Successive chapters will build upon this theme.

Six Steps Towards a New Infrastructure Paradigm

- Promote sustainability, resilience, and carbon neutrality.
- Develop infrastructure plans at the local, state, regional, and federal levels.
- Ensure that life-cycle costs, impacts and maintenance are addressed.
- Encourage innovation and multiple scales for infrastructure provision.
- All stakeholders should be involved in infrastructure planning.
- Equity aspects of infrastructure need continuing attention.

chapter 2

The Emergence of Infrastructure in the United States

In this chapter...

Importance to Local Practitioners. . . . 19
The Colonial Era: 1770–1850 19
The Federal Role 21
Engineers and Plans. . . . 22
The Rise of the Industrial and Scientific City: 1850s–1930 22
Economic and Political Background. . . . 22
Industrial Restructuring and Financing Innovations 23
Infrastructure Systems 24
Planning, Plans, and Engineers from 1850 through 1930. . . . 26
The Depression and World War II: 1930–1945. . . . 27
The Feds Step In. . . . 27
Direct Spending by the Feds. . . . 27
Regulating Private Infrastructure Provision. . . . 28
Concepts Put Forth to Be Implemented in the Post-World War II Period 29
Post World War II: 1945–the 1960s—"Rolling Out the Grid" 30
Pent-up Demand for Growth and Infrastructure. . . . 30
Federal Activity. . . . 30
Planning and Engineering. . . . 31
The Post-Industrial City: 1970s–2001. . . . 31
A Cultural Sea Change. . . . 31
Impact of Globalization on Infrastructure Systems. . . . 32
Federal Involvement in Infrastructure Funding 32
Planning Abandons Infrastructure. . . . 33
Conclusion: Are We Sustainable Yet? 35
Additional Resources 36
Publications. . . . 36
Websites 36

Why Is This Important to Local Practitioners?

Local infrastructure issues today are tied to the emergence of infrastructure as part of the growth and settlement of the United States. Understanding the origins, thought processes, technologies, and financing for each historic era will help local officials to make more informed decisions as they evaluate aging roads, sewers, schools, and civic facilities in older urban areas. Infrastructure rarely goes away—in fact, the current infrastructure is often overlaid on older systems that continue to be serviceable. The traditions associated with the provision of various systems also affect how we plan for infrastructure in greenfield areas. This chapter presents a general history of infrastructure provision in the U.S. The chapter covers five historical periods: the Colonial Era, the Industrial Era, the Depression and World War II, the Post-WWII Expansion, and the Post-Industrial City and the Current Era. Infrastructure has usually been provided during periods of economic expansion, when capital is available for investment and technological innovation occurs. Provision of infrastructure occurred during the boom times of the early nineteenth century, the late 1890s, the 1920s, the 1950s and 1960s, the late 1980s, the late 1990s and 2000 through the bust of 2008.

The Colonial Era: 1770–1850

The earliest public infrastructure projects in the United States probably were the stockades, which protected the first settlements.[1] These were built by local settlers and then, as governments were formed, by members of local and state militia, many of whom were "volunteers." By today's standards, cities were small. Not one city in 1770 had a population exceeding 50,000. By 1860, six had more than 50,000, and only two had more than 500,000.[2]

a, b, c Solid waste being collected by hand into horse-drawn collection vehicles

Individuals, local governments, and some entrepreneurs in the Colonial Era developed streets, roads, and civic buildings on an *ad hoc*, decentralized basis. It was an era of "walking" cities, when most streets were unpaved and funded locally through assessments on property or from general tax revenues. Water systems, initially built by private firms, were replaced by direct government provision at the municipal level as supply failures occurred when fighting fires. Local governments and philanthropists provided parks, civic buildings, and other municipal infrastructure.[3]

Private capital and local assessments built the initial infrastructure of the period, with state financing becoming important only in the early part of the nineteenth century. Infrastructure was primarily the province of cities and individuals within the city. Farm residents dug their own wells and made their own roads.

Provision of infrastructure was tied to the country's agriculture-based economy. As cities became commercial centers for the agricultural hinterland, private interests dominated infrastructure. Those who governed the cities tended to be close-knit patrician clans, a poor man's version of Europe's aristocratic order.[4] Cities and states competed for business, and used infrastructure to make the city more attractive for commercial interests and to open up markets in the hinterland. As cities became more dense, more organized water systems were needed for consumption and fire suppression. Garbage needed to be cleared to make way for commercial vehicles. Considered a necessary part of the economic infrastructure, business interests actively promoted these services.[5]

In 1800, only 16 waterworks had been built to serve the small cities, and all but one were privately owned. The remainder of the population relied upon wells and streams. Initially, waste and wastewater were disposed of privately, or in the streets. In the first quarter of the nineteenth century, cities hired private scavengers to empty privy vaults.[6]

Transportation projects during this era facilitated trade and land development by linking urban areas to extraurban networks. The archetype for infrastructure was the Erie Canal. Established in 1816, the Erie Canal Commission connected the Hudson River to Lake Erie, opening Midwestern markets to New York merchants who sought a competitive advantage over their counterparts in Boston and Philadelphia. Interestingly, the governing body for this enterprise was a commission, with a semi-autonomous board of appointed officials. The canal was financed by special assessments on real estate and consumer goods such as salt. Although it was profit-making, the canal was also expected to serve the public interest.[7]

During this era, privately owned and operated toll roads connecting cities aided development as area land values often increased. However, private owners could not benefit from the increased property values in order to finance the roads; by the 1860s, many were turned over to counties and municipalities. Private capital similarly funded early railroads.[8]

The cost of roads, railroads, and canals to connect the rising commercial centers—as well as water systems for the growing population—eventually exceeded private capital's ability to finance. As a result, states began to issue unsecured bonds, with state bond debts growing from $12 million in 1820 to more than $200 million in 1842. The failure of the U.S. economy in 1837 and again in 1857 led to reduced or nonexistent revenue streams, and defaults occurred. Many state-built canals were abandoned with the rights-of-way turned over to private entities. In response, state legislatures began to enact debt ceilings and requirements for voter approval on the issuance of state government bonds, while simultaneously permitting issuance of bonds by municipalities.[9]

Horse-drawn road construction

The Federal Role

Although vigorously debated during this period, the federal role in providing infrastructure was limited. Among the contested issues was whether state and local governments could "take" land through eminent domain and whether it was constitutional for the federal government to build transportation infrastructure and, if so, whether it was fair.

In 1795, the Supreme Court ruled that the government, when necessary, had the right to appropriate private property. The Supreme Court in 1848 reaffirmed that this right belonged to the government, not to legislators. So, during the nineteenth century, the federal government appropriated millions of acres for railroad, turnpike, and canal systems to be built by private entrepreneurs. In 1806, U.S. Treasurer Albert Gallatin proposed a comprehensive plan for turnpikes and canals.

His plan was followed by many others, but all were rejected on the grounds that intervention might favor one city or state over another in their competition for growth.[10]

Engineers and Plans

In Europe and the U.S. during this period, provision of infrastructure was project-based, using whatever tools, material, and money were available. However, the idea that the new scientific methods could be brought to bear on infrastructure was growing. In 1747, Paris established the first formal college for training road and bridge engineers, and by 1761 efforts were made to standardize the design and construction of London's streets. In 1793, the Society of Civil Engineers was formed in England.[11]

In the USA, which at the time lagged behind its European cousins, those involved in the design and building of infrastructure were "something like engineers," but not necessarily formally trained.[12] In 1794, however, the U.S. Congress established the Army Corps of Engineers as a permanent part of the government. Six years later, in 1800, only 30 persons identified themselves as engineers, but by 1840 the group had grown substantially. In 1848, Massachusetts established the Boston Society of Civil Engineers, the first professional society of its kind in the USA, followed in 1852 by the founding of the American Society of Civil Engineers in New York City. This organization remains the preeminent U.S. professional organization for engineers.[13]

One prominent infrastructure designer of this era was the European-trained Benjamin Henry Latrobe. He designed and built the Philadelphia Waterworks, which consisted of a sophisticated reservoir and distribution system powered by two steam engines.

City planning as a formal profession had yet to emerge, either in the USA or in Europe. Instead, architects performed this function in addition to designing buildings.

The Rise of the Industrial and Scientific City: 1850s–1930

Economic and Political Background

The Industrial Revolution and the century-long wave of immigration to the USA that ended only with the Great Depression brought infrastructure issues to the fore. At first, the pressures of population and industry on infrastructure in the rapidly growing cities resulted in chaotic growth and serious public health problems. In the 1870s and 1880s, thousands of people died due to the spread of yellow fever and cholera. However, the period ended with important technical advances in water, sewer, and telecommunications and widespread confidence in the ability of the scientific paradigm to address infrastructure as well as other issues. Engineering and other development professions in the USA grew and prospered during this era, and local governments actively provided infrastructure for both economic and health and safety reasons.[14]

In urban areas, growth resulting from the transformation of the country's agricultural base to industrial was explosive. While six million people lived in cities in 1860, 40 years later populations had risen to 25 million, and by 1910, 42 million.[15] Growth was facilitated by the invention of the steam engine, which eliminated the need for water wheels, permitting factories to be built anywhere. The steam engine also replaced the horse with the railroad so that waterways were no longer the principal means of transport.[16] In fact, more than half of the largest 150 cities in 1970 were established after 1850, once the railroad network had been fully established.[17]

Industrial Restructuring and Financing Innovations

Two factors permitted cities to provide infrastructure to support population growth and public health needs. First, both the private and public sectors were "restructured" during the last part of the nineteenth century and the first part of the twentieth. On the private side, mergers and vertical integration resulted in the large corporation (which provided railroads, energy, and telecommunications, as well as noninfrastructure services). In the public sector, the Progressive Movement was responsible for the consolidation of city services (streets, water, and civic buildings) into a single municipal corporation which could incur debt. Second, the passage of the 16th Amendment to the U.S. Constitution in 1913, which levied the federal income tax, indirectly resulted in the exemption of interest on municipal bonds from taxation due to a Supreme Court decision that classified securities on debt as property.[18]

The combination of these two factors allowed municipal governments to borrow heavily to finance infrastructure. The first municipal bond was issued after 1812, and by 1932, about $18 billion in municipal bonds were outstanding.[19] The panic of 1873 caused municipalities to default on these debts, prompting legislative action to set limits on debt ceilings for local governments.[20] However, despite the limits, the municipal bond market continued to flourish, pausing only for the Depression and WWII.

The result of widespread municipal financing was the transformation of infrastructure from systems provided by neighborhoods, individuals, and small firms to systems that were standardized and centrally provided at the local level. This transformation applied to privately provided infrastructure (energy and telephony) as well. At the beginning of this period, energy and telephone companies were small, and many competed in a single municipality. By the 1920s and 1930s, however, they were consolidated into large-scale regulated monopolies.

Infrastructure Systems

Transportation infrastructure from 1850 through the 1920s consisted of capital improvements for streets, roads, and mass transit, both within and between cities. Financing and institutional patterns of provision for intercity transportation differed from provision for transportation between cities. Cities could use the increases in property taxes resulting from the increased value to pay for improvements to publicly owned streets and roads in commercial areas. Streets in new residential neighborhoods were built by private developers and deeded over to the municipality for maintenance.

Transportation systems outside cities—both mass transit and highways—relied on user fees. Within cities, municipal transit lines were built and operated by private entities, both entrepreneurs and land speculators. Municipal public transit became more capital-intensive as the technology shifted from horse-drawn car lines to cable and then electrical power. In the 1910s and 1920s, because of over-building and the rise of the automobile, private transit lines became financially shaky. Transit companies in Boston and Philadelphia went bankrupt during WWI. Subways and rapid transit were too expensive for private entrepreneurs, so municipal governments became involved, although actual operations might be privately managed.[21]

Historic photo of refuse collection truck dumping into another truck that will transfer the refuse to a disposal site

The industrial city needed massive quantities of good quality water for public health and fire protection, with public waterworks the most common means of provision. In 1900, all cities but one with population over 300,000 had a municipal waterworks. In smaller communities, privately owned service dominated.

In large cities, the cesspools and privy vaults of the previous era proved inadequate. Sewage disposal was always municipally provided, with concern for a comprehensive, standardized system that would prevent disease. Property taxes financed these systems. Pockets of privately built sanitary sewers were integrated into the overall system. Solid waste provision varied between municipal and private collection during this era.[22]

Private interests often built streetcar lines and water lines to service real estate owned by the principals in the hopes of profiting from the sale of the land. With infrastructure, the land became valuable; without it, the subdivision was just another undeveloped tract. One researcher calls these entrepreneurs "community builders"[23] as opposed to developers, since they built "new towns" complete with infrastructure and housing. Another sees a symbiotic relationship between "place makers," who provided the infrastructure to open up wilderness or farmland areas for development, and "place takers," those who came after and profited from smaller-scale development ventures once the infrastructure was in place.[24]

This era also saw the rise of private electrical and telephone utilities. Electrical companies began as small-scale competitive businesses in the last part of the nineteenth century, but emerged as large-scale, vertically integrated and regulated monopolies. The telephone industry was more unitary from its inception.

Civic infrastructure in this period was built to promote economic development of the city. The "City Beautiful" movement, for example, combined municipal improvements, civic art, and landscape design to bring beauty into grimy industrial cities. Cities began small-scale civic improvements such as paved streets and sidewalks, parks, trees, and public drinking fountains in the latter half of the twentieth century.[25] The first decades of the twentieth century saw the rise of major capital expenditures for monumental civic buildings, parks, and Haussmann-like road systems in downtown areas including Chicago, Cleveland, and San Francisco. Daniel Burnham's City of Chicago Plan was the exemplar, and was built with numerous local bond issues from 1909 through 1931.[26]

Municipalities were also active in building schools, libraries, and concert halls, although philanthropic commissions built the great libraries. Other municipal buildings were also built during this era, many of which contributed visually to their surroundings. A fire station built in Suffolk County, New York, in 1886 is a gentle reminder of this era in public infrastructure.

Chicago Public Library (now Chicago Cultural Center), built 1897

Chicago's Union Station opened in 1925 replacing one built in 1881

Bay Shore, New York, Hose Company No. 1 Firehouse

Tunnel construction

Road construction

Planning, Plans, and Engineers from 1850 through 1930

Just as the crowding, congestion, and disease in the cities eventually produced reforms in sanitary and transportation systems, so unplanned growth and "corrupt" infrastructure provision at the municipal level resulted in the rise of the planning profession and reliance on professionals to guide the growth of the city. At the turn of the twentieth century, the planner's major tool was public works funded by taxing and direct construction. In the first and second decades of the twentieth century, the Supreme Court began to give governments the power to regulate private infrastructure by regulating land use within their jurisdictions. The boom of the 1920s resulted in formalized planning tools to standardize the provision of infrastructure in residential areas—the zoning ordinance, the comprehensive plan, and engineering standards for privately provided infrastructure.[27]

The City of New York first adopted zoning ordinances in 1916. In 1925, the U.S. Department of Commerce, under the leadership of Herbert Hoover, authored the Standard Zoning Enabling Act. This Act was subsequently adopted by all 50 states, most of which had adopted it by 1930.[28]

The concept of a rational comprehensive plan where infrastructure is planned in concert with a vision of future land use emerged in the early 1920s in Cincinnati. Local politicians had made millions on graft from public infrastructure construction and public utility franchises. The ensuing high tax rates resulted in a Citizens Commission that prepared the "Cincinnati Plan." This plan, adopted in 1925, had a 50-year time horizon, and integrated private land use patterns with transportation, community services, and public facilities.[29] It was designed to maximize the efficiency of government infrastructure expenditures and to bypass the corrupt political machine.

While planners worried about infrastructure needs of private residential and commercial development, engineers designed and built the major urban infrastructure that remains today. Herbert Hoover, notably the first professional engineer to be president, was elected in this era. Engineers were key to the development of the railroad, which made possible much of the Industrial Revolution. Successes in public health, bridges, railroads, roads, and the new technologies of electricity and telecommunications resulted in respect and confidence in the civil engineering profession, and in the ability of professionals to make decisions that would serve the best interests of the body politic and business. Robert Moses, who built many bridges, parks, and highways in New York City, is an exemplar of the public entrepreneur during this period.[30]

As the country grew, 85 engineering schools were formed and produced graduates. One aim of the 1962 Morrill Act, which granted land to states to establish state colleges, was to teach engineering. The second half of the nineteenth century saw the rise of the private consulting firm. Although some cities employed municipal engineers, most large-scale infrastructure projects were undertaken by engineers in consulting firms.[31]

The Depression and World War II: 1930–1945

The Feds Step In

Economic failure in the 1930s ended both the U.S. population expansion and the preeminence of the city in funding infrastructure. However, it did not end confidence in the ability of government and science to address economic and social problems. Nor did economic failure end infrastructure investment. President Franklin Delano Roosevelt supported the Keynesian economic policy of public works spending to reduce unemployment and to stimulate the economy. Together with his programs during World War II, which stimulated private sector investment in shipbuilding, aircraft, and munitions, these public programs reinvigorated the economy. They also provided continuity of provision for infrastructure from the economic boom of the 1920s to the postwar era.[32]

Roosevelt took three key actions that affected the provision of infrastructure. First, he and his team pumped money directly into the economy to build infrastructure. Second, he supported a variety of changes to the structure of private markets that indirectly influenced how private and public institutions would provide infrastructure. Finally, he put forward a number of concepts, rejected during his era but adopted in modified form in the postwar infrastructure boom.[33]

Direct Spending by the Feds

One-third of all construction during the Depression was funded by the 1932 Relief and Construction Act, which allocated over $300 million for major public works projects. The Works Progress Administration (WPA), which administered the funds, spent 75 percent on construction. The Public Works

Grand Coulee Dam

Hoover Dam hydroelectric power plant

Administration (PWA) built 70 percent of the schools and 65 percent of the courthouses during the 1930s. Nearly 65 percent of all public construction was federally funded from 1933 through 1938, and accounted for about one-third of all construction.[34]

Many large-scale U.S. infrastructure projects were initiated and or completed during the Depression. For example, the National Industrial Recovery Act funded the Grand Coulee Dam and Columbia Basin Project, which was constructed from 1933 to 1942. This remains the largest concrete structure ever built, and the largest hydroelectric facility in the United States. The Tennessee Valley Authority (TVA), initiated in 1933, resulted in improved productivity and living conditions for farmers in the area; prior to the TVA only 3 percent of the population there had electricity. The Hoover Dam was built by the U.S. Bureau of Reclamation in the 1930s for flood control, water storage, and hydropower generation. The Golden Gate Bridge, which combined leading-edge civil engineering theories with progressive art, was completed in 1937. The American Public Works Association ranks these projects among the top ten public works projects of the century.[35]

Regulating Private Infrastructure Provision

The Depression also gave rise to widespread concern about financial abuses and concentration of power in regional electric holding companies. As a result, regulatory changes were made that furthered the vertical integration of electricity and telecommunications industries while increasing federal oversight. In 1935, the Federal Power Commission began to regulate wholesale prices for electricity marketed across state lines. In addition, the Securities and Exchange Commission (SEC) was given authority to order the electrical companies to simplify their corporate structures and to serve geographically integrated areas. The telephone industry was already functionally integrated on a national scale. However, in 1934, the Interstate Commerce Commission (ICC) became the regulator for this industry as part of the Federal Communications Act.[36]

Regulations which enabled the public authority as a financing tool for infrastructure were developed during this period. A financial bridge was needed until municipalities that had defaulted, or were near their debt ceiling, could rebound. Revenues to retire the bonds issued by public authorities came from the federal government. But, as we shall see in Chapter 3, the public authority was the foot in the door for a financing mechanism for infrastructure that is widely used today.[37]

Finally, creation of the Federal Housing Administration (FHA) mortgage insurance program in 1934 resulted in uniform standards for infrastructure in thousands of suburbs throughout the country for millions of housing units. If a subdivision's streets, sewers, sidewalks, and street lighting did not meet federal standards, the housing units would not receive mortgage insurance. Without mortgage insurance, the bank would not lend money to the buyer to purchase property. Creation of the FHA laid the groundwork for the rollout of billions of dollars in private infrastructure for the postwar period.

The Los Angeles railroad system (street car above) began operation in 1901 and operated successfully through 1945 when it was sold to a group of investors who were later convicted in the "General Motors street car conspiracy." During the 1940s and early 1950s buses replaced much of the rail service which ended completely in 1963.

Concepts Put Forth to Be Implemented in the Post-World War II Period

The concept for the Urban Renewal program in the Housing Act of 1949 was developed during Roosevelt's term. The Act was not implemented due to the outbreak of World War II, but would have provided funds to pay for large-scale "slum removal" and redevelopment of the street grid in the center of older cities. The concept for the interstate highway system was a Depression-era idea, although its genesis is attributed to a trip that Dwight D. Eisenhower took in 1919 when he spent 62 days crossing the country from Washington to San Francisco in an army convoy. This trip convinced him that the country needed a good, fast interstate highway system for national defense. In 1941, the Inter-Regional Highways Committee, appointed by President Roosevelt, produced a report that called for a 32,000-mile interstate system. As a result, Congress passed the Federal Aid Highway Act of 1944, which authorized a system which bypassed the cities. However, implementation was blocked by those who wanted highway development funds to also promote the redevelopment of central cities—"roads fight blight." Consequently, funding for the interstate system was delayed until the 1950s.[38]

Post World War II: 1945–the 1960s—"Rolling Out the Grid"

Pent-up Demand for Growth and Infrastructure

At the end of World War II, a huge pent-up demand existed for development of all sorts. This period became the apex of the rational, comprehensive planned approach to infrastructure, or "rolling out the grid."[39] Although the standardized approach to capital investments had its roots deep in the industrial era and in Roosevelt-era reforms, its heyday was the postwar growth of suburbs and the redevelopment of the central city. Its very success would result in a backlash, and a retreat from physical planning by the planning profession in the 1960s through the 1990s. But during the rollout, the physical plan, uniform standards for the provision of infrastructure, the belief in rational, comprehensive, even regional planning, and the conviction of a rational approach to solve social and physical problems, all were in full swing.

The construction of standardized networks of streets, water lines, sewers, highways, and mass transit systems allowed the industrial system to extend into the far corners of the United States. Mass production and consumption of standardized goods were made possible by distribution to markets via reliable, standardized, national and local transportation systems. Markets could be controlled by the central office of the industry because of the existence of a national and local communication network.[40]

Federal Activity

Curb cuts for postwar subdivision

The Federal Aid Highway Act, which provided a funding mechanism for the interstate highway system, was finally passed in 1956. This Act authorized 41,000 miles of new roads, with most of the construction to take place during the 1960s. This Act also authorized a tax on gasoline and oil—the gas tax—to fund this system. In most cases, the federal government contributed 80–90 percent of the funding, while state and local governments were expected to fund the balance. This macro-grid connected cities, regions, and metropolitan areas. In Los Angeles, additional funds were allocated to its mass transit system. Urban renewal funds were also provided to rebuild central city infrastructure.[41]

Within new subdivisions, FHA standards for infrastructure were the principal template for sidewalks, sewers, roads, streets, and location of energy and telecommunications utility lines. The centrally developed standards were uniform, but capable of being implemented by local developers whose plans were reviewed by thousands of FHA appraisers.

Under this model, infrastructure was to be provided equally and in all locations. Therefore, to complement suburban infrastructure, the Economic Development Administration (EDA) was established in 1961 to ensure public works projects in depressed areas. EDA provided the funds for water, sewers, and roads for industrial parks to stimulate economic activity in these locations. The federal share of public infrastructure spending (excluding public facilities and parks) rose from 17 percent in 1956 to 40 percent in 1977. Coincidentally, this was the height of expenditures for the interstate highway system.[42]

Street construction in subdivision around 1946 to 1948

Planning and Engineering

Many city and county managers were products of the engineering profession during this era of high confidence in planning, engineering, and the role of science to solve social problems in the USA. Planners, engineers, and architects were seen to have the answers to the growth explosion that faced the nation in the wake of World War II. Physical planning dominated the planning profession.

More street construction in 1950s

The Post-Industrial City: 1970s–2001

A Cultural Sea Change

The unbridled exuberance and unfettered growth of the post-war period resulted in a reassessment in the late 1960s and 1970s. The prosperity of the 1960s, the Vietnam War, and the impact of suburbanization led to questions about the value of growth. Birth control, antibiotics, and the increase in the number of service sector jobs meant that women could enter the workforce at their own choosing.

At the same time, the publication of two environmental books, *The Silent Spring* by Rachel Carson in 1962, and Barry Commoner's *Making Peace with the Planet* in 1975, began to affect the next generation. The first Earth Day, on April 22,

Man with jackhammer in urban setting breaking up curbs to widen urban street

1970, focused attention on the importance of action to preserve the environment. However, the rise of information technology and subsequent transformation of the economic base of the country contributed to a new era in infrastructure provision characterized by neglect of the old and fascination with the new.

Although industrial employment in the United States began to decline slowly after WWII, only in the early 1970s was the new economic base named. Called variously postindustrial, the new economy, or globalization, manufacturing began to outsource production to nonunion areas of the USA, to Mexico, and overseas to Asia and India. Corporate research and development functions, as well as the financial sector of the economy, became concentrated in a few "global" cities, such as San Francisco, New York, London, and Tokyo.

The demographic make-up of the country changed dramatically during this period. Immigration, which had come to a standstill with the Great Depression, was reauthorized in 1964. Immigration became an important growth factor for many cities over the next 35 years as the service sector became a stronger component of the nation's economic base. Labor markets saw the rise of female labor force participation rates and the atomization of the family. Single-parent households increased, family size decreased (except for immigrants), and household incomes became increasingly unequal.[43]

The restructuring of the U.S. economy in the 1960s and 1970s resulted in the relocation of manufacturing from the Midwest to the South and West. Population growth in these areas led to a situation similar to that experienced by the Midwest and Northeast in the late nineteenth century. Cities, counties, and regions had no plans or financing in place to provide the infrastructure necessary to support such rapid development.

Impact of Globalization on Infrastructure Systems

During the nineteenth century and most of the twentieth, the role of infrastructure was to connect the country, and to serve household and commercial needs within the cities. The nation's changed economic base caused a shift in emphasis for infrastructure. Airports to connect the global economy became increasingly important. Highways continued to play a key role as "just in time" supply lines, and even Internet retail, relied heavily on trucking to transport goods within the USA.

Federal Involvement in Infrastructure Funding

During the late 1960s and early 1970s, the National Environmental Policy Act (1969), the Clean Air Act (1970), and the Federal Water Pollution Act (1972) were passed by Congress. The result was a shift in federal emphasis on

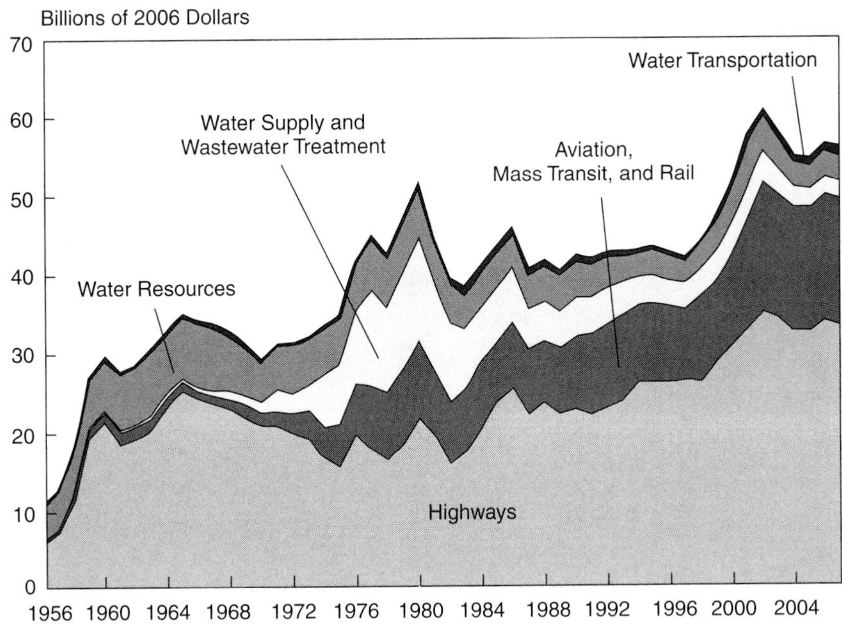

Figure 2.1
Federal capital outlays for transportation and water and wastewater treatment, 1956 to 2007 (est.)

infrastructure away from highways to mass transit, water, and wastewater systems. For example, in 1956, 41 percent of federal expenditures were on highways, and none for water supply and wastewater treatment. By 1976, 36 percent of federal funding went on highways, but water supply and wastewater treatment expenditures had increased to 15 percent of the total. Mass transit was 8 percent[44] (Figure 2.1). (Note that these figures do not include parks and public facilities.)

The 1980s saw a decline in federal infrastructure spending as the ideology of privatization was supported both politically and in academia. In the 1990s, the federal emphasis shifted back to highways and aviation, and local concerns about infrastructure began to include those of the "information highway" and the needs of the telecommunications industry. State and local infrastructure provision emphasized economic development needs, and frequently included financing not only for roads and sewers, but for convention centers and major league stadiums to promote the latest new industry—tourism.[45]

Planning Abandons Infrastructure

The slum clearance programs of the 1950s prompted a backlash against physical planning. Concern about social issues caused many planners to criticize physical planning as being elitist since it was seen to displace the poor from central cities and was insensitive to environmental issues. The comprehensive plan was deemed irrelevant to the issues of social inequality and cultural changes that were occurring. One commentator characterized this era as one where disjointed incrementalism rather than comprehensive planning ruled. Strategic, focused, functional planning became dominant in the postindustrial city.[46]

Therefore, infrastructure planning was rarely addressed comprehensively at the local level. Planners and engineers were well aware of the power of infrastructure to shape urban form, but this did not translate into integrated cross-functional local level planning that included infrastructure. Although T. J. Kent Jr.'s *The Urban General Plan*, published in 1964, kept alive the ideal of integrating land use and infrastructure planning, many local governments did not heed his advice. He recommended that, because utilities such as water, drainage, and sewage disposal influence the land use patterns of a community, they deserved their own section in the general plan—an idea that is finally being acted on today.[47]

In its 1969 Urban Planning Guide, the American Society of Civil Engineers (ASCE) stated that since "civil engineers direct the planning and development of large urban engineering works that form the structure and shape of cities," such engineers will make "outstanding urban planners."[48] In fact, the ASCE went on to say: "the civil engineer is already planning our communities in large measure and will continue to do so." In 1970, Norman Williams, a legal scholar, identified infrastructure investment as one of the three *de facto* systems of land use control in this country—the other two being fiscalization of land use (in which local officials consider the revenue-producing capacities of different land uses), and the "official" land use regulatory system of zoning, subdivision rules, and development permits. Williams concluded that the official system had the least power.[49]

By 1975, however, many experts were concerned that the Federal Water Pollution Control Act amendments, passed three years early were subsidizing suburban sprawl because the Act provided $18 billion in grants to local governments for sewage treatment plants. Critics charged that many communities were expanding these systems without regard to compact growth. Several studies confirmed the fact that infrastructure investments in general, and sewer and transportation investments in particular, were more powerful growth shapers than had been previously thought.[50]

During the 1980s, however, these insights faded from view, partly due to cutbacks in federal funds for land use planning and termination of sewer grants, and partly due to the movement away from planning as strictly a land use activity. For example, the first edition of *The Practice of Local Government Planning*, published in 1979 by the International City/County Manager Association (ICMA) contains an entire chapter on infrastructure planning.[51] In the third edition of the same volume, published in 2000, the word "infrastructure" does not even appear in the index.[52] Similarly, during the 1980s and well into the 1990s, presentations on infrastructure at the annual meeting of planning academics were limited, and sparsely attended.[53]

During the 1980s, several high profile infrastructure failures caused a flurry of interest in the subject. Highlighted by publications such as *America in Ruins* (1981), *Fragile Foundations* (1988), and *Financing the Future* (1993),[54] these studies decried the loss of federal funding and urged massive spending to solve the problem. However, this concern was short-lived and did not translate into integrated planning solutions at the local level.

But times appear to be changing. Ten years ago, in 2002, the APA's *Growing Smart Legislative Guidebook* identified the need to link "the timing, location, and intensity of development … to existing or planned infrastructure."[55] States impacted by rapid growth during the 1980s, 1990s, and the first decade of the 2000s passed legislation to fund local and state infrastructure needs.[56] Infrastructure began to creep back into the lexicon of planners. Today, issues such as climate change, the deterioration and accidents of high profile infrastructure systems, along with calls for investment in infrastructure to promote an economic stimulus have rekindled interest in the role of infrastructure at the local as well as national level.[57] Policy-makers are discovering that managing infrastructure is instrumental in achieving greater social and environmental goals. Perhaps the next decade will see the union of social, environmental, and physical concerns in infrastructure planning and provision.

Conclusion: Are We Sustainable Yet?

At the end of the first decade of the twenty-first century, the economic downturn gave a "time out" for discussions about uninhibited growth, while concern about climate change and employment dominated the public agenda. New growth will continue to be an issue for the USA since the biggest component of population growth over the next decades is likely to be natural increase, not immigration; therefore, infrastructure will continue to be a concern across the country. Maintenance will be a more serious issue in poor economic times. Capital investment in infrastructure likely will be provided for high value activities, such as upper-income residential communities and knowledge-based industrial parks, while infrastructure quality in poor areas will continue to decline. Likely impacts of climate change will affect poorer communities the most. However, the demystification of technical solutions to infrastructure has begun and is likely to continue as the need to reduce energy use and carbon emissions has mobilized a broad spectrum of the community who will insist on applying these metrics to infrastructure investment decisions as well. As noted earlier, sustainable infrastructure design is a key element in stabilizing, mitigating, and adapting to the effects of climate change.

Additional Resources

Publications

Jacobson, Charles D. and Joel A. Tarr. ***Ownership and Financing of Infrastructure: Historical Perspectives.*** **No. WPS 1466. Washington, DC: The World Bank, 1995. Available at: www-wds.worldbank.org/servlet/WDSContentServer/WDSP/IB/1995/06/01/000009265_3961019111943/Rendered/PDF/multi0page.pdf.** A very careful historic analysis by infrastructure function done for the World Bank. Although only available on-line, it is a valuable supplement in this area.

Monkkonen, Eric H. ***America Becomes Urban: The Development of U.S. Cities and Towns, 1780–1980.*** **Berkeley, CA: University of California Press, 1988.** A classic on the development of cities in the United States.

Sbragia, Albert M. ***Debt Wish: Entrepreneurial Cities, U.S. Federalism and Economic Development.*** **Edited by Bert A. Rockman.** ***Pitt Series in Policy and Institutional Studies.*** **Pittsburgh, PA: University of Pittsburgh Press, 1996.** A wonderful history of how cities financed their major improvements.

Websites

American Planning Association: www.planning.org. The website for professional planners.

American Public Works Association: www.apwa.net. A good source for historic data and historic resources for infrastructure in the United States.

American Society of Civil Engineers: www.asce.org. Also a good source for the rise of the civil engineering profession.

chapter 3

Growth, Demand, and the Need for Infrastructure

In this chapter...

Importance to Local Practitioners 37
National Population Trends, Households and Infrastructure Demand 37
New Kinds of Households and Changing Mobility Rates 38
The Aging of America 39
Increasing Ethnic Diversity 40
Rising Incomes and Rising Income Inequality 41
Demographic Trends at the Metropolitan Level 43
Infrastructure and the Economy 45
Does Public Infrastructure Investment Cause Growth? 45
At the Regional and Local Level 46
Importance of Human Capital for Growth 49
Conclusion 49
Additional Resources 49
Publications 49

Why Is This Important to Local Practitioners?

Local and regional officials need to understand the relationship of infrastructure to demand, growth, and needs to make informed decisions about local capital improvement plans and individual projects. Factors that influence the demand for infrastructure are population growth, economic growth, replacement needs, and infrastructure standards. The need to replace old and deteriorated underground pipes, streets, sidewalks, and public buildings is a major factor in local demand, while local efforts to reduce energy use and greenhouse gas (GHG) emissions contribute to the need for a new kind of infrastructure. Infrastructure investment also contributes to economic growth. During times of economic contraction, expenditures for infrastructure may serve as an economic stimulus, while during boom times local officials may use infrastructure to promote business in a particular city or metropolitan area. The economic boom of the 1990s and 2000s saw local officials invest in substantial amounts of networked infrastructure and in quasi-public buildings such as sports arenas, concert halls, and museums.

This chapter begins by describing trends in demographic factors that relate to the need and demand for infrastructure in general. The following section takes a closer look at these trends at the metropolitan level. The section thereafter addresses the relationship of economic growth and infrastructure. The final section of the chapter looks at benchmarks to assess national infrastructure expenditures.

National Population Trends, Households and Infrastructure Demand

The need for infrastructure is triggered by population growth. This growth, from an infrastructure point of view, is reflected in the rate of household formation, which triggers the demand for housing. The kinds and amount of housing constructed and reused form the basis for both on-site and off-site infrastructure in individual jurisdictions. This section looks more closely at past and projected population trends and household characteristics at the national level. The sections following examine the implications of these trends locally.

From 2000 through 2010, the population in the United States increased by 10 percent, and this trend is projected to continue although various analysts disagree about the size. The most recent U.S. Census figures project that the population will increase from 310 million to 439 million people from 2010 to 2050, an increase of 42 percent.[1] This is unlike other industrialized countries where populations are stable or even decreasing. However, the characteristics of the households that will make up the projected population have been and will continue to be quite different from those who built the legacy infrastructure now in place. Four changes will have far-reaching effects on housing choices, and hence on the amount, kind, and location of infrastructure to support these choices. The first change is the rise of smaller households, and fewer households with children. Second, the aging of the population will result in profound changes in residential demand for housing and therefore for infrastructure. The third change is the rising proportion of minority and immigrant households. Finally, although incomes have been rising in the United States, income inequality rose sharply in the past 30 years. These trends are anticipated to continue.

New Kinds of Households and Changing Mobility Rates[2]

Today, U.S. households are smaller, and the proportion of families with children has decreased since WWII. The traditional two-parent family with children has given way to smaller households, many with single parents. In 1950, single-parent households were 4 percent of family households, whereas in 2000 they accounted for 12 percent. Rising incomes and the entry of women into the labor force in the 1970s contributed to higher rates of household formation. At the same time, many single older persons have chosen to live independently rather than move in with their adult children, and households composed of unrelated persons are growing. In 2000, married couples without children were the most common household type, followed by single-person households. Households with children were only 28 percent of the population in 2000.[3] This trend continued through the 2000s.

Average household size decreased from 3.33 persons in 1960 to 2.62 persons in 1990, and from that time through 2010 at a slower rate, reaching 2.58 persons. The economic troubles of the 2000s resulted in a reduction in the rate of household formation as young adults continued to live with parents, or took on additional roommates.[4]

Growing diversity in household types means that traditional per capita multipliers for calculating infrastructure needs may no longer be valid. Housing choices are usually different for those with and without children. In the past, income stratification resulted in homogeneous subdivisions, with a uniform grid of infrastructure. This may not hold true for the future.

Mobility patterns among households also affect the geographic distribution of housing types and need for infrastructure. Young adults are the major interstate movers, moving to areas with strong employment and often away from where they grew up in order to obtain jobs and have children. Therefore, traditional married-with-children families are on the rise in areas with strong growth, such as the South and West. In the North and Midwest, both of which have lost population, single-parent families outnumber the traditional two-parent families. Nationally, in the suburbs in 2000, singles outnumber married-with-children families, and central city population growth reached a high point while the number of households being formed was at a low.

For local officials trying to anticipate infrastructure needs, the mix of households at the local level will impact investments in schools, parks, transportation arrangements, and the provision of water and sewers. Local officials will need to tailor needs assessments to the specific situation for their own locality instead of using national trend information.

The Aging of America

In the past, the distribution of population by age category has been a pyramid, with the oldest categories at a small top, supported by a large number of children at the bottom. Today, people are living longer, and in the next 25 years the increase in the number of households will take place primarily in the older, post-children group.[5]

Three basic age groups can be used to track household formation: 25–44; 45–64; and 65 and older. During the next 25 years, the population is expected to become more equally distributed between these groups, thus creating more of a pillar than a pyramid (Table 3.1).

Demography and Transportation

Changes such as the increased female labor force participation rate and the Baby Boomers reaching driving age have been responsible for the increased demand for infrastructure since WWII. More people own cars and, as a result, vehicle miles traveled and the demand for streets, roads, and highways have increased.

Table 3.1
Changes in proportions of households in three age groups from 2000 to 2025

Age group	2000	% of total	2025	% of total
25–44	85,041	46.7	86,106	37.9
45–64	61,952	34.0	78,416	34.5
65 plus	34,992	19.2	62,641	27.6

Source: M. Farnsworth Rich, "How Changes in the Nation's Age and Household Structure Will Reshape Housing Demand in the 21st Century," in *Issue Papers on Demographic Trends Important to Housing* (Washington, DC: U.S. Department of Housing and Urban Development, 2003).

Housing demographers project that major generational trends will reconfigure housing markets in ways that are particular to the metropolitan level. The Baby Boomers, and the "New Immigrant" generations were responsible for the rapid growth patterns of the past several decades. However, as the Baby Boomers age over the next ten years, they may "rent, relocate and withdraw from the housing market."[6] The Millennials will arrive later into the market, but depending on the preferences of these groups to live in single-family detached or multifamily structures,[7] the availability of land, water, and sewer needs will also be affected. Local infrastructure projections need to be made based on the local distribution of housing types and age categories. See also Table 3.2 for regional projections of households, employment and income through 2025.

Increasing Ethnic Diversity

The past several decades have seen the growth of minority populations—particularly Hispanic, African American, and Asian—to become 30 percent of the U.S. population. Although the rate of household formation, important for predicting the demand for infrastructure, is lower for Hispanics and Asians, it is higher for African Americans and non-Hispanic whites. It is also lower for newly arrived immigrants, although the longer immigrants are in the country, the more their rate of household formation follows that of the native born. Similarly, the majority of white households have no children, whereas the reverse is true for minority households. Not surprisingly, minority households are younger.[8]

Minority households accounted for 58 percent of total household growth between 1996 and 2001. By 2020, minority households are expected to increase from 25 percent to 30 percent. The housing crisis of the first decade of the twenty-first century, however, caused setbacks in homeownership for many black and Hispanic households. The effect was to reduce their median wealth by 50 percent to two-thirds, compared to 16 percent reduction of

Table 3.2
U.S. growth by U.S. Census region, 2000–2025

Measure	Region	2000	2025	2000–2025 growth	
				Number	Growth (%)
Population (#, in 000s)					
	Northeast	53,594	57,223	3,629	6.8
	Midwest	64,393	73,061	8,668	13.5
	South	100,237	127,538	27,301	27.2
	West	63,198	84,328	21,130	33.4
	Total	281,422	342,150	60,728	21.6
Households (#, in 000s)					
	Northeast	19,955	21,431	1,476	7.4
	Midwest	24,773	28,223	3,450	13.9
	South	35,863	46,526	10,663	29.7
	West	22,654	30,519	7,865	34.7
	Total	103,245	126,699	23,454	22.7
Employment (#, in 000s)					
	Northeast	29,964	36,013	6,049	20.2
	Midwest	39,821	50,278	10,457	26.3
	South	54,157	73,179	19,022	35.1
	West	35,448	49,338	13,890	39.2
	Total	159,390	208,808	49,418	31.0
Income (millions of 1992 dollars)					
	Northeast	1,403,731	2,032,287	628,556	44.8
	Midwest	1,507,569	2,287,786	780,217	51.8
	South	2,012,882	3,490,513	1,477,631	73.4
	West	1,426,246	2,541,805	1,115,559	78.2
	Total	6,350,428	10,352,391	4,001,963	63.0

Source: Robert W. Burchell et al., *Costs of Sprawl—2000,* Transit Cooperative Research Program Report no. 74 (Washington, DC: National Academy Press, 2002).

wealth for white households.[9] This will impact the ability of these groups to pay for infrastructure as well as housing.

These differences mean that local officials in areas with strong concentrations of minorities or immigrants must pay particular attention to the local composition of the population when estimating the need for services. Three-quarters of the nation's immigrants live in just six states—California, Florida, Illinois, New Jersey, New York, and Texas.

Rising Incomes and Rising Income Inequality

Increases in household income since WWII have also affected the demand for infrastructure. In the 1980s, household incomes increased as a result of women and boomers entering the labor force, and again in the 1990s as the boomers reached their peak earning years.[10] As incomes rose, demand for larger houses and consumer goods increased, which in turn resulted in more or more expensive on-site and off-site infrastructure. Research done before the housing bust at the end of the last decade notes that the average square foot per household has increased substantially from 1985 through 2005. Generally the larger the house, the greater the consumption of water, energy, and wastewater. As incomes rose, more consumer goods were purchased and discarded, increasing the need for landfill sites, for example.[11]

Rising incomes are also related to increased investment in transportation infrastructure: people purchase faster transportation alternatives such as buying cars instead of taking buses and flying instead of driving. Both roadways and airports become more congested, resulting in demand to expand these facilities. Most analysts feel that future income increases will not affect per capita miles driven, but it will continue to play a strong role in the demand for parks, public facilities, environmental infrastructure, energy use, and telecommunications infrastructure.[12]

Although incomes have been improving on an aggregate basis, serious problems emerge with the widening gap between the top and the bottom, as well as with large pockets of those with no income but substantial service needs. Inflation-adjusted wages rose during the last half of the 1990s, but wages during that decade for the bottom and middle pulled closer together, while the top pulled farther away. This trend continued throughout the 2000s. Labor productivity has grown, but families, especially those with children, are working more hours. In 1998, the average childrearing family worked the equivalent of six weeks more per year than it did in 1989. It was not until 1999 that the poverty rate dropped to the rate at the height of the previous economic cycle in 1989.[13] These rates vary substantially throughout the country, and between and within metropolitan areas.[14]

The geographic segmentation of the poor in the United States has an impact upon the ability of local governments to address infrastructure needs. Although generally the commuting needs of workers are addressed with national, state, or regional programs, the mismatch between the location of the poor and jobs has been and continues to be a serious concern.[15] Water and waste infrastructure are usually provided at the regional level, and often the concern is for building new systems instead of paying for upgrading the old. Frequently all ratepayers are assessed, which means the urban poor pay the same amount for deteriorated services that the more affluent pay for new infrastructure in more recently built areas. Civic facilities such as schools and public safety buildings may be deteriorated in areas with substantial amounts of low-income households since the tax base may not be adequate to upgrade them.

Equity issues are usually missing in the discussion of infrastructure. Frequently the debate revolves around whether middle-class homeowners should pay for services versus landowners or developers. However, exceptions do exist. The location of waste facilities in low-income areas has given rise to the field of environmental justice and increased interest and concern by local governments in rectifying these inequities. In some communities, nonprofits actively pursue commuting alternatives for the poor. Unfortunately, these efforts do not address the systemic equity issues with respect to infrastructure provision.

Demographic Trends at the Metropolitan Level

Ultimately most infrastructure is built locally or at the metropolitan level where demographic trends are diverse, and will continue to be so as the uneven rates of household formation caused by the echo of the Baby Boomers reverberate over the years. The difference in constant dollar household income between different metropolitan areas has increased substantially in the past 25 years. The South, long one of the poorest regions in the United States, has seen household income converge towards the median. Future population growth is expected in only a few of the over 300 metropolitan areas in the USA, primarily in California, Texas, Florida, and the Southwest. That is, 40 of the 3,100 counties in the country will contain one-third of the projected future growth. The South and West combined will account for about 80 percent of future population growth and 70 percent of future employment.[16]

Within the traditional regional divisions, there are also many differences at the metropolitan level. Southern and western metro areas continued to grow the fastest in the 2000s, but had the greatest slowdown in growth from the 1990s. The "bubble economies" had the greatest slowdown in growth in the 2000s.[17] In addition, during the 2000s, the 100 largest metropolitan areas became more distinct from each other during the 2000s, amplifying trends that were on the horizon in the 1980s and 1990s but also responding differently to the market crash of 2008. These metro areas differ dramatically in population size, percentage of minority and elderly, and educational attainment. At the same time the distinction between the central city and suburbs in individual metropolitan areas began to blur as demographic trends became metropolitan, not urban–suburban. High density suburbs have growth and commuter patterns similar to that of their central urban area. White population grew in many urban areas, while a majority of immigrants and ethnic groups lived in the suburbs. Educational attainment levels became more equalized between urban and suburban areas within individual metro areas, and the differences in income fell between these areas as well.[18]

A Brookings Institution study in 2010 grouped the 100 largest metro areas into seven categories that are also useful for analyzing future demand and needs for infrastructure:[19]

- *Next Frontier Metro Areas*: These include nine metropolitan areas with high growth, high diversity and high educational attainment. All but Washington, DC, are west of the Mississippi. They are younger, have a more diverse economic base, their infrastructure is transit-oriented, but incomes and educational attainment are more unequal.
- *New Heartland*: These areas have the high growth and educational attainment as do those above but are not as diverse ethnically, with lower rates of Hispanic and Asian populations. These metro areas include Portland and those in the New South that attracted middle-class migrants during the 2000s.
- *Diverse Giants*: These include the largest U.S. metropolitan areas with lower than average population growth that is intensifying their already dense urban form. These areas exhibit large differences in educational attainment and wages, and have large second-generation immigrant populations.
- *Border Growth Metros*: Characterized by high growth, high diversity but low educational attainment, these metro areas are mostly in the southwestern border states, where immigration caused a housing boom, and subsequent bust.

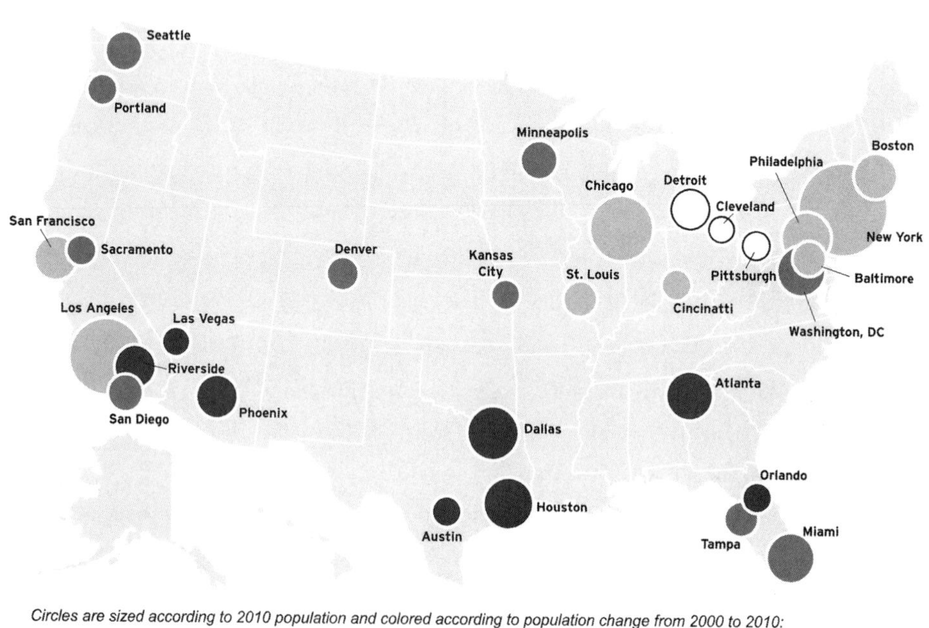

Figure 3.1
Thirty largest metro areas, 2010

- *Mid-Sized Magnet Metros*: These metropolitan areas are similar to the Border Growth Metros, in that they experienced rapid growth and overbuilt housing. They have seniors but otherwise low diversity, and low educational attainment, and rely on automobiles.
- *Skilled Anchor Metros*: Located primarily in the northeast, these metros are slow-growing, and are former manufacturing and port cities that have transitioned to the service economy. Most growth has occurred, however, in the suburbs.
- *Industrial Core Metros*: Lack of in-migration to these low growth metros has kept wage and income inequality growth from rising, but educational levels are low. Population declined overall in the 2000s.

Infrastructure and the Economy

Infrastructure demand is not only caused by growth, but infrastructure investment also causes growth. Business cycles correspond to periods of increased or decreased demand for infrastructure. As economic activity heats up, the demand for employment increases, which causes population shifts within the country. The economic "winners" need more transit, water, schools, and community facilities. In periods of economic decline, local governments often forgo long-term infrastructure investments unless funds were already obligated at the height of the previous boom. In the short term, reinvestment in infrastructure can also be part of an employment or economic stimulus effort.

Does Public Infrastructure Investment Cause Growth?

Generally, it is agreed that infrastructure investment contributes to the growth of a nation, region or city by supporting productivity and employment. One academic put it this way: "core economic infrastructure ... is important in maintaining economic performance" and "infrastructure investments are essential for the functioning of the U.S. economy."[20] Where analysts part ways is about the role of *public* infrastructure. Many empirical studies have been done by economists to determine whether increased public infrastructure investment results in increased economic growth or simply displaces

Spotlight on California

In the past decade, California's population growth slowed to its lowest rate on record, according to the 2010 Census. Still, the state will grow by millions of residents over the next 20 years. By 2025, California's population is projected to reach about 44 million. Annual growth rates are expected to be just 1 percent, similar to growth experienced in the 1990s and the 2000s but substantially slower than in earlier decades. Even so, average annual increases will exceed 400,000—equivalent to adding population equivalent to the city as large as Oakland each year.[1]

The inland areas of California have grown faster than the coastal areas for many decades, but coastal counties are still home to most of the state's population. Projections indicate that the Inland Empire, the Sacramento region, and the San Joaquin Valley will grow faster than other areas of the state. California is still home to more immigrants than any other state, but its popularity as an immigrant destination is dropping: From 1990 to 2007 immigrant growth rates in some other states topped 20 percent per year, more than five times the rate of growth here. Similarly, growth rates in new destinations within the state, such as Riverside and Kern Counties, have soared. There is a decline in the preferences of new immigrants, particularly Latinos, to live in cities with large populations of other immigrants. Simultaneously, widening economic opportunities in new areas have attracted many new immigrants.[2]

Notes

1. Hans Johnson, *California Population: Building for a Better Future* (San Francisco: Public Policy Institute of California, 2012).
2. Sarah Bohn, *New Patterns of Immigrant Settlement in California* (San Francisco: Public Policy Institute of California, 2009).

spending by other levels of government or the private sector. The results are inconclusive.[21]

A controversial study published at the end of 1989 found that public capital investment resulted in higher economic payoffs than private investment, and that 1 dollar of public capital investment led to a 4-dollar increase in overall gross national product. This was due to the high correlation between productivity and public infrastructure. The results were used to argue that the productivity slowdown in the 1980s was due to decreased public investment.[22]

The Congressional Budget Office (CBO), however, after a more comprehensive review, concluded that only a few infrastructure projects could be justified by their contribution to the economy. The CBO noted that most analyses "do not consider the potential impact of shifting funds from low-return projects chosen for noneconomic reasons to projects with higher returns." The CBO noted that public investment in new runways and highways with thicker pavement would create greater dollar benefits if airport and road use were priced based on the vehicle's contribution to congestion, and if commercial truckers would pay taxes based on weight per axle. The CBO review also found that federal spending displaces about 35–75 cents of state and local investment per grant dollar.[23]

More recently, there has been increased discussion about the role of infrastructure in jump-starting economic growth for the country. In the fall of 2010, the U.S. Department of the Treasury laid out an ambitious infrastructure plan to stimulate the U.S. economy. This document noted the arguments above, but concluded that given the dire state of the economy, and the positive impact of infrastructure spending in ARRA, that further stimulus was necessary.[24] In 2011, the head of the Congressional Budget Office testified before Congress that "[o]ver the long run, the nation's output depends on the size and composition of the capital stock, the quantity and quality of the labor force, and the nation's technological progress."[25]

At the Regional and Local Level

Economists such as Alfred Marshall[26] emphasized the importance of external economies in a region—that is, such things as communication and lowered costs of goods transport—as important ways of fostering local growth. Traditional location theory includes a firm's transportation costs in its decision to locate in a particular area.[27] Changes in the nation's economic base in the past 30 years have resulted in changes in the type of infrastructure needed by commercial interests. The restructuring of the economy has changed production processes and labor force requirements. The rise of cheap global transportation and the impact of computer technology have seen the decline of unionized labor, the increase of much lower paid service sector jobs, and the importance of highly skilled professionals in large cities. The increased production of goods and services has resulted in greater use of transportation facilities.[28] Airports and universities, along with a strong local educational system, have become important types of infrastructure for the changing economic base of the country. The availability of an adequate communications infrastructure is also important in attracting high-value employment firms.

Regional economists also refer to the importance of the "commons," which includes infrastructure and an educated labor force, in fostering growth.[29] One author cites the quality of public infrastructure as one of the local determinants of geographic distribution of firms and populations,[30] while another relates the quality of the business environment in a region to the quality of infrastructure.[31] In this sense, public infrastructure forms the basis for increased wealth of the region. Although regional economists have different views on how to ensure the growth of a region, most agree that infrastructure is important in attracting, retaining, and nurturing businesses.

From the 1950s onward, infrastructure has been provided as an in-kind contribution to a business to induce it to locate in a certain city. The types of infrastructure might be the paving of a street next to an industrial park, the provision of street lighting, or the waiving of development fees. Today, economic development strategies include not just bricks and mortar infrastructure, but also infrastructures that place a premium on the development of human capital. In addition, development of fiber-optic networks by some municipalities has been used as a tool to attract what are seen as more high-value industries.[32] For example, the city of Alameda, California, has been able to market office space profitably because of the existence of high bandwidth and high-speed Internet access as a result of a locally owned telecommunications utility.

Despite the inconclusive role of public infrastructure investment noted earlier, economic development specialists

How Infrastructure Affects a Firm's Decision to Locate

Infrastructure can affect a firm's decision to remain in or come to a region in three ways. First, infrastructure can be an input to production. The semiconductor production process, for example, requires vast amounts of pure water, and since the process generates large amounts of hazardous waste, a sophisticated solid waste management system is needed. Second, infrastructure is required to support the region's non-basic industries—the service sector which supports manufacturing. Finally, infrastructure is important in establishing, facilitating, and maintaining linkages and knowledge transfers in a regional economy. The lack of infrastructure causes firms to relocate, shift to a more labor-intensive process, to provide it themselves, or to reduce output.[1]

Notes

1. Kyu Sik Lee, Alex Anas, and Gi-Taik Oh, "Cost of Infrastructure Deficiencies in Manufacturing in Indonesia, Nigeria and Thailand," Working Paper (Washington, DC: World Bank, 1996).

continue to look at ways that regions can improve their competitive advantage. A study comparing the level of public infrastructure investment and private capital investment in Indianapolis, Columbus, and two "new economy stars"—Austin, Texas, and Raleigh-Durham, North Carolina—found strong linkages between local investments in education, street, water, and other public infrastructure, and private capital expenditures, although the strength varied by location.[33] Another study in Florida found that development fees for public infrastructure provision were related positively to employment growth, although employment growth may have caused the need for infrastructure, hence the fees.[34] A transportation literature review in 2003 concluded that investments in public sector transportation resulted in increased output, income, and productivity, and reduced costs.[35]

Although infrastructure such as water and roads can guide exogenous growth, the evidence of whether or not the public infrastructure investment can generate growth is mixed. Sometimes local officials back infrastructure projects to promote economic development that are inappropriate for the local circumstances, result in a net cost to the city, or do not yield the intended effect. The City of Cleveland built a new light rail and a downtown people mover, and subsidized the construction of new office buildings, only to find that the occupants were existing businesses that relocated within the central core.[36] The bidding wars between Fort Worth and Oklahoma City for American Airlines' maintenance operations center (MOC), and between Denver, Indianapolis, Louisville, and Oklahoma City for United Airlines' MOC, are two other examples. The airlines were able to manipulate the governments to further their corporate financial agendas because the cities were outgunned analytically in terms of assessing the fiscal impact of the airlines' proposals and the true benefit to the airline of the cities' proposed subsidies.[37]

Within the metropolitan area, if regional economic forces are strong, the location of growth can be influenced by infrastructure. The presence of infrastructure is often more important than the comprehensive plan in determining where growth will occur within the regional area, or even within a single jurisdiction. The developer of a subdivision, for example, needs three things: good access, water and sewers, and suitable zoning. Of these three items, access and water and sewers are the most important. Development follows roads and sewers. (Zoning can be changed given enough political influence.) Developers rely upon government to provide highways and off-site water and sewer lines. Highway development is too risky for the private sector, especially if the economy suffers a setback and the pace of residential demand slows down.[38]

Importance of Human Capital for Growth

Although debate continues about the role of infrastructure investment in inducing growth, the literature is fairly conclusive about the role of human capital for growth. One respected analyst notes that investment in education (and by extension, educational facilities) reaps benefits in economic growth. The continuing shift from the manufacturing and goods producing sectors to the service sector implies a relatively smaller demand for transportation, utilities, and energy infrastructure (although in absolute terms the demand is there). This shift also means that to remain competitive economically in the global and national economies, metropolitan areas need to continue to invest in the education of their workforces.[39]

Conclusion

The changed economic base of the country, along with the impact of distinct differences between the nation's metropolitan areas means that local practitioners cannot rely upon national trends to decide on infrastructure strategies. Local conditions must be examined separately to determine future needs. Practitioners in declining growth areas must look for funds from a shrinking tax base, while those in areas anticipating growth must look at a longer time frame than usual in local plans to ensure that needed public infrastructure (which has a longer time horizon than private development) is available when needed. The intersection of the date of construction and type of materials used for many pipe, road, and community facilities means that infrastructure in many localities will need to be replaced over the next 30 years, but replacement will depend on local initial conditions. The demand for a more sustainable approach to infrastructure provision may result in increased federal and state incentives or expenditures for innovative infrastructure, but all must be tailored to the individual needs of the jurisdiction.

Additional Resources

Publications

The Brookings Institution. "The State of Metropolitan America: Demographics, Cities, Regions and States," *State of Metropolitan America* no. 2. May 9, 2010. Available at: www.brookings.edu/reports/2010/0509_metro_america.aspx. A thorough overview of past and present demographic and economic trends in the United States for the 100 largest metropolitan areas with individual chapters by pre-eminent scholars. Good for understanding infrastructure demand.

Burchell, R. W., G. Lowenstein, W. R. Dolphin, G. C. Galley, A. Downs, S. Seskin, K. G. Still, and T. Moore. *Costs of Sprawl—2000*. Transit Cooperative Research Program Report no. 74. Washington, DC: National Academy Press, 2002. A projection of population trends and infrastructure costs in the U.S. from 2000 to 2025 done at the beginning of the millennium.

The methods and findings regarding infrastructure costs have not been replicated in the past 12 years. It is well worth looking at by local practitioners as a guide for making local estimates.

Kelly, Eric Damien. ***Planning, Growth, and Future Facilities: A Primer for Local Officials.*** **PAS Reports no. 447. Chicago, IL: American Planning Association, 1993.** A classic publication on the relationship of infrastructure and growth.

Pitkin, John and Dowell Myers. "U.S. Housing Trends: Generational Changes and the Outlook to 2050," available at: onlinepubs.trb/Onlinepubs/sr/sr29pitkin-myers.pdf. An authoritative look at housing trends in the USA. Useful to extrapolate infrastructure demand.

chapter 4

Institutions of Infrastructure: The Providers

In this chapter...

Importance to Local Practitioners . . . 51
Intergovernmental Context . . . 52
- How Infrastructure Institutions Evolved . . . 52
- Who Provides Infrastructure? . . . 52
- Federal Infrastructure Institutions . . . 53
- State Government Institutions for Infrastructure . . . 54
- Local-Level Governments . . . 55
- Regional Institutions . . . 57
- Energy and Telecommunication Services . . . 57

Organizational Structure of the Local General Purpose Government . . . 57
- Planning Department . . . 58
- Public Works Department . . . 60

Special Purpose Districts . . . 61
- Functions of Special Districts . . . 61
- Special District Characteristics . . . 62
- Special Assessment Districts . . . 63
- School Districts . . . 63
- Distinction between Special Districts and Public Authorities . . . 63
- Issues with Special Purpose Districts . . . 64

Regional Infrastructure Institutions . . . 65
- Councils of Governments and MPOs . . . 65
- Renewed Interest in Regional Efforts in the 1990s and 2000s . . . 65

Conclusion . . . 66
Additional Resources . . . 67
- Publications . . . 67

Why Is This Important to Local Practitioners?

Today's local infrastructure must be planned, financed, built, and maintained in a patchwork of overlapping agencies and institutions, many of which have competing jurisdictions, mandates, and goals. At best, this is complex—at worst, it is an environment that seems to preclude effective and sustainable infrastructure provision. Counties, cities, and special districts bear the bulk of the responsibility for infrastructure provision in the United States. Federal and state governments also provide some infrastructure and are the source of many mandates and requirements for local infrastructure. Sometimes this rich stew of institutions seems to work at cross-purposes, often preventing concerted action on infrastructure problems at the local and regional level.

Although the existing network of infrastructure institutions has built a system of infrastructure in the USA that is deemed to be the most advanced in the world, this web of providers no longer seems responsive to today's needs. The old independent ways of planning and financing infrastructure seem to result in sprawl, unsustainable and inequitable solutions in older cities and suburbs, while ignoring important sustainability and climate change issues. The practitioner must be able to make sense of the family of infrastructure actors within the given jurisdiction in order to assess their impact on local land use patterns and local infrastructure plans and programs. This chapter begins with an overview of the intergovernmental system before moving to a more detailed description of how local general purpose governments (GPGs) are organized to plan for and finance infrastructure. Special districts and their role in infrastructure provision are also described. The chapter concludes with a brief discussion of the regional responses to the problem of fragmented infrastructure responsibilities.

Intergovernmental Context

This section of the chapter traces the evolution of infrastructure institutions, and then provides an overview of the roles played by federal, state, local, and private actors in providing infrastructure.

How Infrastructure Institutions Evolved

Growth at the subnational and subregional levels in the United States occurred at the local level based on the ability of entrepreneurs and markets to control the development of land, rather than in accordance with a master plan. The closest thing to a centralized approach to growth in the USA was the deeding of lands by European monarchs to early settlers, and the designation of counties and subsections for homesteading in the 1850s. Local governments were created as they were needed, beginning with provisions for public safety and education, and later to plan, build, and regulate infrastructure development as well as other local services. Hence, we find a wide variety of private and public institutions, authorities, districts, governments, and agencies involved in building, regulating, and operating infrastructure in the United States.

The financial and geographic scale of the infrastructure system influenced the institutional arrangement, as did the timing of the settlement of the area and the culture of the immigrants. Development of railroads and telecommunications occurred at the national level, while water companies had local or regional service areas. School districts and city halls were local and public, while sewer institutions followed the drainage basin contours. The federal government, which had financial capacity to make the investment, developed some large-scale facilities such as dams, while local public authorities developed many bridges, airports, and ports. Municipal governments built the streets and roads.

Who Provides Infrastructure?

Infrastructure is built and maintained by a combination of public and private institutions depending on the function, although by far the biggest players are those in the public sector. In 2004, public agencies accounted for 68 percent of capital expenditures for infrastructure, as contrasted with 32 percent by private agencies. Private expenditures for infrastructure dominate the telecommunications and energy area at 91 percent of the total, and play a smaller role in the provision of community facilities. Most public infrastructure is financed, built, and maintained at local and state levels, a pattern since World War II. From 1990 through 2007, over 80 percent of non-defense investment expenditures were made by state and local governments.[1]

This will be described further in Chapter 11 on the financial context for infrastructure.

Federal Infrastructure Institutions

The federal government provides funds to state and local governments for some infrastructure, and develops major infrastructure projects too large or too risky for states or the private sector. It also mandates certain requirements that influence local infrastructure. The federal government also has its own facilities. It operates dams, waterways, and other floor protection facilities under the perview of the Army Corps of Engineers, for example. Federal capital investments in facilities for the military, federal prisons, and parks, as well as federal office buildings for courts and other federal departments, are part of the federal infrastructure responsibility. For many years, the federal government did not combine these expenditures into a capital budget, nor was the impact of federal infrastructure investments considered as a whole on a geographic area.[2]

For local officials, the location of federal facilities can have both positive and negative impacts, although most are subject to local land use regulations. For example, siting large-scale postal facilities in suburban localities rather than in central cities may be more efficient operationally, but may negatively impact public access. The closing of military bases in the 1990s prompted furious lobbying on the part of state and local officials concerned about job losses. At the same time, the fact that the land which formerly housed the bases was freed up in major urban areas resulted in economic development opportunities that would not have otherwise been available.

The major federal programs for infrastructure funding at the state and local level are transportation, water, wastewater, and hazardous materials. Transportation is administered by the Department of Transportation (DOT) and state-level transportation agencies that usually look to local transportation agencies to plan and administer these programs in their jurisdictions. Enforcement of water and wastewater are regulations of the Environmental Protection Agency (EPA), are delegated to state environmental agencies. A small amount of federal funds for water and wastewater infrastructure is made available to states to help fund local projects.

Historically, the contribution of federal dollars for construction has always been much less than state and local dollars. Between 1956 and 1965, for example, the federal share of total public spending for transportation and water and sewer infrastructure rose from 17 percent to 33 percent, before falling to 29 percent in 1970. The federal share of dollars for these same programs rose to a peak of nearly 40 percent in 1977—mainly because of Congressional actions to allocate funds for wastewater treatment and mass transit programs. During the 1990s, the federal share was about 25 percent.[3] If locally financed schools, libraries, and civic buildings were included in these totals, the contribution of infrastructure funds provided by the feds would be lower still.

Federal government infrastructure funding has primarily been for capital facilities, not for operations and maintenance of the project. State and local governments are responsible even for facilities at the local level built with federal program funds—although not for facilities that house federal government programs, where the federal government operates and maintains the facility.

The influence of federal infrastructure funds on land use patterns in an area has often been disproportionately high compared to the amount spent on infrastructure by the locality. This is because most federal infrastructure funds are targeted for a particular purpose, and decisions have been made about the location of the facilities (highways and sewage treatment plants, for example) by state transportation agencies and local or regional special districts and authorities, rather than by the cities and counties.

The individual programs and specialty areas for infrastructure within the federal government often appear to be separate worlds. Each seems to have its own goals, technologies, special interest groups, types of professionals, training, and professional organizations. The individual infrastructure programs often have trouble relating in a more comprehensive way to each other and to other levels of the intergovernmental system, despite efforts over the years requiring coordination with state and local GPGs.

State Government Institutions for Infrastructure

As with the federal government, the states pass most infrastructure funds on to the next level of government, and are responsible only for providing infrastructure facilities for the programs they operate themselves. States administer the federal transportation and water and wastewater programs, which often means passing the funds on to the regional level. States are responsible for their own set of infrastructure programs, which usually include facilities for higher education, state highways, state jails and court buildings, state parks, and state office buildings. Some states are responsible for aspects of statewide water infrastructure systems and for regulating local and regional activities in their area. Including intergovernmental grants from the federal government, state-level institutions spent $130 billion on infrastructure in 2000. This was 31 percent of the total spent at the state and local level.[4] States are also responsible for setting up the regional and local planning system to address infrastructure needs.

State legislatures also create cities, counties, townships, school districts, special district governments, and government corporations through enabling legislation that grants these entities specific duties and powers.

Local-Level Governments

The bulk of infrastructure is planned, built, and maintained at the local level. General purpose governments such as cities and counties have the responsibility for planning, financing, developing, and maintaining civic infrastructure, such as buildings and facilities involved in running the local government. This can include city administration buildings, police and fire stations, libraries, parks, recreation centers, and fleet facilities. Streets, sidewalks, and the maintenance of sewer collection systems are among the most ubiquitous local infrastructure responsibilities. Special purpose districts (discussed further below) more frequently are involved in water, sewage treatment, ports and airports, energy, and telecommunications infrastructure, although in many large cities and counties these may also be line departments. Local GPGs are responsible for land use planning and development regulation (Table 4.1).

In 2007, the United States had 87,476 units of local government. Of these, 38,044 were general purpose local governments, while the others were

Table 4.1
Responsibilities of general purpose governments and other infrastructure service providers

Functions	Local general purpose government	Other infrastructure service providers
Infrastructure planning	Responsibility for planning for location, timing, and level of service for all infrastructure in jurisdiction	Planning for single system, with different boundaries, data systems, and GIS conventions
Infrastructure financing	Looked to by citizens as major taxing agency	Able to charge fees; also able to put bonds on property tax without GPG approval
Regulating development	Lead responsibility for seeing that land use plans are translated into development regulations, issues permits	State and regional agencies issue specialized permits
Developing infrastructure	Develops own infrastructure	Develops own infrastructure
Managing infrastructure	Manages own and dedicated infrastructure from private development	Manages own infrastructure

Source: David E. Dowall and Jan Whittington, *Making Room for the Future: Rebuilding California's Infrastructure* (San Francisco: Public Policy Institute of California, 2003).

Table 4.2
Number of government agencies in the United States in 1952 and 2002

	1952	2002	Percent change
Counties	3,052	3,033	−0.2
Municipalities	16,807	19,492	15.6
Towns and townships	17,202	16,519	−4.1
School districts	67,355	13,051	−79.9
Special districts	12,340	35,381	184.1
Total	116,756	87,476	−25.0

Source: U.S. Census Bureau, *2002 Census of Governments* (Washington, DC: U.S. Census Bureau, 2002).

special purpose governments, including 13,051 school districts. From 1952 to 2002, the number of special districts, many of which provide infrastructure services, tripled (Table 4.2).[5]

The consolidation of many small school districts from 1952 through 2002 is responsible for the 25 percent decrease in the total number of local government units, despite the dramatic growth in special districts.

Most infrastructure funds are raised and spent by local governments, and, despite the increased use of special districts over the past several decades to pay for many municipal services, the bulk of local infrastructure spending still originates with local GPGs, particularly with cities (municipalities, towns, and townships). In 2000–2001, slightly less than two-thirds of both capital spending and total infrastructure spending was done by city and county governments together (Table 4.3).[6]

Table 4.3
Local government infrastructure and capital expenditures in 2000–2001 by type of local government (in millions)

	County government	Municipalities, towns, townships	Special districts	School districts	Total
Local capital spending amount ($)	26,511	58,883	23,732	42,248	151,374
%	**18**	**39**	**16**	**28**	**100**
Local infrastructure spending* amount ($1)	45,085	136,065	66,768	42,248	290,166
%	**16**	**47**	**23**	**15**	**100**

Note: * Includes capital and operating expenses for environment systems, transportation, public energy, general public buildings, and parking facilities. Does not include libraries or housing and community development functions performed by special districts. Education, parks, hospitals, and corrections includes capital expenditures only.

Source: U.S. Census Bureau, *2002 Census of Governments* (Washington, DC: U.S. Census Bureau, 2002).

Regional Institutions

Within each state, economic and population centers form metropolitan areas. Often they are composed of several cities and counties. Some metropolitan areas have formal government bodies or agencies at the metropolitan level, known as councils of governments (COGs). These 450 regional agencies may plan for single or multiple sectors, and some may operate infrastructure programs. Many special districts that deliver water and sewerage services, or that operate airports or water ports, are regional in nature, although the majority are local.

Energy and Telecommunication Services

Other specialized infrastructure institutions include those that deliver energy and telecommunication services. Most are private corporations, with the rest being municipal energy and telecommunication utilities. Private utilities are national or subnational or regional in nature, while government suppliers are more substate or regional or local in nature. The institutional arrangements and history of the energy and telecommunications utilities are discussed in the individual chapters devoted to those infrastructure systems in this volume (Chapters 27 and 28).

Organizational Structure of the Local General Purpose Government

In 2007, 239 cities nationally had a population of over 100,000; overall there are 19,492 municipal governments and 16,519 towns and townships. Illinois, Pennsylvania, and Texas each have more than 1,000 municipalities, while eight states have fewer than 50. The number of municipal governments has increased by about 15 percent since 1952, while the number of towns declined (see Table 4.2). Most of the increase is a result of new incorporations due to population growth.[7]

Cities (that is, municipalities and some towns and townships) and counties generally have similar sets of programs, and a similar organizational structure. The major difference is that the county administers state programs such as the court system, jails, and health and welfare programs. They may also provide all functions for unincorporated areas within their boundaries. Some cities have their own jails, health, and welfare departments, but these are in the minority.

Although each jurisdiction will have its own unique organizational structure, the core departments responsible for planning and financing infrastructure are planning, public works, and finance. In smaller cities, infrastructure planning and services will probably be part of the public works departments. In larger cities, the various subdivisions of the planning and public works departments may be line departments on their own. Some cities and counties may have a separate capital projects department. The finance department is responsible for issuing bonds for infrastructure projects, and usually administers the budget for the city, although there may be a separate budget office reporting directly to the city or county manager or the mayor. These departments or functions may be known by different names from city to city and county to county. Some larger cities may have a separate budget office, and bonds may be issued by a treasurer's office. A typical organizational structure for a small and medium-sized city may resemble that presented in Figure 4.1.

Large cities and counties might have other enterprise departments, with responsibility for the local port, marina, airport, sports arena, concert hall, football stadium, and hockey arena, all of which have capital needs. Counties and some cities are responsible for public health, which may include running a mental or general health hospital. Some cities and counties are responsible for the infrastructure associated with the criminal justice system, including courts, jails, and the district attorney and public defender's offices. All of these functions include specialized capital infrastructure needs. Some large cities and counties have other governing structures. For example, in Portland, Oregon, each of the city council members has line responsibility for a city department, or bureau. In the City and County of Los Angeles, and in some cities, the police department is governed by a police commission, whose members are appointed by the mayor.

Local GPG functions that are important in the provision of infrastructure are discussed in greater detail below.

Planning Department

The planning department has two major functions and is usually organized accordingly. These two functions are: regulating development projects (zoning

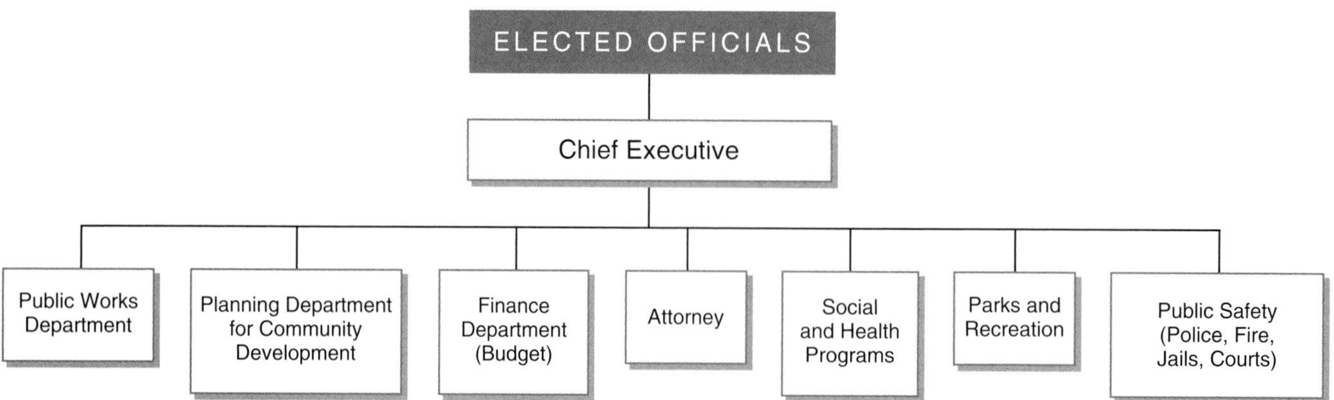

Figure 4.1
Organizational structure of a small or medium-sized city or county

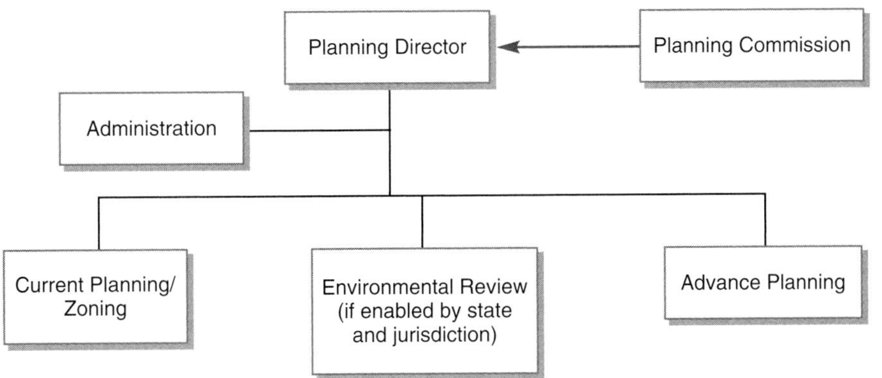

Figure 4.2
Organizational structure of the planning department

or current planning), and preparing the general plan (advance planning). If the state or local government requires an environmental review function, this may be a separate section, or a subsection of the current planning division. The planning department may also be responsible for issuing building permits, but in some localities building permits are the responsibility of the public works department or a separate department. Generally the transportation planner works in the planning department, but this function may also be found in the public works department where the traffic engineer is normally located. The transportation planning and traffic engineering functions can also be consolidated into a separate transportation department (Figure 4.2).

Most planning departments have a planning commission that consists of citizens appointed by the mayor or the elected body of the city or county. The planning commission's responsibility is to provide citizen input to planning decisions, and to provide a forum for hearings and public workshops.[8]

Sometimes the land use planning functions are combined with housing, community development, and redevelopment (sometimes called renewal), in which case the department might be known as the community planning and development department. These other functions may have their own department(s). The community planning and development department may also include responsibility for economic development. Some housing programs are operated by a separate housing authority with an independent board. Alternatively, the housing authority can be a subordinate department or division in a city or county. In some cities and counties, the economic development function may report directly to the mayor or city or county manager.

The redevelopment authority or agency, sometimes also known as an urban renewal agency, is a separate legal entity set up to streamline land acquisition and finance development in blighted areas of the locality. In large jurisdictions redevelopment and urban renewal activities are managed by a separate unit of government, although its governing body is appointed by the chief elected official or elected body of the GPG. In smaller cities and counties, even if the redevelopment agency staff is part of the planning and community development department and the elected officials for the GPG sit on the redevelopment board, it remains a separate legal entity. If redevelopment authorities use federal funds, they are required to have an oversight citizen body with membership from the area.

Public Works Department

In small and medium-sized jurisdictions, the public works department is responsible for engineering, or capital project development of streets, sidewalks, transportation, and facilities, as well as sanitation, solid waste, and water if a special utility district or private utility does not provide these services. The public works department is frequently divided into an operations division and an engineering division which is responsible for the locality's own capital projects (Figure 4.3). In larger cities and counties, many of these functions may be a separate department or bureau with their own engineers and operating staff—and where it is an enterprise function, billing and collections staff as well. Most cities and counties do not have a citizens commission for public works functions, although that is changing as citizens realize the importance of public works, particularly with respect to climate change and sustainability issues. Some jurisdictions will also have a separate capital projects department responsible for nonroutine types of projects such as large buildings or large-scale transportation facilities.

Most of the traditional infrastructure responsibilities are planned and executed by the public works department. The engineering division designs infrastructure facilities owned and maintained by the locality, or is responsible for contracting the design functions out to private consulting firms. The operations

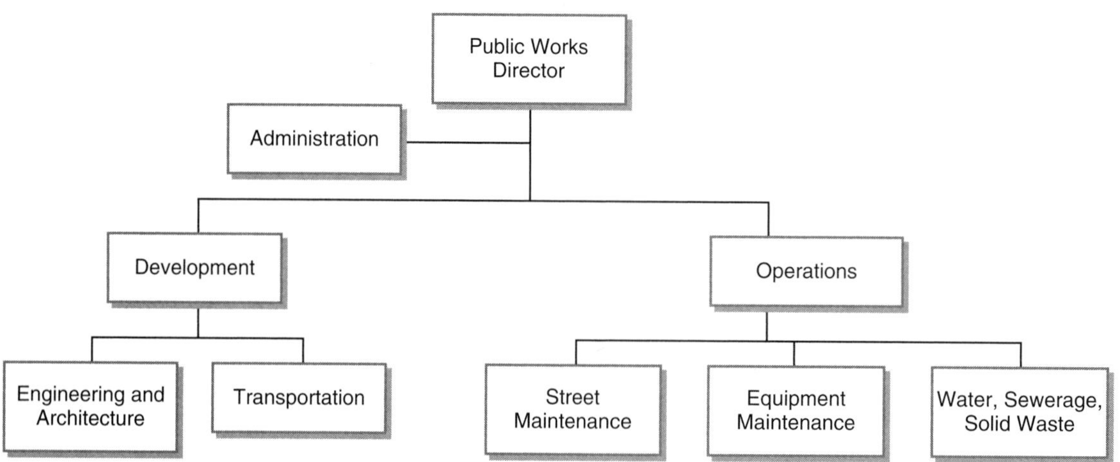

Figure 4.3
Organizational structure of a typical public works department in a small or medium-sized city or county

division usually is responsible for maintaining sanitary and storm sewers, signs, streets and street lights, sidewalks and curbs, and civic buildings. It may be physically located at a separate facility known in many localities as the "Corporation Yard."

Some cities and counties have a separate parks and recreation department. Others place the maintenance portion of parks in the public works department along with the landscape architects who do the parks design work. Some parks departments are separate special regional or area districts.

Special Purpose Districts

Although, as noted above, special purpose districts do not account for the bulk of local infrastructure expenditures, they have been used in the past several decades in high growth areas as a means to quickly build needed infrastructure. In addition, special purpose districts have a long tradition in older areas of being used for "mega-projects" in order to bypass cumbersome local government procedures. Special districts are independent special purpose government units that exist as legal entities separate from the local GPGs. The governing board is popularly elected or appointed by elected officials, and the agency may have the power to levy taxes. A special district has "substantial autonomy"—that is, it is able to determine its own budget and issue bonds without review by another government body.

Functions of Special Districts

All else being equal, states that have passed enabling legislation for special districts have more of them than do other states. Special districts are most prevalent for civic facilities, water, and sanitation infrastructure. Many metropolitan special districts were established to build and operate water and sewer facilities funded by environmental legislation in the 1960s and 1970s.[9] The bulk of special civic facilities[10] districts are for fire services. However, they account for a relatively small amount of the resources. Transportation special districts, small in number, wield large amounts of capital and operating funds, as do the public energy providers (Figure 4.4).

In 2007, over 90 percent of special districts were responsible for a single infrastructure function, although their charter may permit them to take on other functions. Multifunction special districts usually provide water services in connection with fire, sewerage, or other functions. Special districts inside metropolitan areas vary somewhat from this pattern, with more fire, water, sewerage, parks and recreation, and utilities districts, but fewer natural resource, hospital, library, and airport districts.[11]

Although there is a perception that special districts are regional in nature, in 2007, only 20 percent covered two or more counties. The remainder were

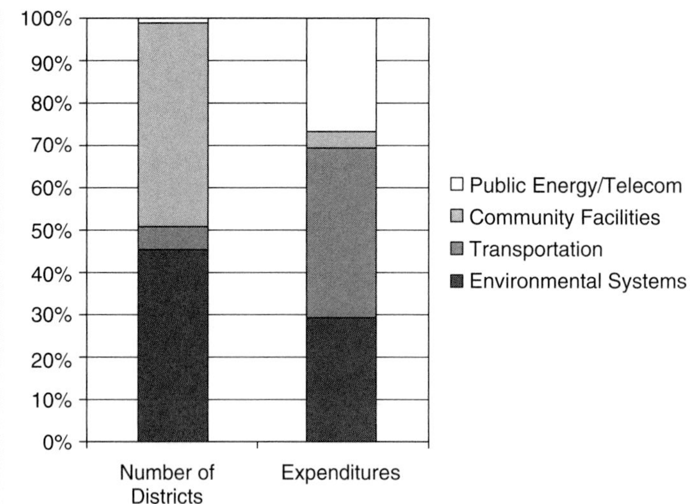

Figure 4.4 **Percentages of special districts by number and by expenditures in FY 2000–2001 for major infrastructure areas**

coterminous with a county (17 percent), a city (13 percent), or were within a county but the boundaries were not the same as the county or any municipality (50 percent).

Special District Characteristics[12]

Special districts have both a public and private character, and as such, some of them are viewed as "government corporations" or government businesses. They have many characteristics of public bodies, including exemption from property taxes, the ability to issue tax-exempt bonds, and the right to enter into contracts and own land. Many have the right of eminent domain to acquire property for capital projects. Nearly all special districts with the power to raise revenue by taxing have elected bodies, but they may also employ user charges and fees. Nontaxing authorities, those that rely primarily on user fees, usually have boards that are appointed by the chief executive or elected board of the parent body. Contrary to the lore about special districts, they are not always self-sufficient financially. In 2007, the majority of the special districts received grants or aid from the federal government and other sources.

Special districts have many characteristics of private businesses. They can be exempt from civil service, government contracting, and pension fund regulations. They can set their own personnel practices, and are often exempt from state and local ceilings on taxing and borrowing. The most important characteristic of a special district for those who create them is the independence from local government and the need to have its budget reviewed and freedom from competing with a broader range of programs for its budget.[13]

Special Assessment Districts

Some states provide for special tax districts that are subordinate units of the city or county to fund infrastructure and other services. These districts can cover the entire jurisdiction or only a portion. These are sometimes called special assessment districts, and can be used both for repaying off bond indebtedness and operating and maintenance costs. These kinds of districts are not included in the special district counts in this chapter since they are not autonomous, independent agencies.

School Districts

School districts are another local government entity. They are usually operated by a locally elected board, and they rely upon the same funding source as the local GPG—the property tax. With respect to infrastructure, school districts are responsible for the construction and maintenance of school buildings. However, most school districts and school boards see as their primary responsibility the delivery of services rather than infrastructure. Consolidation of rural school districts resulted in a substantial decrease in the number of school districts from 1942 through 1972. Since then, school districts have continued to consolidate, but at a much slower rate. However, current educational theory is advocating the effectiveness of neighborhood schools, so perhaps this trend may see a slowing or reversal over the coming years. See Chapter 25 for the relationship of schools and infrastructure planning.

Distinction between Special Districts and Public Authorities

The U.S. Census does not distinguish between special districts and public authorities, although some find the distinction useful.[14] Although exceptions exist, some differences include the following:

Public Authorities are usually formed to take on large, risky, and long-term projects that could be self-sufficient financially. Many were intended to be insulated from elected officials and the normal rules of government. They usually have appointed boards and enjoy more flexibility regarding personnel management (paying higher wages, for example) and contract decisions than the local GPG. Examples include the Palm Beach County Transportation Authority (county), the Delaware River and Bay Authority (bistate), the Maryland Stadium Authority, and the Fargo Municipal Airport Authority. Authorities have been responsible for some of the most complicated and large-scale infrastructure projects in the U.S.[15]

Special Districts (nonauthorities) usually have elected governing boards and may have the authority to tax. Some refer to these as taxing districts. The archetype for this kind of agency is the pollution control district or the fire

Evolution of Special Districts and Public Authorities

Some of the first special districts in the United States can be found in the 1600s, when early settlers provided charters to both private and public corporations to construct canals, roads, bridges, and harbors that would benefit commercial interests. These autonomous institutions were financed by public and private funds.

During the eighteenth and nineteenth centuries, special purpose governments were created to finance and to consolidate the delivery of some regional services. The first instance of this occurred with the City of Philadelphia and its suburbs when special boards were established beginning in 1790 to consolidate prison inspectors, poor relief, health, police, port wardens, and education. These regional special purpose boards delivered services and levied taxes independently from their parent cities. Other localities in New York, Boston, and Chicago followed suit for sanitation, police, parks, and water. At the same time, small area special assessment districts at a local level were used for services such as street lighting and road construction. These districts gained popularity because they were seen as a fair way to tax only those who would benefit from the service.[1]

Special districts were also used to circumvent financing restrictions placed on municipal governments during the end of the nineteenth century. During the Depression era of the 1930s, federal programs were also created to buy bonds from special municipal agencies, or "public authorities" as they were known, to build infrastructure to stimulate the economy. The Port of New York Authority, the Bonneville Power Authority, and the Tennessee Valley Authority were set up as part of these initiatives. By 1946, all but seven states had set up enabling legislation for public authorities to sell revenue bonds, and 25 of the states permitted municipalities to create public authorities by ordinance. Courts paved the way for the proliferation of authorities by ruling that their revenue bonds did not "count" against debt ceilings.[2]

The Depression-era public authorities, really quasi-public corporations, were intended to go out of business once the original debt was retired and their capital inventory turned over to the state or local GPG that authorized them. However, their effectiveness and ostensible removal from the political process made them ideal tools for funding the construction and operations of major infrastructure such as ports, bridges, and tunnels. Therefore, instead of being retired when the Depression was over, hundreds more were created in the post-WWII era of growth.

Notes

1 Kathryn A. Foster, *The Political Economy of Special-Purpose Government* (Washington, DC: Georgetown University Press, 1997).

2 David Perry, "Building the City through the Back Door: The Politics of Debt, Law, and Public Infrastructure," in *Building the Public City: The Politics, Governance and Finance of Public Infrastructure*, ed. David Perry (Thousand Oaks, CA: Sage Publications, 1995).

district. Many view these kinds of special districts as less entrepreneurial and less capable of taking risks than the public authority.[16]

Issues with Special Purpose Districts

The "success" of the special purpose districts has been their ability to bypass some of the restrictions and public scrutiny of the GPG. The primary financial relationship of special districts is with the bondholders and their users. Proponents see special districts as a way of building needed infrastructure and providing other government services such as water, sewerage, and fire prevention and suppression without relying on local taxes and the politics associated with local government.

Critics of special districts view these agencies as a way of circumventing the democratic process, and some even point to the lack of full disclosure about budgets. Special districts are not policy-neutral. Metropolitan areas with many special districts allocate greater shares of their resources to "development and housekeeping" functions than to social service functions. In addition, special districts generally are not more efficient in cost terms than their general government counterparts. The per capita cost of service delivery is higher for special districts than for GPGs.[17]

It is frequently difficult for citizens or elected officials to know how much money is being invested in infrastructure each year for their jurisdiction by special districts, or what is planned for the future. Citizen interest in special districts is minimal and difficult to spark. Special purpose districts, even those with elected boards, do not command the same type of public interest as do GPGs, despite the scale of the funds involved. Public meetings of special districts are sparsely attended, and their financial statements frequently are confusing. In most states no formal way exists for the GPG to change the plans of special districts if they disagree.

From a land use planning point of view, special districts are a two-edged sword. They can provide needed services quickly in the face of growth pressures, but once established, they make it more difficult for the GPG to make coordinated capital investment decisions.

Regional Infrastructure Institutions

Councils of Governments and MPOs

Rapid growth stimulated interest in the development of regional infrastructure and land use planning at three different times in the past century: the 1920s, the post-WWII boom, and the hot 1990s. As automobiles and a booming economy made possible the development of suburbs in the 1920s, businessmen, citizens, planners, and architects were concerned that the regions be developed as a whole. About 15 regional plans were developed in this era, with most being advisory. The New York Regional Plan Association, a private organization, developed one of the most famous regional plans for 5,000 square miles spanning three states: New York, New Jersey, and Connecticut.

In the late 1950s and early 1960s—a period of rapid suburban growth—there was a surge of interest in regionalism surfacing at both the local and federal levels. Five areas, including Portland, Oregon, and Minneapolis-St. Paul, took steps toward reorganizing their local governments to provide cross-functional policy direction.[18]

During the same period, the federal government provided financial incentives for localities to form voluntary councils of governments (COGs) and paid for some regional land use studies. By the year 2000, there were approximately 450 COGs in the U.S. Not all local governments belong to an area's COG, but at least half and usually more do. Some governments may belong to two or more COGs. Local elected officials are usually the majority on the COG boards. With the exception of Portland Metro and Minneapolis, plans implemented by COGs do not have authority over local infrastructure investments aside from federal transportation grants.[19]

Renewed Interest in Regional Efforts in the 1990s and 2000s

The past two decades have again seen renewed activity and interest in regional planning and regional infrastructure efforts. The federal transportation program required areas with populations of over 50,000 to establish a regional agency called a Metropolitan Planning Organization (MPO) with authority to plan and allocate federal transportation funds. The composition of the MPO is determined by the state transportation department. In many states, the COG became the MPO, but in other states they are separate.

Environmental pollution resulted in EPA funding for watershed councils, which are regional in nature. These groups of interested stakeholders use collaborative techniques to solve problems such as rising temperatures in rivers that negatively affect the wildlife, aquatic life, and the rivers and streams themselves.[20]

The global economy also caused policy-makers in the component jurisdictions of many regions to begin thinking of themselves as an integrated metropolitan area[21] and to demand regional responses.[22] Two types of regionally oriented responses have occurred. The first is the restructuring of state planning statutes governing local planning, some of which also created regional planning bodies (as in Florida) or regional land use appeals boards (as in Washington State). Second, many voluntary regional efforts have been organized with varying degrees of success.[23]

In 2008, California enacted policies and statutes to reduce carbon emissions by coordinating transportation infrastructure decisions with land use planning at the regional level using COGs and the MPOs. A recent study questions whether it will be possible to reinforce climate policy with inherently weak regional institutions.[24] Other regional planning efforts have started to emerge around the country, using scenario planning to coordinate growth for a metropolitan area with a focus usually on transportation solutions, and, more recently, on water. Envision Utah, the Blueprint planning efforts in California, and Maryland are particularly noteworthy efforts.[25]

Beyond metropolitan regions, other groups are looking at subnational regions or "megaregions" to analyse infrastructure needs. Megaregions are "networks of metropolitan areas connected by overlapping commuting patterns, business travel, industrial value chains, transportation infrastructure, natural systems and shared historical and cultural characteristics," according to the director of *America 2050*. This organization is pressing for a national strategic investment framework for roads, bridges, transit systems, the energy grid, water infrastructure, and telecommunications. *America 2050* has identified 11 emerging megaregions where over 75 percent of the nation's population will reside by 2050, and where critical infrastructure decisions must be made[26] (Figure 4.5).

Conclusion

The current intergovernmental system, with its multiple and often isolated centers for making infrastructure investment decisions, has been roundly

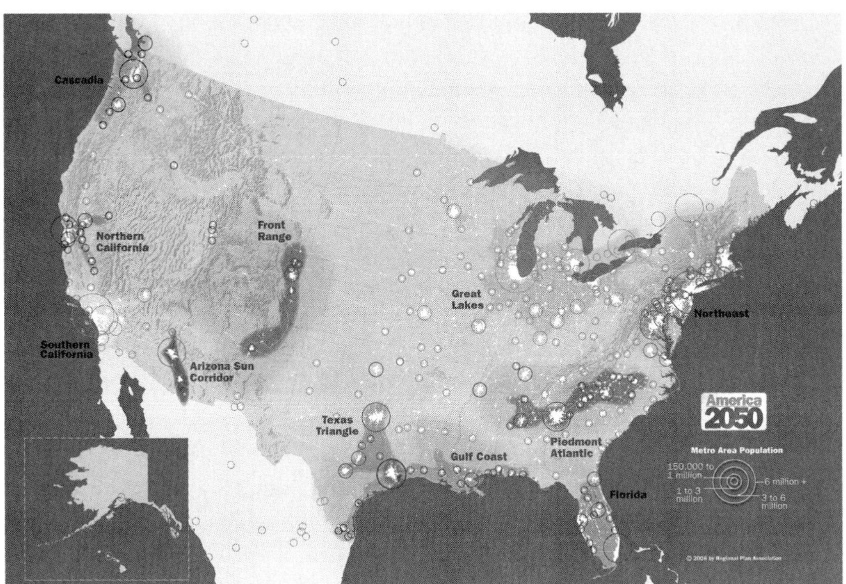

Figure 4.5
Megaregions in the United States

criticized for its lack of coordination across disciplines, across functions and across governments within a region, or even within a single jurisdiction, or a single agency. Some advocate change in government structure, perhaps to home rule for metropolitan regions, or for changes to the regulatory environment at the state and federal levels to encourage long-term cross-functional infrastructure planning at the metropolitan level.

Other thoughtful experts are exploring the "soft approach" to dealing with this conundrum. They urge networking and collaborative solutions as well as reliance upon non-profit organizations to provide the glue for coordinated infrastructure decision-making when governments become too fragmented. At the individual level, the local practitioner can fall back on the locality's own capital improvement budget, and begin to work outward to create a "virtual" planning institution to begin the slow process of working with the patchwork function of infrastructure agencies at the local level.

Additional Resources

Publications

Dodge, William R. *Regional Excellence: Governing Together to Compete Globally and to Flourish Locally*. Washington, DC: National League of Cities, 1996. A practical book on strategies to use in our complex jurisdictional world.

Grigg, Neil S. *Infrastructure Finance: The Business of Infrastructure for a Sustainable Future.* Hoboken, NJ: John Wiley and Sons, 2010. Although written for engineers and others involved in the private sector for large-scale infrastructure projects, this book contains information from a business perspective on institutional arrangements by infrastructure sector not normally found in guides for cities.

Innes, Judith and David Booher. *Planning with Complexity: An Introduction to Collaborative Rationality for Public Policy.* London: Routledge, 2010. A critique of current governmental structures and processes. This volume also provides guidance for dealing with overlapping and multiple jurisdictions through networking and giving voice to non-technical standards.

Selzer, Ethan and Armando Carbonell, eds. *Regional Planning in America: Practice and Prospect.* Cambridge, MA: Lincoln Institute of Land Policy, 2011. A collection of thoughtful chapters on the current state of regional efforts planning infrastructure and land use. Includes case studies as well as an essay on governance institutions that goes into more detail about the patchwork problems of today's local institutions.

U.S. Census Bureau. *Census of Governments.* Vol. 1, No. 1, *Government Organization.* Washington, DC: U.S. Census Bureau, 2007. The tables in this report can provide local practitioners with listing of types of agencies in their jurisdiction. The website also has a handy interactive device to obtain specific information for a particular locality about all its agencies.

PART TWO

Planning and Developing Infrastructure

Chapter 5 (Local Plans and Infrastructure) looks at a variety of plans being prepared today at the local level where sustainable infrastructure can be addressed. Strategic planning and indicator systems are touched on.

Chapter 6 (The Infrastructure Program and its Preparation) describes how to prepare an infrastructure element for the comprehensive plan or a capital needs study to be used as the background to the capital budget.

Chapter 7 (Smart and Sustainable Development Rules) describes the origin of the subdivision ordinance which is the major set of regulations prescribing local-level infrastructure.

Chapter 8 (Developing the Public Infrastructure Project) goes through the steps involved in developing and building an infrastructure project at the local government level.

Chapter 9 (Capital Improvements Plan and the Capital Budget) addresses how infrastructure interfaces with the local budget process. The various budget documents used at the local level for infrastructure are described, including the capital improvements plan (CIP), the capital needs study, and the capital budget.

The final chapter in this part of the book, Chapter 10 (Managing Infrastructure), describes concerns about lack of maintenance of public infrastructure systems. Asset management systems also are discussed.

chapter 5

Local Plans and Infrastructure

In this chapter...

Importance to Local Practitioners . . . 71
Sustainability Planning . . . 72
Origins of Concerns about Sustainability . . . 72
Local Sustainability Plans . . . 72
What Is the Comprehensive or General Plan? . . . 73
Influences on the Development of the Comprehensive Plan . . . 75
The Model Ordinance from the 1920s . . . 75
Environment, Infrastructure, and the Comprehensive Plan . . . 75
Federal Grant Requirements . . . 75
Comprehensive Planning Reform Efforts . . . 76
Status of Local Comprehensive Land Use Planning Requirements Today . . . 76
State Planning Requirements for Local Infrastructure Planning . . . 76
Status of Infrastructure and Capital Planning at the State Level . . . 78
Smart Planning and Sustainability Planning . . . 78
Infrastructure and Strategic Planning . . . 79
Managing for Results . . . 79
Strategic Planning . . . 79
Strategic Planning and the Capital Improvement Plan . . . 80
Climate Change Planning/ Sustainability Planning . . . 81
Adaptation Planning . . . 83
Conclusion . . . 84
Additional Resources . . . 85
Publications . . . 85
Websites . . . 85

Why Is Infrastructure Planning Important to Local Practitioners?

Infrastructure planning is at the center of the two major types of plans that local governments prepare: land use plans and strategic plans. This is due to the interest in sustainability planning as well as climate change plans that are prepared to adapt to the impacts of global warming on the built environment. Infrastructure planning can be a vehicle for managing carbon emissions, energy, and water use as well as toxic and solid wastes. Planning for infrastructure can be a way of assessing and controlling the impact of major investment decisions on the location and density of growth, and on the ecological footprint of the jurisdiction. Planning for infrastructure can also identify life cycle costs that will help prevent the catastrophic failure of older systems.

In the past several years, sustainability issues have dominated the discussions of local government planning. Accordingly, this chapter begins with a brief history of sustainable development before turning to the *comprehensive plan* which is the long-term development plan for the local jurisdiction. The chapter then addresses the strategic plan, which is a short-term plan for the government that guides yearly expenditures and activities. The strategic plan is also the generic category for several types of plans where infrastructure planning is essential: the climate change plan, the emergency preparedness and the adaptation plan. Finally, the chapter talks about the capital needs study and the infrastructure program. This is the nuts and bolts plan for infrastructure which can be a part of the comprehensive program or a strategic plan.

Sustainability Planning

Origins of Concerns about Sustainability

The issues of climate change and going green have renewed interest in sustainable development—a topic that has its roots deep in the past when the Industrial Revolution began to change the urban landscape. Writers of the late nineteenth century were concerned both about the encroachment of industry on nature as well as the poor physical conditions of the inhabitants of the industrial city. The optimism of the first half of the twentieth century quieted many of these concerns, but by the 1960s serious doubts were voiced about the viability of an economic system that relied on ever increasing consumption and non-stop development. The *Limits to Growth* report in 1972 concluded that population and consumption trends would result in the current system crashing in the mid-twenty-first century. Drawing on that, the *Blueprint for Survival* report called for the creation "of a society that is sustainable." In the late 1970s, the World Watch Institute began publishing a series of papers that were highly influential, including the *State of the World* report which is still published annually."[1]

In 1987, the World Commission on Environment and Development, chaired by the Norwegian Prime Minister Brundtland, catalyzed both national and local sustainability concerns. This commission produced a report (reprinted as *Our Common Future*) that contained what is currently the standard definition of sustainable development—"development that meets the needs of the present without jeopardizing the ability of future generations to meet their own needs." In 1992, the United Nation's Rio Earth Summit formally defined sustainable development as that which is "economically efficient, socially equitable, and responsible and environmentally sound." Agenda 21 was the summit's action plan, which contained goals for the sustainable use of land, sustainable energy and transportation systems, and improved human resource development. Agenda 21 encouraged local governments to establish their own plans.[2] The International Council for Local Environmental Initiatives (ICLEI) was formed in 1990 and launched the Cities for Climate Protection campaign to help local governments reduce greenhouse gas emissions and to implement Agenda 21.

Local Sustainability Plans

Local sustainability efforts grew out of the cultural and environmental revolution of the late 1960s and early 1970s. At this time many experimental communities were formed with the goal of reducing their ecological footprint and getting "off the grid." Many of these communities also experimented with new social relationships. A series of highly publicized oil spills in the 1970s and 1980s added to the concern about the sustainability of U.S. consumerism. Some progressive communities initiated sustainability efforts, including local controls on solid waste, toxics, and energy use that were later adopted by some states, and some measures (such as recycling) were incorporated into federal standards.

In the 1990s, a small number of local governments in the United States began to prepare sustainability plans, or otherwise began to adopt aggressive "green" measures. Despite high hopes, many of these plans lacked systematic implementation measures for key systems such as transportation, brownfield revitalization and biodiversity, although leading edge cities were exceptions.[3] More widespread was concern about urban sprawl and how to pay for local infrastructure needed to support the rapid growth of that decade.[4]

Since 2005, when the international Kyoto Agreement on global warming took effect, without the participation of the United States, a veritable cornucopia of state and local actions have taken place to promote local and regional sustainability planning. At that time local mayors banded together to form the Mayors Climate Protection Agreement, which was endorsed by the U.S. Conference of Mayors in that same year. Since then, the "Cool Mayors" have been working with ICLEI, now called "ICLEI—Local Governments for Sustainability," to promote local sustainable programs and published a guide for local communities to sustainability planning.[5] Other state and local efforts are too numerous to mention, save that the trend is towards considering the impact of land use decisions and infrastructure such as transportation and water, on carbon emissions. There is also impetus towards coordinating these decisions between different levels of government.

What Is the Comprehensive or General Plan?

One local tool that can be used for sustainable cross-functional long-range infrastructure planning is the comprehensive or general plan. It is particularly useful for areas experiencing rapid growth. A comprehensive plan is the legal document that lays out the long-term vision for the use of land within the locality's jurisdiction. This plan is often referred to as a "comprehensive plan," although some jurisdictions use the term "master plan" or "general plan." An influential planner in the 1960s, T. J. Kent, Jr., defined the general plan as:

> The official statement of a municipal legislative body which sets forth its major policies concerning desirable future physical development; the published general-plan document must include a single, unified general physical design for the community, and it must attempt to clarify the relationships between physical-development policies and social and economic goals.[6]

This definition still holds today.

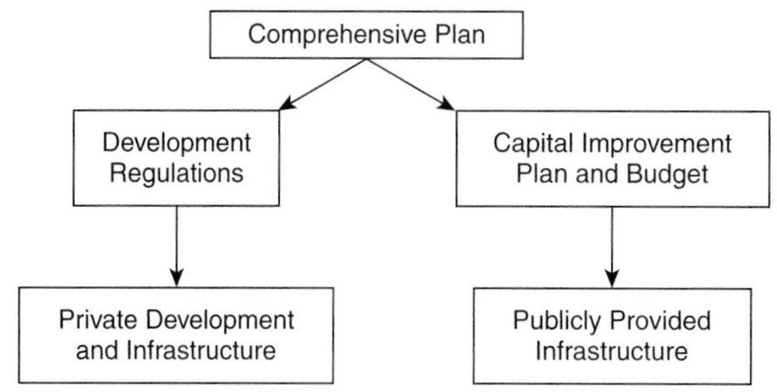

Figure 5.1
Ideal relationship of comprehensive plan to implementation tools

The comprehensive plan is implemented by two sets of documents: development regulations and the capital improvement plan and budget (Figure 5.1). Development regulations translate the land use designations of the general plan into more specific ordinances in order to regulate private market development projects as they are processed and given permits by the locality. They also can provide requirements for the kind and amounts of infrastructure that the private developer is expected to provide for the project.

Preparation of the comprehensive plan is the sole responsibility of the multipurpose or general purpose government (GPG). The comprehensive plan is a document that sets out, at a minimum, the long-range physical plan for the entire jurisdiction. Although other agencies and private interests may significantly influence how development occurs, the GPG is the only government body that has the power to set out the vision of the physical future of the jurisdiction. This is a power entrusted to it by the state, which sets up the legal framework within which the comprehensive plan can be prepared. Typically, state legislation authorizes the GPG to prepare a general plan. In some cases, the states mandate the certain elements, such as land use and transportation. Some states have planning guidelines that may be either prescriptive or advisory. However, most state statutes make local planning optional. As a consequence, most localities do not have comprehensive plans, and if they do, they have not been updated for years.

In this permissive environment, land-use related planning for infrastructure is generally not required, although it is a popular non-mandated element. Aside from the transportation element, those states that do mandate planning do not require the comprehensive plan to have a specific section dealing with infrastructure. Some states provide guidance for an optional public facilities or capital facilities element, or for specific elements such as water, energy, parks, or community facilities. Indeed, most state planning agencies might not agree on the definition of infrastructure used in this book.

Influences on the Development of the Comprehensive Plan

The Model Ordinance from the 1920s

Most state and local planning land use regulations authorizing comprehensive plans originated in two model ordinances developed during the 1920s by the federal government. These ordinances were intended to authorize local comprehensive plans for land use and zoning, as well as infrastructure plans such as transportation and public facilities. However, when the legislation was finally enacted, it confused zoning with land use planning; further, the legislation stated that zoning, the comprehensive plan, and the public facilities (i.e., infrastructure) plan did not need to be adopted at the same time. As a result, most localities prepared zoning ordinances only using the model statute for zoning (the Standard State Zoning Enabling Act—SZEA) as a guide. In many areas, therefore, there is no comprehensive planning tradition for infrastructure. However, all 50 states adopted the legislation delegating land use authority to local general purpose governments. This system worked well for the largely decentralized provision of infrastructure and land use planning until the 1960s.

Environment, Infrastructure, and the Comprehensive Plan

The 1960s and 1970s saw an upsurge of interest in "comprehensive" planning, much of which was spurred by the increased interest in environmental preservation. Some statewide planning programs were enacted that required localities to plan for infrastructure and to ensure that it was available when issuing permits for individual projects. Approximately 20 states enacted environmentally oriented land use laws, and 37 states adopted regional planning programs. In addition, both Florida (in 1972) and Oregon (in 1973) adopted statewide planning programs that required local governments to plan for infrastructure, while Washington and Georgia attempted to do so. Despite these mandates, however, many local governments did not prepare the required plans.[7]

Federal Grant Requirements

The federal government also required local project plans in order for the local governments to be eligible for infrastructure grants for highways, urban redevelopment, and water and sewerage. Ultimately these plans were required to be "comprehensive," that is, to address all the needs for that particular system within the metropolitan area or region. However, these requirements did not consider land use impact, nor were they cross-functional. One analyst notes that federal planning requirements "codified a new form of fragmentation in the planning system" that she calls "vertical regionalism."[8] For example, federal regulations of transportation require the development of comprehensive transportation plans as a condition of ISTEA grant approval.

Comprehensive Planning Reform Efforts

Status of Local Comprehensive Land Use Planning Requirements Today

In 2002, comprehensive planning was required in only 15 states. A publication by the American Planning Association (APA) reported that only 41 percent of general purpose governments in metropolitan areas and 30 percent of those in rural areas had comprehensive plans.[9] In many states that require local comprehensive or general plans, there is no system to ensure that they have been done or contain provisions for infrastructure. For example, in California, a state with mandatory planning, a study done in 2003 found that the general plans of many localities were no longer current.[10] Florida reviews the local element at the state level, but has been considering reducing the state planning staff.

The APA concluded that many state planning statutes "are in dire need of modernization" because almost half of the states still have 1920s vintage state laws on local planning, and most allow local governments to ignore local planning provisions.[11] The APA also noted that the most modernized planning laws are found in the coastal states, which are also the most urbanized. Those states with conditional or no mandates for planning are in the heartland. These states are also those with statutes that have not been updated significantly since their initial adoption in the 1920s or 1930s.[12]

Reform proposals to restructure the local planning process abound. In 1999, about 1,000 state land use reform bills were introduced at the state level, but only 200 were enacted. Reforms fall into three categories: (1) tightening of existing laws and procedures; (2) authorization for innovative land-use controls; and (3) major changes in the overall structure and content of land use statutes. The most important reforms were caused in large part by strong leadership from the governor or legislators, as well as by strong grassroots support.[13] Many high growth states enacted legislation to foster consistency between land use and infrastructure agencies at the local level (Table 5.1).[14]

State Planning Requirements for Local Infrastructure Planning

Currently, many states have some provision for including infrastructure planning in the comprehensive plan, calling for transportation and community facilities (also called capital or public facilities) elements. Oregon, California,

Table 5.1
Significant state planning statute reforms

Year	State	Description/Summary
1985	Florida	Requires all local governments to develop and implement a comprehensive plan on which future land use decisions would be based. These are reviewed by the state. A state report issued in 2001, however, has proposed delegating this responsibility to local governments.
1985	New Jersey	Created the State Planning Commission and Office of State Planning that in 1992 adopted a plan to guide public and private development toward compact, mixed-use projects that make the most efficient use of existing and planned infrastructure. This plan was followed up by a second in 2001.
1988	Maine	Established 10 goals for addressing growth and initially required all local governments to prepare comprehensive plans.
1988	Vermont	Established a system for coordinating land-use planning at the municipal, regional, and state levels.
1989	Georgia	Adopted minimum standards and procedures for local comprehensive planning to help ensure a comprehensive, integrated, and coordinated planning process at the local, regional, and state levels.
1990	Washington	Requires city and county governments to adopt 20-year comprehensive plans that address land uses, housing, capital facilities, transportation, and utilities, to help meet such state goals as reducing urban sprawl and retaining open space and habitat areas. State funding for intergovernmental planning efforts.
1991	Kansas	Established procedural changes for many planning and zoning actions and authorized the use of various new tools and techniques. There is no state role in growth management, and local plans are voluntary.
1992	Maryland	Requires cities and counties to adopt comprehensive plans with certain elements that help fulfill such objectives as concentrating development and protecting natural resources. Also encourages localities to adopt an interjurisdictional element (ICE) in land use plans.
1997	Maryland	Directs new development to municipalities, enterprise zones, and other locally designated priority funding areas by limiting state support for projects to these areas.
1998	Tennessee	Requires local governments to establish growth boundaries and includes penalties for those that do not comply. Municipalities and counties with no metropolitan form of government are required to develop 20-year joint growth plans.
1999	Wisconsin	Establishes consistent and uniform comprehensive planning requirements to, among other things, protect farmland and open space and preserve certain communities. Provides funding to local governments for comprehensive and transportation planning.
2000	California	Requires agency responsible for approving annexations to conduct a service area review (SAR) to ensure that the boundaries of infrastructure providers are consistent and that the most appropriate agency is providing the infrastructure in case of conflicts within the area.
2008	California	SB 375 targets greenhouse gas emissions through land use and transportation planning at the regional level.

Sources: U.S. General Accounting Office, *Local Growth Issues* (Washington, DC: U.S. General Accounting Office, 2000); and Richard T. LeGates, *The Region Is the Frontier* (California State University, Sacramento, 2000), available at: www.csus.edu/calst/government_Affairs/Region_is_the_Frontier.html.

Rhode Island, and Washington require a single transportation element. Some states, such as Florida, require several individual transportation elements including circulation, mass transit, airports, and water ports. Other states, including California, provide that, if a capital facilities element is done, it must be consistent with the general plan.[15] Conspicuously absent from land use planning requirements in many states are provisions for water and sewerage planning at the local level, despite the fact that these are a major determinant of the location of growth and for many localities those facilities will need to be "adapted" to account for declining snow packs, rising ocean levels and the increased incidence of natural disasters due to climate change.

Status of Infrastructure and Capital Planning at the State Level

Infrastructure planning at the state level is uneven. A few states have outstanding policies for their own capital planning and investment, while others have some measure of a capital plan. Most state-level capital plans are not truly cross-functional nor do they target capital expenditures geographically or strategically.[16]

The lack of integrated capital planning by the state, even without the more controversial siting issues, makes it difficult for local governments to forecast the siting of major state facilities, such as jails or university expansions that might have significant impacts upon local land uses. Some states also have infrastructure funds for economic development that are allocated without an overall strategy for where the state wants growth to occur. In some cases, private development is the force driving state investment decisions, rather than the reverse. The result can be a mismatch between state and local infrastructure financing. As the APA *Smart Growth* document indicates, "the timing, location and intensity of development [at the local level] … need to be linked with the existing and planned infrastructure by the state."[17]

Smart Planning and Sustainability Planning

In response to some of the exuberant growth of the 1980s, and concerns about sprawl, various efforts emerged. New forms of sustainability dialogue began to emerge within regional bio-regions. EPA funded efforts for local stakeholders to manage a particular policy area (such as water) against performance standards that are upgraded over time.[18]

The Smart Growth project[19] and New Urbanism[20] were efforts to broaden the scope of conventional planning from organizing the physical growth of a community to looking at the community holistically in all its aspects. Today, smart growth has been married to sustainability planning in many locations. But most climate action plans are not being prepared by planners, and do not take advantage of many of the tools available from the land use and transportation planning regimes.

Therefore, in 2010 the American Planning Association formed the Sustaining Places Task Force in order to "define the role of comprehensive planning in addressing the sustainability of human settlement." They distilled the following principles as criteria for developing for the comprehensive plan from best practices in the U.S. for cities and counties:

Livable built environments	Harmony with nature
Resilient economy	Interwoven equity
Healthy community	Responsible regionalism
Authentic participation	Accountable implementation[21]

The "best practice" today for the preparation of the comprehensive plan is that it is a sustainable development plan; and that infrastructure is thought of from an interdisciplinary point of view and supports the above principles (See also sidebar on Santa Monica's sustainability principles.).

Infrastructure and Strategic Planning

Managing for Results

"Managing for results" and "performance management" are terms used to describe a process of developing a strategic plan for the organization, deploying resources to achieve their goals and objectives, developing performance indicators to measure progress toward the goals, reporting on progress, and making changes as needed to stay on course. A larger context of organizational change and reengineering is built into the concept, but for infrastructure purposes, these terms mean that each project funded by the jurisdiction must help the agency meet its strategic goals.

Strategic Planning

A strategic plan for an organization is a short-term plan that looks at available resources and constraints, and attempts to maximize a specific set of goals and objectives. It can provide an overall framework for infrastructure expenditures. In lieu of an overall vision for the community coming out of the comprehensive land use process, or in addition to, a strategic plan can provide guidance to integrate capital budgets for individual infrastructure systems. That is to say, strategic plans

Santa Monica Sustainable City Plan Guiding Principles

1. The concept of sustainability guides city policy. Santa Monica is committed to meeting its existing needs without compromising the ability of future generations to meet their own needs. The long-term impacts of policy choices will be considered to ensure a sustainable legacy.
2. Protection, preservation, and restoration of the natural environment are a high priority of the city. Santa Monica is committed to protecting, preserving and restoring the natural environment. City decision-making will be guided by a mandate to maximize environmental benefits and reduce or eliminate negative environmental impacts. The City will lead by example and encourage other community stakeholders to make a similar commitment to the environment.
3. Environmental quality, economic health and social equity are mutually dependent. Sustainability requires that our collective decisions as a city allow our economy and community members to continue to thrive without destroying the natural environment that we all depend upon. A healthy environment is integral to the city's long-term economic and societal interests. In achieving a healthy environment, we must ensure that inequitable burdens are not placed on any one geographic or socioeconomic sector of the population and that the benefits of a sustainable community are accessible to all members of the community.
4. All decisions have implications for the long-term sustainability of Santa Monica. The City will ensure that each of its policy decisions and programs are interconnected through the common bond of sustainability as expressed in these guiding principles. The policy and decision-making processes of the City will reflect our sustainability objectives. The City will lead by example and encourage other community stakeholders to use sustainability principles to guide their decisions and actions.
5. Community awareness, responsibility, participation, and education are key elements of a sustainable community. All community members, including individual citizens, community-based groups, businesses, schools and other institutions must be aware of their impacts on the environmental, economic and social health of Santa Monica, must take responsibility for reducing or eliminating those impacts, and must take an active part in community efforts to address sustainability concerns. The City will therefore be a leader in the creation and sponsorship of education opportunities to support community awareness, responsibility and participation in cooperation with schools, colleges and other organizations in the community.

City of Santa Monica, *Santa Monica Sustainable City Plan* (2003).

are a way to ensure that sewerage and water decisions made by one department in a city or county can be coordinated with capital decisions made by other departments, or other jurisdictions.

However, strategic plans are not common at the state and local level. Only 10 percent of states had statewide strategic plans in 2002.[22] Maryland's strategic plan for infrastructure investment as part of its Smart Growth strategy is one exception. At the county level, 25 percent of Maryland's counties had strategic plans.[23] Some large cities, such as Dayton, Ohio, and Phoenix, Arizona, and counties such as Fairfax, Virginia, and Santa Barbara, California, however, do plan strategically for capital projects.[24] However, they represent the exceptions at the local level although considerable progress has been made in the past decade.

An analysis of state and local organizations identified four factors that contributed to excellence in capital planning: vision, strategic planning, the availability of good information, and effective communication mechanisms.[25] For local practitioners, this means:

- *Vision.* Vision means developing a clearly defined mission for the organization, including setting new directions and priorities. Vision is also a process of deciding which areas of the organization should grow and receive greater resources, and which areas should remain stable or receive reduced emphasis.
- *Strategic planning.* Strategic planning is an agency's structured process to translate its vision into general goals and objectives. Strategic planning includes a reassessment of both needs and the current political and economic environment, which can be translated into an annual performance plan that links the outcomes of capital projects to the agency's overall strategic goals and objectives.
- *Good information and data systems.* Good information is essential to support high quality capital planning and decision-making, including asset management and facility inventory systems that track the current condition and value of the existing capital assets and that calculate deferred maintenance, repair, and replacement needs and costs.
- *Communication.* Everyone involved in the capital decision-making process should know the agency's current mission and goals, and what outcomes and results are expected from them personally. When this happens, projects are selected, designed, and implemented to contribute toward the achievement of the locality's strategic goals.

Strategic Planning and the Capital Improvement Plan

The investment in infrastructure made by the local jurisdiction is usually large. Just as the State of Maryland decided to strategically target state investment funds to areas where growth was desirable, cities and counties can do the same with the yearly or biannual budget process. In areas experiencing rapid growth, a comprehensive land use plan may have been recently prepared that included a thorough analysis of infrastructure needs grounded in an awareness of fiscal realities. This would be the basis of the capital improvements plan (CIP).

Another option would be to integrate a strategic planning process into the regular budget process. Leading edge cities do this as a routine matter. For example, in Charlotte, North Carolina, the budget document contains goals and objectives by functional area as well as the traditional line item budget. The CIP budget and capital projects are clearly related to the city's strategy for that year.

Climate Change Planning/Sustainability Planning

Climate action plans (CAPs) are a special kind of strategic plan. A CAP can be a freestanding plan or one that is integrated into the city's budget, strategic plan, and capital improvement program. Climate action plans are often prepared in the Chief Executive's office. Many were begun in response to the ICLEI fourfold challenge: identify the emissions; set goals; implement it; and evaluate it. Localities then identified their emissions by sector, usually beginning with the local government and then moving to the emissions of the entire jurisdiction. See Figure 5.2 for the total carbon emissions in the USA.

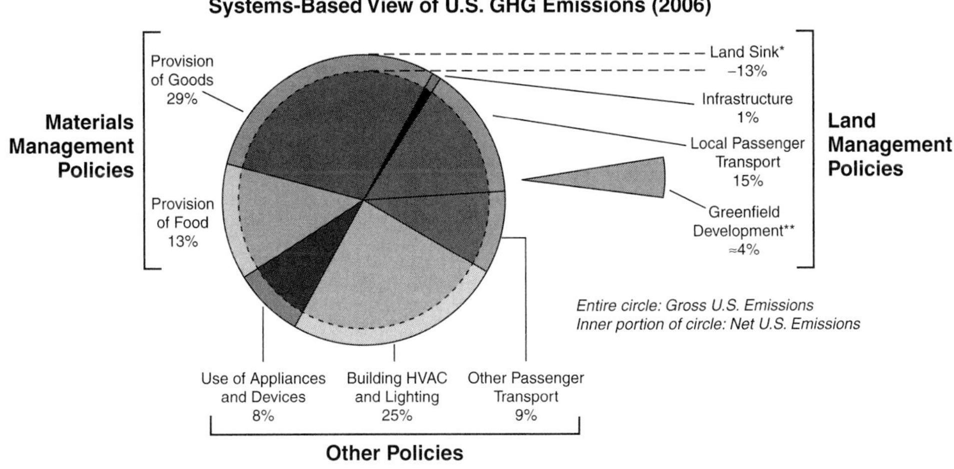

Figure 5.2
Systems-based view of U.S. GHG emissions, 2006. The outer ring represents carbon emissions soaked up by land sinks which erases a portion of total anthropogenic emissions. The inner portion of the pie chart represents net U.S. emissions. Clearing of land for greenfield development is about 4 percent of U.S. emissions but it is not included in the inventory since it may include some overlap with land sink values.

This chart shows where reduction impacts can be made with materials management policies, land managment policies and others.

The local studies would identify areas with greater emissions and generally propose strategies for reducing the emissions to a goal. Some cities then integrated these strategies into the yearly budget cycle, while others are still working to gain acceptance either politically or within the organization. Advanced localities also integrated carbon emission goals into their land use plans. One of the issues with the early efforts is that the CAP process was functionally separate from the land use planners. APA is advocating for planners to become more involved in the CAP process since regulating the built environment is a key part of reducing carbon emissions.

For many localities, mitigating carbon emissions is just one aspect of sustainability. One practitioner exhorts her clients to avoid looking at the city through "carbon goggles."[26] Sustainability plans vary widely in their emphasis. Many are "green-"oriented—that is, primarily focused on environmental outcomes. Others take a broader view, encompassing the three "Es" of the environment, the economy, and equity (or society). Some sustainability plans are strategic or policy plans to influence the full range of the locality's programs, while others serve as the focal point of a land use plan update. There is also a considerable amount of overlap between climate change plans and sustainability plans as they have developed recently in the United States.

A parallel trend is the growth in initiatives to assess, measure, and report on progress towards sustainability. For example, a grassroots movement of citizens and non-profits in Seattle developed a list of sustainable indicators in 1993, which greatly influenced the City of Seattle's 1996 General Plan and prompted some of the regional agencies responsible for infrastructure to "green" their capital investments as well as their operations. ICLEI and the U.S. Green Building Council initiated an effort in 2008 to establish the STAR community index that will work much as LEED did for buildings. The STAR system provides standardized indicators that make it possible to compare progress among localities, and provides a framework for local sustainability plans. A particularly innovative indicators project is the Minneapolis Annual Living

Excerpt from City of Minneapolis 2011 Sustainability Report

For the sixth year in a row, the City is reporting its progress on 26 sustainability indicators. Highlights include:

A Healthy Life

- Once again, fewer children had elevated levels of lead.
- The number of HIV cases was at its lowest level since 2005, and the number of gonorrhea cases was at the lowest level in a decade.

Greenprint

- Transit ridership rebounded, and bike ridership increased.
- Renewable energy installations dramatically increased.

A Vital Community

- The pace accelerated for building and updating affordable housing.
- Violent crime decreased almost 27 percent over the past four years.

Setbacks

The City experienced setbacks too, including continued health disparities among populations of color, a decline in recycling and composting rates, and jobs lost due to the struggling economy.

City of Minneapolis, *2011 Living Well Report*, available at: www.ci.minneapolis.mn.us/sustainability.

Well report that has tracked 26 sustainability indicators for the past six years. (See the sidebar on Minneapolis.)

Adaptation Planning

Although climate adaptation plans can be developed as part of a climate action plan, in the past decade, many jurisdictions that were experiencing more extreme weather events began to develop free-standing climate adaptation plans. Climate change is causing rising coastal waters, droughts, wildfires, more frequent intense storms, along with a variety of other destructive weather patterns. Infrastructure systems are threatened by these changes, but can also be a source of assistance in protecting the built environment. Recently, large coastal cities throughout the world formed an organization called C40/ Connecting Delta Cities to share experiences, plans and solutions for climate risk on a regular basis. ICLEI has also been active in this area for Europe and developing countries. In the United States, cities at risk of flooding have also begun to mobilize around adaptation issues. Table 5.2 is from King County's

Table 5.2
Identification of expected climate change impacts on key infrastructure systems in King County, WA

1. Sectors	2. Planning areas	3. Current and expected stresses to systems in this planning area
Hydrology and water resources	Water supply management	Low water supply during hot, dry summers; managing drought (*current and expected*) Poor water quality during flood events (*current and expected*) Poor water quality during summer (*current and expected*) Aging infrastructure and lack of funding for system upgrades (*current and expected*) Increased demand from population growth (*expected*)
Infrastructure	Stormwater management	Combined sewer overflows (CSOs) during heavy rainstorms (*current and expected*) Lack of funding for system upgrades (*current and expected*)
Transportation	Road operations and maintenance	Road buckling during heat waves (*current and expected*) Winter road maintenance (snow/ice removal) (*current and expected*) Managing brush fires along roadways (*current and expected*) Erosion around bridge footings (*current and expected*)

Source: King County, *Preparing for Climate Change: A Guidebook for Local, Regional, and State Government* (Washington, 2008).

manual for localities to follow in developing an adaptation plan for municipal operations. It notes present and future threats for water and transportation systems.

Adaptation planning should also include health planning, since rising temperatures destabilize existing patterns of disease. Adaptation planning may also include strengthening of the emergency preparedness plan for extreme weather events.

Conclusion

The heart of sustainability is to re-invent existing systems, programs, and procedures to ensure that the new millennium technologies and sensibilities are re-imaged by governments. Infrastructure, because of replacement needs and new ways of conceptualizing these systems that are made possible by advances in technology, can play a transformational role in making cities more sustainable. New York City's path-breaking PlaNYC, issued in 2007, is an example of a plan that looks holistically at the key elements of the city's environment: land, air, water, and transportation infrastructure.

PlaNYC develops policies for city programs, its comprehensive plan, development regulations, and the capital budget in a way that is cross-functional as well as making use of most of the City's planning, development, regulatory, and budget tools. Below are some excerpts that illustrate how integrating planning for multiple infrastructure systems (along with land use planning) can produce better results than looking at them individually.[27]

> The scale, intricacy, and interdependency of the physical challenges we face required a more holistic approach; choices in one area had unavoidable impacts in another … If you seek to solve traffic congestion by building more roads or by expanding mass transit, you make a choice that changes the city. If you care about reducing carbon emissions, that suggests some energy solutions rather than others. If your concern is not only the amount of housing that is produced, but how it impacts neighborhoods and who can afford it, then your recommendations will vary.
>
> The plan outlined here shows how using our land more efficiently can enable the city to absorb tremendous growth while creating affordable, sustainable housing and open spaces in every neighborhood. It details initiatives to improve the quality of our air across the city, so that every New Yorker can depend on breathing the cleanest air of any big city in America; it specifies the actions we need to take to protect the purity of our water and ensure its reliable supply throughout the city; it proposes a new approach to energy planning in New York, that won't only met the city's reliability needs, but will improve our air quality and save us billions of dollars every year. Finally, it proposes to transform our transportation network

on a scale not seen since the expansion of the subway system in the early 20th century—and fund it.

Each strategy builds on another ... encouraging transit-oriented growth is not only a housing strategy; it will also reduce our dependence on automobiles, which in turn alleviates congestion and improves our air quality.

We have also discovered that every smart choice equals one ultimate impact: a reduction in global warming emissions. This is the real fight to preserve and sustain our city.

Additional Resources

Publications

Boswell, Michael R., Adrienne I. Greve, and Tammy L. Seale. *Local Climate Action Planning*. Washington, DC: Island Press, 2012. If infrastructure is being considered in the context of a climate action plan, or an adaptation plan, this is a wonderful guide to preparing these plans with many tips on the major infrastructure systems. Good references at the end of many of the chapters.

Hoch, Charles, Eugenie Birch, et al. *Local Planning: Contemporary Principles and Practice*. Chicago: International City Managers Association, 2009. The latest "Green Book" designed to provide an overview of planning for the generalist. Covers a wide range of planning concerns in a readable manner.

King County, WA, ICLEI et al. *Preparing for Climate Change: A Guidebook for Local, Regional and State Governments*. King County, Washington, 2008.

Roseland, Mark, Sean Connelly, et al. *Toward Sustainable Communities: Resources for Citizens and Their Governments*, 3rd edn. Gabriola Island, BC: New Society Publishers, 2005. A manual for local practitioners about sustainability, and preparing sustainability plans. This is particularly appropriate for local and elected officials who need practical advice about the programs and tools that can be used locally. Individual chapters about key infrastructure systems.

Wheeler, Stephen M. *Planning for Sustainability*, New York: Routledge, 2004. A good overview of the sustainable planning movement with advice about the components of a sustainability plan, including innovative infrastructure approaches.

Websites

American Planning Association: www.planning.org. The American Planning Association is the professional organization for planners. Their website contains many useful planning documents and guides.

American Public Works Association: www.apwa.net. Many good sources on capital improvement planning and sustainability. APWA is the professional association for those working in public works at the local level.

Government Financial Officers Association: www.gfoa. This is the professional organization for budget and financial officers at the local and state level. Its website contains a wide variety of publications on strategic planning and best practices.

ICLEI-Local Governments for Sustainability: www.icleiusa.org/about-iclei. The website where the Toolkit for sustainable planning can be found as well as the latest on the STAR rating system for local governments.

In this chapter...

Importance to Local Practitioners 87
The Infrastructure Program 87
- The Importance of the Infrastructure Program 88
- The Status of Local Infrastructure Programs 89
- The Impetus for the Program 89
- Intergovernmental Considerations 90

Preparation of the Infrastructure Program 91
Organizing the Planning Process 92
- Building the Team 92
- Public Participation Process 93

Analyzing Baseline Conditions 94
- Inventory of Existing Facilities or System 94
- Developing the Condition Assessment for Each System 94

Determine the Goals and Objectives for the System 95
- Developing Level of Service Standards 95

Project Future Demands and Needs 96
- Projecting the Future Population 96
- Translating Future Population into Demand 97
- Sizing and Locating the Facilities 97

Identify and Evaluate Alternatives 98
- Identify Alternatives 98
- Spatial Simulation of Alternatives 99
- Identify Costs and Risks 100
- Identify Financing Alternatives 100
- Identify Other Impacts 100
- Adopt Preferred Alternative 101

Conclusion 101
Additional Resources 101
- Publications 101
- Websites 101

chapter 6

The Infrastructure Program and its Preparation

Why Is This Important to Local Practitioners?

Every local government should have a long-term infrastructure program, whether it is part of a larger comprehensive land use plan, a back-up document to the capital improvements budget for individual systems or a source document for the local sustainability plan. Decisions will be made about infrastructure even without a program. Bringing together the various actors and infrastructure systems allows a jurisdiction to take an informed view about resources and challenges to make better budget decisions and planning decisions.

The chapter begins by describing the infrastructure program, its context and status today. The remainder of the chapter describes how to prepare one. Emphasis is given to organizing the process, identifying the baseline conditions, and projecting future demand.

The Infrastructure Program

The infrastructure program for an individual system is the most frequently prepared infrastructure plan. It is a document that is probably not familiar to many planners, especially those in already built-up areas, since most of them exist in the engineering department of the GPG, where they guide the construction of sidewalks, street repaving, and sewer upgrades, and the construction and major improvements to public buildings and facilities. In a special purpose infrastructure agency, the infrastructure program is the capital or master plan for that function. If it is prepared for a single infrastructure system, it may also be referred to as the capital needs study, since these words signal the intent to integrate portions into the local budget documents.

The infrastructure program may not be a formal document. However, this document, if it exists for a particular system such as sewers or sidewalks and streets, could be the functional equivalent to the "element" in a comprehensive plan for a single programmatic area or individual infrastructure system. In some states, if a local infrastructure program exists, it must be consistent with the locality's general or comprehensive land use plan.

The Importance of the Infrastructure Program

Infrastructure investment decisions will be made even if there is no plan. These decisions are made by the local general purpose government every year as part of its budget process, even if the decision is to allow systems to deteriorate. A systematic long-term program will help improve the impact and efficiency of these decisions. Other reasons to prepare a long-term infrastructure plan and program are:

- *Infrastructure implements the general or comprehensive plan.* Nothing implements the land use element of the general plan more than infrastructure investments.
- *Infrastructure is linked directly to the quality of life in a community.* A long-range infrastructure program allows local government to establish "levels of service" for community facilities so that they will meet the goals of the local government and its residents.
- *Control over timing, phasing, and cost of development.* A long-range infrastructure program allows local government more control over the timing, phasing, and pricing of infrastructure projects. It will also optimize the use of existing facilities as an alternative to facility expansion or new construction.
- *Coordination with other infrastructure providers.* A long-range infrastructure program provides a venue to open discussions with other infrastructure providers that are outside the immediate control of the local government, such as adjacent municipalities, special utility or school districts, state and federal agencies, or private utilities.
- *Provides guidance for changes in infrastructure needs.* Finally, the long-range infrastructure program can include strategies to deal with changes in infrastructure service as systems age and deteriorate, federal or state regulations or best management practices evolve, a community grows (or shrinks), or the nature of supply and demand changes.

The Status of Local Infrastructure Programs

There are several problems with current approaches to long-term infrastructure planning. Infrastructure plans are usually prepared for one particular system, frequently with an eye on the upcoming budget cycle. The quality and depth of the studies for individual infrastructure systems may be uneven at the local level. The timing of the preparation is idiosyncratic. In addition, geographic synergies do not occur. It is one of those local government truisms that sometimes plans are made to pave streets one year, that are torn up the following year as part of a sewer replacement program—despite the fact that the engineers working on the planning for each system can be sitting within shouting distance of each other. The need to replace sewers or sidewalks in a commercial area ideally should be synchronized with a revitalization plan for the affected neighborhood, for example.

Water, wastewater, and transportation infrastructure areas probably have the most complete infrastructure programs at the local level. Similarly, many localities have detailed engineering studies about the condition of their storm sewers along with a plan to replace, enlarge, and build new facilities in order to comply with new National Pollution Discharge Elimination Standards (NPDES). Special districts for water, sewer, and solid waste probably have the most sophisticated capital needs studies since they are usually required to be fully funded by fee revenue. Although they may be developed with a public participation process, they are generally not coordinated with the local government's capital improvement planning process.

Infrastructure plans or capital needs studies for public facilities such as civic centers, parks, public safety buildings, and libraries are probably the least developed because they are usually funded with local taxes or one-time grant money that is not predictable. Instead, although some local public works departments may have a detailed condition assessment for all existing buildings for maintenance purposes, most public facility planning is at the project level.

The Impetus for the Program

The infrastructure program is prepared in response to the need to develop new facilities or to upgrade existing ones. It anticipates a tax being passed, or a realignment of fees to pay for the recommended capital program for a particular infrastructure system, such as a series of upgrades to recreational structures in the locality, the seismic retrofit of school buildings, or a program for upgrading the storm water sewers to come into compliance with new

EPA regulations. For infrastructure systems that rely on fees, the needs analysis is translated into various fee alternatives.

Infrastructure programs are prepared for problematic systems or for a priority issue such as climate change or sustainability. Policy-makers in a developing area will address location of streets and transit stations, and the need for wastewater treatment plants, schools, public safety buildings, libraries, and parks, as well as address critical or sensitive environmental areas.

Existing communities will be concerned about upgrading streets and sidewalks, installing lighting fixtures, perhaps widening sidewalks to create a more pedestrian-friendly commercial area. These communities may also be interested in a full-fledged "redevelopment" of the street grid in a certain part of town. In a city facing economic and community development concerns, infrastructure issues might include redesigning the street grid, or restructuring the plan for sidewalks and street tree patterns. The urban design aspect of infrastructure such as utility boxes, street furniture, and street art will also be important.

Issues such as whether the wastewater treatment plant or water supply will be adequate for future infill can also be dealt with in the infrastructure program. In arid areas, the local government may want to prepare its own water plan even though it may not be the water provider. A locality may also want capital needs studies for energy facilities, solid waste management, or telecommunications if its industrial base has special needs.

When the infrastructure system is run by a special district or special authority, as in the case of many water and sewer facilities, schools, or hospitals, for example, the requirements for the long-range plan are mandated by the rules setting up the district or other legislation. These may vary from jurisdiction to jurisdiction. In many states, there are no requirements for the special district capital plan to interface with the comprehensive plan of the GPG.

Intergovernmental Considerations

In an ideal world, a regional planning agency would plan for transportation, water and sanitary facilities, regional parks, recreational, cultural, and educational facilities. The city or county would then work within the regional perspective to tailor these elements for locally-serving infrastructure. For the most part, regional planning agencies that play such a role are rare.[1]

The ideal infrastructure program is internally, horizontally, and vertically consistent (Figure 6.1). It is internally consistent when implementing regulations and budgets for the jurisdiction carry out the intention of the comprehensive plan. It is vertically consistent when a city or county plan is consistent with a regional plan that is consistent with a higher-level state plan. It is horizontally consistent when a city general plan is consistent with the general plan

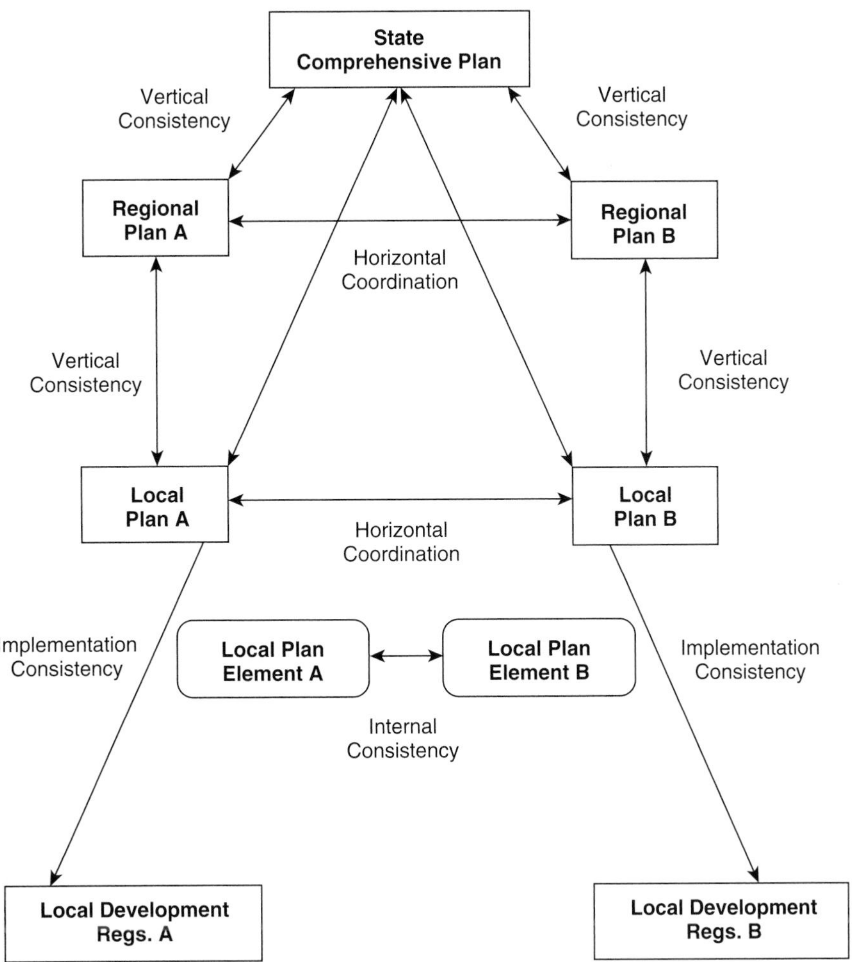

Figure 6.1
Vertical, horizontal, and internal plan consistency at the local and state level

of a neighboring city, or with the infrastructure plan of a special purpose district such as a water or school district. Preparing a cross-functional plan is one way of ensuring horizontal consistency.[2]

Preparation of the Infrastructure Program

Infrastructure investments have several characteristics that significantly alter the way local governments should plan for them. The need for technical experts is more important, the question of boundaries is essential, and the ability to alter demand or need through demand management techniques is extremely important. The ability to make use of innovative technologies to further low-impact development is also key. The organization of the planning process must reflect these differences.[3] The following turns to a detailed discussion of the process, although in practice it may not be as linear as described here.

Organizing the Planning Process

Building the Team

Building the team will start within the local general purpose government (GPG), and then move to include outside agencies. Depending upon whether the impetus for the infrastructure plan comes from a comprehensive or general planning process, or from budget needs, or from the need for a plan for a specific system, the initiating staff person will be a planner, the budget officer, or the engineer. Regardless of how the process is initiated, participants from all three areas within the GPG must be involved early on.

Once the in-house team is identified, others can be identified who should be involved within the local government and from outside agencies. Unlike many elements of land use planning, infrastructure planning requires the support of outside experts for the particular systems. Water and sewer systems require input from civil engineers and the local utility. Power and telecommunications systems need contributions from electrical or communications engineers and other experts. Planning for schools, libraries, jails, and recreation centers requires the involvement of the actual service providers.

Many agencies outside the local GPG often provide infrastructure to a given community. Special districts, state or federal agencies, or private energy or telecommunications firms are examples. Cultivating relationships with these other infrastructure providers is important, especially during the planning process. Representatives of these other agencies should be part of the infrastructure element preparation, at least to be kept abreast of possible changes in their service areas.

Consultants can be hired at the outset to undertake many of the tasks in preparing the infrastructure plan. Specialized consultants from engineering firms are often hired to conduct the inventory and condition assessment, particularly in mature urban areas where the original subdivision maps denoting water and sewer lines may be lost, or where unrecorded infrastructure construction has occurred over the years. Some localities hire engineering firms to develop fee schedules for water and sewer systems, while others hire architectural firms to design large-scale public facilities.

Steps in Preparation of the Infrastructure Plan

Step 1 Organizing the planning process

- Build the team.
- Prepare budget and work program.
- Hire outside consultants.
- Set up the public participation system.

Step 2 Analyze baseline conditions

- Identify service area.
- Assemble inventory of existing systems and facilities.
- Develop the condition assessment for each system or facility.

Step 3 Determine the goals and objectives for the system

- Develop goals and objectives.
- Develop level of service (LOS) standards.

Step 4 Project future demand and needs

- Create population projection.
- Identify infrastructure system needs.
- Address demand management considerations.
- Calculate facility requirements.

Step 5 Identify and evaluate alternatives

- Identify alternatives.
- Create spatial simulation of alternatives.
- Identify costs.
- Identify financing alternatives.
- Identify other impacts (environmental, including eCO_2, social, economic).

Step 6 Adopt preferred alternative

If the state or locality requires an environmental review process separate from the planning process, customarily another set of professionals is hired for this function. Most importantly, however, the local practitioner should not deed over policy decisions to technical consultants. Technical assumptions can hide serious differences in goals and outcomes unless they are discussed.

Public Participation Process

If the infrastructure plan is being prepared as part of the larger land use plan, then the planning commission or local elected body has probably established the hearings and milestones. If an environmental assessment is being done, the state or federal government has its own required hearings and public participation process. In this case, the engineer or planner must ensure that users or other stakeholders in the infrastructure systems who might be overlooked as part of the larger process are included. See the sidebar on collaborative planning for more information on the importance of involving stakeholders in an open process.

If the infrastructure plan is being prepared as part of the budget process, it is likely that the budget officer has already set up the timeline with at least the legally required hearings and notices. The planner or engineer

Collaborative Planning

Preparation of the long-term cross-functional infrastructure plan can benefit from using a collaborative planning approach because most local GPGs do not provide all the infrastructure services in-house. Other agencies at the state and federal level along with special districts may have infrastructure responsibilities that affect the jurisdiction and should be included in the local planning process if at all possible.

A collaborative effort to develop a long-term infrastructure plan could be an interagency, multijurisdictional effort involving all the agencies, each with knowledge of different infrastructure systems and agency constraints and opportunities. In a collaborative process, representatives of those who have a "stake" in the outcome of the planning process are involved, thus, the term "stakeholder." Other stakeholders for the infrastructure plan might include representatives of the consumers, local environmental groups and those private groups that would be affected by the outcome of the plan. The stakeholders must represent all the diverse interests.

Some experts in collaborative planning indicate that broader public participation is not important if all the interests concerning the problem are fairly represented at the table, and if all of the stakeholders have the opportunity to participate equally. Many GPGs, however, have formal citizen participation requirements or open hearings before taking official action. A collaborative planning process can interface with hearing requirements effectively if all the stakeholders were meaningfully involved in the development of the plan prior to the hearing.

The collaborative process is usually long-term, involving face-to-face discussions among stakeholders. The discussion is usually facilitated, and agreements are reached through consensus.

A collaborative planning process involves development of information about the problem that can be trusted by all stakeholders in order to develop a joint vision about the outcome. Sometimes the information is generated by experts who are given explicit instructions about the kinds of data, models, maps, or other information that is desired by the group. The findings of the experts may be subject to challenges by some of the stakeholders, in which case the issue would be further researched. The result should be information that is trusted by the stakeholders.

A good outcome for the collaborative planning effort is an agreement about the problem that brought the group together. The best case is a decision that reflects a shared vision and an innovative solution based on the group's shared knowledge. In collaborative planning, a good plan is one that benefits the interests of each of the stakeholders but also creates benefits for the group as a whole. However, even if the process does not produce an infrastructure plan that can be formally agreed on by all parties, mutual learning will have occurred that will result in individual agency plans being more responsive to the broad array of needs and constraints.

Based on ideas from Sarah Connick and Judith E. Innes, "Outcomes of Collaborative Water Policy Making: Applying Complexity Thinking to Evaluation." *Journal of Environmental Planning and Management* 46(2) (2003); Judith Innes and David Booher, "Metropolitan Development as a Complex System: A New Approach to Sustainability." *Economic Development Quarterly* 13(2) (1999); Judith Innes and Judith Gruber, *Bay Area Transportation Decision Making in the Wake of ISTEA: Planning Styles in Conflict in the Metropolitan Transportation Commission* (Berkeley, CA: University of California Transportation Center, 2001), available at: *www.uctc.net/papers/papersalpha.html.*

in charge of the infrastructure plan may wish to create a more detailed process to ensure that those citizens or stakeholders concerned about the individual infrastructure systems involved have a chance to influence policy decisions at the major steps in plan preparation. Finally, if the infrastructure plan is being prepared for a single function, in response to federal regulations or a deteriorated system, the outcome likely will be a change in fees or require a special bond. In this case, a broad-based stakeholder process is necessary if fees or taxes need to be raised.

Analyzing Baseline Conditions

Analysis of the baseline conditions is the first substantive step in infrastructure planning. To do this, the boundaries must be decided upon for the service area. For infrastructure planning in support of the general or comprehensive land use plan in an urbanized area, these are usually the boundaries of the jurisdiction. In an already urbanized area, there likely will be overlapping service area definitions from outside agencies and the best service area may not conform to political boundaries. Some systems, like water supply and sewage collection, are bound to physical features of the landscape, while others, such as privately provided power and telecommunications services, are governed by market forces.

Inventory of Existing Facilities or System

The next step is to inventory existing facilities. Every infrastructure system has different types and categories of facilities. Transportation facilities include expressways, arterials, and other roads, as well as buses, railways and rail vehicles, and airports. An inventory of solid waste facilities might include transfer stations, landfills, or incinerators, as well as rolling stock. There are even inventory systems for trees on public property. A consultant may be hired to do part or all of this task.

The inventory should include both a listing and an assessment of any infrastructure systems that support the land-use element, and systems over which the local government exerts regulatory authority. The inventory should include at a minimum: the agency that operates the system; the system's service area; both design capacity of the system and current demand on it; and the level of service provided by the facility. Where community facilities are shared, each local government should indicate the portion of the system's capacity allocated to serve its jurisdiction.

Developing the Condition Assessment for Each System

Condition assessment is a description of current state of the system, with an evaluation of performance, an estimate of expected life, facility capacity including surpluses, and deficiencies for each service area. The general assessment discusses ways of addressing any deficiencies that do not include expansion and new construction, such as operational changes, different operating hours, or using multipurpose facilities—in general, ways of optimizing performance of existing facilities. The facility inventory should contain a section for each facility, including maps and tables organized by the relevant analysis unit (such as the service area for water supply, or the drainage area for stormwater management). Both existing conditions and proposed changes should be covered in this map.

Forward-thinking communities consider energy consumption and carbon emissions in their infrastructure planning efforts. For those that do not, however, the condition assessment stage might be a good place to consider these issues along with reduction strategies. Ways of conserving potable water should also be considered in water-scarce states.

Determine the Goals and Objectives for the System

Developing goals and objectives is another key piece of the infrastructure planning process. The individual system chapters in this book address key areas that should be considered but a key consideration is the level of service the community desires.

Developing Level of Service Standards

Level of service (LOS) standards are a way of quantifying the objectives in the plan. By comparing current and future levels of infrastructure provision to the service standard, local decision-makers can identify future needs. LOS standards can be formally adopted into local development regulations for use during project processing to ensure that adequate infrastructure exists to serve new growth. There are two types: technical and policy.

- *Technical Standards.* The first type of standard is the operational or technical requirements, or the standards that a system must maintain to operate properly. For example, the service standard for water supply is usually an acceptable minimum water pressure that usually is governed by fire protection needs.
- *Policy Standards.* The second type of standard is determined by policy. Transportation engineers have rigorously defined level of service standards for transportation facilities; again, local officials must determine which level of service to apply, and where. Schools, jail, libraries, and parks are more difficult to quantify in terms of service standards. Common standards are square foot of classroom space per pupil, or number of books per patron. The choice of service standard must

Needs and Wants

The goal is to identify needs, not wants. Sometimes, the difference is easy to distinguish. Other times, though, the difference between needs and wants is difficult to tell, often confusing the plan developers themselves. Potential solutions to real needs easily become identified with the needs themselves. This confusion is most vividly seen in transportation, where many times decision-makers become seduced by a new technology or practice. Bus Rapid Transit is the name given to a special high-capacity bus system. It approaches the performance of a light-rail system, but at a fraction of the capital cost. It is currently in vogue among transportation professionals, and suddenly, every city in the country needs a Bus Rapid Transit system to address its congestion problem. In reality, what the community needs is a congestion reduction solution, which can be arterial management, demand management, or improvements in current transit performance, as well as Bus Rapid Transit. In other words, separating needs from wants requires critical thinking.

be deliberate. The community should not uncritically accept the recommendations of professional or trade organizations.

Project Future Demands and Needs

Once the baseline condition of the infrastructure system has been identified, and goals and LOS standards set, future infrastructure needs should be projected, mindful that this is an iterative process. This includes replacing existing infrastructure as determined from the condition inventory, as well as meeting future needs.

Projecting the Future Population

The first step is to estimate the future population. In infrastructure planning, "no other calculation is more important in the sizing of the system than the projection of future 'ultimate' population."[4] Most infrastructure service consumption is related to the number of expected households, total and by age cohort, family size, and type of dwelling (single-family detached or multifamily). Three choices must be made:[5]

- *Level of disaggregation*. The first choice considers whether the forecast should be disaggregated by age, by ethnicity or race, or by nativity; and for the foreign born, by duration of residence in the U.S. This choice is a policy decision but also depends upon the availability of data for each subcategory.
- *Household rates*. The second choice the local official must make concerns the household formation rate. The easiest choice is to use the rates measured in the last census. However, most of these trends are changing over time—either rising or falling—and the impact on the sizing of a large infrastructure project may be significant.
- *Home-ownership rates*. The third choice concerns home-ownership rates by age and by ethnicity or race or nativity because this measure distinguishes between single-family detached and multifamily—an important determinant in forecasting transportation, waste, energy, and telecommunication consumption. However, this is also a normative decision requiring a balance between high rates for every group, the realities of the market, and the ability of the locality to provide for low-income housing.

Once the population projection is complete, the simplest way to assess future needs is to determine present use on some unit basis, such as per person, per household or per square foot of commercial space, and then multiply that amount by the expected future increase. This new figure is compared to the current capacity of present infrastructure systems. Any deficiencies are then candidates for capital investment. This method is the traditional supply-side method. It has the virtues of being intuitive, simple to apply, and not data-intensive. And while it is susceptible to error because of some of its assumptions, it often provides a good "first cut" for needs assessment, against which conservation and other demand management programs can be assessed.

Translating Future Population into Demand

One major flaw with the traditional per-capita estimation of need is that the typical response strategy is one of capital-intensive, supply-side measures. Hence, demand management alternatives should be part of the infrastructure planning process. The most obvious benefit is that by addressing a service deficiency with demand management, a community is spared the expense of having to build more supply. The second benefit is that demand management uses less natural resources.

Demand management can be difficult to sell to traditional infrastructure providers because unlike increasing supply, which is generally a technical issue, managing demand often includes behavior modification and nonprice techniques. Engineers and some infrastructure providers are often unfamiliar or uncomfortable with this approach. And since many demand management techniques are not technical, their effectiveness can be difficult to evaluate. Therefore, policy-makers might be reluctant to adopt these tools in a capital improvement plan.

Demand management can also appear on the supply side. A major portion of water use in any metropolitan water system is caused by leaks. Therefore, a leak detection and repair system can significantly reduce water demand. In general, any program that increases the efficiency of the existing system should be evaluated during this stage in the planning process.

Sizing and Locating the Facilities

Another issue that should be addressed in the infrastructure plan is the tendency to oversize facilities. A common engineering philosophy is "better safe than sorry," which means that systems are often designed for acutely large ultimate service populations. In the 1970s, numerous studies of sewer investments found that population projections on which the investments were based tended to be speculative and exceedingly high. Furthermore, these same studies found that, since population and infrastructure investment are linked, these projections tend to be self-fulfilling.[6]

Many public officials do not worry about oversizing, assuming that the capacity will be used some day. Moreover, many communities have enough unused capacity in water and sewerage lines to meet expected demand 100 years in the future, well beyond the expected service life of the facility. Furthermore, these systems require maintenance in the meantime—a local government might spend as much on road maintenance over ten years as it spent on building the road in the first place.

Most infrastructure systems typically do not require large amounts of land. The components of many systems, such as pipes and conduits, share space with other land uses. A water treatment plant, power production plant, or landfill might require on the order of 10 acres. A large school might require 40 to 50 acres. One notable exception is the large commercial airport, which can require up to 50 square miles in area.

Identify and Evaluate Alternatives

Identify Alternatives

The planning team needs to identify a set of alternatives for each system that will meet the needs. This step must be done in the context of land use and other policy goals, and includes incorporating both the management strategies the service standards discussed earlier, as well as considering the larger social goals. For example, a community might decide that it wants to support a multimodal transportation system that caters to low-income, elderly, and handicapped residents. This policy will result in different infrastructure investments than a transportation policy designed to support economic or industrial growth.

Generating alternatives may be an iterative process—going back and forth between alternatives and first-round impacts until a viable set of alternatives is identified for an in-depth analysis of impacts. In states where an environmental impact statement or assessment is required, these two processes can proceed in tandem, each assisting the other. (See the sidebar opposite)

The alternatives may consist of a plan for a physical facility and network along with a series of programs to modify transportation and water consumption behavior, for example. The alternatives may also consist of recommendations about detailed design standards, or changes to the development rules to meet the needs.

Spatial Simulation of Alternatives

As noted earlier, the size of the population depends on infrastructure development just as much as the size of the infrastructure development depends upon the population. When these sizes are translated into spatial alternatives, development patterns resulting from different infrastructure locations and sizes can be simulated so that policymakers can understand the impact of infrastructure decisions. Choices about transportation and water and sewer infrastructure can affect the amount of land consumed in development, just as land development policies and locations can affect the amount of infrastructure required.[7]

The major infrastructure systems that are "drivers" for development are transportation and water and sewer capacity. Although the metropolitan transportation planning agencies usually do not take account of land use impacts of their funding decisions,[8] transportation simulation models were the first to be developed in the 1960s.

Today several types of technologies exist that focus on developing land use alternatives depending upon policy choices about growth and infrastructure. Until recently, most urban development models have focused on transportation planning and development.[9] Smaller areas, such as a single county or a city within an already built-up area, should not ignore the development of different spatial alternatives in preparing a long-range infrastructure plan.

Technology for Community Design and Decision Making

Over the past decade, a family of computer-based tools that enhance decision-making—often called Decision Support Systems—has evolved to the point where it serves as a useful adjunct in community and infrastructure planning. These tools and techniques fall into five categories of decision support that reflect the critical phases of a sustainable community development process for infrastructure:

- *Geographic information systems (GIS) and land use spatial analysis tools.* GISs are information management systems tied to geographic data. Various types of data, such as hydrology, road networks, urban mapping, land cover, and demographic information can be integrated geographically to provide the spatial context for an infrastructure plan. GIS-generated maps and charts can also help nontechnical community members spatially visualize land uses, environmental hazards, transportation access, and other relevant information.
- *Impact analysis tools*, many of which are GIS based, can be used to assess the past, present, and future economic and environmental impacts on infrastructure for a wide range of development projects and policies. Conversely, they can also assess the impacts of a particular infrastructure project as well. By providing both quantitative and visual outputs for a variety of scenarios, impact analysis tools help planners and the public to understand the trade-offs between alternatives and public policy decisions. Tool names include *INDEX*, *Place^{3}s*, *CommunityViz*, and *Whatif?*
- *Community process tools allow greater numbers of people to be involved in a more effective and efficient process.* Electronic meeting systems using technologies such as keypad voting can facilitate very effective large-scale meetings, while Internet resources such as websites and databases can help communities gather and share important data and information to promote better communication.
- *Visualization tools can play an important role in both the process and design phases of a planning project.* These tools allow citizens to experience different alternatives that are difficult or impossible to see in raw data form. Visualization techniques can involve computer imaging using realistic three-dimensional designs and software that allows users to "paint" their own changes to an image during public meetings. These tools can help demystify the "black box" of many engineering assumptions that underlie infrastructure planning.

For more information about specific tools and techniques, visit the PlaceMatters.com database (www.placematterstools.org).

Written by Uri Avin, a principal at PB PlaceMaking and also the Planning Director in Howard County, Maryland and Ken Snyder, of PlaceMatters, where he has pioneered the use of visualization methods.

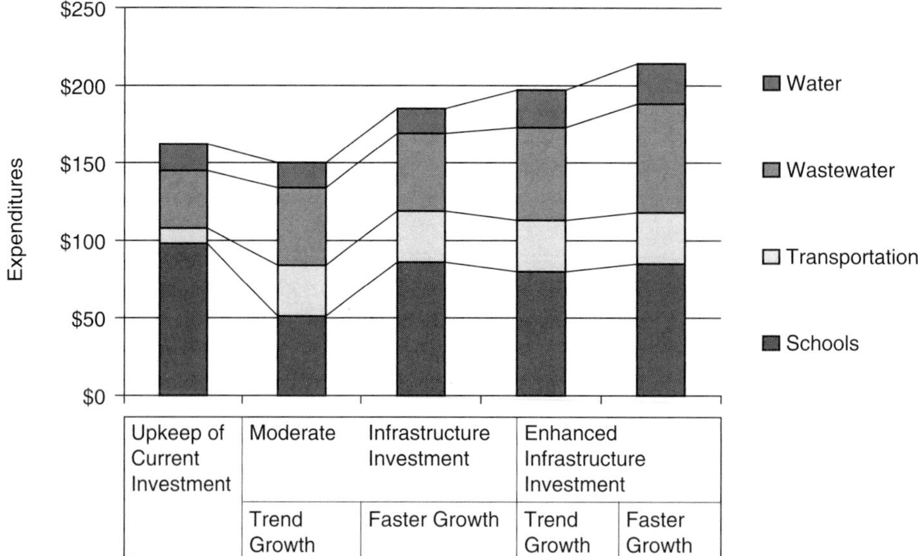

Figure 6.2
Infrastructure costs by assumptions about growth in a comprehensive planning process

Coordination of Infrastructure Plans with Land Use Policy Can Save Local Dollars

Cross-functional long-range planning for infrastructure can result in significant cost savings when combined with land use planning. A report by the American Planning Association, published in 2002, identified a series of needless infrastructure project costs that occurred as a result of improper coordination between infrastructure planning and other development. A sewer district covering two counties in Kentucky paid out over $500 million from 1992 to 2002 as a result of inadequate local zoning policies. Sprawling development in Houston, Atlanta, Dallas-Ft Worth, Miami, and Detroit resulted in transportation expenditures of $2,500 more per year per resident compared to other cities where infrastructure and land use decisions were coordinated. Maine spent $727 million for school construction from 1970 to 1995 even though the number of students dropped by 27,000 during the same period.[1]

The report noted that localities that coordinated infrastructure plans with land use

Identify Costs and Risks

The costs of developing and maintaining the infrastructure system should be identified for the time horizon of the plan. This should include an evaluation of risks due to climate change. See Figure 6.2 which compares costs for several different infrastructure alternatives for a county in Maryland. These alternatives cost out the impact on infrastructure budget needs assuming variations in expected growth whether it will be at the historic trend, or accelerated. Also estimated is the impact of providing a moderate or enhanced level of investment in the system.

Identify Financing Alternatives

Financing, for both capital and operating expenses, is a key component of the infrastructure planning process. Depending upon whether the plan is being prepared as part of a land use plan, for the capital budget, or for a single infrastructure system, this section will be done in broad strokes or minute detail. This topic is addressed at greater length in the subsequent chapters on financing infrastructure in Part III.

Identify Other Impacts

Other impacts of the alternatives including impact on reduction of carbon emissions, should be evaluated.

Adopt Preferred Alternative

Additional information might be required at this point to select the preferred infrastructure alternative or to integrate it into the community's plan.

Conclusion

Long-range infrastructure planning is important for local jurisdictions whether they are growing, stable, or declining. While growing localities may have the funds to pay for a comprehensive plan, the challenge for stable or declining areas will be to find the funds to plan and pay for the improvements. For a jurisdiction where state legislative change is not on the horizon, a useful place to start is with the existing infrastructure plans or capital needs studies. They can be used as the start of a mini "cross-acceptance" process within the jurisdiction, including looking at land use impacts and cross-functional issues. If funding to do this is available for a region, a similar process can be followed.

Additional Resources

Publications

Berke, Philip and David R. Godschalk. *Urban Land Use Planning*, 5th edn. Urbana, IL: University of Illinois Press, 2010. A classic on land use planning with excellent sections on planning for sustainable infrastructure systems.

Kelly, Eric Damian. *Community Planning: An Introduction to the Comprehensive Plan*, 2nd edn. Washington, DC: Island Press, 2010. This book provides an excellent treatment of infrastructure considerations in the development of the comprehensive plan.

Websites

American Planning Association: www.planning.org. Many resources on planning for individual infrastructure systems, such as transportation, water, telecommunications, energy, and parks.

American Public Works Association: www.apwa.org. Many resources for preparing capital needs studies for the local practitioner.

policies expect to save substantial amounts of public and private dollars. Oregon projects $11.5 billion in savings for road expansion costs from 1991 through 2111 because of policies that coordinate transportation planning with other development. Salt Lake metropolitan area cities anticipate savings of $4.5 billion in transportation, water, sewer, and utility funds over the next 50 years by adopting policies that locate growth within the metro area. Similarly, Virginia Beach, Virginia, will save $275 million in infrastructure costs by planning for land use and infrastructure concurrently. Finally, New Jersey's cities and counties estimate that they will save $2.3 billion between 2000 and 2020 in costs for roads, water and sewers, and $160 million a year for schools, because the state specifically plans for infrastructure.[2]

Notes

1 Eric Damian Kelly and Barbara Becker, *Community Planning: An Introduction to the Comprehensive Plan* (Washington, DC: Island Press, 2000).

2 American Planning Association, *Planning for Smart Growth: 2002 State of States* (Chicago: APA, 2002).

chapter 7

Smart and Sustainable Development Rules

In this chapter...

Importance to Local Practitioners . . . 103
Local Responsibilities for Development . . . 103
Development Rules in Growing Areas . . . 104
Subdivision Ordinance and Its History . . . 104
Cluster Development, PUDs, and Green Subdivisions . . . 106
Site Improvements, Dedications, and Exactions . . . 106
Uniform Development Standards . . . 106
Development Impact Regulations . . . 107
Adequate Public Facility Ordinances . . . 108
Level of Service Standards . . . 109
Development Caps Based on Infrastructure Capacity . . . 110
Development Impact Fees . . 110
Development Rules for Urbanized Areas . . . 110
Zoning Ordinances . . . 111
Zoning and Infrastructure . . . 112
Specific Plans, Site Plans, and Redevelopment Plans . . . 113
Environmental Impact Assessments . . . 113
Right-of-Way Regulations . . . 114
Uniform Building Code . . . 115
Infrastructure and the Development Process . . . 116
Preliminary Plat or Tentative Map . . . 116
Detailed Drawings and Final Map . . . 116
"Vesting" Tentative Maps . . . 117
Construction of Public Facilities and Infrastructure . . . 117
Lot-Specific Infrastructure and Connections . . . 117
Building and Occupancy Permit Process . . . 118
Conclusion . . . 118
Additional Resources . . . 118
Publications . . . 118
Websites . . . 118

Why Is This Important to Local Practitioners?

Most development in the United States occurs as a result of investment by the private sector. In the best of all worlds, the comprehensive or general plan and capital budget are complemented by development rules that regulate the provision of infrastructure by the private sector. Development rules are the most ubiquitous planning documents at the local level, and are probably the most well known to local officials and developers, whether they are private or public. Development rules determine the shape, design, and cost of privately provided infrastructure for new development. They are the major mechanism that translates smart and sustainable principles in the general plan into directions that the developer will use in making these goals a reality.

Local officials need to understand the various types of infrastructure rules in order to assess the adequacy of existing development regulations in providing sustainable infrastructure as part of the local permit process. The first section of this chapter concentrates on development rules for infrastructure provision in "greenfield" areas, where streets, roads, civic buildings, and basic utilities must be provided from scratch. The second section discusses development rules for urbanized or already built up areas. The final section of the chapter describes the development permit process.

Local Responsibilities for Development

The local general purpose government (GPG) is responsible for establishing the standards and rules that are used to process local development applications. This is done within the framework of state-enabling legislation and federal constitutional principles. Often local officials play an active role in changing state requirements as well as in establishing local rules. Once established, these rules are used by planners and engineers to evaluate the adequacy of infrastructure provided by private developers as part of a new subdivision, or as part of an individual development project in a built-up area. In an ideal world, the development rules implement the overall vision for the community's land use plan described in the comprehensive plan. In actuality, many localities often "plan by doing" and adopt development rules without an up-to-date comprehensive or general plan.

The local GPG is also responsible for processing the development applications for large-scale infrastructure projects, whether they are being built by the GPG, an external agency, a special district, or a private utility. Development rules enable planners and engineers to evaluate the adequacy, location, and timing of this type of infrastructure as well.

Development Rules in Growing Areas

Subdivision Ordinance and Its History

Subdivision ordinances are a major source of on-site infrastructure requirements in developing areas. They had their origin in rules developed in colonial America, to ensure that land that was bought and sold had a legal paper trail. The subdivision was a legal mechanism for the registration of land ownership. Early versions of the subdivision ordinance focused on rules for recording subdivision plats, or maps, to ensure that the property was adequately described. Initially, recording was voluntary. However, in 1882, Oak Park, Illinois, became the first municipality to adopt an ordinance requiring subdivision maps to be filed with the village.[1]

Local governments then began to require that proposed roads be shown on the maps, but most did not require that these roads be connected to the roads in adjacent subdivisions. This situation changed in the late nineteenth century and first half of the twentieth. Local governments became concerned that new subdivision streets line up with existing city or county streets, and that street capacities were coordinated with the overall municipal or county road pattern.[2]

Similarly, the quality of early subdivision roads was not regulated by local governments and most were private. Instances of substandard roads were common, as were demands from residents that the government take over maintenance responsibilities once the roads began to deteriorate. Eventually, local governments capitulated, but to avoid taking over substandard roads, they began to insert quality standards into the subdivision ordinances.[3]

By the early 1950s, facilitated by federal mortgage insurance requirements, the location and standards for other infrastructure such as water and sewer lines, stormwater drainage systems, sidewalks, street lights, street trees and furniture, telecommunications, and gas and electricity lines were also included in local subdivision ordinances. The installation of physical improvements was often required as a prerequisite to plat approval.[4] By the 1960s, subdivision maps were being used in some localities to control the location of new development by requiring detailed plans for schools, sewers, parks, and community buildings. For example, some localities required that residential development near a sewer line hook up to it, rather than use septic tanks.[5]

Chattanooga, TN, Park

Road construction in Eugene, OR

The subdivision ordinance governs infrastructure requirements for new development in areas without roads, sewers, and water. They specify the requirements for turning raw land into individual parcels or lots. Unlike the zoning ordinance, the subdivision ordinance does not specify the use of the land. Nor do they address the infrastructure needs for a major commercial building in a downtown area, or off-site infrastructure provision for new development. Although many subdivision ordinances often do not address timing issues, they can do so. Usually, they are concerned with the development of the land itself: the layout of streets, lots, and the provision of streets, utility placements, sidewalks, lighting, and water and sewer lines. These ordinances set the standards for streets, sidewalks, water and sewers, energy, and telecommunications provision. They also regulate their location within the subdivision and the street right-of-way.

All states have provisions authorizing subdivision legislation at the local level. In some states, counties or regional agencies, rather than the local government, are responsible for subdivision applications.

Subdivision ordinance standards are a powerful tool to shape both the design and sustainability of the community. They can control whether the community will be walkable by the length of the block and the widths of the street. They influence whether the community will have shared green spaces or whether back yards will be divided up into separate spaces. They can be used to specify how stormwater will be addressed. Lot sizes will impact carbon emissions, and land use patterns embodied in the subdivision ordinance can have a positive influence on reducing motor vehicle emissions.

Cluster Development, PUDs, and Green Subdivisions

As subdivisions proliferated in the post-WWII "roll-out," the limitations of the lot-by-lot approach of the zoning ordinance were challenged. Developers wanted to be able to concentrate development on a smaller portion of the tract, or to be able to mix different land uses on the same parcel. What is now called cluster zoning, or Planned Unit Development (PUD), came into use. Cluster zoning ordinances provide for the same overall density as would be allowed under the existing subdivision and zoning ordinances for the area, but allow the use of more organic street patterns and more natural approaches to drainage.[6] PUDs are still the most common form of zoning in fast-growing areas.[7]

Site Improvements, Dedications, and Exactions

These days, subdivision ordinances contain detailed requirements for on-site infrastructure improvements. In many states, a subdivision map cannot be approved without them. Subdivision ordinances also provide a mechanism for the developer, who usually pays for and constructs the streets, sidewalks, sewers, and water lines, to transfer the title and financial responsibility for maintaining the improvements to the local government.

Cement truck for curbs, gutters and sidewalks as part of street construction

In the case of schools, parks, or a major upgrade to water or sewage treatment infrastructure that may serve a wider regional group of citizens, the subdivision ordinance can require impact fees to pay for them. (See Chapter 14 on development impact fees.) The subdivision ordinance must have a mechanism to apportion the costs among multiple developments as well as a way to collect the fees.[8]

Uniform Development Standards

Uniform development standards are the "guts" of the infrastructure requirements for developing areas. They generally include detailed engineering specifications for the capacity, location, placement, composition, and dimensions for the following infrastructure components:

- streets and roads
- sidewalks, pedestrian pathways, bicycle paths
- traffic control devices and signs, including street name signs
- street and pedestrian lighting requirements
- water mains, connections, and fire hydrants
- sanitary sewers, storm drainage mains, and connections
- utility lines, poles, conduits for energy, and telecommunications
- off-street parking
- landscaping, including drainage and erosion control.

Development standards are usually prepared by the local city engineer, and until recently they were not subject to public scrutiny or debate, despite the fact that the assumptions behind many of the specifications influence the level of service and the way infrastructure is provided. Requiring wide streets, for example, can result in a very different kind of community than permitting narrow streets. Standards for stormwater runoff, landscaping, and drainage will affect whether or not some of the newer wetlands or sustainable site-specific programs for stormwater can be implemented.

Development Impact Regulations

The subdivision ordinance and its variations may ensure that adequate on-site infrastructure is provided for new development. However, off-site or community facilities such as water treatment plants, landfills, regional transportation facilities, or school and library buildings are a more difficult issue. Traditional land use regulations address "the place, not the pace," and often key infrastructure is not provided in a timely manner.[9] In addition, most subdivision ordinances do not effectively address the fact that major transportation, water and waste infrastructure, and school facilities may not be provided by the local GPG.

Therefore, in the last two decades of the twentieth century, as rapid growth engulfed places such as Florida, California, and Washington, DC, a variety of additional development regulations emerged at the local and state levels. These rules are often referred to as growth management tools, but their purpose is mainly to ensure that an adequate supply of infrastructure, both on-site and off-site, is provided for new development in an efficient manner.

These development rules include Adequate Public Facilities ordinances (APFOs), level of service (LOS) requirements, development caps, and development impact fees.

Adequate Public Facility Ordinances

The APFO links approval of the development project to the adequacy of public facilities that serve it. APFOs are also known as "concurrency" requirements, since permits will not be issued until off-site improvements are provided "concurrently" with the impact of the development. The locality adopts an LOS standard for key infrastructure systems, such as water, transportation, or sewage treatment, in order to evaluate the adequacy of infrastructure for the project, to ensure that existing residents will not suffer diminished services if an existing infrastructure system were to be used by more people.[10]

In the 1950s, Clarkstown, New York, was one of the first cities to require public facilities to be in place before a development would be approved. Ramapo, New York's phased growth program adopted in the 1960s, is better known. Under the Ramapo program, if the infrastructure was not adequate, the development could not be approved, although the developer or owner of the property had the option to provide the infrastructure directly. Widespread interest in "concurrency" regulation did not occur until the 1970s, when Clean Water Act requirements resulted in localities refusing to issue building permits until upgraded treatment plants were built.[11]

The best-known of the APFOs was adopted in 1985 by the State of Florida. This legislation required local governments to prepare a comprehensive plan with a capital improvement element. Local governments in Florida were also required to adopt land development regulations with LOS standards for roads, sewers, solid waste, drainage, water, parks and recreation, and mass transit. The model statute designed by the American Planning Association for APFOs is based on the Florida law, with two major exceptions. First, states are responsible for defining the LOS, and, second, for ensuring both uniformity and flexibility across jurisdictions. Criticism arose that in Florida some localities used very stringent standards to preclude growth in their area, and therefore concurrency requirements had discouraged infill and perversely resulted in more development on the fringe.[12]

The Sierra Club's Urban Growth Management Guidelines

The Sierra Club's Urban Growth Management Policy Guidelines call for the following with respect to infrastructure:

> State law should mandate that no development project be approved by a public agency unless a commitment has been made for financing the necessary roads, water and sewer facilities, gas and electricity, parks, open space, schools and other public facilities related to that project. Development should be phased according to short- and long-range Capital Improvement Plans to ensure the timely provision of public facilities and services.
>
> State law should mandate that local governments establish Level of Service (LOS) criteria for major public services/infrastructure—sewer and water, main transit and transportation routes, gas and electricity, parks open space, schools, fire and police—in their community. These should be reviewed annually at a public hearing. LOS criteria should not be used to justify expansion of roads and freeways that serve single-occupancy vehicles.

Sierra Club, *California Urban Growth Management Policy Guidelines* (2001).

The APA model ordinance recommends only five types of facilities for inclusion in an APFO—water, wastewater, stormwater, solid waste, and roads—to meet the basic health and safety infrastructure needs. The *Smart Growth* handbook notes that other facilities, such as parks, recreation, and civic buildings, should be addressed with development impact fees.[13] Most APFOs require that the infrastructure be built concurrently with the impact of the development. Impact fees are one of several funding mechanisms to ensure that infrastructure is built. Additionally, a mechanism is needed to address the impact of development on the school district—frequently the third largest infrastructure cost, after transportation and water and sewers.

Level of Service Standards

Level of service (LOS) standards allow the locality to specify how much of the system is needed in response to the size of the development and the number of new households. LOS standards for water and traffic tend to be performance oriented, while standards for community facilities tend towards per capita measures. Traditionally, highway engineers rank highways and intersections from A to F, based on volume, capacity, and speed. (See Chapter 20 on streets which describes these rankings in further detail.) However, a movement is underway in transportation circles to modify these measures to focus on travel time and person capacity rather than vehicle capacity in order to reduce dependence on the automobile and to increase community livability. In the Portland area, the regional transportation plan permits local governments to forgo the LOS standards if they meet alternative standards such as connectivity, a managed parking supply, and transit service availability, among others.[14]

LOS standards for water measure drinking quality, water supply, and pressure. LOS wastewater standards measure the capacity to collect and treat it. Developing LOS standards for stormwater is more complex since wastewater can be treated by structural, pipe-oriented solutions, or on-site via green methods.[15] LOS standards for parks, libraries, schools, and civic buildings are usually couched in per capita measures of need, rather than standards of performance. Chapter 26 in this volume on parks shows how an LOS standard can be developed by a local government to reflect local preferences for parks and recreation facilities.

Many APFOs do not require an LOS for parks, libraries, schools, and civic buildings because the capacity of these facilities is thought to be more flexible than the networked infrastructure. A hundred students can be added to a large high school by increasing class size rather than increasing the size of the capital facility. Libraries typically welcome more patrons without feeling that their capacity needs to be expanded. Excess capacity in police and fire departments may be easily absorbed by the ability to respond to more calls for service. However, if the new development is significantly distant from the nearest fire or police station or school, a new facility must be built. These needs can be addressed in the APFO, but should also be reflected in the comprehensive or general plan and the capital improvement program.[16]

Development Caps Based on Infrastructure Capacity

Another way to ensure adequate infrastructure is to quantify the development that the existing infrastructure or the physical configuration of the area will support and limit development to that amount. This kind of development regulation allows the community to project the capacity of major infrastructure systems over time, and to limit construction permits to a certain number each year.

An urban service boundary can be based on the determination of the public water and sewer providers about the limits of growth that can be sustained within their service area, given their current ability to provide service and realistic expectations for expansion. Other health and safety systems can also provide an upper limit for development. Sanibel, Florida, determined the upper population limit that the island could house based on the ability of a bridge connecting it to the mainland to evacuate the island in preparation for a hurricane. An urban growth boundary can also ensure that existing infrastructure is used efficiently before expanding—a key motivation behind Oregon's urban growth boundary requirements.

Development Impact Fees

This topic has its own chapter (Chapter 14) in the finance section of this book (Part III). However, since a local ordinance authorizing impact fees is included in local development regulations, a brief discussion is warranted here about the rule itself. A local development fee statute can authorize the local government to assess fees for new development to cover the capital expenditures associated with the development. The fees would cover the cost of a new facility, or the expansion of an existing facility, to service the development.

Most of states that authorize impact fees also require that a local government have a capital improvements program, a facilities plan, or capital budget in place before assessing fees.[17]

Development Rules for Urbanized Areas

Most development in urbanized areas (not including a redevelopment or other specially designated area) takes the form of an individual project. The project is often an expansion of an existing use, or can be a large-scale change of use for a single parcel or group of parcels. Here the project inherits the infrastructure framework of the previous use. Thus the ability of a locality to regulate the provision of infrastructure in an already urbanized area for a development may not rest in the subdivision ordinance since the basic street pattern has already been laid out. Streets are paved. Sidewalks have been built. Water and sewer mains, as well as energy and telecommunications conduits and lines, are already located in the street. Schools, libraries, parks, and other civic facilities already exist. Indeed, most large cities, such as New York and Chicago, have "had little interest in subdivision control."[18] However, a variety of other local ordinances and regulations can be used to address the infrastructure needs of both large-scale and small-scale projects.

Zoning Ordinances

In an urbanized area, where subdivision ordinances may not exist or where they govern development already built and in the ground, the zoning ordinance and the building codes or other special regulations contain the infrastructure requirements.

The zoning ordinance is well known to most developers, since it is the primary determinant of what kind of development can go where in a built-up area. The ordinance itself consists of a written text that becomes part of the municipal code. The ordinance includes a detailed map with a parcel-by-parcel designation of the appropriate use, intensity, and density.[19] Zoning ordinances in the United States had their origin in the Standard State Zoning Enabling Act (SZEA) that was drafted in the 1920s for states to enable localities to adopt local ordinances. All 50 states have adopted enabling legislation for zoning. For the majority, however, zoning was developed without a comprehensive or general plan, and without a long-term infrastructure plan.

The zoning ordinance implements the general plan with respect to buildings and the site. This is accomplished by translating broad land use designations into uniform rules that apply to specific parcels. The major zoning designations are residential, commercial, industrial, and agriculture, although in many localities there can be scores of sub-designations. A typical zoning ordinance specifies the permitted use of the land (or structures on it), the intensity of that use, and the height and bulk, and has a zoning map that designates the zoning category for each property. A zoning ordinance also includes rules for processing the permit application, which include submittal requirements, time frames, and hearing requirements.

Zoning ordinances are not the major set of local regulations that affect off-site infrastructure, if we think of infrastructure as the networked systems like sewers, streets, water, energy, and telecommunication lines. However, any infrastructure system that has a major facility associated with it has to go through the same process that a residential, commercial, or industrial use does. A school, community center, sewage treatment plan, landfill, or telecommunications towers, all must go through the local permitting process set up in the zoning ordinance. Sometimes the land use designation may need to be changed, since in many localities infrastructure uses were not considered when the comprehensive plan was prepared and the zoning ordinance adopted.

Zoning and Infrastructure

Zoning relates to infrastructure in four ways. First, zoning can be used to target growth in the part of town that already has adequate infrastructure. Local governments can do this by rezoning to permit higher densities and mixed-use development in these areas.

Second, zoning ordinances can be modified to address particular infrastructure needs. In some localities, the zoning ordinances contain detailed provisions for street light improvements, landscaping, sidewalks, and underground utilities. Some even contain provisions for impact fees to provide for offsite infrastructure caused as a result of the project.

Third, zoning relates to infrastructure in its traditional way. It regulates the location, intensity, and density of a new infrastructure project, just as zoning regulates the same aspects of private residential, commercial, and industrial projects. When looking for a location for a public building, the GPG's project manager must take into account the height and bulk requirements of the zoning ordinance, absent a strong indication that elected officials will waive these requirements. Similarly, the architect or project manager for a school expansion project, or a new courthouse in the downtown area, must be guided by the local zoning requirements. Zoning regulations apply to developers of small-scale energy plants within a local jurisdiction, as well as to a large-scale sanitation facility or landfill. However, many localities can "zone out" uses like toxic waste disposal sites.

Fourth, zoning ordinances can be used to implement sustainability principles in the general plan, ranging from allowing increased density to reduce energy and water use to requiring water-conserving landscaping.

Height limits can be increased to promote the use of wind turbines. Solar collectors can be included as a by-right accessory use, and over-the-counter permits and fee waivers can encourage the use of energy conservation and on-site production. Water harvesting can be encouraged in local zoning ordinances (and building codes), and landscape credits can be provided for tree preservation and tree planting for carbon sequestration. Green roofs can be required or encouraged to offset CO_2 emissions and help to cool cities.[20]

Specific Plans, Site Plans, and Redevelopment Plans

Zoning regulations usually apply to a single parcel-based project. However, in urbanized areas, sometimes the local government or the developer wants to develop a project that encompasses an entire sub-area in the locality, or perhaps only a block or two with multiple parcels. In either case, many localities have adopted development rules to deal with larger urban projects. In some states the zoning ordinance can designate areas where a specific plan can be used along with a development agreement to ensure that needed infrastructure will be built. The development agreement requires that the locality adopt the enabling legislation to permit its use. Once in place, the development agreement provides a mechanism to negotiate for infrastructure items that otherwise could not be mandated.

Redevelopment plans can also be used for this purpose. A site plan ordinance may be on the books, with some of the more detailed technical development provisions normally found in the subdivision ordinance for developing areas. The local practitioner needs to ensure that rules in the jurisdiction contain adequate protections to provide for needed improvements in infrastructure as a result of the reuse, or increased use, of the area.

Environmental Impact Assessments

In 1969, the National Environmental Policy Act (NEPA) was passed in response to concerns that federal projects were having negative impacts on local communities. NEPA requires an environmental study and a series of hearings for federally funded projects that pass a certain threshold level of impact (including infrastructure impact) on the environment. Since 1969, 27 states have enacted similar legislation for government projects, and five have passed mini-NEPAs for private development.

The environmental review process can identify the impact of the project on schools, water and wastewater treatment plants, and other infrastructure. The review begins with an initial study to see whether further study is necessary. If the project will cause a significant impact upon the infrastructure (and the rest of the environment), a more detailed study is prepared. If significant

Infrastructure Deficiencies Often Mitigated During Environmental Review Process

Infrastructure deficiencies are often identified and "mitigated" during the project review as part of the environmental review process. Savvy developers quickly learned that the best way to get a project through the process was to address the infrastructure impacts (among others) of the project and to receive a negative declaration (an initial assessment finding no significant environmental impacts). In 1990, about 1600 to 1800 full-scale environmental studies were completed by local governments in the USA, compared to about 30,000 negative declarations. At that time, there were as many environmental reports in California as in the entire rest of the United States.[1]

Note

1. Robert Olshansky, "The California Environmental Quality Act: Implications for Local Land Use Planning," *Environmental Assessor* 3(1) (1992).

impacts are found, then the project must be mitigated, or the lead agency must approve the project based on a "statement of overriding considerations." A significant impact on infrastructure may mean that the project is violating certain standards that the community wants met for the system in question.

Mitigation of environmental impacts is probably the most important outcome of the process. Most developers do not like to pay for the preparation of environmental studies and will modify the proposal early on if possible. If not, the project may be modified later on to obtain approval. Local infrastructure mitigations may consist of putting money into a fund for a new stop light or left-hand turn signal to be built by the city. It may also be through providing impact fees to pay for an upgrade to the water treatment plant or local schools.[21]

In 2007, the California State legislature voted to modify its environmental review process to include guidelines to assess greenhouse gas emissions (GHGs) when conducting a CEQA (California Environmental Quality Act) review of a development project. This vote was precipitated by a lawsuit filed by the State's Attorney General against San Bernardino County for failing to consider GHG emissions in its proposed general plan update. The final draft guidelines are available at www.opr.ca.gov.

Right-of-Way Regulations

Another set of local powers or regulations that affect infrastructure concern power over the "right-of-way." The right-of-way (ROW) is either an easement or ownership right that the locality has over land for public streets, sidewalks, and utilities. Early ROWs arose from use, and correspond to the street itself. Today, in new subdivisions, they are created separately and extend past the street to include the median, the sidewalk, and any drainage and snow collection areas.[22]

Until the early 1960s, many cities used alleys behind the houses to locate infrastructure for sewer, water, telecommunications, and power, as well as for garbage collection. Today, however, most of these utilities are in front of the house. Depending on the local ordinances, the city or the private owner may maintain the median in front of the property.

The local GPG has the right to issue a contract, or franchise, to private utility operators, to put infrastructure in the

ROW. This infrastructure can be underground pipes and cables, or poles for telecommunications and electricity, or utility boxes, and may include permission to put wireless transponders on existing city light poles. A bus stop can be put in the row.

These issues are discussed in greater detail in the Chapter 20 on streets and Chapter 28 on telecommunications.

Construction of local bus stop, Eugene, OR

Uniform Building Code

The Uniform Building Code was developed in response to concerns about health and safety in tenements at the end of the nineteenth century. The first building code in the U.S. was published in New York City by the Tenement Housing Commission in 1901. It contained standards for space, fire protection, and the provision of plumbing. Building codes were subsequently adopted across the country in the early part of the twentieth century.

Today, the term "building code" refers to the panoply of local codes that regulate construction of new development, usually consisting of electrical, plumbing, mechanical, and building requirements. Although originally codes were generated locally, today they are based on model codes prepared by the national associations of professionals in the field, including architects, engineers, and building officials. Over 60 agencies set standards for types of material that can be used in construction:[23]

- *The Uniform Building Code* is developed by the International Conference of Building Officials, and addresses the structural elements of the building. It also has provisions for earthquake safety, hurricane and tornado safety, and fire safety.
- *The Uniform Plumbing Code*, promulgated by the International Association of Plumbing and Mechanical Officials, covers everything from proper diameter of the drain opening, to the correct piping material, to the required water pressure for a particular type of toilet.
- *The Uniform Mechanical Code*, developed by the International Association of Plumbing and Mechanical Officials, deals with issues of indoor air quality and HVAC system design.
- *The National Electric Code* discusses electrical safety, and is published by the National Fire Protection Association. This association publishes a series of other codes, with topics including the fire resistance of wall materials, means of egress during emergencies, and the proper location of fire alarms and fire extinguishers.

Innovative Building Codes, Standards and Rating Systems

Within the past several years, a plethora of standards and rating systems designed to regulate greener buildings and communities has come onto the stage. LEED v4 is in its final stages of development at the time of this writing. The International Green Construction Code has been published while the California Energy Code and CAL Green are undergoing important updates. The Massachusetts Stretch Energy Code has been adopted in more than a third of its communities. A certification system for sustainable infrastructure is available online at the Institute of Sustainable Infrastructure's website.

Building codes are local creatures, adopted, modified, and enforced by local officials. The national codes are recommendations that local governments may choose to adopt in part or in whole, with or without modification. They are not intended to limit developers to certain materials if the local building official determines that others will do as well (and can convince the local legislative body to adopt these provisions as part of the local code). This flexibility can be especially important with green building practices and conservation methods which are comparatively new and may not be addressed in local ordinances.

Infrastructure and the Development Process

Preliminary Plat or Tentative Map

The first step in the designation of infrastructure in a greenfield subdivision begins when the developer submits a preliminary map (plat) that shows the layout of the streets, lots, public utilities such as water, sanitary, and storm sewers, and private utility placement for cable, gas, electricity, and telecommunications. Contour lines are shown along with the major physical features of the site, such as streams, ponds, and forests. The application is submitted to the planning department, which circulates it to affected agencies and other City staff for comments. The application then goes to the planning commission for its initial review.[24]

At this point, the adequacy of the on-site and off-site infrastructure is reviewed. Utility easements must be in place, sites must be dedicated for parks, and "green" infrastructure improvements noted. Specific infrastructure issues of concern to the planning commission are:[25]

- whether sites have been reserved for public facilities such as parks, schools, or other civic needs;
- whether utility easements for electricity, gas, and cable are indicated;
- whether stormwater is treated on-site or not;
- whether transportation easements are adequate.

Detailed Drawings and Final Map

The second phase of the subdivision application process includes preparation of the detailed maps for the development. The developer hires a surveyor, engineers, and design firms to prepare the detailed final plat or map. This legal document fixes the location of lots and streets, and will create the legal land title record for locating the lots. It is also the legal document that serves as the basis for designating streets, sewers, sidewalks, and other infrastructure items that the city or county will ultimately maintain.[26]

In addition to the final map, the developer submits detailed engineering drawings of the street profiles, and drawings that show water and sewer lines, location of street lights, fire hydrants, and other infrastructure. Construction specifications required by the locality are incorporated. Compliance with the zoning ordinance and the comprehensive plan is also determined. Usually, the final documents return to the planning commission, and sometimes the elected officials of the local GPG, for final approval. Sometimes final recordation is delayed until the developer actually constructs the public improvements.[27]

"Vesting" Tentative Maps

In some states detailed drawings are created earlier so that the developer can lock in the requirements and general configuration prior to the final map. The planning commission approves with some conditions, which are then added to the package. Final approval in this case is usually administrative.

Construction of Public Facilities and Infrastructure

The third phase of the infrastructure process in a greenfield development is the construction of the on-site public facilities and infrastructure by the developer. The local engineering or public works department may inspect the streets and sidewalks frequently during construction, since they ultimately will be deeded over to the local GPG for maintenance. (Off-site infrastructure is handled separately, since it may be funded from several development projects.)

Lot-Specific Infrastructure and Connections

The last step is to connect the public infrastructure on the residential or commercial site itself. In cases where the infrastructure—streets, sidewalks, and sewers—have been built prior to the units, the individual lateral connections on the private property must be made. Once the construction of the unit begins, these lateral connections are constructed by the developer, and include telecommunications connections, sewer and water lateral, stormwater lateral, and wires or pipes for electricity and gas. Local building codes contain the requirements for these connections.

Principles for Assessing Adequacy of Local Development Rules

- Local rules should ensure that development does not occur without adequate infrastructure.
- Development should generally pay its own way for infrastructure, with provisions for subsidizing high housing costs or constructing smaller units to ensure a portion of the development is affordable.
- The community should have a mechanism to direct new development to areas with adequate infrastructure.
- The subdivision ordinance, or site plan, and the technical specifications (development rules) should reflect current best practices for environmentally friendly, carbon-neutral, and cost-effective infrastructure.
- These documents should also reflect any "hardening" required to withstand natural disasters, or the decision by a locality to "withdraw from a high threat area.
- The local GPG should take the initiative to work with special districts (parks, schools, water, and sewer) and regional agencies on making sure that development rules in the metropolitan area integrate local and regional needs, and are consistent with each other.
- Development standards and LOS requirements along with the building code should be reviewed to see if they need to be brought in line with current community values.
- The cost of planning studies should be included in infrastructure fees.

Building and Occupancy Permit Process

Application for the building permit is made after the locality issues its subdivision approvals and use permits. The occupancy permit is issued once the locality has ensured that code requirements for infrastructure (and the other codes as well) are followed. This is done by reviewing the plans before construction begins; and then inspections by local officials during the construction process.

Conclusion

Development rules for infrastructure are the responsibility of the local jurisdiction. No one right mix fits all localities. They need to be tailored to the condition of the existing infrastructure and development pressures. Usually, the issue of whether the locality has adequate development rules for infrastructure arises in response to a problem. A comprehensive assessment of infrastructure development rules is expensive, and must be balanced against the number of new projects and the impact that the rules will have. Development regulations are also one of the key places that changes can be made to reduce carbon emissions and otherwise ensure that the community is moving towards sustainability.

Additional Resources

Publications

Duerksen, Chris, Erica Heller, James Van Hemert, et al. *Sustainable Community Development Code: A Code for the 21st Century, Beta Version 1.1*. Denver, CO: Rocky Mountain Land Institute. A framework for a set of development regulations for a more sustainable community including various infrastructure systems.

Kelly, Eric Damian. *Community Planning: An Introduction to the Comprehensive Plan*. Washington, DC: Island Press, 2010. Contains good sections on development regulations.

Meck, Stuart. *Subdivision Control: A Primer for Planning Commissioners*. Chicago, IL: APA Planners Press, 1996. Basic guide for policy-makers.

Morris, Marya. *Smart Codes: Model Land-Development Regulations*. PAS Report # 556. Chicago, IL: American Planning Association, 2009. A guide to model smart growth ordinances that includes sustainable infrastructure regulations.

Websites

Department of Energy: www.energycodes.gov. The latest with respect to energy-efficient building codes at the federal and state levels.

State of Washington: www.depts.washington.edu/trac/concurrency/lit_review/lit_review.html. Website with an extensive literature on concurrency containing abstracts of levels of service standards used by different communities in the United States.

chapter 8

Developing the Public Infrastructure Project

Why Is This Important to Local Practitioners?

Although some on-site infrastructure is built by private developers in the course of developing subdivisions or individual projects, most off-site and other large-scale capital improvements, including the replacement of older infrastructure in the right-of-way, are built and financed by public agencies. About half of this is the responsibility of the local general purpose government (GPG), while the remainder is the purview of special districts for water, sewer, and the local school district. Indeed, annual expenditures for the capital budget can account for a substantial part of the GPG's annual budget. It is likely that most local practitioners will be involved with the development of an infrastructure project sometime in their careers. Many will be project managers, while others will be on the project team, or will review the project as it is permitted. Therefore, an overview of how to develop a public infrastructure project is important.

This chapter focuses on the development of public buildings, but the process is similar for the development of other infrastructure facilities as well.[1] Differences will be highlighted in the chapters devoted to the individual systems. This chapter begins with an overview of the infrastructure development process (Figure 8.1) and an assessment of current local limitations. It then describes the five stages in developing a public project: (1) developing the concept; (2) organizing the project; (3) financing and site selection; (4) design, permits, and site acquisition; and (5) construction. The chapter concludes with current issues with public infrastructure construction.

In this chapter...

Importance to Local Practitioners . . . 119
Phase I: The Concept . . . 120
Phase II: Organizing the Project . . . 121
The Design Team . . . 122
Environmental Consultants . . . 123
The Finance Team . . . 123
The Site and Right-of-Way Acquisition Team . . . 123
The Construction Management Team . . . 123
Phase III: Financing and Site Selection . . . 124
Financing . . . 124
Selection of the Site . . . 124
Phase IV: Design, Permits, and Formal Site/ROW Acquisition . . . 126
Preliminary Design . . . 126
Life-Cycle Cost Analysis . . . 126
Design Review and Value Engineering . . . 127
Discretionary Permit Process . . . 127
Site Acquisition and Eminent Domain . . . 128
Phase V: Construction Contract and Building Permits . . . 130
Finalize Design Contract . . . 131
Prepare Detailed Plans and Specifications and Plan Review . . . 131
Enter into the Construction Contract . . . 132
Phase VI: Construction and Project Closeout . . . 133
Who Monitors Construction? . . . 133
Local Agency Inspections: Building and Engineering . . . 134
Project Close-out . . . 134
Conclusion . . . 134
Additional Resources . . . 135
Publications . . . 135
Website . . . 135

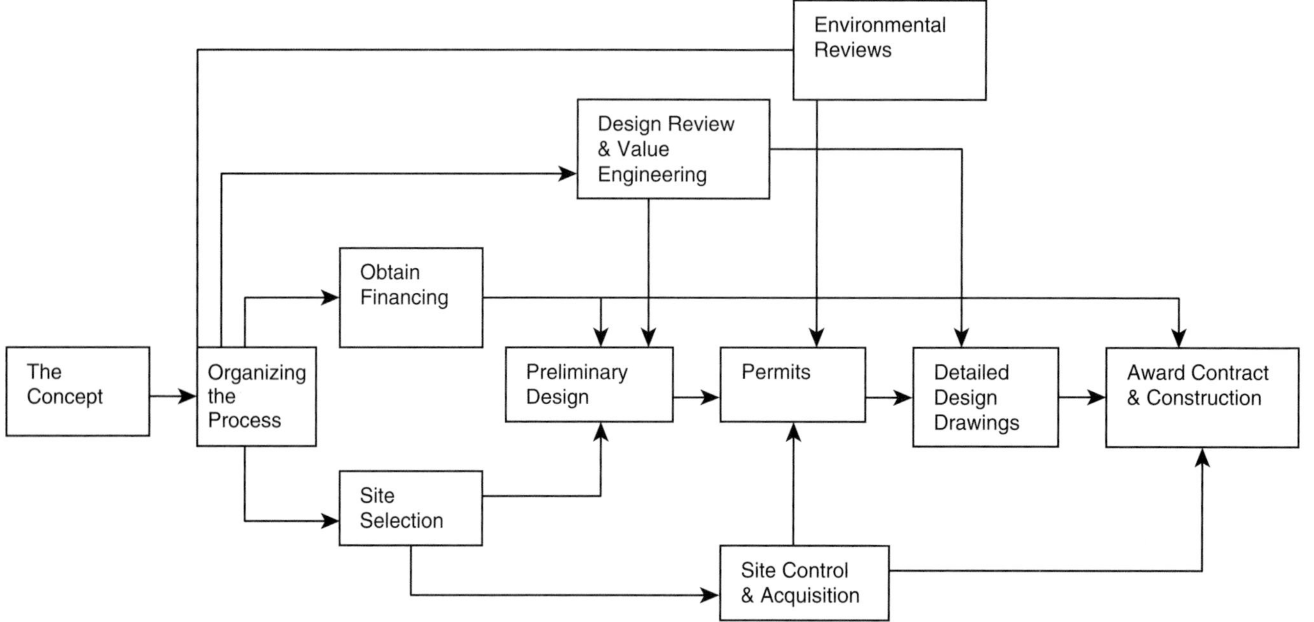

Figure 8.1
Overview of infrastructure project development process

Phase I: The Concept

The concept phase is the most fluid, and for the large-scale public infrastructure project, the most important. At the concept stage, the preliminary budget and time line are developed. These provide a baseline that can be used to assess tradeoffs for making the project larger or smaller, or to do more or less quickly. They should be regarded as temporary, and subject to constant revision.

At the concept stage, a temporary project team is formed with representatives from finance, planning, and the department with operational responsibility for the project area (Figure 8.2). In a city or county, this would mean representatives from the finance, planning or community development, and public works departments.

If this is the first time in several years that the jurisdiction has proposed a project of this size, the composition of the original core team will probably change once the jurisdiction proceeds to the next stage. The first team should think of itself as temporary—putting together a work program to get staffing, financing, and the political go-ahead for the project. However, specialized consultants may also be needed to participate in the preparation of the work

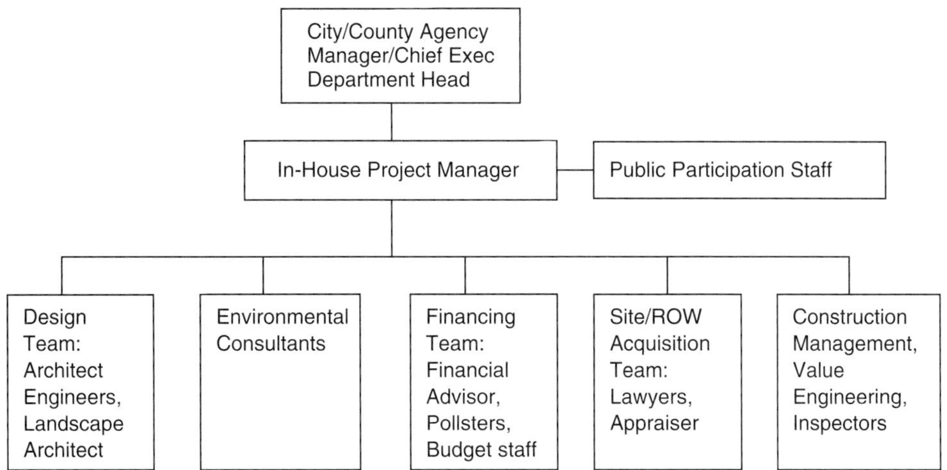

Figure 8.2
Organization chart of an idealized project development team

program for the rest of the project—a specialist in jail construction or civic center design, or a library, park, or large-scale street or sewer specialist, for example. A short-term contract can be let. If a consultant is hired, he or she may be asked to stay on through the next stage.

At this stage, the task is to lay out the range of possibilities for size, budget, and issues, so that the policy-makers can proceed to the next stage—that of deciding to look for financing and a site and/or right-of-way. The equivalent of the sophisticated private sector market analysis for the typical public sector infrastructure project—perhaps a civic center, recreation center, school, parking garage, or public safety building—is the poll taken of the taxpayers to determine whether they would vote for the bond issue to pay for the project, and at what size. Quasi-public buildings, such as convention centers, museums, and sports arena, are usually developed in concert with the private sector, and extensive market feasibility studies are done to target the commercial development that usually accompanies them. A refined concept for the project should come out of this phase along with the decision to hire the skeleton of the permanent infrastructure development team.

Berkeley Public Library, New Addition

Phase II: Organizing the Project

The next step is to prepare a work program, working budget, and the plan for the development team that is tailored to the specific needs of the jurisdiction and project.

A decision must be made as to whether the project will be designed in-house, or whether consultants (architects, engineers, and other specialists) will be hired to develop the design or some combination. The local agency will be influenced in its decision by the source of the funds and its rules for the design and construction process. State laws also prescribe contracting procedures, and some will not permit local governments to let a design/build contract.

If the infrastructure project is large and complicated, the locality would do well to hire a staff person with development experience from the outside as the overall project manager. Project managers for large-scale infrastructure projects are often known as "public entrepreneurs" and should be a salaried, in-house employee even if they are hired just for the project. They usually are paid more than other planning or engineering staff because they have skills that are generally in short supply—including a personality to make the project happen in the face of conflict.

The Design Team

If the infrastructure project is a building, the head of the design team is the architect. Depending upon the size and location of the project, a series of engineers will be part of the team. The responsibility of the structural engineer is to make sure the facility is structurally sound and that the mechanical systems are designed properly. The mechanical engineers are responsible for the HVAC (heating, ventilation, and air condition systems) and plumbing, while electrical engineers design the energy and telecommunication systems. Engineers and their firms bear legal liability for the plans and specifications they develop.

Civil engineers may be part of the team—every building needs streets, water, sewer, and other basic infrastructure systems. Geotechnical or soils engineers determine the site's bearing capacity and can make recommendations for fill or soil replacement so that the structure is stable. These are the specialists who conduct seismic safety, percolation, and compaction tests on the soil. Environmental engineers determine whether any hazardous materials are on the site.

A transportation engineer will advise on parking, street improvements, and traffic signals. The transportation consultant can play many roles for the large-scale infrastructure facility such as developing parking for bicycles and cars, or transit plants. The presence of a higher density development may result in changes to the transportation system. A landscape architect will do the site planning and produce the master plan for all landscaping and hard surfaces. All these professionals have legal liability for their work.

In the case of a series of smaller projects, such as a street reconstruction, sidewalk repair, or sewer replacement project, the design may be done by in-house civil engineers. It makes sense for the locality to hire the staff permanently if the project will last for longer than five years, and if the projects are routine. If the projects are unique, or if the financing source is not stable, it makes more sense to contract for the design.

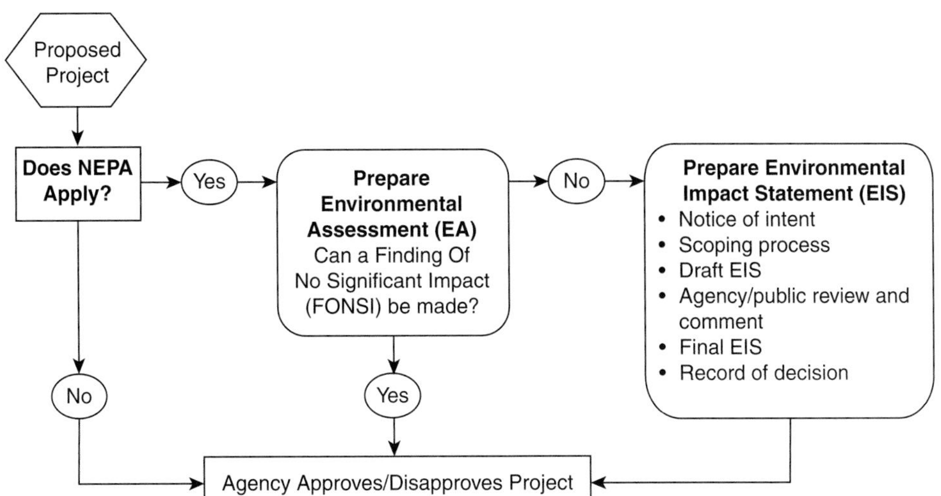

Figure 8.3
Flow chart of the environmental review process

Environmental Consultants

The environmental consultants may be in charge of the environmental review process (Figure 8.3). These firms may also be of assistance in understanding regulations about hazardous materials and wetlands, and can help the project manager understand the types of permits needed beyond that of the city or county.

The Finance Team

If the project is a large one for the GPG, the finance director, or budget officer or his or her representative, will likely be part of the team. Depending upon the type of project, the project manager might wish to hire an appraiser or a cost-estimating firm to provide a range of costs to help in framing the concept of the project, and then to cost out the project in advance of bidding.

The Site and Right-of-Way Acquisition Team

The assistance of lawyers skilled in site acquisition and environmental regulatory matters is also important. The GPG's legal counsel needs to be part of the team from the beginning. It is usually the city attorney or county counsel who makes the call whether to hire specialized counsel for the project or to do the work in-house.

The Construction Management Team

The project manager can be in charge of construction management. If the project is quite large, a separate firm expert in these matters should be hired for the day-to-day on-site construction management. The building department needs to be alerted that the project will be coming through if it is a non-right-of-way project, and therefore will need building, electrical, plumbing, and other permits. The engineering staff who review right-of-way matters and issue permits should also be notified about the project.

Phase III: Financing and Site Selection

Once the policy-makers and management have decided that it is worthwhile to pursue the infrastructure project further, a more permanent project management team is assembled to firm up the financing and site selection.

Financing

Bridge in Chattanooga, TN, part of waterfront redevelopment area.

It is rare that the locality will have the financing in place for a large-scale infrastructure project when it first is conceived. Once in a while the availability of a grant triggers the project. For small-scale sewer, sidewalk, or street reconstruction projects that are let out to bid once a year, it is likely that funding is in hand. However, these smaller projects can be thought of as pieces of a larger infrastructure project that also had to be initiated, sized, and an on-going source of funds secured.

As the revenues are tabulated, the core team might also hire the services of a professional estimator to provide an overall picture of what various levels of funding will achieve. Project managers usually assume, accurately enough, that the project will not be abandoned midstream for cost overruns. In addition, at the beginning of the project when the funds are not in hand, there is enormous political pressure to low-ball the project cost, particularly if the project is a controversial one, such as the Boston Tunnel or the San Francisco Airport expansion.

Picnic table near wildlife center in Berkeley

Selection of the Site

The location of the project for a public infrastructure facility can be controversial, particularly if the public agency or local government does not own the land. In a mature urban area, the location for a new school, toxic waste site, baseball field, court house, electrical peaker plant, stadium, or water treatment plant, or even a bike trail, may be the subject of heated political debate. Sometimes at this point the GPG may hire a consultant to do a site reconnaissance, to list vacant sites and sites presently for sale that meet criteria given to the consultant by the jurisdiction. This might include estimated cost of purchase, square footage, present zoning designation, owner, and the cost of demolition if the property has a structure on it. A bare minimum analysis would be a list from a realtor. A more in-depth analysis consists of cost estimates and preliminary height and bulk zoning constraints for the available sites.

The following factors should be considered in selecting the site for the infrastructure facility:[2]

- *Physical features of the site*: What is the size of the site or parcel? What are the soils and topography and hydrology of the site? Is it in a flood plain? Are the soils stable enough to support the projected use?
- *History of the use of the site.* What has the site been used for? Was it a gas station, or the site of a dry cleaner, thus requiring expensive hazardous material cleanup? Re-use of industrial sites always has this concern, especially for urban parks.
- *Zoning.* What are the legal constraints on the use of the site? Can it be used for a water treatment plant? A school? What restrictions are there on density and setbacks? Will variances be needed? For what? What are the adjacent land uses? Are these compatible with the needs of the public facility?
- *Local attitudes.* Even if the infrastructure meets the zoning (and it is a rare development that does not need a variance or two), will there be significant opposition to the infrastructure facility on this site? Is there sufficient political will, or is there a community group or political coalition willing to see this library or convention center built here?
- *Infrastructure.* Even infrastructure needs infrastructure. Will the sewer lines and treatment facilities support the expanded use if the project is a school, a public safety building, or a convention center? Will the water supply be adequate? If this is a large-scale facility, will the cost of retrofitting the existing lines outweigh the benefits of the site?

Sometimes site selection is integrated into a public design process. In this case, a land use or architectural firm specializing in site design is hired to develop several different land use configurations. This is done publicly, in order to assist policy-makers in developing a consensus about the finished project, so that when site selection actually occurs, the public has helped to craft the final result. Even so, there still may be opposition to the acquisition of certain sites.

Phase IV: Design, Permits, and Formal Site/ROW Acquisition

When the financing and site selection looks secure, the process of developing more detailed plans for the site can be initiated. The formal site acquisition process and the local permit process can begin. Usually these tasks await determination of financing, but it is not unknown for a locality to pay for preliminary design work, the project manager's salary, and the initial round of consultants right up to the decision to start construction. This would be about 15–20 percent of the total project cost.

Preliminary Design

For large-scale building projects a design competition may be warranted. A design competition can be used to select the design team for the more detailed design work.

For projects in the right-of-way (ROW) that do not typically require a use permit, preliminary designs (for sidewalks, sewers, street trees and storm sewers, and street reconstructions) go right to the funding agency for approval and then to the engineering department of the local GPG. The engineering department usually prepares the detailed designs, or contracts with an engineering firm. Frequently, ROW projects go right to detailed plans and specifications used for bidding (see below).

Life-Cycle Cost Analysis

Life-cycle analysis, also known as life-cycle cost analysis, is the consideration of all agency expenditures and user costs throughout the entire service life of an infrastructure system, not just the initial capital investments. (It should not be confused with life-cycle assessment, a term used in the consumer product sector and environmental community to refer to the process of evaluating the effects that a product has on the environment over the entire period of its life, including raw material extraction, manufacture, packaging, and marketing.) Life-cycle analysis is not a simple cost comparison, but can offer sophisticated methods to determine and demonstrate the economic merits of the selected alternative in an analytical and fact-based manner.

Life-cycle analysis can help infrastructure providers answer questions such as:

- Which design alternative results in the lowest total cost to the service provider over the life of the project?
- Will facility replacement be better than facility preservation?
- What are the user-cost impacts of alternative maintenance strategies?[3]

Design Review and Value Engineering

Value engineering refers to the process of hiring a second engineering firm or architect and engineering firm to go over the detailed plans and specifications, eliminating the more expensive tasks or design features to reduce the size of the construction contract.

A process that encompasses this step but also looks at the relevance of the design itself is called a facility design review (Table 8.1), which is a more extensive process. These kinds of reviews can be done at the beginning or end of the design phase, or after construction bids are received, if they are well over the funds available, or throughout and after the process. These kinds of reviews have been lauded for their cost effectiveness, but they are sometimes criticized for eliminating the aesthetic amenities of the project as well.

Discretionary Permit Process

When the preliminary design has been completed, the discretionary permits can be applied for, and the environmental review process can be initiated.

Table 8.1
Key questions when preparing a facility design review

Topic	Key question to be addressed
Owner satisfaction	Does the constructed facility meet the owner's expectations as originally defined by the project scope definition or statement of work (i.e., performance characteristics, architectural statement, level of quality, cost, schedule, and any relevant owner-published standards and/or policies)?
Sound professional practice	Is the approach taken in each of the specialty areas (architectural, civil, mechanical, and electrical) commensurate with professional standards?
Code compliance	Does the design comply with all applicable codes, such as fire protection, life safety, and access?
Architectural statement	Is the overall presentation representative of established architectural standards?
Sustainability	Does the design minimize greenhouse gas emissions and energy and water use and use environmentally friendly materials?
Value engineering	Are there any less expensive methods or materials that could be used in the design without impacting project quality or performance (or life-cycle costs)?
Bidability	Are the construction documents sufficiently clear and comprehensive so construction contractors will have no difficulty developing an accurate bid with minimal allowance for contingency?
Constructability	Does the design impose any unnecessarily difficult or impossible demands on the construction contractor?
Operability	Does design of the facility operating systems ensure ease and efficiency of operation during the facility's useful lifetime?
Maintainability	Does the facility design allow for easy and cost-effective maintenance and repair over the useful life of the facility?
Life-cycle engineering	Does the design represent the most effective balance of cost to construct, cost to start up, cost to operate and maintain?
Post-occupancy evaluation	Could any unexpected difficulty have been avoided by a different design approach?

Source: U.S. General Accounting Office, *Role of Facility Design Reviews in Facilities Construction* (Washington, DC: U.S. General Accounting Office, 2000).

Use permits. Non-right-of-way projects—that is, anything located on a parcel, such as a stadium, a parking garage, water treatment plant, telecommunications tower, police department, or school—need a zoning or use permit. Typically, a design at this stage shows height and bulk, and contains a site plan and elevations—enough for the local planning body to determine what types of permits are needed. Specific requirements are contained in the local zoning ordinance.

Specialty permits. If state or federal grants are involved, they will have their own permits that must also be applied for, with their own application requirements. Jails, schools, water and sewer treatment plants, and solid waste facilities usually have their own permitting agencies at the state level. Telecommunications and energy providers are regulated by state public utility commissions, and facilities must meet their specifications and permit requirements.

Environmental review. If the project has a significant impact upon the environment and is located in a state with an environmental review process; or if it meets the federal NEPA standards, the preliminary design must present several alternatives that can then be evaluated, including the no project alternative. The federal process consists of three phases: deciding whether NEPA applies; preparing the environmental impact assessment; and then preparing a finding of no significant impact (FONSI) or and environmental impact statement (EIS). State environmental review processes follow the same general process.

Site Acquisition and Eminent Domain

The public entity has selected the site and secured or identified financing. It is now time to begin acquisition of the site and any rights-of-way needed for the project. In the best of all possible worlds, a site may become available for sale, and the local general purpose government will buy it. Local governments are usually required by their charters, or by state law, to pay no more than fair market value for the site. Again, in the best of all possible worlds, the seller is cooperative and the GPG is able to gain access to the site to obtain soils borings to determine the condition of the soil, and whether there are toxics on the site. The appraiser finds that the site is priced at the fair market value. The real property acquisition officer of the local government (often in the finance department), negotiates successfully with the owner since toxics (in the best of all possible worlds) are not a problem, and the site is acquired at the appraised value.

In the real world, this does not always happen. Site acquisition can be a thorny process. The ideal site may not be for sale. Necessary right-of-way access routes for pipes or gas lines may pass through the backyards of the most recalcitrant of neighbors. An owner originally willing to sell the perfect parcel for a transformer station may be pressured by irate neighbors and friends to take the parcel off the market. The last remaining sites needed for a regional bike path system may pass through the backyard of a private school that publicly and repeatedly refuses under any circumstances to negotiate with the city, and even attempts to get a law passed at the state level to keep the city from getting the site through eminent domain. Or, once municipal, county, or special district interest is shown for a site, the realtor announces that the price has been doubled.

All of these situations can be resolved if the local officials have the political will to go through condemnation—that is, a process of forced acquisition through eminent domain. "Eminent domain is the power of the government to take private property for a public purpose without the owner's consent." When eminent domain is used to acquire property (and even when it isn't) the local government must pay just compensation or the fair market value of the site. Fair market value can be thought of as the highest price that a seller, with full knowledge of the desired use, would accept being under no duress to the seller, and that a buyer would agree to, also being under no duress to purchase.[4]

Transit stop and Bus Rapid Transit, Eugene, OR

Bike lane, and bike box in Eugene, OR

Given that it is not possible to know in advance whether best case or worst case will occur during site or right-of-way acquisition, it is prudent to take precautions so that the jurisdiction is protected if it needs ultimately to go to court and a jury trial to set the value of the land. This is why many large public authorities with substantial amounts of development have their own in-house site acquisition teams. Most cities, counties, school, and park districts will not find themselves in this situation, and they may have to use an outside lawyer, appraiser, and other staff during this process.

The first few steps of the site acquisition process are the same, whether site acquisition will proceed smoothly, or whether it will be turbulent. The first step in condemnation proceedings is to negotiate in good faith with the owners or his or her representative. To negotiate with some degree of acumen, an arm's length appraisal is required. Care must be taken in the appraisal process to ensure that if the amount the property is valued at goes to court, the owner will not be able to accuse the public agency of influencing the appraiser.[5]

If things are going well, it is at this point that the owner's representative will allow access to the site for soils testing, boring, and hazardous material appraisal. If the owner resists the condemnation, and will not grant access to the site for the tests, the formal condemnation process must be started to obtain a court order for access to the site. The site does not need to be condemned before this occurs. The formal condemnation process provides for a "resolution of necessity," that the agency adopts, indicating why the site is needed for a public purpose. This is usually enough both to allow access to the site, and also to show site control for the use permit process referred to previously.

If the agency still wants the site after the results of the testing, the remainder of the condemnation process is followed. If significant hazardous materials are found on the site, the appraised value of the site may be affected. See Richard G. Rypinski's book, *Eminent Domain: A Step-by-Step Guide to the Acquisition of Real Property*, and consult your agency's own attorneys, for specifics on how to ensure that this process will stand up in court.

We have talked about a free-market negotiation, and also a hostile property owner. There is a third option—a willing owner who wants the tax benefits that condemnation provides him or her. In this case, the costs of a court battle are avoided. The basic condemnation process is followed. This is called a "friendly condemnation."

A further consideration that should be addressed at the beginning of the process is whether the infrastructure project is being financed in whole or in part by a federal or state grant, or is a type of infrastructure (such as a jail or school) that is subject to regulation by a state, regional, or federal agency. These grants often contain provisions that regulate how sites are to be acquired. These must be taken into account from the beginning. The agency's lawyers should be involved from the beginning of the process. There may also be a need to have a special liaison for the project from the GPG to the funding agency or to the state or federal agency that regulates the use of the facility to ensure that site acquisition and the condemnation process meet their requirements as well.

Phase V: Construction Contract and Building Permits

When discretionary land use permits and any variances have been acquired, and control of the site is in hand, the steps towards construction can occur. They include:

- Finalize design contract.
- Prepare detailed plans and specifications and plan review.

- Value engineering (if not done earlier).
- Enter into construction contract.

Finalize Design Contract

For large-scale projects where the design work is contracted out, the design contract for the detailed plans and specifications, and perhaps construction management, must be finalized. The initial contract with the architect or the engineering firm may have included a two-step process, with a commitment only for preliminary design work to provide for the possibility that a community would not have been able to secure permits or financing to proceed further. Many grants and financing arrangements require that someone be professionally responsible for the quality of the design documents. This can be the architect, the engineer, and, if the work is done in-house, the city or county engineer who typically certifies infrastructure plans. (See sidebar for key elements in the design contract.)

Prepare Detailed Plans and Specifications and Plan Review

Whereas the previous steps for a public building or large-scale infrastructure project were very public, the preparation of the detailed plans usually occurs out of the public's eye. The private firm, or in-house engineer or architect, prepares the detailed plans and specifications that are then submitted to the local building department for review and approval. This department ensures that the fire marshal (for toxics matters) and the engineering department review the plans. Other specialty reviewers may be needed, depending on how complicated the project is.

If the project is in the right-of-way and is not a building, the detailed plans and specifications are either certified by the contract firm or by the city or county engineer. They can also be submitted to another firm or to the design review team for value engineering (to ensure that the most cost-effective plan has been prepared) if this has not already been done.

Once the discretionary permits have been received (use permit and environmental review), the detailed plans and specifications are prepared by the design team. Whereas the

Elements in the Design Contract

- Responsibility for errors and omissions insurance.
- Indemnification responsibilities.
- How attendance at public hearings will be compensated.
- For buildings, whether interior design is included.
- How disputes will be resolved: mediation, arbitration, or litigation.
- Ownership of plans should belong to the public agency.
- How much cost-estimating or budgeting work will be required.

U.S. General Accounting Office, *Role of Facility Design Reviews in Facilities Construction* (Washington, DC: U.S. General Accounting Office, 2000).

Local building construction project, Eugene, OR

agency might want to take a chance on funding the preparation of the initial design, the detailed plans and specifications, which are usually the construction bid documents, are extremely expensive to prepare for large-scale projects. They are often not prepared until the issuance of the bond or receipt of the grant. The plans and specifications for facilities and buildings must receive a building permit. As noted in Chapter 7, the term "building permit" actually covers a wide range of permits, including plumbing, electrical, and mechanical, as well as actual construction.

Enter into the Construction Contract

Most local agencies have a "purchasing" unit, often located in a larger administrative department that handles the formal bidding process. The bid documents will be available at this office (and on-line in some cities), which will also ensure that the bids are submitted promptly. Most state laws provide (as do federal grants) that construction contracts go to the lowest responsible bidder. Every firm that meets the bid requirements is assumed to be qualified to carry out the contract. Although the local engineering department may be unhappy about the quality of the "lowest responsible bidder," unless it can document its case to the city attorney or county counsel's satisfaction, the locality must accept the bid and enter into a contract. Elements of a good case might contain instances of default on other recent contracts in the area, or a proven history of construction complaints from other jurisdictions.

Most RFPs (requests for proposals) or bid packages for governments contain a provision that indicates that the agency may choose not to award the contract. This is to protect the agency in the event the financing does not come through, or if there is a problem with formal close of escrow for the site or the permits. Similarly, most construction bids from contractors contain a caveat that their figures are good only for 30 days, or another specified period.

Usually, the construction contract is let to a general contractor who then subcontracts with specialty subcontractors. The GPG and the general contractor sign one contract, and the general contractor signs individual contracts with the subcontractors.[6] The agency should determine whether it would be appropriate to split the contract up into several smaller contracts in order to permit small or minority contractors to participate.

Federal construction contracts are obligated to abide by the provision of the Davis Bacon Act, which requires that the contractor and all of his or her subcontractors pay prevailing wages.

Most government construction contracts require that the contractor provide a bond to guarantee completion of the project. The bond company evaluates the financial and management capacity of the construction firm before issuing the bond. There are two types of bonds. One is a completion bond, where the insurance company guarantees that the project will be completed, even if the contractor defaults or goes bankrupt. A payment bond provides for the insurance company to pay all valid claims—which means that the insurance company will step in to pay the subcontractor if the general contractor does not.[7]

Phase VI: Construction and Project Closeout

It is extremely important that the local agency have someone in charge who is experienced and capable to monitor construction. Construction management is a very technical field, with sub-specialties for the various types of infrastructure—jails, highways, airports, tunnels, and the like.

Renovation of the California Palace of Honor

Who Monitors Construction?

During the design phase, the agency needs to decide how construction will be monitored. This is important not only to ensure that the contractor is accurately translating the plans and specifications into a physical reality, but to allow for progress payments to the contractor. Most construction contracts provide benchmarks for payments tied to milestones of progress, with 10 percent withheld until project completion and closeout. Therefore, a representative of the local agency must inspect construction and certify compliance with design specifications. This responsibility can be given to a member of the in-house team such as the architect or engineer in charge of the design group. It is customary for those who created the design to certify that construction is proceeding in accordance with the plan.

Typical structural steel erection

A specialty firm can also be hired to assist in monitoring construction. For larger jobs, it has been common to rely on the architect or engineer who designed the project. These days, in addition to the construction manager (the representative of the agency), there may be a "clerk of the works" or an on-site engineer who acts as liaison between the design team and the contractor. The design team may also have specialists or specific members, who conduct technical reviews of different components of the construction.

If the infrastructure project is a mega-project, or a very technical one, like an energy plant, a telecommunications center, a jail, or a concert hall, the design firm may be involved from the beginning of the design process all the way through to contractors and final payout.

For small-scale jobs, the in-house engineer or architect will design the project and monitor construction.

Local Agency Inspections: Building and Engineering

In addition to the agency's construction management team, for non-ROW projects, the local building officials will also inspect the facility during construction. For large-scale projects, the local agency may wish to establish a dedicated team, to be paid out of the project's funds.

For right-of-way projects, the engineering department is in charge of the inspections, and for ensuring that the roadway is adequately repaved once dug up—whether it is by contractors for the city or county, or for private utilities.

It is also important to think about the "as-builts"—about how the locality will record the location of the infrastructure into the local GIS, and how construction drawings will be modified to reflect actual construction, and where they will be retained in the agency.

Project Close-out

Project close-out is the paperwork part of the project that is easy to ignore, but essential to complete. This consists of confirming that all payments have been made, and that all bond or grant funds have been drawn down. Any reports required from granting agencies must be filed. Sometimes there are post-audit requirements for projects by state or federal agencies. Sometimes a post-occupancy review is done to develop a lessons learned document.

Conclusion

Developing an infrastructure project can be a challenging task. The engineering department is a good place to start, since most agencies do routine capital improvements and the staff there can provide advice on setting up a development team. The redevelopment agency's staff also has considerable experience with development and can provide advice to the GPG. The planning and community development department may also have experience in housing construction or rehabilitation. Although a major infrastructure project has many technical challenges, the overall development process is generally the same for most projects. Indeed, the local government will have to decide where the project management is located, since all of these departments may want lead responsibility. It is important to remember to be a "smart buyer." The more expertise available to the project from in-house staff, the better the project will be. Infrastructure development can be frustrating, exciting, but ultimately rewarding on the day the doors of a new facility open, or the water runs to a new development as predicted, or the stormwater in a 100-year event is absorbed by state-of-the-art green infrastructure.

Additional Resources

Publications

Bass, Ronald E., Albert I. Herson, and Kenneth M. Bogdan. *The NEPA Book: A Step-by Step Guide on How to Comply with the National Environmental Policy Act*. Point Arena, CA: Solano Press, 2001. Good for lawyers and planners who are trying to shepherd a project through the NEPA process.

Goodman, Aluin, and Makarand Hastak. *Infrastructure Planning Handbook*. Restan; VA: American Society of Civil Engineers, 2006. A thorough guide geared towards engineers.

Flybjerg, Bent, Nils, Bruzelius, and Werner Rothengatter. *Megaprojects and Risk: An Antomy of Ambition*. Cambridge University, Press, 2003. An appraisal of all the factors that cause large scale infrastructure projects to have cost overruns and revenue short falls.

Miles, Mike E., Bayle Berens, Mark J. Eppli, and Marc A. Weiss. *Real Estate Development*. Washington, DC: Urban Land Institute, 2007. This is an excellent resource for local practitioners doing their first project. The focus is on private real estate, which would assist those doing quasi-public buildings, but also suitable for public buildings paid through taxes.

Ortolano, Leonard. *Environmental Regulation and Impact Assessment*. New York: Wiley & Sons, 1997. Recommended for the local practitioner who will contract out or prepare the environmental assessment. Very detailed with an emphasis on the scientific basis.

Peca, S.P. *Real Estate Development and Investment*. Hoboben, New Jersey: Wiley and Sons, 2009. For the private secter practitioner.

Peisner, Richard and David Hamilton. *Professional Real Estate Development* (3rd edn). Washington, DC: Urban Land Institute, 2012. Although geared toward the private developer, a helpful guide for local city and country practitioner.

Rypinski, Richard G. *Eminent Domain: A Step-by-Step Guide to the Acquisition of Real Property*. Point Arena, CA: Solano Press, 2002. A guide to the intricacies of eminent domain that many infrastructure projects must consider at some point. Emphasizes legal aspects.

U.S. General Accounting Office. *Role of Facility Design Reviews in Facilities Construction*. Washington, DC: U.S. General Accounting Office, 2000. An excellent overview on how to be cost-conscious during the development process. Geared towards building construction, but the principles apply to networked facilities as well.

Website

Urban Land Institute: www.uli.org. Contains many useful publications on the development process.

chapter 9

Capital Improvements Plan and the Capital Budget

In this chapter...

Importance to Local Practitioners 137
Local Capital Budgeting Today 137
- Status of Capital Improvements Plans and Budgets 137
- Technical Tools Need Improvement 138
- Lack of Routine, Comprehensive Assessment of Infrastructure Needs 139
- Comprehensive Plan and the CIP Often Are Not Related 140
- Local Governments Do Not Control All the Local Infrastructure Funds 140

Capital Improvement Planning Documents in the Agency 141
- Capital Improvements Plan 141
- Organization of the CIP 143
- Capital Needs Study 143

The Capital Budget 145
- What Is the Capital Budget? 145
- Relationship of the Capital Budget to the Operating Budget 145
- Why the Capital Budget Should Be a Separate Document 146

GASB 34 and Infrastructure Budgeting 147
Developing the CIP and the Capital Budget 148
- Organizing the Process 148
- Identifying Projects and Funding Options 149
- Selecting Projects 150
- Prepare and Recommend Capital Plan and Budget 150
- Adoption of a Capital Budget 150

Conclusion 151
Additional Resources 151
- Publications 151
- Websites for Selected County and City CIPs 152

Why Is This Important to Local Practitioners?

Local practitioners need to be familiar with the capital improvements plan (CIP) and the capital budget, the principal documents used to implement the jurisdiction's infrastructure plan. The CIP and the capital budget enable the locality to translate its long-term land use plan, or shorter-term strategic plan, into specific infrastructure projects. Strategic planning during the budget process ensures that these investment decisions are most effectively deployed to achieve the agency's overall goals and objectives, perhaps with a geographic focus.

This chapter begins with an overview of capital budgeting at the local level today. It then describes the necessary elements of the capital plan documents used at the local level including the CIP, the capital needs study, and the capital budget. The impact of Government Accounting Standards Board Rule 34 on infrastructure budgeting is discussed, followed by a section on how strategic planning can produce a more effective CIP and capital budget. The chapter concludes by describing the major steps in the preparation of the CIP and the capital budget.

Local Capital Budgeting Today

Status of Capital Improvements Plans and Budgets

Multiyear capital improvements plans and capital budgets are common in the USA, and their use is growing. A survey of state governments at the end of the 1980s found that 42 of the 50 states had capital budgets[1] while a 1992 survey of cities with populations over 75,000 found that 73 percent prepared

Capital Budgeting as Practiced by Local Governments

One expert classifies local jurisdictions into four categories, according to their relative sophistication in capital budgeting:

- *Project-specific capital financing and little debt.* Multiyear capital planning occurs but it is project-specific. Capital financing comes mostly from annual revenues and operating fund balances or grants and state loans. The procedures and forms used for capital budgeting are the same as for the operating budget.
- *Special procedures for capital projects and basic debt instruments.* Although there is no jurisdiction-wide capital planning or financial forecasting, multiyear planning occurs for equipment replacement, some utilities, and perhaps streets. Capital financing is from operating revenues, earmarked reserves, and basic debt instruments. Special procedures and formats are used for capital projects, but the process is part of the general operating budget context.
- *Strategic planning, asset management systems, and special budget procedures.* Multiyear and jurisdiction-wide capital planning and financial forecasting occur, and the jurisdiction uses strategic planning principles. Condition assessment and cost information comes from asset management systems. Capital financing comes from a wide variety of sources. Special budgeting procedures and formats are used for capital projects in a separate budget and authorization process. Coordination with land use plan occurs.
- *Interjurisdictional coordination for capital facilities planning.* Same as above, but all infrastructure providers in the jurisdiction are involved in capital planning. Can be regional. Comprehensive asset management systems use the same location and investment conventions. Coordination with land use plans occurs.

Based on John A. Vogt, *Capital Budgeting and Finance: A Guide for Local Governments* (Washington, DC: International City/County Management Association, 2009).

capital budgets, and 61 percent prepared the capital budget as a separate document.[2]

Capital budgeting at the local level is likely to be relatively unsophisticated compared with the federal government and the private sector. The federal government follows key principles and practices for effective capital decision-making (Figure 9.1). Although local governments recognize that capital planning and budgeting are important, the quality of the process can be uneven. Local practitioners often struggle with how to improve the quality of the budget given the necessary political overlay. Capital budgeting frequently has no more than a paper relationship with the comprehensive plan. Strategic planning is not commonly done as part of the budgeting process for infrastructure. Condition assessments are spotty. Deferred maintenance is still the rule. These problems are discussed in greater detail below.

Technical Tools Need Improvement

A 1987 survey of finance officers from a sample of cities with populations over 50,000 found that the most popular methods of evaluating capital projects were the less sophisticated quantitative methods, and that nonquantitative methods predominated.[3] This approach may be appropriate in many cases. For example, although parks and community facilities compete with each other for scarce general fund tax dollars, the "time value" of the expenditure is evaluated in the political arena. A new library or a replacement for the city hall may have been long anticipated and alternative uses for the funds vigorously debated in newspapers and council meetings. The technical challenge for the local practitioner is to resist the pressure to underestimate the costs, and to provide nimble analyses of the land use impact of different scenarios as appropriate.

Quantitative models for evaluating capital investments such as Internal Rate of Return (IRR), or Net Present Value (NPV), and cost–benefit ratios are designed to consider the time value of money when

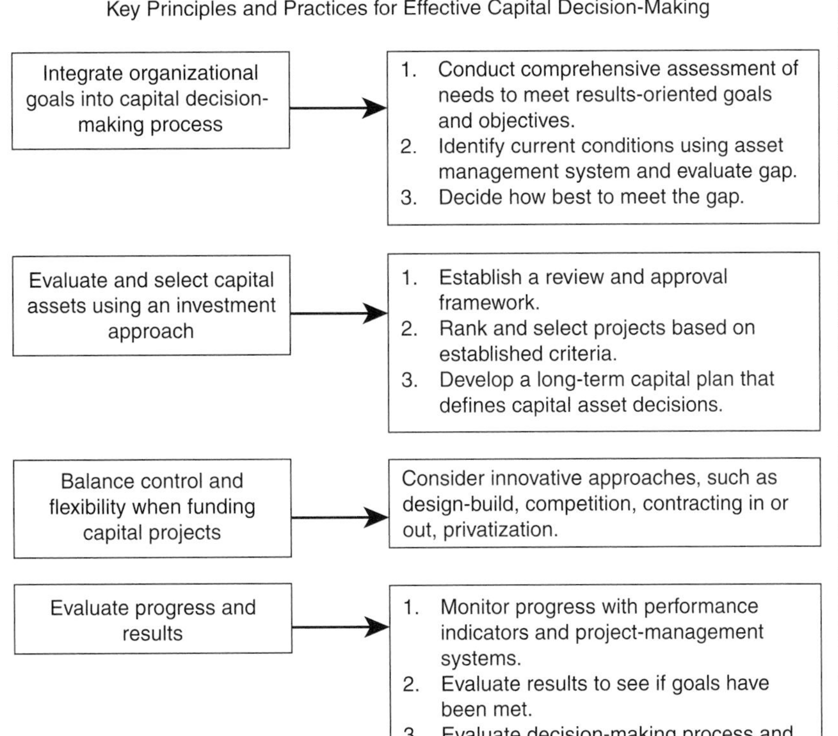

Key Principles and Practices for Effective Capital Decision-Making
Integrate organizational goals into capital decision-making process
1. Conduct comprehensive assessment of needs to meet results-oriented goals and objectives.
2. Identify current conditions using asset management system and evaluate gap.
3. Decide how best to meet the gap.
Evaluate and select capital assets using an investment approach
1. Establish a review and approval framework.
2. Rank and select projects based on established criteria.
3. Develop a long-term capital plan that defines capital asset decisions.
Balance control and flexibility when funding capital projects
Consider innovative approaches, such as design-build, competition, contracting in or out, privatization.
Evaluate progress and results
1. Monitor progress with performance indicators and project-management systems.
2. Evaluate results to see if goals have been met.
3. Evaluate decision-making process and improve it.

Figure 9.1
Federal government guidelines for capital decisions

comparing alternative projects within a particular functional area. These models are best used in areas where funding is more predictable, and where the alternatives are more alike. Capital expenditures for water, sewers, and transportation projects, mainly relying on fees, are more appropriate venues for sophisticated quantitative analyses.

Local practitioners should invest in tools and training to project population growth under various growth scenarios. They should be able to respond quickly to requests from policy-makers to "model" the effect of one capital scenario compared to another.

Lack of Routine, Comprehensive Assessment of Infrastructure Needs

Despite recent improvements in this area, the lack of a routine, quantified, and comprehensive condition assessment for existing infrastructure, as well as solid projections for future needs, remains a concern for local decision-makers. This concern is being actively addressed today by the transportation, water, and wastewater utilities under the rubric of asset management (discussed in Chapter 10).

A jurisdiction, however, need not spend scarce resources to gather existing data on conditions and to make incremental yearly improvements. Government Accounting Standards Board (GASB) Rule 34, discussed later in this chapter, has also been touted as a requirement that may assist local jurisdictions in this area. However, the rule may result in another set of figures to explain to local policy-makers unless these estimates are carefully integrated with those of individual facility plans.[4]

Comprehensive Plan and the CIP Often Are Not Related

The capital improvement program process can be an important part of implementing the comprehensive plan, and indeed, some capital budgets are organized into sections that parallel the "elements" in the comprehensive land use plan. In the first five years of the new millennium, growth pressures and infrastructure replacement needs have resulted in the emergence of a new generation of facilities plans linked closely to land use designations in cities like San Jose, Los Angeles, and Seattle. Some states (Wisconsin is a good example) have also begun to require that all infrastructure service areas and land use planning areas be consistent.

However, many jurisdictions still regard the long-range comprehensive plan as suggestive. In some jurisdictions, the plan is not current; in others, the goals and policies are too general to guide the construction of capital facilities. Other comprehensive plans and zoning ordinances may permit densities and uses that do not reflect the current availability of infrastructure. Real fiscal constraints may not be integrated into the planning process. So, planners and plans may be seen by local budget and public works officials in many jurisdictions as irrelevant to the highly specific nature of the CIP process.[5]

Local Governments Do Not Control All the Local Infrastructure Funds

Although 80 percent of infrastructure is provided at the local level, a significant amount of capital funds is budgeted by special districts that are accountable to an elected board that may not feel compelled to coordinate closely with the general purpose governments (GPGs). Special district capital budgets include spending within the jurisdiction of the GPG, but many kinds of special districts have no requirements to coordinate capital planning efforts, although this situation has been changing recently.

Ideally, the capital improvements plan should be coordinated at the regional level, since the major facilities for the two infrastructure systems most influential for growth—transportation and water and wastewater—need to be planned at that level. Solid waste disposal and many park and cultural facilities can be regional in nature. The regional plans should "flow" into the statewide plan, and should also provide guidance for local CIPs and capital budgets. The special purpose districts, many of which operate at the county level, should be required to have their capital plans and budgets integrated into this overall plan, perhaps even approved by local cities or counties.

Capital Improvement Planning Documents in the Agency

Capital Improvements Plan

The capital improvements plan or program (CIP) is a plan that assesses the infrastructure needs in a jurisdiction against its overall goals and objectives, using a multiyear planning horizon, and then evaluates and prioritizes specific capital projects that the jurisdiction will fund. It usually includes a narrative and a series of tables that show which projects will be built, when, where, and how much they will cost. Ideally, it is based on a capital needs study that identifies long-term needs on a system-by-system basis, and a strategic plan for the jurisdiction that prioritizes capital investments based on fiscal realities. The following features should be considered.

Timeframe. The accepted wisdom is that the CIP should encompass at least five years. Fairfax County's CIP (this Virginia county is a leader in "best practices" for the CIP) is for five to seven years.

Designation of percentage of revenues to capital. Some communities designate a fixed percentage of revenues to capital, so that when the CIP is prepared, the players know how much money is available. In other communities, one-time revenues are used and as many of the top priority projects as can be funded are built. In still other communities, the CIP provides the context for a bond issue.[6] The governing body can adopt explicit principles, which should be flexible enough to allow for underfunding of infrastructure during hard economic times.[7]

Debt affordability limit. State law, the local charter, or voters impose limitations on the level of debt financing for many local governments. The constraint may be a requirement to go to the voters if the debt amount exceeds a certain threshold. Sometimes this constraint acts as a real restriction, but in other cases the limit is higher than fiscal prudence warrants, and the locality must generate its own limit. Bond rating agencies appreciate this type of analysis as an indicator of sound financial management and long-range planning. See Table 9.1 for how one county determined how much debt can be undertaken for capital improvements.

Long-term operating budget plan. Lengthening the time frame of the operating budget to include the out years' maintenance and operating costs for new infrastructure might be desirable. A typical capital project that

Table 9.1
Fiscal year 2004 debt affordability projections for Anne Arundel County, MD

	FY2004	FY2005	FY2006	FY2007	FY2008	FY2009
New authority, recurring ($)	70,000,000	70,000,000	70,000,000	70,000,000	70,000,000	70,000,000
New authority, one time ($)	3,590,000	5,000,000	4,500,000	4,000,000	3,600,000	0
Total authority	73,590,000	75,000,000	74,500,000	74,000,000	73,500,000	70,00,000
Debt service as % of revenues (9%)	7.5	7.9	8.0	8.1	8.1	8.0
Debt as % of est. full value (1.50%)	1.45	1.46	1.46	1.46	1.45	1.43
Debt per capita ($1,000)	1,120	1,161	1,197	1,227	1,251	1,272
Debt to personal income (3.0%)	2.5	2.8	2.8	2.8	2.7	2.7
Debt service ($)	69,130,062	73,078,363	77,153,875	81,679,012	85,128,443	87,767,261
Debt at end of fiscal year ($)	560,294,704	585,480,861	607,784,370	626,852,152	643,763,835	658,667,090
General fund revenues ($)	886,600,000	926,100,000	965,600,000	1,006,500,000	1,049,200,000	1,094,500,000
Estimated full value ($000)	38,731,000	40,087,000	41,490,000	42,942,000	44,445,000	46,001,000
Population	500,440	504,100	507,580	511,060	514,540	518,020
Total personal income ($000)	19,718,000	20,605,000	21,232,000	22,501,000	23,514,000	24,572,000

Source: Anne Arundel County, MD, 2004.

involves significant operating budget impact will take about three years from design to construction completion. This time lag in the impact on the operating budget impact creates a significant problem. Policy-makers can easily indicate that the funds will be available when needed but not truly deal with the operating impact. A commitment to a long-term operating budget plan would force acknowledgement of the long-term costs.

Funded and unfunded needs. Practitioners disagree about whether the CIP should contain the total capital needs—whether they can be funded or not, or just those that will be funded. We think it is more helpful to present both the CIP projects as well as unfunded needs, a practice followed for state transportation plans which have a CIP and an unfunded project work plan (UPWP), each with a different time horizon. See Figure 9.2 for a simple way one city provided a visual representation of the capital expenditures and needs for sidewalks.

Definitions of a capital project. What is defined as a capital project differs among jurisdictions. Usually, all long-lived infrastructure is included (water and sewers, streets, parks, buildings), along with equipment such as fire trucks, radios, police cars, telecommunications equipment, furniture, and computers. The capital budget may also include a land purchase as a capital project. The charter of Anne Arundel County in Maryland defines a capital project as:

> 1) any physical public betterment or improvement and any preliminary studies and surveys relative thereto; 2) the acquisition of property of a permanent nature for public use; and 3) the purchase of equipment for any public betterment or improvement when first constructed.[8]

Although ordinary maintenance and operations of infrastructure may not belong in the capital budget, deferred or capital maintenance that rehabilitates or extends the life of an infrastructure asset can be included in the capital budget. In addition, the dollar amount cutoff for a capital project differs among jurisdictions, ranging from $1,000 to $250,000. Equipment and vehicles must meet a threshold dollar amount to be considered a capital project.[9]

Organization of the CIP

Ideally, the CIP and its projects are organized into major categories that correspond to the major elements of the comprehensive plan: housing; economic development; environmental resources (water, sewer); open space and recreation; public facilities; and transportation. In some jurisdictions, projects are categorized by organizational entity—a practice to be avoided in order to get managers out of their organizational silos. The public is usually more concerned about the service and not who provides it. More important is maintaining a similar format over time so that tracking historical information such as relative shares of spending is easier for the public.[10]

Capital Needs Study

Ideally, the individual projects in a CIP are abstracted from what the City of Phoenix calls a capital needs study. This long-term assessment of the system needs for replacement and

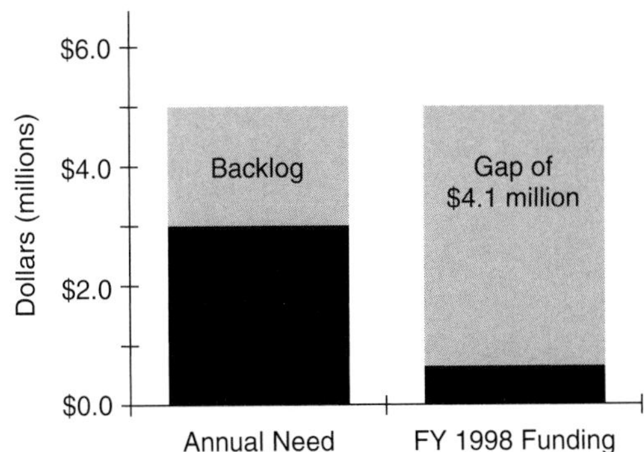

Figure 9.2
FY1998 sidewalks gap in Berkeley, CA

The Finer Points of Defining a Capital Project

Capital improvements are defined in many ways. For example, although technically maintenance because of its infrequent nature and large expense, street resurfacing is a prime candidate for a capital improvements plan. In addition, while engineering, architecture, and landscaping studies are not physical improvements, when necessary to build a large infrastructure project, they too are contained in a CIP and can be funded by bond financing. A flexible definition can sometimes help solve a sticky problem, but care should be taken not to stretch the definition of a capital project too far.

Adapted from Robert A. Bowyer, *Capital Improvements Programs: Linking Budgeting and Planning* (Chicago: American Planning Association, 1993).

City of Los Angeles Uses Report Card Approach to Jump-Start Infrastructure Planning Process

In 2002, the Los Angeles City Council passed a resolution designating the city engineer as infrastructure manager for the city. At approximately the same time, the American Society of Civil Engineers published a report card on the condition of the nation's infrastructure. Local ASCE chapters and the City of Los Angeles soon followed with their own report cards. Los Angeles evaluated seven elements of its infrastructure—bridges, stormwater, streets and highways, street lighting, wastewater collection, wastewater treatment, and water supply. Grades ranged from a D+ for streets and highways to B+ for the wastewater system, with an average grade of C–.

Shortly after the report card was published, the mayor convened a task force of 40 experienced professionals from the infrastructure stakeholder community. The task force used a 10-year horizon to prepare a comprehensive expenditure and funding plan to address the most critical infrastructure investment needs. Fifty-five recommendations for increased funding and improved maintenance practices were proposed.

Not surprisingly, the task force identified a $3.9 billion funding shortfall. Surprisingly, Los Angeles actually did something about it. Revenue-generating agencies increased fees to pay for infrastructure improvements, and a $500 million bond issue was passed to pay for stormwater improvements. The report card served its purpose—to get decision-makers talking and doing something about the infrastructure of the city. (See Table 9.2 for the funded and unfunded street maintenance needs in the City of Los Angeles as of 2003.)

Table 9.2
Estimated annual expenditure needed to eliminate street maintenance backlog over 10 years in the City of Los Angeles, 2003 (in millions)

Street functional classification	Annual cost for rehabilitation ($ millions)	Annual cost for maintenance ($ millions)	Ten-year need ($ millions)
Primary arterial	30.6	6.8	37.4
Secondary arterial	23.8	5.2	29.0
Collector	17.3	4.5	21.8
Residential	42.7	18.0	60.7
Total	114.4	34.5	148.9

Source: City of Los Angeles, CA, *L.A. Infrastructure Report Card*, 2003.

Written by Vitaly Troyen, former city engineer with the City of Los Angeles and former Deputy Public Works Director in San Francisco, currently with Parsons Brinkerhoff.

modernization, as well as new growth, includes buildings and the networked infrastructure.

A capital needs study consists of a comprehensive inventory of existing facilities for the system, as well as an assessment of its condition and a schedule for repair and replacement. It also includes a detailed plan for new facilities. When the program is funded by fees, such as water, sewer, or solid waste, it may also contain a budget and an analysis of alternative fee structures as well as a risk analysis.

Although the development of the capital needs study for many infrastructure systems may occur outside the regular CIP process, these efforts form the basis of the project requests from individual departments. The projects should then be translated into the capital improvement plan, ideally using a strategic plan and an understanding of the overall fiscal health of the community.

Capital needs studies have several issues. First, they may not exist. Conducting an in-depth condition assessment is expensive. When they do exist, they may be based on a technical assessment of needs inconsistent with the strategic plan or comprehensive land use plan of the locality. Finally, some studies may not use the same population projections, funding constraints, or even GIS conventions as the others. The capital planning and budget process can encourage the use of a common format, budget, and data conventions for all capital needs studies. The process can also ensure that a systematic process exists for infrastructure condition assessments and needs studies.

The Capital Budget

What Is the Capital Budget?

The capital budget contains the planned reserves and expenditures for the first year (or first and second year) of the capital improvement plan or program. The major difference between the capital budget and the capital improvement plan is that the capital budget is a legal document, usually adopted yearly by the elected officials. The capital plan contains the first year's projects as well as projects in subsequent years for which funding may not yet be obtained or authorized. Since the CIP is not a legally binding document, it can change in the "out" years.[11]

Generally, a capital budget contains a summary narrative document along with summary charts of revenues and expenditures. The capital budget also contains a description of each project, as well as five-year expected revenues and expenditures. Some localities also include maps that show the location of the project. See Table 9.3 for the summary table of Santa Barbara County's capital improvement budget from 2009/10 through 2014.

Relationship of the Capital Budget to the Operating Budget

The capital budget is usually a separate budget document from the operating budget. The operating budget contains expenditures for annual or routine services of the jurisdiction, such as police and fire salaries, salaries for

Table 9.3
Five-year CIP for Santa Barbara County, CA, through fiscal year ending June 30, 2014 (in thousands of dollars)

	FY 2009–2010			
Expense category	**Funded**	**Unfunded**	**FY 2010–2014***	**Total**
Land, buildings and facilities ($)	5,725	18,576	231,806	256,107
Major equipment	3,013	1,194	8,931	13,138
Major improvement to building facilities	5,817	3,927	86,897	96,641
Transportation	20,273	33,845	285,351	339,469
Water resources	12,226	6,069	112,367	130,662
Resource recovery & waste management	6,948	0	36,897	43,845
Major maintenance	250	0	3,138	3,388
Total	$54,252	$63,611	$765,387	$883,250

Note: *25% funded.
Source: Santa Barbara County, CA, *Proposed CIP*, 2005.

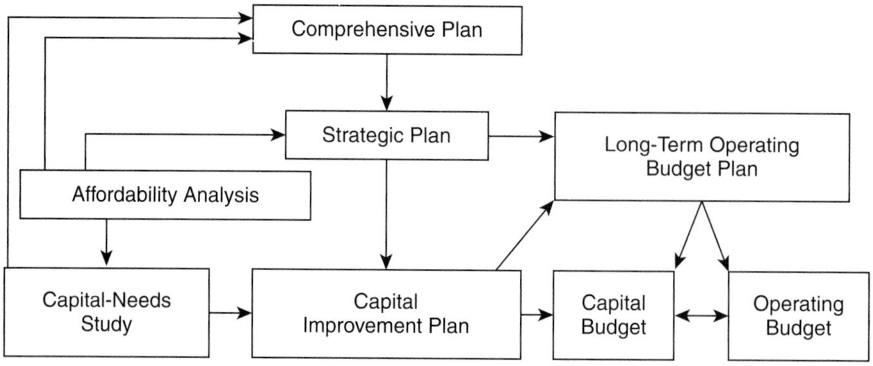

Comprehensive Plan
Strategic Plan
Long-Term Operating Budget Plan
Affordability Analysis
Capital-Needs Study
Capital Improvement Plan
Capital Budget
Operating Budget

Figure 9.3
Relationship of key plans and budgets for infrastructure purposes

recreation department staff, and the local permit center. The operating budget is the legal document that authorizes the expenditures by department to meet the annual goals and objectives of the jurisdiction. Figure 9.3 details the relationship of the key plans and budgets for infrastructure purposes.

The capital budget, on the other hand, contains detailed information about the design, costs, financing, and schedule for infrastructure projects for the coming year. Just as the elected body's approval of the operating budget authorizes recurring expenditures, its approval of the capital budget authorizes each infrastructure project.

A commitment to a long-term operating budget plan is necessary to ensure that capital budget commitments to build a facility are carried through with proper long-term maintenance and upkeep, and that adequate operating funds are there when the doors open or the water or sewage runs through the new pipes.

Why the Capital Budget Should Be a Separate Document

Capital expenditures have several characteristics that merit a separate budget: their long life, their infrequency, and their size. Once capital expenditures are dedicated, they cannot be changed easily. Infrastructure projects take a long time to build, and the benefits are not obtained until the project is completed. Therefore, the budgeting process for annual expenditures is not appropriate, and something which reflects the "lumpy" nature of the capital expenditures is required.[12]

Additionally, infrastructure projects are frequently funded by one-time special funds, and separating these funds from the operating budget ensures that they are used appropriately. Further, the process to develop the capital budget differs markedly from that used for the operating budget. A list of projects must be prepared, ranked, and compared against each other. As projects are built, or as funds become available, new projects are added to the list, while programs and services funded by an operating budget generally are the same from year to year.

Finally, the time frames for the capital and operating budgets differ. All the activities funded by an operating budget occur in one or two fiscal years, while the projects in a capital budget often take three to five years to complete, or may be part of a larger program envisioned over a 20- to 30-year period. The funds required for a particular project may be uneven—small initially, and larger in the out years as construction proceeds. A capital budget provides for separate accounting for the projects, which permits financial oversight.[13]

GASB 34 and Infrastructure Budgeting

A major infrastructure issue for local governments has been the continuing deterioration of infrastructure in older jurisdictions. Until the past decade, there were no consequences for local governments that chose to ignore long-term infrastructure asset management issues when preparing their budgets. However, in June 1999, the Government Accounting Standards Board (GASB)[14] put forward Rule 34, known generally as GASB 34. Today, GASB standards are used by external auditors to evaluate the adequacy of the city's financial statements, which are particularly important for bond ratings.

Prior to GASB 34, most localities omitted infrastructure such as streets, sewers, and telecommunications from annual financial statements (government buildings had already been included as assets in the financial statement). Projects were funded when they were built, and "expensed" at that time. Because infrastructure projects have long lives and were thought to retain their economic value for many years without an annual decline, and because there were no tax advantages as for the private sector, they were not depreciated. Infrastructure was treated as a sunk cost and was often ignored by local decision-makers until a replacement was required.

Since these assets did not appear on the locality's financial statement, budget officials exerted no pressure to appropriate maintenance funds. The unreported assets and their worsening conditions could dwarf the size of capital assets that actually were included in the financial statements. No early warning system existed in the budget to alert local officials to the potentially astronomical costs of replacing key infrastructure systems. GASB 34 was enacted to address this problem.[15]

GASB 34 required all state and local governments to set a value on all infrastructure assets as part of their budgets by 2003 for newly built or recently acquired infrastructure, and to report on that value each year.[16] Preexisting infrastructure assets were not required to be valued until 2007. The rule applies to over 83,000 local jurisdictions and to state agencies, public universities, school districts, water districts, and other special districts.[17]

GASB 34 regulations do not cover three infrastructure components normally found in a CIP: buildings, land, and equipment. Land is not included because it does not depreciate in value. The regulations do not apply to buildings because these facilities were already supposed to be included in the locality's financial statement which may or may not have occurred. Nothing, however, prevents a local government from including buildings and using the same standards of reporting and depreciation.

Developing the CIP and the Capital Budget

The five steps in developing the CIP and capital budget include: (1) organizing the process; (2) identifying the projects; (3) selecting the projects; (4) formatting and presentation of the material; and (5) adoption and implementation.[18]

Organizing the Process

Designate the lead department and the CIP committee(s). The process begins by a decision to prepare a CIP and capital budget, and the designation of the lead department. The lead department begins the CIP process by setting up a committee from the major departments with infrastructure—including public works, community development (for infrastructure), and police and fire (for rolling stock). Two permanent CIP committees might be appropriate, the first consisting of the department heads (or their deputies) for those with the dedicated revenue sources, as well as the planning director. This committee may be headed by the CEO or the deputy. Working under the direction of the senior committee is a second group, composed of staff who do the work—usually planners and engineers.[19]

In most cities and counties, the CIP and capital budget are prepared by the public works and finance departments, and receive close scrutiny by the elected officials. Land use planners, typically not involved, should be, since decisions made about public facilities during the CIP process significantly affect the growth of the jurisdiction.

The choice of lead agency often reflects the type of capital plan the jurisdiction desires. If a major policy change is needed to address issues of serious growth or decline, the planning department and planning commission will play a significant role. If the emphasis is on maintenance and replacing existing infrastructure, then more involvement from the public works department is appropriate. Usually, however, the finance department or budget office takes the lead.

Develop the process, the forms, the criteria, and the schedule. The next step is to create the overall budget schedule, budget instructions, and forms for the departments. A rough idea of the amount of unearmarked money available should also be developed.

The budget form should have a section for future operating costs that includes the number of staff, annual cost, additional capital equipment expenditures needed, and all other operating costs. Some jurisdictions also require estimates for depreciation. The County of Santa Barbara, California, for example, uses four forms: the first describes and justifies the project; the second is a summary; the third is the impact of an individual project on the operating budget; and the fourth is a program area summary.[20]

The criteria used to allocate the funds should be based on the policies that the jurisdiction has established to guide its growth, or which are generally perceived to be consistent with the values of the elected officials. If the locality is going through a strategic planning process, the results should be reflected in the criteria. In a built-out community with little new residential or commercial growth, the most important criterion might be the maintenance of existing infrastructure. Other objectives might be: to promote health and safety; to support retail development; to encourage new development; to increase the city's tax base; to improve operating efficiency; to encourage private investment; or to encourage conservation of resources.[21]

Choose citizen and stakeholder involvement process. Involvement of citizens and important stakeholders (other infrastructure providers in the area) should be developed. In some localities a public works commission will take the lead, while in others the local planning commission or a budget committee will do so. The local charter may dictate a specific role for citizens in the planning and capital budgeting process in some jurisdictions.

Identifying Projects and Funding Options

CIP projects should be drawn from the capital needs studies for the individual systems. They can also be drawn from other planning documents in the community. Even if these documents do not exist, however, it is possible to put together a good capital improvements budget in a more incremental fashion.

Good project proposals are the heart of a CIP. In a typical CIP preparation process, one department will already have an excellent multiyear program and be in a position to develop its capital projects immediately. There will probably be a department that is not so sophisticated. In the absence of a formal capital needs study, the department could undertake an informal survey of its capital needs, using in-house staff. The project request forms should be developed to accommodate both types of situations.

Selecting Projects

Once all the project proposals have been submitted, the next step is to evaluate the projects to determine individual worth, and to establish priorities among projects.

Although the textbooks indicate that for best results the evaluation of all projects should be done without regard to the funds available, in actuality, this approach works only with a surplus of general funds or with enterprise funds, where raising fees is possible. The ability of a project to leverage funds from state or federal sources can also be considered. The first level of review is against the criteria established at the beginning of the process, either numerically or qualitatively. Projects can then be divided into priority groups—those which are urgent and for which efforts should be made to find funding; those which should be done as funds become available, and so on. Projects with funding from enterprise funds should also be evaluated.

The next step is to make project selections. Projects can be deferred. Projects that must be coordinated can be assessed and scheduled in common. A project can be funded at a lesser amount than requested, or approved contingent upon obtaining other funds and grants. Projects can be grouped, and a bond issue proposed. Finally, a project may be deemed unlikely to be funded in the near future.

Projects competing for general funds—that is, money that is not earmarked for a special purpose—generally receive the most scrutiny. Local sewer, water, and transportation projects funded from development fees, outside grants, or enterprise funds are frequently merely ratified by the CIP. However, these projects should be evaluated against each other within a life-cycle context, and they also should be reviewed for consistency with the comprehensive plan, the sustainability plan, or other community goals. The City of Phoenix has a process for bid-leveling, to spread the bidding of construction contracts throughout the year so that they are not all issued in one quarter.

Prepare and Recommend Capital Plan and Budget

The program is then compiled and presented for approval to the decision-making body—usually the elected officials. A series of hearings, workshops, and other outreach efforts should follow to ensure that all stakeholders and interested parties can comment.

Performance indicators and project development milestones should be developed for the capital plan for subsequent reporting purposes.

Adoption of a Capital Budget

A project can be approved in one of three ways as part of the capital budget. The first and most common is to adopt the CIP as part of the operating budget, whether it is annual or biennial. This method funds the projects only a year or two at a time, although the projects are fully planned out to completion. One drawback of this system is that the project comes up for "approval" each year by the elected officials.

A second method is to adopt the capital budget with the entire amount for all the projects approved in that fiscal year, regardless of whether it will be spent that year or not. Although this is how the federal government and many state governments operate, it is unusual at the local level. The carryover funds for capital projects from one year are put into the next year's budget and approved again by the elected officials.

A third method is to approve the bond measure that both funds and authorizes the project(s). However, the annual or biennial appropriations for the project are usually still required by state and local regulations.[22] Once approved, the development, design, and construction process can begin.

Conclusion

The capital planning and budgeting process is one of the most important ways that the local practitioner can improve the quality and impact of infrastructure investments. The capital budget represents the public's investment funds, and can be the focal point for sustainable infrastructure planning. Ensuring that capital projects are strategically focused and targeted allows each dollar to work toward local strategic goals. In many localities, the process may be fragmented, and key decisions are hidden as technical assumptions. In other localities the capital budget may be only a compilation of projects developed rather than supporting the overall goals of the community.

The importance of the locality's overall strategy for infrastructure investment cannot be emphasized enough. Capital expenditures can play a critical role in guiding the future of the jurisdiction but only if they are coordinated under an overall strategy. Cross-cutting teams for the CIP budget process as well as an active citizen participation process can help facilitate this process.

The linkage of the CIP with the general plan will protect the jurisdiction in the event of litigation regarding impact fees or growth management restrictions. The capital budget is also an important growth management tool.

Additional Resources

Publications

Beaudet et al. *A Capital Planning Strategy Manual.* Denver, CO: American Waterworks Association Research Foundation and the American Waterworks Association, 2001. Capital planning and budgeting from the perspective of the water and wastewater industry.

Bowyer, R. A. *Capital Improvements Programs: Linking Budgeting and Planning.* Chicago: American Planning Association, 1993. A classic document for the planner seeking to understand the relationship of budgets and planning.

Government Finance Officers Association. "Planning and Budgeting for Capital Improvements." In *Budgeting: A Guide for Local Governments.* Chicago: Government Finance Officers Association, 1993. Classic overview of the capital improvements process in a book that contains a wide range of budget advice. A good investment. Check out their website for new publications (www.gfoa.org).

Massachusetts Department of Revenue. *A Practical Guide for Implementation of Government Accounting Standards Board Statement # 34 for Massachusetts Local Governments.* Massachusetts, 2001. A very readable document explaining how GASB 34 works and how local financial officers should implement it.

Robinson, S. "Capital Planning and Budgeting." In *Local Government Finance*, ed. J. E. Petersen and D. R. Strachota. Chicago, IL: Government Finance Officers Association, 1991. Another classic article on capital budgeting from GFOA.

U.S. General Accounting Office. *Leading Practices in Capital Decision Making.* GAO/AIMD-98-110. Washington, DC: U.S. General Accounting Office, 1998. A good document for the local practitioner that includes good practical advice.

Vogt, A. John. *Capital Budgeting and Finance: A Guide for Local Governments*, 2nd ed. Washington, DC: ICMA, 2009. A thorough discussion of the topic. Highly recommended for local practitioners involved with the capital budget. Sections on types of financial strategies and long-range forecasting are detailed and informative, with many local examples.

Websites for Selected County and City CIPs

Most of these websites have documents available for downloading that contain local policies, procedures, and sample documents as well as the capital budget or advice for local capital budget preparation. Some of these are recognized by the Government Finance Officers Association and other organizations as having the "best practices" for a CIP:

Anne Arundel County, MD: www.aacounty.org/Budget/index.cfm.

Fairfax County, VA: www.co.fairfax.va.us/gov/dmb/CIP.htm.

King County, WA: www.metrokc.gov/budget.

Phoenix, AZ: phoenix.gov/BUDGET/bud03cip.html.

Santa Barbara County, CA: www.countyofsb.org/cao/budgetresearch/cip2005.asp.

chapter 10

Managing Infrastructure

In this chapter...

Importance to Local Practitioners. 153
Problem of Deteriorating Infrastructure 153
Need for Asset Management Focus on Infrastructure 154
Causes of Maintenance Backlogs 155
Asset Management 156
Linkage with the Organization's Goal-Setting Process 157
Information Management . . . 157
Setting Investment Priorities for Infrastructure Maintenance 161
Reporting and Evaluation . . . 164
Performance Measurement 164
Performance Indicators 165
Benchmarking 166
Executive Management Reporting Systems 167
Conclusion 170
Additional Resources 170
Publications 170
Websites 170

Why Is This Important to Local Practitioners?

Although developing an infrastructure project can be quite exciting, often with political and media attention, once the ribbon is cut, the public's attention then turns to other areas. This lack of interest can be reflected in the annual budget process, and infrastructure may suffer from lack of maintenance and deteriorate over the years. Research has shown, however, that the cost of consistent maintenance of infrastructure compared to replacement after system failure is considerably lower. Further, attention to maintenance and other life-cycle issues during the budget process and project design phase can ensure lower total life project costs for the public and users. Maintaining infrastructure also is important for public safety and environmental protection, and, more practically, it can reduce claims and lawsuits over poorly maintained facilities.

This chapter begins by describing the problem that deterioration will pose for our infrastructure assets during the next few decades. It then addresses two recent reforms—performance measurement and comprehensive asset management. It concludes with the basic steps needed to create an executive management reporting system to monitor infrastructure management.

Problem of Deteriorating Infrastructure

Infrastructure systems pass through a number of stages during their lifetimes: planning, design, construction, operation and maintenance, refurbishment or repair, and disposal. Infrastructure systems require much more than construction or installation; they require lifelong maintenance, the quality of which can in part determine how quickly the end of adequate service arrives. Public officials and infrastructure providers must be aware of this concept of "service life" if they are to make sound infrastructure decisions, both during initial system selection and design and throughout its service period. Life-cycle analysis can save or at least maximize scarce public resources. This type of life-cycle analysis is not to be confused with the life-cycle analysis (LCA) which calculates the total carbon emissions footprint of a particular product from cradle to grave.

Need for Asset Management Focus on Infrastructure

The value of the existing infrastructure inventory in the USA is significant as noted in previous chapters. Deferred maintenance backlogs for infrastructure exist at every level of the system. The huge expenditures of funds for the Interstate Highway System built during the 1950s through the early 1980s were for capital—not maintenance and operations. Federal expenditures for water and wastewater pipes and treatment plants in the 1970s and 1980s also focused on capital needs, while local agencies were responsible for operations and maintenance funded by local user fees.

Operating expenses for infrastructure (excluding parks and civic buildings) have grown dramatically at state and local levels since 1976 (Figure 10.1). Operating expenditures by the federal government have been a relatively minor proportion of all the funds spent during this period. In contrast, in the mid-1990s, operating expenditures for state and local government were 87 percent of total public infrastructure spending.[1]

Most localities have significant backlogs of deferred maintenance that they know about, and even more that they have not yet quantified. More importantly, in most cities, infrastructure maintenance does not come close to being fully funded. Even localities with strong capital planning systems may not fully fund ongoing repairs and preventive maintenance. For example,

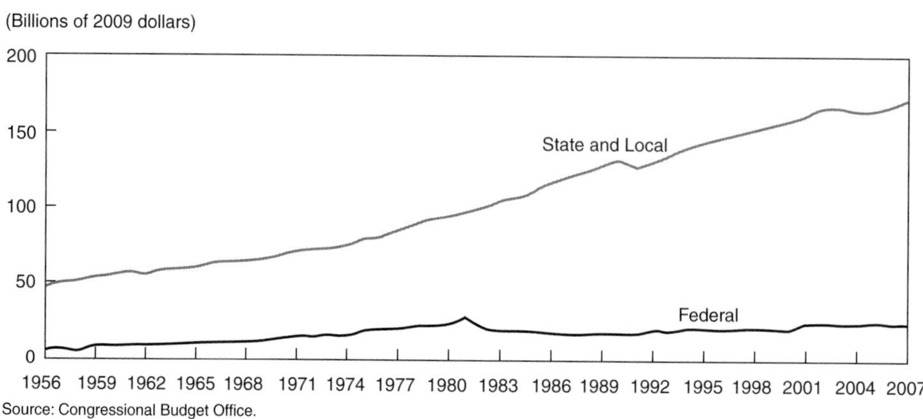

Figure 10.1
Public spending on the operation and maintenance of transportation and water infrastructure from 1956 to 2007

in 2000, Baltimore, Denver, Detroit, and Dallas had infrastructure maintenance needs far beyond their financial capacity to address. The City of Los Angeles estimates that 80 percent of what is needed to maintain existing infrastructure is not budgeted.[2] Today, budget problems at the state and local levels are making it even more difficult to maintain infrastructure not supported by fees.

Examples of infrastructure failures continue to be reported in the trade journals and mainstream media. For example, in 1992, the Chicago River flooded downtown Chicago due to a small break in an abandoned underground tunnel which would not have happened had $10,000 been allocated for a routine patch to the pipe. This caused the evacuation of over 200,000 people and caused over $1 billion in losses.[3] Still, some cities and counties are committed to full maintenance. The City of Phoenix has a five-year facilities management plan with a database system that updates the replacement schedule. Street conditions are monitored every two years, and adequate maintenance is budgeted. Fairfax County, Virginia, also places a high priority on infrastructure preservation in its budget process.[4]

Causes of Maintenance Backlogs

Lack of attention to maintenance has four causes. First, investment in infrastructure during the past 150 years has followed the major business and political cycles, as well as the migration and growth of population, from north and east to the south and west of the country. Given the expected life of these systems, many are now due or overdue for replacement or major rehabilitation. Most urban water supply systems were installed between 1860 and 1900, and sewers between 1890 and 1910. Big urban street (hard surface roads) construction programs were undertaken from 1910 to 1930. Many schools were built in the 1930s. More recent investments in transportation and water systems (see above) occurred in the space of a few decades.

Second is the political dimension of infrastructure. Politicians gain recognition with new capital expenditures, but not for maintenance of existing systems. The needs of social programs in the 1960s shifted many urban budgets away from infrastructure.[5] Third, the institutional structure of the

Landscape Asset Maintenance

At the micro level, maintenance is essential to ensure that capital expenditures for design and landscaping do not deteriorate. "Proper landscape maintenance is one of the most critical aspects in assuring the long-term viability of landscape designs," noted Willson McBurney, the landscape architect responsible for designing two central Florida transportation landscapes. He notes that "Money spent for designing a landscape plan, buying trees and plants, and putting them in the ground is too often wasted because the vegetation fails to thrive. The key to reversing this scenario can be found in what is often the least-valued step in the process: maintenance."[1]

Note

1. Willson McBurney, "Keep It Green," *American Public Works Association Reporter* (2004).

engineering profession traditionally has not rewarded or recognized maintenance as much as development.[6] Maintenance is the responsibility of local public works departments, but competing priorities, often dictated by the agency's governing boards, can change priorities away from maintenance.[7]

Finally, the lack of systematic and inexpensive approaches for appraising the condition of infrastructure facilities may contribute to the lack of focus on maintenance.

Today little disagreement exists about a serious shortfall in funding for maintenance of public infrastructure in the older cities. Conservation, preservation, and demand management are touted along with planned strategic maintenance to address infrastructure management.

Two reforms offer guidance to the local practitioner. The first is "asset management," specifically targeted at infrastructure by transportation and water officials at the federal level, but adaptable at the local level for all infrastructure systems. The second is performance management, around since the 1960s, which is an essential element of the performance budgeting movement that enjoyed a boom in the 1990s and is promoted today as a way of holding government accountable.

The following sections address these two reforms, highlighting elements useful for local GPGs in managing infrastructure.

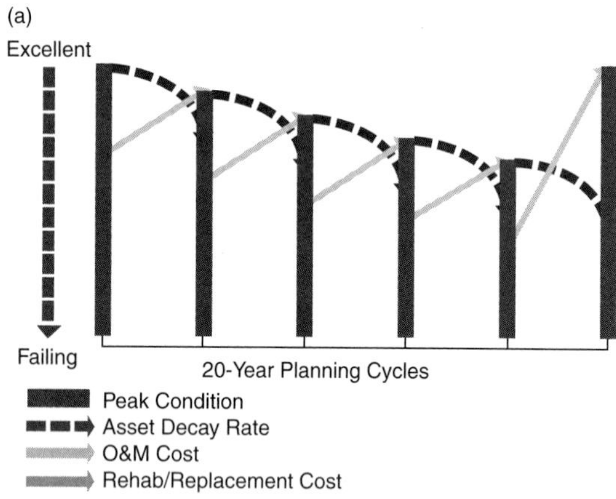

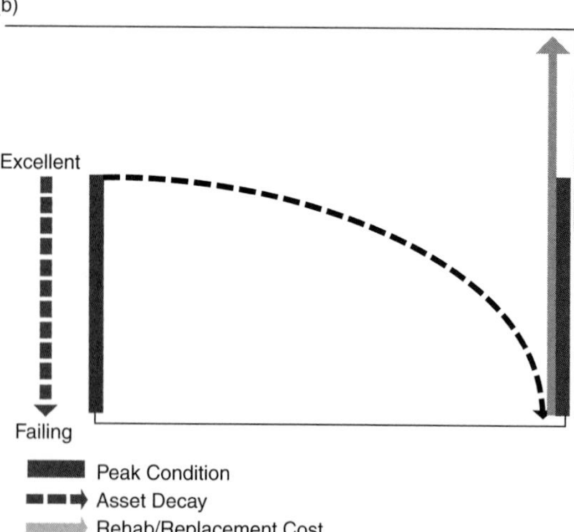

Figure 10.2 a, b
Comparing a planned approach to system maintenance with "run to deterioration"

Asset Management

Asset management is "a systematic approach to managing capital assets in order to minimize costs over the useful life of the assets while maintaining adequate service to customers."[8] It is a process of allocating resources for infrastructure that is strategic and proactive as opposed to tactical and reactive. Preservation of existing systems to prevent deterioration is emphasized, rather than waiting for problems to occur. An asset management approach also seeks to integrate decisions between various infrastructure systems and subsystems, so that investments in repair or maintenance for bridges, for example, are compared to streets and highways.[9]

Asset management uses a planned approach to system replacement and maintenance needs rather than "run to deterioration." See Figures 10.2a and 10.2b for a graphic representation of how these approaches differ.

Asset management has four key elements: (1) linkage with the organization's goal-setting process; (2) rich information management tools that also support the requirements of GASB 34; (3) increased use of sophisticated analytical software; and (4) strong performance monitoring and evaluation systems.

Linkage with the Organization's Goal-Setting Process

The first element of an asset management system is a clear set of performance expectations for the system that is linked to the organization's overall mission and goals. These can be developed during the budget process. Based on these expectations, performance measures are developed to guide the asset management systems process. Regardless of the size and budget available for the infrastructure system, this aspect of asset management is essential.

Asset management can be integrated into existing strategic planning or performance budgeting and measurement efforts. In fact, asset management guides published by EPA and the Department of Transportation incorporate the data management and analysis techniques unique to asset management into the larger "managing for results" framework.

Information Management

Information management forms the heart of asset management. Many local governments already have some systems in place for tracking infrastructure assets. Computerized maintenance management systems (CMMS) or work order management systems (WMS) are often used to organize the daily, monthly, or annual work of the public works crews. These systems track sewers, street trees, medians, parks, sidewalks, and often public buildings in order to manage system failures, repairs, citizen complaints, and preventive maintenance.

Most GPGs will develop individual data systems for the unique purposes of their particular functional area. Transitioning to an integrated system can be expensive and laborious, whether it involves making connections between existing systems, or purchasing new software systems with standard identifiers and conventions for the jurisdiction. All the systems can be integrated together physically (an expensive and

WIIFM (What's in it for me?)

Everyone seems to agree that asset management is a "good thing." If that is true, why do so few cities implement it? Consider the human aspects:

- *It's a lot of work.* Inventorying what you have, defining how you will measure condition, and then measuring the condition of each asset takes a lot of staff time—with no extra staff.
- *Managers are afraid to report bad news.* What will the political reaction be when a manager reports his assets are poorly maintained? Only high risk takers are willing to give themselves a "D" or an "F" in a dramatic attempt to get more funding.
- *Regulatory agencies get upset* when the asset they are regulating (wastewater, stormwater, etc.) is reported to be substandard.
- *Asset management does not produce funding.* Too often, an asset management report gets a one-day headline about how terrible things are, but no new funding. Managers are not willing to waste resources on what is seen as a wasted effort.

To make asset management work, officials must find the WIIFM for the manager. WIIFM can include: direction from a strong mayor; justifying a proposed fee increase; building public support for a bond issue; and reporting problems which can be attributed to prior administrations. Whatever the reason, asset management will not happen until a WIIFM is identified for the manager.

Written by Vitaly B. Troyan, former city engineer for the City of Los Angeles and Deputy Public Works Director for the City of San Francisco.

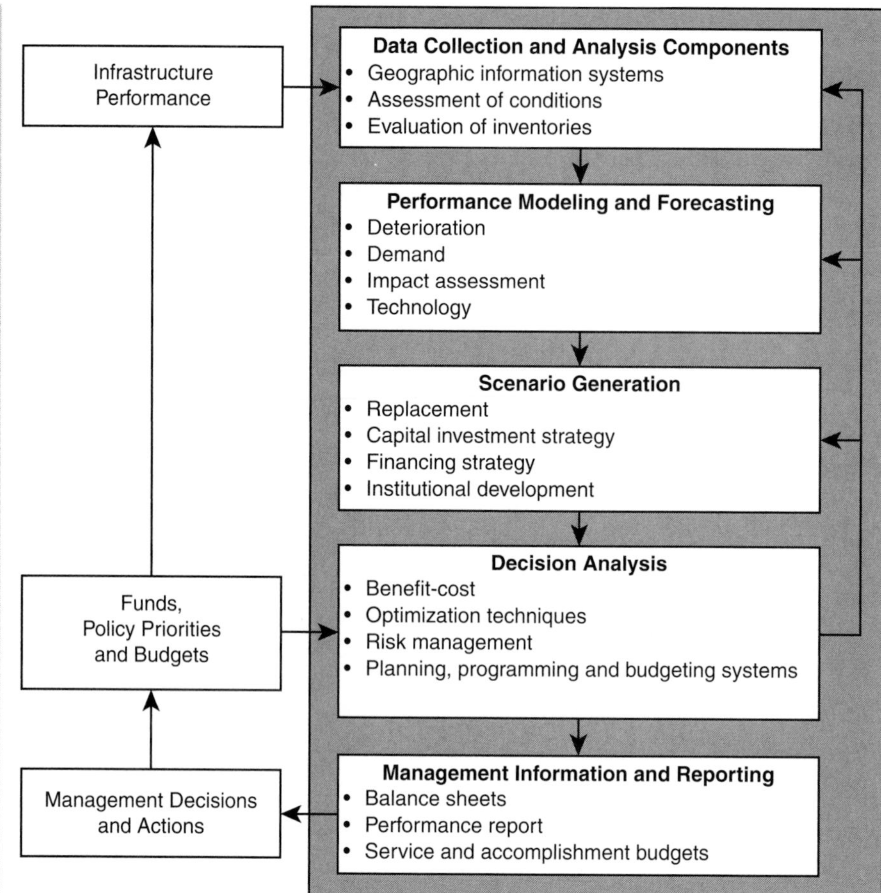

Figure 10.3
Asset management system and relationship to budget

difficult process), or virtually, using common data definitions for location, and uniform criteria for investment decisions. The "virtual" system can be set up as part of the budget preparation and reporting process (Figure 10.3).

Some of the more sophisticated infrastructure data systems for individual functional areas have software that prioritizes sub-elements of the system for capital investment and repairs used for streets in the form of pavement management systems (PMS) since 1970 and for bridges (PONTIS) since the late 1980s. However, the advent of the personal computer in the early 1980s allowed PMS to operate on desktops for smaller agencies without mainframe computers (see sidebar on PMS in Chapter 20 on streets). The federal government requires the use of a pavement management system for all local streets as a condition of receiving federal funds for federal highways in the municipality's jurisdiction.

Three kinds of information are necessary for asset management: (1) an inventory of the infrastructure; (2) condition assessment information; (3) and dollar cost or value. Sometimes these are combined into a core data system, but they can also be analyzed even if the underlying systems are distributed between multiple functional organizations, as long as common identifiers and investment conventions exist.

Infrastructure inventory. An inventory of an infrastructure system can be contained in a simple database or Excel file. These days, however, localities are moving toward installing automated systems. A core asset database can be compiled from maintenance management systems and the geographic information system if the locality has one. The level of detail in these systems is far greater than that needed for making capital investment decisions.

The database for asset management should contain elements for the individual systems that are common to the supporting databases. It may contain the following data elements: identifier, description, location, system, quantity, size, construction materials, installation or construction date, original cost, and service life. It should also have fields for condition assessment, rehabilitation, and renewal schedules.

Data for an infrastructure inventory can be transcribed from "as-built" project records or taken from existing systems as noted above. It can also be obtained from visual observation or windshield surveys. Photos or video logs can be used for roads, sewers, and large water lines, although reducing this information to elements in a database can be time-consuming. Ground-penetrating radar and other similar tools also can be used to obtain inventory information. GPS locating has been a boon to field data collection for some utilities.[10]

Condition assessment information. The purpose of condition assessment is to identify components of an infrastructure system that are not performing adequately, determine the cause of the deficiency, predict when failure might occur, and determine what types of remedies should be employed.[11] Historically, condition assessments have been incomplete. They can also be highly subjective. The most useful systems are those that have standardized criteria and can be quantified so that they can be integrated into an overall condition index. See Table 10.1 and Figure 10.4 for a simple condition assessment used by the City of Los Angeles for street lights.

Condition assessment information for the database can be collected routinely, or can be the subject of a one-time

Getting Started

Getting started on a plan and beginning to invest in renewal is more important than developing a completely accurate and fully comprehensive database. Communities have used the following strategies to implement an asset management system:

- *Begin with part of the system.* Rather than conducting a comprehensive inventory of one infrastructure system or all the systems, one community began with a subsection of a wastewater system. Another began with all of the city's streets.
- *Use selective surveys.* Survey representative portions of an infrastructure system to obtain a preliminary assessment of condition deficiencies. Include more detailed information and information about less critical systems as time and resources allow.
- *Begin with the oldest parts of the system with obvious deterioration.* A mid-sized city in the mid-Atlantic area began with a detailed analysis in the neighborhoods with the oldest infrastructure, while conducting an overall valuation study for the entire city to guide additional investments in the system.

Adapted from Association of Metropolitan Sewerage Agencies, *Managing Public Infrastructure Assets to Minimize Cost and Maximize Performance* (Washington, DC, 2002).

Table 10.1
Street light report card for the City of Los Angeles in 2003

Category	Definition	Total number of streetlights
A Very good	Not in need of energy-efficiency upgrade. Safe and efficient multiple circuit. Maintainable and reliable. Less than 30 years old (HPS and MH).	107,561
B Fair to good	Candidate for energy-efficiency upgrade. Safe and efficient multiple circuit. Less than 30 years old (MV and Inc).	17,355
C Poor to fair	Candidate for energy-efficiency upgrade. Expensive to maintain/unreliable. 30–40 years old (any light sources).	41,594
D Very poor	Candidate for energy-efficiency upgrade and system replacement. Dangerous high-voltage series circuits. Not maintainable/unreliable. Over 40 years old (any light source).	29,222
F	Unlit streets	69,000 (not in total)
Total		195,732

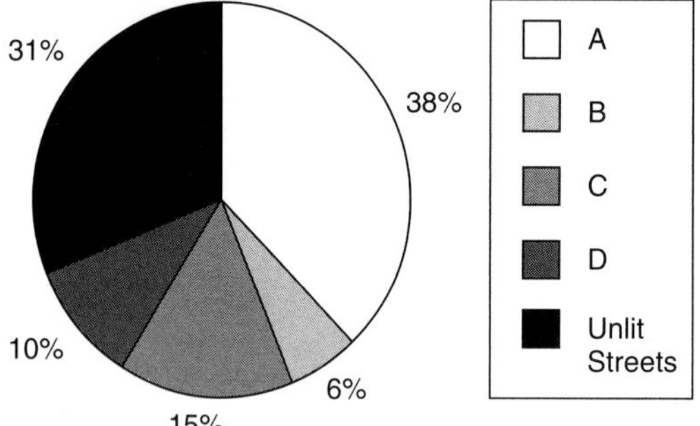

Figure 10.4
Street lighting grades from LA City report card

study for the purposes of making a specific investment decision. A wide variety of techniques are available to assess the condition of a facility. The use depends upon the purpose and the budget of the agency. They can range from "nondestructive" methods to ones where samples of the structure are taken to determine whether the bearing walls of a building or bridge will fail during an earthquake. Predictive models can also be used if the system cannot be observed directly, as with sewer pipes or the interior of a bridge pylon or a support in a building. See Figure 10.5 for how Edmonds, Washington, visually maps the conditions of its road inventory.

If the locality is using the GASB 34 modified accounting option, the condition assessment must be based on an up-to-date inventory, and the methods used must be documented and replicable. A measurement scale is needed and a minimum acceptable standard established by administrative rule of the jurisdiction. As noted in Chapter 9, GASB 34 requires a condition assessment every three years.[12]

Cost or value information. Information is also needed on the original cost of the system (its historic value), replacement or rehabilitation cost, and operating and maintenance cost. Many times the original cost is not available. An estimate can be made by historically deflating the replacement cost. Or the reverse can be done—the replacement cost can be estimated by escalating the original or historic cost to today's dollars. A variety of cost indices exist to assist this process, including the *Engineering News Record*, which has a construction and building cost index that is published monthly. The Bureau of Labor Statistics also produces both the producers' and consumers' cost indices.

Other sources that can assist in valuing the asset if the original records are not available include construction cost estimating guides, manufacturer quotations, and cost data from similar facilities. These are often combined with the expected service life. See Tables 10.2 and 10.3.

The value of the asset includes not only the original cost, but all costs connected with placing the system into service, including engineering costs, legal fees, right-of-way fees, and installation costs.

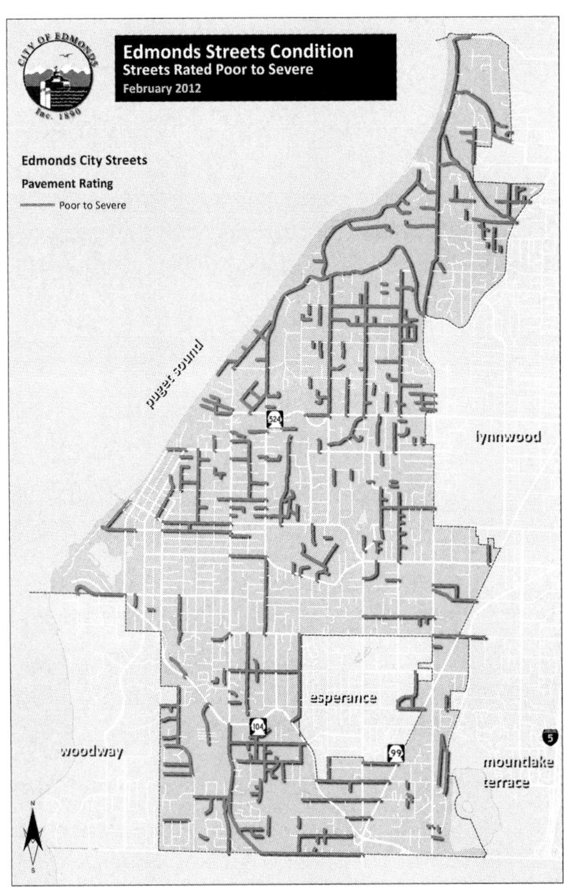

CITY OF EDMONDS

Figure 10.5
Street assessment of Edmonds, Washington, in 2012

Setting Investment Priorities for Infrastructure Maintenance

Another element of a comprehensive asset management system is the use of analytical tools, many drawn from the

Table 10.2
Infrastructure cost measures

Cost type	Example
Initial costs	Planning and design costs Construction costs Debt service
Recurring costs	Operations costs, Maintenance costs, Depreciation
Other costs	Fund source Expenditure timing Source availability timing

Source: Committee on Measuring and Improving Infrastructure Performance, National Research Council, *Measuring and Improving Infrastructure Performance* (Washington, DC: The National Academies Press, 1996).

Table 10.3
Expected service life of various infrastructure facilities

Infrastructure type	Expected service life (years)
Airport terminal	50–70
Airport runway	40–50
Public buildings	30–100
Electrical transmission/telecomm lines	50–100
Water lines	50–100
Streets	10–40
Sewer lines	50–100
Landfill	Up to 250
Electrical power plant	20–50

Source: Ronald W. Hudson, Ralph Haas, and Waheed Uddin, *Infrastructure Management* (New York: McGraw-Hill, 1997).

field of economics, to set investment priorities for existing infrastructure maintenance, repairs, and replacement. The objective of these tools is to allow the local practitioner to prioritize investment or budget alternatives according to the relative economic efficiency of each. Life-cycle and risk assessment are two techniques that can be useful.

Life-cycle cost analysis. Life-cycle cost analysis looks at initial or "first costs" as well as maintenance and rehab costs over the life of the project to identify the mix of local activities that will achieve the desired standards at the lowest long-term cost. In comparing investment alternatives, differences in the cost of installation, operating efficiency, maintenance, and repair frequency are evaluated. This technique is the principal ingredient in bridge and pavement management systems. It can also be used for a stand-alone evaluation of design alternatives for a major infrastructure investment.

In the transportation field, the chief performance measure traditionally used to evaluate investment options was "speed" or "volume to capacity ratio" into investment decisions. In addition, different transportation modes often were not evaluated against each other at budget time. Today, growth management and livability concerns have prompted a new generation of performance standards that include "mobility," "accessibility," "livability," and "sustainability." A livable community also looks at the automobile as one of several travel choices, and is concerned about fossil fuel consumption and air pollution.[13]

The use of life-cycle cost analysis is accepted in making transportation investment decisions but has been limited primarily to pavement design, and does not include user costs.[14] Issues such as the following appear to stand in the way:

- Selecting the discount rate.
- Quantifying nonagency costs such as user costs.

- Obtaining credible supporting data.
- Projecting costs and travel demand.
- Estimating salvage value and useful life.
- Estimating maintenance costs and effectiveness.
- Modeling asset deterioration.

Risk assessment. Risk assessment is also a factor for some infrastructure investment decisions. It is becoming increasingly important given the stresses upon infrastructure caused by climate change. The goal is to identify those systems or portions of systems that will have the most adverse impact upon the community if they fail so that the jurisdiction can focus on these first. Risk assessment is used to determine how critical the particular system is for the users or how critical a particular portion of the infrastructure system is for overall operations of the system. The likelihood that the system will fail, as well as its consequences if it does, are assessed and included as costs.

The organization needs to identify the types of risk events that might impact the infrastructure. These can include natural events, such as floods, tornados, or earthquakes. Risks can also include those posed by actors external to the system, such as a shortage of gas or oil in the event of geopolitical factors or earthquakes. Physical failure risks include the failure of the system itself through aging, or through design or operational considerations.

These risks then need to be analyzed in terms of their consequences to the jurisdiction. These can include repair costs, loss of revenues or service, loss of life, injury or other health impacts, property

Looking at Life-Cycle Needs of a Landscape Investment

A senior landscape architect points to five "lessons learned" in working with landscape contractors to make sure that life-cycle maintenance issues are addressed for new trees and plants put in by a contractor so that they remain healthy for many years to come:

1. *Upgrade the Role of Maintenance.* Those who will maintain the landscaping should be involved in the design.
2. *Integrate Maintenance into Project Plans.* Maintenance costs must be calculated into the overall project costs.
3. *Face-to-Face Meetings with Bidders and Selected Contractor Prior to Work.* Explain project goals in meeting with all bidders and again with selected contractor after award, and shortly before construction of landscaping begins.
4. *Detailed Project Specifications.* Consider whether to prescribe the number of visits to each landscaping site to inspect and repair irrigation system operations.
5. *Adequate Supervision of Maintenance Activities Following Installation.*

Adapted from Willson McBurney, "Keep It Green," American Public Works Association Reporter (2004).

Importance of Life-Cycle Costing

The high up-front costs of capital acquisition often dominate the capital improvement planning process. It is important, however, to evaluate capital improvement alternatives relative to the blend of capital and life-cycle costs and the expected useful life of the asset. For instance, it may cost $1 million to construct a 36-inch sewer using a 4-inch compacted gravel bed and $5 million to build the same line using an 8-inch gravel bed. Over time, however, the higher maintenance costs and shorter useful life related to the first design would more than make up for the difference in up-front cost. Other life-cycle costs that may affect the cost include the risk of harm to human health or the environment, or the risk of private or public property damage in the event of failure.

Adapted from U.S. Environmental Protection Agency, *Fact Sheet: Asset Management for Sewer Collection Systems* (Washington, DC: U.S. Environmental Protection Agency, 2002).

Table 10.4
Infrastructure reliability measures

Measure	Description
Demand peak indicators	
Peak-to-capacity ratios	If the peak demand approaches system capacity, breakdown is more likely
Return frequency, mean time to failure	Longer time between breakdown or failure of some sort is better
Design safety factors or planning contingency factors	Infrastructure systems often have safety or contingency factors built in to account for uncertainties

Source: Committee on Measuring and Improving Infrastructure Performance, National Research Council, *Measuring and Improving Infrastructure Performance* (Washington, DC: The National Academies Press, 1996).

damage, and failure to meet regulatory requirements, as well as loss of respect in the community. Some measures of reliability for the system are shown in Table 10.4.

Once an organization has completed a risk assessment, it might focus on the most critical assets for a detailed inventory of condition instead of inventorying all the facilities, as well as other strategies to prevent or mitigate the identified areas of vulnerability.

Reporting and Evaluation

The fourth element of an asset management system is a routine process of reporting and evaluating the infrastructure system against performance measures already established. Different levels of the organization have different responsibilities for asset management, and consequently different information needs. It is beyond the scope of this book to outline all of these needs. However, the establishment of an overall reporting framework for the local jurisdiction is critical to preserving scarce infrastructure assets. The Executive Reporting section of this chapter describes several aspects of a system.

Performance Measurement

Performance management grew to prominence at the federal level in the 1990s as governments sought to "reinvent" themselves. Performance management was seen as a way of promoting government accountability. Many state and local governments also adopted performance budgeting and tried to link service accomplishments to budget decisions.

Performance measurement is systematic monitoring and evaluation against indicators of performance. Performance measurement collects, analyzes, and reports information on resources used, the work produced, and the results achieved.

Regular Measurement of Progress

Harry Hatry, performance indicator guru, notes that "Regular measurement of progress toward specified outcomes is a vital component of any effort at managing for results."

Harry Hatry, *Performance Measurement: Getting Results* (Washington, DC: Urban Institute, 1999).

Performance Indicators

A key element of performance measurement is the performance indicator, a measurable data item that shows how well the jurisdiction is doing with regard to its goals and objectives. For infrastructure, most indicators are suitable for monitoring the services delivered once the major capital investment in the system has occurred. They can be used for solid waste services, transportation improvements, water delivery, wastewater collection systems, parks, and facilities maintenance.

The indicators should address quality, cost, cycle time, customer satisfaction, and sustainability objectives, as well as other outcomes. A few local governments use productivity measures that combine efficiency and effectiveness into a single indicator.[15] Some common performance indicators are:

- *Workload or output measures*. These provide information on what is produced by the government, or how many units of service were provided. They can measure how many work orders were processed, how many miles of sewage pipes were cleaned, how many foot lane miles were repaved, how many tons of garbage were collected, or measure units of preventive maintenance services. They do not track the quality of the services, or how efficiently they were provided. They provide information about the demand or need for a service, but do not by themselves assist in more effective infrastructure management. Table 10.5 shows some traditional infrastructure workload output measures that have been used for several decades.
- *Cost-efficiency measures*. Cost-efficiency measures relate outputs to the resources consumed. Examples are the unit cost per foot lane mile of repaving, the cost per ton of garbage collected, the unit cost for a preventive maintenance service, or the number of staff hours per street tree trimmed or street light replaced. These measures combine workload information with information on the dollar costs or staff hours expended.
- *Process-efficiency measures or cycle-time measures*. These include turnaround time, such as the percent of work orders receiving response within 24 hours, or the average length of time taken to repair a sewer main break. These measures enable managers to assess the delivery of infrastructure services.
- *Impact or outcome measures including customer satisfaction*. These indicators are the most closely linked to the objectives of the agency or government. Sometimes they are known as quality-of life indicators, or quality-of-service indicators, because as noted above, infrastructure is built and maintained in order to improve people's lives. These indicators can include the quality of effluent, levels of traffic congestion, the number of trips and falls for a sidewalk repair program, and other

Table 10.5
Infrastructure output measures

Infrastructure system	Performance measure(s)
Water supply/Wastewater treatment	Gallons of water treated
	Number of service connections
	Leakage/loss rate
	Conformance with environmental regulations
	Number of main breaks
Wastewater treatment	Sewered area
Solid waste management	Solid waste tons collected
	Recycled material diversion rates
Energy	Megawatts produced
	Production efficiency
Public buildings	Library books per patron;
	Cell space per inmate
Transportation	Revenue hours of transit service
	Lane miles of highway
Telecommunications	Bandwidth capacity

Source: Committee on Measuring and Improving Infrastructure Performance, National Research Council, *Measuring and Improving Infrastructure Performance* (Washington, DC: The National Academies Press, 1996).

indicators related to the purpose of the infrastructure. They can also include the satisfaction of citizens, ratepayers, or recipients of the local service.[16]

- *Sustainability indicators*. These are a variant of impact measures which are used to evaluate the community's progress toward sustainability. Currently cities are measuring reduction in carbon emissions and energy use. Other examples include the Eutrophication Index, used in the Netherlands to measure the release of phosphates and nitrogen compounds, while an index of wild salmon runs through local streams is used in Oregon to assess the health of the watershed.

The list of additional resources at the end of this chapter contains several references that describe how to develop good indicators or performance measures, both generally and for specific infrastructure systems.

Benchmarking

Benchmarking is another aspect of performance measurement. Once indicators have been developed, they can be compared to each other over time within a single jurisdiction. The Government Financial Officers Association (GFOA) calls this "process benchmarking." Indicators can also be the basis for another popular management initiative, continuous process improvement (CPI), often called corporate-style benchmarking, as managers try to continuously improve their performance. Benchmarking has also come to mean comparing the indicators of performance in one jurisdiction with those in another. GFOA calls this "performance benchmarking." These external benchmarks are examined to find best practices in other jurisdictions that can help to improve service delivery. Another objective of external benchmarks is to use them as a target. GFOA calls these "strategic benchmarks."[17]

Three efforts are currently underway in the United States designed to assist local government practitioners in benchmarking services. The first is the ICMA Center for Performance Measurement, which collects performance and cost data for police, fire, emergency medical, and neighborhood services, including solid waste, highways and road maintenance, parks, and facilities management. A group of 44 cities and counties developed the indicators to be measured. An annual survey is the basis of an annual report for use by other cities and counties to compare service delivery.[18] The other two efforts are regional, one in the Northwest, where 27 cities share performance information, and one in North Carolina sponsored by the state with 14 participating cities.[19]

Oregon's benchmarks have tracked 272 indicators of well-being since 1991, while "Sustainable Seattle," a nonprofit organization, published an urban sustainability report in 1993 with 20 indicators, and has since been updated.

Executive Management Reporting Systems

Monitoring progress against the performance indicators developed during the budget process is the single most important way that managing infrastructure can be tied to the locality's management systems. Performance information can also be used by managers throughout the year to identify problems with attainment of the yearly targets and to reallocate resources to solve operating problems. The underlying systems can also be used by line managers to plan workloads by the week or month.

A very simple system that can be used even with the most diverse set of individual asset management systems, and even when some systems have no databases at all, is an executive management monitoring system. This consists of periodic face-to-face meetings about the organization's

Two Steps in Benchmarking

The first step in benchmarking is to compare two performance data points with one representing a benchmark. The second step is to analyze the difference between these two points and to make recommendations for improving the output, the outcome, or the efficiency of the delivery of the service.

Adapted from Janet M. Kelly and William C. Rivenbark, *Performance Budgeting for State and Local Government* (Armonk, NY: M. E. Sharpe, 2003).

Advice on Benchmarking

Remember that … (1) no benchmarks and data measuring them are ever perfect; (2) direct comparisons to another jurisdiction or to an absolute standard should be done only to find red flags; (3) small differences should be ignored; and (4) never expect an exact comparison to another jurisdiction.

Adapted from Richard Fischer, "An Overview of Performance Measurement," *Public Management* (1994) online at the ICMA Center for Performance Measurement, available at: www.icma.org.

performance in delivering infrastructure services. The system is supported by performance reports drawn from existing asset management systems or other databases. The effectiveness of this process depends more on the consistency with which managers meet over a common framework than the complexity and sophistication of the underlying systems.

Reporting system. Two types of activities are performed by local governments—services delivered on a recurring basis, and projects undertaken and completed. Services are measured with a performance indicators system, whereas projects have a beginning, middle, and end, and are tracked with a project management system. Both kinds of reports are prepared on a monthly, six-week, or quarterly basis, and exception reporting forms the backbone of discussion during face-to-face meetings. These two systems are an essential part of the monitoring process for the capital plan and budget, and are relatively new at the local level.

These reports rely for information on existing asset management systems and existing reporting systems, although when such do not exist, an original effort may be developed for the executive management reporting system. These reports can also be incorporated into the official budget docouments.

Project management systems track the progress of capital infrastructure projects. (A project management system also tracks other "lumpy" local efforts that have a beginning, middle, and end, such as the completion of an audit, preparation of a zoning ordinance, development of a fee schedule, or preparation of a management study.)

Local GPGs, or even special districts, with many small capital improvements projects can use software systems designed to both plan the work over a year period and then report progress on a weekly or monthly basis. These systems provide for the entry of financial and cost information which is useful for consulting firms. However, many local governments that have centralized accounting systems will probably need to modify this software so that the expenditure information is consistent with that in the budget office.

Performance indicator systems were discussed in detail above. In reporting, many indicators and their targets can be placed on one page, and a space can be held for comments describing why a target was not met.

Periodic face-to-face meetings. Although department and division heads may monitor progress toward key goals on a weekly or monthly basis, at the executive level, six weeks is about the right amount of time to follow up on actions from the previous meeting. Another alternative is to rotate departments so that each department meets only six times a year. Key to the success of the meeting is to have the appropriate support staff present. The chief executive convenes the meeting, which is attended by the heads of finance,

Performance Measures Need to Be Part of an Overall Management System

"The hardest part of a performance measure system is not in derivation of indicators but in application of results. The problem is that unless a government ties its performance measures meaningfully into its management systems—unless those measures are something more than decorations for the budget document—any performance management program will fail, and probably deserves to."[1]

Note

1. Janet M. Kelly and William C. Rivenbark, *Performance Budgeting for State and Local Government* (Armonk, NY: M. E. Sharpe, 2003).

purchasing, and affirmative action, the personnel director, and possibly legal counsel.

The team goes through the status of the projects division by division. If there are more projects than time allows, they can be addressed on an exception basis. The usual kinds of problems that prevent projects from meeting their milestones concern hiring, litigation or permits, or procurement. Having the lead department head in the meeting for these functions assures that some problems can be resolved on the spot, or appropriate action can be developed for follow-up. A staff person takes the lead on action minutes, which are then tracked for the following meeting.

Performance measures are also reviewed during the meeting. Usually these include workload or efficiency measures. Outcome measures are more appropriate for the yearly budget review since they change more slowly in response to government action.

Implementing an executive management reporting system. If the system does not already exist, start-up should begin as part of the budget process. Since this will require a change in the way the government entity is managed, managers should be involved from the beginning in the development of the system for the locality. The two-tiered budget committee set up for the capital budget noted in the previous chapter should develop the forms and reporting instructions.

For project management reporting, each project should have its own form with a brief project description, milestones, and budget information. In addition, each program area should have a page where key indicators of performance are developed and reported on. A mechanism should be developed for collecting the data for reporting. The mechanism can be manual, or can rely upon existing asset management or computerized maintenance management systems, as well as the accounting system. Generally, the budget office or the chief official's office assigns a staff person responsible to take the lead on assembling the reports and for staffing the function.

The executive management reporting system should be jurisdiction-wide. It is a critical tool to ensure effective management of infrastructure assets (as well as other programs), and flexible enough to use highly sophisticated measures and systems or simpler but no less effective reporting systems. In some jurisdictions, elected officials may be invited to attend the meetings if they are interested, or to gain access to the reports. Performance against the project milestones and performance indicators for infrastructure (and for other programs) can be used in personnel evaluations as well.

Improving the Performance Measurement Process

- Address service quality and outcomes explicitly when reviewing services and programs during the budget process.
- Set a target for each performance indicator and assess progress regularly with written reports and face-to-face meetings.
- Include indicators of both "intermediate" outcomes and "end" outcomes in the performance measurement process.
- Report and meet on performance at least quarterly if not monthly.
- Calculate key breakouts of the data for each indicator, perhaps by geographic area, or suborganizational unit.
- Ask program managers to provide explanatory information when targets are missed.
- Incorporate outcome-related performance requirements into contracts wherever feasible.
- Use service quality and outcome progress information as part of the internal personnel evaluation process.

Adapted from Harry Hatry, Craig Gerhard, and Martha Marshall, "Eleven Ways to Make Performance Measurement More Useful to Public Managers," in *Public Management* (1994), online at the ICMA Center for Performance Measurement, available at: www.icma.org.

Conclusion

Today's sophisticated information technology enables local practitioners to manage infrastructure assets in a way only dreamed of even 10 years ago. Careful investment in a computerized asset management system, coupled with implementation of an executive management system, will repay the local taxpayers tenfold. When individual asset management computer systems are linked up with GIS, they can also be a valuable resource for land use planners during the preparation of the comprehensive plan and during the permitting process for individual development projects. Care needs to be taken, however, to ensure that a locality's individual systems are internally compatible with respect to geocoding, software, and analytical conventions.

Additional Resources

Publications

Ammons, David N., Erin S. Norfleet, and Brian T. Coble. *Performance Measures and Benchmarks in Local Government Facilities Maintenance*. Washington, DC: International City/County Management Association: ICMA Center for Performance Measurement, 2002. Contains a wealth of indicators suitable for many levels in the organization for managing facilities and buildings.

Hatry, Harry P. *Performance Measurement: Getting Results*, 2nd ed. Washington, DC: Urban Institute Press, 2007. This is the authoritative guidance on developing and reporting performance indicators for governments, by the guru of indicators.

Kelly, Janet M. and William C. Rivenbark. *Performance Budgeting for State and Local Government*. Armonk, NY: M. E. Sharpe, 2003. A thorough and interesting book on what performance budgeting is, its status at the local level, and how to do it.

Websites

American Public Works Association: www.apwa.net. The professional organization for public works officials. The website has a wealth of information on infrastructure asset management, including publications and links to over 60 chapters around the USA and Canada.

American Society of Civil Engineers: www.asce.org. Many public works officials also belong to ASCE. The website has helpful resources about asset management in general as well as for specific infrastructure systems.

International City/County Management Association: Center for Performance Measurement: www2.icma.org. This website has many helpful resources for performance measurement including statistics and current data for selected cities and counties that can be used as a benchmark.

U.S. Dept. of Transportation Federal Highway Administration: Office of Asset Management: www.fhwa.dot.gov/infrastructure/asstmgmt. Asset management information for transportation systems.

U.S. EPA: www.epa.gov/npdes/pubs/assetmanagement.pdf. This site links to asset management information for water and wastewater systems.

PART THREE
Financing Infrastructure

Almost a third of a trillion dollars is needed annually to build local infrastructure to accommodate the anticipated growth in population and to replace and upgrade older systems over the next 30 years. This annual need is large even when compared to the $1.9 trillion in federal budget outlays in 2001 or the $10.8 trillion for the GDP in that same year. Even if only the rate of expenditures in 2001 is considered, a hefty $167 billion, financing capital infrastructure projects by local governments, requires the ingenuity and best skills of the local practitioner.

Part III is designed to assist the practitioner in deciding upon the best way to raise these funds, whether as part of a long-range infrastructure plan, as part of the capital budgeting process, or simply to finance an individual project. Each of the five chapters looks at a different aspect of financing issues and tools.

Chapter 11 (The Financial Context for Infrastructure) briefly describes the major infrastructure financing tools before discussing revenue sources for local governments. This chapter also goes into detail about the traditional source of local infrastructure financing—the property tax.

Chapter 12 (Bonds and Borrowing) outlines the different kinds of bonds, the actors in the bond financing process, and the steps in issuing a bond. Bond financing strategies are addressed.

Chapter 13 (User Fees and Public Pricing) describes various types of user fees and the issues associated with each. The steps used to calculate fees are presented, along with how to decide whether or not user fees are appropriate. The steps in setting up a special district to be supported by fees are outlined briefly.

Chapter 14 (Exactions and Impact Fees) is concerned with the growth of a newer form of financing that has been used to finance off-site infrastructure for new development in the past three decades. The pros and cons of using impact fees are discussed, as is a collaborative process with developers being assessed for these fees. How to develop impact fees is described.

The last chapter in this part of the book, Chapter 15 (Competition and Privatization) addresses various privatization strategies, ranging from asset sales and contracting out, to introducing competition in the provision of infrastructure services. The pros and cons of each approach are discussed, along with a description of the incidence of privatization. The steps necessary to undertake such a process at the local level are outlined.

chapter 11

The Financial Context for Infrastructure

In this chapter...

Importance to Local Practitioners. 173
Infrastructure Expenditures 174
Historic Trends on Capital Expenditures. 174
Public Versus Private Spending. 175
Capital Spending by Level of Government 176
State Versus Local Capital Spending on Infrastructure. 176
Capital to Operating Expenditures. 179
Revenue Sources of Local Government. 179
Property Taxes. 180
Funds: General, Special, and Enterprise. 182
Federal or State Grants and Loans 182
A Local CIP Example 184
How Much Should We Spend on Infrastructure? 185
National Benchmarks 185
Local Benchmarks 186
Conclusion 188
Additional Resources 188
Publications. 188
Websites 188

Why Is This Important to Local Practitioners?

Local practitioners need to understand the financial context at the national, state, and local levels in order to prepare capital budgets, to develop project budgets, and to prepare local infrastructure plans. This includes understanding the relationship of the amount spent at the national level to the Gross National Product (GNP), as well as the amounts spent over time by individual infrastructure systems at various levels of government and the private sector. In addition, local practitioners should have some background information on the terminology and sources of revenues for governments so that they can make informed decisions about where to go to finance an infrastructure program or project. Although previous chapters have discussed the preparation of the capital improvement budget, and subsequent chapters in this section of the book go into detail about individual revenue sources, this chapter provides the context for various financing mechanisms for infrastructure.

The chapter begins with trends in infrastructure financing by the federal government for key programs. The amounts by infrastructure system are described nationally before drilling down to expenditures by system at the state and local level. The chapter then turns to a discussion of the property tax, which is the traditional source of funds used to pay off infrastructure bonds. Finally, some criteria are developed for assessing how much should be spent on infrastructure at the local level.

Impact of Density on Local Infrastructure Costs[1]

A five-year effort sponsored by the Federal Transit Administration developed a model to compare the impact of population growth projections over the next 25 years for all 3100 counties in the United States under controlled and uncontrolled (sprawl) growth scenarios. The controlled growth scenario allocated a portion of the projected development to already developed counties, or to areas as close as possible to already developed counties. The objective was to reduce growth to 75 percent of the sprawl-growth threshold for rapidly growing counties. Receiving counties were allowed to accept growth only to that same level, with the exception of the Northeast, where a stricter limit was put in place to ensure that growth was not being sent to declining urban locations.

This research found that sprawl would occur in about one quarter of the nation's counties. However, the study also found that growth could be redirected from lesser developed to more developed counties almost 60 percent of the time. That is, almost 60 percent of the counties could reduce the total expected increase in employment and population growth by one quarter by redirecting it to contiguous areas that were not growing as fast. Most of these counties were located in the West and South.

The costs of water and sewer infrastructure, transportation, and other services were compared for directed and undirected growth alternatives. Significant cost savings were found to occur with more dense development alternatives. Overall, the U.S. could save about $12 billion in water and sewer costs with more compact development. The reduction in costs for local roads (lane miles) was almost 12 percent for a total savings of $110 billion. The South, which is projected to have the greatest amount of absolute growth in the next 25 years, would be able to save $5.5 billion in water and sewer costs if denser alternatives were followed.

Note

1. Robert W. Burchell et al., *Costs of Sprawl—2000*, Transit Cooperative Research Program Report no. 74 (Washington, DC: National Academy Press, 2002).

Infrastructure Expenditures

Unfortunately there is no existing analysis of infrastructure expenditures that covers all the systems, that is also historic, and that disaggregates by level of government. The following therefore looks at several sources of analysis and data, concentrating predominately on capital spending rather than operations.

Historic Trends for Capital Expenditures

Looking at trends in constant dollars for all public investments, federal, state and local, from 1995 to 2010, the amount steadily increased each year, although state and local expenditures declined in 2009 and 2010 (Table 11.1).

When a longer time period and only water and transportation are examined, a slightly different picture emerges. From 1956 to 2004, capital spending increased by 1.7 percent in inflation adjusted amounts, or from 1987 by 2.1 percent a year. This increase (quite large in actual dollars) was caused by the increase in capital spending by localities to accommodate post-WWII growth. However, capital spending as a percent of GDP for these programs has been relatively flat since the 1980s[1] (Figure 11.1). Between 2003 and 2007, constant dollar spending on transportation and water infrastructure decreased by 6 percent due to increases in cost of materials that outpaced the increase in the amount spent in current dollars.[2]

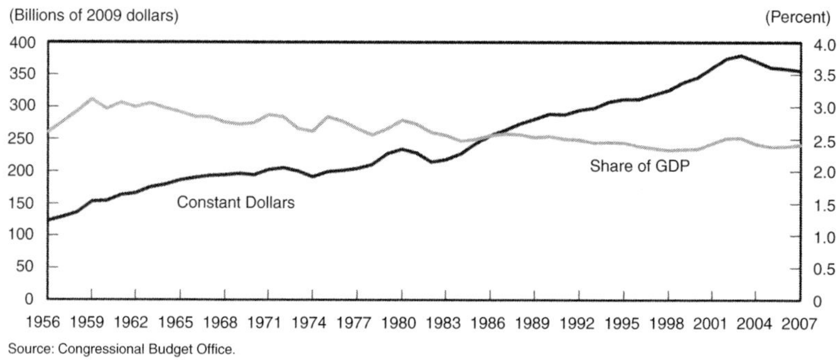

Figure 11.1
Total public spending on transportation and water infrastructure in constant dollars and as a share of the GDP, 1956 to 2007

Table 11.1
Investment expenditures on infrastructure from 1995 to 2010 in chained (2005$) millions for federal and state and local governments

	Federal non-defense			*State and local*			
Year	*Investment expenditures*	*Total expenditures*	*Investment expenditures as % of total*	*Investment expenditures*	*Total expenditures*	*Investment expenditures as % of total*	*Investment ratio of state/ local to federal*
1995	27,103	227,526	11.9	201,907	1,183,618	17.1	7.4
1996	29,340	225,714	13.0	211,133	1,211,140	17.4	7.2
1997	28,988	231,921	12.5	227,125	1,254,306	18.1	7.8
1998	31,981	233,749	13.7	233,711	1,303,762	17.9	7.3
1999	34,774	238,692	14.6	252,697	1,361,832	18.6	7.3
2000	31,643	244,355	12.9	266,578	1,400,145	19.0	8.4
2001	31,337	255,497	12.3	279,740	1,452,302	19.3	8.9
2002	34,171	273,915	12.5	289,443	1,500,641	19.3	8.5
2003	34,579	281,713	12.3	292,241	1,499,670	19.5	8.5
2004	34,432	284,574	12.1	289,753	1,497,074	19.4	8.4
2005	36,252	287,253	12.6	281,553	1,493,582	18.9	7.8
2006	39,095	296,550	13.2	286,445	1,507,180	19.0	7.3
2007	39,544	294,205	13.4	288,319	1,528,100	18.9	7.3
2008	42,450	313,338	13.5	290,963	1,528,083	19.0	6.9
2009	44,083	333,788	13.2	285,329	1,514,167	18.8	6.5
2010	50,395	357,653	14.1	274,299	1,486,963	18.4	5.4

Source: National Income and Product Accounts, 2012

Public Versus Private Spending

Public agencies play the dominant role in capital expenditures for transportation infrastructure (94 percent), and for environmental infrastructure associated with water, wastewater, and solid waste disposal (92 percent). Not surprisingly, public agencies have the lion's share of the capital expenditures for community facilities such as correctional facilities, parks, and other civic buildings. Private capital is a substantial portion in two of the subcategories, however, with 58 percent for solid waste disposal and 100 percent of the investments associated with freight railroads. By far the largest infrastructure sector is transportation, followed by the private utility sectors of telecommunications and energy. Capital expenditures for community facilities are close to the amount for the telecommunications and energy sectors combined (Figure 11.2). (In national discussions, community facilities are not often included but this area is a major concern for the local practitioner, see below.) The environmental sector is the smallest component.

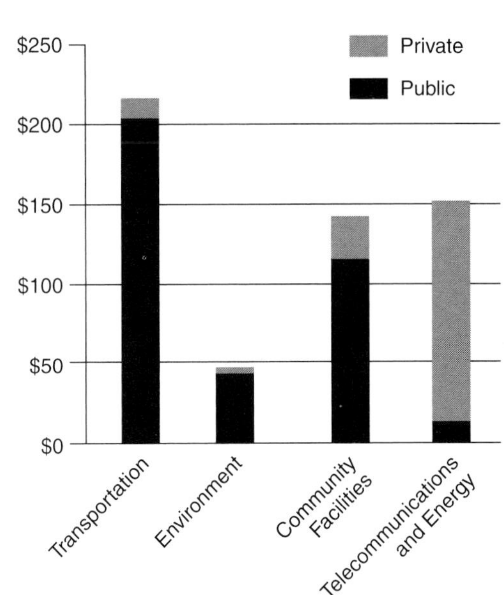

Figure 11.2
Breakdown of capital spending in 2004 by public–private institutions

Within the transportation sector in 2004, capital expenditures for streets, roads, and highways formed 61 percent of the total. Mass transit accounted for 14 percent, while passenger rail outlays were only 6/10ths of a percent.[3]

Capital Spending by Level of Government

Although much of the media attention goes to expenditures on infrastructure by the federal government, in actuality most public infrastructure spending takes place at the state and local level. Federal expenditures on infrastructure have been confined to those projects with a national interest such as interstate commerce, security, or an overriding health issue. Except for a brief period in the 1970s when state and local capital expenditures declined sharply in response to historically high interest rates, and when federal capital spending for water and sewer and transit projects peaked, relative shares have been fairly constant from 1956 through 2004. Federal "dominance" in capital spending was only for traditional public works projects, and only from 1976 to 1984.[4]

Another way of looking at this is to examine the ratio of local and state government expenditures on infrastructure to those at the federal level. From 1995 through 2010, the percent of total expenditures varied by year for the federal government with a low of 11.9 percent in 1995 to a high of 14.1 percent in 2010. State and local percent of total expenditures ranged from 17 percent to a high of 19.5 percent in 2003. Until 2008, state and local governments invested 7 to almost 9 times that of the federal government. This changed in 2008, when the ratio dropped sharply, reaching 5.4 in 2010. However, generally over time, local and state expenditures for infrastructure are a relatively constant proportion of total expenditures—varying between 17 and 19 percent, although this depends on growth pressures and the economic cycle. Federal non-defense investment expenditures for infrastructure as a percent of its total expenditures dipped in the year 2000, rising again to a high of 14 percent in 2010, no doubt as a result of stimulus funds[5] (Figure 11.3).

State Versus Local Capital Spending on Infrastructure

Overall, local governments spend over twice as much for capital outlays for infrastructure than do state governments.

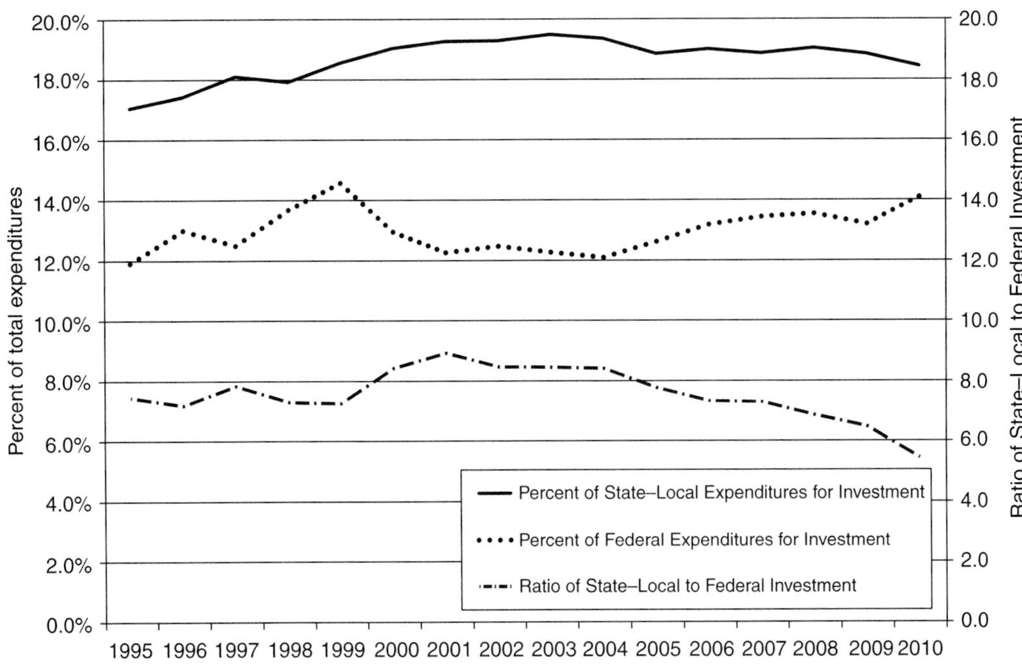

Figure 11.3
Investment ratios, federal compared to state and local

In 2009, local capital outlays were 69 percent of the total for state and local governments. This was the result of predominately local spending for community facilities and the environmental infrastructure where local governments outspent states by a factor of almost 3 for the former and 9 for the latter. Community facilities is the largest sector for capital expenditures, accounting for almost $127.5 billion in 2009 followed by transportation at $85.7 billion, $42.2 billion for the environmental systems, and $21 billion for publicly owned energy and telecommunication systems (Table 11.2).

Community facilities capital spending includes most non-networked facilities. Capital spending for education was the largest component of this sector in 2009, with higher education expenditures dominated by the state, and local for primary and secondary education. Capital expenditures for parks and hospital were also substantial, again, primarily a local matter.

Transportation capital spending is largely a state affair, with 63 percent of the total. The state is the dominant player for highways local streets, but airports, ports, and transit are dominated by local-level provision of capital facilities. Despite the fact that state spending for highways is over twice as much as at the local level, local level capital outlay for streets and highways at $11.2 billion was a substantial amount.

Capital expenditures for water supply, sewers, solid waste, and natural resource protection occurred mostly at the local level, with 90 percent of the total in 2009. It should be remembered that many solid waste facilities are privately owned and operated and do not show up in this data.

Table 11.2
Capital outlays by state and local governments in 2009 (in billions of dollars)

Description	*State*	*Local*	*Total*	*Percent of Total*
Direct general expenditure	605.6	1,291.8	1,897.8	100.00
Capital outlay	106.1	184.4	290.5	15.31
Other direct general expenditure	499.5	1,107.4	1,607.3	84.69
Percentage	31	69	100	
Transportation				
Highways	61.8	23.9	85.7	4.52
Transit	2.6	11.2	13.8	0.73
Air transportation (airports)	0.2	2.1	2.3	0.12
Sea and inland port facilities	0.2	0.4	0.6	0.03
Subtotal	64.8	37.6	102.4	5.39
Percentage	63	37	100	
Environment:				
Natural resources	3.2	3.8	7.0	0.37
Water supply	0.1	13.4	13.5	0.71
Sewerage	0.5	18.7	19.2	1.01
Solid waste management	0.3	2.1	2.5	0.13
Subtotal	4.1	38.0	42.2	2.22
Percentage	10	90	100	
Community Facilities				
General public buildings	0.6	1.7	2.3	0.12
Parking facilities	0.0	0.3	0.3	0.01
Parks and recreation	1.1	10.3	11.3	0.60
Hospitals	3.0	6.2	9.2	0.48
Correction	2.1	1.7	3.8	0.20
Higher education	25.9	5.4	31.3	1.65
Elementary & secondary	1.6	67.7	69.3	3.65
Subtotal	34.3	93.3	127.5	6.72
Percentage	27	73	100	
Energy and Telecommunications				
Electric power	3.6	15.2	18.7	0.99
Gas supply	0.0	2.3	2.3	0.12
Subtotal	3.6	17.4	21.0	1.11
Percentage	17	83	100	

Source: Adapted from U.S. Census Bureau, *State and Local Government Finances by Level of Government and by State: 2008–2009* (Washington, DC: U.S. Census Bureau, 2011). Water, public buildings, and parking are estimates.

Capital spending for publically provided energy and telecommunications was also a local rather than state matter although we do not have figures for the latter. Again, it must be remembered that public capital spending in this sector is overshadowed by that of the private sector.

Capital to Operating Expenditures

Infrastructure is capital-intensive. For most systems, operating expenses are around 50 percent or less of total expenditures. The U.S. Census of Governments of State and Local Finances done in 2007 (published in 2010) found that for state and local governments combined highways and airports had the highest percentage of capital to total expenses with 58 percent and 49 percent respectively. This means that operating expenses, that is to say, their operating expenses were less or about half total yearly expenditures. Sewers were also fairly capital-intensive, with capital expenditures ranging from 45 percent in 1990 to 40 percent in 2008. Water and mass transit followed suit. Solid waste management expenditures on capital by state and local governments during those years were about 10 percent.

Revenue Sources of Local Government

Local practitioners must finance capital outlays and maintenance for infrastructure in the context of the overall revenue picture of the jurisdiction. Local governments receive revenue from the federal government as well as state governments; from taxes; from utility revenues (earmarked by fees to pay for particular services); from liquor stores and trust funds for pensions and unemployment insurance (Table 11.3). In 2009, state and local governments together received 22 percent of their funds from the federal government, 53 percent from taxes, 16 percent from user fees and charges[6].

Taxes are required revenues exacted for public purposes. They are the largest source of revenues for state and local governments. The major categories of taxes are property taxes, income tax and sales taxes, and federal general revenue tax sharing, although the latter is more important for states than for cities and counties. Property taxes are the most important revenue source for local governments accounting for 74 percent of tax revenue in 2009. State governments were most dependent on sales taxes and income taxes which accounted for 83 percent of their revenues in that year. (See Figure 11.4 for combined sources of revenue.)

General revenues (as opposed to user fees and charges, and impact fees) can be spent on whatever the local politicians decide. These are also called discretionary funds since their uses are not prescribed for a particular function. The following addresses first property taxes, other general

Table 11.3
Sources of Revenue for State and Local Governments in 2009 (billions)

Description	Combined	State	Local	Local % of Combined
Revenue from federal government	537	476*	61	
General Revenues				
Taxes	1,271	716	556	44%
Property	424	13	411	97%
Sales and gross receipts	434	345	89	21%
Individual income	271	246	25	9%
Corporate income	46	39	7	15%
Other taxes	97	73	25	25%
Charges and miscellaneous fees	605	285	321	53%
Utility revenues (water, energy, transit)	144	17	127	89%

Source: U.S. Census Bureau Data from Table A1. State and Local Government Finances by Level of Government and by State: 2008-2009 Issued October 2011. Pension Trust Funds and Liquor Store Revenues Excluded.

* 470.7 from the federal funds is suballocated to local government.

revenues and finally special purpose grants. Charges, fees and other ways of financing infrastructure are dealt with in subsequent chapters.

Property Taxes

Since colonial times, property taxes have been the most important general revenue source for local governments. Today, the property tax is being supplanted by other general revenue sources such as sales and income taxes. The property tax provided two-thirds of the revenue of local governments prior to the 1930s, compared to one-fourth today.

The property tax declined as a total share of local revenue in two eras: the Depression through WWII, and in the 1970s. In the 1930s and 1940s, property values declined (resulting in less property taxes), while state and federal government grants for highways, education, and welfare became important revenue sources for local governments. The share of local general revenues attributable to property taxes stabilized for the next 15 years until the mid-1960s. At that time, although federal and state aid increased, local governments began to employ user charges and other non-property taxes. The incidence of

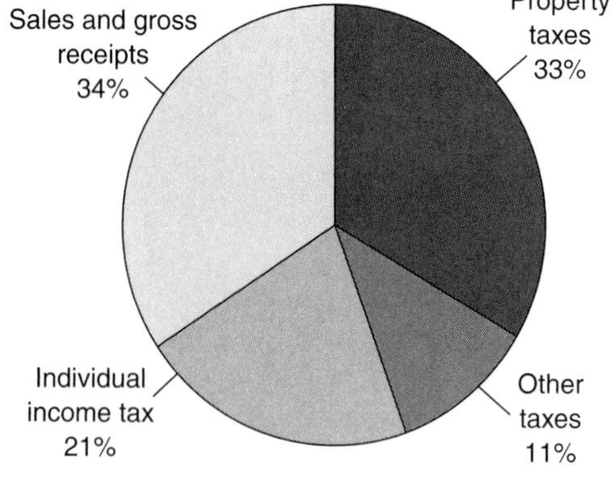

Figure 11.4
State and Local Government Tax Revenue 2009

special districts, which rely primarily on user charges, began to increase rapidly. Similarly, the tax revolts of the 1970s placed new restrictions on property tax collection, which contributed to a declining share. Towards the end of the millennium, the property tax share of local general revenues again stabilized.[7]

The property tax is a local tax that is universal at the local level. School districts receive about 40 percent of the total amount levied, and it accounts for about 80 percent of their total revenues. Cities and counties combined account for another 40 percent of the total amount levied, although property taxes comprise only half of all municipal taxes, and three-quarters of county taxes.[8] Special districts and authorities rely very little on property taxes; instead, they employ user charges and fees. The property tax remains the principal source for general obligation bonds for municipal buildings and schools.[9]

Until the 1990s, most states required that all property be taxed at the same rate. However, in the past two decades at least 15 states have divided properties into different classifications and rates. Owner-occupied residential and farmland have the lowest tax rates, and business and utility properties have the highest. New York has a ratio of 5:1 from the highest to the lowest rate, but most states with differential rates have a smaller range.[10]

Property taxes today are primarily ad valorem (percent of value) taxes applied to the assessed value of real property, that is, land and improvements. Taxation of personal property (such as automobiles) in the U.S. used to be more common, but this has been declining—from about 16 percent in 1961 to only 10 percent in the mid-1980s. Each state has its own regulations that govern how property will be assessed. Local governments then set the tax rate (or sometimes the portion of the tax base that will be taxed) that will be applied to the assessed value.[11]

The county usually is responsible for assessing the value of the property, and for collecting and disbursing the tax to the various taxing agencies. To raise the property tax, the district or government desiring the funds will go to the voters. Depending upon the state, this may require a simple majority, a super majority, or two-thirds of the registered voters.

The property tax bill will specify the various components that make up the total. These elements may be related to the elections held to approve each tax increase. Some property tax bills show the amount that goes to the school district, the city, the county, and special districts. See Figure 11.5 for a sample property tax bill from DeKalb County in 2011.

Funds: General, Special, and Enterprise

During the past 50 years, alternatives to the property tax, such as sales tax and income taxes, have become more important for local governments. Sales taxes, since they generally accrue to a locality in proportion to the amount of retail activity in the area, have been criticized for encouraging retail and discouraging residential land uses.[12]

Property taxes declined by almost 40 percent as a percentage of general local revenues from 1968 to 1998, as their share declined from 27 to 17 percent. All general revenues increased in constant dollars during that period by 175 percent, with property taxes increasing only 72 percent (Table 11.4).

The area of general revenue with the largest increase (these figures do not include user charges or grants) is the individual income tax, which increased 368 percent. Sales tax and income taxes are generally not used for infrastructure *per se*, because they are a major share of discretionary income at the local level. Instead, infrastructure services can usually be quantified into unit costs that can be charged back to the user.

Since general funds can be used for any local purpose, most of the discussion during the budget process revolves around how these funds will be allocated. However, in many local governments, general funds may make up only one-third of a municipal budget. The rest (aside from intergovernmental grants) will consist of "Special" or "Enterprise" funds. Most governments in the United States are required to use fund accounting that separates the budget into individual funds depending upon the source, each with its own rules about what money is included and how it can be spent. Each fund must balance its revenues and expenditures, and surpluses in one fund cannot be used to bail out another fund.[13]

Federal or State Grants and Loans

Taxpayers both inside and outside the benefit area pay for federal or state grants or loans (which have interest rates lower than the locality would have to pay in the municipal bond market). Most studies indicate that the availability of federal and state grants has a strong substitution effect—that localities do not assess themselves if funds are available from other sources. An advantage from the local point of view is that the funds do not have to be

2011 DEKALB COUNTY REAL ESTATE TAX STATEMENT

PAY ONLINE AT www.yourdekalb.com/taxcommissioner
OR BY PHONE AT 404-298-4000

CLAUDIA G. LAWSON
TAX COMMISSIONER

OWNER	TAXPAYER JOE
CO-OWNER	
PARCEL I.D.	06 280 02 034
PROPERTY ADDRESS	123 MAIN STREET
TAX DISTRICT	50 DUNWOODY
PIN.	1234567

APPRAISAL VALUES AND EXEMPTION INFORMATION

LAND	125,000	40% TAXABLE ASSESSMENT	140,000
BUILDING	225,000	BASE ASSESSMENT FREEZE	140,000
MISCELLANEOUS		NET FROZEN EXEMPTION	0
TOTAL	350,000	EXEMPTION CODE	H1F

YOUR TOTAL TAX SAVINGS FOR THIS YEAR IS 1,363.25 A REDUCTION OF 585.25 IS DERIVED FROM YOUR LOCAL CONSTITUTIONAL HOMESTEAD EXEMPTION(S). THE HOST CREDIT OF 778.00 IS THE RESULT OF AN ADDITIONAL HOMESTEAD EXEMPTION FUNDED BY PROCEEDS FROM THE HOMESTEAD OPTION SALES TAX.

COUNTY GOVERNMENT TAXES

Levied by the Board of Commissioners: representing **24.73%** of your tax statement

TAXING AUTHORITIES	TAXABLE ASSESSMENT	X MILLAGE	= GROSS TAX AMOUNT	- FROZEN EXEMPTION	- CONST-HMST EXEMPTION	- HOST CREDIT	= NET TAX DUE
COUNTY OPNS	140,000	0094300	1320.20		94.30	563.91	661.99
HOSPITALS	140,000	0008800	123.20		8.80	52.62	61.78
COUNTY BONDS	140,000	0008700	121.80		0.00	0.00	121.80
UNINC BONDS	140,000	0009400	131.60		0.00	0.00	131.60
FIRE	140,000	0027000	378.00		27.00	161.46	189.54
TOTAL COUNTY TAXES							1166.70

BOARD OF EDUCATION - SCHOOL TAXES

Levied by Board of Education: representing **67.63%** of your tax statement

TAXING AUTHORITIES	TAXABLE ASSESSMENT	X MILLAGE	= GROSS TAX AMOUNT	- FROZEN EXEMPTION	- CONST-HMST EXEMPTION	- HOST CREDIT	= NET TAX DUE
SCHOOL OPNS	140,000	0229800	3217.20		287.25	0.00	2929.95
TOTAL SCHOOL TAX							2929.95

STATE & CITY TAXES, AND OTHER CHARGES

Levied as applicable by State, City, or County: representing **11.47%** of your tax statement

TAXING AUTHORITIES	TAXABLE ASSESSMENT	X MILLAGE	= GROSS TAX AMOUNT	- FROZEN EXEMPTION	- CONST-HMST EXEMPTION	- HOST CREDIT	= NET TAX DUE
CITY TAXES	140,000	0027400	383.60		167.40	0.00	216.20
STATE TAXES	140,000	0002500	35.00		.50	0.00	34.50
CITY STLIGHT	100 FRONT FEET	.33	33.00				33.00
DEKALB SANI	UNIT RATE	1	265.00				265.00
CITY SWTRFEE	UNIT RATE	1	48.00				48.00
			265.00				
TOTAL STATE, CITY AND OTHER ASSESSMENTS							596.70

TOTAL PROPERTY TAXES

	TOTAL MILLAGE	GROSS TAX AMOUNT	- FROZEN EXEMPTION	- CONST-HMST EXEMPTION	- HOST CREDIT	= NET TAX DUE
TOTAL DUE	0407900	6056.60		585.25	778.00	4693.35

PLEASE DO NOT FOLD, STAPLE, OR CLIP REMITTANCE COUPONS TO PAYMENT

MAKE YOUR CHECK PAYABLE TO:
DEKALB COUNTY TAX COMMISSIONER
P.O. BOX 100004
DECATUR, GA 30031-7004

DUE DATE
NOVEMBER 15, 2011

PLEASE NOTE THAT A 5% PENALTY WILL BE IMPOSED IF THE AMOUNT SHOWN IS NOT PAID BY THE INSTALLMENT DUE DATE

PARCEL I.D.	06 280 02 034
TOTAL ANNUAL TAX	4693.35
INSTALLMENT AMOUNT DUE	2346.68

SECOND INSTALLMENT

0212345671000003015697000000603138 8

TAXPAYER JOE
123 MAIN STREET
DUNWOODY, GA 30338

RETURN THIS COUPON WITH YOUR SECOND PAYMENT

ENTER AMOUNT PAID

PLEASE DO NOT FOLD, STAPLE, OR CLIP REMITTANCE COUPONS TO PAYMENT

MAKE YOUR CHECK PAYABLE TO:
DEKALB COUNTY TAX COMMISSIONER
P.O. BOX 100004
DECATUR, GA 30031-7004

DUE DATE
SEPTEMBER 30, 2011

PLEASE NOTE THAT A 5% PENALTY WILL BE IMPOSED IF THE AMOUNT SHOWN IS NOT PAID BY THE INSTALLMENT DUE DATE

PARCEL I.D.	06 280 02 034
TOTAL ANNUAL TAX	4693.35
INSTALLMENT AMOUNT DUE	2346.67

FIRST INSTALLMENT

0212345671000003015697000000603138 8

TAXPAYER JOE
123 MAIN STREET
DUNWOODY, GA 30338

ENTER AMOUNT PAID

RETURN THIS COUPON WITH YOUR FIRST PAYMENT

Figure 11.5
2011 DeKalb County real estate tax statement

Table 11.4
State and local government general revenues by source from 1927 to 1998 (in millions of nominal dollars)

Fiscal year	Total	Property taxes	Prop. taxes as % of total	Sales/ gross receipts taxes	Individual income taxes	Corp. net income taxes	Fed. gov't revenue	All other
1927	7,271	4,730	65.1	470	70	92	116	1,793
1938	9,228	4,440	48.1	1,794	218	165	800	1,811
1948	17,250	6,126	35.5	4,442	543	592	1,861	3,685
1958	41,219	14,047	34.1	9,829	1,759	1,018	4,865	9,699
1968/69	114,550	30,673	26.8	26,519	8,908	3,180	19,153	26,118
1978/79	343,236	64,944	18.9	74,247	36,932	12,128	75,164	79,821
1988/89	786,129	142,400	18.1	166,336	97,806	25,926	125,824	227,838
1998/99	1,434,464	240,107	16.7	290,993	189,309	33,922	270,628	409,505
% change 1968/69–1998/99	175.8	72.4	–37.5	141.7	368.1	135.0	211.2	245.4

Source: U.S. Census Bureau, *U.S. Statistical Handbook*, 2001.

raised locally. However, there may be strings tied to the grant or loan, or a need for matching funds. The availability of the grant or loan may make the locality spend matching funds on a priority it did not previously have. In addition, federal and state grants and loans usually do not pay for the operating costs of a facility.

The principal disadvantage of relying on grants and loans for infrastructure construction is that they are erratic. It is difficult for a locality to ramp up with the internal capacity to quickly use these funds, and then to ramp down when the funding cycle is concluded. It often takes a year to hire a new staff person in local government, from the time the need is officially felt to the employee's start date. In addition, except for transportation, these funds are often scarce. However, this may be changing as infrastructure is looked at as an economic stimulus. Localities should designate a staff person to scan for federal and state infrastructure grants, because if the grants support existing priorities, they can enable a locality to build something they otherwise could not.

A Local CIP Example

See Figure 11.6 to see how Fairfax county VA proposed to use a variety of revenue sources and revenue mechanisms to fund its capital improvement program for 2013.

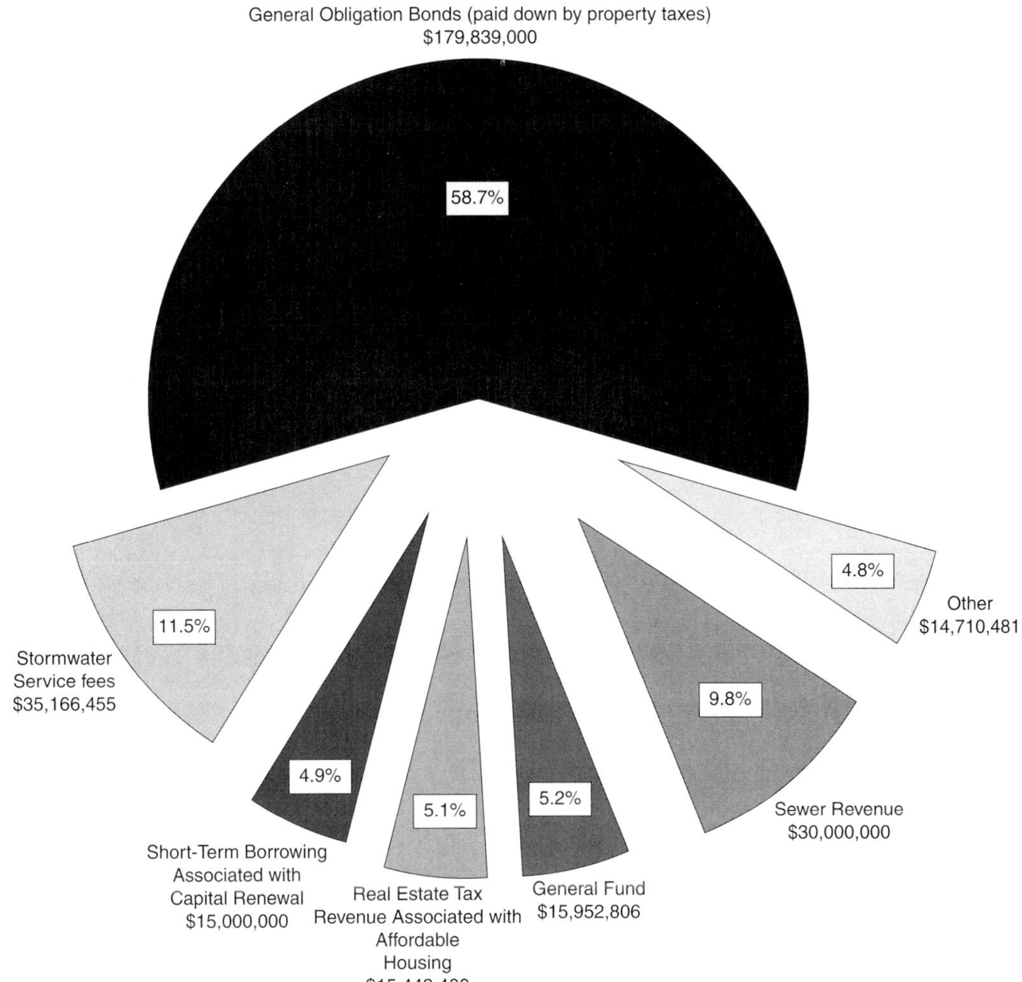

Figure 11.6
Fairfax County, Va., CIP revenue sources, 2013

How Much Should We Spend on Infrastructure?

National Benchmarks

National analysts evaluate how much should be spent on infrastructure by comparing expenditures to the past; to other countries, as a percentage of GNP and as a percentage of government expenditures (in the case of public provision). In some sectors, analysts have aggregated local growth and replacement needs to come up with a total figure. Some analysts also compare present-day expenditures to past expenditures either by industry or on an aggregate basis as an indicator of need.

An analysis by the Lincoln Institute of Land Policy that included public and private expenditures on transportation (air, road, transit, rail, water), the networked utilities (energy, water, sanitation and sewers, telecommunications), irrigation and flood control, and solid waste but not civic facilities, noted that in 2004, the U.S. spent only 2.6 percent or $302 billion. These authors suggest that an additional $103.5 billion per year spent on infrastructure by the federal government would be "economically justified" and the total amount as a percent of GDP would rise to 3.5 percent.[14] Many developed countries spent 4 percent of their GDP on infrastructure.

The ASCE Report Card described in Chapter 1 estimated that $3.3 trillion was needed for a five-year period of 2009 through 2014 which would be $66 billion a year, well under the Lincoln Institute number. This estimate was based on engineering estimates aggregated from local needs studies.

These analyses were intended to inform the national infrastructure debate to assess whether federal subsidies are needed. However, although private sector infrastructure was included, many local community facilities were not, despite the fact that they are a major capital cost item for the local government.

Local Benchmarks

Local governments do not use capital expenditures as a percent of GDP to judge the amount of infrastructure expenditures. Instead, needs drive much of the discussion, along with the ability of the jurisdiction to afford the amount.

Our estimate of local needs found that to meet growth and replacement needs with local funds would cause expenditures to rise by 34 percent although in some categories, such as transportation, water and wastewater, the increases are well over that amount. Replacement costs for existing infrastructure are two to three times those needed for new growth, particularly when the additional revenues to local government caused by the growth are taken into account.[15]

Is it possible for local governments to raise this amount? One way of assessing reasonableness is to see whether the amount could be borrowed. Rating agencies for local governments use two benchmarks: first, the total amount of bonds outstanding per capita compared to per capita income should not exceed 15 percent. Another benchmark is 10 percent of local assessed valuation.[16] For the U.S. as a whole in 2001, the total amount of bonds outstanding per capita compared to per capita income was 17.8 percent, an amount slightly over the benchmark of 15 percent used by the rating agencies.[17] Similarly, for the U.S. as a whole, the amount of debt in 2001 was 13.5 percent of the total assessed market value of real estate compared to the 10 percent benchmark. At the local level for two counties in 2004, local debt benchmarks (discussed in greater detail in Chapter 12 on bonds) indicate that money for this infrastructure could be borrowed (Table 11.5). But could taxes or fees be raised to pay off the bonds?

Our calculations showed that the fees/taxes needed to accommodate the 34 percent increase in capital spending by local governments to meet local needs for new growth and replacement would result in a 4.9 percent

Table 11.5
Local criteria for assessing infrastructure expenditure increases

	Debt as % of assessed market value	GO bond debt per cap/per capita income	Debt service as % of operating revenues
Fairfax County VA, benchmark	3	n/a	10
Fairfax County VA, actual 2004	1.45	n/a	8.30
Anne Arundel County, MD, benchmark	1.50	3	9.00
Anne Arundel County, MD, actual 2004	1.45	2.80	7.80
Benchmarks used by bond rating agencies	10	15	20
US Total 2001	13.5*	17.8**	n/a

Notes: *Using Residential Assessed Value.
**All types of bonds.
Source: Compiled by the authors.

increase, while a smaller increase that would not meet local needs would bump up these rates by 1.4 percent annually. Actual capital outlays by local governments in 2001 were 13.9 percent of all expenditures. This amount would increase to 15.3 percent with the 10 percent increase and to 21.1 percent with the 34 percent increase—not altogether impossible and close to some of the historic figures presented earlier. As a percent of the national GDP in 2001, the increased local capital expenditures would rise from 1.4 percent to 1.5 percent and 2.1 percent respectively. Even adding private, state, and federal capital expenditures to the local figure, the total is in the range of the 4 percent figure noted above as a reasonable minimum percentage for U.S. infrastructure expenditures (Table 11.6).

Table 11.6
Comparison of annual capital needs estimates at the local level

	2001 Actual	With 10% increase	With 34% increase*
Annual capital outlays ($ billions)	147	162	223
As % of local government outlays	13.9	15.3	21.1
Ratio to federal annual outlays	7.9	8.7	12.0
As % of GDP	1.4	1.5	2.1
2001 Gross Domestic Product ($)	10.8 T		
Total federal budget 2001 ($)	1.8T		
Annual tax or fee increase needed (%)	n/a	1.4	4.9

Note: *Based on actual needs analyses.
Source: Authors' calculations and BEA Economic Indicators Report, 2002.

Conclusion

Some localities may be more affected than others by infrastructure needs. In addition, the actual increases needed by sector vary greatly. For community facilities and local road replacements that are funded primarily by taxes, local politicians may choose to defer replacement and concentrate on new facilities where the cost of new construction can often be charged to the new development itself. For airports, ports, and telecommunications and energy facilities that rely on fees, the public has come to reluctantly accept rapidly escalating fees and charges. In sum, although from an economic perspective the needed capital costs at the local level are affordable, affordability varies by locality and sector. A variety of revenue sources are available to meet these needs.

Additional Resources

Publications

Bland, Robert L. and Irene S. Rubin. *Budgeting: A Guide for Local Governments.* Municipal Management Series. Washington, DC: International City/County Management Association, 1997. A useful overview for the local practitioner who wants to know more about fund accounts and budget processes.

Vogt, John A. *Capital Budgeting and Finance: A Guide for Local Governments.* 2nd ed. Washington, DC: International City/County Management Association, 2009. Comprehensive overview of property taxes, enterprise accounts, and other items useful to understanding the financial context in which the capital budget is prepared. This guide is highly recommended.

Websites

Infrastructure Finance Alternatives: Council of Development Finance Agencies: www.cdfa.net.

Lincoln Institute of Land Policy, *Fiscal Dimensions of Planning*: www.lincolninst.edu/subcenters/teaching-fiscal-dimensions-of-planning/materials/kotval-mullin-fiscal.pdf. Although initially designed to provide materials for planning professors to use to teach fiscal matters to planning students, this contains many documents on fiscal matters useful for the local practitioner.

National Conference of State Legislatures: www.ncsl.org.

chapter 12

Bonds and Borrowing

In this chapter...

Importance to Local Practitioners. 189
General Facts about Bonds 189
How Big Is the Bond Market? 190
Who Issues Municipal Bonds and What Are the Proceeds Used For? 191
Who Buys Bonds?. 192
Types of Bonds 192
Tax-Exempt GO Bonds 193
Revenue Bonds. 194
Lease-Financing Bonds (Certificates of Participation). 195
Tax-Exempt Bonds for Geographically Defined Areas 196
Taxable Bonds, Including Build America Bonds 197
Actors in the Bond Issuance Process 197
Government Agencies 197
Outside Experts and Professionals 198
Defaults and Credit Worthiness 199
Bond Ratings. 200
Factors Rating Agencies Consider 201
Key Steps in the Bond Issuance Process 202
Who Should Pay? 203
Sizing the Bond. 203
How Will the Issue Be Sold? 204
Preparation of the Bond Documents 205
Approval of the Issue by the Governing Body 205
Conclusion 205
Additional Resources 206
Publications. 206
Websites 206

An earlier version of this chapter appeared in White and Kotval, 2012-see Additional Resources

Why Is This Important to the Local Practitioner?

Most infrastructure projects and programs in the United States are too large to finance out of local revenues. In fact, most state and local-level infrastructure is built with funds borrowed from investors by issuing a "bond" or a promise to repay over a certain length of time. Debt financing has built streets, schools, hospitals, parks and art museums, along with sidewalks, city halls, sports arenas, water purification plants and much more. Deferring the cost of a facility that is needed now, but which will last for many years has enabled the United States to grow over the past 200 years. Most governments have good credit ratings, and investors are willing to lend the government money (compared to a corporation) because of its security. The municipal bond market is big business indeed in the United States, and local practitioners involved with infrastructure need to be familiar with bonds. Most local practitioners involved with infrastructure will certainly run into the question of whether and how to borrow funds to make his or her project a reality.

This chapter begins by defining bonds, the size of the bond market, and provides information on who issues bonds and who buys them. It then moves to a discussion of different types of bonds: general obligation (GO) bonds, limited liability, or "revenue," bonds, lease-based bonds, geographically based bonds, and taxable bonds. The actors in the process are defined before going into the bond rating process. Lastly, the process of issuing a bond at the local level is described.

General Facts about Bonds

A bond is an interest-bearing certificate that an organization issues in order to borrow money. A bond is a loan between the borrower, who is called the issuer, and the lender (investor). Bonds are similar to a promissory note—a promise by the issuer to repay the investor the principal of the loan by the end of a fixed period of time, which can be anywhere from 1–40 years, plus interest. Bonds are often contrasted with stocks, which are shares in a company sold to the public to raise money. Stockholders are therefore subject to both the ups and downs of a company. In the corporate world, bonds are considered a more secure form of investment than stocks because they must be paid back before stockholders are issued dividends.[1]

Both corporations and state and local governments can issue bonds, but governments cannot sell shares in the government entity to raise money the way a corporation does. Instead, governments issue municipal bonds, the term used for loans entered into by both local and state governments as well as special districts and authorities. These loans or bonds are repaid with taxes or revenues from user fees, exactions, leases, and, more recently, with intergovernmental grants.

Municipal bonds are also called tax-exempt bonds, because the interest received by the investor is not subject to the federal income tax. In some states, the interest is not subject to state or local income taxes either. By contrast, the interest received from corporate bonds is taxable.

How Big Is the Bond Market?

There really is no such thing as a bond market that resides in a specific location, like the stock markets. Instead, all transactions occur between individual buyers and sellers. This occurs because bonds must be tailored to the individual requirements of the project being financed, as well as to the different state laws and to specific sectors (e.g., health care, public power). There are over 1.5 million different types.[2]

The first debt issuance in what was to become the United States occurred in 1751, in Massachusetts. In the 1800s, as the country expanded west, state and local debt was a major source of financing for canals, roads, and railroads. In the 1900s as New York, Chicago, San Francisco and other major cities grew, debt financing helped fund transportation, water, sewer lines, ports, hospitals, and housing.[3]

The total value of the bonds issued in a year depends upon economic conditions, population pressures, and federal and state policies.[4] In the 1950s, capital spending at the state and local level was financed primarily by general obligation bonds, paid by taxes. In the 1960s and 1970s, federal grants were added to the mix to fund a variety of local capital projects. In the 1980s and 1990s, in addition to taxes, user fees became an important source of funds for local bonds. The 1986 Tax Reform Act eliminated many tax-exempt activities, leading to a decline in yearly municipal bond issuances as did economic downturns in 1993/94 and the dot.com bust in 2000/01.[5] (Figure 12.1).

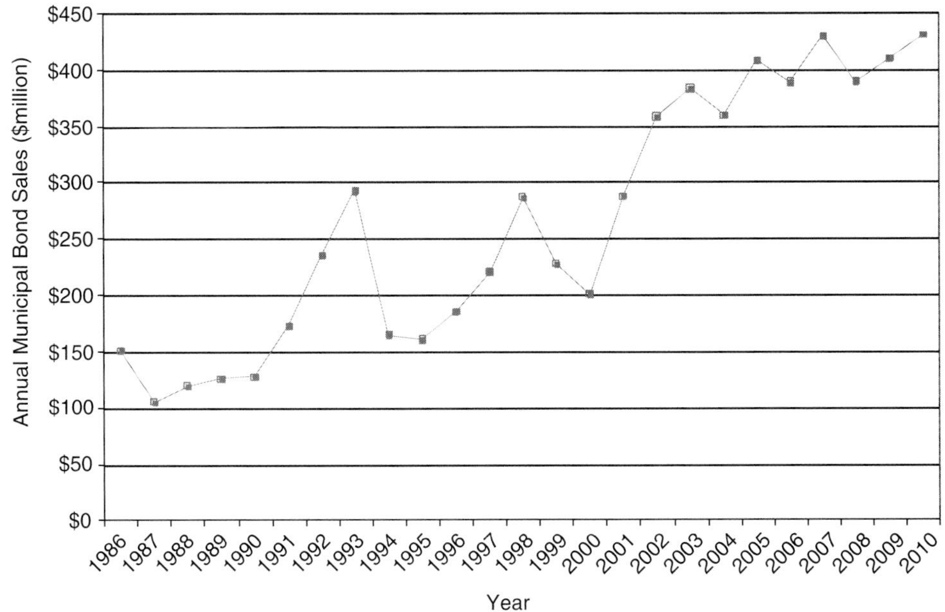

Figure 12.1
Dollar volume of annual municipal bond sales from 1986 through 2010

Who Issues Municipal Bonds and What Are the Proceeds Used For?

Municipal bonds are issued by state and local governments, school districts, nonprofit organizations such as hospitals and universities, and by special authorities and special districts. There are over 55,000 different issuers of municipal bonds in the United States, and they range from very large, such as the states of California and Illinois, to very small rural school districts.[6] Generally, authorities have been the largest borrowers, followed by special districts, then cities, and then colleges and universities.[7] In 2010, special districts (including school districts) and authorities issued $293 billion of new bonds, while cities and counties issued $86 billion. States issued only $55 billion (Table 12.1).

Municipal bonds are issued to fund a wide variety of projects, including non-infrastructure expenditures, such as student loans and pollution control, and the more typical education projects, utility facilities, economic and industrial development projects, and health facilities. Although the amounts in each category fluctuate depending upon economic vagaries, the relationships have been roughly the same over time, with education and health (including both hospitals and nursing homes) eliciting the largest amounts of bond funding, followed by transportation, industrial aid, and finally water, sewer, and flood management utilities.[8] In 2010, 24 percent of new issuances were for education projects, 17 percent for transportation, and 5 percent for municipal utilities (Table 12.1). Although the majority of bonds financed with

Table 12.1
New security issues, state and local governments ($ billions), selected years

Type of issue, issuer, or use	1990	1995	2000	2005	2010	
					N	% of issues
All issues, new and refunding[1]	122.9	145.7	180.4	409.6	434.5	
By type of issuer:						
State	15.0	14.7	19.9	31.6	55.5	12.8
Special district of statutory authority[2]	75.9	93.5	121.2	298.6	292.9	67.4
Municipality, county, or township	32.0	37.5	39.3	79.4	86.2	19.8
Issues for new capital	97.9	102.4	154.3	223.8	284.7	
By use of proceeds:						
Education	17.1	24.0	38.7	71.0	66.8	23.5
Transportation	11.8	11.9	19.7	25.4	48.6	17.1
Utilities and conservation	10.0	9.6	11.9	9.9	13.2	4.6
Industrial aid	6.6	6.6	7.1	18.6	47.8	16.8
Other purposes	31.7	30.8	47.3	60.6	89.9	31.6

Notes:
1 Par amounts of long-term issues based on date of sale.
2 Includes school districts.
Source: U.S. Census Bureau, *Statistical Abstract of the United States 2012* (Washington, DC: U.S. Census Bureau, 2012).

municipal bonds are public, under certain circumstances (discussed later) the bonds may be used for private purposes associated with industrial development.

Who Buys Bonds?

Tax-exempt municipal bonds have always been attractive to wealthy corporations or individuals in the United States. In 1980, investments by individual households or their proxies accounted for about 34 percent of all investment in municipal bonds, but this figure had doubled by 1999 to almost 75 percent. This coincided with the shifting of household investment to mutual funds, which invest in municipal bonds. By contrast, commercial banks were the largest holders of municipals in 1980 at about 37 percent, but this dropped to 7 percent by 1997, since tax reform permitted banks to deduct only 80 percent of interest on tax-exempt securities. At the end of 2011, the household sector and mutual funds held 66.7 percent of municipals, while banks and insurance funds accounted for another 23.9 percent (Figure 12.2).

Types of Bonds

Although every bond issue is different, they can be put into some general categories. There are basically three major types: General obligation bonds which are paid back or retired by taxes and Revenue Bonds which are paid by

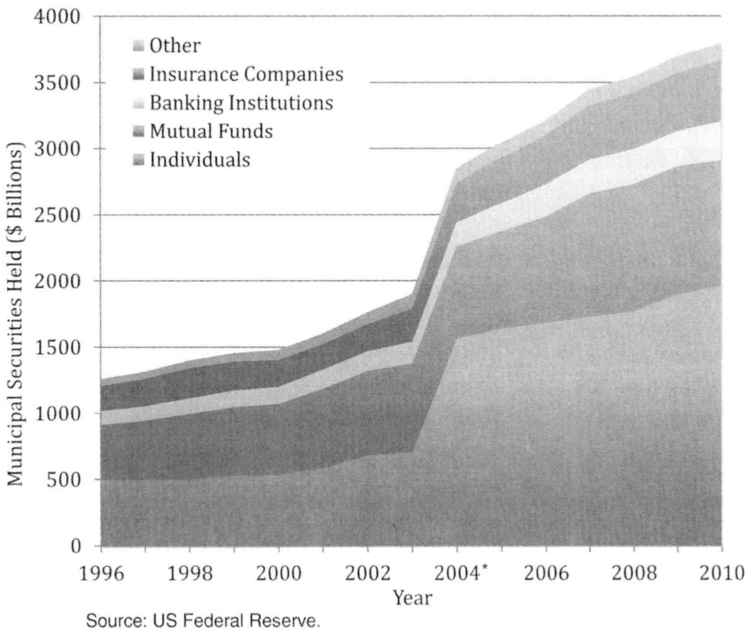

Figure 12.2
Holders of U.S. municipal securities from 1996 through 2010 ($ billions)

revenues from a project. Revenue bonds may be retired with fees from utilities, but also through fees or payments from leases or special districts. Taxable bonds can be used by a locality but these are rare, with the exception of the Build America Bonds (BAB) program described below.[9]

- General obligation bonds
- Revenue bonds
  - Leases/certificate of participation
  - Geographically based bonds
- Taxable bonds (including Build America Bonds)

Tax-Exempt GO Bonds

GO bonds are backed by the full faith and credit of the issuing government. This means that the government is obligated to use its unlimited taxing power to repay the debt. GO bonds are subject to some restrictions. First, local governments in 44 states have constitutional or statutory limits on the amount of GO debt they can incur, while cities in 40 states have limits on the amount of interest they can pay. Second, 42 states require voter approval of GO bond issues. In some states, two-thirds approval is needed at the local level for a bond issue (other areas require only a simple majority). In Oregon, a supermajority is required: more than 50 percent of the registered voters must turn out to vote, and, of these, 50 percent must approve the issuance of the bond.

Although GO bonds were the original mechanism local and state agencies used to finance large-scale capital improvements, in 2010 these bonds formed only 34 percent of all new issues, having been overtaken by

Table 12.2
U.S. municipal bond issuance[1] from 1996 through 2010: general obligation and revenue ($ billions)

	GO		Revenue		
	Amount	% of total	Amount	% of total	Total
1996	64.6	34.9	120.6	65.1	185.2
1997	72.4	32.8	148.3	67.2	220.7
1998	93.7	32.7	193.1	67.3	286.8
1999	71.0	31.2	156.5	68.8	227.5
2000	66.6	33.2	134.3	66.9	200.9
2001	101.7	35.4	186.0	64.7	287.7
2002	125.7	35.2	231.8	64.8	357.5
2003	142.1	37.1	240.6	62.9	382.7
2004	129.6	36.0	230.1	64.0	359.7
2005	144.2	35.3	264.0	64.7	408.2
2006	114.8	29.7	272.0	70.3	386.8
2007	131.0	30.9	293.5	69.1	424.5
2008	110.2	28.6	276.2	71.5	386.4
2009	154.9	38.2	251.9	61.9	406.8
2010	146.9	34.1	283.4	65.9	430.3

Note: 1 Excludes maturities of 13 months or less and private placements.
Source: Thomson Reuters.

revenue bonds. GO bonds generally have a lower interest rate than revenue bonds (Table 12.2).

Revenue Bonds

Revenue bonds rely upon user fees or dedicated revenue sources from the proposed capital facility to repay investors. They are also called limited-liability bonds since they do not rely upon the taxing power of an agency for repayment, and because they are not repaid through taxes, they are not legally part of a government's debt ceiling and do not require voter approval. Sometimes these are referred to as special obligation (SO) bonds. They generally carry a higher interest rate because they are less secure than GO bonds. However, they are still tax-exempt.

Since 1990, limited-liability bonds have been the security of choice for local governments, special districts, and authorities and in 2010, over 60 percent of all new issues for state and local governments were of this type. Revenue bonds are the workhorse of single-purpose special districts formed to build and operate infrastructure such as water and sewage plants or solid-waste facilities. They are also used to pay for convention centers, sports facilities, and parking garages.[10]

As part of deregulating the energy and telecommunications sector, federal rules were changed to permit certain private companies to issue tax-exempt revenue bonds as well. These are called private-activity bonds (PABs).

Lease-Financing Bonds (Certificates of Participation)

Lease-financing bonds, sometimes called certificates of participation, are a category of revenue bonds issued by the municipality secured with lease payments from the local jurisdiction. The term "certificate of participation" refers to the way a capital lease is marketed to investors. Certificates of participation are used for large leases (over $1 million). The capital lease is divided into certificates of participation that are sold publicly. For smaller leases, the debt can be privately placed.[11]

Since leases are an obligation on the part of the municipality or special district to make rental payments, not a commitment to service a debt, they are not subject to the limitations placed on debt by state and local laws. This enables the local agency to issue the bonds without voter approval and the large majorities required for GO bonds.[12] Schools, public buildings, hospitals, and even parks and transportation facilities have been built using lease-financing bonds. They can be used to finance prisons, courthouses, convention centers, or similar projects when political support is insufficient for a GO bond. The local government usually enters into an agreement with a developer, a nonprofit, or a joint-powers authority to build a facility and then lease it back to them with a long-term lease. Bonds to finance the capital facility are issued at a tax-exempt rate, with the rental stream and interest rate determining the size of the issue, and hence the size of the facility to be built. See Figure 12.3, which illustrates the flow of funds in Philadelphia that was used to pay for a new airport terminal.

Certificates of participation can also be used to purchase fire trucks, police vehicles, computers, and telecommunications equipment using city funds to enter into the lease agreement: general funds, gas taxes, sewer funds, or even community development block-grant funds.

On the Use of Certificates of Participation for City Buildings

The County of Santa Barbara has used Certificates of Participation (COPs) as a primary means of financing capital needs that do not have a fee revenue stream to secure bonds such as roads, bridges and flood control projects. These include several public safety buildings, a children's clinic, the county elections facility and the lining of a landfill. COPs are lease-financing agreements in the form of securities that may be issued and marketed to investors as tax-exempt debt which are an obligation of the General Fund.

The County states: "Issuing COPs is a method of leveraging public assets in order to finance other new assets. *By entering into tax-exempt lease financing agreements, the County is using its authority to acquire or dispose of property, rather than its authority to incur debt.*"

Santa Barbara County Capital Improvement Program and Budget, 2009–2014 (italics by the authors).

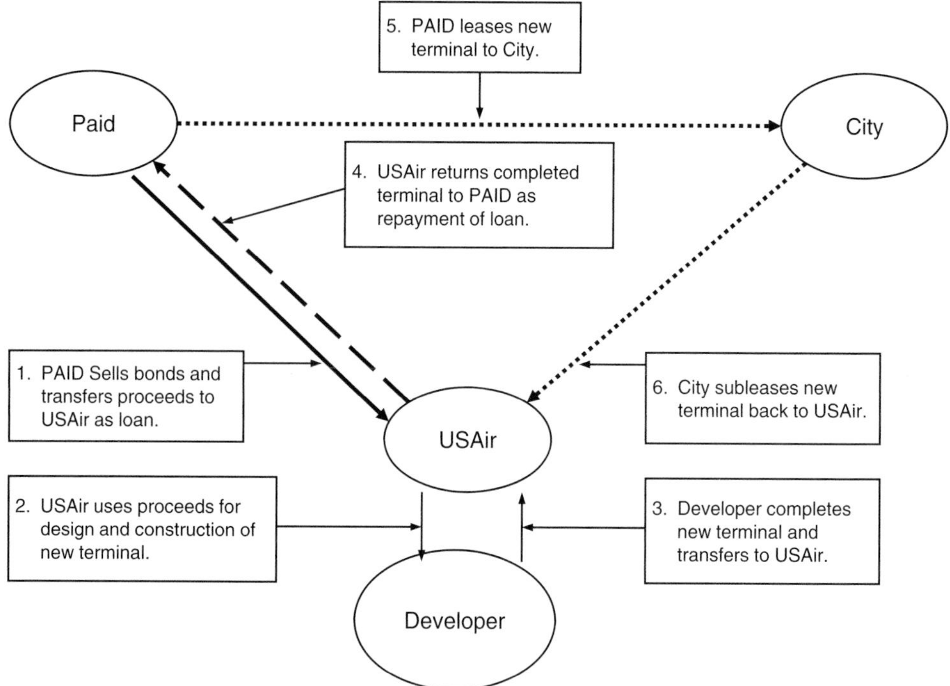

Figure 12.3
A complicated sale and leaseback bond issuance in Philadelphia

Street Lighting and Special Assessment Districts

In California, many localities used to rely on citywide special assessments to replace property tax revenues lost as a result of Proposition 13 (a 1978 amendment to the California Constitution that lowered property tax rates and made future tax increases more difficult to pass). One of the most popular special assessment districts was the street lighting and landscape assessment district. Local governments were able to use assessments for capital and maintenance activities for streetlight replacement and tree maintenance in the right-of-way, as well as for park improvements and maintenance. However, the rise in local expenditures caused by these districts led to the passage of Proposition 218 in 1996. Among other reforms, this initiative required increases in citywide assessments to be put to the voters instead of being adopted legislatively by the local government.

Tax-Exempt Bonds for Geographically Defined Areas

Another kind of revenue bond can be issued based on fees or property tax increments from a particular geographic area. These are another type of limited-liability bond since they are not secured with the full faith and credit of the taxing power of the jurisdiction. Instead, they are repaid with a tax or a fee assessed for the purpose of the project, or taxes projected to be received as a result of the investment. Three different types of geographic areas are discussed below.

Special assessment districts. If the benefit of this project can be linked with a particular geographic subsection of the locality, such as curbs and gutters for a specific neighborhood, the local government may consider establishing a special assessment district to raise the revenue to repay the bond. Assessment districts are not independent of the government that creates them.[13]

Special assessments are similar to property taxes, since the amount assessed is related to the value or size of the property. The assessment can be a function of the property's value, or its street frontage, or any other mechanism that relates the cost of the capital improvement to the benefit received by the property owner. To determine whether or not to levy the assessment, all the property owners are polled, and if a majority concur, then the assessment is made mandatory for all properties in the area. Assessment districts have been used by many localities to fund federally required stormwater improvements. The cost per parcel might be a function of the amount of its impervious surface.

Redevelopment and tax increment bonds. Redevelopment agencies were originally established to provide financing for "slum clearance" and infrastructure provision for blighted areas in the large urban areas. The funds to pay off the bonds they issue come from the difference between the property tax that would have been collected in the area and the increased revenue the area is able to provide thanks to the rising property values attributable to the redevelopment. How this actually works varies from state to state.

Business improvement districts. The past decade has seen the rise of business improvement districts (BIDs) in older urbanized areas. The impetus often comes from local merchants who are trying to revitalize an area. Since they are voluntarily assessed, usually by a simple majority of area property owners, the funds can be used for whatever was put in the assessment district formation, including operating costs and maintenance items. They are typically used for items like pedestrian lighting, street furniture, curbs, gutters, and paving for a business district. The proceeds can be used to issue a bond or to pay for improvements on a pay-as-you-go manner.

Taxable Bonds, Including Build America Bonds

The American Recovery and Reinvestment Act of 2009 included a new program for municipal bonds called Build America Bonds, or BABs. Localities began issuing bonds through this program in April of 2009, and the program expired on December 31, 2010. This program provided a reimbursement of 35 percent of the interest on the bond to the locality. The program was enthusiastically received, with BABs accounting for 27 percent of all municipal issuances in 2010.

Actors in the Bond Issuance Process

Actors in the bond issuance process can be divided into two broad groups: government agencies, on the one hand, and outside experts and professionals, on the other.

Government Agencies

The major actors in the bond issuance process within the government agency are usually in the finance department or budget office. However, someone from the economic development department, the redevelopment agency, or even the planning department may be the project manager for a large-scale capital project such as a convention center, a downtown revitalization, or a green infrastructure project. He or she will be responsible for convening the team charged with issuing the bond and will work closely with the finance and budget offices on the issuance.

Because of the highly specialized nature of redevelopment law and financial reporting, often the redevelopment staff takes the lead in managing their own bond issuance process. However, even though the redevelopment agency is a different legal entity than the city or county, the bond issue's impact on property taxes affects both entities, so the bond issue needs to be coordinated with the appropriate general-purpose government.[14]

Outside Experts and Professionals

Since issuing a bond is usually a major event for general-purpose governments—and even for special-purpose districts and authorities—a team of outside experts is usually hired or is put on retainer to complement the local staff. Often the finance department has many of these experts already on retainer to assist with other financing activities. The following outlines the roles of the major players.[15]

Bond counsel. Ever since the late nineteenth century, when some issuers of bonds for railroads defaulted on their obligations by asserting that the bonds had not been properly authorized, bond issuers and underwriters have included a bond counsel opinion with the issue that states that all the relevant laws and regulations have been complied with. The bond counsel prepares the bond proceedings, confirms their tax-exempt status, and drafts key financing documents. Today, a bond issue is not marketable without a positive opinion by bond counsel.

Financial adviser. Local governments that do not routinely issue long-term debt will want to have a financial adviser. The financial adviser helps the local government understand the amount of debt it can afford and decide whether or not debt is the most appropriate strategy for the desired purposes. The financial adviser can develop requests for proposals (RFPs) to hire the other professionals and can help the locality prepare for the bond rating process. He or she also advises the local government about whether to accept the bids in a competitive bond sales process or assists in the negotiated sales.

This professional should also be hired at the beginning of the bond issuance process. In the past, financial advisers were compensated as a fee based on the size of the bond issue, but this provides an incentive to make the issue larger. Instead, an hourly, flat fee, or a retainer method should be used.

Underwriter. The underwriting firm buys the debt from the issuer and resells it to investors. If the firm is successful, it holds the loan briefly and is able to sell it for more than it paid for it. If the issue is strong, the local government can sometimes pressure the underwriter to waive the risk portion of its fee. The underwriter can be an investment banker, a securities dealer, or a consortium to help spread the risk.

Consulting engineer. Revenue bonds, including special assessment districts, usually require the opinion of an outside engineer about the feasibility of the projected revenues. This study is used to reassure all parties that the revenues will be adequate to cover the debt service as well as operating and maintenance costs.

Polling expert. To assist in determining the size or focus of a bond issue, a locality may contract with a polling firm to conduct a voter survey. Many firms specialize in polling voter attitudes, and although their primary expertise may be political campaigns, many can quickly and easily provide results on the feasibility of a prospective bond measure.

Rating agencies. Three national firms rate local government and corporate debt—Standard & Poor's Corporation (S&P), Moody's Investors Service (Moody's), and Fitch IBCA. The local government will pay a fee to one of these firms to rate its debt.

Defaults and Credit Worthiness

A default occurs on a bond when a payment is not made on time. Since the Great Depression, when there were almost 5,000 defaults recorded on municipal bonds, through the mid-1970s, there have been very few defaults. Between 1972 and 1984, there were 21 municipal bankruptcies and from 1981 through 2000, there were 156 filings of Chapter 9 municipal bankruptcy. Historically, however, most defaulters recover and continue to make debt payments, or otherwise satisfy the creditors.[16] Recovery rates on defaulted municipal bonds are fairly high at almost 70 percent of par.

There are three categories of default risk. The lowest risk category is traditional tax-backed or revenue bonds, which has an historic default rate of 0.24 percent (for comparison, triple-A-rated global corporate bonds have a default rate of 0.43 percent). The second category is bonds for essential functions that are not fully protected from the economic cycle or competition, including hospitals, private colleges, airports, and toll roads. Here the default rate is 0.70 percent (comparable to double-A-rated corporate bonds, which have a default rate of 0.76 percent).

The third risk category of bonds comprises projects that compete against private-sector activities or that have volatile revenue streams. These include industrial development revenue bonds, tribal gaming bonds, local multifamily housing, nursing homes, and retirement communities. Their default rate was 0.65 percent (comparable to triple-B-rated corporate bonds, which have

Table 12.3
Cumulative default rates of municipal and corporate bonds by bond rating category, as of February 2010

Bond rating		Moody's		S&P	
Moody's	**S&P**	**Municipal (%)**	**Corporate (%)**	**Municipal (%)**	**Corporate (%)**
Aaa	AAA	0.00	0.05	0.00	0.65
Aa	AA	0.03	0.54	0.32	1.20
A	A	0.03	2.05	0.25	2.91
Baa	BBB	0.16	4.85	0.37	7.70
Ba	BB	2.80	19.96	2.07	19.33
B	B	12.40	44.38	7.70	33.14
Caa-C	CCC-C	11.60	71.38	41.76	52.93
Investment-grade only		0.06	2.50	0.25	3.76
Noninvestment-grade only		4.55	34.01	6.75	27.82
All		0.09	11.06	0.33	11.38

Note: Moody's data reflect the average ten-year cumulative default rate from 1970 and 2009. S&P municipal data reflect the average cumulative 23-year default rate from 1986 to 2008. S&P corporate data reflect the average cumulative 15-year default rate from 1981 to 2008.
Source: Moody's, S&P, LPL Financial.

a default rate of 3.97 percent). Specifically, industrial revenue bonds had a cumulative default rate of 14.62 percent, multifamily housing, 5.72 percent, and nonhospital health care facilities, 17.03 percent. These latter three types of bonds accounted for only 8 percent of bonds issued, but for 56 percent of the defaults.[17]

Overall municipal bond default rates have been between 0.9 percent and 0.33 percent, while corporate default rates have been over 11 percent[18] (Table 12.3).

Bond Ratings

Even though default is a remote possibility, investors want to know whether they are buying bonds from a locality that will repay the loan. Rating agencies are used to establish the rating for a particular bond issue, and this in turn sets the interest rate that the jurisdiction will have to pay on the funds. A bond rating is a significant expense for the issuer. One agency notes in its literature that a rating costs from $1,000 to $350,000. Some issuers may get two ratings to make it easier for the underwriter to sell the bonds, while others may not get one at all. According to one source,[19] a third of all newly issued debt was rated, while another found that between 6 and 13 percent of bonds issued during the 1990s were rated.[20]

The rating firms use letter grades to rate long-term debt. The four highest categories are called "investment grade," because many banks and municipal mutual funds are prohibited from investing in debt that is rated lower. Lower-rated bonds are often called "junk bonds" and are viewed as speculative. Most issuances by local governments are investment grade.

Factors Rating Agencies Consider

The factors ratings agencies consider vary depending on the type of bond. Below, the factors relevant for GO bonds, revenue bonds, lease financing, and geographically based bonds are considered.

GO (tax-backed) bond rating factors. The process for rating GO bonds evaluates the risk for the debt over the life of the issuance. The rating agency typically looks at the political mood of the jurisdiction, existing debt, the condition of the local economy, municipal financial health, and the management capacity of the jurisdiction. The rating agency may evaluate the jurisdiction's per capita income and its ability to address infrastructure needs. Local educational levels may be used to indicate whether the economic base of the locality will be able to compete in the twenty-first century. The long-term impact of pension requirements for local governments also is scrutinized carefully because of the long-term increases locked in place during the late 1990s, when local governments were flush.[21]

One of the key steps is an analysis of the existing local debt supported by the same tax base. The level of debt as a percentage of the local government's budget is scrutinized, with 10 percent or more raising questions. Rating agencies also use debt per capita (including debt from overlapping jurisdictions) to assess the ability of local residents to support debt. Additionally, rating agencies look at local debt as a percentage of real estate market value in the jurisdiction. The average range of total debt as a percentage of market value is 2–5 percent, with above 6 percent considered high, and 10 percent a credit problem.[22]

Revenue bond rating factors. In the case of bonds issued by organizations, a rating agency is primarily concerned about whether or not the enterprise will produce the revenue to repay the loan. Therefore it looks at the viability of the overall organization, as well as the specific capital facility that is being financed. The rating agency will want to know that the organization, if it is fully funded by user fees, will have more than enough funds to repay the loan after considering needs for operating and maintenance. This is called "coverage ratio." In the past, ratios of net available funds to the debt service of the bond were about 1.5–2 times the debt service. These days, they are about 1.25, lower if the project is strong and higher if it is a weaker project. The rating agency will require, among other information, five years of audited statements, a rate study, an engineering report, and lists of customers by class, along with overall economic information for the area, as above.[23]

Rating factors for lease financing (certificates of participation). Since the primary security for repayment in the case of lease financing lies with the facility that is being leased and the creditworthiness of the lessee, the rating agency will look at these. This will include much of the information noted above for the GO bond. The rating agency will look at how essential the facility is to the local government. For example, jails, schools, and water and sewer facilities are deemed to be essential and receive higher ratings, while parks and recreational facilities are thought of as less essential and (unfortunately) receive lower ratings. The rating agencies will also look at whether there was opposition to the project locally and the local government's attitude toward debt repayment. Some local governments can pledge additional collateral or establish reserve funds for repayments to mitigate these concerns.

Geographically based bonds. For special assessment district bonds and redevelopment districts, ratings agencies look at the wealth in the specific area. For special assessment districts, they look at the collection practices of the issuers. Redevelopment (urban renewal) bonds are evaluated on the history of tax base growth in the area, the powers of the redevelopment agency, state laws, and the wealth and general credit of the jurisdiction creating the redevelopment area.[24]

Use of cooperative bond pools and credit enhancement devices. Sometimes when the issuance is a small one, or when the locality may not have a competitive bond rating, it may join with another local government that is issuing a bond to take advantage of shared overhead costs. If there is a regional council of government that offers a program for small issuances made by its member jurisdictions, the locality can participate in that. Sometimes states will do a large bond measure to provide low-cost loans for infrastructure for local jurisdictions. For revenue bonds, the jurisdiction can pledge tax revenues. Bond insurance can also be purchased, although the financial collapse of 2008–2009 resulted in all but one of the bond insurance agencies going out of business. Use of credit enhancement devices is currently on the wane. During 2004–2008, 41 percent of new issuances used them, but only 20 percent of the issuances in 2009 and 10 percent of issuances in the first quarter of 2010 had credit enhancement.[25]

Key Steps in the Bond Issuance Process

Two types of considerations govern whether the locality chooses pay-as-you-go or borrowing. The first concerns values, and the second concerns financial capacity of the agency and the political and economic climate in the jurisdiction.

Who Should Pay?

At the heart of the values question is who should pay. Those in favor of new debt argue that users should pay for the facilities—the pay-as-you-use argument. Since most infrastructure is long-lasting, borrowing today and paying back the debt over time permits this to occur. On a practical basis, often the ability to borrow makes a large capital project feasible for the local government. On the other hand, some argue that borrowing burdens future generations with debt and infrastructure decisions that may not be appropriate. Those who support pay-as-you-go note that paying for a facility out of current revenues means that there are also no interest costs.[26]

Sizing the Bond

The size of the bond is one of the most important decisions in the bond issuance process. This determination is often a complicated process, involving a series of triangulations between project needs, funds available for the project from all sources, and the increase required in the revenue source that will be used to pay off the bond. The size is not dictated by the need alone. The size of a GO bond, for example, may be determined by the political feasibility of obtaining a positive vote from the elected body. The size of a revenue bond is determined by the ability of the facility being financed to raise revenues. In addition, very few jurisdictions permit the city or county manager, or CEO of the special district, to proceed without approval from the elected officials during the capital facility process.

Determine the demand for and cost of the project. The first step is to determine what the need or the demand for the project is. Usually, several alternative configurations are developed—high, medium, and low. Quite often the locality will contract with subject area specialists to help develop the cost and revenue estimates. Depending upon the locality, for streets, street improvements, and public buildings, a public participation process may be used, as well as a design competition.

Determine the impact on the local tax bill. Once agreement on the costs or alternatives has been reached, the budget office, working with the finance department and the financial adviser, identifies the financial impact on the average citizen over time. Every agency with revenues on the property tax bill is contacted to determine if they contemplate issuing bonds or raising their assessments in the near term. This information is analyzed to show the impact of different bond levels over time and often compared to equivalent data from other jurisdictions.

Table 12.4
U.S. municipal bond issuance[1] in 2000, 2005 and 2010 by type of sale ($ billions)

	Competitive		Negotiated		Private placement		
Year	N	(%)	N	(%)	N	(%)	Total
2000	48.6	24.2	146.0	72.7	6.2	3.1	200.8
2005	76.1	18.6	330.3	80.9	1.8	0.4	408.2
2010	73.1	16.9	357.2	82.5	2.8	0.6	433.1

Note: 1 Excludes maturities of 13 months or less.
Source: Thomson Reuters.

How Will the Issue Be Sold?

By far the most common way to market the bond is through a negotiated sale. In 2005, local and state governments issued $330.4 billion of bonds using this form, or 81 percent by dollar volume, compared with only $76.1 billion through competitive bidding and $1.8 billion by private placement (Table 12.4). GO bonds are more likely to be competitively bid than revenue bonds, which are overwhelmingly negotiated. A similar pattern is seen for 2010.[27]

Competitive sale. At a competitive sale, bonds are sold at an auction to the underwriting firm that provides the best bid on true interest or net interest cost of the bonds. This used to be called an advertised sale. The date, time, and place where the bids for the bonds will be opened are advertised. The bids are then opened, and bonds awarded to the underwriter, who then sells the bonds to investors. It is the easiest way to sell, but is more risky for the underwriters, who actually own the bonds until they are turned around. Many state statutes require GO bonds to be sold through a competitive bidding process.

Negotiated sale. As the municipal bond market has tilted more towards limited-obligation bonds and/or revenue bonds, negotiated sales have become more common. For a negotiated sale, the underwriter is chosen before the sales date of the bonds, usually through an RFP process. The underwriter, using its sales force, then drums up interest in the bond so that, when the day of the sale arrives, there will be ready buyers. Some believe that with negotiated sales, underwriters are really acting more like brokers than underwriters who are paid to take risks. Negotiated sales are useful when the issue is quite complex or unusual and needs special explaining. They are also useful when the bond market is unsettled.[28]

Private placement. One alternative to both the competitive and negotiated sale is private placement, where the issuer goes directly to the investors, bypassing the underwriter. Without competition, however, administrative and interest costs may be higher, and the locality may be open to charges of cronyism. To avoid this, several banks or private investors can be approached in a competitive bid process, a common practice in some parts of the country. Another problem is that, unlike publicly sold debt, for which an active secondary market exists, private placement is less liquid. Private investors may buy directly because they intend to hold the debt to maturity, and they may demand higher interest rates.[29]

Preparation of the Bond Documents

The SEC requires a preliminary official statement (POS) and an official statement (OS) for public sales of municipal bonds. The POS is used to market the debt and is prepared before the sale, while the OS is prepared after the sale, and includes sales results such as the interest rates and prices. For GO bond sales, these documents contain information on the general economic condition of the jurisdiction issuing the bonds, while the OS for a revenue bond also includes information about revenue sources and operating information. The documents for revenue bonds, or nontax sources of revenue, can sometimes run to hundreds of pages. If a capital lease is involved, the POS and OS contain information about the specific property described. They are prepared by the team of technical experts described earlier.[30]

Approval of the Issue by the Governing Body

Generally, the agency's elected or governing body must approve the bond documents. Depending on how controversial the capital facility is, hearings may be required before the elected body, which is usually the entity that votes to place a bond issue on the ballot, if required.

Conclusion

Bonds are the chief mechanism that local governments use to borrow funds. Debt repayment may come from a variety of sources, including property taxes and revenue generated by the project. Today's bond market is complex, but a large cadre of professionals is available to assist local practitioners once the decision has been made to go forward with the project or program. Borrowing to make needed large-scale public projects happen has been the backbone of expansion in the United States, and although the occasional default makes the news, most municipal debts are quite secure. An important issue for the locality is to consciously be aware of the lifetime of the project and the projected life of the debt.

Additional Resources

Publications

Petersen, J. E., and T. McLoughlin. "Debt Policies and Procedures," in *Local Government Finance: Concept and Practices*, ed. J. E. Petersen and D. R. Strachota. Chicago, IL: Government Finance Officers Association, 1991. A classic chapter still timely today that goes over the basics.

Temel, J. W. for The Bond Market Association. *The Fundamentals of Municipal Bonds*. New York: Wiley and Sons, 2001. This is the handbook for bond professionals and has more detailed information on the bond market.

Vogt, J. A. *Capital Budgeting and Finance: A Guide for Local Governments*. 2nd ed. Washington, DC: International City/County Management Association, 2009. Another good source for local practitioners about bonds and the capital budget. Written for those working in or for cities and counties.

White, S. B., and Z. Kotval. *Financing Economic Development in the 21st Century (2nd edn)*. New York: ME Sharpe, 2012. Many chapters about financing infrastructure as part of economic development at the local level, including the use of funds in a redevelopment area.

Websites

Bond Buyer: www.bondbuyer.com. A subscription service that provides excellent daily, weekly, and monthly reports and has a good data archive.

The Securities Industry and Financial Markets Association (SIFMA): www.sifma.org. This organization issues a brief quarterly research report that provides analysis and data—both quite readable.

chapter 13

User Fees and Public Pricing

In this chapter...

Importance to Local Practitioners . . . 207
Definitions of User Fees and Public Prices . . . 207
Fee Use in the United States . . . 209
Fee Use, Taxes, and Total Local Revenues . . . 209
Use of Fees by Infrastructure Function . . . 210
Relative Levels of Fees . . . 211
Why and When Should We Charge Fees? . . . 212
Functions of Prices . . . 212
Types of Infrastructure Services Appropriate for Public Pricing . . . 213
Criteria for Determining Type of Fee or Public Prices . . . 213
Public Pricing Concepts and Types of Fees . . . 214
Less Than Average Cost, Including No Cost . . . 214
Average or Full Cost Pricing . . . 215
Marginal Cost Pricing . . . 216
Types of Fees Used at the Local Level . . . 217
Issue of "Crowding" in Public Pricing . . . 218
Congestion Pricing . . . 218
Demand Management Pricing . . . 219
Pricing to Finance Expanded Capacity . . . 220
Public Pricing to Address Equity Concerns . . . 220
Setting a Fee: Basic Steps . . . 221
Setting Up a Special District . . . 222
Conclusion . . . 224
Additional Resources . . . 224
Publications . . . 224
Websites . . . 225

Why Is This Important to Local Practitioners?

User fees and public pricing have been the fastest-growing revenue source for infrastructure in the past three decades. As the "tax revolt" spread across the United States in the late 1970s and 1980s, local governments found that they could charge fees for many infrastructure services. General tax revenues were then freed up to mitigate the impact of caps and other restrictions on general taxing authority in order to preserve other public services. User fees are also an important revenue source to finance revenue bonds issued to build new infrastructure or to replace or rehabilitate the old. In addition, user fees can be used in a strategic manner to regulate demand so that both natural resources and the useful life of existing infrastructure facilities can be conserved.

This chapter begins by defining user fees, before looking at their use in the United States. These include congestion pricing, conservation pricing, peak period, and equity considerations in pricing. The functions of public prices are described, including their role in demand management. The chapter then addresses when public prices are appropriate, and the criteria that should be used to calculate them. Finally, the basic steps in setting user fees and in starting a special district are outlined.

Definitions of User Fees and Public Prices

Many definitions are associated with user charges or fees. Some authors include exactions and fees assessed during the project development process as part of the general discussion of user fees and charges. We will not follow that convention in this book. Instead, Chapter 14 is devoted to the special case of exactions and impact fees that are assessed as part of new residential or commercial development. We do this because developing pricing strategies for infrastructure services has a different set of constraints, objectives, and methods than when local governments require fees for infrastructure from developers, even if both ultimately are used to pay for the same type of infrastructure.

The formal typology government finance officers use for fees includes user charges, utility charges, special assessments, and regulatory fees:[1]

- *User charges and fees.* User charges or fees are prices charged by governments for consumers of publicly provided infrastructure (or other) services. The benefits of consumption are clearly received by an individual, household, or group, but there may be additional benefits to others as well. An individual may avoid paying the user fee by not consuming or acquiring the good or service. Examples of user fees include charges for each can of garbage picked up, or use of public recreational fields, golf courses, and swimming pools.
- *Utility charges.* Utility charges or "rates" are user fees for infrastructure services that the community has either chosen to provide publicly or have provided privately, but where the federal and state government regulates the enterprise to some extent. Such goods and services often have economies of scale in their production—that is, the cost per unit declines as more is produced over large quantities. These are primarily electricity, gas, water, and waste utilities. The local practitioner

Table 13.1
Sources of revenue for state and local governments from 1980 to 1996 emphasizing fees and user charges

	1980		1996		1980		1996	
	Total (millions)	**% of total**	**Total (millions)**	**% of total**	**Per capita**	**% of total**	**Per capita**	**% of total**
From federal government	83,029	18.4	234,891	15.5	367	18.4	886	15.5
From state and local sources	368,509	81.6	1,278,742	84.5	1,627	81.6	4,822	84.5
General revenue taxes	223,463	9.5	689,038	45.5	986	49.5	2,598	45.5
General revenue user charges and miscellaneous	75,830	16.8	298,892	19.7	335	16.8	1,127	19.7
Water supply user charges	6,766	1.5	25,433	1.7	30	1.5	96	1.7
Electric power fees	11,387	2.5	35,476	2.3	50	2.5	134	2.3
Transit system fees	2,397	0.5	6,889	0.5	11	0.5	26	0.5
Gas supply system fees	1,809	0.4	3,795	0.3	8	0.4	14	0.3
Liquor stores	3,201	0.7	3,732	0.2	14	0.7	14	0.2
Insurance trust revenue	43,656	9.7	215,487	14.2	193	9.7	813	14.2
Total state and local revenues	451,537	100.0	1,513,633	100.0	1,993	100.0	5,708	100.0

Source: U.S. Census Bureau, *Statistical Abstract of the United States: 2001* (Washington, DC U.S. Census Bureau, 2001).

will typically only be involved in setting or commenting on fees for local utilities.

- *Special assessments.* Special assessments are charges to a property or a business in a specific geographic area based on the cost of public services rendered, but distributed to each payer according to an estimate of the benefit received. Although identifiable individuals receive benefits, participation may not be voluntary.[2] Examples of special assessments can include charges for public street lighting, residential street maintenance, or power line undergrounding in the area in which the payer maintains their home or business.
- *Regulatory fees.* Regulatory fees include licenses, franchises, and permit fees, such as those for a land use application, a permit to use the right-of-way, or a building permit. These kinds of fees are authorized under the police power of the agency, instead of its taxing authority. Regulatory fees are generally set to cover the administrative costs of regulating the activity, for example, the costs of issuing the permit. Some localities and states permit fees for these activities to be used to generate revenue as well as to cover the cost of the regulation. Many states, however, including California, prohibit certain fees from exceeding the cost of administering the regulation.[3]

The U.S. Census uses a category called "current charges" when it reports on user fees at the local and state level. This category includes assessments, but excludes utility charges and regulatory fees (licenses and permit fees). We deal broadly with the issue of user fees and public prices as they concern infrastructure provision. Therefore, utility pricing and user fees are both discussed in this chapter, while exactions and development fees are addressed in Chapter 14. This chapter also includes a discussion of fees or rates that are used for special assessment districts.

Fee Use in the United States

Fee Use, Taxes, and Total Local Revenues

Local level fee use has been increasing since the end of World War II. In 1957, the ratio of user fees to tax revenues was 40:100, by 1977 it had increased to 45:100, and by 1987 this figure was 64:100.[4]

Looking at the period of 1980 through 1996 (see Table 13.1), we find that most categories of user charges have grown faster than local taxes. Taxes dropped as a share of state and local revenues from 50 percent to 45 percent from 1980 to 1996, while general revenue user charges increased from 17 percent to 20 percent. Water fees increased slightly as a total share, while electric power fees decreased slightly. Transit and gas revenues, never a large proportion of state and local revenues, remained constant during this period.[5]

The increase in the use of fees by governments was related to several factors that coincided in the late 1970s and 1980s. During this period, there was much discussion about downsizing government and making it more like the private sector in terms of accountability, customer service, and pricing. Restrictions were placed on the ability of local governments to raise general purpose taxes, but roll-backs and "take-aways" were rare, Proposition 13 being an exception. Federal grants as a percentage of total state and local revenues decreased during this period, and cities and counties increased their use of fees substantially. In California, user fees went up from 4 percent of local revenues to 23 percent within three years after the passage of Proposition 13. However, the incidence of local general purpose taxes continued to increase, albeit more slowly.[6]

Use of Fees by Infrastructure Function

The increase in user fees in the United States differed by infrastructure function. First, increased spending occurred for functions that have always relied on fees. These include airports and government-operated gas and electrical utilities. Second, some functions rapidly increased the proportion of their funding from fees. Sewers, hospitals, and educational facilities are in this category. Third, recreation and waste management services have always relied to a large extent on funding from fees, but the increase has not been that great.[7]

Cities had the greatest number of fee increases for operating costs for parks and recreation, and for permit fees for development. Counties used fees for health, parks, and development fees. Parks and libraries saw political resistance to user charges. Recreation departments were under the most pressure to adopt user charges, but the recovery rate compared to the total cost is only about 25–66 percent.[8]

More than any other city or county department, the public works department relies upon user charges or fees for revenue. Sanitary sewer charges are the most likely to be fee based, and have close to 100 percent cost recovery. This is probably because the fees can be added to the water bill. Solid waste collection has followed a similar path, particularly since the use of high technology devices for billing and tracking weights have allowed localities to be able to charge by the amount of waste generated.

Solid waste and sanitary services can be set up as a public enterprise with their own funds within a local government, or can be operated by a special district. Fees for storm drainage have been resisted at the local level since it is often difficult, although not impossible, to establish a link between "use" and the fee. Streets and traffic services have about a 50 percent cost recovery. Bridge and tunnel tolls and airport landing and departure fees have also been levied with some success.[9]

User charges related to individual properties (that is, assessments) have also been successful politically and have the highest-cost recovery rate. These kinds of fees are the easiest to collect since they are part of the property tax bill and the services are not discretionary. In that respect, assessments are not like public prices or user fees. In some states, user charges related to properties meet local definitions of taxes and are subject to more stringent adoption procedures than other user fees.

Relative Levels of Fees

Another way of looking at fees is to compare relative levels. Although information is not available for all kinds of user fees, the Institute of Public Utilities at the University of Michigan compiled BLS data on fee levels for infrastructure services that are commonly thought of as utilities. Their analysis reveals that the four functions where fees have grown the most in the past 20 years have been cable TV, garbage collection, water and sewer maintenance charges, and local telephone service. Within the past four years, gas prices have also increased rapidly; the price of electricity has also climbed, but at a more moderate rate. Both interstate phone service and cell phone service fees have declined, and Internet access prices appear to be on the downturn as well (Figure 13.1).

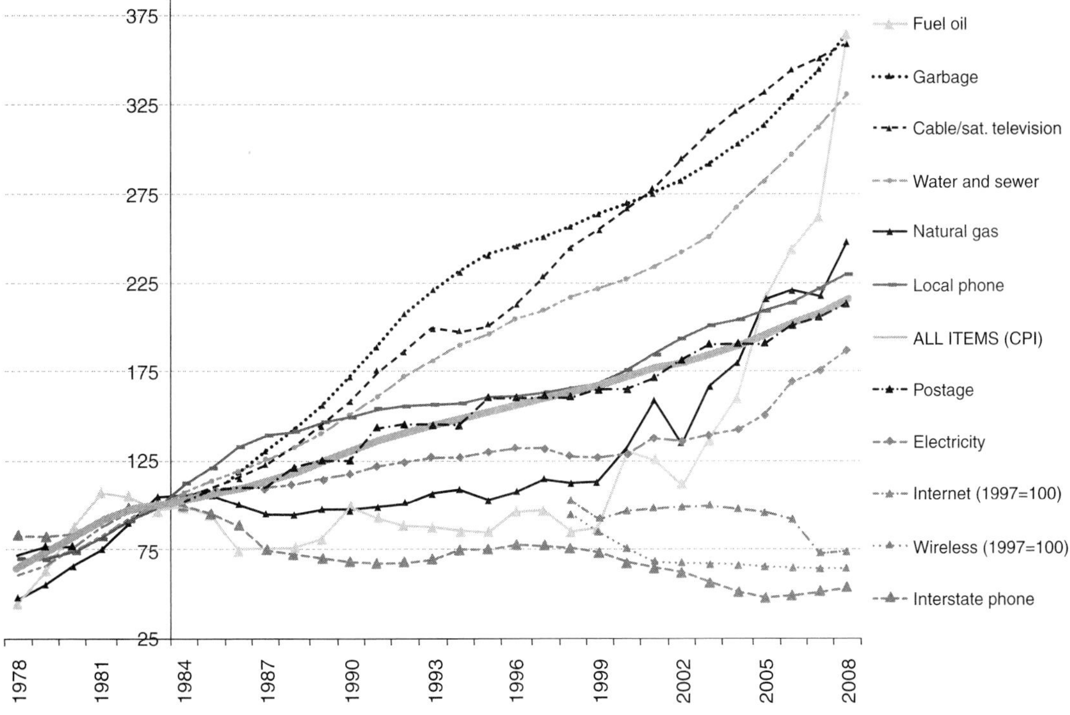

Note: The index is set to 100 for 1982–1984 except for internet and wireless services, where the index is set to 100 for 1997.

Figure 13.1 **Trends in consumer price index for public utility services, 1970–2007**

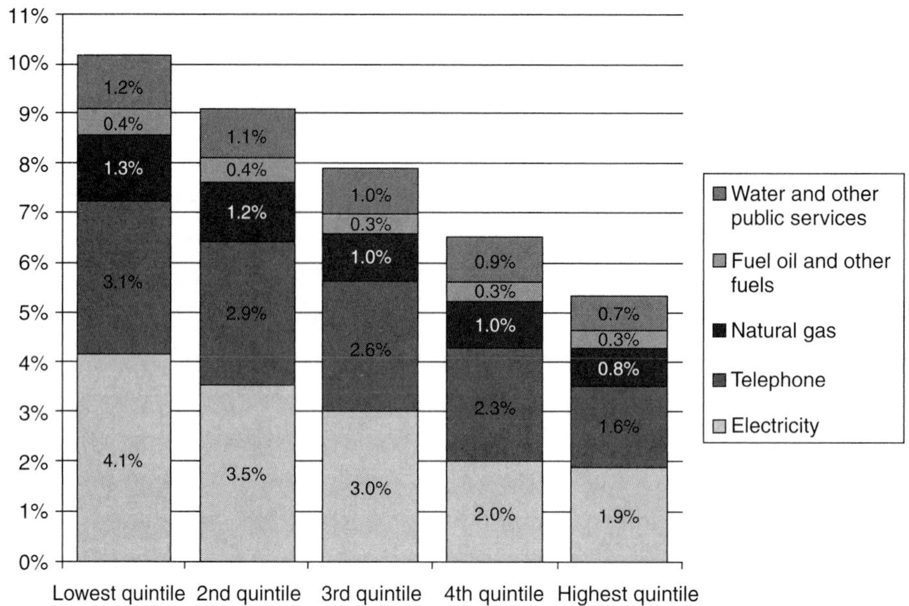

Figure 13.2
Consumer expenditures on utilities by income quintile in 2007 (all consumers %)

The average four-person household in the United States spends between 6 and 7 percent of their total expenditures on utilities and this figure has been constant since the mid-1980s despite the increase in fees and rates during this period. In 2007, this figure was 6.7 percent. Within overall household utility expenditures, the highest percentage was for electricity (37 percent), telephone (33 percent), and natural gas (15 percent). This same household spent only 13 percent, however, on water and other public services.

Unfortunately, when the data is disaggregated by income quintile over time, a pattern of "persistent disparity … and persistent regressivity" exists. Higher income households spend more for utilities than do lower income households, but it is a smaller proportion of their overall total expenditures. (See Figure 13.2.) Lower income households spend less, but expenditures for utilities take a bigger bite out of their total. In addition, recently, the proportion of income for lower income households spent on utilities has risen more rapidly than for others. The highest increases were for electricity and for water.[10]

Why and When Should We Charge Fees?

Functions of Prices

The key rationale for fees is that those who consume ought to pay the cost, so that they do not consume more than has value to them. Fees prevent wasteful consumption and excessive use of infrastructure services and natural resources. Thus fees have long been the basis for financing public utilities.[11]

In the private market, prices have three functions. First, they signal what a service costs so that consumers can judge whether the value they derive from more consumption justifies the expense. Second, as prices rise, they signal producers to provide more of the service. Third, they allow a customer to "vote" on the way a particular producer is responding to a need—rewarding firms that are more customer service-oriented with their purchases.

Although these three principles do not translate exactly to infrastructure, where a single monopoly utility or government agency usually provides the service, the concepts can be modified to serve the public purpose. Prices can be used to raise revenues to cover the cost of service in a way that links use to the amount paid. Prices can also be used as a signal to government about the right size of an infrastructure facility. They can also be used for income redistribution, where one type of user pays more than another. Finally, fees can be used to regulate demand and conserve scarce resources. This is called demand management. If a service or resource is free, or if citizens pay little or nothing additional to use more of the service or resource, then people think the use is unlimited and may consume more than is desirable for the public good.[12]

Types of Infrastructure Services Appropriate for Public Pricing

Types of infrastructure services that lend themselves to public pricing or user fees are natural monopolies—water/waste, energy, and telecommunication—or congested public goods—recreation centers, highways, and mass transit.[13]

Three conditions need to be met before deciding whether to use fees as a public price. First, it must be possible to identify the "customer"—that is, the party or individual receiving and paying for the infrastructure service. Second, it must be possible to exclude people from the benefits of the service. Third, it must be possible for a household to decide whether or not to "purchase" the service, or to at least vary the amount consumed.[14]

Criteria for Determining Type of Fee or Public Prices

Once the decision is made to attempt to recover some or all of the cost of service provision, the local official needs to decide on the type of fee or rate structure. There are as many different types of fees and price structures as there are people to think them up. Determining the kind of fee is often a balancing act—trying to optimize several criteria, some of which may be working in opposite directions. The following discussion is drawn from Hanemann's "Price and Rate Structures," 1998.[15]

Three basic criteria should be met when determining the kind of public price or user fee to set for infrastructure services. Fees should generate revenue, allocate costs fairly and accurately by use and user, and provide the desired incentives from the public agency's point of view:

- *Revenue generation.* The rate or fee structure should provide revenues sufficient to allow the agency providing the infrastructure to operate on a self-sustaining basis to the extent that the community does not wish to subsidize the function. The fee or rate structure should provide a stable and predictable stream of revenue over time and over changing circumstances. Administration costs should be minimized, and it should not be more trouble to collect a fee than the revenue warrants.
- *Allocation of costs.* The fee or rate structure should apportion costs (including social costs) of infrastructure services fairly and not arbitrarily between individuals or classes of users. Inadvertent cross-subsidies of one group to another, such as existing development subsidizing new development, should be avoided. Explicit policy goals to subsidize some group may warrant cross-subsidies. Some communities set commercial fees higher than residential, for example. Children and youth are usually charged lower fees for the use of recreational fields or facilities than adults. Lower rates are sometimes charged for low income residents or the elderly than for the general population.
- *Incentives for socially best use of the infrastructure.* The price or fee structure should also encourage the efficient use of the facility or service both in the present, as well as in the future. It should also encourage conservation, and it should be easy for the users to understand, so that heavy or light use telegraphs the correct signal back to the agency's policy-makers.[16]

Public Pricing Concepts and Types of Fees

Three major concepts are behind user fees: less than average cost, average cost, and marginal cost pricing, and several different types are used to implement these concepts.

Less Than Average Cost, Including No Cost

Until recently, the most common pricing option for infrastructure services has been provision at less than the average cost—and sometimes at zero or no cost to the consumer. This has been done for many reasons. First, many policy-makers and the electorate believed that all citizens should have access to basic services, and that their use should not be deterred by having to pay a fee that may be unaffordable. This has resulted in a political tradition at the local level of providing many services at no cost to the user or for a nominal sum.[17]

Sometimes grants have been received for the capital cost of the facility, and policy-makers reason that it is "only fair" to pass on the good luck of the jurisdiction to the users. For some services, such as bridges, streets, and parks, the marginal cost of providing service is seen to be zero. In addition, for some services, it is hard to exclude persons from receiving the service.

Collecting fees and billing users is also inconvenient, costly, and sometimes politically unpopular. Historically, most general purpose governments are not set up to do this well. Although recent advances in information technology, accounting, and software systems have made the process easier, for many communities, the administration of a new billing and collection function at the local level should be carefully considered before making the decision to charge for a service. For example, audits in the City of Berkeley in the mid-1990s revealed that more than $15 million for sewer and refuse fees had not been billed or collected. More recent audits in Portland, Oregon, found that telecommunications providers under contract were not paying full fees to use the city's right-of-way. This is not uncommon throughout the United States, although many localities have not performed such audits.

Special districts that rely for their full funding on fees do not have the choice of whether to use fees. They typically invest adequately in the internal mechanisms needed for billing and collection.

Average or Full Cost Pricing

Average cost is the total cost to provide the infrastructure service divided by the quantity provided. Under this method, the user charge is set at a level that covers all the costs associated with the provision of the services. This includes operations and maintenance, overhead, and costs associated with the capital facility such as bond payments. The budget is based on the total cost of the facility divided by the number of people to be served. This is the method most commonly used by local governments, special districts, and assessment districts. Some cities deduct from total costs the amount of any outside grants received that help to defray costs.

Average cost pricing is the easiest to implement and explain to the policy-makers and residents. It also ensures that all the costs of the infrastructure service are covered by the fee. However, it also results in a higher price and a lower quantity of services than consumers would pay for if the service could be privately provided.

Marginal Cost Pricing

The marginal cost is the cost of producing another unit—the additional cost incurred by the agency to provide another unit of the infrastructure service. Marginal cost pricing is based on the cost of producing an additional unit of the service, or the last unit to be produced, while average-cost pricing is based on the cost of producing all units. If costs are decreasing as output increases, average cost will be higher than marginal cost.

The logic of marginal cost pricing rests on the theory that a consumer who knows the true cost of the unit to be consumed can decide whether the benefit justifies using one's resources to finance the additional cost. A wide array of institutions have evolved that approximate marginal cost pricing—impact fees for new development, using special districts to finance services to new users, varying prices by length of route or time of day in public transit, or by the amount or weight of trash picked up ("pay as you throw"), or recreation fees to pay a facility's operating costs.

There are three problems with marginal cost pricing. First, it is difficult to calculate, because detailed information on both historical and projected capital costs is required. Most local governments hire outside experts to provide some of this information. Marginal cost pricing also requires detailed, timely information on actual use, which may require an expensive data collection system.[18] In fact, marginal cost pricing is calculated primarily by utilities and by academic studies of public services. In practice, most communities calculate marginal cost as the average cost in a geographic area where service is extended, or the average cost of the additional capacity needed to handle the higher demand at peak usage.

The second problem is that, when the average revenue is less than the average cost, an operating deficit will result. With most goods and services, the marginal cost is greater than the average cost, and usually the private market provides the service because there's profit to be made. With a monopoly such as an infrastructure service, the average revenue is less than the marginal cost, and this can result in an operating deficit. Financing the deficit through taxes can also create distortions by encouraging more than optimal use. In addition, when a public agency or utility relies on taxes or grants to finance deficits, the incentives for efficient management decline. In fact, deficits needing taxes or outside finance may become much larger than the one inherent in the difference between marginal and average costs.[19]

The third problem is getting public support for pricing differentially across space and time, or for charging prices that yield a "profit" to a public agency.

The latter will occur with marginal cost pricing when marginal costs are rising and are greater than average costs. However, it is hard to justify or even explain the advantages of marginal cost pricing to the public.[20]

Types of Fees Used at the Local Level

The discussion about average cost and marginal cost pricing translates to the real world of fee schedules in a variety of ways. The following discusses many of these local solutions. The word "tariff" is used instead of "fee" or "rate" in some situations, such as the private utilities.

Flat rate. A flat rate fee is one where all users pay the same amount regardless of quantity consumed.

Base rate with two-part tariff. This is an innovative way of charging close to the marginal cost, while allowing full cost recovery for the infrastructure service. A recurring flat rate is charged for the minimum level of service provision. Thereafter, use is charged according to the actual costs of producing an additional unit—that is, the marginal cost. Real-time information on use is needed. This method is used by the water/waste providers and by telecommunications and energy utilities to preserve the efficiency characteristics of marginal cost pricing and also to recover the total cost. Because of the need to have highly accurate and timely information in order to calculate these fees, water, energy, and telecommunications have very sophisticated billing and information systems.

Two-part tariff with block rates. Sometimes the local agency does not have real-time or even monthly information on the quantity of use. Instead, sewer rates may be tied to water use rates, which may be available only from special computer runs obtained once a year from the water agency. Or the amount of solid waste disposed of by a household can only be approximated by the size of the container used. In these cases, the infrastructure agency can try to approximate exact use data by categories of rates, either by type of user or by some other category. Increasing block rates are those where for the higher levels of use the consumer pays a higher rate when his or her use passes a designated threshold. This pricing scheme can be used for conservation purposes. (See below.) Decreasing block rates can be used for high

Flat Rates, Variable Rates, and Lumpy Costs

Another way of looking at the difference between average cost and marginal cost pricing is to distinguish between flat rates and rates that vary by use. A flat rate is what historically was charged by water and sewage companies, for example, based on average cost. Current bridge tolls, which do not distinguish between high and low use periods, or the number of people in a car, for example, are also flat rate fees. Rates that vary by use are similar to marginal rates. The development of a marginal rate, or fee, is constrained only by the inventiveness of the budget analyst and the availability and frequency of information on use.

If there were no "lumpy" costs, average cost and marginal costs would be the same. But this is not the case for public infrastructure. Average costs and marginal costs diverge because of the "lumpy" nature of building the capital facility (the "fixed" cost in the short run, to use economists' terms) as compared to the more routinely budgeted operating and maintenance costs ("variable" costs). Overhead and administrative costs may be "lumpy" as well, for example, when it comes time to hire another analyst to deal with the billing function.

Controversies over Increasing Block Rates

Increasing block rates were implemented briefly during a drought in the late 1990s by the East Bay Municipal Utility District (EBMUD). EBMUD provides water to homeowners in the Berkeley/Oakland Hills, in northern California, where lot sizes are smaller, the climate is wetter, and where many homeowners had drought-resistant, low water-use gardens. Voters in these areas supported the increasing block rates. EBMUD also provides water to homeowners in the Lafayette, Moraga, Orinda, area, where lots are larger, conventional grass lawns predominate, and where the climate is much drier in the summer time. Voters in this area were incensed when increasing block rates caused enormous water bills. The utility ultimately rescinded this pricing scheme. The rates, however, did promote conservation and decrease water use, thereby creating an unanticipated problem. Water use fell to the point where the utility's original revenue projections failed to be met. Today the utility is known as a pioneer in water conservation strategies and its rate increase in 2011 was an across the board increase of 6% to cover decreased water use.

volume users. The more of an infrastructure service used, the lower the price per unit. The theory behind this is that economies of scale results in lower costs as the amount of the service consumed increases.

Capital costs covered by taxpayers, and maintenance and operation by users. A variation of the two-part tariff is when the capital cost of the facility, such as a pool, recreation center, bridge, or sanitation plant, is paid for with a general obligation bond issue that is paid off with an increase in the property tax for all tax payers. However, operating and maintenance costs come from fees charged to the users. The fees may be apportioned by quantity of use, or may be a flat fee based on the average cost of maintenance and operations. Another example would be a regional landfill, where the voters authorized a bond issue for a regional plant debris facility, which is shown on the tax bill, but the households are charged a flat fee per plant debris container. Or, the voters authorized a bond issue to develop a landfill, but the operations are paid for by a fee charged to each city or private company based on the weight of the total garbage disposed of there—the "tipping fee."

Issue of "Crowding" in Public Pricing

Sometimes the use of an infrastructure facility by many consumers at one time will negatively affect the use for all. Swimming pool, energy, and water use can increase dramatically during the summer; streets, intersections, airports, and bridges become congested during the rush hour and the holiday season. Internet connection times slow in early evening hours. Cell phone use during the day can overload the system. Infrastructure planners can manage peak demand in three ways: pricing to reduce demand during peak use; encouraging conservation (by reducing demand) through price and non-price strategies; or building extra capacity and charging additional to cover the cost.

Congestion Pricing

Congestion pricing, which can ration the use of a fixed infrastructure resource during peak periods, has been commonly used in the telecommunication field for years. All of us have probably experienced trying to make decisions about how many "anytime" minutes we want, compared to "night and evening minutes" for cell phones. Long-distance phone calls on weekends have always been cheaper than during the business week. The airplane industry also employs a variation of congestion pricing, with reduced fares and mileage clubs for non-business travel periods, both on a daily and seasonal basis. More recently, the same concept has been put forth as an alternative for reducing traffic congestion on streets and highways during peak hours. The concept of congestion pricing has been fairly controversial for publicly provided infrastructure services. Some see it as a relatively easy way to smooth the use of scarce resources—streets, highways, and bridges—and encouraging the use of mass transit. The theory is that some commuters could choose to pay high prices for peak hour individual vehicle use, while others would switch to mass transit, or travel in off-peak hours. Critics argue that congestion pricing would discriminate people with limited income, who may not have the ability to choose. The right to travel is seen as an entitlement that benefits the entire society.

Sometimes the argument is framed as pitting those who favor markets (conservatives) against those who favor government (liberals), but the underlying issues defy categorization. Efforts to implement congestion pricing are often linked with provision of better and less expensive options for the poor, such as more bus runs in peak periods.

Demand Management Pricing

Until recently, the supply-side approach and cost-based fees dominated infrastructure planning. Planners would forecast population growth, and then engineers and planners would apply technical standards to determine how much infrastructure to produce. This process was used for major transportation projects, water and sewage treatment plants, and refuse collection. The experts could ultimately link the amount of the infrastructure to accident rates, or incidence of death or disease.[21]

Pricing, which can easily manage demand, is one of many tools in the local practitioner's kit bag. Shifting demand by time of day or geographic location can contribute to the efficient use of existing capacity and the reduction of needed supply. Another form of congestion pricing that attempts to

What Is a Fair Price for New Users If All or Part of a Capital Facility Already Exists?

Sometimes infrastructure service will be provided to a new user where the capital facility already exists. This might be a new recreational program using the existing equipment and facilities. It might also be new water service that can be handled with the existing treatment plant, but where new mains and laterals are needed. The concept of allocating indirect costs can be used to include the facility and equipment maintenance costs (as distinct from the day-to-day operating costs) and other equipment that has already been purchased. It may have been paid for by one-time funds, or already allocated to existing users. Whether or not these costs are included in the fee basis is usually a political decision, depending upon the service.

meet demand without adding to capacity is increasing block rates, which is discussed elsewhere in this chapter. Other non-price policies and programs used for conservation and demand management are discussed in the individual infrastructure system chapters.

Pricing to Finance Expanded Capacity

For some forms of infrastructure—particularly those for energy, telecommunications, and water—meeting peak load needs without bringing about negative impacts from overuse is important. The expectation is that all customer demand will be met. Once the decision has been made to provide the infrastructure service to meet peak demand rather than rationing use, pricing strategies to finance capital facility expansion must be developed. One strategy is peak load pricing where capacity costs are absorbed by those who cause the peaks.[22]

The pricing analyst must first determine whether the peak is random or systematic. A random peak is illustrated by the need to provide water for fire use. A fire can occur at any time—in the midst of high water use as well as at night. Pricing strategies to pay for infrastructure to meet this demand are called capacity charges. These are fixed charges not based on volume, but on the amount of money needed to increase system capacity to respond to an emergency.

A systematic peak (such as seasonal use) is treated differently for pricing purposes. A volume charge affects all customers equally. This is because, whether the use is large or small, it contributes to the need to have increased capacity for the season or other time period.

Traditionally, public infrastructure providers use the same pricing schedule for periods of high and low demand. This can result in over-consumption during peak hours and in the long run may cause overdevelopment of the facility. It can also result in peak users being subsidized by off-peak users, since they are paying for costs for which they were not responsible. Peak pricing strategies are complicated to administer and usually require more information than what is needed for uniform rates.[23]

Public Pricing to Address Equity Concerns

Often, governments are concerned that low-income customers or those with other special needs might be precluded from using an infrastructure service because of high fees. To address this issue, means-based and discounted fees or fares have been developed for children, the elderly, and the disabled. While such discounting or assistance is justifiable on equity grounds, price variation can also be used for a wide array of groups where the equity issue is less clear. Discounts for new first-time customers—or for bulk or high-volume, longtime, and business users—fall into this shadow category, and intense lobbying for these fee subsidies often occurs at the political level.[24]

Since delivery costs can vary significantly among groups of users, charging the same fee for the same service may provide hidden subsidies. Charging a standard rate for water use, transit, and other types of infrastructure or public service has often resulted in significantly subsidizing one group of users at the expense of others. Cost-based fees have the additional advantage of helping avoid such subsidies when fees are set on the basis of actual user cost.[25]

Setting a Fee: Basic Steps

The basic elements in the development of a user fee are straightforward. However, the details of obtaining the data, selecting the fee strategy, and actual billing may be extremely complicated. The following are the basic steps into which the issues of the previous sections can be integrated (based on the recommendations of the Government Financial Accounting Association as presented to the General Accounting Office in 2001):[26]

- *Adopt formal policies about fees and charges.* The first step for the locality is to formally specify the factors to be taken into account when pricing infrastructure services. This includes whether the local government will recover the full cost of the service, and what circumstances would prompt the government to charge more or less than actual costs. Some states preclude charging more than the actual costs, while utilities generally must balance their budgets.
- *Calculate the full cost of providing the service.* This includes direct and indirect costs. Direct costs are staff salaries and benefits, supplies and materials, capital facilities and equipment, depreciation of equipment, and any other costs attributable to service delivery. This should also include the cost of using capital facilities. Equipment and facility costs can be cash purchases, the cost of borrowing (paying debt service on a bond, for example), and maintenance and operations. Indirect or overhead costs are associated with payroll processing—accounting, computer support, and other administrative services.

Developing a Fee

The first step in developing the fee is to project future capital needs for the infrastructure service, obtain detailed operating and maintenance costs, and generate annual revenue requirements. The second step is to allocate costs among types of users—this can be contentious if subsidies are ending or costs are increasing. The third and fourth steps involve calculating the unit cost of service by user and, from that, generating annual user fees. For large-scale projects, the locality may hire an engineering firm to conduct special studies for setting fees.

Issue of Indirect Costs

One issue for infrastructure pricing is how to allocate indirect costs. Indirect costs include administrative costs such as legal, budget, and personnel, including a share of the county chief executive's salary or city manager's salary if the infrastructure function is administered by the general purpose government. Administrative functions in a city may be paid for from the general fund, and not charged back to specific services, using the argument that the added cost to the government of providing the infrastructure service does not require additional overhead. When it does, however, then these costs should be included as part of the fee basis. Even without a general fund problem, canny budget managers may have already shifted the general funds out and replaced them with charges against the special fund for the infrastructure.

- *Allocate the costs among users.* Decisions must be made about cross-subsidies and whether some users generate more costs than others, and accordingly should be charged more. Demand management decisions can be made at this point.
- *Review and update fee schedules periodically.* The cost of providing infrastructure services may vary from year to year due to inflation, staffing cost increases, and the need to replace particular infrastructure. If the cost of providing the infrastructure service increases during the year, particularly in a general purpose government, the cost is de facto shifted to the property tax or other revenue sources. In addition, if yearly or frequent adjustments to the fees are not made gradually, then substantial one-time increases may be needed, which are more difficult for the legislative body and the infrastructure service user.
- *Public review and participation.* The policies, procedures, and other aspects of the fee setting process should be made public, and should be subject to public comment and input.

Setting Up a Special District

In planning for infrastructure that will be supported by fees or public prices, sometimes it is necessary to set up a special district. There are two types of special districts: the classic independent district with its own elected board; and the dependent assessment district.

Local and state regulations govern the formation of a dependent assessment district, and the practitioner should consult the city or county counsel for assistance. Since a new agency is not being formed, the steps consist of deciding upon the geographic area, costing out the service, deciding whether the assessment is by parcel, street frontage, or another method, providing notice to the property owners, and legislative action.

Setting up an independent special district is more complicated, since, in effect, a free-standing unit of government is being created. Most independent special districts are for water, sewer, or other environmental infrastructure services. While cross-functional coordination can be difficult, an independent district funded by the user fees is an effective way to provide these services. The formation of special districts is not recommended, but the local practitioner may need to know the critical elements involved.

The basic process of setting up a district consists of a petition from the affected landowners to the designated government body, which either approves or denies the request. The landowners then elect the district's board, frequently business associates of the developer, who serve through the construction phase of the improvements. Thereafter, public elections for the board occur. The necessary steps (based on Porter et al., *Special Districts: A Useful Technique for Financing Infrastructure,* 1992), are as follows:[27]

- *Identify state requirements for establishing a special district.* The first step in the process is to understand the requirements of the specific state for special districts. The local practitioner may seek the advice of the county or city legal department in this task, as well as look at any information on the state's website. Many fast-growing states, such as Florida and Washington, have manuals and a wealth of other information about special districts and the rules and regulations surrounding their formation and operation.
- *Plan for the special district.* The next step is to put together the plan for infrastructure improvements. This includes setting the boundaries and identifying affected landowners. It also involves projecting demand by forecasting the population to be served and preparing an analysis of existing infrastructure. A financial analysis is made to determine the feasibility of the fees to be charged and the services performed. Alternate fee scenarios (described above) must be developed, along with an assessment of the likely property owners in the district who will want to join.
- *Petition of property owners.* A petition must be signed by a certain percentage of property owners, depending upon state regulations. Some states specify all owners, while others require at least 25 percent. State law determines the petition's contents. This may include the boundary description, a detailed property map of the district, its proposed name, initial designation of board members, the nature of the infrastructure improvements, and the timetable for completion. The petition is then presented to the appropriate governing body for approval. In most states this is the city or county, and in California, Oregon, and Washington, approval by a boundary setting agency is required.
- *Public participation and approval.* The agency receiving the petition usually sets a hearing and requires appropriate notice to be sent to the public. Some states have fast track processes if all the landowners

have signed the petition, while other states require the same hearing procedure to be used as for local ordinances. After the hearing, the local government body approves or disapproves the petition.

- *Election of the governing board.* Once the special district has been approved (and it may have an interim board), the election is noticed. Voting in a special district election can be based on one person one vote, or one acre one vote, depending on state requirements.
- *Hiring the administrative staff.* It is in the interests of the local general purpose government, the developer, and the actual or potential residents to ensure that professional staff is quickly hired to implement the infrastructure construction process, including administering the user fees and issuing any bonds that were envisioned in the plan. This staff also then operates and maintains the infrastructure.

When setting boundaries for a special district, long-term land use issues should be considered. California, for example, recently passed legislation authorizing the local agency for boundaries to review the decisions. Whether this will reduce overlap and provide local governments with more control over land uses is not yet known.

Conclusion

Constraints on general tax revenues have caused many local governments to charge fees for infrastructure services. However, user fees or public prices are also an effective tool to manage demand for services, and to see that infrastructure facilities are not overbuilt. Where local governments have not yet taken a comprehensive look at their revenue situation to see whether some infrastructure services should be provided with fees, it may warrant the cost and the inconvenience of setting up billing systems. Although special districts, once established, make it more difficult for city or county governments to manage land uses, in some cases they are warranted. If the local practitioner is charged with setting up such a district, it would be wise to ensure that boundaries will reflect the larger public interest about land use and development, as well as the immediate infrastructure purpose.

Additional Resources

Publications

American Water Works Association. *Principles of Water Rates, Fees, and Charges.* 5th ed. Denver, CO: American Water Works Association, 2000. This manual provides detailed guidance on how to evaluate and select water rates, charges, and pricing policies. It is a bit pricey, but may be worth it.

Hanemann, W. Michael. "Price and Rate Structures," in *Urban Water Demand Management and Planning.* Edited by Duane D. Baumann, John J. Boland, and W. Michael Hanemann. New York: McGraw-Hill, 1998. An excellent discussion of the issues regarding types of fees that are useful for water and sewer systems. Applicable to other infrastructure systems as well.

Zorn, C. Kurt. "User Charges and Fees," in *Local Government Finance*. Edited by John E. Petersen, and Dennis R. Strachota. Chicago, IL: Government Finance Officers Association, 1991. A good basic description in an excellent volume for the local practitioner who wishes more information on local government financing for infrastructure.

Websites

Government Finance Officers Association: www.gfoa.org. This website has numerous practical publications as well as links to best practice sites at the local level. Highly recommended for the practitioner who wishes to go further in the area of fee setting and public prices.

Institute of Public Utilities: www.ipu.msu.edu. Great website for those interested in setting utility fees. Publications and training opportunities.

State-specific websites for special districts. These contain local regulations and requirements. Most have links to other resources helpful to the local practitioner seeking local resources. For example, Special District Association of Oregon, California Special District Association, Special District Association of Colorado, Washington Association of Sewer and Water Districts, Florida Department of Community Affairs Special District Information Program, Florida Association of Special Districts.

chapter 14

Exactions and Impact Fees

In this chapter...

Importance to Local Practitioners. 227
History of Impact Fees and Exactions 227
Colonial Times Through the 1960s 227
More Recent Developments. 228
Definitions of Exactions and Impact Fees 229
Land Dedications 229
Development or Impact Fees. 230
State Enabling Legislation. 231
Amount of Impact Fees 232
When and Why Are Impact Fees Used? 232
Developing an Impact Fee Program 234
Which Services Are Appropriate for Development Fees? 236
Elements in a Development Impact Fee Program. 238
Policies Governing Development Impact Fees. 238
Preparation of the Impact Fee Schedule 240
Administrative Procedures for the Impact Fee Program 243
Conclusion 244
Additional Resources 245
Publications. 245
Websites 245

Why Is This Important to Local Practitioners?

Over the past three decades, exactions and development impact fees have become highly visible as a way to finance the infrastructure needed as a result of new development. Exactions encompass a broad range of local infrastructure requirements, including impact fees and land dedications, which are assessed during the permit approval process.[1] The term "impact fee" itself refers to local government charges for offsite capital facilities for water, sewers, sidewalks, and street improvements. Sometimes exactions and impact fees are used to mitigate the development's social or economic impacts. These fees are distinguished from user fees, a broader term used for all fees that local governments charge for public services. Exactions and impact fees help ensure that new development has the infrastructure needed for a healthy city or county.

This chapter begins with a history of their use, before defining the wide range of terms that apply to exaction and impact fees. Reasons for using impact fees are presented, followed by key legal and policy considerations a locality must address in a development impact fee program—including the rationale nexus, proportionality and double taxation issues, and who pays. The chapter also outlines the elements of an impact fee program, including how to develop the fee schedule and other critical administrative issues.

History of Impact Fees and Exactions

Colonial Times Through the 1960s

Cities have required developers to provide land for parks and open space since colonial times. The federal government, as it released or sold land for private use, required that portions of the land be set aside for public uses.

In 1789, the Northwest Ordinance mandated set-asides of land for town halls, churches, and schools. Most federal ordinances of the early nineteenth century did the same.

By the twentieth century, local governments required that developers or landowners provide the land for streets, although the improvements such as sewers and streets were usually constructed at public expense—meaning with tax funds. Many real estate speculators subdivided land far from urbanized areas, anticipating that those who bought the lots would request and eventually receive city services, among them basic infrastructure.

The Great Depression found cities unable to extend service to every home and every street in a subdivision. However, the adoption of the U.S. Commerce Department's model subdivision statutes by states and local governments across the country in the first part of the twentieth century resulted in local regulations requiring developers to construct on-site infrastructure such as streets, sidewalks, water mains, sewer lines, and other utilities.[2]

After World War II, many states authorized legislation for localities to require dedication of land for parks and schools from new subdivisions, or to require in lieu fees. According to one commentator, these were the first efforts of using fees to mitigate the impact of the development on local infrastructure.[3] However, until the late 1960s, most local off-site infrastructure was financed by local taxes or federal grants, with exactions or impact fees limited to on site improvements, land dedication, and in lieu fees.

More Recent Developments

In the last quarter of the twentieth century, continuing population growth, property tax limitations, and reduced federal spending on infrastructure have resulted in the expansion of the role of impact fees and exactions—both in kind and amount. Impact fees were first used in the 1950s and 1960s as a way of funding water and wastewater facilities.[4] In the 1970s and 1980s, the number of localities using them, the services funded, and the proportion of these fees to total development costs and local government budgets all exploded.[5]

High growth municipalities in states such as California and Florida began to require that development pay its own way for new off-site infrastructure needs. This included upgrades to sewer and water treatment plants, civic parks, schools, fire and police stations, and other public facilities not located on the project site. In some localities, impact fees financed social services such as day care[6] and housing for low-income families.[7] Between 1979 and 1988, at least ten jurisdictions adopted programs requiring new commercial development to fund low-income housing, day care centers, and employment training programs, while another six adopted voluntary programs.[8]

In the early 1970s, many of these fees and exactions were "negotiated" between the local officials and the developers as part of the subdivision approval process, without many guidelines or standards. As the practice evolved, courts forced localities to link the amount of the fee to the proportionate share of the impact caused by the development. Many states adopted enabling legislation, and more standardized methods were used to calculate them. Today, some localities use complex formulas and computer programs to calculate the panoply of fees involved.[9]

Impact fees can also be assessed on projects where land subdivision is not involved. In Florida, for example, thousands of acres of land had already been platted years before development. Exactions and impact fees are also used for development in a built-up area whether it is infill or redevelopment of a particular parcel.

Today's impact fees have usually been adopted in an open process, and apply to a wider array of infrastructure services than did the first generation of impact fees, including civic facilities as well as the traditional utilities. Some localities also exempt impact fees for low-income housing projects.

Definitions of Exactions and Impact Fees

The term exactions is a broad one that includes development impact fees, in-kind contributions, and dedications of land levied or negotiated as part of the permit process. The definitions below are divided into land and site improvements and development fees.

Land Dedications

Dedications of land for new development may include the right-of-way, land for drainage, streets, sidewalks, and easements for other infrastructure. Dedications of land may also be required by local ordinances for bicycle paths, transit facilities, parklands, schools, and other civic facilities, depending upon the size of the development. Land swaps between the locality and the developer to provide for parking, bike paths, or other amenities may also occur. Land dedications may be required without payment, or the land may be purchased from the developer at fair market value.[10]

On-site infrastructure improvements may also be required from the developer. As discussed in previous chapters, subdivision ordinances usually include requirements for the design and construction of streets, sidewalks, sewers and storm drainage, energy, and telecommunications—the networked infrastructure. In an urbanized area, the developer of an infill project may be asked to repave the street and sidewalk damaged by construction adjacent to the project, or to provide public amenities in the right-of-way near the property.

Development or Impact Fees

The American Planning Association defines impact fees as "fees upon new development projects to cover capital expenditures by the governmental unit on the infrastructure required to serve the new development."[11] They are usually assessed as a condition of the issuance of a building permit, occupancy permit, or subdivision approval. The legal basis for imposing these fees comes from the local government powers to regulate growth and provide adequate facilities and services. The terms "impact fees," "development fees," "in lieu fees," and "mitigation fees" are used interchangeably. Other terms that may be used are "capacity fees," "facility fees," and "system development charges (SDCs)." Local and state laws regulate how these fees may be assessed, and also determine their definitions. Merely designating a fee as such does not make it legally binding.[12]

- *In lieu fees* are payments by a developer instead of dedicating land within the development or providing necessary off-site infrastructure improvements. In lieu fees are charged when the development is too small for land dedication, or perhaps to pay for the purchase of land elsewhere when there is not appropriate land on-site for the required infrastructure. In an urbanized area, the development may be required to make in lieu payments for infrastructure improvements that are needed as a result of infill development. These fees could pay for a traffic signal, for example, or a new left-hand turn lane, an upgrade to the sewer or water system, or the paving of a previously unpaved street. Such fees can only be used for the designated purpose, and should be deposited into a special account.
- *Linkage fees* are fees to mitigate secondary impacts of commercial and industrial projects, such as the need for housing and day care. They are based on a broader interpretation of the police power of the locality than early subdivision ordinances, with the rationale being that employees of these projects should be able to live in the jurisdiction. They are most common in California and Massachusetts, in areas with high housing costs.
- *Mitigation fees* refer to impact requirements under NEPA and state and local environmental impact assessment processes.

Jurisdictions with Linkage Fees

- Berkeley, California
- Concord, California
- Cupertino, California
- Los Angeles, California
- Livermore, California
- Menlo Park, California
- Oakland, California
- Petaluma, California
- Pleasanton, California
- Napa, California
- Napa County, California
- Sacramento, California
- San Diego, California
- San Francisco, California
- Santa Monica, California
- Sunnyvale, California
- Washington, DC
- Miami, Florida
- Key West, Florida
- Atlanta, Georgia
- Hawaii County, Hawaii
- Shreveport, Louisiana
- Boston, Massachusetts
- Cambridge, Massachusetts
- Somerville, Massachusetts
- Detroit, Michigan
- Philadelphia, Pennsylvania
- Seattle, Washington

Jennifer Evans-Cowley, "Development Exactions: Process and Planning Issues," Lincoln Institute of Land Policy Working Paper (2006), available at: www.lincolninst.edu/pub/pubdetail.aspx?pubid=1177.

- *Connection fees* are charges a locality, water or sewage company, or utility company levies to hook up the development project to the larger network. These revenues can only be used for the specific purpose designated, and are prevalent in older, built-up communities to help pay for upgrades to the capital cost of the sewage treatment plant, for example. They are not usually put into a special capital fund for the infrastructure service, but they should be.[13] They are sometimes called "tap fees."

State Enabling Legislation

Although states were initially cautious about using impact fees, by 2005, 27 states had passed enabling legislation authorizing local governments to set impact fees for new development (other than for water and wastewater), with Texas being the first state to do so in 1987. Florida, another early user of impact fees, does not have authorizing legislation per se.[14] Florida localities used development impact fees for many years without authorizing legislation. Despite litigation based on whether they were an appropriate use of the police power of local government, the authority of cities and counties to adopt impact fees is solidly established in case law.[15] Ohio's use of impact fees was also validated by case law, based on home rule authority of individual jurisdictions.

Impact fees are the most common in the South and West, especially Washington, Oregon, California, Arizona, and Florida—the states where most of the growth in the U.S. has occurred over the past three decades (Table 14.1). Impact fees are rare in the Midwest and Northeast.

Most state enabling acts codify the principles developed in case law set by early users of impact fees: rational nexus and rough proportionality.

Table 14.1
States with enabling legislation for impact fees

State	Year of adoption	State	Year of adoption	State	Year of adoption
Arizona	1988	Indiana	1991	Rhode Island	2000
Arkansas	2003	Maine	1988	South Carolina	1999
California	1989	Montana	2005	Texas	1987
Colorado	2001	Nevada	1989	Utah	1995
Florida	2006	New Hampshire	1991	Vermont	1989
Georgia	1990	New Jersey	1989	Virginia	1990
Hawaii	1992	New Mexico	1993	Washington	1991
Idaho	1992	Oregon	1991	West Virginia	1990
Illinois	1987	Pennsylvania	1990	Wisconsin	1993

Source: Clancy J. Mullen, *Annual Impact Fee Survey* (Texas: Duncan and Associates, 2011), available at: www.impactfees.com.

The acts range from general statements of principles (Arizona, Arkansas, Maine, Vermont, and Wisconsin) to what one expert calls California's "exhaustive, confusing and conflicting" statute. Most states prohibit impact fees from being collected before the certificate of occupancy, although one third permit collection at any point in the development process. Most also require that the revenues be spent within 5 to 15 years or else be refunded to the developer. Half of the statutes permit waivers of impact fees for affordable housing or economic development, but half of those require the local government to reimburse the impact fee fund. A few states lock in the fee schedule at the time the subdivision map receives final approval (platting), while a few also have requirements for updating the fee schedule.[16]

One issue that many states are still struggling with is whether and how to reduce impact fees if additional development makes use of the capital improvements paid for by the impact fees. Some states require that consideration be given to this issue.[17]

Amount of Impact Fees

California's impact fees are at least two to three times the amounts of the next highest states for single family residential unit if water and sewer fees are exempted—followed by Florida, Maryland, Oregon, and Connecticut, according to a "non-random" survey by one consulting firm.[18] The largest fees are for roads, parks, schools, and water and wastewater. A breakdown of these fees by type of expenditure shows that for the United States in 2011, the average fee in the 283 jurisdictions included amounted to $11,908 for a single family dwelling unit, and about $7,028 for a multifamily unit. Figures for retail, office, and industry are provided in Table 14.2 by costs per 1000 square feet.

In 1999, residential development fees in California ranged from $4,000 to $60,000 per unit, and accounted for about 10 percent of the median price of a home. Capital facilities accounted for 80 percent of the fees for subdivisions and 86 percent for infill development. The remainder were fees for the permit processing.[19] In 1997, development fees for infrastructure in one upper-income area in California accounted for 19 percent of the mean sales price of a home.[20]

From 2003 through 2007, non-utility fees increased by 77 percent while construction costs increased by 19 percent. The increases were the largest in Florida, where school fees were increased aggressively. Utility fees increased at a modest rate.

When and Why Are Impact Fees Used?

Before the last quarter century, the traditional view towards financing infrastructure was that both existing residents and the incoming population should

Table 14.2
Average impact fees by type and land use, 2011

Facility ($)	Single family (unit)	Multifamily (unit)	Retail (1,000 ft²)	Office (1,000 ft²)	Industry (1,000 ft²)
Roads	3,228	2,201	6,066	3,367	2,077
Water	3,838	1,452	683	624	638
Wastewater	3,692	1,736	749	701	764
Drainage	1,423	765	1,033	873	999
Parks	3,089	2,341	**	**	**
Library	431	321	**	**	**
Fire	523	388	420	373	260
Police	393	311	414	266	183
General government	1,538	1,192	598	613	398
Schools	4,646	2,558	**	**	**
Total nonutility*	8,422	5,612	6,431	4,090	2,741
Total*	11,908	7,028	6,710	4,484	3,212

Notes: * Totals do not represent sum of average fees, since not all jurisdictions charge all types of fees.
**Rarely charged to nonresidential land uses, with the exception of school fees in California.
Source: Clancy J. Mullen, *Annual Impact Fee Survey* (Texas: Duncan and Associates, 2011), available at: www.impactfees.com.

bear the costs of growth. The logic was that existing residents had benefited from infrastructure put in by the generations that preceded them, so they in turn should help pay the way for infrastructure for newcomers by using taxes that everyone in the jurisdiction pays.

The modern view is that only beneficiaries should pay. Part of this shift is due to the pragmatic need to obtain the funds. There is also a reluctance of the electorate these days to approve increases in general purposes taxes such as the property tax, or sales and income taxes.[21] Some also find that fee use is associated with lower levels of capital spending, and may act as a pricing mechanism that regulates the amount of infrastructure demanded by developers.[22] Others note the decrease in federal funds for water and sewer.[23]

Local governments may have other political and institutional rationales for adopting impact fees on new development to finance major capital facilities. The following discussion is based on Frank and Downing (1988):[24]

- *Shift development costs from tax roles to new revenue source.* Given the difficulty of getting political approval for bond financing through increased taxes, impact fees are a way of financing infrastructure by increasing the sales price of the home, which is then financed by the mortgage market. The cost of the infrastructure levied as impact fees becomes part of the house price to be financed. This prevents the new residents from being at the mercy of the existing residents and having to spend time and resources on getting political approval after they move in for essential services. As has been well documented, capital facilities in most localities have not been well maintained over the years, and need upgrades, but many projects have been delayed to hold down tax rates increases resisted by existing voters. It is the rare locality that does not have an unfunded liability in this area. The introduction of a new revenue source, albeit if only to pay the proportionate share caused by the new development, has been attractive to local officials and politicians. The temptation to load as many costs as meet the "rational nexus" test below is often hard to resist.

- *Concurrency, or timely availability of new facilities.* A second reason to charge impact fees is to line up funds to ensure that the infrastructure facility is ready when the new development is occupied. Florida's concurrency policy requires that all localities stop issuing development permits where there is insufficient capacity in streets, sewers, water systems, and solid waste.
- *Use as quasi-market prices.* With impact fees it is possible to reflect spatially differentiate costs and to provide an incentive for development to choose the less costly locations. A recent study found that the use of impact fees was associated with lower levels of capital funding, thus lending credence to the argument that these fees act as a quasi-pricing mechanism and reduce the amount of capital investment demanded by the market, acting through the developers.[25]
- *Discrimination.* Setting the impact fee high enough to raise the cost of residential housing so that low-income families, who cost more to serve, are excluded, may motivate some local officials. Similarly, officials and/or developers may wish to take advantage of a hot real estate market to deliberately market towards the high end, where the full proportionate share of the capital facilities can be charged.[26]
- *Placate anti-growth advocates.* Those who oppose unrestricted growth in a community often question whether the costs of new development should be subsidized by the taxpayers as a whole. They argue that growth should pay its own way. To the extent that local officials feel they need to be sensitive to this concern, impact fees are a way of ensuring that growth is charged for its proportionate share.

Developing an Impact Fee Program

Ideally, development of the impact fee program should be linked to a long-term infrastructure plan created as part of the general plan update. In this case, the impact fee methodology and the enabling legislation would be addressed as part of the implementation of the general plan. The public

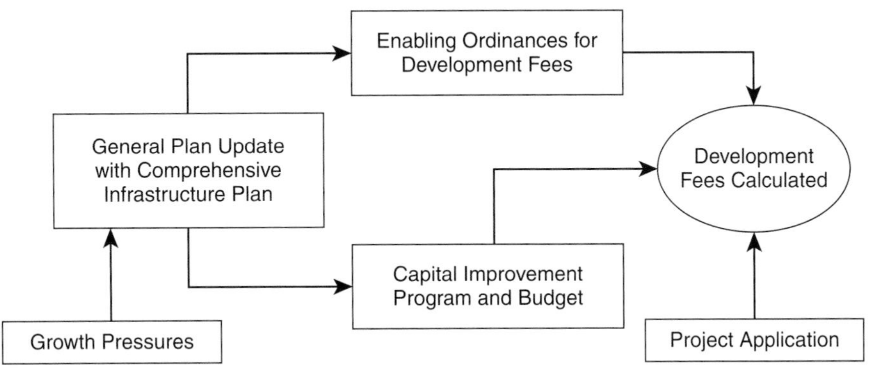

Figure 14.1
Developing an impact fee program as part of the general plan update

participation process set up for the general plan would be used. Specific revenue needs arising from new development would be included in the capital improvement plan and budget (see Figure 14.1).

In the real world, however, rapid growth in urbanized and greenfield areas may not be met by a general plan update because it is viewed as a long and expensive process. Instead, local officials may be faced with a situation where development fees need to be found fairly rapidly to pay for offsite infrastructure, such as an upgrade to a sewer or water treatment plant. Or, the fees already in place may be proving to be inadequate to fund the needed expansions, and general taxes are not a viable option.

If it is not possible to update the general plan, local officials must set up a stand-alone process to develop or update the locality's development fee program and policy (see Figure 14.2).

Although it is technically possible to establish and adopt an impact fee program without a broad-based public involvement effort, this is not wise. The more input that a locality has on the development of policies, procedures, and the assumptions used for the calculations, the less likely it is that there will be legal challenges, from taxpayers, developers, or environmental groups.

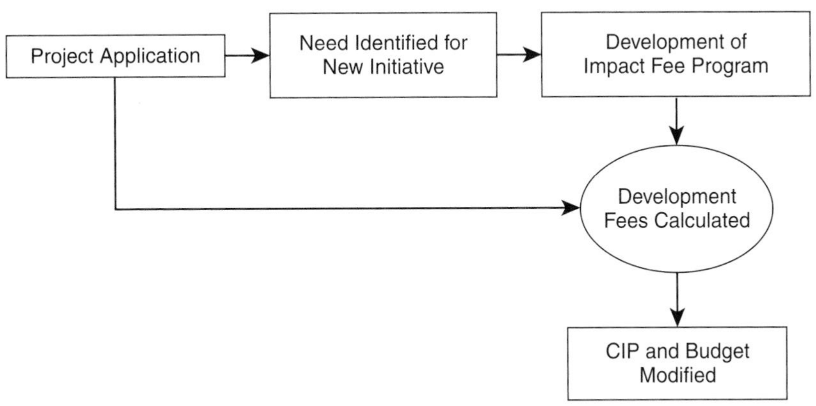

Figure 14.2
Developing an impact fee in response to a specific project development application

Which Services Are Appropriate for Development Fees?

The local practitioner needs to decide which services to include in an impact fee program. Although the local context may override more theoretical considerations, public finance experts provide the following advice based on both fairness and efficiency criteria. The following is based on a classic analysis by Douglass Lee:[27]

- *Water, sewer, and other pipeline infrastructure.* Impact fees are most appropriate for capital improvements for water, sewer, and drainage improvements because the costs are paid by those receiving the benefits. In fact, water, sewer, and other drainage improvements are one of the most frequent uses of impact fees. Special assessment districts, however, are a more equitable tool if it is possible to enlarge existing smaller districts and reassess the fees. Practically speaking, the organizational politics of changing boundaries may tip the scales back in favor of impact fees.
- *Transportation improvements.* Transportation or traffic impact fees are also commonly assessed in the development process. Impact fees will tend to moderate the amount of improvement built because the demand for this type of infrastructure will drop as costs increase. If the cost of improvements is too great, some localities will choose greater congestion instead of the improvements. In this case (and depending upon who will suffer the congestion), the jurisdiction may want to partially subsidize the improvements with taxes paid by the general population.
- *Public buildings with exclusive service boundaries.* If a bounded service district can be defined, the capital improvements for fire, police, parks, libraries, and other municipal or public buildings are suitable for impact fees.
- *Schools.* Public finance experts indicate that impact fees are not really appropriate for schools on a fairness criteria since they are, in effect, taxes levied on the newcomers for benefits that are felt throughout the community. Although some have argued that education is not a "merit" good, most citizens regard education as important for society as a whole. Following that argument, community wide taxes would be more appropriate. Indeed, perhaps this feeling, along with the high cost of providing new schools in a growing community, explains the authorization for this type of impact fee in only seven states. Impact fees for schools are found only in Florida (authorized by case law), California, Washington, and Maryland (where the state legislature has provided specific authority for impact fees for selected counties).[28]

School impact fees are quite large compared to other types of infrastructure (see Table 14.2), slightly less than half the total for a single family unit in the

United States, and higher than every other category of infrastructure, including water, sewer, and road, three other capital intensive efforts. They are usually also levied only on residential development. One expert attributes the lack of more states authorizing impact fees for schools to the relative power of the homebuilding associations in the face of high costs.[29] However, if this is so, the homebuilders are joined by public finance theorists in their opposition.

Table 14.3 shows where impact fees are authorized for different infrastructure functions by state. All of the 27 states with authorizing legislation

Table 14.3
Facilities eligible for impact fees by state in 2011

State	Roads	Water	Sewer	Stormwater	Parks	Fire	Police	Library	Solid waste	School
Arizona (cities)	•	•	•	•	•	•	•	•		
Arizona (counties)	•	•	•		•	•	•			
Arkansas (cities)	•	•	•	•	•	•	•	•		
California	•	•	•	•	•	•	•	•	•	•
Colorado	•	•	•	•	•	•	•	•	•	
Florida	•	•	•	•	•	•	•	•	•	•
Georgia	•	•	•	•	•	•	•	•		
Hawaii	•	•	•	•	•	•	•	•	•	•
Idaho	•	•	•	•	•	•	•			
Illinois	•									
Indiana	•	•	•	•	•					
Maine	•	•	•		•	•			•	
Montana	•	•	•	•	†	•	•	†	†	†
Nevada	•	•	•	•	•	•	•			††
New Hampshire	•	•	•	•	•	•	•	•	•	•
New Jersey	•	•	•	•						
New Mexico	•	•	•	•	•	•	•			
Oregon	•	•	•	•	•					†††
Pennsylvania	•									
Rhode Island	•	•	•	•	•	•	•	•	•	•
South Carolina	•	•	•	•	•	•	•			
Texas (cities)	•	•	•	•						
Utah	•	•	•	•	•	•	•			
Vermont	•	•	•	•	•	•	•	•	•	•
Virginia*	•									
Washington	•				•	•				•
West Virginia	•	•	•	•	•	•	•			•
Wisconsin (cities)	•	•	•	•	•	•	•			•

Notes: † Can be imposed by super-majority vote of city council or unanimous vote of county commission.
†† School construction tax up to $1,600 per unit authorized in districts with populations up to 50,000 (NRS 387.331).
††† Development tax of up to $\$1.00/ft^2$ for residential and $\$0.50/ft^2$ for nonresidential may be imposed by school districts.
*Impact fees may be imposed on by-right residential subdivision of agriculturally zoned parcels for a broad array of facilities under certain circumstances.
Source: Clancy J. Mullen, *Annual Impact Fee Survey* (Texas: Duncan and Associates, 2011), available at: www.impactfees.com.

provide for roads. The next most frequent authorization is water and sewer. Interestingly, impact fees for parks are authorized in slightly more states than fire and police stations. Libraries are rarely able to assess capital costs as part of new development.

Elements in a Development Impact Fee Program

Given the highly charged nature of the development process, it is understandable that much litigation followed the initial use of exactions and impact fees. The essential legal issue revolved around whether impact fees are a tax or a regulation—and whether or not a locality has the authority to impose the fee. Courts eventually found that exactions can be imposed under both the taxing and the police powers of local government. Most localities impose such fees based on their police powers to regulate development to further public health, safety, and welfare of the community.

The locality is in a better position legally, if the state has authorizing legislation for localities to charge such fees.[30] Texas and California are among states with such legislation. The locality is also in a better situation if there is a local development impact fee program.

A good development impact fee program has three elements: general policies, the fee schedule, and administrative mechanisms to ensure the program operates properly. The following addresses these in greater detail.

Policies Governing Development Impact Fees

The local plan should address both legal and policy issues. These include issues of rational nexus, proportionality, double taxation, impact of the fees on the local economy, and relationship to the general plan.

- *Rational nexus.* Impact fees for shared infrastructure costs must meet the "rational nexus" test. Based on a well-known legal case, *Nollan v. California Coastal Commission (1987) 483 U.S. 825*, there must be a linkage between the dollar amount assessed to the development, or the amount and location of land

CIP and Impact Fees in California

A study conducted at the height of boom time found that most capital facilities fees in California are determined using an average-cost methodology, based on historic or projected capital costs and current or projected population. The study also found that the link between development fees and long-term capital improvements programming was weak to non-existent.[1]

Note

1. John Landis, Michael Larice, Deva Dawson, and Lan Deng, California Department of Housing and Community Development, "Pay to Play: Residential Development Fees in California Cities and Counties," 2001, available at: www.hcd.ca.gov/hpd/pay2play/pay_to_play.html.

dedicated, to the development itself.[31] For example, when the City of Berkeley was planning a 172-acre parcel that was prime waterfront property, it could not just zone the entire site as a park, as the majority of council members at the time wished. Instead, the city had to ensure that land dedicated for a park was related to the profit-making uses of the parcel so that the designation would not be ruled a "taking."[32]

- *Rough proportionality requirement.* A locality also cannot charge a development fee for the entire cost of a facility that will be used by others—the impact fee must be proportionate to the use by those in the new development. The requirement for "rough proportionality" was established by *Dolan v. City of Tigard (1994) 512 U.S. 374.* If the exaction is in the form of funds, generally most localities require that the funds be segregated, so they cannot be used for purposes other than which they were designed. And the amount cannot exceed the cost of the facility. In some early exaction efforts, some developers paid, in essence, a fee to develop, not a fee for infrastructure. Usually, however, developers are better negotiators than cities, and this was not a common phenomenon.[33]
- *Double taxation issues.* Another issue concerns potential double taxation. When impact fees were first used, the major concern of developers was that new development would not bear the brunt of the cost for a new facility that would serve both new and existing development. As both localities and developers have become more sophisticated, another fairness issue has emerged. If impact fees are paid to cover the cost of a capital facility, but the locality will also issue a bond for the facility to be repaid with property taxes, the residents may be paying for some portion of the facility twice. Developers argue that these costs must be calculated and figured into the impact fee assessed at the initiation of the project.[34] The weakness of this argument is that it ignores the disproportionate share of the infrastructure to the exaction payers. This is a fair argument only if a special assessment district is created for the new residents to pay over time. If all citizens had paid exaction fees, there would be no equity issue.[35]

Who Pays?

The jurisdiction needs to consider who will actually be paying the development impact fees. Costs are almost totally passed on to the buyers and renters if exactions and impact fees have been in effect for a long period of time, and if they are uniform throughout the market area. One reputable scholar indicates that buyers of new homes bear about 75 percent of the burden of infrastructure costs levied as impact fees, while the remaining quarter falls on the owner of the undeveloped land.[1] However, there are four situations in which this is not true. (The following is based on Altshuler and Gomez-Ibanez.[2])

- In the short run, owners and developers absorb some of the cost—that is, the fee cuts into profit margins and expected land values.
- If only one community in the area has the development fees, then the owners of the undeveloped land will bear most of the costs. However, communities that buyers regard as comparable usually have similar attitudes towards growth and so fees and exactions become similar over time.
- In a hot real estate market, where developers have a hard time keeping up with demand, the price of housing is determined by the temporary shortages. Development fees and exactions may not be able to force the prices higher, and therefore they cut into the windfall profit of the developer.
- Finally, if building permits are limited, the developers and land owners will bear the cost of exactions—but this is because of the limited building permits not the level of development fees.

Notes

1 John Yinger, "School Finance Reform: Aid Formulas and Equity Objectives," *National Tax Journal* 51, no. 1 (1998).

2 Alan Altshuler and Jose A. Gomez-Ibanez, *Regulation for Revenue: The Political Economy of Land Use Exactions* (Washington, DC: Brookings Institution Press, 1993).

Market Test of Impact Fees

Impact fees differ from the user fees discussed in Chapter 13 in that they are paid for by the developer, who is the intermediary between the consumer and the local jurisdiction. The developer will attempt to pass the impact fees on to the buyers of his development to the extent the market allows. Hence the test for the size and nature of the impact fee is whether or not the developer is willing to pay for the development of the infrastructure necessary to support the development. If not, perhaps the development needs to be scaled back, built somewhere else, or the local government needs to pay for the infrastructure through general fund revenues.

- *Influence of impact fees on local economy.* Other local policy issues concern whether the fees will be a damper on local jobs, and whether the ability to levy development fees will cause the jurisdiction to change its policies about land use. Some argue that impact fees are a damper on the market, and that they negatively impact economic development, including investment and job growth, since they function as a tax on capital. However, research looking at the impact of Florida's impact fees from 1993 to 1997 found "modest but significant positive effects of impact fees on job growth," when controlling for property taxes, location, and demographic factors.[36] This likely varies from jurisdiction to jurisdiction, however.
- *Fiscalization of land use issues.* Just as localities shifted existing local services that could pay for themselves off general revenues and into special fees, so too, some localities are accused of going after the new development which results in the largest amount of impact fees, (among other revenues) with the smallest service requirements.[37] Initially, some argued that impact fees could be a tool to impose discipline upon sprawl.[38] However, research in Illinois found that impact fees are more a reflection of the strength of development demand.[39]
- *Relationship to general plan.* Ideally, there should be an up-to-date local comprehensive or general plan that outlines large-scale infrastructure needs. In built-up areas, off-site infrastructure needs for permitted uses may not be predictable in advance of the permit application, so policies guiding ad hoc assessments should also be specified. If the locality does not have a capital improvements program, this is an ideal time to set one up as part of the impact fee program.[40]

Preparation of the Impact Fee Schedule

For the developer and the financial officer, the fee schedule is the heart of the impact fee program. Most impact fee programs develop fees by land use categories and have at least four categories (depending upon the detail of the zoning ordinance): residential single-family detached; multifamily residential; commercial; and residential. The following discussion relies on Burchell, Listokin, and Dolphin (1994).[41]

The local practitioner should go through the following steps for each major land use category:

- *Develop population indicators for each land use.* Estimate the population profile of the anticipated growth for each land use category. The number of households (or people, students, and employees) for each anticipated use over the time frame (25–40 years) should be projected. This should be converted to a per unit basis for residential, and into 1,000 square foot increments for non-residential. Sometimes the per acre factor is used. Demographic multipliers for the number of persons per type of structure can be generated from the American Housing Survey for the area, which is updated every two years, or the Public Use Microdata sample from the Census 2000. Non-residential multipliers are available from national industry reports, and are also presented in Burchell, Listokin, and Dolphin (1994), noted above.[42]
- *Identify infrastructure facilities needed for projected growth.* This is often done by interviewing the providers, analyzing existing CIP documents from the affected agencies, and sometimes by conducting special studies for a particular infrastructure area. In addition, level of service standards (discussed in the chapters on specific infrastructure systems in this book) can be used to help triangulate in on the size of the needed facilities. The four major categories of infrastructure presented in this book should be addressed, including streets, sewer and water, police and fire, schools, drainage, phone, electricity and gas, and sanitary services including solid waste collection and disposal.
- *Identify cost of infrastructure needs.* Sometimes cost figures can be obtained from the local providers, and other times a special study needs to be done, or a cost estimator hired for specialty estimates. Non-monetary costs such as the value of lost habitat, as well as the effects of urban runoff, should be addressed in some manner. Both capital and operating costs need to be factored in. Life-cycle costs should be addressed. Assumptions about the financing mechanisms for the

Principles for Designing Impact Fees and Exactions

- Each exaction or fee should be designed to meet service needs directly attributable to the project bearing the cost.
- Where facilities are to serve more than a single development, costs should be allocated in proportion to services rendered.
- Facilities funded by impact fees and exactions should be elements of a long-range comprehensive local plan for infrastructure services.
- Where facilities are financed by a combination of tax and impact fee revenues, care must be taken to ensure that project occupants, who pay taxes like everyone else, are not double-billed.
- Impact fee revenues should be segregated until used, and should be expended in a timely fashion (generally, within five to six years) for the purpose originally designated.

Adapted from Wes Clarke and Jennifer Evans, "Development Impact Fees and the Acquisition of Infrastructure," *Journal of Urban Affairs* 21, no. 3 (1999).

capital investments need to be laid out. Some geographic locations require more expensive capital facilities. Hilly areas, and those near a flood plain, may have higher unit costs for sewers, streets, and other components of infrastructure.

- *Identify timing of infrastructure provision.* The costs of infrastructure provision may vary by the order in which development occurs. This should be addressed.
- *Identify cost per unit of capacity.* First, calculate the total capacity of the new infrastructure needed by the new population indicator. For residential uses, the new capacity can be expressed in terms of households or new residents and students, while for nonresidential, the capacity is framed in employees. Special studies must be done for industrial uses. Then, the cost per unit of capacity can be calculated. If the development will not be paying double with future taxes on the same infrastructure, then this represents the development impact fee for that use.
- *Include administrative costs for program.* The practitioner should ensure that the costs to administer the program are reflected in the fees. This ranges from about 2–5 percent of the funds collected (excluding the engineering costs). As with the bulk of the fee, this must be documented and justified as to both the nature and amount.
- *Double taxation and fiscal impact adjustments.* If the locality has decided to adjust the development impact fee due to the fact that residents or commercial uses will be paying property taxes in the future on bonds used to raise funds for the needed infrastructure, or to adjust the fee downward due to other revenues to the locality, these adjustments are done at this point.
- *Adjustment 1.* The total infrastructure cost is calculated for each land use by multiplying the capital cost per person, student, or employee by the number anticipated for the land use category.
- *Adjustment 2.* Then property tax revenues and the portion for the bond are extracted, after accounting for the difference between one time fees and annual expenditures.
- *Adjustment 3.* An annual equity credit might also be calculated for future property taxes for the project so that the new development does not pay twice for infrastructure.

The fee schedule and program should contain procedures to deal with cases that arise in urbanized areas such as the demolition of a structure and replacement with a more intense land use and building expansions that will require more off-site infrastructure, and idiosyncratic industrial uses. For expansion projects, the fee should reflect the increment in use. Credits against impact fees may be provided for in the program under certain circumstances, as well as waivers. A waiver for infrastructure impact fees for low-income housing development is common among localities with impact fees, as is a provision for a waiver for "overriding public purpose." Some states require that the policy behind the exemption be included in the general plan, or in another local policy document adopted legislatively. Some allow the waiver for low-income housing, but pay the difference from the general fund.

Administrative Procedures for the Impact Fee Program

Although the fee schedule and its basis are probably the most scrutinized aspect of the impact fee program, the local practitioner should not overlook the administrative procedures that will actually implement it. Timing, handling of the funds, and the accounting systems need to be part of the program as well. This discussion is based on Nicholas et al. 1992:[43]

- *Timing*. The issue of timing should be addressed in the program. A program should not be put into effect until the locality has set up its administrative procedures for processing applications and for accounting. The program should provide for applications already in the pipeline. The fee is usually (but not always) assessed at the time of the building permit, which is when the development application is deemed vested in most states. However, developers may balk at this since they prepared their financial pro forma based on earlier fee schedules. Depending upon the severity of the need for the infrastructure, the locality may decide either to forego or insist upon the impact fees.
- *Timeframe*. Similarly, the issue of how long fees may be retained by the locality should be addressed. Many experts agree that they should be expended within a five-year timeframe. However, this is not always feasible, and if so, the justification must be clearly laid out. For example, the County of San Diego developed a plan for capital improvements for a 20-year period, outlining when each improvement would be built. Impact fee contributors are not deprived of benefits, since the improvements will be built when needed. In addition, if delays in construction occur, fees need not be refunded. Sometimes growth rates are lower than expected and the total needed for a new facility is not accumulated as projected.
- *Collection, accounting, enforcement, and refund procedure*. Mechanisms for handling the funds must be outlined. Staff responsibility for collecting, accounting, enforcing, and refunding the fees must be designated. This is usually the building department, but it could also be the planning or finance department. Funds for specific project must be earmarked for that use. Sometimes, in a built-up area, exactions or

Impact Fee Standards

- The imposition of a fee must be rationally linked (the "rational nexus") to an impact created by a particular development, and to the demonstrated need for related capital improvements pursuant to a capital improvement plan and program.
- Some benefit must accrue to the development as a result of the payment of a fee.
- The amount of the fee must be a proportionate fair share of the costs of the improvements made necessary by the development, and must not exceed the cost of the improvements.
- A fee cannot be imposed to address existing deficiencies except where they are exacerbated by new development.
- Funds received under such a program must be segregated from the general fund and used solely for the purposes for which the fee is established.
- The fees collected must be encumbered or expended within a reasonable timeframe to ensure that needed improvements are implemented.
- The fee assessed cannot exceed the cost of the improvements, and credits must be given for outside funding sources (such as federal and state grants, developer-initiated improvements for impacts related to new development, etc.) and local tax payments that fund capital improvements.
- The fee cannot be used to cover normal operation and maintenance or personnel costs, but must be used for capital improvements, or, under some linkage programs, affordable housing, job training, child care, etc.
- The fee established for specific capital improvements should be reviewed at least every two years to determine whether an adjustment is required, and similarly, the capital improvement plan and budget should be reviewed at least every five to eight years.
- Provisions must be included in the ordinance to permit refunds for projects that are not constructed, since no impact will have manifested.
- Impact fee payments are typically required to be made as a condition of approval of the development, either at the time the building or the occupancy permit is issued.

Source: Policy Guide on Impact Fees, American Planning Association, 1998, pp 3–4.

in lieu fees for off-site improvements are levied as part of the conditional use permit, not as part of an impact fee program per se. In this case, the building department has to be directed and staff trained to review for this type of fee as well as impact fees. If no procedures have been set up, it is likely that the fee will not be collected, or it will be "forgotten." If there are multiple localities involved in collecting impact fees, procedures must be developed to make sure that the funds do not drop through the cracks between the agencies. If the intended infrastructure was not built, procedures should address how the monies will be refunded, and whether interest will be paid.

- *Level of accounting and geographic subdivisions.* Some localities initially set up accounting systems for impact fees at geographic levels that precluded an effective construction project, for sewer replacement, for example, or road improvements. The level of accounting should preserve the "nexus" requirements, but be large enough so that excessive amounts of overhead do not result in building small infrastructure projects. Larger localities divide up accounts into different geographic zones to ensure that the development is paying for the infrastructure services it receives. The American Planning Association recommends service areas where the impact fees are both assessed and spent. This can be a "powerful tool" in ensuring that the constitutional requirements are met.[44]

Conclusion

Development impact fees and exactions are one of the most powerful tools that a locality has to ensure that new development pays its own way. Although a program of this sort results in the costs being passed on to newcomers through higher housing prices, most local policy-makers find it otherwise difficult to raise taxes to cover these costs. In developing an impact fee program, local officials should try to balance the need for affordable housing against the need to have adequate infrastructure to serve new residents. In addition, the local practitioner should not forget to include adequate administrative measures for the program, including adequate staffing. The local practitioner should also try if at all possible to tie the impact fee program to a long-range capital improvements or infrastructure plan, since this will be less costly and more likely to be upheld in court if challenged.

Additional Resources

Publications

Abbott, William W., Peter M. Detweiler, M. Thomas Jacobson, Margaret Sohagi, and Harriet A. Steiner. *Exactions and Impact Fees in California: A Comprehensive Guide to Policy, Practice, and the Law*. Point Arena, CA: Solano Press, 2001. A thorough guide for the local practitioner with emphasis on California laws.

Burchell, Robert W., David Listokin, and William R. Dolphin, "Shared Infrastructure Costs." In *Development Assessment Handbook*. Washington, DC: Urban Land Institute, 1994. A methodology for calculating off-site infrastructure costs. Also a generally fine handbook about the development process despite its age.

Meck, Stuart. *Growing Smart Legislative Guidebook 3: Model Statutes for Planning and the Management of Change*. Chicago, IL: APA Planners Press, 2002. Suggested legal language for state and local development impact fees.

National Association of Home Builders of the United States of America. *Impact Fee Handbook*. Phoenix, AZ: Development Planning and Financing Group, 2008. This handbook is also available on-line as a downloadable pdf at: www.nahb.org/infrastructurefinance.

Nicholas, James C., Arthur C. Nelson, and Julian C. Juergensmeyer. *A Practitioner's Guide to Development Impact Fees*. Chicago, IL: APA Planners Press, 1991. Although written over a decade ago, this is still a comprehensive look at development impact fees written in a practical manner. Its descendant, *A Guide to Impact Fees and Housing Affordability*, published by Island Press in 2008, contains instructions on how to design an equitable impact fee that accounts for variation in the size and selling price of a home. See also Impact Fees: Principles and Practice of Proportionate-Share Development Fees by Arthur Nelson, published by the APA Press in 2009.

Websites

www.impactfees.com. An absolutely first rate website sponsored by Duncan and Associates with every single link you will ever need for finding out what impact fees are, how to calculate them, what the latest techniques and court cases are, along with a variety of consultant contacts.

American Planning Association: www.planning.org. This also has numerous publications about development fees: www.planning.org/policyguides/impactfees.html.

National Association of Home Builders: www.nahb.org/infrastructurefinance.

National Association of REALTORS: www.realtor.org/library/library/fg805.

U.S. Department of Housing and Urban Development: www.huduser.org/publications/affhsg/impactfees.html.

chapter 15

Competition and Privatization

In this chapter...

Importance to Local Practitioners . . . 247
Privatization in the Past 30 Years . . . 248
Resurgence of Interest in Privatization . . . 248
Privatization: Costs and Benefits . . . 249
Status of Privatization Today in the United States . . . 250
Forms of Market-Based Approaches . . . 252
Full Privatization and Divestiture . . . 252
Public–Private Partnerships . . . 254
Contracting Out . . . 255
Re-Engineering: Internal "Competition" . . . 256
Planning Considerations . . . 257
Considering Private Involvement . . . 257
Evaluating the Privatization Project . . . 258
Implementing the Privatization Project . . . 259
Institutional Issues about Privatization . . . 260
Conclusion . . . 262
Additional Resources . . . 262
Publications . . . 262
Websites . . . 263

Why Is This Important to Local Practitioners?

The use of private capital to build and operate major infrastructure systems has a long history in the United States and abroad. Many important infrastructure systems in the USA—such as water, sewer, telecommunications, and energy—began as small-scale private enterprises at the beginning of the twentieth century. Involvement of the private sector is an important consideration for local practitioners. Decisions about which agency should provide new infrastructure may need to be made during the review of a subdivision application or a use permit application for a major project. Institutional arrangements and their implications for financing when the capital improvement plan and the capital budget are developed need to be addressed. Issues about privatization and competition may also arise during the preparation of an environmental impact report, when lower cost alternatives are needed to mitigate or minimize negative effects on existing infrastructure from a large-scale project. The local practitioner may also be involved in the decision about whether to develop a special purpose agency to undertake a large-scale infrastructure project.

This chapter begins with a brief summary of recent privatization events and a précis of the private–public debate. It then surveys the use of these efforts in the United States, before turning to a description of the four major privatization strategies: full privatization and divestiture, public–private partnerships, contracting out, and reengineering within the agency. A process to be used to decide whether to consider some form of privatization or contracting out at the local level is outlined, along with steps to be followed if the locality decides to go down this path. The chapter concludes with recommendations for changes in state and local rules that local officials might wish to pursue in order to make a market-based approach to service provision easier.

Privatization in the Past 30 Years

Many government services were contracted out in the United States prior to the resurgence of interest in privatization. In fact, many infrastructure services were originally provided by the private sector in the nineteenth century. Public provision came only when the private sector was found to be corrupt (the good government movement), or when, in the case of water and some municipal energy utilities, the private sector did not provide adequate standards. Solid waste collection, street cleaning, street construction, large-scale architectural and engineering services, equipment maintenance, and tree trimming historically have also been performed by the private sector under contract to the local government.[1]

This system seemed stable until about 30 years ago when a combination of ideology, growth pressures, new technologies, and a poor U.S. and world economy resulted in increased pressures to privatize. Many efforts both here and in Europe were taken during the last quarter of the twentieth century to privatize government-owned and government-operated infrastructure and to de-regulate privately owned and operated monopolies. Although the focus of this book is on infrastructure in the United States, the global context deserves mention because the issue of the privatization of infrastructure receives much of its current impetus from the international scene.

Resurgence of Interest in Privatization

The issue of whether private firms or government should provide infrastructure services is part of the larger debate about the effectiveness of markets versus regulation and the role of government and business. This debate, never completely quiet, heated up again in the early 1970s when one of the leading advocates for market provision from the Chicago School received a Nobel Prize. Thereafter, in the 1970s and 1980s, both Margaret Thatcher and Ronald Reagan were strong proponents of reducing the size of government, privatizing many government services, and de-regulating the private sector.

At the height of the debate about privatization, the argument was ideological—the hidden hand of the market versus the public interest-oriented government.[2] In Europe, privatization of government-owned infrastructure agencies was part of a larger effort to privatize other government-owned businesses such as Volkswagen, Lufthansa, and Renault, in addition to the oil and telecommunication companies. One author estimates that from 1985 to 1998, over $100 billion in government assets were sold to private investors.[3]

However, as the global economy weakened, privatization of public infrastructure also occurred for more pragmatic reasons. In New Zealand, the government faced huge deficits in the mid-1980s. The government there owned and operated almost 13 percent of the nation's economy, including the entire telecommunications industry, all wholesale electricity distribution, the ports, the rail system, the only national airline, the only two television channels, and half of the commercial forest land. In 1987, these were all converted to public corporations. By 1995, most had been sold to the private sector.[4] Deficits were reduced by layoffs and subsequent reorganization of the enterprises, which resulted in increased productivity and lower costs.

Developing countries also look to the private sector to mobilize the capital and technical resources necessary to build transportation infrastructure, energy plants, and transmission and distribution systems, as well as telecommunications services. Here the issue is less the institutional arrangements of who will design and build the infrastructure, and more how to raise funds. However, many developing countries have unstable capital markets. Therefore, in these countries:

> The issue is not whether to privatize urban infrastructure or enter into some form of joint venture between the public and private sector or to contract a service out. The question is how to create stable accessible markets of domestic capital, so that local governments will not be forced to continue to fully build, finance and operate public works.[5]

Today, both in the USA and in developing countries, local officials faced with rapid growth look to the private sector as a way to mobilize capital and organize quickly to build new infrastructure facilities. Changes in the capital markets have decreased the difference between private and public borrowing. Traditionally, federal and state grants or tax-supported municipal bonds have been the lowest-cost capital sources. However, over the past decades, pension and insurance fund managers have become more familiar with infrastructure financing and are willing to accept longer-term debt. More efficient construction methods by the private sector can reduce the overall project cost; and finally, both state and the federal government have come up with new low-cost financing programs for which the private sector is eligible. For example, TIFIA (the Transportation Infrastructure Finance and Innovation Act of 1998), provides funds to write down up to 33 percent of financing charges for major transportation projects undertaken by the private sector.[6]

Privatization: Costs and Benefits

Governments are attracted to privatization strategies for infrastructure provision for three reasons—to raise funds, to reduce costs, and to increase the effectiveness of service delivery. First, when it sells the asset entirely, the government receives cash. Second, the promise of cost savings is also an important motivator. Private industry is generally thought to be more efficient than the public sector in addressing labor issues. They are also thought to have more innovative organizational structures and production strategies. Private industry is generally regarded as more open to the development and use of new technologies, which can improve productivity, and thereby cut costs.[7]

Private firms can frequently build large-scale capital projects faster using design-build techniques, more efficient procurement mechanisms than state or local governments, and more sophisticated project management mechanisms. Outsourcing to private firms for construction can also permit the government to avoid staffing "ramp ups," and then having to lay off government staff who often have developed political clout during their five- to seven-year tenure with the agency. Finally, oddly enough since a century ago the reverse was true, private firms are thought to have greater access to capital. Some of them may even invest in the project directly.[8]

Those who are opposed to privatization generally argue that private industry may sacrifice quality for profits. Others argue that unless care is taken, instead of an unresponsive, costly public monopoly, the only change may be to an unresponsive, costly private monopoly.[9] Some also indicate that certain infrastructure, such as water, telecommunications, and energy, are so important to the public that even though the government may operate them less efficiently, this is worth having the control and the long-term institutional capacity in the public sector.

The local official should be aware that many of the arguments regarding the public versus private provision have an ideological stance. However, it should also be noted that the debate, once quite polarized between those who supported "markets" and those who supported "government" has quieted somewhat as alternative ways have been found to introduce demand management, performance accountability, and cost-minimizing measures into government organizations.

Status of Privatization Today in the United States

A survey done by an accounting firm of local governments in the mid-1980s found that about one-third of the respondents had contracted with the private sector to provide roads, bridges, and tunnels, and another 30 percent contracted out street lights and wastewater collection and treatment. A sizeable number had privatized solid waste and resource recovery facilities. Municipal buildings and garages were turned over to the private sector in another 20 percent of the localities. Local governments mentioned solid waste, garages, jails and stadiums, convention centers, and recreational facilities as likely candidates for further privatization.[10]

A study for the International City/Country Managers Association (ICMA) using longitudinal data for cities and counties from 1982 to 1997 found that the areas of public works and public utilities that are part of the local general purpose government had the highest incidence of contracting with for-profit private firms. Almost a quarter of these services were provided by the private sector in 1997. The figure was higher for strictly public works activities and for parks and recreation activities, and was lower for activities that are sometimes referred to as public utilities.[11] The researchers also found that richer suburbs favor privatization, while higher-cost metro and rural communities tend to favor public provision.[12]

Research on U.S. cities from 1992 through 2007 show a rise in mixed public–private delivery and the decline of complete contracting out (Figure 15.1). Early enthusiasm for market competition was tempered by the challenges of contract management and increased transaction costs in that area for the locality. In 2002, direct public delivery accounted for 60 percent of local government services, up from 54 percent in 1992 and 1997. In 2007, public delivery fell back to 1997 levels. However, this period saw a rise in intergovernmental contracting, from 17 percent to 20 percent. The infrastructure services with public delivery over 60 percent were street cleaning, water and wastewater management. Refuse collection continues to have high levels of for-profit contracting. Only about half of the contracts are monitored by local government, being the highest among suburban governments, "where citizen interest is lower and more homogeneous, and competition is high. Contracting levels are lower among urban governments, where citizen interest is highest and internal economies of scale are present (making contracting less economically beneficial)."[13]

The picture is not monolithic, however. Governments seem to contract in and contract out for infrastructure services overtime. New contracting out for services by local governments dropped from 1997 to 2002 from 18 percent to 12 percent. Similarly, reverse contracting, or the re-publicizing of services was on the increase (see Figure 15.2).

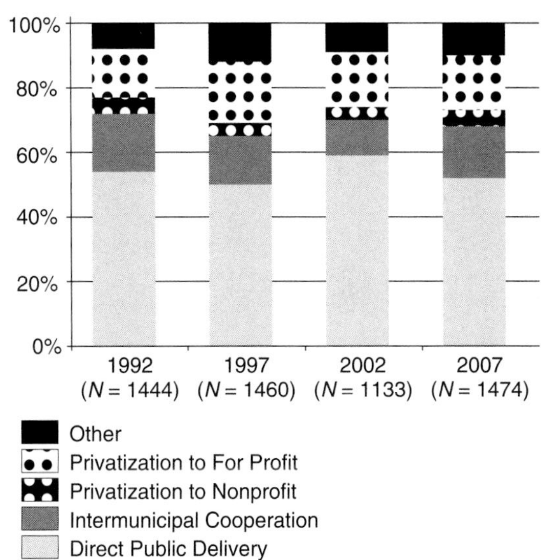

Figure 15.1
Composition of local government service delivery, 1992–2007

Dynamics of Local Government Service Delivery, 1992 to 2002

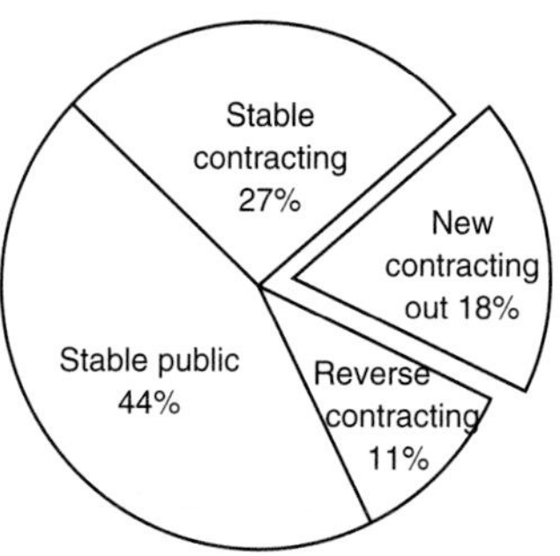

1992 to 1997 Survey

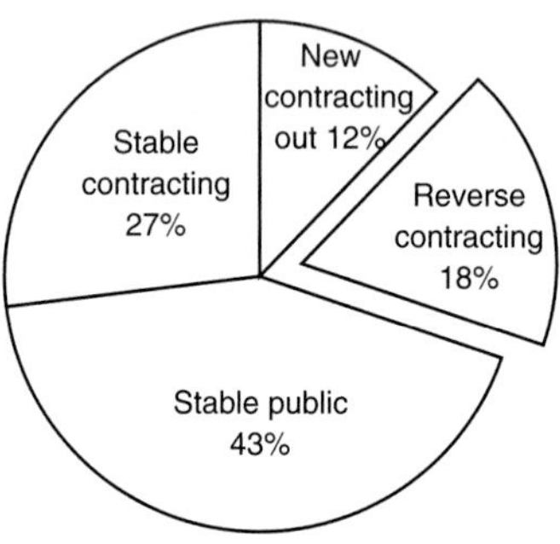

1997 to 2002 Survey

Source: Hefetz, Amir and Mildred Warner, 2007. "Beyond the Market vs. Planning Dichotomy: Understanding Privatisation and its Reverse in US cities." *Local Government Studies*, vol. 33, No 4, page 555–572.

Figure 15.2
Dynamics of local government service

Internationally, the primary candidates for infrastructure privatization have been transportation, water, telecommunications, and energy. In the United States, interest has been more focused on the privatization of water and sewer, and in many places solid waste services are delivered by private providers. There has also been an ongoing debate about the private provision of transportation facilities and services.

Forms of Market-Based Approaches

Four major categories of strategies are usually addressed as part of the market-based approach to infrastructure provision: (1) full privatization; (2) public–private partnerships (restructuring with partial privatization); (3) contracting out; and (4) re-engineering within the government agency. There are, however, as many strategies as there are people to think them up and politicians to debate them. The following discusses the four general categories with an emphasis on describing the range of options open to local officials.

Full Privatization and Divestiture

Full privatization is where the infrastructure service responsibility is transferred to a private or nonprofit entity, and the capital asset is sold or becomes part of an investor-owned utility. The purchasing company pays a sales price. This is also called divestiture. Additional terms and conditions may apply to rate increases and performance goals. Some instances of this approach have resulted in the utility being regulated, while only a few are totally free of controls. Public transport is the only instance of no regulation following divestiture. However, in most cases of full privatization, at the same time that the asset was sold, a regulatory body was formed to monitor prices and service quality. Similarly, before divestiture, the agency is usually restructured, either by converting the public department to a stock corporation and or by streamlining operations and cutting costs to make it more attractive for sale.[14]

Full privatization of infrastructure systems has been limited to only a few countries and a few instances.

In England, for example, several municipal bus companies and the nationally owned long-distance bus company were privatized. In addition, in 1989, water and sewer provision was privatized. Ten public companies covering all of England and Wales were formed from existing public agencies, and then sold to investors. To monitor the private companies, a regulatory office, the Office of Water Services, was established. Two bridges in Dublin, Ireland, were financed and constructed privately. In the late 1990s, Great Britain's Private Finance Initiative (PFI) encouraged the public sector to purchase private sector services for electricity, gas and water, prisons, hospitals, and schools, instead of providing them publicly. New Zealand sold off assets worth 14 percent of its Gross National Product between 1987 and 1991; the United Kingdom sold assets worth 12 percent of its GDP between 1979 and 1991.[15]

The United States has privatized fewer existing public infrastructure agencies in the past 20 years than its European counterparts. This is primarily because the major energy and telecommunications infrastructure, provided publicly in Europe and developing countries, was already privatized. Some exceptions are the privatization of Conrail and the Great Plains Coal Gasification Plant.[16] In addition, in the 1990s, some smaller water systems were sold to existing, nearby investor-owned utilities. Only one sale of a federally subsidized municipal wastewater facility has occurred in the USA—in Ohio in 1995. Birmingham, Alabama, proposed to privatize its wastewater facility, but the issue went to the voters in 1998 and was defeated.[17]

Some of the benefits of divestiture are to transfer risk to the private sector, which some feel is better able to calculate risk, estimate demand, and control costs. Other benefits include the ability to raise capital more easily and be more "nimble" than the public sector in dealing with labor issues, and the ability to develop new technologies and to be responsive to customers. Full privatization is intended to reduce risk and failure for the public sector.

On the other hand, an asset sale and full divestiture may result in less community control over quality and the supply of public goods. Full and even partial privatization, as will be discussed below, usually requires the governmental entity to play an active role in monitoring, which is not cost-free in terms of staff. There may be other costs not quantified at the time of the sale. Asset sales are generally quite controversial, since public employees fear losing their jobs and customers fear degradation of the service. This would not, however, be the case if the private infrastructure services were being proposed for a new subdivision.

Public–Private Partnerships

Partial privatization of the provision of infrastructure services can take many forms, and is often referred to as a public–private partnership. Generally, some form of public–private partnership is used when a new facility must be built. Planners in a local government dealing with the impact upon a utility of a particular development might also raise these ideas, but generally, decisions to even explore these alternatives must come from the chief executive officer or the budget staff. Planners working for a developer or for a firm charged with doing the environmental impact report on a development proposal that has a substantial impact upon a utility should, however, consider raising alternatives such as these if new capital capacity is needed.

Although there are many terms for public–private contracts or relationships—among them, concessions, leasing, franchising, and public–private partnerships—the basic idea is that the government contracts in some manner for the private sector to build the capital facility. Various forms of ownership and financing agreements are available—even some profit-sharing arrangements—but there is no standard recipe. The following terms have been used after the fact to characterize the arrangement reached by the government agency:

- *Design–Build–Operate (DBO)* is the term used where the government has a long-term contract with a private company to design the facility, build it, and then operate it. The government generally finances this through the use of public debt.
- *BOOT (Build–Own–Operate–Transfer)* is where a private firm finances, designs, builds, and operates the infrastructure under a franchise from a government entity. The private entity receives the revenues. At the end of the contract period, the facility reverts to the government. This can also be referred to as BOT (Build–Operate–Transfer).
- *BTO (Build–Transfer–Operate)* is when the company builds the facility, then transfers it to the government but continues to operate it.
- *Design–Build* is when the same entity designs the structure and then builds it. One contract covers both activities.

Similar approaches can be used for existing facilities in need of renovation. The governmental entity can lease a substandard facility to a private operator with a contract that provides for upgrading of the facility and subsequent operations and maintenance. Obviously this can be done for a facility not in need of capital infusion, but generally the motivation behind these efforts is either cost containment, or to raise enough capital to make substantial capital improvements.

Some of the more interesting examples of this strategy come from abroad, and their influence is being felt domestically. In France, water systems are publicly owned, but 75 percent are operated by private companies who receive long-term (25–30 years) contracts to finance, build, and operate the facilities. This system has been in effect for over 100 years, and is recognized for technological innovation, although from time to time there have been allegations of corruption and some environmentalists have noted the poor water quality in the rivers receiving the effluent. In 1984, the provision of water in Paris was restructured so that the facility could be modernized without increasing taxes. The central facility was vested in a company where 70 percent of the shares are owned by the city, and the remaining 30 percent divided between two private companies who also have the concession, or franchise, for water distribution. The workforce was reduced by 40 percent through reassignment and attrition, while the accounting and technology were overhauled and modernized.[18]

Contracting Out

There are two general approaches to contracting out. The first is for all of the maintenance and operation of an infrastructure utility—such as contracting for the maintenance and operations of a port, bridge, water treatment facility, school, or plant debris recycling plant. The second is contracting for some subset of the tasks performed by the utility. The following discussion applies to both categories.

First, it should be noted that although much of what was described in the discussion about building a new facility could be thought of as "contracting out," at the local level in the United States, this term means transferring work from in-house staff to another entity, usually a private firm, but it could also be to a non-profit or other government agency. This can be for the entire maintenance and operation of the infrastructure service (the core function), or for a particular aspect of the service. The ICMA studies referred to earlier indicated that government service provision is dynamic, and that there is a great deal of contracting out, and contracting back in as conditions change.

Contracting out is usually done to save money, and it is usually done for the same reasons that many corporations in the 1990s outsourced specialized services. Either they did not have the services in house and they did not want to develop the capability; or, they had the function in-house but could obtain it less expensively and with higher quality from another firm specializing in the same. One of the reasons that contracting out can be less expensive is that private firms have the ability to hire temporary staff to whom they do not pay pensions or, in many cases, health benefits. Even if the staff in the private firm is "permanent," they usually are not unionized, and wages may be less than in the public sector.

Examples of contracting out for specific functions might include billing—local governments are notoriously poor at billing and collections. Engineering and architectural services for the design and detailed contract specifications have long been contracted out by local engineering departments for roads, sewers, and large-scale public facilities construction, since these expenditures may be lumpy and so do not justify permanent employees. The economic downturn in the early 1990s in the United States saw many local governments contracting out janitorial work, tree trimming, and other services for which performance standards could relatively easily be specified and turned into a contract where monitoring was straightforward.

At the local government level, there may be considerable disagreement about what is a "core" function that should remain in-house. Although the re-engineering gurus suggest that only the policy or steering function is "core," employees in affected jobs usually disagree, and are often in a position to apply considerable political pressure on the city council, the county board of supervisors, or the governing board of the special purpose district. Therefore, in infrastructure agencies where employee pressure could not be ignored, a phenomenon called "managed competition" has arisen. Public utility managers and their staff are encouraged to submit their own bid packages to see if they can compete with the private sector.

There are significant issues on both sides of the fence with respect to managed competition. Many argue that private bidders have an unfair advantage because they do not pay union wages, and do not offer the same health and retirement benefits as do local infrastructure agencies. Bid specifications can, however, be adjusted so that the government agency is not penalized for these costs. Some argue that this discourages the transition of the infrastructure system to a more cost-effective model that is more equitable for all tax payers, not just those who work for the utility. Others respond that managed competition is inherently rigged in favor of the local agency due to the failure to include the risk borne by the private firm and the taxes paid by the firm.

One successful example of managed competition occurred at the Charlotte-Mecklenburg Utility Department in North Carolina, where a city team put together a proposal to cut costs by nearly 50 percent. This bid came in 19 percent lower than the closest private bid, with projected savings of \$4.2 million.[19]

Re-Engineering: Internal "Competition"

Another way in which departments or agencies responsible for infrastructure can become more competitive without privatizing is through internal restructuring. Advice on how to do this has given rise to a plethora of books and government efforts that go well beyond infrastructure provision. At the heart of all these efforts is the ability of an organization to voluntarily give up old ways of doing business, in favor of increased efficiency and effectiveness.

Local governments that have been the most successful in restructuring have found ways to provide incentives for employees to develop strategies to cut costs and or be more customer-oriented. Sometimes these incentives are a guarantee of another job if the employee recommends eliminating his or her task or program. Sometimes the incentive is a pot of money for which employees submit proposals for innovative ideas that improve productivity or customer service. Cities such as Indianapolis, Phoenix, Arizona, San Jose, Charlotte, NC, and several in Florida are exemplars in this area.

Other re-engineering techniques involve the formation of cooperative agreements with other government agencies. One of the ICMA studies previously noted found that the use of inter-governmental cooperation agreements for all services increased from 12 percent in 1992 to 15 percent in 1997. These may be to obtain economies of scale in purchasing, for example, or to avoid having to duplicate services. If one government agency has an efficient and effective equipment maintenance function, such as the city or the county, it might be cheaper for the school district to enter into a cooperative agreement with the other agency rather than to establish its own function. For example, the City of Berkeley obtained an EPA grant to establish a fueling station for city vehicles using propane gas. It contracted with the local school district to service their propane gas-powered vehicles as well.

Along similar lines, local governments commonly use statewide purchase orders to purchase vehicles, since the state government has stronger purchasing power and can negotiate a cheaper price for the vehicle. Cities with increased need for water often enter into cooperative agreements with neighboring cities to purchase their water, rather than rebuild their own plant.

Finally, creating the infrastructure department as its own cost center or enterprise is a way to begin the internal discipline of seeing the utility as a cost-conscious entity that should live on its revenues.

Planning Considerations

Considering Private Involvement

Private provision should be examined if the infrastructure project is extremely large or complex. Although a century and a half ago size and complexity made public provision necessary, this was partially due to unstable capital markets and boom/bust cycles where even the public sector defaulted. Today, the local government needs to carefully consider privatization, particularly in light of the global financial meltdown of 2009. Although most private infrastructure providers maintained stable service provision, continuity of service is an important consideration, as is the protection of public assets.

Another situation where private involvement should be considered is when the service has been provided by the agency for a long period of time, and there are numerous complaints about service quality and the cost of provision. If a major rehabilitation is required, and public funds are not readily available, private involvement could be a possibility. (See Figure 15.3 for a summary of factors that make for a successful privatization effort.)

Evaluating the Privatization Project

Once the decision has been made to consider privatization, a more formal assessment should be undertaken. The following conditions outlined in a 1997 GAO report on the efforts of states to privatize may be useful in guiding the assessment process at the local level.[20] This discussion is also drawn from the report of the Roundtable on California's Experience with Innovations in Public Financing for Infrastructure held in December 2000.[21]

- *Gearing Up the Institution.* The degree of political and organizational support for embarking on a changed service delivery system should be evaluated. If this does not exist, the change probably will not work. The government agency sponsoring the project needs to have the authority and mandate to be able to clear the way for the private sector through the various governmental hurdles. Therefore, both the organization and the governing board need to recognize the need

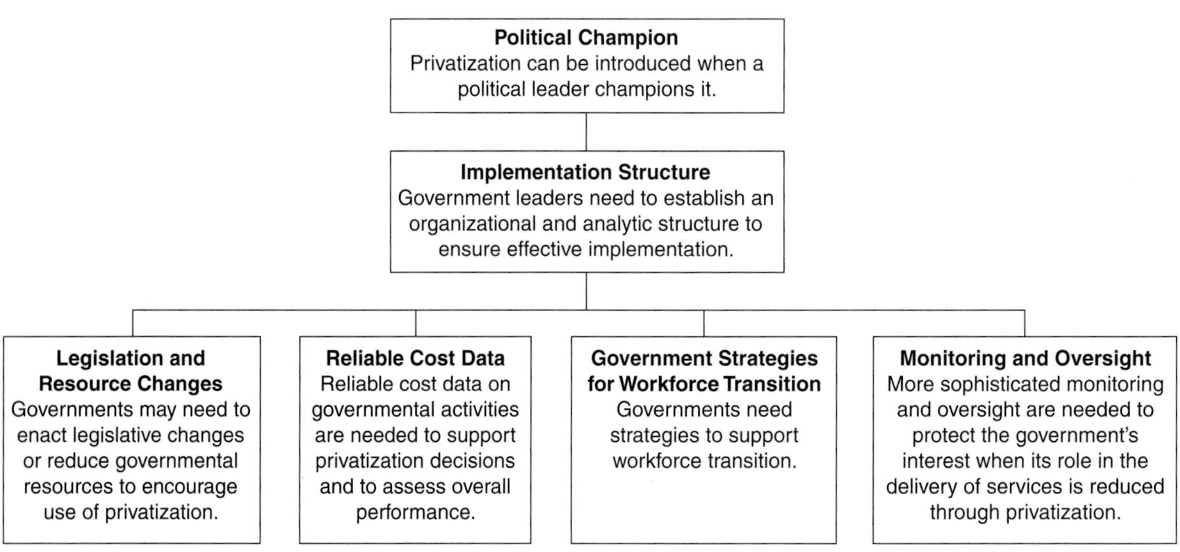

Figure 15.3
Factors necessary for a successful privatization effort

for privatization. Closely related, is whether there is widespread public support for the project itself. Trying a new method of service delivery will have its ups and downs, and sustained public and political support may be needed over a long period of time.

- There also needs to be a *champion for the change*, and the ability of the political structure to protect that champion.[22] The GAO study found that in the six governments they surveyed (five state and one local), the governor and mayor played major roles in introducing and leading the privatization efforts.
- *Stability of the government.* For developing countries, perhaps the most important requirement for privatization efforts is the stability of the governmental system, its markets, and currency. Although the U.S. capital markets are the most stable in the world, they are also subject to the shocks of the business cycle, as the events of 2008 and 2009 revealed. An expansionary cycle, where demand is soaring and would seem to go on forever, always comes to an end, and with it reduced revenues and the need to retrench. Although the government may step in with a strong infrastructure program to act as a counter-cyclic balance, this may not always be possible.

Implementing the Privatization Project

If the initial assessment of the political and institutional context proves positive, the agency can then move to take the next steps to implement the project outlined below.

- *Lead organization designated.* The lead organization should be designated to staff the effort .
- *Legislative changes enacted to support privatization/competition effort.* The lead agency needs to see that legislative changes necessary to implement the efforts actually happen. The GAO study found that legislation also signaled to the organizations involved that the politicians were serious about making these changes.
- *Cost out the impact of alternatives.* Having complete cost information for all the alternatives is a necessary element for services. It should be noted that most local accounting systems have a limited capacity to provide accurate data in this area. Therefore, most local officials will have to use forecasts and estimates to develop an assessment of whether privatization will reduce costs. Studies after the fact can be ambiguous about the extent to which privatization efforts have saved money.
- *Involve the affected workforce.* The workforce should also be involved in the process. Some local efforts have stalled based on the opposition of the unions and management, for example, but when the effort is

made to involve them (and when they realize they cannot change the decision to proceed but only how it is done), the result has been a productive attitude.

- *Effective contract monitoring function.* Another important element of success, which both the General Accounting Office (GAO) and the International City Managers Association (ICMA) studies found essential, is the ability to establish an effective contract monitoring function. Unhappily, the ICMA study shows that very little of this is done, and the GAO study echoes this as a concern by top officials. Establishing such a function requires retraining or hiring new staff experienced in contract management, which is altogether a different function from supervising in-house construction or maintenance staff. Frequently, in-house supervisors have risen up through the organization and may not have strong financial or contract management skills. Managing contracts also requires setting performance standards—usually where quantity and quality can be quantified. This is generally easier in the infrastructure area, but not altogether something that can be taken for granted.

Institutional Issues about Privatization

Five institutional issues need to be addressed to permit greater use of public–private partnerships in the United States—or to improve public sector efficiency—while protecting the public interest. (The following is based generally on Price, 2002.[23]) These include revisions to state contracting and procurement provisions, more sophisticated accountability systems for federal and local reviews, institutionalization of life-cycle costing, comprehensive planning at both the regional and state level, and continuing education on private provision.

- *Design-build enabling legislation needed.* The present design-build-bid contracting rules at the state level were developed as a reaction against fraud and corruption in the letting of government contracts. These rules effectively preclude design-build projects, a common organizational arrangement for large-scale infrastructure projects overseas. Some states have begun to authorize design-build contracts on an experimental basis, but this needs to be expanded. In California, in 1999, design-build was authorized for transit agencies, and for other large projects in some counties. A joint powers agreement has also been used in some places where one of the partners could obtain, or already had on the books, authorization to use design-build for construction projects. A word of caution here, however—design-build requires sophistication on both sides and is not suitable for localities with limited development experience and expertise. Design-build legislation should be coupled with capacity-building efforts at the local level to enable local agencies to use these mechanisms.[24]

- *More sophisticated (and efficient) accountability systems at the local level for development.* Local review processes for development have been built up over time, and often are not efficiently or consistently implemented. Although they were effective in the pre-information age to ensure that developers would follow health, safety, and aesthetic requirements, they are prime candidates today for a more performance-based approach. The element of uncertainty and risk these regulations pose for large-scale investment in public infrastructure in built-up areas is enormous. Yet, similarly, these regulations protect the interest of the locality, and cannot and should not be summarily sent to the trash heap. One step in the right direction was taken in Silicon Valley, where the electronics sector was instrumental in assisting the local governments to revise (streamline) their processing procedures and to adopt uniform building codes (but not zoning requirements) across the jurisdictions in the area. This was necessary because the product cycle for some businesses was six months or less, often the time it would take to get a building permit. Until the streamlining took place, some businesses just ignored the requirements and built without permits. The private sector feels that a similar constraint exists at the federal level for design-build efforts. The National Environmental Protection Act prohibits environmental contractors who prepare an impact statement from being involved in the design and construction stages of the project. While this avoids conflicts of interest in the review process, it also results in a loss of continuity of staff in the project, according to developers.[25]
- *Institutionalize life-cycle costing requirements.* Institutionalization of life-cycle costing requirements for public–private infrastructure ventures would protect the public agency and the private individual from irresponsible contractual arrangements. This should include requirements to adequately cost out maintenance responsibilities, as well as "exit" strategies for the private firm, particularly in costing out the value of the asset.[26]
- *Comprehensive infrastructure planning at the regional level.* Comprehensive planning at the regional level, at a minimum for water, sewage, transportation, and networked utilities, would provide an agreed-upon framework for growth and replacement. Therefore, these issues would not be raised for the first time when the public–private partnership is being discussed.
- *Capacity building at the local level.* Finally, the capacity at the local level to undertake these efforts is limited. Current educational and

experience requirements do not qualify many to be the project manager for the government in a complex and sophisticated arrangement—nor to even supervise someone with the requisite training. Public agencies and universities need to recognize this and make arrangements for appropriate in-service training as well as in the academic requirements for civil engineering and related professions.

Conclusion

In some respects, the entire discussion of public versus private for infrastructure services is artificial, driven by vested interests on one side of the debate or the other. If it does not actually provide the services, the government structures the capital markets, institutions, and regulatory environments that set standards for service. In the United States, where stable capital markets exist, the question is whether those with capital, whether it is the private investor or the taxpayer, are willing to commit the massive amount of resources required for the construction and operation of a large-scale facility.

Also at issue is the ability of the electorate to face up to the large-scale infrastructure needs of rapid growth. A similar point can be made about the need to retrofit, upgrade, and replace the systems in older urban areas. Without this awareness and commitment, it will not be possible to develop the institutional framework to ensure that new systems are effectively and efficiently designed. This means that the major infrastructure systems—transportation, water, and utilities—are located in accordance with a reasonable plan, and that they are developed and built in the most cost-effective manner.

Additional Resources

Publications

Delmon, Jeffrey. *Public–Private Partnership Projects in Infrastructure.* New York, NY: Cambridge University Press, 2009. An easy-to-read guide on public–private partnerships that contains the basics for someone interested in doing such a venture. Contains references to more in-depth guides.

Dowall, David E. "Rethinking Statewide Infrastructure Policies: Lessons from California and Beyond," *Public Works Management and Policy* 6, no. 1 (2001): 5–17. Useful policy advice for the state practitioner that can also be used at the local level. Set in California but the lessons are universal.

Osborne, Peter, and Peter Plastrick. *Banishing Bureaucracy.* New York: Addison-Wesley, 1997. An overview of the re-engineering movement for government. Also has practical techniques and strategies that can be used by the local practitioner to evaluate when and how privatization and competition can be used for infrastructure provision among other government services.

U.S. General Accounting Office. *Privatization: Lessons Learned by State and Local Governments*, Washington, DC: U.S. General Accounting Office, 1997. Although GAO only looked at five local governments and one state, this document presents a very good overview about what succeeds and what doesn't when the government is trying to "privatize" or become more competitive.

Websites

International City and County Managers Association: www.icma.org. The ICMA sponsors a survey every five years on local government delivery of services, whether they are public or private. This is reported on in the *Municipal Yearbook*. The last survey was done in 2007. The avid practitioner should monitor their website for the results of the 2012 survey.

Reason Foundation: www.reason.org. This website has a wealth of publications about private provision of government services by type.

Section TWO

The Major Infrastructure Systems

PART FOUR

Environmental Systems

The so-called environmental systems, those infrastructure systems concerned with the protection of human (and increasingly environmental) health, were among the earliest to be provided by local government in this country, and are generally the first considered when one reflects upon the term "infrastructure." They are also, according the ASCE report card, in the worst shape: water and wastewater systems both earned a D- in 2009.

Chapter 16 (Water Supply) begins with a description of water supply, treatment, and distribution systems, as well as water sources. It continues to a description of the institutional actors, and concludes with a discussion of conservation.

Chapter 17 (Wastewater and New Paradigms) contains a history of sewers and wastewater treatment and a description of the system elements. The institutions that provide and regulate wastewater treatment are outlined. Innovations in decentralized and distributed systems are described. The chapter concludes with a discussion of fees and financing and planning issues.

The next chapter, Chapter 18 (Stormwater and Flooding), addresses the increasingly important topic of managing the runoff from rain and snow, from the standpoint of pollution, water supply, urban design and adaptation planning.

The last chapter, Chapter 19 (Solid Waste Management), opens with a history of waste management, leading up to current practices and institutions. It then goes on to describe the technical and operational details of waste management policy. It closes with a discussion of local planning and programs.

chapter 16

Water Supply

Jeff Loux, Adam Leigland, and Vicki Elmer

In this chapter ...

Importance to Local Practitioners. 269
History and Current Issues. . . . 270
Supply and Distribution of Water. 270
Water Treatment. 270
The Fifth Paradigm. 271
Current Issues 272
The Elements of the Water System 275
Water Sources. 275
Water Transportation 278
Water Treatment. 279
Water Distribution 280
The Institutions of Water 281
Who Provides Water?. 281
Who Uses Water? 282
Who Regulates Water?. 283
Water Planning 284
Key Element of the Water Planning Process: Projecting Demand 284
Financing Water 285
Considerations for Local Government Land Use Planners 286
Influencing the Planning Process of the Water Planners. 286
Long-Range Land Use Planning 287
Review of Development Projects 288
Demand Management and Conservation. 289
Cost–Benefit Analysis of Conservation Options 290
Reduction of Water System Water Losses and Leaks 290
Public Education Programs . . 290
Indoor Residential Water Conservation. 290
Outside Water Use Conservation. 291
Conclusion 292
Additional Resources 292
Publications. 292
Websites 293
Key Federal Regulations 293

Why Is This Important to Local Practitioners?

Water infrastructure is going through a major shift from traditional "gray" highly centralized structures and institutions to decentralized practices and designs that integrate water into the fabric of the city in eco-friendly ways. This new approach moves away from silo-based thinking to one where multiple scales permit site, neighborhood, city, and regional solutions that integrate water, energy, waste, transportation, and food production with the rest of the built environment. Many in the water industry are working towards the "City of the Future" where water infrastructure contributes to the resilience of the city and where water engineers work side by side with planners and architects to develop new urban water solutions.

Part of this shift is in response to climate change, where water is a visible manifestation of the effects of global warming. Population growth and fluctuations in water supplies caused by climate change will strain the capacity of existing water sources, while new water sources will be increasingly difficult to access. The high cost of energy, the need to reduce carbon emissions, and the challenge posed by emerging contaminants such as pharmaceuticals and heavy metals in our water supply also require a new way of thinking about water infrastructure. The era of "water development" has been replaced by the era of "water management and reuse" where jurisdictions will struggle to stretch existing supplies in innovative and environmentally positive ways.

This chapter is limited to discussing water supply. It begins with a history of water provision and a discussion of these issues before turning to a description of how the systems of water supply, treatment, and distribution work. Water sources and accompanying issues are addressed before turning to the institutional context of water, including users, providers, and regulators. The responsibilities of the water agency are discussed before turning to the role of the land use agency with respect to water. The chapter concludes with the elements of a conservation and demand management program.

History and Current Issues

Supply and Distribution of Water

The concept of collecting, storing, and distributing water for human use is as old as civilization. Perhaps the most famous water system of ancient times is that of the Romans. Beginning in 330 BC, the Romans built vast aqueducts for transporting water from the mountains to their population centers—the longest carries water for 59 miles. Experts have calculated that at the height of the Imperial period, when the population of Rome was about one million, the water distribution system delivered over 250 gallons of water per person per day.[1] By comparison, in 1995, the per capita indoor residential water use in the United States was 70 gallons per day.[2]

The modern municipal water system, as we know it in the United States, originated in the late 1800s.[3] At first, waterworks were just as likely to be privately as publicly owned. However, as time went on, more and more systems came under public ownership. This was due to several factors, including the widespread failure of private systems to provide equitable and high-quality service, the increasing fiscal health of municipalities, and stronger regulations. Today, 82 percent of water systems in the USA are publicly owned.[4]

Water Treatment

Modern formal water treatment was first aimed at the aesthetic qualities of water, such as color and odor. In 1835, however, in London, Dr. John Snow determined that cholera was waterborne and in the late 1880s, Louis Pasteur developed the germ theory of disease, which explained how water transmitted diseases. Soon, water treatment focused on these new public health aspects. Filtration to remove turbidity (cloudiness) and the widespread use of chemical disinfectants, principally chlorine, significantly reduced the outbreak of waterborne diseases.[5] Today, 64 percent of U.S. community water systems disinfect with chlorine. However, chlorine is losing favor as concern grows over hazardous products formed during its use. Alternatives such as ozone and chloramine are taking chlorine's place.[6]

One Water Management

In the past decade water industry experts have advocated a new approach to water management with water services infrastructure integrated into the design of the city. The need to reduce GHG's and to adapt water infrastructure to the new future created by climate change has resulted in a more urgent driver for the changes that have been advocated by the New Urbanists for several decades. This vision includes a compact urban form that integrates nature with the city, and also integrates water, energy, and waste services. The result has been a transformation in approaches taken by leading-edge designers and the elites responsible for urban water infrastructure systems.[7] Water planning used to be the purview of engineers in the utility or water department but increasingly architects, urban designers, and planners are becoming involved at the building and site level. Leading-edge water practitioners envision an infrastructure system with regional, local, and site-based delivery systems.[8] Some are also calling on the water industry to recreate itself as a water-based, clean power industry.

Some authors call this the Fifth Paradigm[9] looking at the evolution of water systems since the beginning of civilization. Others use a more detailed framework, to describe the evolution of water infrastructure and the city (see Figure 16.1). Although a new concept for many in the United States, the conviction that our water infrastructure must change dramatically has gained currency in Europe and Australia. Recently, officials from the professional water associations in the United states have formed a water alliance to promote the concept of One Water Management.

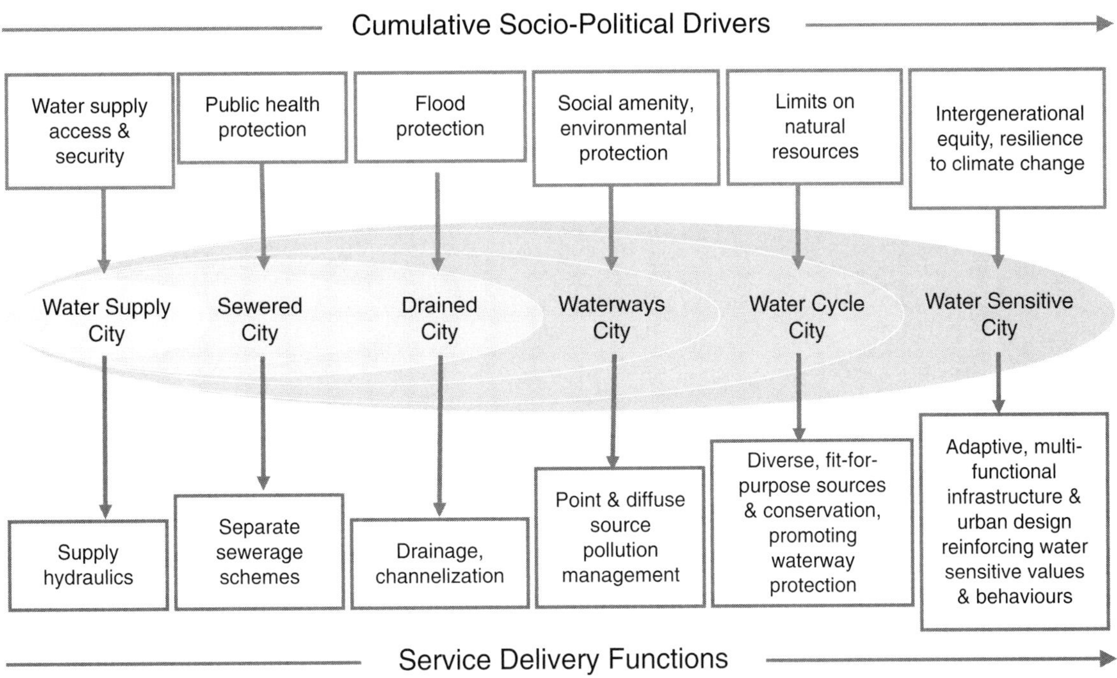

Figure 16.1
Changing the water paradigm

Current Issues

The issues facing water suppliers depend upon not only local and regional hydrologic conditions but also are heavily impacted by climate change and concerns about carbon emissions throughout the life-cycle of water provision.

Climate change. During the past 100 years, increasing amounts of hydrocarbons have become part of the atmosphere due to burning fossil fuels and volcanic activity. Hydrocarbons do not "rain out," but stay in the atmosphere indefinitely. Hydrocarbons block heat from leaving the planet, resulting in higher temperatures. In North America, average temperatures have risen 1 to 3 degrees in the past decades, and are projected to rise anywhere from 6 to 8 degrees more during the twenty-first century.[10] (See Figure 16.2.)

The result is a critical problem for water systems that rely upon "storing" precipitation in the form of snow in high mountainous areas for use during the summer months. In addition, as is described further in Chapter 18 on stormwater, weather patterns are becoming more extreme. Wet areas become wetter, and dry areas are prone to extended droughts. Traditionally wet areas may experience more rainfall, and yet have intense dry periods. Traditionally arid areas may also experience intense short-lived storms that stretch the

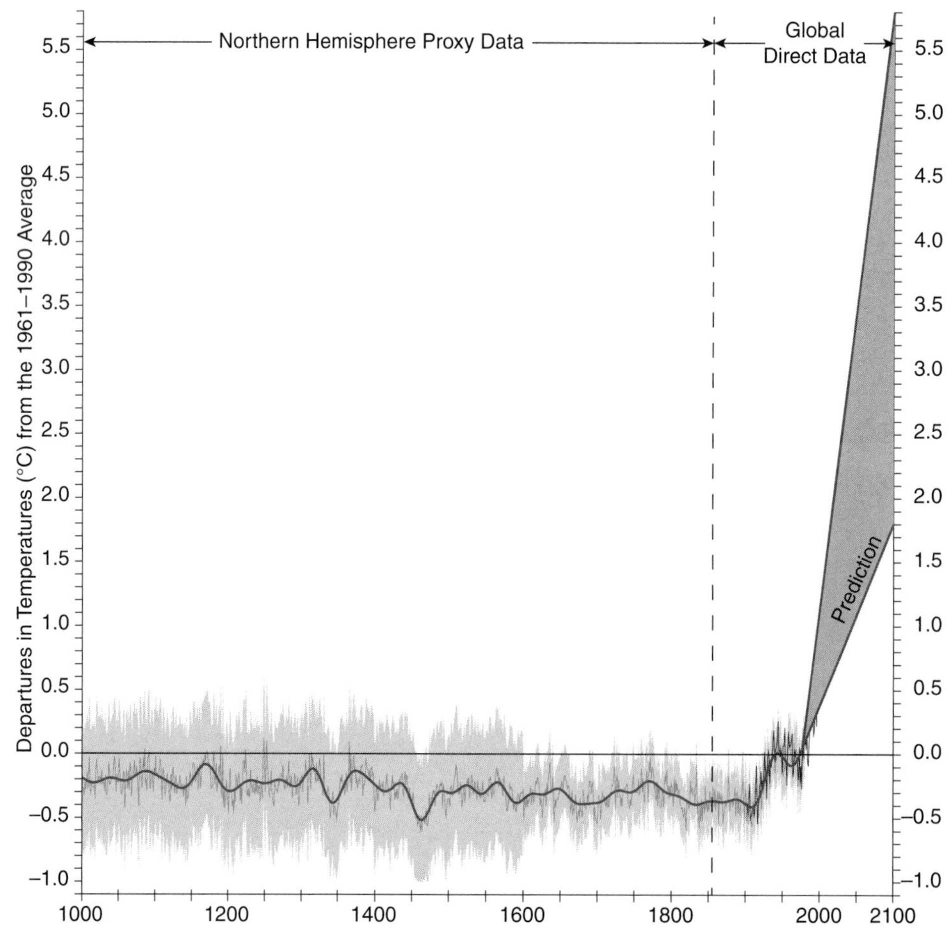

Figure 16.2
Variations in the Earth's temperatures from 1000 to 2000 AD and predictions for 2100

capacity of the infrastructure. For the water planner, uncertainty abounds.

In those areas with growing demands, matching future supplies to needs is becoming increasingly difficult. The "easy" water sources have either been appropriated or are presently in use. Environmental regulations make new surface water diversion, storage, and treatment projects more difficult, costly, and time-consuming to bring online. Meeting increased demands without major new water projects or radically new ways of providing water has been a challenge for many communities, and changing precipitation patterns will only make this more difficult.

Energy (carbon emissions) and water. The connection between water and energy is one that has only recently been on the radar for water planners. The water industry is both a consumer of energy while also being an energy producer with hydropower. Nationally, around 2–3 percent of power generation is used for water supply and treatment, and in drier parts of the country the percentage is much higher. For instance, in California, the water sector is the state's largest energy user, accounting for about 19 percent of total electricity consumption.[11] Drinking water and wastewater plants are typically the largest energy users in local governments, accounting for 30–40 percent of total energy. Energy as a percentage of drinking water system operating costs can be as high as 40 percent. This figure is expected to increase by 20 percent in the next 15 years due to population growth and more stringent drinking water regulations.[12] At the same time, however, hydroelectricity accounts for 6 percent of the nation's power.[13]

Surface water requires much more treatment than groundwater, and hence greater inputs of energy. Furthermore, surface water is often transported over longer distances and the pumping requires electricity. Water-consuming end uses, like water heating, dishwashers, and the like also require significant amounts of electricity and energy.

The collection, distribution, and treatment of drinking water and wastewater in the U.S. release 116 billion pounds of carbon dioxide (CO_2) each year—as much global warming pollution as 10 million cars on the road.

Water is also used to produce energy. The average amount of water used per kilowatt-hour of electricity production in the United States has decreased from 63 gallons in

Water and Growth Are Linked in the Arid West

In the arid west, the ability to develop reliable water sources was a major driver of settlement patterns and population growth. In California, the early missions were sited to take advantage of what little "native" water was available. The Gold Rush of the 1850s was fueled in part by the massive quantities of Sierra river water available for hydraulic mining operations and to transport slurries of minerals. The growth of the farming industry in California depended upon groundwater extraction technology and irrigation canals. The growth of both Los Angeles and San Francisco depended upon importation of surface waters from as far away as 300 miles.[1]

Note

1. Norris Hundley, *The Great Thirst: California and Water* (Berkeley, CA: University of California Press, 2001).

1950 to 25 gallons in 2009 and is projected to decline further by 2030 to less than 12 gallons by 2030. However, the total amount of water used nationally for energy production has increased due to increased consumption. It is estimated that total U.S. primary energy consumption will increase by 14 percent between 2008 and 2035, while demand for electricity will grow by 30 percent.[14]

Institutional disconnects between water providers and land use planning. Land use planning is the responsibility of the local general purpose government, which may have little or no control over the water agency. Local governments typically plan for future population growth for 20 years at the most. Water projects are usually planned 20–30 years in advance of need, and sometimes water planning has an even longer horizon—climate change is making some planners look to a 90–100-year time horizon. Water projects are a key determinant of growth. As a result, water providers often find themselves debating issues of land use and growth, and not exclusively water. Integration of land use and water use planning is likely to be an increasingly important issue, particularly in water-short areas where drought cycles and imported water are of concern.

Environmental water need. The need for new water sources often leads to conflicts between environmental protection, recreational uses, and the potential consumptive water use. This is especially acute in the arid west and southwest, where rivers and streams have highly variable flows, and where prolonged droughts are common. The battles over how much water can be taken out of the system for consumptive use and how much water needs to remain for aquatic health have raged for several decades in the western states.

Aging infrastructure and redevelopment/intensification. For older communities, a key issue is the age, condition, and capacity of the physical infrastructure supporting the water system. Facilities for water diversion, treatment, conveyance, storage, and distribution may have become obsolete, be in need of repair, or be incapable of current capacity requirements. The annual needed rate of replacement will peak in 2035 at 2 percent, which is four times the current rate of replacement. Nationally, this amounts to $250 billion in 2000 dollars over the next 30 years. For an individual utility, the fee increases needed to address replacement may be politically difficult.[15]

Emerging contaminants. An issue facing nearly every urbanized watershed (and many rural watersheds) in the USA is the discovery of pollutants in drinking water. Hundreds of chemicals from industrial and commercial uses such as dry cleaners and gas stations,

Groundwater Contaminants

In California, MTBE (methyl tertiary butyl ether) has shown up in wells across the state. MTBE was required to be added to gasoline in the 1980s to improve air emissions; it turned out to be a potentially carcinogenic compound, and has moved into many drinking water wells, rendering them unusable, and necessitating extensive remediation. Similarly, perchlorate, a byproduct of the jet and rocket fuel industry, has shown up in valuable drinking water aquifers. Near Sacramento, California, a perchlorate-contaminated system has taken out nearly half of the groundwater production of a private drinking water company. The analysis, modeling and remediation costs exceed $50 million annually and the complete aquifer clean-up will take over 240 years.[1]

Note

1. Douglas H. Benevento, "Remarks," in Second National Drinking Water Symposium (National Regulatory Research Institute, 2003).

as well as from agricultural uses, are being detected in water supply wells. When a well becomes contaminated by such chemicals, it is typically taken out of production. The purveyor must find an alternate water source, often at high cost in terms of infrastructure or environmental impact. In addition, trace amounts of pharmaceuticals and heavy metals are being found in surface water supplies.

Rural areas and small community systems. In isolated rural areas and small community systems where growth and change may not be at issue, questions of adequate water quality and the relation between water service and wastewater disposal may be of concern. Small systems do not have the economies of scale that larger utilities have for treatment, source development, and capital financing.[16]

The Elements of the Water System

A water supply system is composed of the water source, the treatment facility, storage facilities, and the conveyance and distribution system. The water source can be a river, a lake, wells, or some innovative sources that are starting to be used in drought areas (such as reclaimed water). Water storage is typically a surface reservoir, but increasingly water providers are taking advantage of nearby groundwater basins to store and retrieve water when needed. Water treatment is usually done at a centralized treatment facility although for some well systems, wellhead "in situ" treatment has become common. Distribution is accomplished with pipe networks, pumping stations, storage facilities, connections, and water meters.

Water Sources

Municipal water can come from a variety of sources. Reservoirs, natural lakes, rivers, deep wells, and shallow wells are common sources, and all have unique attributes and implications for raw water quality. If water comes from a lake or river, it is called surface water. If it comes from a well, it is called groundwater. Only 0.6 percent of the world's water is fresh and available for human use. Of that amount, 97 percent is groundwater. In the United States, 79 percent of community water systems draw from groundwater sources and 21 percent from surface water. However, surface water serves 67 percent of the population.[17] Both groundwater and surface water sources form a watershed.

Water Infrastructure Trends

- Large rate increases will be needed to replace and upgrade water pipes and facilities.
- Many water utilities will have funding difficulties.
- Regulations and economics will result in the consolidation of the smaller water providers.
- Raw water sources will be impacted by environmental concerns.
- Small treatment units and point-of-delivery devices will become important.
- Conservation and water recycling will increase.

Steve Albee, "Future Water Trends," in *Water Conference* (Washington, DC: U.S. Environmental Protection Agency, Office of Water, 2003).

Sobrante Ozonation Facility, El Sobrante, CA

Exchequer Dam in Merced, CA

Table 16.1
The world's water

Water sources	Water volume (mi³)	Water volume (km³)	% of freshwater	% of total water
Oceans, seas, & bays	321,000,000	1,338,000,000	—	96.54
Ice caps, glaciers, & permanent snow	5,773,000	24,060,000	68.6	1.74
Groundwater	5,614,000	23,400,000	—	1.69
Fresh	2,526,000	10,530,000	30.1	0.76
Saline	3,088,000	12,870,000	—	0.93
Soil moisture	3,959	16,500	0.05	0.001
Ground ice & permafrost	71,970	300,000	0.86	0.022
Lakes	42,320	176,400	—	0.013
Fresh	21,830	91,000	0.26	0.007
Saline	20,490	85,400	—	0.007
Atmosphere	3,095	12,900	0.04	0.001
Swamp water	2,752	11,470	0.03	0.0008
Rivers	509	2,120	0.006	0.0002
Biological water	269	1,120	0.003	0.0001

Source: Igor Shiklomanov, "World Fresh Water Resources," in Peter H. Gleick (ed.), *Water in Crisis: A Guide to the World's Fresh Water Resources* (New York: Oxford University Press, 1993).

The Water Cycle

The water cycle is the process where water evaporates into the air, becomes part of a cloud, falls to the earth as precipitation, and evaporates again. This repeats in a continuous cycle. Water is warmed by the sun and changes from a solid to a liquid to a gas. Precipitation (rain or snow) creates runoff that travels over the surface of the ground and fills lakes and rivers. It also percolates through the soil to refill underground aquifers. Some places receive more precipitation than others. These areas are usually close to oceans or large bodies of water that allow more water to evaporate and form clouds. In colder, mountainous areas, as the water vapor freezes, it condenses into snow that falls on the peaks and is stored there until all or part of it melts in the spring (Figure 16.3).

In the past ten years, desalination has also become a water source but it is not viewed as a sustainable solution. See Table 16.1 for a comparison of freshwater sources to all other sources of water on Earth, including the oceans and the atmosphere. The amount available for human consumption is a small portion of the total.

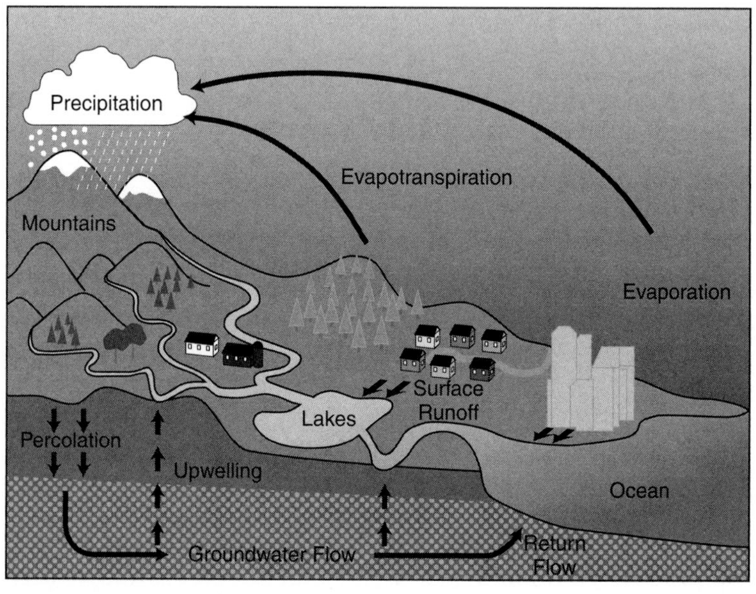

Figure 16.3
The water cycle

The American Waterworks Association, modified by the authors.

Surface water. A source of surface water can be a river, stream, or lake. It is diverted directly from a stream, river, or canal and stored in natural lakes, reservoirs, and sometimes in aquifers. Water is generated from snow or rainfall, and then diverted from these storage areas throughout the year as needed. There is usually an intake structure in the river, lake, or reservoir that connects to a nearby storage facility and/or to a water treatment plant. Sometimes water is collected from alluvial deposits next to streambeds. Water is filtered through the soil to an "infiltration gallery" and thence to the treatment plant.

Groundwater. Groundwater supplies the majority of drinking water in the United States, but only 21 percent of all uses. It is found in underground basins and aquifers that are replenished primarily by precipitation. Groundwater is found underneath 40 percent of the land surface and may be tapped into by many individual wells, as well as large district well fields. Aquifers often consist of many interconnected layers of materials with varying degrees of storage capacity, permeability, and connectivity. Permeable layers are often interspersed with impermeable clay layers. Groundwater can serve as a supplement to surface water and if the groundwater basin is replenished in wet years, this system can work well. However, increased use may result in an "overdraft" condition for the groundwater, when more water is being taken out than is flowing in.

Overdrafting has occurred in aquifers throughout the United States, in some places for over 100 years.[18] Water levels in aquifers have dropped between 300 and 900 feet during the past 50 years. Portions of the Mississippi River aquifer have dropped 100 feet in the past 100 years, while aquifers in the San Joaquin Valley in California, and in Baton Rouge, Louisiana, have dropped over 200 feet. Groundwater levels in Chicago and Milwaukee have declined by 900 feet[19] (Figure 16.4).

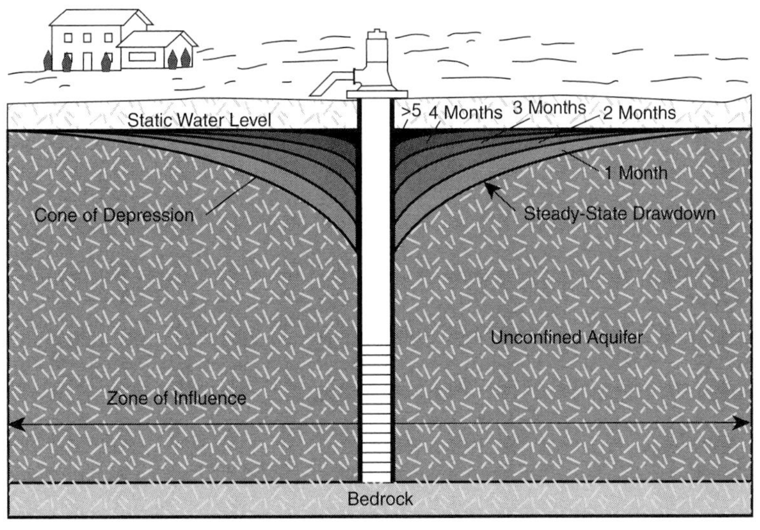

Figure 16.4
Water table drawdown from pumping and time to recovery

Conjunctive use. This is an increasingly important tool to accommodate more demand without increasing withdrawals from streams. The water purveyor withdraws water from the stream or river when it is plentiful and does not impact aquatic resources or affect recreational uses. During lower flows, the purveyor switches over to groundwater and "gets off the river." This allows the water to remain in the river or stream to protect aquatic resources and fisheries. This is also called in-lieu recharge—using groundwater instead of surface water and vice versa.

Desalination.[20] In 2000, 15 percent of all water withdrawn was saline compared to a relatively minor amount in 1950. Desalination is the process of removing salt, other minerals, or chemical compounds from impure water, such as brackish water or seawater, for the purpose of creating potable water. Depending on the technology used, the final water product is generally of high quality.

However, desalination is expensive because of the corrosion-resistant capital facilities required, the large amounts of chemicals needed, and the high energy requirements. The costs substantially exceed those for traditional water sources in most areas. Desalination also results in the need to dispose of large volumes of brine (waste salts), typically by diluting them back into the ocean.

The energy required for seawater desalination is higher than the energy expected to be required for reusing treated wastewater effluent or interregional transfers.[21]

Water reuse and rainwater harvesting. These sources of water are discussed in Chapters 17 and 18 on wastewater and stormwater.

Water Transportation

From the source, water is transported to the municipal region. The transport mechanism can be a channel, a pipe, a flume, a tunnel, an aqueduct, or a combination. The choice of a particular method depends on topography, cost, and amount of water to be carried. Water usually arrives at the water treatment plant in channels open to the atmosphere, flowing under the force of gravity only.

Water line under construction

Many large metropolitan regions have to reach far into their hinterland for adequate water supplies. New York City gets its water from the Croton Reservoir, 38 miles away, as well as from the Catskill/Delaware Watershed, which is 125 miles away in upstate New York (Figure 16.5). San Francisco gets its water from the Sierra Nevada, 150 miles to the east. Los Angeles gets part of its water from the

Figure 16.5
New York City's water supply system showing the Catskill and Croton watersheds

Owens and Mono valleys through a 338-mile aqueduct, and part of its water from the Colorado River, also several hundred miles to the east. If pumping is required to transport the water, it will add both to the cost of the water and the carbon emissions for the water supply.

Water Treatment

Water is treated so that it is free of chemical and microorganisms that can harm humans. It can also be treated to provide adequate quality for industrial uses. Surface water might have more bacteriological content than groundwater, but less mineral content. All surface water must be treated according to the Safe Drinking Water Act and any additional state or local health standards. In addition, certain industrial processes might demand a much higher degree of purity than human consumption requires.[22]

Some groundwater-based systems also use treatment facilities. There are technologies that allow "in-situ", wellhead treatment of groundwater for naturally occurring contaminants in the soil or aquifer, such as arsenic or selenium.

Water Distribution

From the treatment plant to the household, the water is distributed under pressure. The pressure is provided either by pumps, or more commonly, elevated storage tanks, such as the water tower on a ridge at the edge of town. One of the main needs for pressure is water for fighting fires, which may result in substantial additional costs to store and distribute water. One advanced concept for water distribution systems is a dual system—one for drinking water and the other for fire fighting. This might be less expensive since drinking water pipes are smaller and require less pressure than those for fighting fires. The concept might also improve drinking water quality since it would reduce the amount of time it is "stored," thereby reducing the opportunity for the growth of pathogens, as well as the need for excessive disinfectants.[23]

Actual distribution is through a network of interconnected pipes and pump stations (Figure 16.6). The network can either be a grid system or a branching system determined by the topography, street layout, and type of development. Water lines vary in size, depending on the application, and can be anywhere from over 18 inches in diameter for the large mains to ⅝ of an inch for domestic service lines.[24]

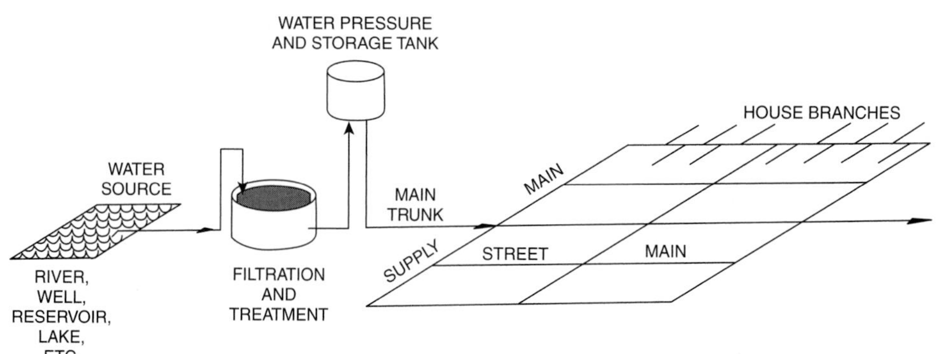

Figure 16.6
A typical urban water distribution system

The Institutions of Water

Who Provides Water?

In contrast to many other countries, water provision in the United States is provided by a decentralized system of multiple agencies, both public and private.[25] Of the 170,000 public drinking water systems in the United States, only 54,000 serve more than 25 people a day all year round. EPA calls the latter "community" water systems. The rest are systems that serve campgrounds, schools, and other small users.[26]

Small systems that serve less than 10,000 people form 93 percent of all community water systems, but the remaining 7 percent serve 81 percent of U.S. customers. In sheer numbers, most water supply systems are privately owned—53 percent of systems nationwide in 1994. Public ownership dominates the total amount of water provided in the United States, although there is a continuing interest in privatization.

Organizational arrangements for publicly owned systems. A water agency can be a retailer, a wholesaler, or a combination of the two. Many of the large western water districts are wholesalers, meaning they purchase water from the federal and state government and sell it to local entities such as cities, smaller water districts, or private water companies. Some large districts are both wholesaler and retailer, selling some water to cities and districts along its pipelines and canals and some directly to customers. Some systems integrate state or federal government agencies, water districts, jurisdictions, and even private companies in the "supply chain."[27]

The local water agency may control all or some of the different links in the water supply chain, including: holding the actual water rights; owning and managing the dams, reservoirs, and aqueducts; owning and managing the treatment plants; and the distribution system. The water agency may also own the land that forms the watershed. Local or regional water districts might own and operate the conveyance systems such as canals, pumping stations, and pipelines. The local general purpose government might sell the water to the customer and own and operate the terminal storage reservoirs, treatment facilities, and distribution lines.[28]

Staffing and governance. Local water districts or agencies are usually governed by an appointed commission or an elected board unless they are part of a city or county agency. The commission or board sets general policies. The day-to-day operations are managed by a professional staff, usually dominated by managers, engineers, and technicians. However, planners are becoming more frequent in water agencies as policy issues regarding water become more urgent.[29]

Privatization and water. Investor-owned water facilities account for about 14 percent of total water revenues and for about 11 percent of total water system assets in 1995. The National Association of Water Companies estimates that the total amount of water provided by private water companies in the United States has been around 15 percent since World War II. Experts conclude that continued public ownership and operation is the most probable future for most water utilities. They note that attempts to privatize water utilities will likely spur the public systems to improve their performance.[30] Small utilities perhaps have the greatest potential for consolidation, privatization, or coordination with another provider, given that 86 percent of them are within five miles of another system.[31]

Who Uses Water?

Water use is typically broken down into eight categories: public supply, domestic, commercial, irrigation, livestock, industrial, mining, and thermoelectric power cooling. Total water use in the United States in 2005 was about 410,000 million gallons per day. The states of California, Texas, Idaho, and Florida accounted for more than one-fourth of all water withdrawn in the United States in 2005. Irrigation accounted for over half as 53 percent of the total withdrawals in California were for irrigation while 28 percent were for thermoelectric power. Most of the withdrawals in Texas were for thermoelectric power—43 percent, and irrigation at 29 percent. Most of the water withdrawals in Idaho were for irrigation (85 percent) while thermoelectric power accounted for 66 percent of the water withdrawals in Florida.[32] See Figure 16.7 for a graphic representation of withdrawals in 2005 with and without thermoelectric

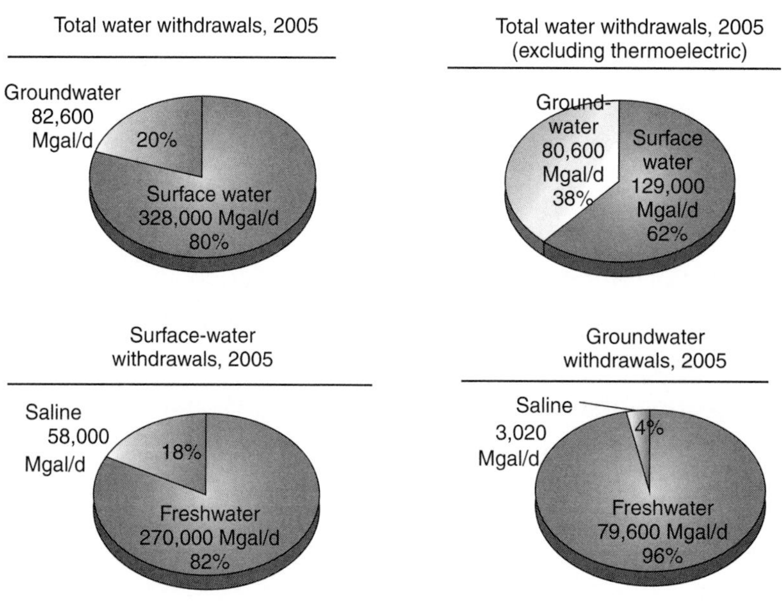

Figure 16.7
U.S. water withdrawals in 2005

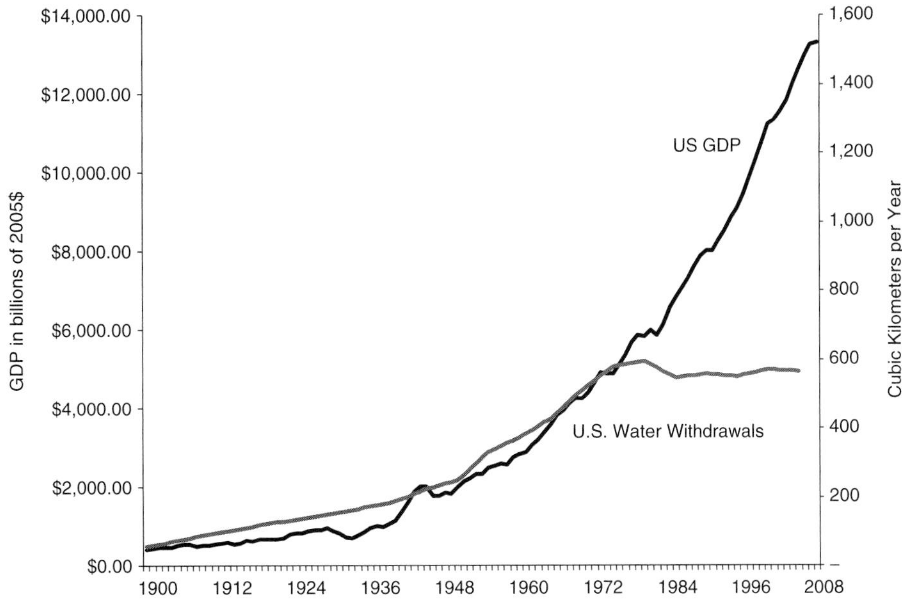

Figure 16.8
U.S. GDP and water withdrawal, 1900–2008

uses factored in. Both saline and freshwater withdrawals are shown and divided between ground and surface waters.

During the past 20 years, per capita daily use has declined after rising steadily from 1900 onward. Similarly, the gallons of water per dollar of GDP has also declined (Figure 16.8). Whether this is attributable to conservation measures or the outsourcing of many manufacturing processes and food production is a source of debate.[33]

On average, in 2000, residential use accounted for about 11 percent of total water consumption. The two largest water use categories are for thermoelectric power (47 percent) and irrigation (33 percent), not counting the "environmental" use.[34]

Direct per capita urban water use in the USA tends to be between 150–500 gallons per capita per day. Residential consumption is divided between 42 percent for indoor purposes and the 58 percent used outdoors. Another important "use" is to maintain the existing ecological systems and environment. In California, typical calculations suggest that the environment "uses" about 40 percent of the available water.[35]

Who Regulates Water?

The major legislative mandate for water are the Clean Water Act of 1972, and the Safe Drinking Water Act of 1974. The Environmental Protection Agency has the lead to develop regulations and provide resources to local water departments and agencies. Enforcement has been delegated to most states, and within the larger states again to counties. Water rights are regulated at the state level. Local governments also play a role in regulating water use through their police powers and land use authority. (See p. 293 for a list of federal legislation pertaining to water.)

IRP as defined by the American Waterworks Association

Integrated resource planning (IRP) is a comprehensive form of planning that encompasses least-cost analyses of demand-side and supply-side management options as well as an open and participatory decision-making process, the development of water resource alternatives that incorporate consideration of a community's quality of life and environmental issues that may be impacted by the ultimate decision, and recognition of the multiple institutions concerned with water resources and the competing policy goals among them. IRP attempts to consider all direct and indirect costs and benefits of demand management, supply management and supply augmentation by using alternative planning scenarios, analyses across disciplines, community involvement in the planning, decision-making and implementation process, and consideration of other society and environmental benefits.

American Waterworks Association, "Total Water Management", Mainstream (June 1994).

Water Planning

Water districts (or departments, boards, or agencies) were at one time focused only on providing high quality reliable infrastructure and service to whatever new demands were requested of them. However, in the early 1990s, the industry began promoting the concept of integrated resource planning (IRP).[36]

Integrated resource planning calls for demand management strategies to be considered along with other supply alternatives during the planning process. In its ideal form, the process includes a wide variety of stakeholders, such as the multiple institutions that are involved with local water and wastewater services. It is a process that would be quite familiar to the land use planner.[37]

During the same time period the concept of total water management began to gain currency in the industry. Total water management looks at water as a limited rather than unlimited resource. More recently, the ideal water planning process has been expanded to include asset management concerns, and capital budgeting.[38]

Key Element of the Water Planning Process: Projecting Demand

Water planning follows the same steps outlined in an earlier chapter for an infrastructure plan. A key aspect, however, that warrants further discussion is projecting demand.[39]

Population projections are the most commonly used because the data are easy to acquire and because of historical precedent. Total population estimates are multiplied by per capita water demands (which can be calculated based on existing usage) to arrive at future projections. The limitation of population-based projections is that they are only estimated accurately every ten years by the U.S. Census. They are also averaged over all land uses, and are difficult to track for individual development projects. Separate studies are needed to project the impact of conservation or water recycling savings.

Socio-economic models use sophisticated computer simulations to project jobs and housing growth and changes in consumption patterns and trends. These models tend to be more precise than population-based projections, but are costly to develop and are not easy for the public to understand. Like population-based projections, it is often difficult to link the demand to a geographic area or development proposal.

Land use-based projections are the most publicly accessible, most directly linked to the land use planning process, and the most easily disaggregated for detailed analysis of water demands and conservation savings. The benefits of a land use approach is that it is accurate, allows for detailed water system planning, is easily updated (especially if GIS is used), and can enable planners and the public to see exactly which land use proposals are covered by water supplies in the planning documents. The disadvantages of this approach are that it is time-consuming upfront, and that community land use plans are often not a complete representation of a community's growth.

Financing Water

The water supply industry is the most capital-intensive industry in the nation, even more so than the power and telecom industries. Water suppliers invest more than $3.50 for every dollar of annual revenues received from customers.[40]

Historically, municipal water supply agencies have relied on user fees for operating revenue and bonded indebtedness for capital investments. Today, most local water agencies meter use and charge by volume.

Full cost pricing. In the past, water has been underpriced, that is, subsidized by other levels of government. In addition, water prices have not reflected the full social and environmental cost of provision. Water, sewerage, and solid waste disposal together account for a relatively small share of the average household budget—less than 0.8 percent of total expenditures. By comparison, electricity accounts for 2.4 percent and telecommunications for 2.1 percent. In fact, on average, a four-person household spends about the same amount each year on cable television and tobacco products as on water services.[41] Today, most economists advocate full cost pricing where all costs, including operations, maintenance,

Rate Shock in Colorado

Moving to a full cost of service rate structure may result in "rate shock." For example, the city of Longmont, Colorado, increased its fee structure by about 40 percent in 1989, which according to one official was "pretty painful and not particularly well received." However, the rate was designed to accumulate funds for a modern water treatment plant without another increase until 2009. The city's program, also includes conservation measures and development fees for new growth.[1]

Note

1. Ari M. Michelson, J. Thomas McGuckin, and Donna M. Stumpf, *Effectiveness of Residential Water Conservation Price and Nonprice Programs* (Denver, CO: AWWA Research Foundation, 1998).

and capital costs are recovered with subsidies for low-income households where necessary.

Fee structures.[42] The uniform rate, declining block rates, and the flat rate do not promote conservation. Uniform rate structures charge the same price per unit for all residential customers. The uniform rate and declining block rate structure usually has a base fixed cost used for capital facilities, while the variable fee is used for operations and maintenance. Another type is a flat rate, which is usually used by smaller or rural agencies without water meters. In 1996, 39 percent of water agencies still used a uniform rate, and 33 percent had declining block rates, but the number using flat rates to charge for water had declined to 4 percent.

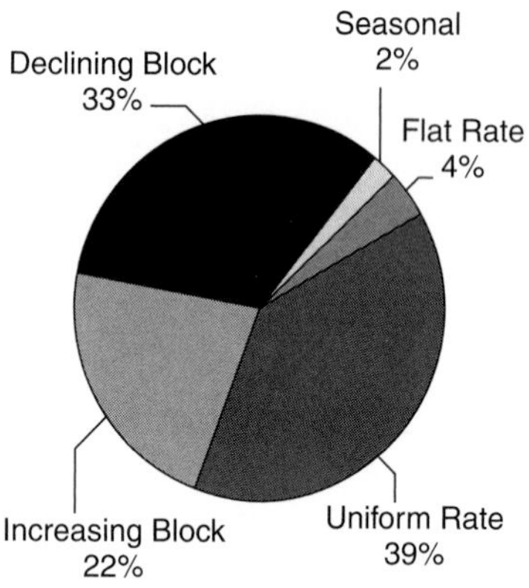

There are four pricing strategies or fee structures that can be used to promote conservation: (1) increasing block rates; (2) time of day pricing; (3) water surcharges for use higher than average; and (4) seasonal rates.[43] Over one-fifth of the water agencies in the United States used increasing block rates while another 2 percent employed seasonal price structures[44] (Figure 16.9). Rate structures differed by residential, commercial, and industrial use (Figure 16.10).

Using conservation rate structures can reduce water demand by 7–20 percent. However, it may be difficult for a utility not facing an imminent crisis to switch from a uniform rate structure to an increasing block rate, or "inclining" rate structure. If the agency does not meter water use, this is a good first step toward conserving water and reducing demand.

Figure 16.9
Percentage of agencies using different fee structures in 1996

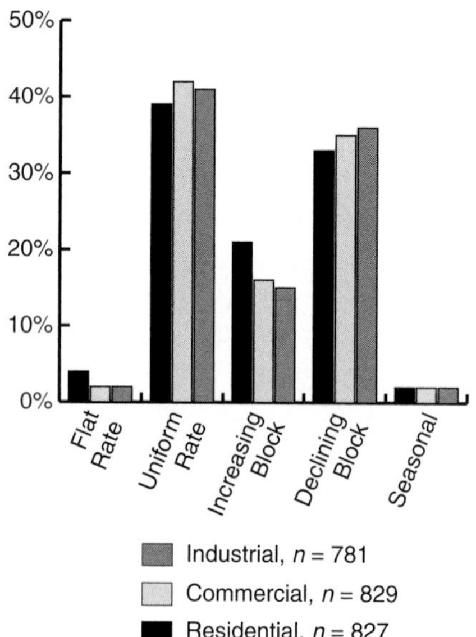

Figure 16.10
Water rate structures by residential, commercial, and industrial use

Considerations for Local Government Land Use Planners

There are several roles that land use planners can play with regard to water supply. The first is to coordinate with the local provider to ensure that the water district's plans reflect the development and conservation vision of the local general purpose government, and that its water facilities are consistent with the local land use plan. The second is to be sure that the GPG's long-range or comprehensive land use plan is linked to adequate water quantity and quality. The third is to ensure that when specific development proposals are made, there will be adequate water supply and site-specific infrastructure at the time they are needed for future users.

Influencing the Planning Process of the Water Planners

In full-service cities (and some counties), the city is responsible for water supply and for land use and growth decisions. In these cases, the planning department needs to coordinate with the public works or utilities department—sharing data, working cooperatively on project review, and keeping consistent land use plans and water supply plans.

Coordination is more difficult when the land use agency and the water district are separate. In either case, however, the city or county land use planners need to share data, plans, and information on comprehensive plans and development proposals. Information exchange can be challenging when very small water purveyors are involved who have not had the staff or budgets to conduct extensive planning studies. Private water companies also are not subject to the same public disclosure requirements and may not have completed the same level of long-range planning and reliability analyses as a public agency has.

Reviewing an application or proposal for a new water facility. The local general purpose government has the responsibility for reviewing the application of the water department or agency as it builds individual water projects. If the project is a large-scale treatment plant or a new reservoir, it is likely the project will go through the use permit process. If the project is part of the collection or distribution system in the right-of-way, the review may be non-discretionary and performed by the engineering department.

Location of a water treatment facility is likely to be set largely by the location of the water source, cost, and engineering feasibility issues. Satellite or cluster treatment plants that are managed by the central utility but located close to the user are becoming more common.

Long-Range Land Use Planning

The community's general plan or comprehensive plan is the blueprint for future growth and development. For many communities, especially in the arid west and southwest, water supply is a factor in how much the area can grow, where it might grow, and at what densities. Land use planners can also prepare a water element as part of the community's comprehensive plan. A water "element" can be a single planning document or chapter in the general plan where all aspects of the local hydrologic cycle can be described and policies adopted for water supply, groundwater and source watershed

Key Elements of a Water Planning Process

Although the context and values of the policy-makers will determine the scope and issues dealt with in the planning process, today's planning process for water infrastructure should address:

- growth needs;
- renewal and replacement needs;
- compliance with federal, state, and local regulations;
- sustainable supplies;
- stakeholder goals;
- energy use of alternatives;
- changing climate conditions;
- adaptation needs.

Legislation Requiring Linkage between Land Use and Water Supply

Five western states have passed legislation that requires demonstration of adequate water supply when approving residential subdivisions. *Arizona's* law, known as the "Assured Water Supply program," was passed in 1995; it prohibits new subdivisions in the Phoenix, Tucson, and Prescott areas without use of "renewable" (i.e. surface) water. *Colorado* passed the Subdivision Act of 1972 (SB 35), requiring all counties to adopt subdivision regulations, including a requirement that "adequate evidence that a water supply that is sufficient in terms of quality, quantity, and dependability will be available." *Nevada* passed the 1973 Subdivision Act (NRS 278.335), which requires state-level approval of all tentative subdivision maps for water supply. *New Mexico* passed the 1995 Subdivision Act (HB 1006), requiring all counties to adopt subdivision rules with water supply requirements meeting certain criteria. In 2001, *California* passed Senate Bills 221 and 610, designed to address water supply adequacy for large subdivisions (over 500 units) and other projects at earlier stages of planning. Each of these laws is designed to ensure that adequate long-term water supplies exist prior to the approval of new housing projects. In all cases, except California, there is a state-level review of the water supply adequacy. In California, the findings are left up to local jurisdictions. In some cases, "long term" is defined: Colorado requires a 100–300-year supply of water, and in New Mexico a 30–100-year supply must be assured. Some of the statutes affect all residential subdivision regardless of size: in Arizona 100 percent of all subdivisions in the affected water-short areas are reviewed by the state, while in California only subdivisions over 500 units are subject to the legislation.[1]

Note

1. Based on Ellen Hanak, *Water for Growth: California's New Frontier,* Public Policy Institute of California, 2005; and Ellen Hanak, "Show me the Water Plan", Golden Gate University Environmental Law Review (4)1.

protection, wastewater, drainage and runoff controls, and water quality.

Review of Development Projects

The local general purpose government has the responsibility of ensuring that adequate water is available for new developments such a subdivision, or a large-scale change of use in a built-up jurisdiction.

When planners review area master plans, specific plans, and individual development proposals, they need to ensure there is a reliable, sustainable source of water, and that the infrastructure for treatment and delivery will be in place when development occurs. While it may seem obvious that development applications should not be approved without adequate supplies of water, in several states this did occur, prompting new regulations at the state and local levels.

What to look for in water supply analysis for new development. The type of questions that should be asked during the permit review process by land use planners include:[45]

- Are all current sources fully "on line" and producing water?
- Are there dry year limitations with any sources, water rights or environmental limitations, or anticipated dry year cutbacks or delivery issues?
- Are there potential water quality risks (especially for groundwater) that might affect the use or availability of a source?
- When planned water sources are included in a water supply assessment, are they fully funded and permitted?
- If agricultural water transfers are part of the water supply reliability estimate, have all legal and permit hurdles been cleared?
- What is the nature of the transfer and for how long is it available?
- Has the water supplier conducted a risk assessment of the water supply portfolio?
- What "margin of safety" has the water supplier considered in each type of hydrologic year?
- Does the water supplier have a plan for a prolonged drought?

- What type of cutbacks or auxiliary sources (such as agreements with neighboring districts) are planned for?
- Are all of the local facilities in place, or at least financed, to treat, store, distribute, and deliver water?
- What share is being financed by the development community and what share is being financed by future rate payers?

A current list of planning applications and development proposals showing the status of each major project, its size and proposed land use mix, location, required permits, and specific water needs can serve as a tracking system for land use planners, water planners, other infrastructure providers, and the public.

Demand Management and Conservation

Today, most forward-thinking water and land use planners advocate a comprehensive demand management program that includes a combination of regulations, incentive programs, technological advances, and public education. Some call it a "soft path" that complements the "hard path" of centralized capital investments in water infrastructure. The soft path looks for decentralized solutions in human behavior and new technologies—often "green"—that will improve the productivity of water use. This approach applies not only to residential customers but also to industrial and commercial uses.[46]

Demand management and conservation strategies (often called water use efficiency programs) are the first step to take when more water is needed in the jurisdiction. For example, the Southern California region has added over 3 million water customers in the past 20 years, yet has not increased its total water supply.[47] Today, conservation and water efficiency programs are mainstream ideas.

California's Water Supply Availability Statutes: SB 610 and SB 221

Two California statutes require information about water supply availability prior to approving certain development proposals. Senate Bill 610, passed in 2001, requires that the land use agency include a water supply assessment for any development project of over 500 residential units, or the equivalent of commercial, industrial, or other development, as part of the environmental assessment process. Within 30 days of the request by the land use agency, the water agency must evaluate whether there is an adequate supply of water to serve the project, taking into account present and planned growth. If the water supply is insufficient, the water provider is required to document how it will develop new water sources or otherwise meet the overall water demands. If there are documented water deficiencies, the local land use agency can still approve the project, but it must ensure that water supply information is provided as part of the record. A similar statute (SB221) emphasizes the responsibility of the land use agency to find water that is adequate for the development, including finding that the water provider has valid entitlements (water rights), has the funding, and has all permits and approvals for the source before approving the development project.

Water Conservation in Colorado

The State of Colorado's Water Conservation Act of 1991 requires that retail water agencies that supply 2000 acre feet of water a year to commercial or residential customers adopt water conservation plans with the following as suggested activities:

- water-efficient fixtures and appliances;
- low water-use landscapes and efficient irrigation;
- water-efficient industrial and commercial water-using processes;
- water reuse systems for potable and non-potable water;
- distribution system leak repairs;
- public information programs about water use efficiency;
- water rate structure to promote conservation;
- local land use and building regulations to promote water conservation;
- incentives, including rebates, to customers using conservation methods.

Cost–Benefit Analysis of Conservation Options

Many utilities use a demand analysis model to evaluate different mixtures of conservation activities. This is called an *end use analysis.* One set of authors who have used this tool in a variety of settings both in the USA and in Australia reached the following conclusions about general results:

- Proper plumbing fixture requirements save large amounts of water/wastewater.
- The majority of program benefits come from deferring new supply/treatment projects.
- Measures that reduce peak demand can produce high benefit–cost ratios.
- Technology to reduce demands significantly is readily available.
- Demand reductions through conservation of 10–20 percent over 20–30 years is possible.[48]

Reduction of Water System Water Losses and Leaks

A good first step in a conservation program is to address water loss, which can account for as much as 10–15 percent of water "use." One Massachusetts water agency discovered that 30 percent of water use was due to leakage.[49] Agency water loss can be caused by evaporation from a storage area, seepage from an unlined channel, or leaks in a distribution system. Agencies have adopted on-going leak detection and repair programs and have made efforts to reduce evaporation in storage facilities. Other programs can include "audits" for residential facilities to identify leaks in private lines. A San Francisco Bay Area water agency estimates its leakage rate was about 15–20 percent in 2008, with a significant portion caused by leaky toilets.

Public Education Programs

Public media messages, internet ads, printed, and other materials that increase public awareness and encourage conservation, and strong school education programs have all been used by local governments and water agencies. For example, the City of Albuquerque set up a massive public education program when it learned that the local groundwater supply was going to run out much earlier than expected. The city's goal, according to their mayor, is to become "the most water-conscious city in the West."

Indoor Residential Water Conservation

As noted earlier, indoor residential water use accounts for 42 percent of total water residential water consumption. The plumbing fixtures and appliances used inside the house therefore are a prime target for conservation. In the

City of Albuquerque, toilets use 26 percent of total water, while showers account for another 17 percent, faucets 16 percent, and the clothes washer 22 percent.

Many localities have developed ordinances that prohibit wasteful urban water use practices and can be enforced by the water agency or the general purpose government. Local ordinances can also require the use of low flow plumbing fixtures for all new construction and rehabilitation as part of the building code.

Water agencies often get grants or loans to offer low cost access to new technologies that are low water use, such as low flow toilets, low water use washing machines, and low flow spray and rinse systems for restaurants. Rebate and retrofit programs can be sponsored by local governments for remodeling, using water-efficient fixtures. The City of San Diego requires homeowners to retrofit or certify that a home has low-flow fixtures when it is sold.[50]

Outside Water Use Conservation

Many local governments have focused on the reduction of irrigation for conservation measures since, as noted previously, it accounts for 52 percent of residential water use. Although studies show that urban landscapes are underwatered by some 58 percent, few would argue that when plant selection, irrigation use, and related landscape choices do not reflect local conditions, water can be "wasted."

Land use ordinances that promote smaller lot sizes are one way of addressing the problem. Other landscaping provisions run the gamut. In the arid west and southwest, native, drought-tolerant, and water-conserving landscaping has taken on a high profile as a means to conserve supplies. When the designer matches the landscape choices to the water available (and seasonality of the water), per capita water demand is significantly reduced.

Local governments can also require that all major landscapes for new and redeveloped construction must use water conserving landscape technologies. The State of California passed legislation in 1991 and again in 2009 that requires developers to submit plans that demonstrate the outdoor landscaping will use 70% or less water than a "typical" lawn. This is to be accomplished by many of the technologies

Residential Irrigation Facts

- Homes with in-ground sprinkler systems use 35 percent more water outdoors than those who do not have an in-ground system.
- Households that employ an automatic timer to control their irrigation systems used 47 percent more water outdoors than those that do not.
- Households with drip irrigation systems use 16 percent more water outdoors than those without drip irrigation systems.
- Households who water with a handheld hose use 33 percent less water outdoors than other households.
- Households who maintain a garden use 30 percent more water outdoors than those without a garden.
- Households with access to another, non-utility, water source displayed 25 percent lower outdoor use than those who used only utility-supplied water.
- On average, homes with swimming pools are estimated to use more than twice as much water as homes without swimming pools, everything else held constant.

U.S. Geological Service, "Irrigation Water Use", available at: ga.water.usgs.gov/edu/wuir.html.

described in this chapter such as native planting, hydro-zoning, mulching, computer controlled irrigation recycled water use, rain water harvest, storm water capture and similar measures. All cities and counties in the state are required to have a local ordinance to enforce these requirements.

Computer-controlled irrigation scheduling, hydro-zoning, mulching, and other tools are also available. The irrigation industry, hard hit by the drought in many states, is developing new technologies such as SWAT (smart water application technology) that uses smart controllers, soil sensors, and evaporation-transpiration sensors to target scarce water supplies.

Many water agencies perform home or business "audits" to help people modify their systems; many agencies supply water rebates for landscape modification or provide technical assistance to improve landscape water performance. Finally, some localities impose temporary ordinances to require consumers to reduce specific water use. Clauses can be included in the covenants of a development to prohibit in-ground sprinkler systems, drip irrigation systems, or automatic sprinkler timers—factors that have been shown to increase water consumption.

Conclusion

The water industry is currently at a crossroads. Rising energy costs, the impact of climate change on water supply, and concern about the carbon emissions of a water transport and distribution system that was developed in an era of abundant water and energy are causing the industry to radically rethink basic infrastructure decisions. Whereas water provision used to be considered merely a supply problem to be dealt with by the water agency, land use and climate change considerations have become increasingly important. In addition, demand management techniques and water efficiency concerns have catapulted the water engineer into a new world of economists, planners, landscape architects, lawyers, and statisticians.

Additional Resources

Publications

Baker, Lawrence, ed. *The Water Environment of Cities*. New York: Springer, 2009. A series of articles that cover emerging institutional practices for holistic water planning at the local and regional level.

Baumann, Duane D., John J. Boland, and W. Michael Hannemann. *Urban Water Demand Management and Planning*. New York: McGraw-Hill, 1997. Although this work is over ten years old, the basic message of demand management and how to pay for water is still current today. A classic.

Johnson, Karen E. and Jeff Loux. *Water and Land Use: Planning Wisely for California's Future*. Point Arena, CA: Solano Press, 2004. This book provides sections on how to plan for water and land use both from the water agency's point of view and that of local government.

Vickers, Amy. *Handbook of Water Use and Conservation: Homes, Landscapes, Businesses, Industries and Farms.* Amherst, MA: Waterplow Press. 2001. The best guide to water conservation practices around. For the individual or business.

Websites

American Society of Civil Engineers (*Journal of Water Resources Planning and Management*): www.ojps.aip.org/wro.

American Water Works Association: www.awwa.org.

Association of Metropolitan Water Agencies: www.amwa-water.org.

Bureau of Reclamation, U.S. Department of the Interior: www.watershare.usbr.gov.

State Water Resources Control Board: www.swrcb.ca.gov.

U.S. Environmental Protection Agency (Office of Water): www.epa.gov/watrhome.

U.S. Geological Service: www.usgs.gov.

Key Federal Regulations

Early water regulations, passed in the late 1800s and early 1900s, focused on development of water resources. Later regulations, arising mostly in the late 1960s and early 1970s, focused on water quality and environmental protection.

Rivers and Harbors Act of 1899

This Act gives the U.S. Army Corps of Engineers authority over navigable waterways in the United States. Of particular importance is Section 13, which prohibits the discharge of refuse material into any navigable water without permission.

Reclamation Act of 1902

The Reclamation Act created the Bureau of Reclamation, which is responsible for constructing irrigation systems in the western states.

Federal Water Power Act of 1920

The Act created the Federal Power Commission, which had authority to permit hydroelectric facilities and to regulate interstate sales and transmissions of electricity. In 1986, the Act was amended so that the Commission, now named the Federal Energy Regulatory Commission, would consider environmental and recreational issues in water planning as well.

Federal Water Pollution Control Act of 1948

This was the first environmental law passed by Congress. It required states to identify water bodies that were polluted beyond tolerable levels, and then to identify the polluters and suppress the polluting discharges. This law was difficult to enforce, however, and led to the Amendments of 1972.

Federal Water Pollution Control Act Amendments of 1972

This law changed the focus of water quality control from regulating the pollutant levels in the water body to regulating the levels in the effluent at the source of the pollution.

Safe Drinking Water Act of 1974

Prior to enactment of the Safe Drinking Water Act (SDWA), only standards for communicable waterborne diseases were in force. The SDWA expanded these national health-based standards for drinking water to both naturally-occurring and man-made contaminants that may be found in drinking water.

Clean Water Act of 1977

The Federal Water Pollution Control Act was amended again in 1977, and was renamed the Clean Water Act. The Clean Water Act authorized grants for planning, research, and wastewater treatment plant construction. It also enacted a system of water pollutant regulations.

Safe Drinking Water Amendments of 1986

These amendments established a wellhead protection program which also protects the recharge area of aquifers by setting contaminant limits in the recharge water and by regulating underground injection wells.

Water Resources Development Act of 1986

This law authorized several large-scale water projects, and also implemented many hazard mitigation programs, such as the Dam Safety Program. The Dam Safety Program allocated federal money to states to implement dam safety initiatives.

Water Quality Act of 1987

The Clean Water Act was reauthorized in 1987 as the Water Quality Act. This Act included provisions to address non-point sources of pollution.

Safe Drinking Water Amendments of 1996

The 1996 amendments expanded the SDWA to require that all community water systems prepare and distribute annual reports about the water they provide, including information on detected contaminants, possible health effects, and the water's source. States must also provide information on their sources.

chapter 17

Wastewater and New Paradigms

Why Is This Important to Local Practitioners?

The construction of wastewater collection systems and treatment plants in the United States was one of the great public health success stories of the millennium. Today, however, many question the ability of the current system to meet the challenges of climate change, urban growth and replacement needs, and the prospects of global disease transmission. This model uses vast quantities of clean water as transport, does a poor job of reusing nutrients, is increasingly energy-intensive, disperses contaminants, and disrupts natural ecological cycles. Solutions to these problems are now possible given recent technological advances in treatment and remote sensing technologies, but the exact form and institutional configuration are under development. In the past, sanitary sewers were called the skeleton of the city, with one author noting that "[t]he planning process, operating through a professional planning staff and/or a citizen-based planning and zoning board, should direct the laying of the [sewer] pipe; if it does not, the laying of the pipe is likely to direct the planning of the community."[1] The "next generation" of wastewater technology still has major implications both for land use and development patterns and for building design.

This chapter begins with a history of sewers and wastewater treatment and the current issues. It then describes the elements of existing centralized systems and their institutions and regulations. New directions for sanitary sewer infrastructure are addressed. The chapter concludes with a discussion of fees and financing.

In this chapter ...

Importance to Local Practitioners . . . 295
History of Wastewater Collection and Treatment . . . 296
- Early Sewers . . . 296
- Wastewater in the United States . . . 296
- The Rise of Wastewater Treatment Facilities . . . 297

The Institutions of Wastewater . . . 298
- Who Regulates Wastewater? . 298
- Who Provides Wastewater Collection and Treatment Services? . . . 299

The Elements of Municipal Wastewater Systems . . . 299
- Wastewater Collection Systems . . . 300
- Wastewater Treatment . . . 301

Current Issues . . . 303
- Emerging Contaminants in Surface and Groundwater . . . 303
- Sewer Systems Overflows . . . 304
- Infiltration and Inflow . . . 305
- Wastewater System Replacement Needs . . . 306
- Rural On-site Treatment . . . 307
- Eutrophication and Peak Phosphorus . . . 308
- Carbon Emissions and Energy Use . . . 310

New Directions for Sanitation . 310
- Water Reuse or Reclamation . 310
- Centralized Water Reclamation . . . 311
- Centralized Biosolid (Sludge) Reuse . . . 312
- Energy and Wastewater . . . 312
- Decentralized and Distributed Wastewater Management Systems . . . 313
- Zero Emissions Wastewater . . 314
- Natural Treatment Systems . . 317

Planning for Wastewater . . . 318
- Wastewater Planning and Land Use . . . 318
- Key Steps in Wastewater Planning . . . 319

Financing Mechanisms . . . 322
- Financing Sanitary Wastewater . . . 322

Conclusion . . . 323
Additional Resources . . . 324
- Publications . . . 324
- Websites . . . 324

History of Wastewater Collection and Treatment

Early Sewers

As long as humans began congregating in large settlements, the collection and disposal of human waste have been a concern. Many cultures had persons whose job it was to collect and dispose of human waste. Others devised passive infrastructure systems to carry away "night soil," while the elites in many pre-industrial societies had sophisticated systems for carrying away human waste. For example, in 2100 BC, the Minoan civilization of Crete had a sewer system that rivals many of the most advanced systems today. The early plumbing engineers took advantage of the steep grades on this island to devise a drainage system with lavatories, sinks, and manholes. The king's palace at Knossos had a main sewer constructed of masonry that linked four large stone shafts arriving from upper floors. The drainage system consisted of terracotta pipes, about 4–6 inches in diameter. Rainwater from the roofs and the courts and the overflows from the cisterns were carried down into buried drains of pottery pipe. The pipes had tapered connections, which caused a jetting action and prevented the accumulation of sediment.

The collection and disposal of sanitary sewage for urban dwellings using pipes is relatively new concept. Prior to 1800, human wastes were collected in cesspools and privy vaults, the environmental solution. Prior to 1833, there were no water closets in the United States, but once running water became available in private dwellings, this modern convenience became indispensable. This in turn led to an increase in the production of sewage. In Boston, a city of 180,000, there were over 14,000 water closets in 1864. By 1880, over one-third of urban households nationwide had water closets.[2]

Many cities had storm sewer systems, but local laws prohibited their use for human waste. At first, the increased flow from water closets went into the existing on-site cesspools and privy vaults, but they did not have the capacity for large amounts of water. Overflows were common. In the 1830s, Edwin Chadwick, an English lawyer-turned-sanitarian, developed the idea of "water carriage" of human waste to avoid this problem. Chadwick promoted the use of what he called "an arterial-venous system" to bring potable water into the house which was then used to carry human waste, or "liquid manure," through public sewer lines out to agricultural fields.[3]

Wastewater in the United States

It took 40 years for the idea to catch on in England and longer to cross the Atlantic, but by the 1880s, most U.S. policy-makers had endorsed Chadwick's innovation. Between 1870 and 1920, the percentage of the urban population in the United States with sewerage increased from 50 to almost 90, while the number of communities with some sort of sewer system increased from 100 to 3,000 in the same period.[4] Two obstacles stood in the way of universal sewer installation. The first was the cost. Like water supply services, wastewater is very capital-intensive. Costs to lay the networks of sewer pipes for private entrepreneurs were prohibitive. In addition, sewer systems were seen as instruments of public health, which was a government responsibility. Therefore, the wastewater industry developed almost entirely as a public enterprise.

The second obstacle concerned the debate between combined sewers and separate sewers. A combined sewer carries both stormwater and sanitary sewage in the same pipe. A separate system has two pipes, often running in parallel, that carry stormwater and sanitary sewage to the point of disposal. Combined sewers arose naturally as private citizens made connections to existing storm sewers to empty their water closets. However, as noted above, these ad hoc combined sewers worked poorly, so in Europe, many cities installed new sanitary systems and let the old storm sewers revert to their original purpose.[5]

American municipalities, however, did not have the resources to install parallel systems, so combined sewers became the accepted method. Memphis, Tennessee, was the first major city in the United States to install a separate system for sanitary sewage. After several epidemics of yellow fever in 1873, 1878, and in 1870, in which over 7,000 people died, national public health and engineering experts pioneered a separate sewer system for sanitary waste in this town. This concept soon caught on in small towns, where surface drainage of stormwater was possible. In larger built-up areas in the Northeast and Great Lakes regions, combined sewers remained the preferred system. A few older West Coast cities such as San Francisco and Seattle also have combined sewer systems.[6]

The Rise of Wastewater Treatment Facilities

Until the end of the nineteenth century, wastewater was still dumped untreated into local water bodies since it was thought that running water purified itself. Later, cost factors were an issue. As a water supply expert said in 1914:

> It is … cheaper and more effective to purify the water and to allow the sewage to be discharged without treatment, so far as there are not other reasons for keeping it out of rivers … $1 spent in purifying the water would do as much as $10 spent in sewage purification … The water works man therefore must, and rightly should, accept a certain amount of sewage pollution in river water, and make the best of it.[7]

Dilution was not a permanent solution. Many cities found their water sources increasingly contaminated due to this practice. But by the early 1900s, downstream cities that drew their water from rivers where upstream cities dumped their sewage saw rising typhoid death rates in the 1880s and 1890s. At first, it was considered cheaper to treat the incoming water than the outgoing sewage.[8] In Chicago, the water intake "cribs" in Lake Michigan which were built in the mid-to-late nineteenth century became contaminated from sewage discharging from the Chicago River. As a result, the flow of the river was reversed, a project that was completed in 1900.

Even with the rise of primary and secondary treatment, throughout the first part of the twentieth century, surface water sources around urban areas remained polluted with municipal and industrial sewage. For example, Manhattan's West Side sent 300 million gallons of raw sewage a day into the Hudson River in the 1960s. And by 1965, as a history written later by the EPA notes:

> Lake Erie was dying. The Potomac River was clogged with blue-green algae blooms that were a nuisance and a threat to public health. Many of the nation's rivers were little more than open sewers and sewage frequently washed up on shore. Fish kills were a common sight. Wetlands were disappearing at a rapid rate.[9]

Accordingly, federal legislation was passed in 1972 and again in 1987 to address these problems.

The Institutions of Wastewater

Who Regulates Wastewater?

The U.S. Environmental Protection Agency (EPA) and its Office of Wastewater Management have the responsibility for ensuring compliance with the 1972 Clean Water Act and its amendments. The EPA has delegated responsibility for implementing the National Pollutant Discharge Elimination System (NPDES) program and permits to all but a few states.[10]

The NPDES program was one of the central features of the 1972 Act. The purpose of this permitting program was to regulate point source pollution into surface waters in the United States. Point source pollution is that which is discharged from a specific single source, such as a wastewater treatment plant, or an industrial plant. These facilities are required by the 1972 Act to obtain a permit (the NPDES permit) that specifies both the quantity and concentration of regulated pollutants. The Act was modified in 1987 to cover non-point sources which are a problem during storms. This is discussed further in Chapter 18 on stormwater. More than 200,000 wastewater sources are regulated under the NPDES program.

The bulk of the enforcement of the 1972 Act has been delegated to the state level. Most states have a Department of Environmental Quality or a Division within a larger natural resources agency that is responsible for implementing this program. Many states also have a health department that may share enforcement powers with regard to water quality.

Who Provides Wastewater Collection and Treatment Services?

In the United States, there are more than 15,000 public wastewater treatment facilities and 23,000 privately owned treatment facilities. The latter are primarily used for treating wastewater from industrial plants and commercial operations. Organizational arrangements are myriad, reflecting the decentralized way in which local governments addressed the need to protect human health and then the natural systems. Many water supply agencies are also full service wastewater agencies. If the wastewater agency is a part of the local general purpose government, the agency is likely responsible for the wastewater treatment plant and the interceptors and the laterals. However, if the wastewater agency is a special district, perhaps at the county or regional level, then it may only be responsible for the treatment plant and the major interceptors. Local general purpose governments are usually responsible for the wastewater collection systems in their jurisdiction that connect to the interceptors.

Local governments also play a key role in local land use and building regulations for sewers. Private property owners are generally individually responsible for their own pipes.

The Elements of Municipal Wastewater Systems

A municipal sewer system is essentially a water distribution system in reverse. Although advances have been made in the treatment of sewage over the past 150 years, the basic system design for capital facilities has remained virtually unchanged for 4,000 years. Today's centralized municipal wastewater system consists of a network of buried pipes for collecting sanitary sewage from households and businesses, pipes for collecting stormwater, and a treatment plant[11] (Figure 17.1).

This system is designed to ensure that waterborne substances—pollutants—from residences, industry, and agriculture that may negatively affect humans and the environment do not get into surface or groundwater sources. Pollutants are defined by the EPA as:[12]

- *Conventional pollutants.* These include human wastes, food wastes from a garbage disposal, and gray water such as laundry and shower water. These kinds of pollutants are usually carried away from a home, business, or industry by the sanitary sewer. They are biodegradable. A pollutant that can cause disease in humans is called a pathogen.

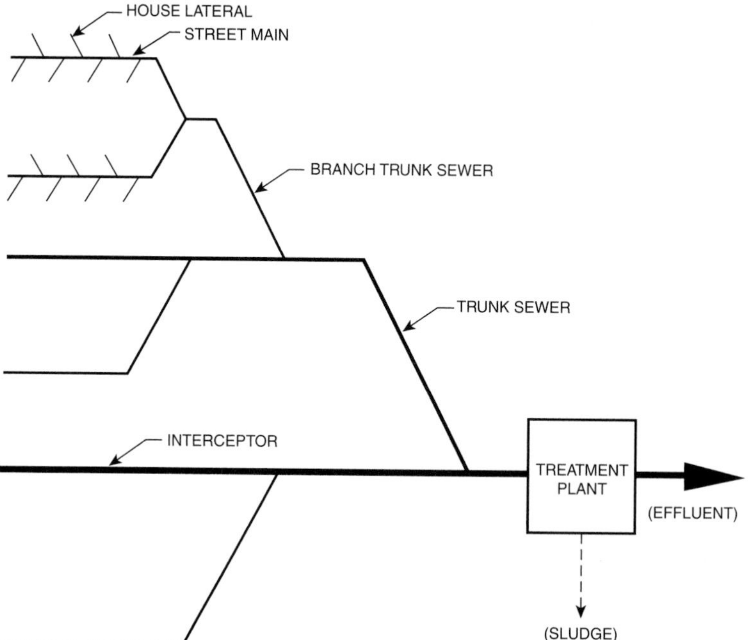

Figure 17.1
Sewer system conceptual schematic

- *Toxic pollutants.* One hundred toxic substances have been defined by the EPA as harmful to animal or plant life. These are organics (pesticides, solvents, polychlorinated biphenyls—PCBS, and dioxins) and metals (lead, silver, mercury, copper, chromium, zinc, nickel, and cadmium). Toxic pollutants can come from households, government, and industry. They are usually regulated with pre-treatment programs and not discharged into the sewer system.
- *Nonconventional pollutants.* These are anything else that requires regulation, and may include nitrogen/nitrates and phosphorus, which in small amounts are beneficial but in large doses lead to disruption of aquatic plant and animal life. Thermal pollution may fall into this category.

Wastewater Collection Systems

Wastewater collection systems are underground pipes for sanitary waste and stormwater. As noted earlier, these systems of pipes can be separate or combined. The pipes leading from individual houses or buildings are called laterals (Figure 17.2). These connect to sub-mains, which generally run down the street. These in turn connect to trunk lines or *interceptors*, which are large lines that terminate at the wastewater treatment plant. Laterals are usually 6 inches in

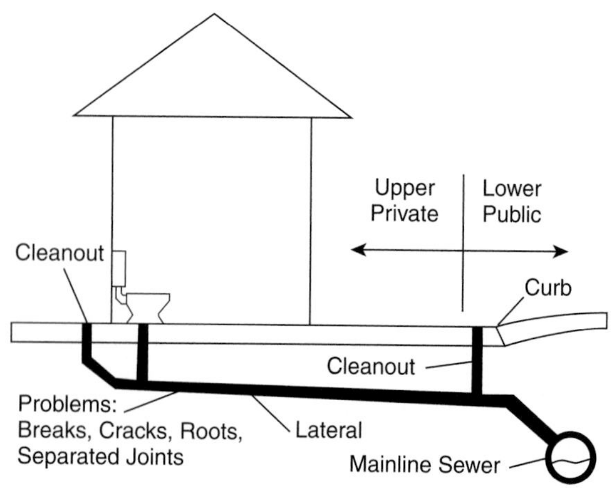

Figure 17.2
Building sewer system connection detail

diameter and made of clay, concrete, or PVC. Sub-mains are usually 8, 10, or 12 inches in diameter. Trunk lines range from 15 to over 27 inches in diameter.

Sewer systems have been designed to flow by gravity wherever possible. Furthermore, sewer lines are designed to flow half-full to ensure ventilation. These two facts make the physical configuration of sewer lines important. Whereas pressurized water lines can have any alignment, sewer lines need to have a specific slope, usually about 0.5–2 percent. Engineers indicate that flows within sewer lines need a velocity of between 2 and 10 feet per second, fast enough so that solids are not deposited, but slow enough that grit in the water does not scour the pipes and cause leaks.

Manholes, also called *utility access ports* (UAPs), are located at any changes in direction, pipe size, or slope, or any time two lines intersect. For sub-mains, manholes are usually located about every 400 feet.

Sometimes, gravity alone is insufficient to carry the sewage. This occurs when the interceptor must traverse a great distance, or when the topography prohibits it. In this case, a pressurized sewer line called a force main is used. Sewage flows into what is called a lift station where pressure is added. Lift stations are prone to failure, particularly during storms. However, sometimes they are necessary in areas where elevation changes abound (Figure 17.3).

Artistic manhole covers

The Crescent City water pollution control facility, a complex of three wastewater treatment buildings

Wastewater Treatment

The wastewater treatment plant (WWTP) is where the sewer pipes converge. This is a centralized facility, usually located near a large body of water, where the sewage is treated before being discharged into the receiving water bodies. This process removes disease-carrying organisms, harmful chemicals, and excess minerals.

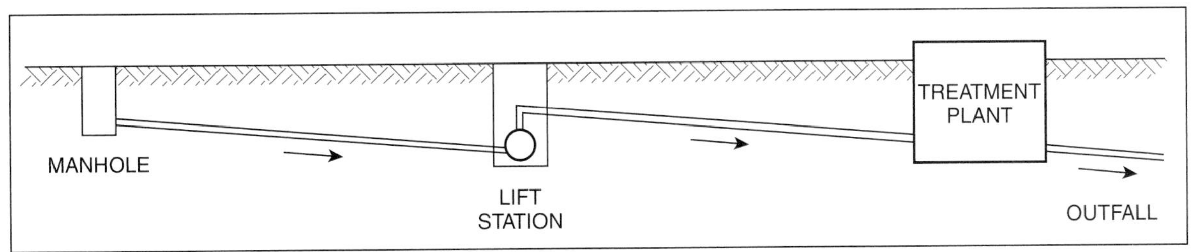

Figure 17.3
Sewer system lift station detail

Sewage treatment plant

There are two basic stages in the treatment of wastewater: primary and secondary. In the primary stage, solids are allowed to settle out of the water in what is really a physical process. In the secondary stage, biological processes actually attack the pathogens in the water. The two most common secondary treatment techniques are the trickling filter and the activated sludge process. Regardless of which process is used, treatment is not complete until the treated wastewater has been disinfected with chlorine. Chlorine is fed into the water to kill whatever bacteria are left, and to reduce odor. Done properly, chlorination will kill more than 99 percent of the harmful bacteria in the effluent. However, since after treatment the wastewater is dumped into a natural body of water, where it becomes harmful to wildlife, many states now require the removal of excess chlorine before discharge to natural waters in a process called dechlorination. Alternatively, some treatment plants use ultraviolet light or ozone instead of chlorine for disinfection.[13]

Tertiary treatment. Some treatment plants are also doing tertiary treatment. Tertiary treatment is a general term for any process that removes wastewater constituents that are not removed in primary or secondary treatment. Usually, these constituents are location-specific chemicals that the local community has chosen to remove or is required to remove. For example, a certain industrial practice might introduce high levels of mercury into local wastewater. The wastewater treatment plant could then use a tertiary treatment process to remove the mercury.[14]

Pre-treatment. Publicly owned sewage treatment plants are generally designed to treat domestic sewage. Many industrial and other non-residential uses have wastewater that is toxic and where a discharge of this sort into the sewers would interfere with the biological processes at the treatment plant or make it impossible for the plant to recycle its used water and solids. Other impacts are more dramatic. In Louisville, Kentucky, hexane was discharged into the public sewers, resulting in an explosion in 1981 that destroyed 3 miles of sewers and caused $20 million in damage.

Current Issues

Significant challenges face the current system. Surface water still suffers from serious pollution from known sources, while emerging contaminants in both surface and ground water are of deep concern. The urban sanitary sewer infrastructure is old and suffers from spills and overflows into surface waters. Although progress has been made in addressing rural sanitary sewer needs, untreated sewage from failing septic tanks affects the watersheds and waterways for all. Finally, climate change concerns and rising energy prices focus new attention on sewage collection and treatment systems.

Emerging Contaminants in Surface and Groundwater

Wastewater treatment systems were originally designed to ensure that pathogens capable of spreading infectious diseases were removed from the waste stream before being discharged into the surface water. However, as society has begun to use more and more chemicals, a wider variety of contaminants are being found in waste streams and in both ground and surface water supplies. One researcher found that of the 30,000 chemicals used commercially in the United States and Canada, about 400 are not broken down in the environment, and 75 percent of them have not even been studied.[10] These are called "emerging" contaminants because their health effects are only now being assessed. Some of these substances include industrial solvent stabilizers, fuel oxygenates (MTBE and TBA), disinfection byproducts, other pharmaceuticals (antibiotics/drugs), personal care products, pesticides/herbicides, and flame-retardant chemicals. Researchers have found concentrations of the latter, used for plastics and fabrics, in human breast milk, and have also determined that their concentrations have doubled every five years for the past quarter century.

Some of the emerging contaminants can come from wastewater treatment plants, which do not remove substances such as estrogen excreted in urine from birth control pills, anti-depressants, pain killers, antibiotics, steroids, and fragrances. Other types of chemicals can make their way into the water supply through agricultural and urban runoff, and from untreated sewage leaks from deteriorated pipes. Low impact development practices such as rainwater

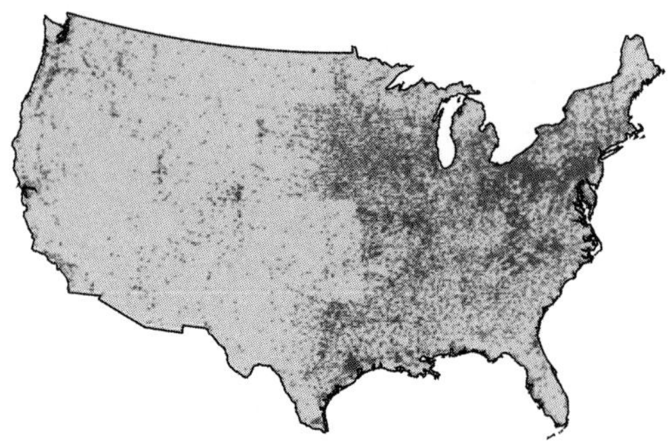

Figure 17.4
Map of U.S. sanitary sewer overflows

Figure 17.5
Sewer overflow, Rhode Island

harvesting and the deliberate infiltration of stormwater and rainwater back into groundwater supplies without treatment may also result in increased concentrations of heavy metals.[16] Although technologies exist to remove many of these contaminants, they can be expensive and the risk to the public must be balanced against the cost.

Sewer Systems Overflows

In times of little or no precipitation, most sewers work well. The contents of the sewers, both combined and separate, are mostly sanitary sewage and the water required to transport it to the treatment plant. But during a storm, the total volume of wastewater arriving at the treatment plant may exceed its capacity and can be discharged untreated into surface water bodies such as lakes, rivers, estuaries, or coastal water. This is known as either a combined or separate sewer overflow (CSO or SSO).[17]

The EPA estimates that there are between 23,000 and 75,000 SSO events a year, discharging anywhere from 3–10 billion gallons annually. They estimated that it would take $41.2 billion in 1995 dollars to abate CSOs. Local figures are somewhat higher. Planners in Fort Wayne, Indiana, estimated that local communities would need to spend a total of $1 billion over 15 years to solve the CSO problem in the Maumee River Basin.[18] Atlanta estimates that it will take $3 billion to solve its CSO problem.[19] These are only two of the many CSO communities. (See Figure 17.4 for a map of sewer overflows in the United States.[20])

The 2004 Report to Congress on CSOs by the EPA indicates that almost 60 percent of the CSO communities with permits have implemented the required measures and that CSO discharges of untreated wastewater had dropped to 850 billion gallons from more than a trillion gallons in 1994 when the CSO policies were put into effect.[21]

During an SSO or CSO, untreated sewage is released from manholes onto city streets and sometimes onto private properties (Figure 17.5). Untreated sewage can also back up into basements, causing property damage and threats to public health. Ultimately it can make its way into surface water used for drinking and recreation.[22] Because sewer overflows contain untreated human and industrial wastes as well as surface runoff, many pollutants may end up in surface water. These may adversely affect drinking water and aquatic habitats, resulting in shellfish harvesting restrictions, beach closures, and the death of fish.

In 1998, more than 1,300 residents of the Brushy Creek area of Austin, Texas, were infected with cryptosporidia parvum after 167,000 gallons of untreated sewage spilled into the creek in the summer of that year.[23] SSOs have a variety of causes, such as severe weather, poor system operation and maintenance, and vandalism.

Infiltration and Inflow

A problem related to SSOs is called "infiltration and inflow." When sewer lines crack or break, sewage leaks out. Another problem occurs when stormwater enters the sewer pipes either from the ground—called infiltration—or water is discharged into the pipes from other sources—inflow. Inflow and infiltration (I/I) drive peaks in sewage flows, and therefore are determinants of infrastructure size and cost. Infiltration can be caused by damaged pipes, joints, or connections, or leakage of manhole walls. Inflow can be caused by roof leaders, cellar, yard, or other area drains, including sump pumps being connected to the sanitary system. It can also be caused by cross-connections from storm sewers and combined sewers or by illicit connections.

The EPA notes that some cities attribute 60 percent of their sanitary sewer overflows to problems from the laterals (Figure 17.6). The service lines or laterals that connect the home to the public sewers are generally the responsibility of the homeowner, depending upon local rules about the portion in the public right-of-way.[24]

The EPA found that over 80 percent of sanitary sewer overflows could be improved by such measures as periodic sewer line cleaning (see Figure 17.7, a vactor truck pumping out obstructions in the city's sewer) and repair or by reducing the amount of inflow and infiltration to reduce the peak flows during storms. This was confirmed in 2008 by a study done in the State of Washington that compared I/I practices between four counties. Data showed that more than 90 percent of I/I occur on homeowner properties, manholes as well as trunk, lateral, and side sewers. Those counties that addressed all four reported 80 percent reductions in I/I.

Failure of Sanitary Sewer System in Milwaukee

Massive and persistent dumping of untreated sewage into Lake Michigan by the Milwaukee sewer agency during a series of storms in May and June of 2004 gave it the ranking of the worst sewage polluter in the Great Lakes Region. Although Milwaukee has a deep tunnel for a storage area, it was already 65 percent full when the storms hit that year.[1]

Note

1. Steve Schultze and Marie Rohde, "Sewer Fixes Costly and Complex, EPA Says," *Milwaukee Journal Sentinel* (June 4, 2004), available at: www.jsonline.com/news/metro (accessed December 19, 2004).

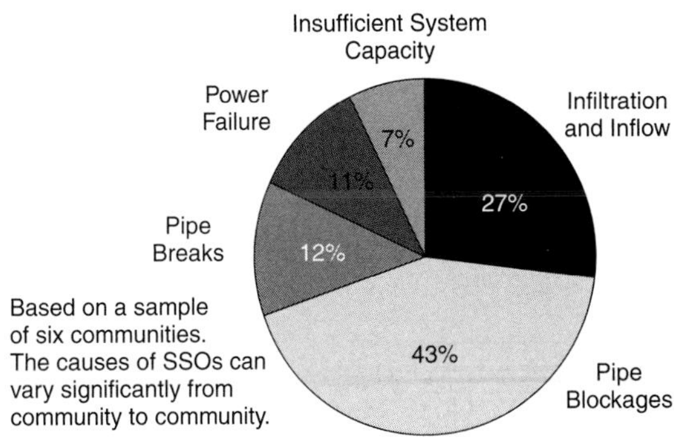

Figure 17.6
Estimated occurrence of sanitary sewer overflows by cause

Figure 17.7
Vactor pump truck keeps sewer lines pumped out

One county that focused on manholes only achieved just 23 percent reduction in comparison.[25]

Wastewater System Replacement Needs

The amount of funds needed during the next 20 years to replace the aging "legacy" wastewater systems and to upgrade them to meet regulatory standards ranges from $331–450 billion or, if financing is included, from $402–719 billion in 2000 dollars. This is the equivalent of anywhere from $16–25 billion annually. These figures include pipe replacement and the rehabilitation of treatment plants.[26] One official estimates that the entire network of sewer pipes valued at $1 trillion will need to be replaced over the next 30 years.[27] These figures do not include the costs of providing new facilities for population growth, which is an additional 25–30 percent. They also do not include new requirements for stormwater.

Aging pipes. The age and quality of sewer pipes currently in the ground vary substantially at the local level. For example, parts of the St. Louis system were built before the Civil War. There is no national inventory of the amount or age of sewer pipe for wastewater systems.[28] Most of the pipe construction followed increases in population and suburbanization, with an 1890s boom, a World War I boom, and a Roaring Twenties boom, although most of today's wastewater pipes were installed after the World War II. If replacement does not occur, deterioration projections indicate that 44 percent of the nation's sewer pipes will be in poor or very poor condition by 2020.[29]

Treatment plants. Most wastewater treatment plants were renovated or built in the 1970s. In the next two decades about 22 percent of the plants will need upgrading or replacement, the rest will come due after 2020. The beginning of this replacement wave will occur at the same time that many of the post-WWII pipes will need replacement. As the EPA Gap Analysis stated, "The typical system could experience a very significant bump in expenditures over a very short period of time to accommodate replacement of old pipes, new pipes, and plant structures."[30] Wastewater needs over the next 20 years are generally agreed to be substantially larger than those for water.[31]

Infill need. The smart growth emphasis on infill development will accelerate the need to upsize or replace existing pipes and increase the capacity of treatment plants. Replacement needs will cause increases in fees. In some East Coast cities, pipe replacement costs will fall on a smaller and poorer population and progressive fee schedules may need consideration. In some older cities, the per-capita replacement value of mains is more than three times higher than in the newer areas because of population declines. Although the cost of collecting and treating wastewater for a residence is only a fraction of that spent for other utilities, the need to triple or even increase the rates by as much as eight times in some jurisdictions will focus attention on these decisions.[32]

Rural On-site Treatment

Twenty-one percent of the population in the United States is not connected to a centralized treatment facility and instead use septic tanks or package units, with 95 percent of these being septic systems. EPA estimates that annually 10–30 percent of all septic tanks fail. When this occurs, pathogens may leach into surface and groundwater and lead to viral and bacterial illness. Industrial facilities may also have an on-site treatment unit specially designed for their particular discharges. See Figure 17.1 for a schematic of a septic tank and leach field.[33]

Many states have passed ordinances restricting the use of septic tanks in new developments, and have required local

Alligators and Sewers

A persistent piece of folklore is that New York's sewers are home to alligators. The story goes that baby alligators were brought home from Florida by returning tourists and were dumped down the toilet when their owners tired of their new pets. The tale further goes on to say that some of these alligators are blind albinos who have lost their sight due to the darkness. Herpetologists, however, agree that NYC sewers are too cold and full of bacteria for alligators to survive. The source of this urban legend appears to be the documented capture of an 8-foot alligator down an East Harlem manhole in 1935, but even at the time it was thought the reptile had come from the Everglades and had been dumped in the sewer shortly before it was "found."[1]

Note

1. About.com, "Alligators in the Sewer: An Urban Legend," (2004), available at: urbanlegends.about.com.

Useful Life of Components of the Wastewater System

Component	Useful life
Collection Pipes	80–100 years
Treatment Plants (concrete structure)	50 years
Treatment Plants (mechanical and electrical)	15–25 years
Force Mains	25 years
Pumping Stations (concrete structure)	50 years
Pumping Stations (mechanical and electrical)	15 years
Interceptors	90–100 years

U.S. Environmental Protection Agency, *The Clean Water and Drinking Water Infrastructure Gap Analysis* (Washington, DC: U.S. Environmental Protection Agency, 2002).

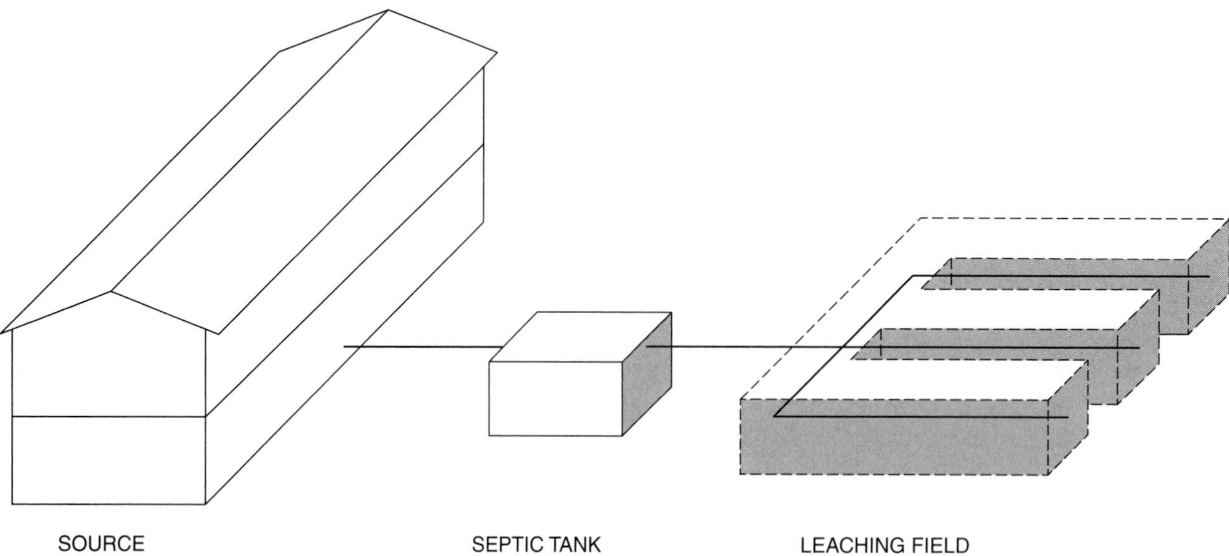

Figure 17.8
Private septic system schematic

sewer agencies to develop programs to phase them out or restrict their use. One former planning director recommends that group septic tanks be used so that there are enough funds to pay for a dedicated maintenance contract.[34]

Eutrophication and Peak Phosphorus

Phosphorus is an essential nutrient for all plants and animals and critical for the world's food supply. It is a component (along with nitrogen and potassium) of commercial fertilizers. Over the past 50 years, over half a billion tons of phosphorus has been mined globally for fertilizer. Although these fertilizers have made possible the Green Revolution that met the need for food as a result of population growth, fertilizer waste products have ended up in oceans, streams, and lakes around the world, causing algal blooms, eutrophication, and dead zones.

World phosphorus supplies are finite and are being depleted. Experts predict "Peak Phosphorus" by 2030, after which the demand will exceed the supply. Five countries control 90 percent of the world's phosphate rock reserves. The largest are found in China, the United States, and Morocco/the Western Sahara. The U.S. Geological Survey estimates that the U.S. reserves will last for 50 years at the current rate of extraction, which is increasing at 2 percent per year. At a rate of 3 percent increase (which has been exceeded in recent years), they will last less than 45 years.[35]

There is no substitute for phosphorus in the food chain, but unlike oil, it can be recaptured and reused. In addition, there are many opportunities for efficiencies. During the process of mining, fertilizer and food production, and consumption, 80 percent of the phosphorus is lost. For example, worldwide, around 3 million tons of phosphorus is produced every year in human feces. More than half is generated in cities and most ends up in receiving water bodies. Similarly, almost half the phosphorus in harvested crops is lost in food processing or food waste. These are areas where efficiencies can occur. In addition, recovering phosphorus (and nitrogen) from wastewater/feces is important not only for phosphorus recovery but also to reduce eutrophication. Changing eating habits towards a more plant-based diet will assist, as will food and organic waste recovery.[36] See Figure 17.9 which shows the increasing use of mined phosphates from 1900 to the present and then models the impact of alternative courses of action.

Until recently, wastewater treatment plants that process domestic sewage in the United States have not been required to divert phosphorus and nitrogen from the waste stream. Additionally, agricultural runoff, a major source of phosphorus and nitrogen, is not regulated in the United States. (In Europe, all

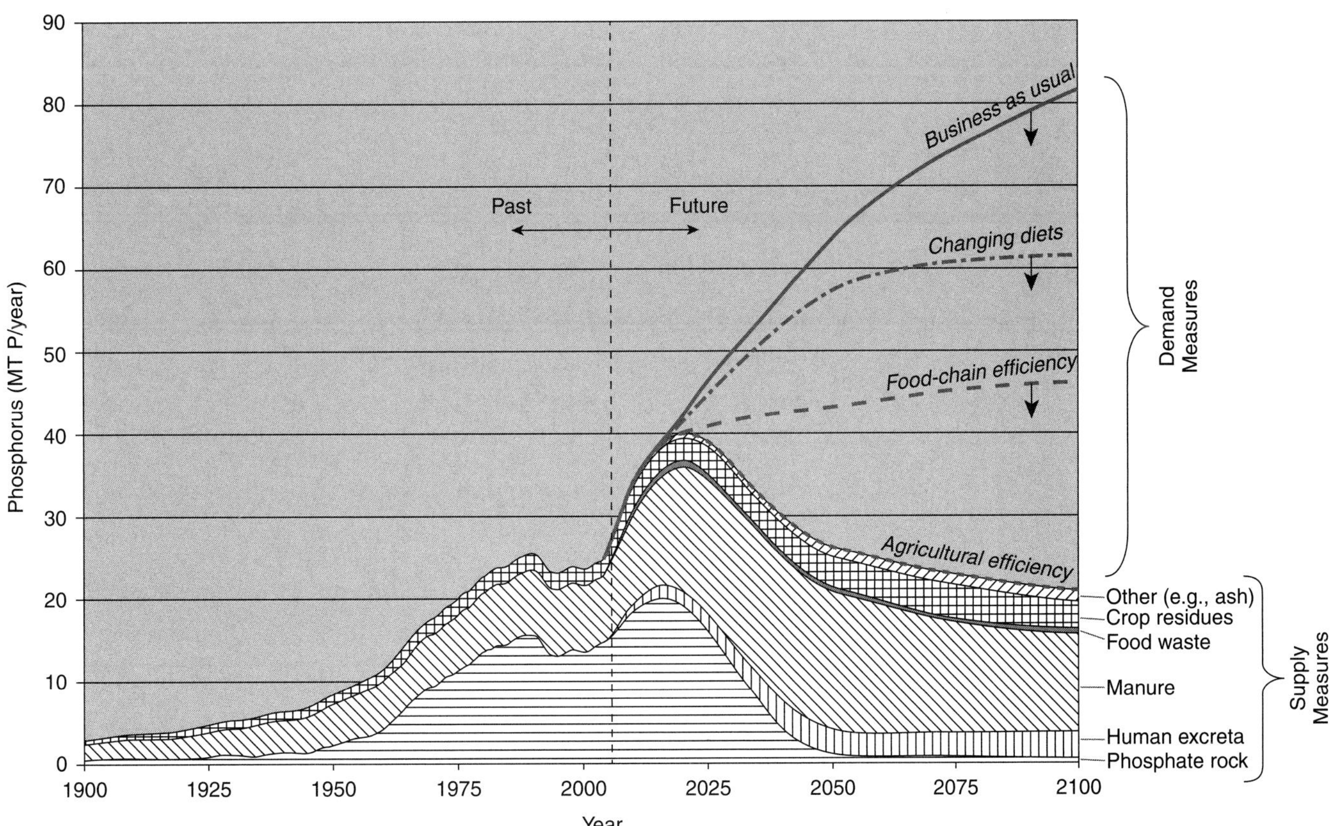

Figure 17.9
Global sources of phosphorus over time and alternatives to avoid peak phosphorus

discharges of nitrogen and phosphorus are regulated.[37]) However, this is changing, and many wastewater treatment plants (WTTPs) today are under pressure to reduce these discharges.

Carbon Emissions and Energy Use

Wastewater treatment plants also use 21 billion kilowatt hours annually, and this is anticipated to increase as the need to treat emerging contaminants becomes more of an issue. Energy is used throughout the wastewater treatment process with pumping and aeration operations taking the largest share. Stricter standards for the treatment of biosolids as well as aging wastewater collection systems which require higher pumping and treatment costs at the central plant will also require more energy and hence carbon emissions.[38]

Wastewater treatment processes also produces greenhouse gases. At the plant, anthropogenic CH_4 (methane) is produced and when the effluent is discharged into surface water, N_2O (nitrous oxide) is also produced. Septic tanks also produce methane as a byproduct of their process. Both emissions are more powerful than CO_2 (carbon dioxide).[39]

The Water Environment Research Foundation notes that the energy contained in wastewater and biosolids is ten times greater than the energy needed for treatment. They have set a goal of assisting wastewater utilities to become energy-neutral, and even energy-producing.[40]

New Directions for Sanitation

There is an emerging consensus among wastewater experts that we need to rethink our definition of waste. Indeed, "wastewater" consists of valuable resources including water, carbon, nutrients, and trace metals useful for industrial processes. This new definition implies

> a commitment to use solid and liquid waste to create energy, reduce greenhouse gas emissions, conserve water and recover nutrients … [and requires] … shifts in technology and in social processes … [including] … regulatory, institutional, and financial arrangements to shift behaviors of all players, from water authorities to consumers.[41]

At the same time, architects and designers are coming at this problem from a site- or project-specific perspective. The result is a cornucopia of innovations by utilities at the system level and by architects at the site and building level, some of which are detailed below.

Water Reuse or Reclamation[42]

If properly treated, municipal wastewater can be recycled and the water reused for a wide variety of potable and non-potable uses. The use depends upon the degree of treatment. Wastewater that has undergone primary treatment is suitable for surface irrigation on orchards and some crops. Secondarily treated wastewater can be used for groundwater recharge of non-potable aquifers, industrial cooling processes, wetland habitat, and a wider variety of crop and on-contact irrigation such as median strips. Finally, tertiary-treated water can be used for applications that involve bodily contact such as golf course and park irrigation, or fountains and decorative lakes.

Recycled and treated water can also be used for direct or indirect potable reuse. Indirect potable reuse (IPR) is recycled water that is purposely discharged into either groundwater or surface water that ultimately supplies a public drinking water system. IPR projects include groundwater spreading, where recycled water is released into open basins and the water seeps into the groundwater basin; groundwater injection where recycled water is pumped directly into the groundwater; and reservoir augmentation where treated recycled water is added directly to a water reservoir. Direct potable reuse refers to water used for drinking immediately after treatment.

Centralized Water Reclamation

Most water reuse in the United States in urban areas is associated with a centralized wastewater treatment plant that delivers using a second set of pipes for reclaimed water. The City of Irvine, California, has been delivering recycled water to commercial buildings for over a decade with "purple pipes." The water can be used for toilet flushing and other non-potable uses. The City of San Francisco has recently established a requirement for all new development to have a second set of pipes to be able to use recycled water.

Orange County's groundwater replenishment system

The Orange County Groundwater Replenishment System (GWRS) is a state-of-the-art facility that takes the treated effluent from the local wastewater treatment plant, treats it again with a three-step advanced process of microfiltration, reverse osmosis, and ultraviolet light with hydrogen peroxide, a process that destroys bacteria, protozoa, viruses, pharmaceuticals, and any trace organics and inorganics so that it is potable. The potable water is then percolated back into the groundwater table, where

Orange County staff check out the reverse osmosis connections for the GWRS

Close-up of a rack of microfiltration membranes

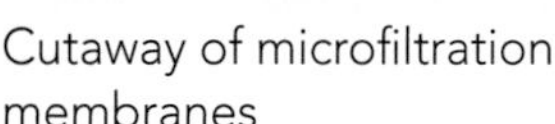

Cutaway of microfiltration membranes

It tastes like water because it is water

it is withdrawn by 19 local water departments and special districts before it is blended with imported water supplies, treated with chlorine, and distributed for potable purposes. The Water Reuse Foundation estimates that municipal wastewater recycling is growing at a rate of 15 percent a year.

Centralized Biosolid (Sludge) Reuse

"Biosolid" is another name for treated sewage that can be used to amend local soils and sometimes as an energy source. In the U.S. there are 16,583 wastewater treatment facilities producing over 64 pounds of sludge per person, every year, which is about 6.5 million metric tons of "dry solids"—sewage sludge with the water squeezed out of it—annually. Sludge management can account for about 30 percent to half of the operating cost of a utility.[43] The Water Environmental Research Foundation notes that wastewater treatment plants consume 21 billion kilowatt hours annually, but that sewage contains ten times the energy needed to treat it. The technology exists to recover that energy and thereby reduce the agency's dependence on conventional electricity.

Currently, 45 percent of sludge produced by wastewater treatment plants is incinerated or goes to landfills, 49 percent is treated and used in land applications, and only 6 percent is reused for other purposes—such as energy production. The King County Wastewater Agency, which serves the Seattle, Washington, area, treats a portion of its effluent so that it can be used on-site for landscape irrigation. In addition, biosolids and digester gas are recycled for fertilizer and power generation. The agency has set up agreements with local farmers who use the biosolids as fertilizer, and currently the demand exceeds the supply.

Energy and Wastewater

One of the most exciting efforts in the wastewater field today is the use of the heat in sanitary wastewater for energy. The premier example of this is the False Creek Energy Center in Vancouver, Washington, which is linked to a wastewater pumping station. Solids are filtered out.

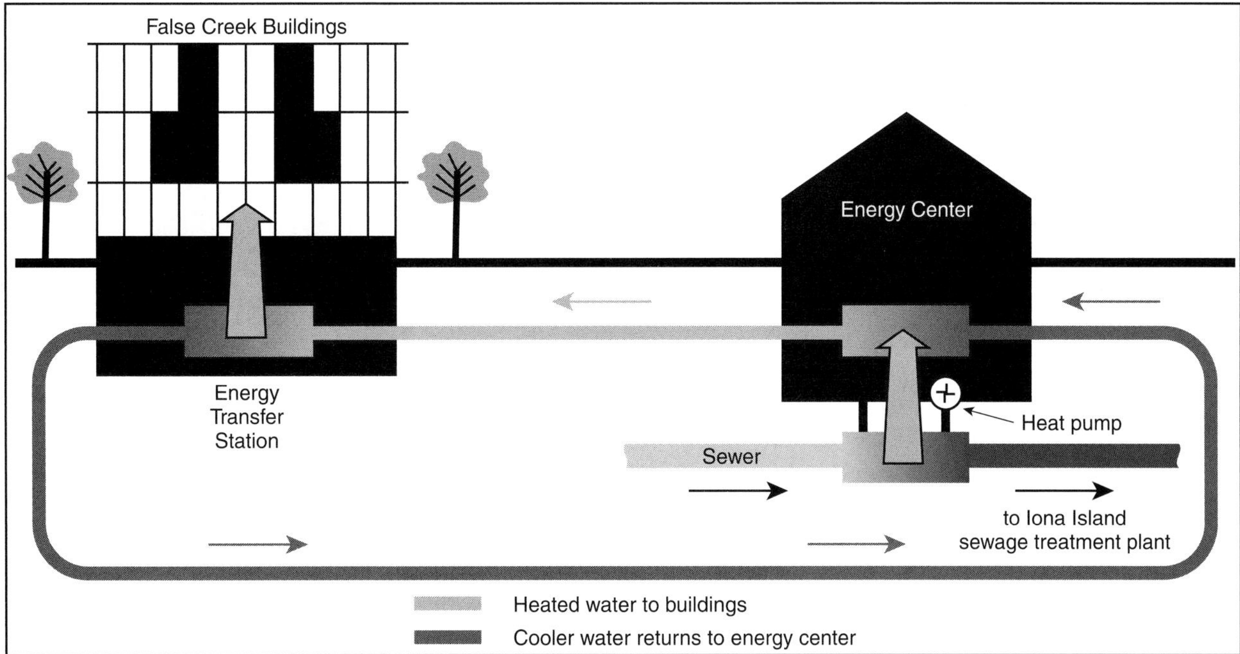

Figure 17.10
False Creek Energy Center diagram

Heat pumps transfer the heat energy from the wastewater to a hot-water distribution system. After the heat is recovered, the water and solids are recombined and pumped to the WWTP. The ratio of sewage to heating and cooling is 3:1—sewage from 300 apartments provides 90 percent of the heating and cooling for these units[44] (Figure 17.10).

Other wastewater utilities are also experimenting with innovative energy production schemes involving biogas. The East Bay Municipal Utility District powers its recycled water treatment plant by incinerating sludge and food scraps from the local solid waste program. The development of Hammarby in Stockholm uses wastewater to produce methane gas that is used to power city buses. Similarly in Finland, municipal outdoor toilets provide feces that also produce methane for local buses.

Decentralized and Distributed Wastewater Management Systems

A variety of new technologies have appeared on the scene in the past several years that make possible on-site or decentralized and satellite systems. These include the use of watertight butt-welded plastic piping, small-diameter effluent collection systems, and collection of source-separated waste streams. In addition, new techniques have been developed to repair and rehabilitate existing collection system infrastructure that do not require digging up the streets. New technologies for remote control and monitoring permit centralized management of distributed systems. Finally, the decreasing size of treatment

Membrane Bioreactors and Package Plants

Membrane bioreactors (MBRs) is a technology that was first developed to process wastewater on cruise ships. MBRs are capable of filtering out pathogens, viruses, and other contaminants that conventional wastewater treatments do not. They have a smaller footprint than conventional methods because they combine biological treatment with filtration and often do not require separate primary treatment. MBR systems are simple to operate and take up very little space. Decentralized and on-site wastewater recycling in urban areas becomes feasible with this technology—both for the developed and developing worlds.[1]

Membrane bioreactor side stream configuration

Certain high value buildings in the United States have begun to use MBRs in small-scale package plants to recycle gray water on-site, such as the Frank Lloyd Wright Fallingwater Home in Chicago, the Bank of America Tower, and the Solaire building in New York City.

Note

1. S. Judd, *The MBR Book: Principles and Applications of Membrane Bioreactors for Water and Wastewater Treatment* (Oxford: Elsevier, 2006).

facilities as well as costs have been made possible by technological advances in membrane bioreactors (MBRs), an alternative for the treatment train described earlier in the chapter. (See sidebar on MBRs.)

Distributed or satellite systems are used typically for servicing dispersed small wastewater flows and small community-scale systems independent of a centralized wastewater system. The wastewater is collected, treated, and reused at or near the point of generation. Recently the term has come into vogue with both the environmental and the engineering community because decentralized systems are considered to be both more cost-effective and environmentally friendly, although proper operations and maintenance can be an issue.[45]

In urban areas, distributed or satellite systems refer to smaller-scale treatment facilities built as part of a larger centralized system of treatment and collection. They are generally built to minimize transport costs of sewage and stormwater. Package units make this type of system possible. A package unit is a small, all-in-one, turnkey unit, installation of which requires only a concrete slab, a power source, and a connection to the wastewater stream. A package unit performs all the functions of a municipal wastewater treatment plant, but on a much smaller scale. The units are typically about the size of a standardized shipping container, and can serve about 500–1,000 people (or about 150–350 homes). They can also be linked together to serve large populations. They require active (though not necessarily skilled) maintenance, and can be quite expensive to both purchase and operate.[46] The advances in MBRs now make possible either satellite treatment plants, or on-site treatment. See Figure 17.11 for a schematic for a distributed wastewater recycling system.

Zero Emissions Wastewater

Attempts have been underway in Europe and Asia for the past decade to develop "zero emissions" technology for wastewater that can be used on-site or on a highly decentralized basis. This concept considers water and waste as a system that can provide high quality recycled water, safe fertilizers, and energy.[47] This is called variously "resources management," "Eco-San," DESAR (Decentralized Sanitation and Reuse), ecological waste management, or sustainable sanitation. The aim

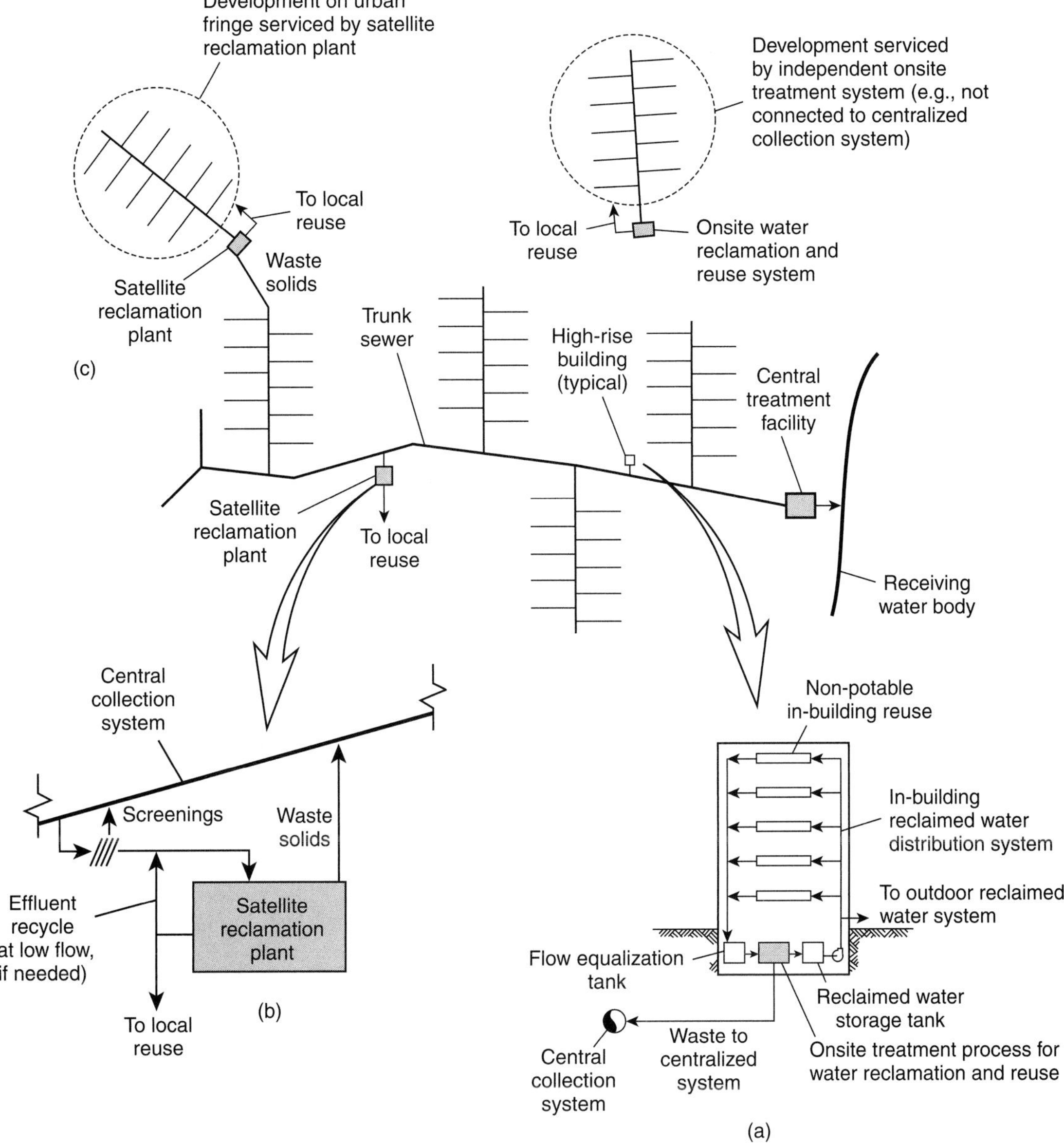

Figure 17.11
Schematic for a distributed wastewater recycling system

of these efforts is to mimic the natural water and wastewater cycle.[48] These efforts use source separation of the wastewater into black, gray, and yellow wastewater to recapture and reuse the water and nutrients. Treatment processes are targeted therefore more precisely to the type of wastewater (Figure 17.12).

Wastewater can be characterized as black water (usually defined as water from toilets and the kitchen), gray water (from showers, baths, and washing machines), and yellow water (urine) (Figure 17.13). The health danger of wastewater comes primarily from fecal matter, while urine contains most of

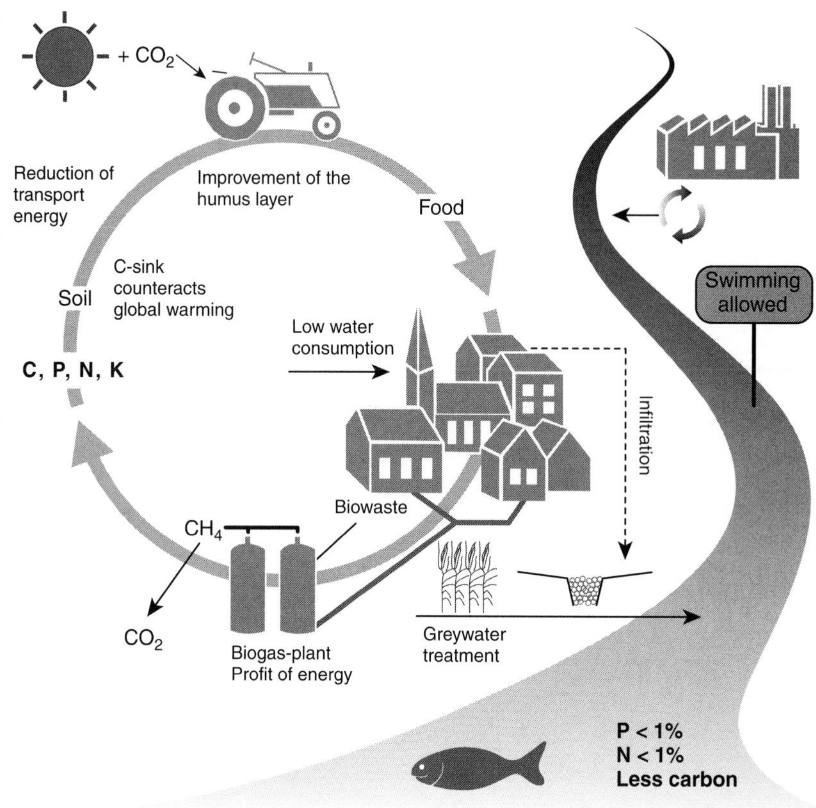

Figure 17.12
The ideal Carbon–Phosphorus–Nitrogen–Potassium cycle as promoted by Eco-San

the nutrients excreted in sewage: 80 percent of the nitrogen, 50 percent of the phosphorus, and 70 percent of the potassium.[49] One component of some of the pilot efforts in Europe is separation toilets (also called no-mix toilets, sorting toilets, or separating toilets). These separate black water from the urine (Figure 17.14).

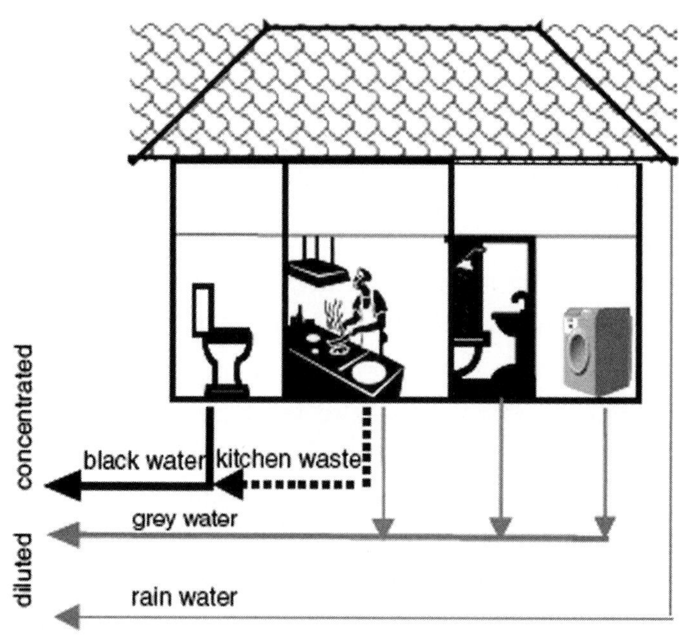

Figure 17.13
Household wastewater types

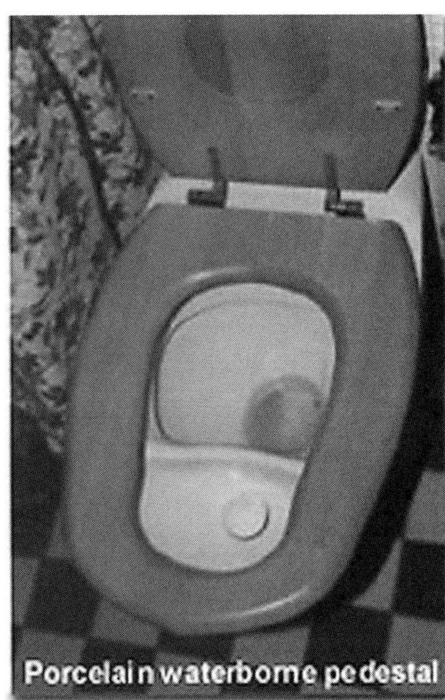

Figure 17.14
Three types of urine-diverting toilets

Gray water can be reused on-site for irrigation and toilet flushing.[50] Black water, including food wastes, is used for biogas and energy production or for soil amendments. Urine is used as fertilizer. The low tech solutions rely on natural processes. The advanced system pilots use membrane bioreactors (MBRs). Pilot schemes in Switzerland, Germany, the Netherlands, and China range from a fourplex to an eco-town of 430,000 where black water and food waste are combined to produce biogas.

Natural Treatment Systems

Natural treatment systems are both an old and a new approach to wastewater treatment. These include the Swedish "ecological villages" of Toar and the Ostra Tom secondary school.[51] In addition to the DESAR source separation approach noted above, these projects use oxidation ditches, aerobic ponds, and both open and underground wetlands. This is an area of considerable interest today both in the United States and abroad because of wetlands' important role in carbon sequestration and their ability to naturally treat wastewater. Constructed wetlands can be designed for storm or sanitary wastewater treatment, and for surface or subsurface flows. They are attractive in rural areas where land costs are low because they can provide secondary or tertiary wastewater treatment with low energy costs. In urban areas of the United States, where land costs are higher, they are more useful for tertiary treatment and to act as buffers for wet weather effluent flows both from storm sewers or from combined sanitary and stormwater sewers.[52]

Source separation of urine from gray and black water in the United States has not occurred except with the use of waterless urinals. However, the urine has not been reused. Schools in California recycle gray water for irrigation. On-site black and yellow water reuse in urban U.S. areas using natural processes is rare. The Living Machine@, ECO Machines, and Solar Aquatic Systems are some exceptions but they are used primarily for small-scale developments and individual buildings.

Planning for Wastewater

Wastewater Planning and Land Use

Many wastewater departments and agencies are becoming quite sophisticated in their planning activities. Whereas a decade ago, planning for sewer treatment and collection facilities was driven primarily by engineering standards, today a new generation of wastewater planners are engaging in public participation processes and actively seek to link facilities to land use plans and community values. In addition, conservation and pollution prevention are becoming mainstream in the field of sanitary engineering.

The state-of-the-art publications for wastewater capital and strategic planning call for explicit coordination with the local general purpose government and many states have mandated such coordination.

At the city, county, and regional level, wastewater planning also looks to the local land use requirements. The county of Spokane, Washington, developed a plan in 2002 to protect the Spokane-Rathdrum aquifer, consistent with the land use policies of the county's comprehensive land use plan that called for wastewater services only within its Urban Growth Boundary unless necessary to protect the aquifer. To prepare the plan, some of the existing sewer service areas designated by the various wastewater utilities had to be changed to comply with the land use plan.[53]

Many local land use plans are not quite as sophisticated with respect to wastewater planning. However, sewers are powerful determinants of where growth in an area will go. They are often called growth shapers because they lower the costs of development. Studies have shown that sewered vacant land ("sewered" means access to an interceptor or trunkline) in suburban communities is worth from two to four times as much as equivalent unsewered land in the same communities.[54] In shaping development, a city's choice of sewer investment can be more important than the city's land use plan. A large interceptor that is run out to the new office park south of town will suddenly make all the land it traverses attractive to development, even if that land is identified elsewhere as open space or agricultural.

Table 17.1
Sewer-related land use policies and implementation mechanisms

Policy	Mechanism	Governmental level	Time horizon for planning
Sewer moratorium	Ban or quota on new sewers, connections, building permits, subdivision approval, rezonings	State or local	Two years or less
Refusal to sewer	Refuse to provide sewer to rural fringe	Local	n/a
Environmental review	Use environmental review to highlight sewer impacts	All	n/a
Facility design	Deliberately design facilities to encourage types of land use, such as smaller treatment plants	Local, county	20 to 50 years
Facility sizing	Limit excess capacity in facilities to preserve future land-use options	Local, county	20 to 50 years
Urban services boundary	Establish an urban service boundary beyond which no sewers will be expanded	County, regional	10 to 20 years
Staging	Stage sewer projects to promote certain development patterns	Local, county, regional	5 to 10 years
Water reuse	Local building code requires dual pipes	Local	5 to 10 years
Energy conservation	Local policies innovative design	Local	5 to 10 years

Source: Richard D. Tabors, Michael H. Shapiro, and Peter P. Rogers, *Land Use and the Pipe: Planning for Sewerage* (Lexington, MA: Lexington Books, 1976).

The courts have viewed access to sewers as a public health matter rather than a local police power or growth management matter, which means without careful planning policy-makers can be limited in their ability to control growth by sewer investments. The local general purpose government, however, has a variety of mechanisms to ensure that sewers are guided by the land use vision of the community rather than vice versa. Local policy-makers can decide where and, just as importantly, when to expand municipal wastewater service. See Table 17.1 for some varied land use policies that can be used by the local government for sewers.[55]

Key Steps in Wastewater Planning

It is likely that the local agency will hire an engineering firm to do the detailed wastewater plan. However, the local practitioner will need to set the overall policy and objectives for the process. This section highlights some key steps.

Content of the wastewater plan. The areas that must be addressed during the planning process are:

- growth and expansion;
- renewal and replacement needs;
- compliance with federal, state, and local regulations;
- community objectives, including water recycling and nutrient re-use;
- financial strategy.

It is important that a high-level strategy framework be put in place to help guide the planning process. Is the community committed to looking at wastewater as a resource? Is the community ready to support adequate maintenance of the system? Is there a balance between the needs for growth and the need to replace or renew the existing system? What are the financial constraints? What is the 50–100-year picture for the total system? What are the regulations that need to be adopted by the local general purpose governments to further conservation goals?

Wastewater service area definition. Even though wastewater collection service areas are constrained by topographic conditions, it is possible to preclude certain areas in the drainage basin from service, and it is also possible to accept sewage from an adjacent drainage basin by pumping. Precise delineation of sewer service areas is a significant variable for encouraging, discouraging, or shaping growth within a developing community. For example, the King County Wastewater Plan for 2000–2030 defined the boundaries of its service area as the service areas of the 31 local agencies that send their flows to the county's wastewater system for treatment (Figure 17.15). Spokane,

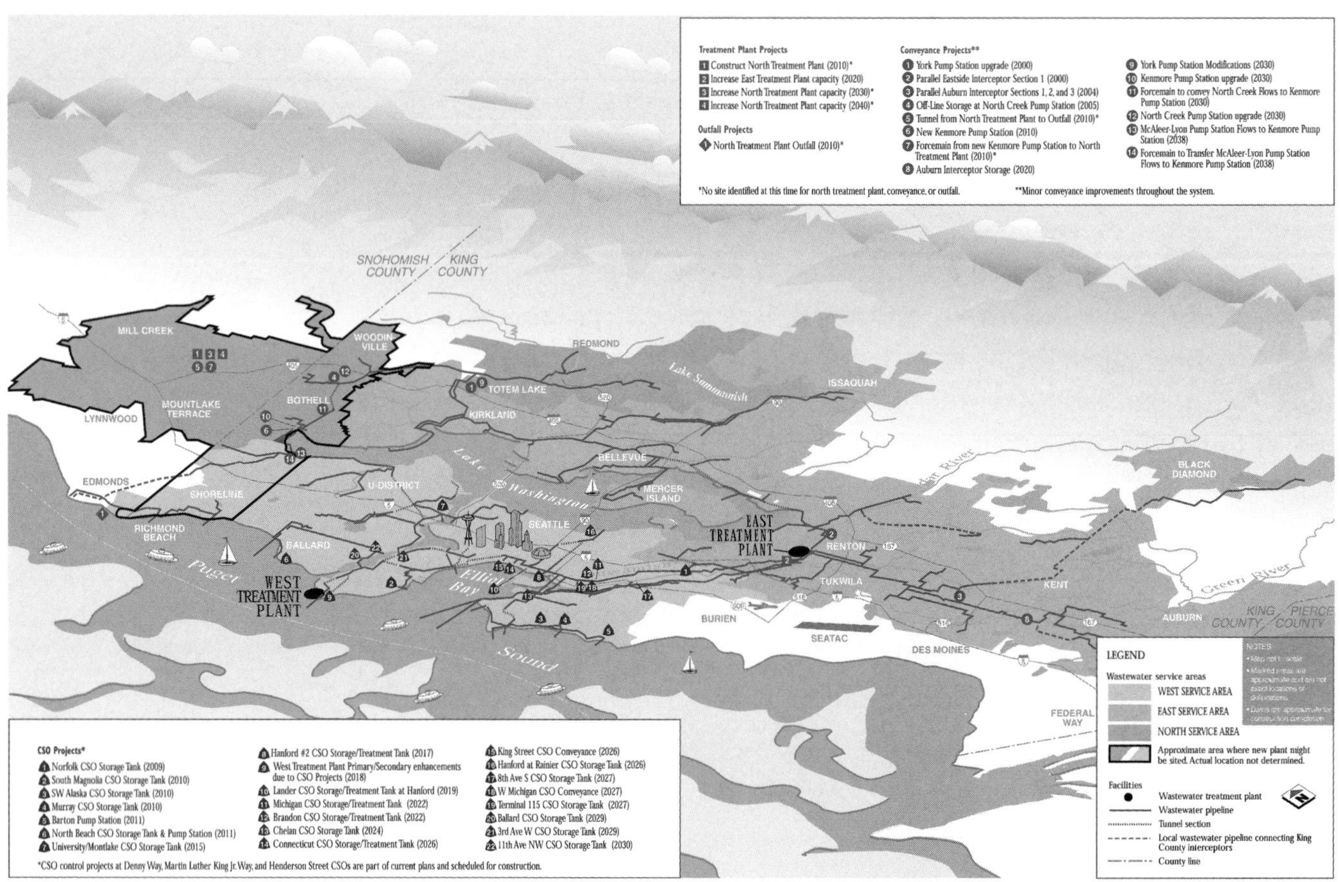

Figure 17.15
Regional wastewater services plan, 2000–2030

Washington, adjusted the boundaries of some of the collection agencies in order to stay within the urban growth boundary.

A simple method for determining a service area is to draw a line around the top of the drainage basin served by the current sewer system. The geographic area within that line should receive the highest priority for future growth. Any system expansion beyond that line will require pumps, new treatment plants, or other capital-intensive solutions. The capacity of the current treatment plant must be kept in mind during this process.[56]

Condition assessment of existing facilities. The capacity of the wastewater treatment plant is the primary determinant of the existing capacity of the sanitary sewage system. The capacity of a treatment plant is usually expressed in million gallons per day (MGD); one MGD is the amount generated by about 3,500 households. Sanitary wastewater generation depends directly on the amount of water used. About 40–85 percent of water used enters the sanitary sewer system, depending on climatic conditions and the exact pattern of usage and the condition of the collection pipes.[57] The amount of stormwater depends on the climatic conditions of the area.

In addition to assessing the capacity of the treatment plant, the condition of the collection lines should also be assessed. Infiltration, illicit connections, and sanitary sewage discharges during rainfall periods should be evaluated as part of the planning process. This kind of assessment is usually a major project and should be linked to an asset management system.

Future sewer and treatment plant needs. Future needs are calculated based on population and household projections by land use type as discussed in Chapter 16. The same methodology can be used for sewage flows. In addition, today, sophisticated sewer models are available to local planners that integrate parcel information and land use information with the existing sewer system and a GIS system. They can be used to identify problems in existing systems by modeling the impact of storms on the system, and to model different sewer configurations for new growth areas.[58]

Location of the wastewater treatment plant. Wastewater treatment plants require careful thought regarding their location. Depending on the capacity, they can require 10–20 acres or more. A plant with a capacity of 50 million gallons per day, which is about the size needed to serve a population of 150,000–200,000, needs about 12–13 acres.

Since they rely on gravity, they are placed at the lowest point in the service area near a body of surface water in which to discharge their effluent. If the receiving body of water also serves as the source of drinking water for the community, the wastewater plant should be downstream of the water treatment plant. If recycled water is to become a viable source for landscape irrigation or industrial use, the more that potential "users" are in proximity to the plant or outfall lines, the greater the chances of making it cost-effective. These include golf courses, industrial parks, major open spaces and parks, and other potential recycled water users.

Financing Mechanisms

Sanitary sewer system capital and operating costs are usually paid for with user fees, although in the past two decades, impact and development fees have been assessed to pay for new infrastructure.

The Fee Development Process

The basic process for fee development is:

- Assess revenue requirements.
- Conduct cost of service analysis.
- Develop fee structure.
- Recommend, adopt, and implement.

Financing Sanitary Wastewater

Fees. Most sewer agencies rely upon on-going fees from existing residents and businesses to pay for operations, maintenance, and capital replacement and upgrades. It is hard to meter sanitary sewage, so generally sewer fees are based on a percentage of water use, and they often show up on the water bill.

Over the past 20 years, EPA guidelines for wastewater charges suggest that if sewerage charges exceed 2 percent of median household income, or if water and sewerage together are greater than 4 percent of median household income, this may be a hardship for some families. At the beginning of the fee setting process, household income should be projected to obtain an envelope of funds that are available for the sewer system. Comparisons of existing and projected costs with neighboring or similar jurisdictions should be done.

Using fees for equity and land use conservation. Sewer system fees can have both land use and equity effects. Research has shown that the costs of providing sewer (and water) service are higher in areas that are further away from existing centers, and area with larger lot sizes. If water and sewer costs are charged to the development, there is an incentive for the developer to minimize these costs. In addition, where average cost pricing is used for both new development and replacement of existing infrastructure, those in more compact areas subsidize the costs of those in less dense areas. Fortunately pricing strategies exist to overcome these problems.[59]

Paying for capital facilities for wastewater. During the 1970s and 1980s, the federal government funded a construction grant program that provided over $60 billion to construct publicly owned and operated wastewater treatment projects. The grant program ended in 1987 and was replaced with the Clean Water State Revolving Loan Fund (SRF) which consists of seed money for states to capitalize state loan funds. In 1998, the program had over $26 billion in assets and $3 billion distributed in new loans each year.[60] Today, sewer capital investments are generally repaid through development impact fees, which are sometimes included in connection fees, the per-dwelling unit charge associated with hooking up to the sewer. Sometimes these are charged to the developer as a block.

Development impact fees. There are three approaches used for development impact fees for wastewater facilities. The first is a system buy-in or reimbursement approach for new development based on existing facilities and costs. Under the second approach, the new development pays for the cost of new facilities. The third is a combination of these two. All three approaches are used for waste and stormwater today, and are guided by the enabling state legislation or case law.[61]

The heart of any fee structure is the documentation of the methodology. Optimally this is based on a detailed capital improvement plan so that the costs of the infrastructure can be allocated to the development. To do this, an indicator is selected (often called a scaling measure) to distinguish between different classes of users so that customers who are larger or use the infrastructure more intensively pay the costs of the capacity required to serve them. For water and wastewater, this data element or indicator can be the water meter size, the PFU (the number of plumbing fixtures), or any other measure readily available that is a reasonable surrogate for the development's share of the capital costs.[62]

Reclaimed water fees. Reclaimed water rates pose a special issue because reclaimed water usually costs more to produce than water from conventional sources and it can be of lower quality. However, use of reclaimed water reduces the demand for potable water and may even help lower costs when it is used to recharge groundwater supplies. Reclaimed water rates set by water utilities are generally based on some percentage of the potable water rate. Some agencies price reclaimed water below the potable water rate to encourage its use.[63]

Conclusion

One of the most important issues today is the technology of sanitary sewers. Just as stormwater runoff has ventured into the area of natural processes and away from traditional brick and mortar solutions, increasingly the field of sanitary engineering is turning to sustainable solutions. Decentralized treatment facilities for sewage will increasingly be important

Connection Fees in Boise, Idaho

The connection fee is made up of the trunk fee, the treatment fee, and the interceptor fee. Each of these is paid by new and existing development at the time of connection.

- *Trunk fee*—This fee recovers the cost of over-sizing sewers to accommodate large geographic areas and to recover part of the cost of sewers in existing neighborhoods.
- *Treatment fee*—This fee recovers the capital costs of the treatment plant.
- *Interceptor fee*—This fee is for properties in the Southwest Sewer Planning Area and will recover the cost of constructing a 42-inch diameter, 5-mile-long interceptor sewer to the West Boise Treatment Facility located below Hewlett-Packard. Commercial connection fees for unmonitored users are calculated per plumbing fixture based on the estimated strength of the discharge. Monitored users pay on actual production of sewage.

Adapted from City of Boise website.

alternatives to the traditional treatment at the end of the pipe. Source control, source separation, and recycling of urine and feces are issues that need to be explored, particularly given the increasing shortage of pollution-free water sources for consumption, and the high costs that the United States faces to replace the existing system. Local regulations that promote water recycling and the reuse of biosolids should be actively promoted by the local practitioners. In addition, the growth-inducing effects of sanitary sewers and treatment facilities should be explicitly addressed in the local land use plans and development policies, so that the vision of the community guides the placement of these facilities.

Best practices in the water and wastewater field are to plan holistically for the entire water cycle, but often professionals are stymied by the myriad of local and state agencies that can prevent this. The local practitioner can play a pivotal role in requiring and supporting the emergence of a new infrastructure that looks at wastewater as a resource rather than a nuisance. The stakes are high, given the challenge of emerging contaminants and the continuing deterioration of the existing pipes and plants.

Additional Resources

Publications

Feiden, Wayne M. and Eric S. Winkler. *Planning for Onsite and Decentralized Wastewater Treatment*. Planning Advisory Service Report Number 542. Chicago, IL: American Planning Association, 2006.

Melosi, Martin V. *The Sanitary City: Urban Infrastructure in America from Colonial Times to the Present*. Baltimore, MD: The Johns Hopkins University Press. 2000.

Novotny, Vladimir S., Jack Ahern, and Paul Brown. *Water Centric Sustainable Communities: Planning, Retrofitting, and Building the Next Urban Environment*. Hoboken, NJ: John Wiley & Sons, Inc., 2010.

Sarte, S. Bry. *Sustainable Infrastructure: The Guide to Green Engineering and Design*. Hoboken, NJ: John Wiley and Sons. 2010. Chapter 8, Integrated Water Management, has detailed information on designing that views "waste as a resource, not waste product."

Websites

American Public Works Association (APWA): www.apwa.net/ResourceCenter/Category/Water-and-Sewers. This has a good listing of websites that address water and sewer issues.

National Association of Clean Water Agencies (NACWA): www.nacwa.org. This is one of three professional organizations representing sewerage agencies, with concerns about both urban and environmental issues. Their website has the latest in policy issues.

Water Environmental Research Foundation: www.werf.org. This organization recognizes the importance of a multidisciplinary approach to wastewater management that integrates the engineering side with that of city officials, including planners and developers. They have a wealth of resources on their website and a good search engine.

chapter 18

Stormwater and Flooding

In this chapter ...

Importance to Local Practitioners . . . 325
History of Stormwater Infrastructure . . . 326
Early Responses to Rainfall, Storms, and Flooding . . . 326
Industrial Era Stormwater Management . . . 326
End-of-the-Pipe Controls . . . 327
The Fifth Paradigm and Stormwater . . . 327
Elements of the Stormwater System . . . 327
The Institutions of Stormwater Infrastructure . . . 330
The Federal Level . . . 330
The Role of the EPA . . . 330
Local Governments . . . 332
Current Issues . . . 333
Stormwater Runoff . . . 333
Climate Change and Flooding . . . 335
Sea Level Rise . . . 337
Planning for Stormwater . . . 338
Maintaining Pre-Development Stormwater Patterns . . . 338
The "Rational Formula" . . . 338
New Development Planning . . . 340
Reducing the Level of Pollutants . . . 342
Automobile-Related Measures . . . 342
Other Pollution Prevention Strategies . . . 344
Stormwater Financing . . . 345
Local Stormwater Fees . . . 345
Low Cost Loans from the State under the Clean Water State Revolving Fund . . . 347
Conclusion . . . 347
Additional Resources . . . 348
Publications . . . 348
Websites . . . 348
Key Federal Regulations . . . 348

Why Is This Important to Local Practitioners?

Stormwater infrastructure policy and practice have come full circle in the new millennium. Stormwater infrastructure was originally designed to address flooding in urban areas, but during the past two decades the emphasis has been on addressing pollutants from stormwater runoff. Federal regulations have required large-scale "gray" infrastructure solutions to retain stormwater at or close to the central treatment plant until pollutants have been removed. Now leading-edge cities are looking at the costs of this massive infrastructure and are promoting green infrastructure options and resilient city developments to slow down the path of stormwater. The new vision for stormwater infrastructure is also due to global warming, which causes more severe storms, higher ocean levels and potentially serious damage to existing urban settlements. In drought areas, a fresh look is being taken at stormwater as an urban water source. These factors have brought together hazard mitigation, pollution control, and water supply professionals in search of a more decentralized approach that involves urban design, zoning, and building regulations and emergency response planning as well as distributed and natural approaches to stormwater control. Adaptation planning in response to increased and more severe weather events ties into the "one-water" paradigm described in the previous chapters.

This chapter begins with the history of stormwater infrastructure and then moves to more detail about the current issues. The elements of the stormwater system are outlined before moving to the regulatory and institution environment. Best practices of the new approach to stormwater are addressed before stormwater infrastructure financing is touched on.

History of Stormwater Infrastructure

Early Responses to Rainfall, Storms, and Flooding

Civilizations have been designing systems to collect and control stormwater for thousands of years. In many civilizations, the stormwater was collected for eventual use in agriculture or as drinking water. The Nabotaean civilization of present-day Jordan built an elaborate gutter system that channeled rainwater into cisterns located throughout the city. However, as cities became larger and denser, systems purely for the purpose of draining stormwater developed. Greek houses had gutters that emptied into drainpipes that in turn emptied into channels in the street. The Romans followed suit with the Cloaca Maxima, which is a 600-meter-long tunnel that drains the Forum in Rome. It was built in 600 BC and is still in use today. Yet, although some streets and areas were paved in these communities, hydrologically speaking, large amounts of pervious surface were also available and infiltration of stormwater into the groundwater table was not a problem. In Europe, it was not until the Middle Ages that cities had primitive sewer systems.[1]

Industrial Era Stormwater Management

Stormwater infrastructure was initially designed to address the volume or sudden quantities of runoff and to move it as quickly as possible away from where it could do damage. The in-migration to cities caused by the Industrial Revolution resulted in covering natural pervious surfaces with buildings, and paved streets of asphalt and concrete. Major cities filled in their wetlands and floodplains for development. The groundwater table dropped in many cities because of the lack of recharge by precipitation, and because of sump pumps draining away water so that basements, foundations, underground garages, and the like could be built. Urban streams were straightened, lined, and ultimately covered to reduce flooding.[2]

In the post-World War II era, often stormwater infrastructure was built only after flooding occurred. An underground drainage system of pipes and catch basins became common in the USA to transport storm flow downstream. Frequently, however, cities and developers employed these devices without regard to the cumulative downstream effect, which was frequently disastrous.[3]

Floods are one of the most common hazards in the United States. Flood effects can be local, impacting a neighborhood or community, or very large, affecting entire river basins and multiple states. Most flood control has been viewed primarily as a local matter with federal flood control efforts focused on the large-scale infrastructure such as the dams and levees built by the Army Corps of Engineers. Until recently, the focus was on volume control not pollution.

End-of-the-Pipe Controls

Pollution is another consequence of stormwater runoff. This issue reached a critical mass in the United States at the end of the 1980s. Since then the same "large-scale collection and treatment at the end of the pipe" paradigm that was used for sanitary sewage has been applied to stormwater discharged into receiving water bodies. Called "non-point pollution" to distinguish it from pollution coming from wastewater treatment plants, this approach calls for localities to use detention basins, ponds, or tunnels both at the site level and at the end of the stormwater pipe to sequester the storm flows until they could be treated, or discharged harmlessly. The culmination of this approach is the so-called "deep tunnel" concept. The Chicago Tunnel and Reservoir Plan (TARP) facility has a length of 109 miles and 33 feet in diameter and cost $3 billion in 1990 dollars. The deep tunnel in Milwaukee, Wisconsin, was built at a cost of $1 billion in 1990 dollars, and is 46 kilometers long but even that did not prevent a serious outbreak of disease following a storm in the 1990s. Currently, there are more than 47 tunnel projects in the United States and overseas. Large tunnel projects in operation in the United States include Austin, Boston, Cleveland, Houston, Detroit, Los Angeles, Portland, New York, Minneapolis, and San Francisco.

The Fifth Paradigm and Stormwater

Concern about the cost, energy, and carbon emissions impacts of the deep tunnel "gray infrastructure" requirements intersected with the call by environmentalists to maintain or restore the natural hydrology of the watershed and to reduce pollutant loads in water bodies. The result has been a transformation in the national regulatory environment to permit decentralized stormwater management solutions. Called variously, "green infrastructure" or low impact development (LID) in the United States, or SUDS (Sustainable Urban Drainage System) in Australia, natural stormwater solutions are an integral part of the "Fifth Paradigm", in which water management is integrated into the design of the city.[4]

Elements of the Stormwater System

Stormwater is water or snowmelt that comes from precipitation. When it rains, or when snow melts, some water is absorbed by the ground but the rest forms puddles or makes its way to rivers, lakes, and other surface waters, carrying elements it has picked up on its journey. This is called surface runoff. Forests, wetlands, and grasslands trap rainwater and snowmelt and allow it to slowly filter into the ground (infiltration). In these natural situations, runoff reaches surface waterways gradually and is usually stripped of many of the foreign substances it had been carrying (Figure 18.1). As discussed further in the

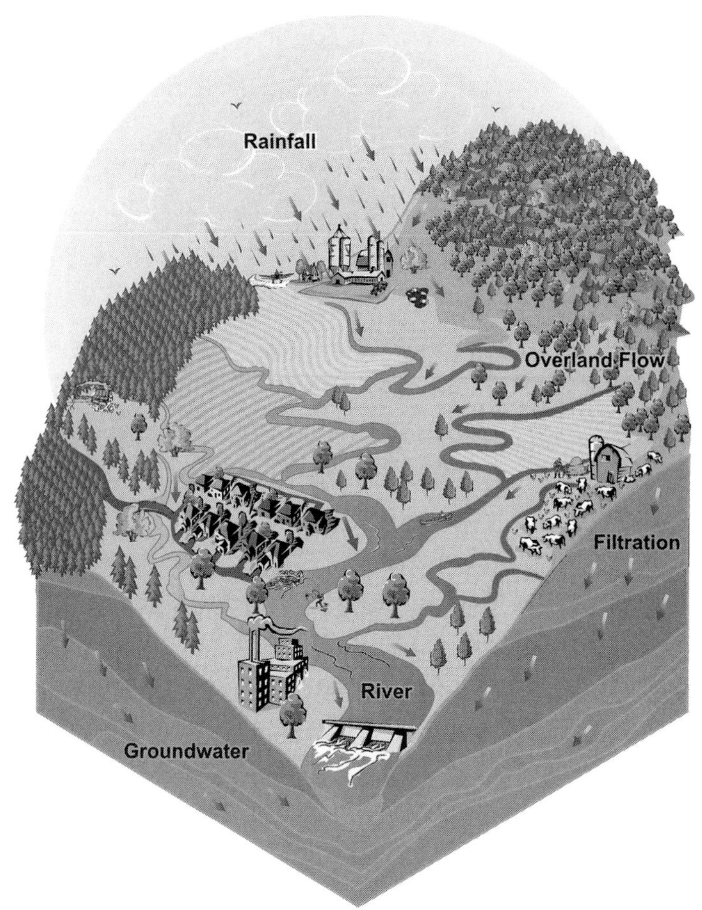

Figure 18.1
Components of a watershed

chapter, urbanization has caused this natural cycle to be disrupted. Urbanization has also required stormwater infrastructure to protect citizens and property from the adverse effects of storms and flooding.

The stormwater infrastructure system is composed of two subsystems: minor and major elements. The minor elements are designed to remove stormwater from areas where people walk, bike, or drive. They include curbs, gutters, street inlets, underground culverts, ditches, channels, and swales (a grassed depression). Municipal stormwater sewers as well as combined sewers are part of this system (discussed further below). Storm sewer systems follow a pipe hierarchy similar to that for sanitary sewers. Lateral connections often drain the stormwater from a roof and property to either a separate storm sewer, or in the case of older and larger cities, a combined sanitary and stormwater sewer. To collect the stormwater from the streets, storm sewers have inlets, which are the grated openings often found at street corners. These lead to sub-mains in the street, which in turn lead to trunk lines. However, there are other kinds of sewer systems in use.

Municipal separate storm sewer systems (MS4). MS4s are sewer systems where the pipes carry only stormwater and not sewage, and are "separate" from the sanitary sewers. EPA regulations apply to them since they often carry polluted stormwater runoff that is discharged into local water bodies. They are built to carry runoff from the 5- to 10-year storm runoff. In a city with separate systems for storm and sewage, the stormwater sewers discharge directly into surface water bodies.

Combined sewer systems (CSS). Combined sewers are remnants of the nation's early infrastructure and are typically found in older communities. Combined sewer systems serve over 880 communities and more than 43 million people in the United States. A combined system carries domestic, commercial, and industrial wastewaters and stormwater runoff through a single pipe system to the treatment plant, where it is treated and discharged into the surface water. Combined sewer systems can have large storage facilities (the "deep tunnels" referred to above are one type of storage facility) to store sewage until plant capacity is restored. Combined sanitary and storm sewers have a capacity of six times the flow in dry weather.

On-site storage and small detention facilities may also be parts of the minor element of a stormwater system. It can also include parking lots and anything that transmits water from the site, including culverts and drainage ditches. Street gutters and drains are typically designed for 10-year peak flows, while floodways and ponds are designed for the 50-year or 100-year peak flow.[5]

The major elements of a system serve flood needs and emergency flows. They include creeks and rivers, lakes, ponds,

Lowering a drainage pipe at Sandlake Galloway Road Project

Drainage pipes at Sandlake Galloway Road Project

A stormwater culvert under construction

marshes, estuaries, and the ocean. They are evaluated in terms of fairly infrequent events, such as a 100-year storm. (This has a 1 percent chance of occurring in any given year, a 5-year storm has a 20 percent chance, and a 2-year storm a 50 percent chance.) The major element is referred to as the receiving water component.

Wetlands. Wetlands are natural elements of the stormwater control function that have only been appreciated within the past few decades. Wetlands are natural sponges that trap and slowly release surface water, rain, snowmelt, groundwater, and flood waters over the floodplain. The combined storage and slowing of the flow lower flood heights and reduce erosion. Preserving and restoring wetlands, together with other water retention, can often provide the level of flood control otherwise provided by expensive dredge operations and levees.[6] Wetlands also sequester carbon.

The Institutions of Stormwater Infrastructure

Stormwater management has historically been a local government responsibility. Today, however, both the federal and state governments are involved. The following begins at the federal level.

The Federal Level

The U.S. Army Corps of Engineers is one of the oldest actors in the stormwater arena. It has the responsibility for federal flood protection. It builds and repairs dams and levees, keeps major streams and waterways clear, and supports state and local agencies in addressing flood management. In 1968 the National Flood Insurance Program also became involved. As a condition of the insurance, localities were required to map their flood plains and to adopt hazard mitigation policies, such as restricting building on flood plains. The Federal Emergency Management Agency (FEMA) was then created in 1979 to coordinate federal disaster-related programs. It currently also manages the flood insurance program which today has a community rating system (CRS) program which is a voluntary program for localities that go beyond minimum requirements. In return, there are discounted insurance rates. Finally, the Agriculture Department's National Resources Conservation Service funds some flood control projects that affect watersheds.

The Role of the EPA

The Environmental Protection Agency (EPA) has the bulk of the federal responsibility for regulating state and local stormwater programs. The NPDES permitting program (discussed in Chapter 17 as it applied to discharges from municipal wastewater treatment facilities) was expanded by Section 319 of the Water Quality Act of 1987 to cover non-point source pollution that often occurs as part of stormwater runoff. This section requires states and local governments to monitor and control pollution from diffuse sources. Section 402(p) of this Act expands the scope of the NPDES permit to include discharges from stormwater systems, as well as construction and industrial discharges.

Combined sewers systems (CSS) have been subject to EPA regulations and required to obtain an NPDES permit since the 1972 Act was passed. This is to prevent illicit discharges and pollution from entering receiving water bodies. They fall under the "point source" category since CSS carries sewage and stormwater to the centralized treatment plant. Municipal Separate Storm Sewer Systems (MS4) have only been required to obtain an NPDES permit for their discharges since 1990. Utilities in Phase I operators (for medium and large cities and some counties) were required at that time to prepare a stormwater management plan (SWMP) as part of the permitting process. As of 1999, utilities in Phase II cities (smaller MS4s in urbanized areas), were also required to develop a plan and obtain a permit.

Requirements for stormwater programs as part of the SWMP include: public education and outreach; public participation and involvement; illicit discharge detection and elimination; construction site runoff control; pollution prevention; and pollution remediation (or "treatment").

Recently, the EPA has begun to recognize the need to plan on a holistic basis for wastewater and stormwater.[7] In a draft strategy,[8] the EPA notes that integrated planning will help localities to make better infrastructure investments, and will also "facilitate the use of sustainable and comprehensive solutions, including green infrastructure." The objective of this approach is to integrate the NPDES permitting process for these functions, regardless of the actual agency administering the program; and to involve all stakeholders in the process.

The 1972 Clean Water Act also requires all states to specify total maximum daily loads (TMDLs) for each pollutant for water bodies within their jurisdictions that do not meet quality standards. Implementing regulations were not written until 1985 and again in 1992. They were modified again in 2000, but Congress prohibited the EPA from spending money to implement them. Current EPA regulations on TMDLs still call for states to submit lists of impaired and threatened water-bodies and to develop TMDLs for them.[9] States are currently completing their TMDLs and putting into effect those that have been completed.

A TMDL is the maximum amount of a single pollutant that a water body can receive and still meet water quality standards. The TMDL will include allocations of pollutants from point sources and non-point sources, as well as from natural background conditions. It also considers seasonal variations. The TMDL approach steps away from saying that a particular level of effluent is acceptable and instead focuses on the pollutant loads that will not impair the water body, given its particular characteristics and ecology. The TMDLs are critical in watershed planning since they are a tangible goal or criteria that can be used for assessing alternate programs.[10]

Local Governments

Local government agencies are generally responsible for stormwater management and flood control. Many states require that the land use element of the plan must identify areas that are subject to flooding, and that it must be coordinated with other local agencies that deal with water or flooding. As part of the general plan requirements, the land use element of the plan must identify areas that are subject to flooding and this should be coordinated with other local agencies involved with water or flooding or reclaimed land.

There is no general pattern across the country for which particular agency has the responsibility for flood management. County flood control districts and county water agencies often are the principal agencies but in many areas this responsibility rests with cities, counties, reclamation districts, levee districts, drainage districts, resource conservation districts, and community service districts. Construction and maintenance of the "minor system" are often the responsibility of the local general purpose government (GPG) because they are also responsible for the streets.

A stormwater utility usually is set up as a special assessment district so that it can generate funding for its infrastructure expenses. Users within the district pay a fee for drainage plans, maintenance and upgrading of existing storm drain systems, flood control measures, and sometimes capital construction projects. In 2003, the American Public Works Association (APWA) estimated that about 500 communities had established a stormwater utility, with many concentrated in Florida, Washington, California, and Oregon.[11] Some estimate that by 2010 there will be more than 2,500 such entities.[12] Other localities use development impact fees to pay for the cost of the stormwater management facilities.[13]

Many of the smaller Phase II MS4 communities referred to above have been aggressive in establishing stormwater utilities so they could pay for federal requirements. Early

Santa Barbara County's Flood Plain Management Program

Through Santa Barbara's Flood Plain Management program, the County reviews proposed subdivisions and single building permit applications for elevation above the 100-year flood level. Proposed development is checked for conformance with the Flood Plain Management Ordinances, setback from major watercourses, adequacy of drainage plans, regional drainage planning, and protection of existing development. This review is intended to prevent future flood hazards from being created in developing areas and to eliminate the need for constructing future expensive flood control facilities.

Individuals within the unincorporated areas of the County receive a discount on their flood insurance rates. Benefits include the prevention of losses in flood-prone areas and reduced need for public emergency response during storm activity. The floodplain maps developed by the Federal Emergency Management Agency for Santa Barbara County and the seven cities are a valuable tool in regulating development in floodplain areas.

Adapted from Santa Barbara County's Flood Plain Management Program.

adopters in Florida and California saw local elections for stormwater utility districts going down to defeat and many attribute the passage of a measure in California in the mid-1990s that required voter approval before assessing fees in general to rapid increases in stormwater fees. However, the use of a dedicated enterprise or utility supported by fees for stormwater programs is now becoming more accepted throughout the U.S.

Current Issues

Stormwater Runoff

Stormwater runoff was not considered a serious issue for many years. Today, there are two main concerns. First, both urban and agricultural runoff, primarily as a result of rainfall and storms, is the major cause of pollution for 40 percent of water bodies in the United States that have quality problems.[14] (See Table 18.1.) Second, uncontrolled stormwater runoff can cause flooding, threats to the built environment, and waterway erosion, which in turn results in more pollution and destruction of the natural environment. Agricultural and mining runoff are also issues.

Faster and more intense runoff. Stormwater runoff in urban areas causes changes to the pattern of water runoff. In urban areas, the slow and natural process does not occur. Much of the urban landscape such as roofs, streets, and parking lots are impermeable, and they do not let the rainfall or snowmelt slowly percolate into the ground. Water remains above the surface and runs off in large amounts. Storm sewer systems that collect and channel the runoff result in higher hydraulic efficiency, that is to say, increased speed of the runoff reaching the water body. (Hydraulic efficiency refers to the rate of energy loss in the water as it travels through the stormwater system.) These factors combine to increase peak flows—they become "flashier." Creeks that would normally see a certain level of storm flow every two years experience the same flow several times in one year.

Table 18.1
Causes of pollution in different types of surface water bodies

Rank in order of severity*	Rivers and streams	Lakes, ponds, and reservoirs	Estuaries
Area	3.7 million miles	65,625 mi^2	90,000 mi^2
Polluted (%)	35	45	44
1	Agricultural runoff	Agricultural runoff	Municipal point source
2	Hydromodification**	Hydromodification**	Urban runoff
3	Urban runoff	Urban runoff	Atmospheric deposition

Note: * Severity is evaluated by percent of area polluted.
** This includes the heating of water through receiving plant cooling water, elimination of bank shading, as well as frequent changes or reductions in the flow from irrigation or climate change.

In addition, because not as much water is percolating into the ground, dry water flows are also decreased. The net result is lower water depths during nonstorm periods, higher-than-normal water levels during wet weather periods, increased sediment loads both on site and further downstream, and higher water temperatures. Stream banks erode and widen and streamside vegetation is damaged.[15] (See Figure 18.2.)

Urban runoff pollution. Sediments and solids constitute the largest volume of pollutant loads to surface waters in urban areas. These include sediment from development and new construction; oil, grease, and toxic chemicals from automobiles; nutrients and pesticides from lawns and gardens; and road salts from de-icing. When runoff enters storm drains, it carries many of these pollutants with it. This polluted runoff is often released directly into the water without any form of treatment. Dissolved metals may also negatively impact aquatic life.[16] In addition, during wet weather, conventional pollutants contained in municipal wastewater may also overflow into the surface water.

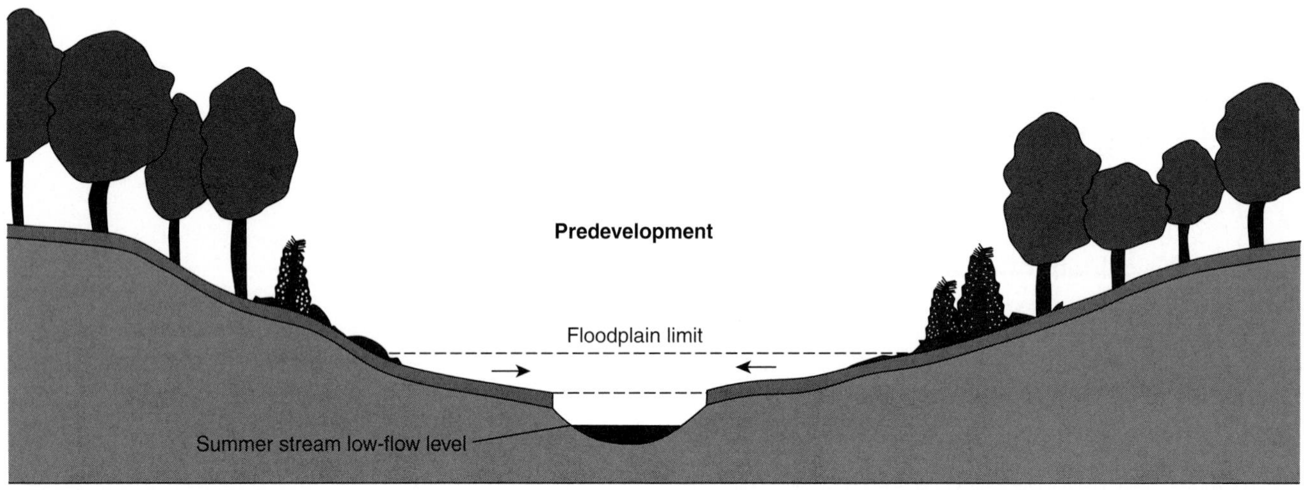

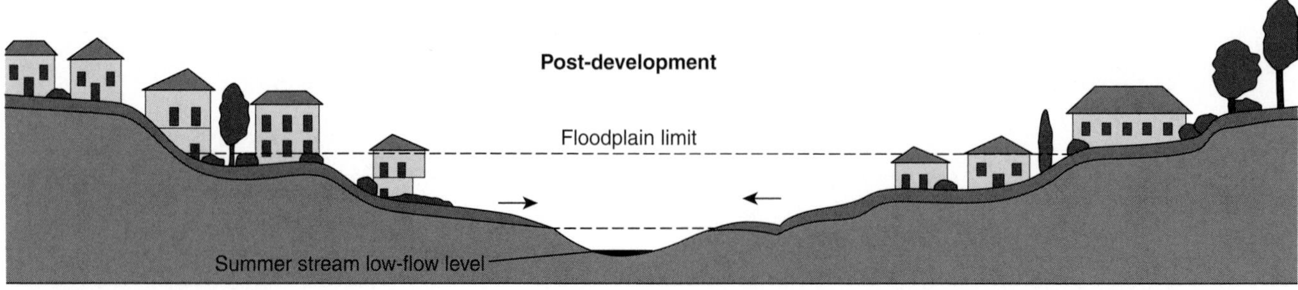

Figure 18.2
Development effects on river flow

Agricultural pollution. In rural areas, animal feeding operations (AFOs) are significant producers of polluted water. There are 450,000 animal feeding operations in the United States, such as hog, cattle, and poultry farms. A smaller number are classified as CAFOs (concentrated animal feeding operations) where the animals are confined, and therefore their waste products become concentrated. If not managed, animal waste can cause serious water pollution, including nutrient enrichment of surface waters, fish kills, and contaminated drinking water.

Many "factory farms" do not have adequate plans for storing and disposing of large volumes of animal waste. Hog farms may use open-air lagoons and spray-fields to store manure that can contaminate groundwater and streams.[17] The CAFOs are classified as point sources under the NPDES program and must apply for a permit. However, as of 1998, fewer than 25 percent of the 6,600 CAFOs had a permit.[18] In addition, farming that relies extensively on irrigation and fertilizers can lead to significant nitrate problems in groundwater as well as salting problems in the soil.[19]

Mining. Current and past mining activities also produce acid mine drainage wastewater, particularly in northern Appalachia. EPA estimates that there are 200,000 abandoned mines nationwide, with between 2,000 and 10,000 active mines. Some of these produce acid mine wastewater as well as heavy metal discharges into local streams and waterways. Runoff from inactive and active mines is more difficult to control since mining is regulated by many different state and federal agencies. Recent efforts have focused on a watershed approach for mining.[20]

Climate Change and Flooding

Climate change has had and will continue to have a significant impact on stormwater infrastructure needs. It has altered weather patterns so that some areas will suffer from more severe and more frequent weather events, including flooding. Climate change is resulting in rising coastal waters as well as more extreme weather events that exacerbate the pollution and water volume runoffs and which also are threats to the built environment.

Sea levels are predicted to rise as polar icepacks melt. Current predictions project an annual increase of 10–40 percent in river runoff in some areas, with a decrease of 10–30 percent in dry areas.[21] This means that most treatment plants will be at risk of flooding; that wastewater infrastructure must be sized to account for more severe events. The cost to adapt water and wastewater infrastructure has been estimated at $448 billion to $944 billion.[22] Wetlands, which sequester carbon and process methane, carbon dioxide, nitrogen, and sulfur, will also suffer. Ocean warming, caused by climate change, has resulted in an increase in hurricane intensity and destructiveness

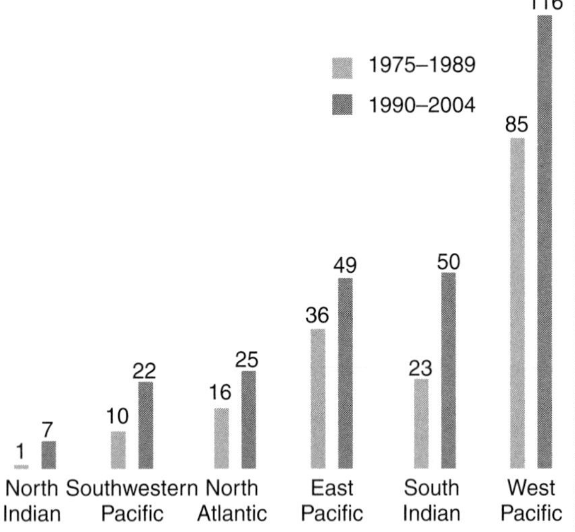

Figure 18.3
Change in number of Category 4 and 5 hurricanes by ocean basin for the 15-year periods, 1975–1989 and 1990–2004

over the past three decades.[23] See Figure 18.3, which shows the rise in Category 4 and 5 hurricanes around the world since 1975.

In the United States, the past several decades have seen an overall increase in rainfall and extreme weather events. From 1958 through 2008 on average, precipitation has increased 5 percent, with significant subnational variations. For example, average precipitation declined by over 40 percent in southern Arizona, and by 15 percent in Georgia, and by 20 percent in Southern Carolina. During that same period, rainfall increased by over 20 percent in North and South Dakota, and by 15 percent in Minnesota and New York. Decreases in annual precipitation have occurred in the Southeast except during the fall season, and in the Southwest except in the spring[24] (Figure 18.4).

Big storms are also becoming a problem in the Midwest, which results in major floods. This is the area's most expensive regularly occurring natural disaster. Between 1961 and 2011 in the Midwest (Illinois, Indiana, Iowa, Michigan, Minnesota, Missouri, Ohio, and Wisconsin), the largest storms (considered as 3 inches of rain or more in one day) have increased by 103 percent. Less severe storms (ranging between 2–3 inches of rain in one day) have increased by 81 percent, and moderate storms (1–2 inches per day) increased by 34 percent.[25]

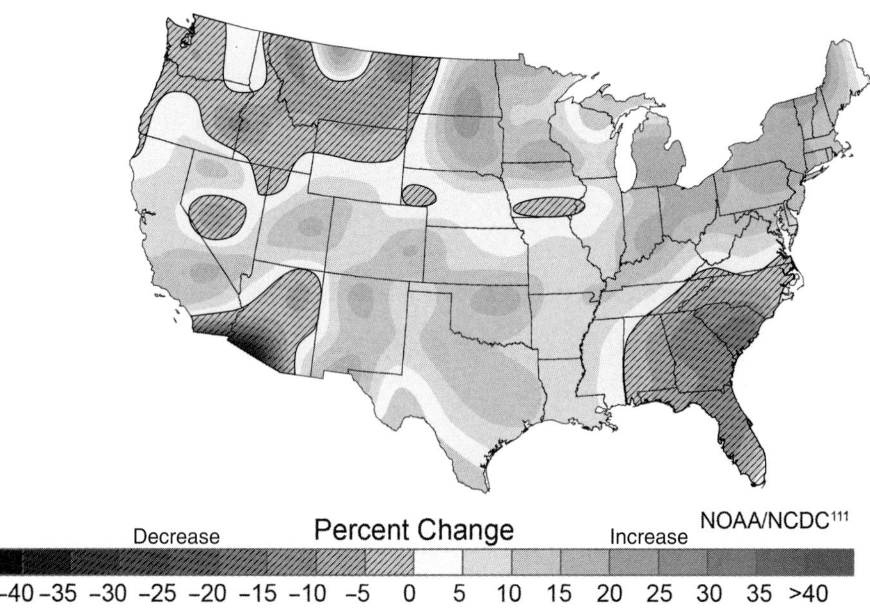

Figure 18.4
Observed change in annual average precipitation, 1958–2008

While U.S. annual average precipitation has increased about 5 percent over the past 50 years, some regions have seen decreases of anywhere from give to forty percent in this fifty year period.

At the same time, there has been and will continue to be an increase in both the frequency and intensity of extreme weather events. Northern areas generally will become wetter, while southern areas will become drier. From 1958 through 2008, there was a substantial increase in the heaviest 1 percent of all daily events across the United States. The highest rate of increase, 67 percent was in the Northeast, with the smallest in the Southwest at 9 percent.[26] This means that regional and inland flooding will continue to be an issue for stormwater planning. During the twentieth century, inland flooding caused billions of dollars of property damage, hundreds of deaths, and untold damage to ecological systems.[27] For example, in 1993, $20 billion in damage occurred in the Mississippi River Basin due to a long period of excessive rainfall.

Sea Level Rise

Climate change is also resulting in sea level rise in most parts of the United States (with Alaska and the northern part of the State of Washington being an exception). In the past two decades, the annual rate of increase in the sea level has doubled.[28] Scientists anticipate that sea levels could rise by as much as 6.25 feet by 2100 if greenhouse gas reduction efforts are not effective. Satellite measurements taken over the past decade, however, indicate that the recent rate of increase is significantly higher than the average rate for the twentieth century. Projections suggest that the rate of sea level rise is likely to increase during the twenty-first century, although there is considerable controversy about the likely size of the increase due to uncertainty about the three main processes responsible for sea level rise: thermal expansion, the melting of glaciers and ice caps, and the loss of ice from the Greenland and West Antarctic ice sheets (Figure 18.5).

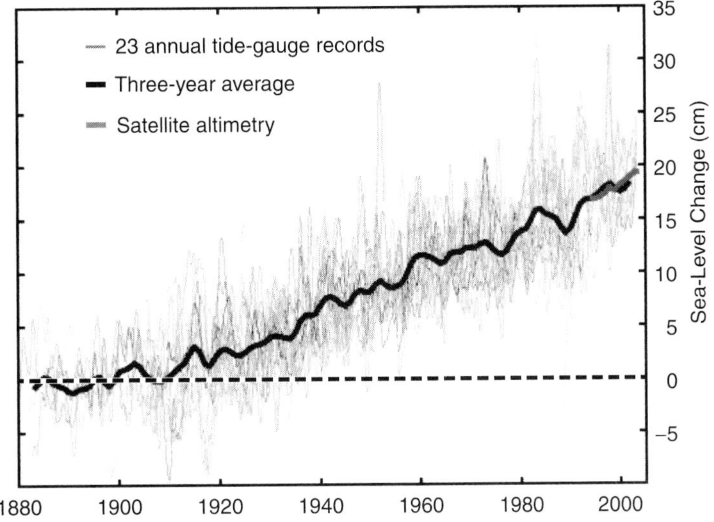

Figure 18.5
Sea level rise, 1880–2010

Planning for Stormwater

In the past three decades, planning for urban stormwater has evolved from being primarily concerned with adequate drainage and flood control provided by public works infrastructure to a more integrated approach that has added concerns about mitigating stormwater flows on-site, treating stormwater pollutants, and protecting and restoring natural drainage channels. As the change in paradigm occurred and strategies expanded to include natural and biological methods, so did the actors involved. Originally the purview of engineers, today land use planners and others work together in processes that result in behavioral changes by institutions, land owners, and individuals, as and the more traditional public works projects.[29]

Maintaining Pre-Development Stormwater Patterns

Most of the programs and "best practices" that seek to mimic pre-development stormwater patterns rely on changing the patterns of development and its design. The development community has mobilized to produce many innovative and attractive design changes that can be used to ensure that new development minimizes its impact upon water runoff patterns. Many of these practices are concerned with reducing the size of the urbanized area and the impermeability of the urbanized area and other methods that slow the speed of urban runoff. This type of approach is often called "low impact development" or LID. Green Infrastructure is a complementary system. Both are discussed further below. First, however, we will look at some of the factors that influence stormwater hydraulics and runoff through an explication of the rational formula.

The "Rational Formula"

Although the "rational formula" has been criticized for contributing to the construction of oversized drainage systems, the concepts behind it are useful to understanding how stormwater runoff impacts the environment. It was used from 1851 until about 1980 in the design of most drainage systems all over the world.[30] The rational formula calculates the peak runoff rate from the area of rainfall catchment area, the permeability of the surface, and the intensity of the rainfall.[31]

The formula is:

$$Q = CIA,$$

where:

Q = peak runoff rate, in cubic feet per second (cfs).
C = a dimensionless coefficient that captures the permeability of the surface.
I = rainfall intensity, in inches per hour.
A = area under consideration, in acres.

The rational formula assumes that rainfall intensity and the coefficient C are the same over the entire area A. Multiplying Q by the length of the storm gives the total runoff volume.

The runoff coefficient for a watershed (C) is the fraction of rainfall on that watershed that becomes stormwater runoff. It has a value between zero and one. The factors affecting the value of C for a watershed are the land use, the soil type, and the slope of the watershed. (See Figure 18.6 for a graphic representation of runoff, evapotranspiration, and infiltration for three different land use categories.)

There are many published sources for the values of the runoff coefficient, C, including local land use agencies. Large areas with permeable soils, flat slopes, and dense vegetation have the lowest C values. Small areas with dense soils, impervious surfaces, steep slopes, and little vegetation have the highest C values. See Table 18.2 for some sample values. These values differ from area to area, and are usually developed for local conditions.

In the past, communities would size their facilities to capture the first half-inch of runoff, believing that 90 percent of the pollutants would be addressed, since pollutants are the most concentrated in the first part of a

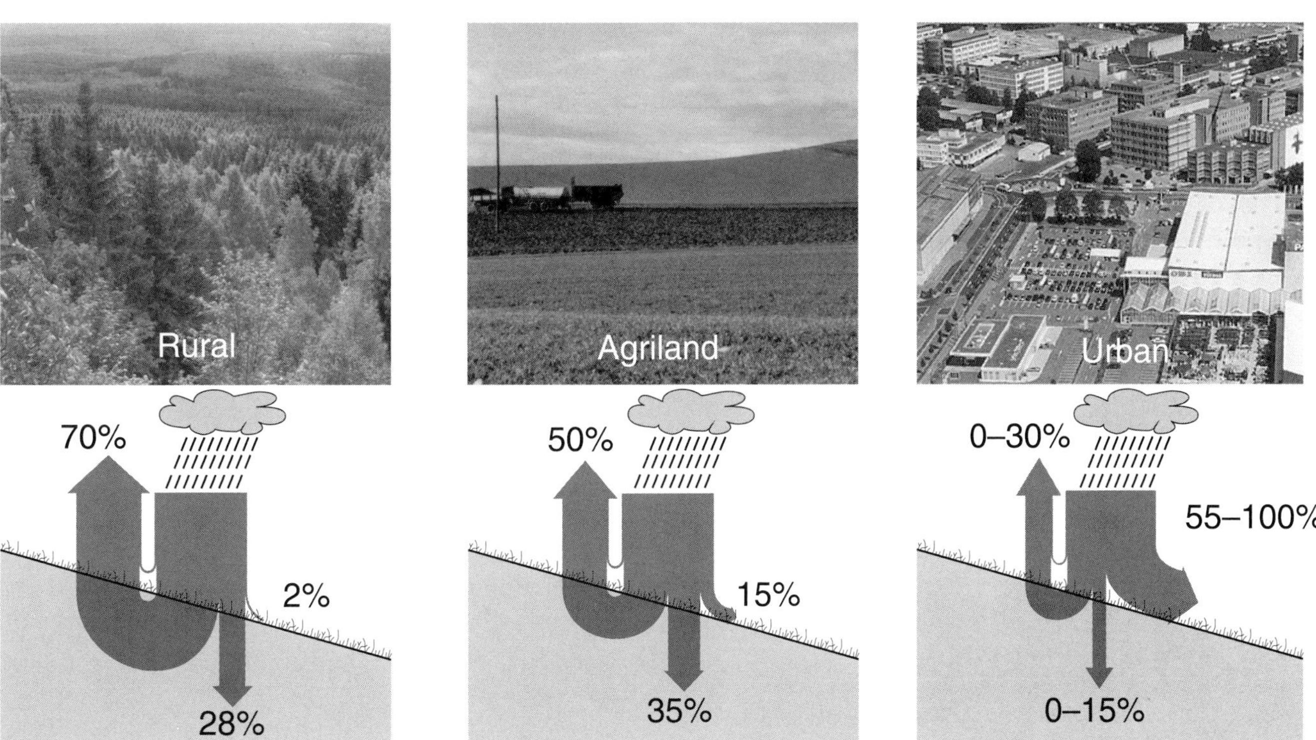

Figure 18.6
Stormwater runoff by major land use category

Table 18.2
Runoff coefficients for types of land uses and soil in Knox County, Tennessee

		Soil Group			
		A	B	C	D
Slope		2–6%	2–6%	2–6%	2–6%
Rural	Forest	.11	.14	.16	.20
	Meadow	.22	.28	.35	.40
	Pasture	.25	.34	.42	.50
	Farmland	.18	.21	.25	.29
Residential	1-acre plot	.26	.28	.32	.35
	½-acre plot	.29	.32	.35	.38
	⅓-acre plot	.32	.35	.38	.40
	¼-acre plot	.34	.37	.40	.42
	⅛-acre plot	.37	.39	.42	.45
Urban non-residential	Industrial	.85	.86	.86	.86
	Commercial	.88	.89	.89	.89
	Streets & ROW	.77	.82	.85	.91
	Parking Lots	.96	.96	.96	.96

Source: Adapted from *Knox County Tennessee Storm Water Management Manual*, section on the Rational Method. Available at: www.brighthub.com (2010).

storm (the first flush effect). Studies, however, have shown that this was not an accurate method for highly impervious surfaces. Today, engineers size storm facilities with the goal of capturing and treating 80–90 percent of the stormwater pollution, rather than treating stormwater volume.[32]

New Development Planning

There are two complementary approaches that are being used today to manage stormwater runoff with distributed and decentralized controls and designs: LID (low impact development) and green infrastructure. LID's goal is to mimic a site's predevelopment hydrology by using design techniques that "infiltrate, filter, store, evaporate, and detain runoff close to its source," instead of relying solely on large-scale end-of-the-pipe facilities at the bottom of the drainage area (Figure 18.7). LID addresses stormwater through small-scale landscape practices and design approaches that preserve natural drainage features and patterns. (Figure 18.8)[33]

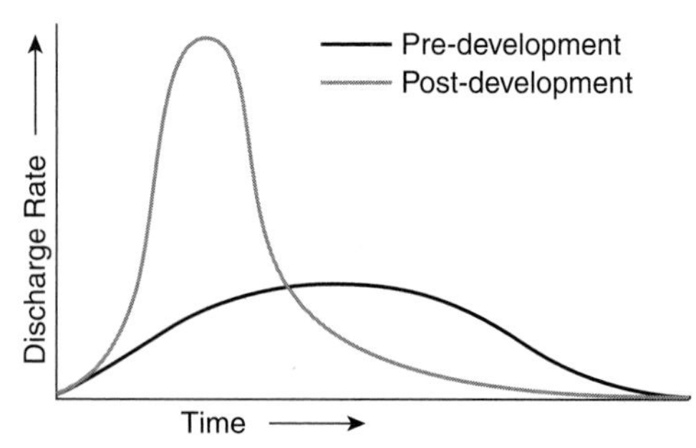

Figure 18.7
Pre- and post-development hydraulics

Green infrastructure. Green infrastructure runs the gamut from green roofs, green walls, rain gardens, and pervious pavers at the site and building level to systems of parks and open spaces

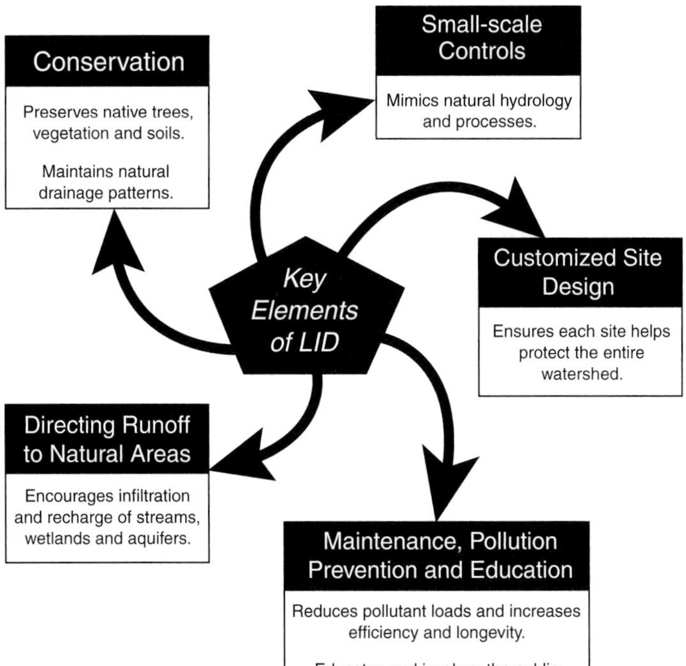
Conservation
Preserves native trees, vegetation and soils.
Maintains natural drainage patterns.
Small-scale Controls
Mimics natural hydrology and processes.
Key Elements of LID
Customized Site Design
Ensures each site helps protect the entire watershed.
Directing Runoff to Natural Areas
Encourages infiltration and recharge of streams, wetlands and aquifers.
Maintenance, Pollution Prevention and Education
Reduces pollutant loads and increases efficiency and longevity.
Educates and involves the public.

Figure 18.8
Key elements of low impact development (LID)

at the neighborhood, city, or regional level. These are systems that address stormwater runoff using plants, soils, and microbes and the natural cleansing tools of the environment. On the regional scale, green infrastructure consists of open spaces and natural areas including forests, wetlands, and floodplains. At the site scale, green infrastructure is similar to LID and consists of site-specific management practices to absorb and infiltrating rainfall[34].

Design process. During the design process for new development, the area of developed land in the watershed should be limited, since less developed land has lower C values. Zoning for higher densities in part of the watershed can reduce the amount of paved, impermeable surface in a metropolitan area and in the area of new development. Local requirements can also be tailored to reduce the area of impermeable surfaces such as parking lots, rooftops, and driveways.

Bio-retention areas such as detention and retention ponds are other alternatives. Detention ponds, as the name suggests, are basins designed to detain stormwater to gradually release it through an outlet in a way that mimics the natural rate of water flow. This flattens and spreads the peak stormwater volumes. Retention ponds are similar to detention ponds, except they do not have a positive outlet. Instead, they rely on evaporation or infiltration into the ground to release the water. Often, natural or constructed wetlands serve this purpose. Buildings and their sites are increasingly being designed to slow down the rush of stormwater. Such efforts include making driveways and parking lots more pervious, the use of swales instead of storm sewers, rain gardens, and tree planting.[35]

Local development rule changes. Measures to increase the permeability of development as part of local development rules. These include use of permeable materials and green storage facilities to keep volumes and speeds down. For example, using concrete pavers instead of asphalt to pave a parking lot can reduce the C value by as much as 50 percent. Similarly, using grass swales—natural or earthen "gutters"—instead of concrete to collect and channel stormwater will reduce the C value. Filter strips are level vegetated strips of land, whereas swales are concave.[36] The City of Chicago has initiated a program to retrofit its paved alleys with permeable concrete or porous asphalt. Stormwater will then filter down into the groundwater table instead of ending up as polluted runoff.

Reducing the Level of Pollutants

The ideal strategy for reducing pollutants in stormwater runoff is to prevent them from getting into the formal collection system—an approach called source control. This requires stronger legislation for producers. However, there are many things that local practitioners can do. See Table 18.3 for a menu of actions.

Automobile-Related Measures

One of the biggest sources of urban runoff pollution is the automobile. Oil, grease, antifreeze, brake dust, and tire particles all find their way into stormwater. In some areas, especially along the U.S. coast, polluted runoff from roads and highways may be the largest source of water pollution (Figure 18.9). For example, about 75 percent of the toxic chemicals that reach the Puget Sound in the State of Washington are carried by stormwater that runs off paved roads and driveways, rooftops, yards, and other developed land. Therefore, any measure that reduces car use will also decrease the pollutant levels.

Similarly, road management policies that call for graveling roads instead of salting them during winter months will also decrease pollutants. Regular and fastidious street sweeping will remove many of the non-point source pollutants before they enter the water. These are, however, generally the larger particles. Sweeping does not remove the smaller and dissolved particles. Modern street sweepers have new and improved brushes that can pick up particulates like asbestos and copper from automobile brake drums.

Table 18.3
Planning tools for water quality protection

Planning tool	Function
Comprehensive master planning	• Provides a coordinated approach with consideration of water quality protection
Stream and wetland setbacks	• Places limits on development activities within certain distances of streams and wetlands • Maintains natural flood- and erosion-control functions • Use protection zones o Based on public health and safety services o Establishes minimum setback widths o Details permitted and prohibited structures and uses o Allows for nonconforming structures and uses o Includes variance section to maintain buildability and provide flexibility
Conservation design	• Allows for partial conservation of natural resources while development occurs • Allows homes to be closer together while remainder of development area is in open space • Requires understanding of site's soils, topography, and natural features
Distributed stormwater management	• Implements stormwater controls at the source of impacts as site features • Potentially reduces stormwater infrastructure costs • Promotes public participation and public education in pollution prevention and maintenance of stormwater management practices • Satisfies NPDES Phase II stormwater requirements • Application for retrofit projects

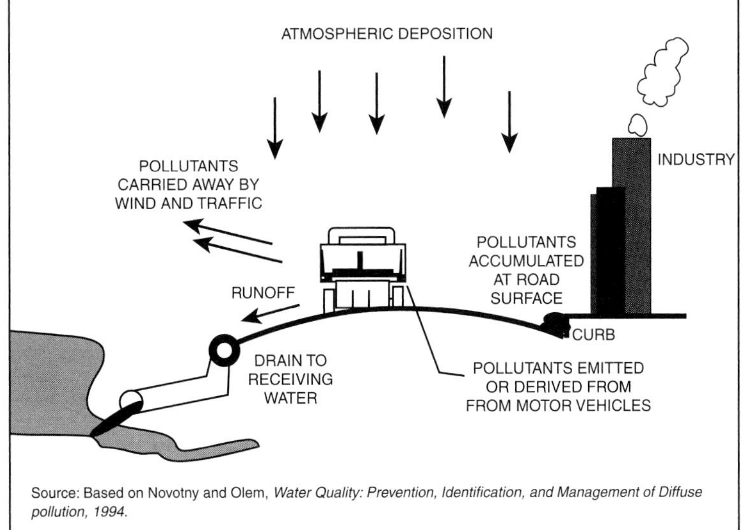

Source: Based on Novotny and Olem, *Water Quality: Prevention, Identification, and Management of Diffuse pollution, 1994.*

Figure 18.9
Pollutant accumulation on impervious street and highway surfaces

This program may be implemented in coordination with a parking ban on certain days of the month to ensure that street sweepers can get to the gutters on a regular basis to remove the trash. Parking ban programs can often require a great deal of public education, however, and they are not always popular.

Road and highway systems can quickly convey polluted runoff during wet weather events to nearby waterways. Road surfaces can carry both land-adjacent and road-vehicle pollutants including heavy metals from tires, brakes, and engine wear, and hydrocarbons from lubricating fluids. If these pollutants are not properly controlled, they can impair waters, causing them to no longer support the water's designated uses and biotic communities.

Transportation authorities are responsible for maintaining stormwater systems along streets, roads, and highways by managing the quality and quantity of stormwater discharging to the waters via those systems. Transportation MS4s management differ in significant ways from traditional MS4s (e.g., cities and towns). For example, linear transportation systems often stretch for many miles, and cross numerous waterways, watersheds, and jurisdictions. They also discharge stormwater and associated pollutants that originate outside of the transportation right-of-way. There are no local ordinances to regulate these discharges.

Other Pollution Prevention Strategies

Public education. Schools can conduct education projects that teach students how to prevent pollution and keep water clean. In addition, educational outreach can target specific enterprises, such as service stations, that have on-site opportunities to control runoff. Many communities have implemented storm drain stenciling programs that discourage people from dumping trash directly into storm sewer systems.

Reduction in pesticide use. Ordinances or clauses in covenants can be implemented that manage either the amount of lawn or the maintenance of lawns in order to reduce pesticide and herbicide use. This applies to the maintenance of public space such as parks and cemeteries as well.

Managing construction. Managing construction is important in fighting NPS pollution. Construction site runoff is one of the largest sources of sediment loading. Many communities have ordinances in place that require the control of runoff and erosion during construction.

Design approaches. Grassy swales, pervious materials, and wetlands not only reduce the destructive results of higher stormwater flows discussed above, but they can also provide natural water treatment in less developed areas. In built-up areas, sand trenches can be placed along the perimeter of a parking area to collect, store, and filter the many small storms that compose

80 to 90 percent of the storm flows. Examples of other source control measures include rooftop storage tanks, green roofs, ditches, or small areas that only serve a parking lot or neighborhood.[37]

Stream and creek restoration. Another component of the water-energy-nature cycle is stream and creek restoration. This is important in reducing the speed of stormwater flows, promoting infiltration, along with increasing wildlife diversity and the amenity value of the natural waterway. These efforts, important to a zero emissions, natural cycle approach, must be thought of at a district or city-wide level.

Retrofitting existing storm drains. Existing storm drains can be retrofitted with a storm drain outlet separator or filters that traps and removes sediments and hydrocarbons from the stormwater.[38] These resources can also be used for discharges from industrial and commercial areas. Some regulators are concerned that storm drain filters by themselves do not effectively remove pollutants.

Stormwater Financing

Municipal bonds financed by taxes or user fees are the primary way that traditional stormwater infrastructure is financed. Only about 1,500–2,000 of the roughly 52,000 water utilities in the United States are large enough to issue their own bonds, however.[39] These are supplemented by state revolving loan funds, or other low interest loan funds at the state and federal levels. Green infrastructure solutions, and distributed innovative systems may find this more difficult and turn to private equity for financing.

Local Stormwater Fees

Stormwater fees are in their infancy. Many of the major stormwater control systems such as detention or retention ponds found in large parking lots are paid for by the private sector, either as a direct cost of construction or through development impact fees. Some localities pay all or part of stormwater management costs with general funds. Many of the Phase I communities (the larger cities and counties) began with low rates, but the smaller Phase II communities presently coming on line are being aggressive with rate structures since many do not have large general fund budgets, and there is also a level of understanding about the stormwater requirements that was not present for the earlier cities.[40]

Most stormwater fees are based on the volume of runoff generated by different land use types using the rational formula described above. Some also calculate the percentage of impervious surface on the parcel.[41] A survey of 200 utilities indicated that about half of fees are based on impervious area, one-quarter on a combination of impervious and gross area, and the

Pollutants of Concern in Stormwater

The six pollutants of concern in urban stormwater runoff (listed in the permit for the city or county) are:

- *Trash*. Usually associated with residential and industrial use.
- *Hydrocarbons*. Oil and grease loads, often from streets and parking lots.
- *Sediments*. Soil erosion, particularly from agricultural areas.
- *Nutrients*. Fertilizers from landscaped and agricultural areas (phosphates, ammonia, and nitrates).
- *Pathogens*. Common in first-flush rainfall events.
- *Toxics*. Heavy metals such as copper in brake dust, cadmium, mercury, and lead. Loading factors can be developed for each one that relates fees by land use to its own mitigation or remediation program.

Grant Hoag, "Developing Equitable Stormwater Fees," *Stormwater*, no. 1 (2004).

rest on intensity of development. Advances in information technology permit localities to merge land use, customer billing, and tax assessor data to develop more accurate and targeted fees. Dallas, Texas, determined the runoff coefficient for each land use type and applied it to the size of the parcel in the assessor's files. San Antonio, Texas, calculated the impervious areas on each parcel to develop their stormwater fees.[42]

Another approach is emerging that assesses the costs to remediate specific pollutants produced by different types of land uses. This approach develops a loading factor for the six pollutants of concern in stormwater (see sidebar on pollutants of concern in stormwater) by major land use categories and applies them to remediation-specific costs. The results are quite different from using runoff volumes (however apportioned). Peaking factors may also be developed since the treatment plant has to be large enough to treat the maximum flow. Table 18.4 shows the differences in percentages of land use, stormwater volume, and pollutant loads for a city of 3,000 acres. It also shows the resulting annual square foot fees that were generated, based on pollutant loads and remediation costs. Note that the industrial and agricultural areas of the city composed 23 percent each of the area, but 34 percent and 36 percent of the pollutant loads. Single family uses compose 20 percent of the land, but account for only 12 percent of pollutant loads.[43]

Table 18.4
Stormwater fees by land use category for a city of 3,000 acres

Land use category	Parcel area (%)	Storm water volume (%)	Pollutant loads (%)	Fees per square foot (annual)
Multifamily	3	4	4	$1.54
Single family	20	20	12	$6.83 per yr (flat fee)*
Landscaped	3	4	2	$0.61
Undisturbed	20	12	1	$0.12
Industrial	23	40	34	$2.73
Commercial	3	6	4	$2.38
Institutional	5	9	7	By specific type
Agricultural	23	5	36	$0.96
Total (3,000 acres)	100	100	100	

Stormwater fees and development fees have been subject to a great deal of litigation. The case law to date supports the ability of local governments to assess these fees if they are developed logically, and if costs for past and future investments are allocated fairly. As the information technology and local databases improve, it is likely that fee methods will become both more targeted and more complex.[44]

Low Cost Loans from the State under the Clean Water State Revolving Fund

Clean Water State Revolving Fund (CWSRF) programs provided more than $5 billion annually in recent years to fund water quality protection projects for wastewater treatment, non-point source pollution control, and watershed and estuary management. Nationally, interest rates for CWSRF loans average 2.2 percent, compared to market rates that average 4.5 percent. For a CWSRF program offering this rate, a CWSRF-funded project would cost 19 percent less than projects funded at the market rate. CWSRFs can fund 100 percent of the project cost and provide flexible repayment terms up to 20 years.[45]

These funds can be more actively managed at the state level to make more money available for infrastructure projects by using a portfolio approach.[46] Instead of using the revolving loan fund money received from the federal government to make loans directly to water and wastewater utilities, states can issue bonds, using the federal money as security—also called "equity." The state bond proceeds can then be loaned to the utilities for infrastructure projects. Using general obligation bonds instead of revenue bonds makes it possible for the state revolving loan fund managers to invest the "equity" in a broader class of assets that can generate more returns to the fund—similar to funds managed by educational institutions and government pension funds.[47]

Conclusion

Climate change is resulting in increased precipitation and more intense storm events. Our current stormwater infrastructure system was designed to rapidly convey water away from urban areas, but this has only resulted in increased downstream flooding and erosion of natural waterways. Combined and separate storm sewers are not sized to meet the increased volumes, but even if they were, urban runoff pollution from automobiles and other sources is causing serious disruption to natural systems. Sewer overflows will be increasingly common. We need to move from "*fast conveyance infrastructure systems*" for stormwater, to decentralized "*storage-oriented, slow release and soft treatment*" systems.[48] Major flooding needs to be addressed by these measures, but also by systematic land development and redevelopment policy changes that integrate stormwater challenges into the design of the city.

Additional Resources

Publications

American Rivers, the Water Environment Federation, the American Society of Landscape Architects and ECONorthwest. *Banking on Green: A Look at How Green Infrastructure Can Save Municipalities Money and Provide Economic Benefits Community-wide* (2012). A report on the savings and benefits that green infrastructure can provide to municipalities.

Blakeley, Edward, and Armando Carbonell, eds. *Resilient Coastal City Regions: Planning for Climate Change in the United States and Australia.* Cambridge, MA: Lincoln Institute of Land Policy, 2012. Nine case studies of how high carbon emitter cities, also facing major flooding risks, have responded proactively.

Howe, Carol and Cynthia Mitchell, eds. *Water Sensitive Cities.* London: International Water Association Publishing, 2012. A thoughtfully edited group of essays by leading-edge thinkers in the water industry that shows the way to integrating water infrastructure planning with other city services.

Meyer, Han. *Delta Urbanism: The Netherlands.* Chicago, IL: American Planning Association Planners Press, 2010. An overview of how one low-lying country has developed the policies, tools, technology, planning, public outreach, and international cooperation needed to save their populated deltas in the face of climate change.

Sarte, S. Bry, *Sustainable Infrastructure: The Guide to Green Engineering and Design.* Hoboken, NJ: Wiley and Sons, 2010. A guidebook to green design that looks at stormwater solutions holistically.

U.S. EPA Region IX and the California Dept of Natural Resources. *Climate Change Handbook for Regional Water Planning.* www.water.ca.gov/climatechange/CCHandbook.cfm. The handbook uses the State of California's Integrated Regional Water Management (IRWM) planning framework to analyze climate change impacts in order to plan for water adaptation and mitigation.

Watson, Donald and Michele Adams. *Design for Flooding: Architecture, Landscape, and Urban Design for Resilience to Climate Change.* Hoboken, NJ: John Wiley and Sons, 2010.

Websites

American Rivers: www.americanrivers.org/library/reports-publications/going-green-to-save-green.html.

Environmental Protection Agency's Stormwater website for Best Practices: www.epa.gov/npdes/stormwater/menuofbmps. A wealth of information for local practitioners.

The International Stormwater Best Management Practices (BMP) Database project website features a database of over 300 BMP studies, performance analysis results, tools for use in BMP performance studies, monitoring guidance, and other study-related publications.

Water Environmental Research Association Website for Livable Communities: www.werf.org/liveablecommunities/index.htm. The Water Environmental Research Foundation conducts studies on innovative best practices for the wastewater industry. They are leaders in the "one water" effort and their case studies of best practices for stormwater address the institutional context as well as design issues.

Key Federal Regulations[49]

In Chapter 16, the important federal water regulations were listed. Some of them are relevant to sewers and non-point source pollution as well. This section expands on the regulations that are important for wastewater.

Rivers and Harbors Act of 1899

This Act gives the U.S. Army Corps of Engineers authority over navigable waterways in the United States. Of particular importance is Section 13, which prohibits the discharge of refuse material into any navigable water without permission. This provision is absolute: polluters are liable even if they pollute accidentally or ignorantly. In 1973, the Supreme Court ruled that this applies to any foreign substance, regardless of its effect on navigation. This in effect turned the Rivers and Harbors Act into a powerful piece of environmental regulation.

Federal Water Pollution Control Act of 1948

This was the first environmental law passed by Congress. It required states to identify water bodies that were polluted beyond tolerable levels, and then to identify the polluters and suppress the polluting discharges. It also included a program for interstate exchange of technical information and formation of multi-state pacts to deal with pollution problems effectively. This law was practically impossible to enforce and poorly funded, however, and led to the Amendments of 1972.

Federal Water Pollution Control Act Amendments of 1972

This law set its goal as "to restore and maintain the chemical, physical and biological integrity of the Nation's waters." This objective translates into two fundamental national goals: "eliminate the discharge of pollutants into the nation's waters, and achieve water quality levels that are fishable and swimmable".[50] This law changed the focus of water quality control from regulating the pollutant levels in the water body to regulating the levels in the effluent at the source of the pollution. It established the National Pollutant Discharge Elimination System (NPDES), which issues permits to municipalities and industries that discharge wastewater. It included a program for construction grants for wastewater treatment facilities. In general, this Act required the federal government to take the lead in water pollution control.

Clean Water Act of 1977

The Federal Water Pollution Control Act was amended again in 1977, and was renamed the Clean Water Act (CWA). In general, the CWA delegated permitting, administrative, and enforcement aspects of the law to state governments. The law gave the EPA the authority to set effluent standards on an industry basis (technology-based) and continued the requirements to set water quality standards for all contaminants in surface waters. The CWA makes it unlawful for any person to discharge any pollutant from a point source into navigable waters unless a permit (NPDES) has been obtained under the Act.

Water Quality Act of 1987

The Clean Water Act was reauthorized in 1987 as the Water Quality Act. This Act included provisions to address non-point sources of pollution. This Act focused on toxic substances, authorized citizen suit provisions, and funded sewage treatment plants (publicly operated treatment works, or POTWs) under the Construction Grants Program. Section 319 of the Water Quality Act established the Non-point Source Management Program.

chapter 19

Solid Waste Management

In this chapter ...

Importance to Local Practitioners . . . 351
History of Solid Waste Management . . . 351
  Early Solid Waste Management . . . 351
  Solid Waste and Cities . . . 352
  Need for Systematic Disposal Methods . . . 352
  Sanitary Landfills and Incineration . . . 353
Waste Management Today . . . 354
Elements of the Waste Management System . . . 356
  Definition of Waste . . . 356
  Solid Waste Collection . . . 357
  Solid Waste Facilities . . . 358
The Institutions of Solid Waste . . . 361
  Who Generates What Kinds of Waste Today? . . . 361
  Who Collects and Disposes of Solid Waste? . . . 362
  How Is Solid Waste Management Financed? . . . 363
  Who Regulates Solid Waste Management? . . . 365
Planning for Waste Reduction, Recycling, and Reuse . . . 366
  The Three "Rs" and Zero Waste . . . 366
  Planning for Zero Waste . . . 367
  Needs Assessment . . . 368
  Developing the Local Program . . . 369
  Emerging Programs and Concepts . . . 370
Building or Expanding Capital Facilities . . . 376
  Siting and Designing Landfills . . . 376
Conclusion . . . 379
Additional Resources . . . 380
  Publications . . . 380
  Websites . . . 380
Key Federal Regulations . . . 380

Why Is This Important to Local Practitioners?

The management of solid waste is one of the United States' most protracted sustainability concerns. Making room for the 250 million tons of solid waste generated per year is a serious issue.[1] In addition, landfills account for over one-quarter of methane gas emissions—one of the most potent of the greenhouse gases that contribute to global warming. Land use, energy, and environmental policies are challenging current disposal methods.

Local practitioners can play an important role in this critical infrastructure area since most solid waste programs are designed and implemented at the municipal or county level. Solid waste programs are a key element in local sustainability and climate change plans. Landfills and other facilities also need to be considered in local land use plans and included in the capital improvement plan and budget.

This chapter starts with a history of waste management, leading up to current practices and institutions. It then goes on to describe the technical and operational details of waste management policy. The final two sections address planning and programs for local jurisdictions.

History of Solid Waste Management

Humans have been generating solid waste as long as they have been consuming plants and animals. The methods for disposing of solid waste have clearly changed over time, reflecting the nature and volume of the solid waste generated and the prevailing attitudes toward solid waste.

Early Solid Waste Management

Ancient human settlements disposed of animal bones, plant residues, and other debris into what archaeologists now call *middens*. Today, these middens are valuable sources of information about ancient cultures. The Ohlone people of the San Francisco Bay area left shell mound middens over 200 feet long and 30 feet high. Archaeologists now think that these middens had spiritual as well as practical value.[2]

The Ohlone middens are a particularly dramatic example of early solid waste management. Most cultures merely disposed of solid waste in nearby piles. When humans were sparsely settled and when the composition of the solid waste was primarily food waste, management was not an issue since it could be fed to animals or left to decompose. The industrialization of the 1700 and 1800s led to increased urbanization and consumption, and solid waste disposal became a more serious and complex issue.

Solid Waste and Cities

It was not until the 1880s that the refuse problem began to receive widespread public notoriety. Sanitarians, already concerned about water quality and sanitary sewage, began to turn their attention to this new danger to human health, the so-called "third pollution." The public, who saw the benefits of publicly managed water and sewer services, embraced the idea of a publicly managed solid waste management effort. Boards of health or health commissions were created and given power over waste collection and disposal. These boards were usually composed of elected officials; very few boards had medical or technical members.

Early disposal methods were crude. Most cities simply dumped their collected wastes into nearby bodies of water, or, if water was not handy, onto vacant lots outside city limits or near undesirable neighborhoods. In 1880, 40 percent of cities dumped their waste on land or buried it, 22 percent used it as fertilizer or animal feed, 10 percent dumped it in water, and 1 percent burned it. New York City dumped over one million cartloads of garbage into the ocean in 1886.[3]

Need for Systematic Disposal Methods

As America shifted from a producer to a consumer society from 1880 to 1920, waste generation skyrocketed. Mass production techniques gave rise to new products like gum, razors, and tin cans that had never been made in the home and were cheap enough to be disposable. Rising incomes and declining prices allowed ever-increasing consumption. Pittsburgh saw a 43 percent increase in garbage between 1903 and 1907. In 1905, the average American generated 860 pounds of waste a year.[4]

The early disposal methods quickly became inadequate, and public health officials struggled with new ways to dispose of the mounting refuse. In 1895, Colonel George E. Waring, Jr., New York City's Street Cleaning

Commissioner and a civil engineer, tackled the problem in a systematic manner.

Waring started a public education campaign to encourage citizens to throw away less, to separate their waste at the curb (into garbage, rubbish, and ashes) to facilitate collection, and not to litter (which was acceptable public behavior at the time). He professionalized the street cleaning crews. He redesigned the garbage barges so they dumped waste more efficiently and farther from shore. Finally, he built rubbish-sorting and garbage-reduction plants, where waste would be sorted and salvageable materials, such as rubber, tin, or grease, could be picked out and resold. The profits from this were used to offset collection costs. Waring's impact was large. First, he demonstrated that the solid waste problem had many fronts, not just that of disposal. Second, he assured the public that the municipal government was the best body to handle solid waste. Finally, and perhaps most importantly, he was key in passing the responsibility for solid waste from public health officials to civil engineers, where it has resided ever since.[5]

The civil engineering community approached the problem pragmatically, with the goal of minimizing cost and maximizing efficiency, particularly with respect to collection and disposal. The engineers also focused on technological solutions, ignoring issues of waste generation, consumer behavior, or public participation.

By default, land disposal—dumping on vacant land—became the primary disposal method. This had problems so using the inorganic material as fill or reclamation material became an alternative. Davenport, Iowa, used refuse to build levees along the Mississippi, for example.[6] The idea was popular in California where between 1860 and 1960, the San Francisco Bay decreased in size from 680 square miles to 430 square miles as communities dumped millions of tons of garbage and dredging material to generate new land for development.[7]

Sanitary Landfills and Incineration

The concept of "sanitary" fill arose in Britain in the 1920s. The method was based on using engineering techniques to control the putrefaction of organic wastes in an open dump. Alternating layers of garbage and dirt were used. In 1934, Fresno, California, opened the first modern sanitary landfill in the United States which employed the first "cut and cover" method, in which a huge trench or hole was dug and subsequently filled with the alternating layers of waste and soil. The idea caught on quickly. By 1945, over 100 cities had adopted the sanitary landfill. It replaced incineration and open dumping as the disposal method of choice for local communities.[8]

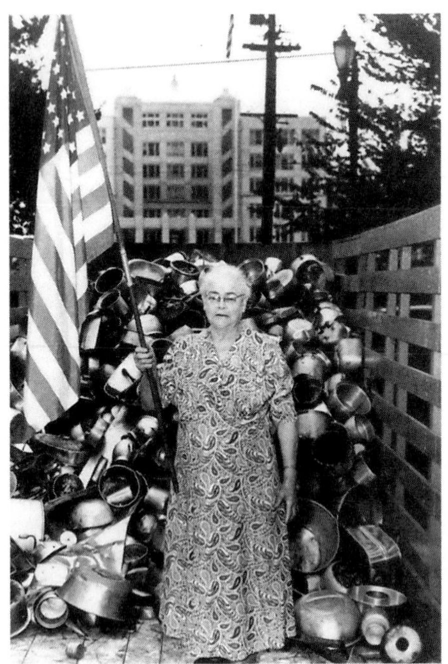
Mayor Carrie Hoyt in 1944 standing with "recyclable" metal for the war effort in front of the Old City Hall, Berkeley, California

World War II caused more "recycling" and salvage in the United States. Metals in particular were salvaged for use in the war effort. However, the post-war economic boom, coupled with the end of the urgent need for wartime supplies, caused the American public to move away from recycling. Rapid post-war growth resulted in less land available for landfills. In addition, many landfills caused contamination in nearby water sources, while others had methane gas explosions, and uncontrolled fires. In 1965, the Solid Waste Disposal Act was passed, which required environmentally sound methods of waste disposal. These forces contributed to an interest in incineration as a replacement for landfills.[9]

Private industry in the 1970s took advantage of new techniques and economies of scale to build large "waste-to-energy" plants, capable of burning up to 1,000 tons of unsorted waste a day. The EPA and many state officials touted incineration as the best method to relieve landfill capacity.[10] Despite the hoopla, incineration is not a prefered method in the USA due to smoke, odors, and toxic by-products. The issue of other types of "hazardous waste" began to receive attention. The result was the landmark passage of the Resource Conservation and Recovery Act (RCRA) in 1976. Its primary focus is hazardous waste management, but it also established minimum federal technical standards and guidelines for state plans to promote environmentally sound disposal methods, maximize the reuse of recoverable resources, and foster resource conservation.

Waste Management Today

The RCRA was not as successful in dealing with normal solid waste as it was with hazardous waste. Therefore, in 1990, Congress passed the Pollution Prevention Act. This Act established the current set of priorities for solid waste management at the federal level: the preferred way to manage solid waste is first to practice source reduction, then recycle and compost, and if absolutely necessary, finally to combust waste at a waste-to-energy facility or place it in a sanitary landfill (Figure 19.1). Sadly, although recycling is one of the environmental success stories of the past two decades with recycling rates increasing from 6.4 percent in 1960 to 34.1 percent in 2010 (see Table 19.1), per capita waste generation rates have also increased over that same period in the United States. In 1960, each person in the U.S. generated 2.7 pounds of solid waste per day. In 2010, that figure was 4.4 pounds (Table 19.2). By contrast, European countries such as Germany, Italy, Spain, Switzerland, the Netherlands, and the United Kingdom generate around 2 to 3 pounds per person per day. Similarly, the two Bs—"burning and burying," or

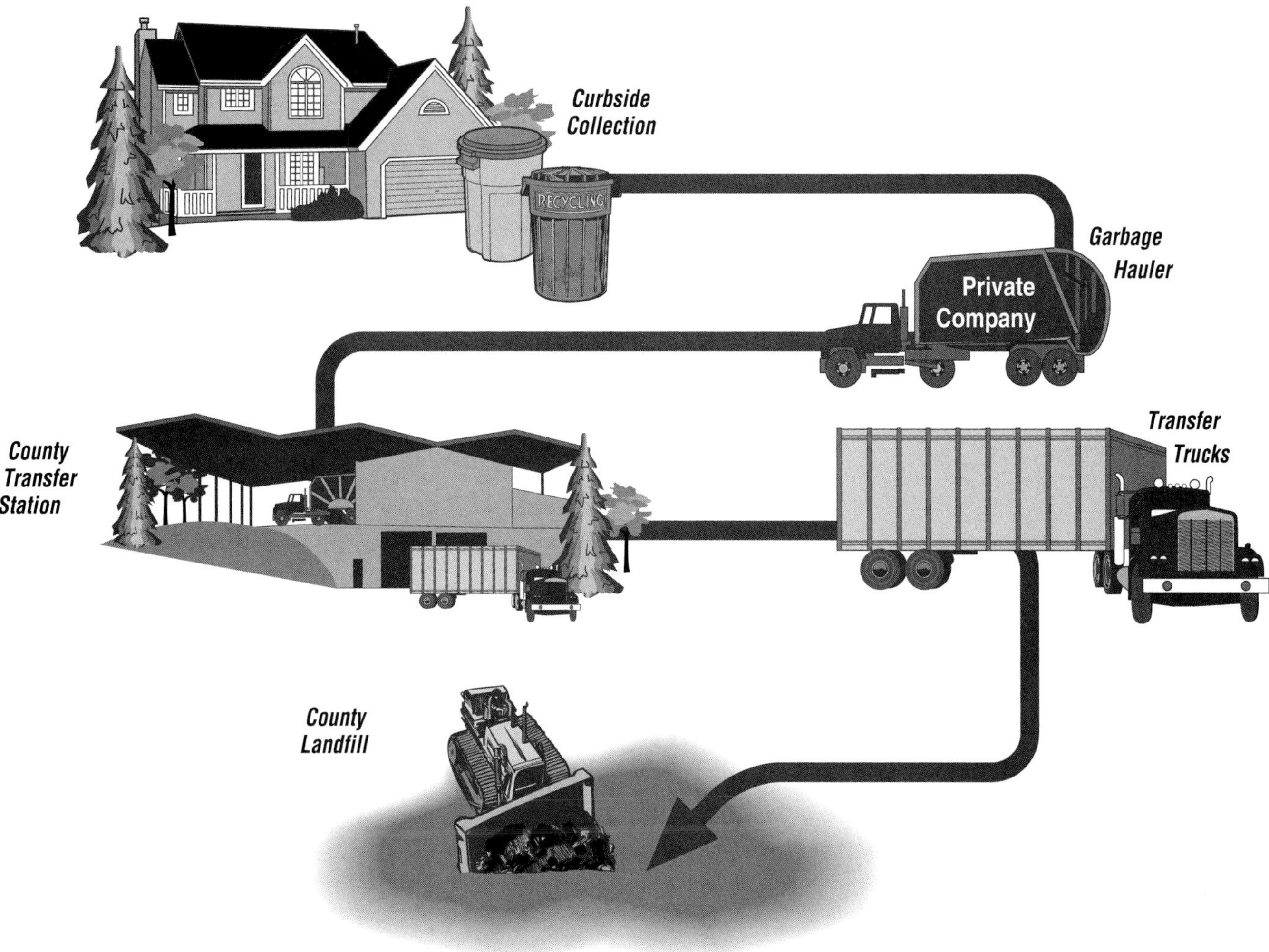

Figure 19.1
Journey of solid waste from the curb to the county landfill

Table 19.1
MSW recycling rates

Year	% of total waste recycled
1960	6.4
1970	6.6
1980	9.6
1990	16.2
2000	29.1
2010	34.1

Source: U.S. Environmental Protection Agency, *Municipal Solid Waste in the United States* (Washington, DC: U.S. Environmental Protection Agency, 2010).

Table 19.2
Solid waste generated in the United States from 1960 to 2010

Year	Total waste generated (million tons)	Per capita generation (pounds/person/day)
1960	88.1	2.7
1970	121.1	3.3
1980	151.6	3.7
1990	205.2	4.5
2001	229.2	4.4
2010	249.9	4.4

Source: U.S. Environmental Protection Agency, *Municipal Solid Waste in the United States* (Washington, DC: U.S. Environmental Protection Agency, 2010).

incineration and landfilling—remain the primary disposal methods. Even though the number of landfills in the United States is steadily decreasing—from 8,000 in 1988 to 1,908 in 2010, the total capacity is staying relatively constant. New landfills are simply much larger than in the past. In 2002, there were 107 incinerators, with the capacity to burn up to 96,000 tons of waste per day.[11] Despite these doleful figures, local commitment to waste diversion from landfills and incineration has increased throughout the USA, Canada, and the world, and many localities are diverting over 50 percent of their waste stream. More recently, sparked by concerns about greenhouse gases, a variety of cities throughout the world have pledged to achieve zero waste. "Zero Waste means designing and managing products and processes to reduce the volume and toxicity of waste and materials, conserve and recover all resources, and not burn or bury them."[12]

Elements of the Waste Management System

Definition of Waste

Solid waste. The term *solid waste* encompasses a wide range of different products.[13] *Garbage* is food wastes: animal, fruit, and vegetable residues. It excludes human waste. *Rubbish* or *trash* is composed of such materials as paper, cardboard, plastics, wood, glass, and cans. *Ashes* are what remain after the burning of wood, coal, and other materials. *Construction and demolition wastes* include such things as concrete, stones, plaster, plumbing parts, and shingles. *Special waste* includes street sweepings, roadside litter, dead animals, and abandoned vehicles. *Treatment-plant waste* includes the sludge and residue from water and wastewater treatment plants. The term *refuse* is often used to describe solid waste generated by a household and so includes garbage and rubbish. RCRA defines solid waste as any solid, semi-solid, liquid, or contained gaseous materials that have reached the end of their useful life and can be discarded. Solid waste can come from industrial, commercial, mining, and agricultural operations, as well as households.

The EPA and Definitions of Hazardous and Solid Waste

On October 7, 2008, the EPA issued a final Definition of Solid Waste (DSW) that was intended to make it easier and more cost-effective to recycle hazardous material by exempting hazardous wastes designed for recycling from RCRA's safe-handling and reporting requirements. Activities that are most affected are metals and solvent recycling. In 2009, the Sierra Club, joined by other environmental groups, filed a petition with the EPA protesting the exemption. This led to a revised proposed DSW in 2011 that is still in public comment. The EPA has committed to take final action on this proposed rulemaking on or before December 31, 2012.

Solid Waste Collection

The mechanism of waste collection is a familiar one. Once a week or so, usually early in the morning, a garbage truck comes to the residence or business to empty garbage cans. Garbage trucks can carry about 5–7 tons of garbage and have compactors built in, which gives them greater capacity.

Types of trucks. *Front loader* trucks are used for commercial and industrial garbage and recyclables. The truck has automated forks on the front which the driver slips into grooves in the dumpster. The dumpster is lifted over the truck, flipped upside down, and the waste dumped into the truck's hopper where it is compacted with a packer blade into the rear of the truck. *Rear loader* trucks usually serve residential areas. Sometimes the carts are lifted and emptied mechanically, and sometimes the work is done manually. Once inside, the garbage is compacted against a moving wall towards the front of the truck. These trucks need a driver and one or two workers for loading the waste. Automated side loader trucks are being seen more frequently these days in residential areas as they pick up the 32-, 64, and 96-gallon rolling carts. These trucks can be operated by one person.

Many communities also have curbside separation of recycling, and the same or a different truck and crew will pick that up. Similarly, many communities have separate collection of plant and yard waste. Since this waste is able to be composted, it often goes to a special processing facility.

Curbside source separation. In the early days of recycling, households were required to separate the recyclable materials into separate bins on the curb. This practice

Disposal of Household Hazardous Waste (HHW)

Generation of some HHW is inevitable but it cannot be discarded along with other solid waste. For this reason, programs or facilities must be established that collect and dispose of HHW. Some communities have exchange areas for unused or leftover paints, solvents, pesticides, cleaning and automotive products, and other materials. Local businesses can also serve as collection points for HHW. Some local garages, for example, may accept used motor oil for recycling, while paint stores can accept unused oil-based paint for use in graffiti abatement programs. Kinkos and Office Max, among other stores, accept used printer toner for recycling. The local electronics store can be asked to dispose of used batteries, one of the most toxic elements in a landfill.[1] On a broader scale, a high priority program for local governments should be to reduce HHW generation to begin with. This can mean many things: using or encouraging the use of substitute products that do not contain hazardous chemicals; using hazardous products more judiciously; or, recycling the products at the end of their lives. Local land use planners should be aware that the size of residential yards has a direct relationship to the amount of pesticides used. In addition, generation of automotive wastes like used oil and antifreeze will be reduced in more compact communities where walking and biking options exist.

Note

1 J. A. Tarr, *The Search for the Ultimate Sink Urban Pollution in Historical Perspective* (Akron, OH: University of Akron Press, 1996).

Automated garbage pick-up truck

has changed in many communities and presently households have a single recyclables bin, and separation is done by workers at a central site. This permits a single truck to serve more households and reduces injuries to workers. As central-site separation technologies improve and become cheaper, source separation into "wet" and "dry" streams only will dominate. All dry waste, recyclable and nonrecyclable alike, is placed in the same receptacle. This is known as "single-stream."

If the community is small, the garbage truck might go straight to the landfill or incinerator at the end of its route. More usually, the truck will go to a transfer station or a materials recovery facility (MRF), where the waste is sorted, compacted and often recyclaldes recovered. From the transfer station, the waste goes on to final disposal at either a landfill or an incinerator. Some waste travels across state borders to a landfill or incinerator; New York, Maryland, and New Jersey were the top three exporters of waste in 2004. Pennsylvania is the largest waste importer, most of which ends up in landfills.

Solid Waste Facilities

As mentioned earlier, the two most common means of solid waste disposal are sanitary landfills and incinerators. Recycling and resource recovery centers are becoming increasingly more common, as are material recovery facilities (MRFs).

Sanitary landfills. A sanitary landfill is in principle a simple thing. Solid waste is placed into an excavated trench, where a bulldozer spreads and compacts it. At the end of the day, soil is placed over the compacted waste, in a layer of 6–12 inches deep. This layer is known as the daily cover. The next day, more waste is placed on top of the previous day's daily cover, and the process is repeated. This way, the landfill gradually gains height. When filled, the landfill is capped with clay or an impervious membrane. The final earth layer is called the final cover, and it is 2–5 feet thick (Figure 19.2).

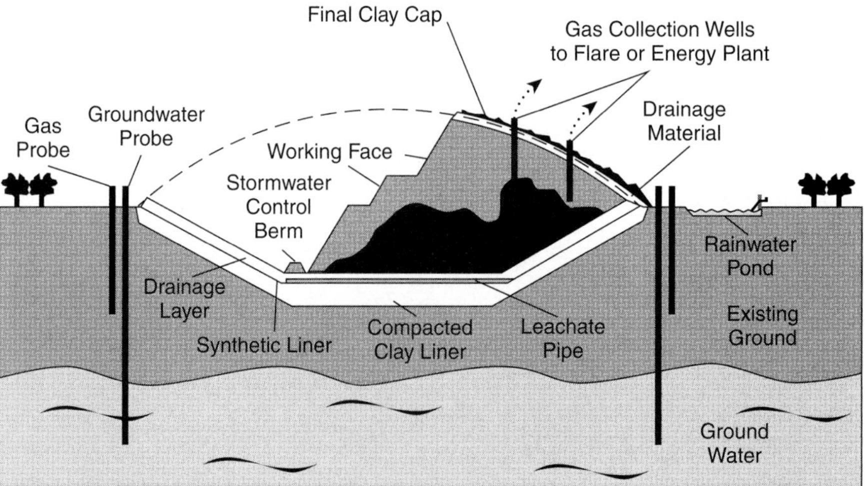

Figure 19.2
Cross-section of a landfill showing the suite of containment and monitoring devices required for closure and beneficial reuse

Sanitary landfills generate *leachate*, which is water that becomes contaminated by the wastes in the landfill. Leachate must be managed. Leachate and the elements of the landfill are discussed in greater detail later in this chapter.

Landfills also generate *methane*, and although current regulations require methane to be "flared," or used for energy, in 2010, landfills were the third largest source of methane emissions in the United States, at 16 percent. On a more positive note, despite increases in the amount of material going to landfills, net methane emissions from landfills decreased by 27 percent from 1990 to 2010 as a result of combusting landfill gases.[14]

Bulldozers moving trash around at a landfill

Incineration. Incineration reduces solid waste by up to 90 percent by volume and 75 percent by weight. It also helps to destroy bacteria and germs, and reduces the amount of waste to be landfilled. After efficient combustion, only residual ash remains for disposal. However, imperfect combustion may result in the creation of micropollutants such as dioxins, which have deleterious environmental effects. Also, only 20 percent of the U.S. municipal solid waste incinerators are able to recover recyclables before shredding and burning the rest (Figure 19.3).

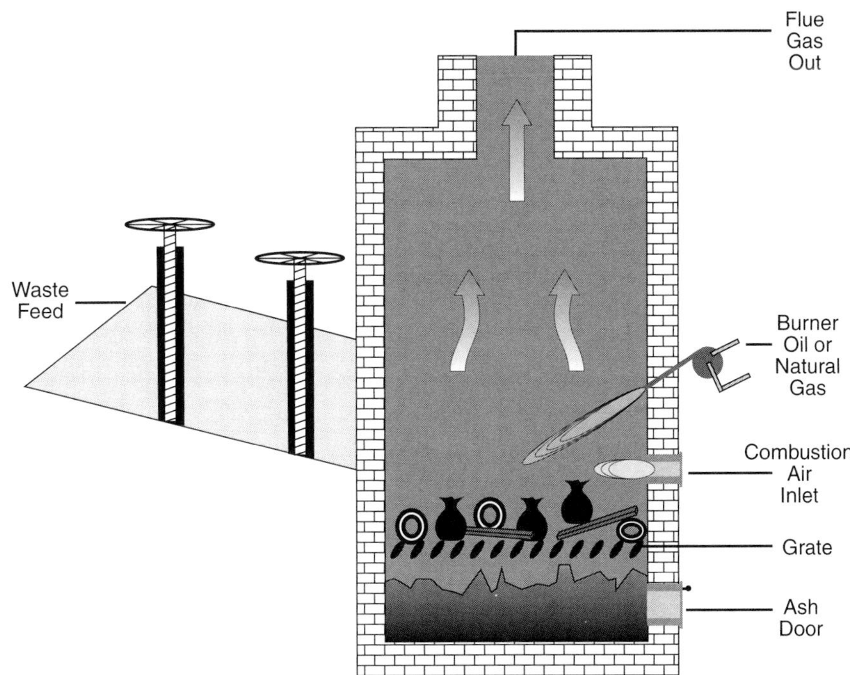

Figure 19.3
Solid waste incinerator

Waste-to-energy (WTE) plants are another form of incineration, but here the energy in solid waste is recovered and turned into electrical energy. One pound of waste can keep a 60-watt light bulb burning for five hours. The WTE plant is characterized by highly controlled combustion supported by extensive air pollution control and ash management systems. In a typical process, the waste is first sorted to remove any large noncombustible material. From there, it enters the combustion area, where it is burned at a temperature of about 1800°F. The hot gases from this combustion are used to generate steam, which in turn is used to generate electricity. (See the section on combined-cycle plants in Chapter 27 on energy.) The gases then go through a scrubbing process, which cools them, neutralizes any acids, and removes particles. In 2008, 13 percent of municipal solid waste (MSW) was burned to produce energy, a percentage that has remained constant since 1990.

Materials recovery facility (MRF). An MRF is a facility where waste that was not sorted at curbside is loaded onto a conveyor belt and hazardous waste such as poisons, paints, and chemicals are removed. Recyclables are separated out and put into bins so that they can be sold. Anything that is not hazardous and cannot be reused or recycled is sent to the solid waste landfill. A "dirty" MRF receives all waste from residences and commercial front end-bins. "Dusty" MRFs receive material from commercial and industrial debris boxes, focusing on certain more productive commercial routes for recyclables and construction & demolition (C&D) debris.

Recycling and resource recovery centers. Recycling and resource recovery centers are places where unwanted materials can be reused and resold. These centers may be private enterprises, separate from the agency that collects and disposes of solid waste, or they may be somehow attached to the normal disposal facility. Such centers may be simply private businesses, or may also be community-based service organizations. A resource recovery park usually is the co-location of reuse, recycling, compost processing, manufacturing, and retail businesses in a central facility. Resource recovery parks also sell or give away reused and recycled-content products.

Resource recovery parks can be quite successful. One such park located in San Leandro, California, recovered 15 percent of Alameda County's total waste materials (Table 19.3).

Table 19.3
Material recovered at San Leandro Resource Recovery Park

Material	Tons
Wood/green waste	119,505
Curbside recyclables	84,532
Dirt	15,881
Concrete	7,275
Paper	3,706
Appliances	1,998
Scrap metal	1,690
Tires	1,554
Total	236,141

Source: California Department of Resources Recycling and Recovery, 2002.

However, resource recovery parks often need help from local government, in the form of amenable zoning, promotion, and financial assistance.

The Institutions of Solid Waste

Who Generates What Kinds of Waste Today?

In 2010, U.S. residents, businesses, and institutions produced almost 250 million tons of solid waste. Residential waste for both single and multifamily homes is about 55–65 percent of municipal solid waste. Schools and businesses, including hospitals, accounted for 35–45 percent.[15]

The mix of wastes produced by a community is known as its *waste stream*.[16] The composition of the waste stream in the United States has changed over the years, as shown in Table 19.4. Glass and metals make up a smaller percentage of the total than they once did, reflecting the success of source reduction and diversion efforts and changes in materials packaging. Yard waste has dropped dramatically from about 23 percent in 1960 to 13 percent in 2010, while glass declined from 7.6 percent to 4.6 percent. However, plastics and rubber, leather, and textiles now make up a larger percentage. The proportion of paper has remained relatively unchanged.[17]

Households also produce large amounts of hazardous waste: leftover household products that contain corrosive, toxic, ignitable, or reactive ingredients. These are collectively known as *household hazardous waste* (HHW) and include products such as fluorescent lamps, paints, cleaners, oils, batteries, and pesticides. Americans generated 1.6 million tons of HHW in 2000. The average home can accumulate as much as 100 pounds of HHW in the basement and garage and in storage closets.[18]

Table 19.4
Municipal waste stream composition for the USA from 1960 to 2010

Material	1960	1970	1980	1990	2000	2010
Paper	34.0	36.6	36.4	35.4	37.2	28.5
Glass	7.6	10.5	10.0	6.4	5.4	4.6
Metals	12.3	11.4	10.2	8.1	7.8	9.0
Plastics	0.04	2.4	4.5	8.3	10.7	12.4
Rubber, leather, textiles	4.1	4.2	4.5	5.6	6.7	8.4
Wood	3.4	3.1	4.6	6.0	5.6	6.4
Food	13.8	10.6	8.6	10.1	11.2	13.9
Yard	22.7	19.2	18.1	17.1	12.0	13.4
Other	1.6	2.1	3.2	3.0	3.2	3.4
Total	100.0	100.0	100.0	100.0	100.0	100.0

Source: U.S. Environmental Protection Agency, *Municipal Solid Waste in the United States*, 2010.

Who Collects and Disposes of Solid Waste?

Today, the solid waste management industry is a mix of the public and private sector, of small and large firms, of horizontal and vertical integration. In 1999, an estimated 27,028 organizations operated in the industry. Almost 56 percent of these organizations were public sector entities, while 44 percent were private companies. Of the total, approximately 15,500 (57 percent) provided collection and hauling services only and did not own or operate any solid waste disposal or process facilities such as landfills or incinerators. The remaining 11,500 organizations operated the estimated 15,700 solid waste management facilities in the USA. The private sector owned 53 percent of the solid waste facilities and the public sector owned 47 percent.

In total, the industry managed approximately 545 million tons of waste. Of that total, about 374 million tons were landfilled; 31 million tons were incinerated; and 140 million tons were recycled. The private sector handled 69 percent all the solid waste recycled, incinerated, or landfilled, while the public sector handled 31 percent.

Most private firms are small—a single landfill, or a fleet of garbage trucks. However, several large national firms dominate in some areas. Waste Management, Incorporated (WMI) is the nation's largest waste disposal and garbage hauler, with revenues of $13 billion and a customer base of over 20 million. Republic Services serves markets in 40 states and Puerto Rico.

A 2002 study that explored differences in communities that chose public or private provision found that although government provision of residential waste collection and recycling dominates in central cities, franchises and contracts play a greater role today than in the past two decades. The private market is active in commercial recycling while local governments are largely responsible for residential waste and recycling. The processing of recyclables is private. Communities with large populations and dense populations are likely to have government-provided services for both waste and recycling—the "economies of density." In addition, local jurisdictions in states with mandated recycling are more likely to provide these services publicly.

Flow Control Decision

In 2007, the U.S. Supreme Court supported the right of two counties in New York State to pass flow control ordinances requiring that all solid waste generated within their jurisdictions go to a public solid waste facility run by a bicounty waste management authority created by the state legislature. *United Haulers v. Oneida–Herkimer Solid Waste Management Authority.* The ordinances were challenged by a trade association of private haulers, but the court ruled that since a public facility was involved, any burden on interstate commerce was overridden by the authority's waste management goals. Previously, in 1994, the U.S. Supreme Court had ruled a flow control ordinance passed by a municipality was unconstitutional since the city had hired a private contractor to build the facility. *Carbone v. Clarkstown.* Between 2007 and today some haulers have continued to challenge local flow control ordinances in lower courts, but generally the municipal governments have prevailed. Flow control ordinances can be important to ensure that bonds issued to build a facility can be paid off through a steady revenue stream.

A public sector waste management organization can be a department of the local government. The City of Houston, for example, has a Department of Solid Waste Management. It can also be a special district. The Cuyahoga County, Ohio, Solid Waste District is a dependent district. Palm Beach County, Florida, and Alameda County, California, have county Solid Waste Authorities, which is an independent district that serves many jurisdictions within the county. The Solid Waste Agency of Lake County, Illinois, for example, provides technical and business expertise to help municipalities manage their solid waste contracts.

Many times, the public and private sector complement each other. In Houston, for instance, the city Department of Solid Waste owns the landfill and recycling facilities, as well as collects the waste from about 67 percent of residences and businesses. Private firms collect from the remaining 33 percent. Since this portion tends to be large businesses, private firms collect about 65 percent of total waste. It is important to remember that even if a local government does not actually perform waste collection or disposal, it will often have an internal staff or government agency responsible for managing the franchise with the private firms.

How Is Solid Waste Management Financed?

Solid waste agencies generally raise revenue through user fees. These user fees come in various types. The most common type is a monthly or yearly fee charged for waste collection. This fee often shows up as a separate line item on some other bill like the property tax bill or the water bill. Typical rates might be $75.00 a year for a single-family residence, and $50.00 a year for apartments. Commercial establishments usually pay based on the size of disposal container they use.

The second type of user fee is called the tipping fee. This is the fee charged for actually disposing of the waste at the waste facility. The specific tipping fee structure varies from facility to facility, but generally, waste disposal is charged on a per-weight basis, and can range from $12 per ton in Western states to over $75 per ton in the Northeast. Different types of waste are priced at different amounts. For example, a landfill in North Carolina charges $29.50 per ton of municipal solid waste, but $59 per ton for construction and demolition waste.

Tipping Fees

A *tipping fee* is charged by a waste facility to those disposing of waste. The name comes from the fact that many dump trucks had to tip the bed of the truck backward to unload garbage.

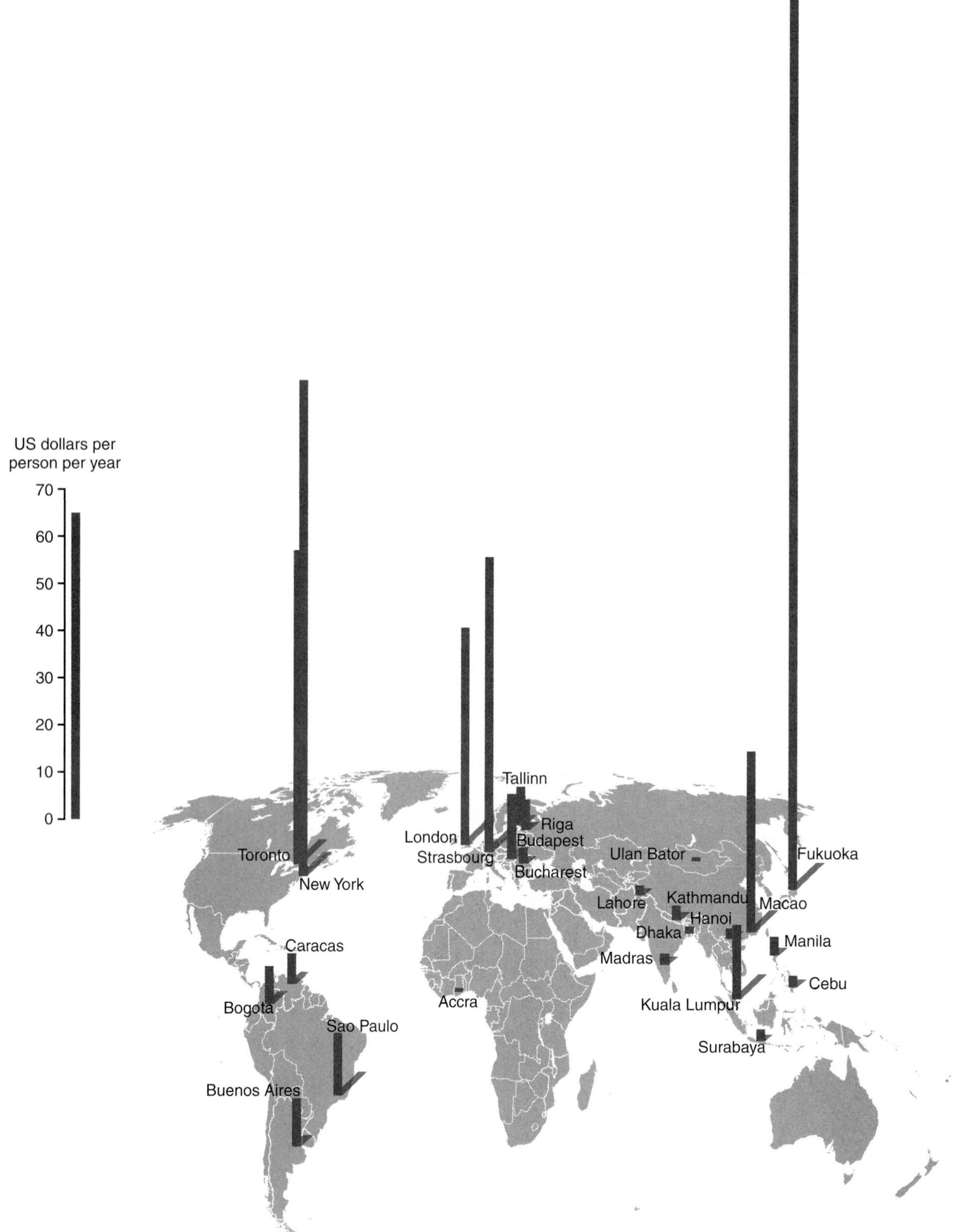

Figure 19.4
Per capita fees in selected cities around the world compared to the average in the United States

Individuals who visit the landfill are charged a flat fee of $20 per visit. Since the organization that collects solid waste is not necessarily the organization that operates the disposal facility, the tipping fee is separate from the collection fee.

Fees are thought to regulate demand. However, inclining single-family residential rates are not that effective in reducing per capita waste disposal. Households usually only change service levels when reminded by the collection agency and otherwise do not seem to pay that much attention to garbage fee levels. However, progressive rates for commercial customers do have an impact on reducing disposal. Unless the local government has a mandatory recycling ordinance in effect, most businesses must make the decision to recycle and, therefore, cost considerations for collection can play a major role in diversion.

In many ways, the current state of solid waste management and finance is amazingly successful. On average, it only costs a municipality about $70 to $80 to collect and dispose of one ton of solid waste. A household may only spend $100 a year on waste collection. Considering the labor, capital, and transportation costs associated with the whole process, these figures are quite low compared to other infrastructure services.

Who Regulates Solid Waste Management?

Solid waste management is regulated in varying degrees at all levels of government. The regulations touch on all aspects of waste management, including waste generation, disposal facility location and operation, and collection fees.

Federal regulation. The Environmental Protection Agency is the federal agency charged with regulating solid waste management. These regulations (described earlier in this chapter) are designed to ensure public health and safety, and apply to final disposal of the waste. The EPA has also established standards for location, liners, operating procedures, groundwater monitoring, landfill gas emissions, and closure and post-closure care of landfills. The EPA also has emission standards for incineration smoke and the final disposal of the ash.

Hazardous waste management is subject to a host of federal regulations noted in the earlier sections of the chapter, including the very definition of hazardous waste. Hazardous waste generation, storage, transportation, and disposal are all subject to federal standards.

State regulation. The extent and nature of state regulation of solid waste management vary from state to state. Generally speaking, states set standards in addition to those of the federal government for solid waste collection and disposal facilities. Many states operate registries in which solid waste facilities must register and provide annual reports.

In addition, many states establish standards for the business practices of solid waste management firms. Fees, service areas, and fleet composition are all aspects that states may control. Many state agencies also have programs to educate solid waste generators and to encourage pollution prevention, recycling, and other waste minimization practices. Finally, many states where disposal options are expensive, or where there are concerns about global warming, are pursuing mandatory recycling ordinances or outright bans on certain materials in landfills. These restrictions are discussed further in the section on regulatory options for reuse and reduction.

Local regulation. In most jurisdictions, solid waste collection and disposal is a service mandated by the local government that requires solid waste generators to pay to have the waste collected and disposed of. This is an exercise of the local government's police power to protect the health of its citizens. The local government also has the franchise authority to permit private solid waste firms to collect the waste. Local regulation can be used to promote local sustainability and climate-change goals.

Planning for Waste Reduction, Recycling, and Reuse

The Three "Rs" and Zero Waste

Many factors are coming together today to reinforce the "three Rs" approach to waste management: reduce, reuse, or recycle. The need to use less energy to transport waste to landfills; reduced availability of land for landfills; public opposition to incineration; the greenhouse gases produced in breaking down waste; and a changed attitude about waste—that it constitutes a resource rather than a nuisance—all contribute to a planning environment that moves toward zero waste. This environment is producing a greater commitment to reducing or diverting waste, and interest in working with the private sector to take more responsibility for consumer product waste.

Several decades ago, many thought that no more than 25 percent of the waste stream could be diverted from landfills and incinerators. Today, however, the national recycling rate is over 30 percent, and some local governments have diverted 60 percent and more from their waste stream, while some hospitals and schools have reached 90 percent (Figure 19.5). San rancisco, San Jose, Long Beach, and Los Angeles all divert more than 60 percent of their total waste from city landfills through recycling, green waste, and composting programs and some have adopted diversion rates of 75 percent. Nonetheless, the early adopter communities are finding it difficult to push recycling significantly past 60 percent.

Against this background, the "Zero Waste" movement arose in New Zealand in 1996, followed by a host of other cities and states in the U.S. and

Pollution Prevention

A special form of source reduction is called pollution prevention. This is the active identification of equipment, processes, and activities that generate excessive wastes or use toxic chemicals, and then somehow changing the equipment, process, or activity to reduce the waste or toxics produced. This could include substituting the raw material, or modifying the production process to be more efficient or use energy-efficient equipment.

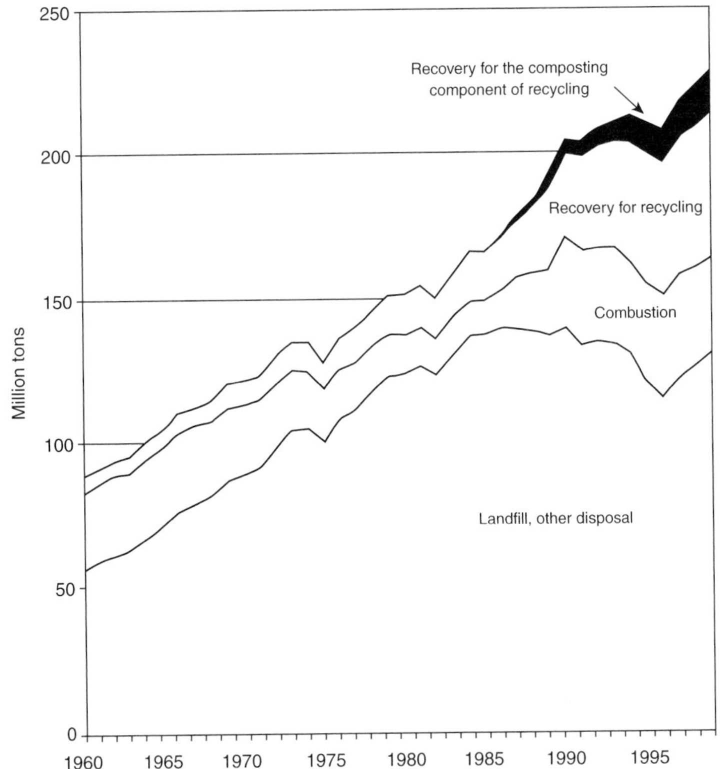

Figure 19.5
Tonnage of solid waste disposal methods from 1956 to 1999 highlighting the growth of composting as an important diversion method

around the world. The concept is that waste is a resource rather than a nuisance. Zero waste proponents consider municipal solid waste a result of poor product design, inefficient markets, and tax subsidies for waste and virgin materials that are sending the wrong signals to the consumers and producers. Zero waste strategies emphasize diversion and producer responsibility for upstream waste reduction and product disposal—"cradle to cradle" responsibility.

Planning for Zero Waste

Planning in this environment involves setting waste diversion goals—usually an overall goal accompanied by detailed targets by individual waste stream—and then developing a strategy and a series of programs to achieve the goal (see the sidebar on the next page for suggested goals from Earth Resource, including a diversion rate goal). A diversion rate is "the percentage of its total waste that a jurisdiction diverted from disposal at landfills and transformation facilities through reduction, reuse, recycling programs, and composting programs."[19] For many agencies, goals are developed by the

The Three "Rs"[1]

Reduce

Source reduction, often also called waste prevention, refers to any change in the design, manufacture, purchase, or use of materials or products (including packaging) to reduce their amount or toxicity before they become municipal solid waste.

Reuse

Reuse is closely tied to source reduction. Reusing products decreases both the amount of post-consumer solid waste, but also reduces the amount of waste created in the production of those products. Simple ways to reuse include using cloth napkins, using refillable pens or rechargeable batteries, and turning empty jars into leftover food containers. It also includes donating or selling old or unwanted clothing and other items. Reuse is considered better than recycling.

Recycle

Recycling is the best-known technique for reducing the amount of solid waste generated. Recycling means reusing materials and objects in their original or changed forms rather than discarding them as wastes. Recycling turns materials that would become waste into materials that can be reprocessed for reuse.

Note

1 Reduce and Reuse: www.epa.gov/osw/conserve/rrr/reduce.htm.

Suggested Zero Waste Goals and Principles for Landfill and Incineration

We divert more than 90 percent of the solid wastes we generate from landfill and incineration from all of our facilities. No more than 10 percent of our discards are landfilled. No mixed wastes are incinerated or processed in facilities that operate above ambient biological temperatures (more than 200 degrees F) to recover energy or materials.

Source: www.earthresource.org.

voters or by policy-makers. Once the goal has been set, the agency needs to figure out how to achieve it. The following discussion outlines some general steps, although the process should be tailored to each agency's own situation.

Needs Assessment

The first step is usually a needs analysis. This consists of a condition assessment and inventory of facilities and equipment as well as a waste characterization study.

Solid waste facility inventory. This will include the rolling stock (the garbage trucks), the transfer stations, the landfills and incinerators, and the recycling centers. The capacity of the disposal facility needs to be assessed. The condition inventory will also include the trash receptacles since they are usually provided to the household or business.

Collection route analysis. The waste collection routes and their adequacy should also be assessed once a year during the budget process. This is important since routing directly affects the size of the truck fleet and the labor force. Solid waste generation has seasonal aspects: Christmas, spring cleaning, fall leaves, and months of heavy moving (usually May, June, and July) all bring higher average waste quantities. Labor over-hires or temporarily increased frequency may be more cost-effective than investing in expensive new rolling stock. This is also important because waste generation rates may vary throughout the community. Higher incomes usually mean more waste. Certain manufacturing processes generate more waste than others, and office parks generate larger amounts of paper waste. These factors can impact the design of the truck routes as well as the need to expand or contract the service at transfer stations, recycling centers, and indeed the ultimate disposal site. Some localities use GIS to assist this process.

Waste characterization study. The waste characterization study identifies existing levels of waste generated by type of waste. Often this is identified by generator segment—such as single family, multifamily, commercial—and mode of delivery to the transfer station—roll-off and self-haul—which contain waste from both commercial and residential. If the plan is being prepared for a solid waste district, the waste stream characterization study may be produced for the district as a whole and then separated out by municipality and or county.

Business-related studies. Special studies on the local business sector are also important to determine practices to be targeted for waste prevention. One such study is a *weight-based disposal* study of waste from commercial routes "sampled" for composition and correlation with business type. The needs analysis is used to identify the areas where the largest gains in diversion could be obtained. Usually organics (food waste and garden waste) as well as construction debris are the largest component of the waste stream. An analysis of the markets for recyclables might be appropriate to develop revenue stream estimates for a buyback program.

Evaluation of current program. Other studies done to help set the baseline might include a *performance audit of current programs* and a survey of other likely programs of interest to the locality. The website for Seattle's 2007 Waste Management Plan has more detail on the methods they used to develop their plan, as well as a comprehensive listing as of 2006 of recycling and recovery programs used across the county (www.seattle.gov). If a locality is just beginning to put together a recycling and diversion program, the EPA website has many different fact sheets about starting a program as well as linkages to other sites.

Developing the Local Program

Developing the local program can be a very complicated task for which the locality will hire a consultant, or the locality can tweak an already existing plan with an additional program or two. It can be limited to diversion programs, or it can include consideration of new or expanded capital facilities. Regardless, programs can be looked at in terms of: tonnage diverted; environmental benefits but small tonnage; or costs. Another set of factors to be considered are local preferences for intervention, which can be: (1) providing services that can be voluntarily used to increase waste diversion; (2) providing financial incentives for waste diversion; (3) incentivizing the private sector to develop infrastructure for diversion; or (4) bans and mandatory diversion regulations. The City of Seattle characterizes its solid waste program as moving away from government provision of disposal to prevention and producer responsibility (Figure 19.6). The consumer is at the center also with links to all four quadrants.

Program scenarios need to be assessed from a technical and risk perspective, particularly with respect to participation rates and to select the overall action plan. Another set of considerations are the groups of users or producers. Table 19.5 is a summary plan for a waste management

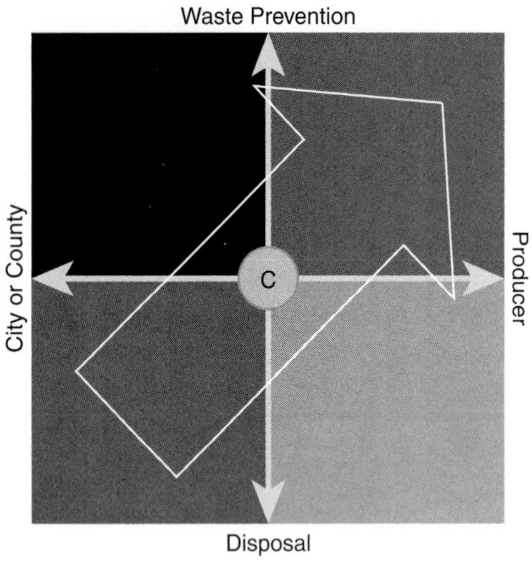

Figure 19.6
Desired policy movement in solid and hazardous waste toward more prevention and producer responsibility

Table 19.5
Overview of a waste management program

Program	Material	Sectors
Green Building	Unpainted wood Inerts Cardboard	Roll-off Self-haul
Business	Mixed paper Cardboard Food waste Film plastics	Commercial Roll-off Self-haul
Organics	Food waste Yard waste Other paper	Residential Commercial Roll-off Self-haul

Source: Alameda County Waste Management Authority, 2008.

district showing the linkages from three program areas to the materials in question and then to the manner of delivery of the waste.

Translating the Program into Tonnage Targets. The program should have the estimated number of tons that will be diverted by each program for the locality. If the plan is being developed by a county or regional district, tonnage targets are also set for each level of government. In this case, the subordinate levels of government will have their own collection systems, transfer stations, and diversion programs. Table 19.6 shows tonnage targets for a countywide waste management authority by waste stream category in order of rank by producing sector. Targets are shown for the central authority and for member agencies.

Evaluating Waste Diversion and Recycling[20]. There is no single measure for diversion or for recycling that is used throughout the United States. The states of California, Oregon, and Washington include construction and demolition debris in their definitions of materials that can be counted towards a diversion rate. There are other materials that are also worth recycling, such as concrete, asphalt, batteries, oil filters, tires, and wood, that are not included in the generally accepted definition of municipal solid waste. With this caveat, Table 19.7 shows some recycling, recovery, and diversion rates from selected west coast states and cities in 2010.

Emerging Programs and Concepts

The voluntary cans, bottles, newspaper, and cardboard programs are familiar to many. Plant debris programs have also grown over the years. For more information on how to plan these programs, see the EPA website. Some emerging programs and concepts being considered by leading-edge communities are discussed below.[21]

Product stewardship and extended product responsibility (EPR). These terms denote an environmental management strategy that assigns responsibility for minimizing the product's environmental impact throughout all stages of the product life cycle (cradle-to-cradle) to whoever designs, produces, sells, or uses it. Cradle-to-cradle impacts include energy, water, and materials use; greenhouse gas and other air emissions; toxic and hazardous substances; materials

Table 19.6
Diversion targets for a county by sector in tons by ranking in the waste stream

Rank	Material categories	Single-family	Multi-family	Commercial	Roll-off	Self-haul	Total (tons)
1	Food waste	78,274	25,708	57,429	21,708	1,612	184,717
2	Wood-unpainted	2,970	2,443	22,624	70,232	38,465	136,741
3	Other paper	48,448	16,277	39,979	15,298	2,495	122,485
4	Plant debris*	16,939	8,558	14,806	11,388	57,692	109,393
5	Wood-painted	2,853	1,587	14,134	30,335	36,442	85,357
6	Uncoated corrugated cardboard	8,737	4,384	24,827	29,412	9,249	76,602
7	Composite bulky items	1,394	1,564	5,258	32,915	34,396	75,538
8	Mixed plastics	12,569	4,461	20,453	25,216	10,599	73,294
9	Film plastics	21,378	7,086	21,276	14,894	2,124	66,753
10	Other ferrous metals	3,484	2,177	12,589	29,711	18,274	66,238
11	Crushable inerts	2,289	752	7,847	20,160	25,449	56,503
12	Mixed paper	17,414	5,556	12,970	14,820	3,210	53,969
13	Newspaper	19,417	6,846	10,776	3,705	1,446	42,189

Note: *Composite of three categories: leaves and grass, branches and stumps, and pruning and trimmings.
Source: Alameda County Waste Management Authority, 2008.

Table 19.7
Comparing actual performance of Washington, Oregon, and California for recycling, recovery, and diversion (%)

	Recycling	Recovery	Diversion
Seattle 2010	53.7		
Washington 2010	48.7		54.3
Portland 2010		57.9	
Oregon 2010	50.0		
California 2010			65
United States 2010	34.1		

Source: Compiled by the authors.

recovery and waste disposal; and worker safety. The current waste management system costs are usually borne by local governments—product stewardship or EPR shifts costs to those responsible for creating it.

Product stewardship or EPR programs can be mandatory or voluntary, and often take the form of "take-back" programs where a private infrastructure is established (reverse-distribution) to recover end-of-life products. EPR programs are funded in a variety of ways, including advanced disposal fees collected at the time of product purchase, end-of-life disposal fees at time of disposal, or with charges incorporated in the purchase price of the product.

The State of California is considering an EPR system for five categories of products: major appliances, batteries, electronics, mercury-containing light bulbs, and paint.

Mandatory recycling and bans. Bans and mandatory recycling ordinances have similar intents: to reduce or eliminate the disposal of certain materials. *Material bans*, starting with hazardous materials (such as lead-acid batteries, medical waste, asbestos, and mercury), have been around since the 1980s, but are increasingly being considered these days by state and local governments for other wastes that could easily be diverted. Two models exist: the first is a statewide ban with local enforcement, while the second is a purely local or regional requirement. Bans can either prohibit materials from the landfill, or from being collected. Bans can target particular recyclables, such as glass, aluminum cans, magazines, cardboard, compost, wood waste, tires, and oil. California bans tires, used oil, and electronics from landfills, but does not ban cardboard or yard debris. North Carolina also bans oyster shells! In many areas, local districts have instituted their own bans. Chapel Hill, North Carolina, banned non-residential cardboard from its landfill in 1996.

Mandatory recycling ordinances have been used by some agencies since 1989 and are gaining more favor today. These ordinances generally specify the material to be recycled and an enforcement mechanism. Addison County, Vermont, implemented mandatory commercial and residential recycling in the early 1990s, while Gainesville, Florida, also adopted mandatory commercial recycling. More recently in 2006, Seattle began mandatory recycling of paper, cardboard, aluminum, and glass. San Diego adopted mandatory recycling in 2007 for all properties generating over 6 cubic yards of solid waste per week. These ordinances apply to the "point of collection" and specify the materials to be recycled and the enforcement mechanism.

For more specifics on how to implement mandatory recycling programs and facilities bans, see the Alameda County Waste Management District's recent program assessment on their website at www.stopwaste.org.

Organics: food scraps and composting. Decomposable matter, also called "organics" (yard trimmings and plant debris, food scraps, wood waste, and paper and cardboard), accounts for over two-thirds of the municipal waste stream. Organics in landfills produce the methane gas which contributes to global warming, as well as resulting in the transformation of mercury from discarded batteries into a toxic gas and the production of carcinogenic compounds. In the past decade many successful large-scale plant debris or yard trimming programs have been implemented. From 1990 to 2000, recovery of yard trimmings went from 12 to 57 percent, and as a percent of the waste stream dropped from 23 percent in 1960 to 13 percent in 2006.

Composting (with or without food waste) is emerging as a "best practice" alternative to the traditional landfill since it has the potential to both reduce landfill size and methane emissions. In 2006, there were approximately 3,500 community composting programs. One facility in Quebec is producing compost that has an organic content of almost 70 percent. Local districts and cities in New York and Massachusetts have feasibility studies underway for large-scale composting facilities that will sell the compost.

Food waste is the next hot topic in solid waste. In 2010, it accounted for 14 percent of the waste stream but less than 3 percent was composted. The rest (33 million tons) was thrown away, making food waste a significant component of solid waste reaching landfills and incinerators. EPA and others are promoting the joint composting of yard trimmings and food scraps. Local governments in the San Francisco Bay Area have begun to provide curbside collection of homeowner food waste for composting at municipal or commercial facilities. Homeowners can also compost food waste in their back yards, and even in kitchens. One organization, "COOL" (Composting Organics Out of Landfills), advocates source separation of residential and business waste into three streams: compostables, recyclables, and residuals. Food waste can also be processed with biosolids from sanitary sewers to produce energy (Figure 19.7).

Wet/dry collection and processing. Closely tied to the increased attention to organics is consideration of changing the current "three-stream" solid waste collection scheme (plant debris, recyclables, and garbage/putrescibles) to a two-stream approach: *wet* and *dry*. *Wet* materials include plant trimmings, clean wood, food scraps, soiled paper, dog droppings, and

Figure 19.7
Truck delivers food waste to EBMUD biodigester for production of energy that is used on site

disposable diapers. *Dry* materials include cans, bottles, paper, plastics, and small scrap metal. The advantage of this collection approach is that materials such as newspapers, paper, and cardboard, which are highly recyclable, stay dry.

An analysis by the EPA found that participation rates for food scrap programs in the Bay Area were only 20–40 percent of eligible households, whereas participation rates in Toronto and the UK were 60–80 percent. The higher rates were attributed to every other week collection of "dry" waste—while "wet" garbage was collected every week.

Electronics recycling.[22] Although discarded electronic equipment is less than 2 percent of the waste stream, it is the source of large quantities of highly toxic heavy metals such as lead and mercury. For example, one cathode ray tube (CRT) television can contain four pounds of lead. Despite the fact that there has been a substantial increase in recycling tonnages for electronics, the recycling rate is only somewhere between 13 and 20 percent. These recycling programs are usually funded and operated by local governments or non-profits. The EPA has tried to promote the reuse and recycling of cathode ray tubes by changing regulations and working with private industry on a voluntary basis. Both California and Massachusetts have banned discarded cathode ray tubes from municipal landfills, and electronics are high on the list for inclusion in EPR programs. Fluorescent light bulbs save 75 percent of the energy used by an equivalent incandescent light bulb, but many types contain mercury and are banned from landfills in some states (Figure 19.8).

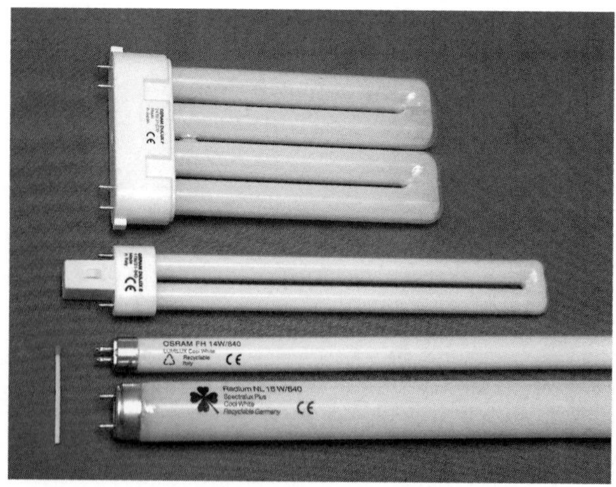

Figure 19.8
Fluorescent light bulbs

Regulatory and land use tools. Local governments have many regulatory tools to help them minimize waste and reduce carbon emissions associated with waste. For instance, if individual manufacturers and businesses are forced to publicly disclose their solid waste generation and efficiency of material use, it could affect consumer behavior and in turn force waste generators to produce less waste. Although the federal government has disclosure standards, a locality may wish to pass more detailed and fine-tuned requirements. Environmentalists in Eugene, Oregon, were successful in achieving passage of an initiative to require toxics disclosures from private firms. Similarly, retail products could be labeled with environmental information in the same way that food is labeled with nutritional information.

Local officials can specify recycled-content products in the design and construction of public buildings and in other procurement contracts. The building industry is one of the largest sources of solid waste; if recycled-content material and material-efficient procedures are designed into buildings or specified in contracts, significant waste reduction can result. Local practitioners can also specify quality standards on things like compost or mulch, to guarantee a certain, reliable level of quality. This in turn will encourage greater use in agriculture, where current uncertainty in compost and mulch quality is an obstacle.

Local governments can also prohibit certain products in order to promote sustainability. The cities of Los Angeles and Berkeley, California, for instance, have ordinances that require all purchases involving wood products to be free of old-growth wood. Oakland, California, passed a measure to change the local tax treatment of raw materials for goods manufactured in the city. This approach could also be used to influence product content. The local practitioner should consult with the city or county attorney to determine what state laws permit in these categories before going further.

On a larger scale, local practitioners can promote the incorporation of "green" building standards into building and land use regulations. This is discussed in greater detail in Chapter 24 on public and quasi-public buildings, but in addition to design considerations, they can also address the generation of construction and demolition waste.

Thrift stores. Thrift stores are remarkably efficient venues for reusing post-consumer goods. Goodwill stores, for instance, re-sell almost 99 percent of what they receive in donations. And they are not marginal operators, either; Goodwill and Salvation Army stores generate over $14 billion a year in combined revenue. Unfortunately, but not surprisingly, thrift stores often fall victim to zoning regulations. In some cases, local citizens see thrift stores as magnets for undesirable people and so agitate against them in the form of enacting zoning changes. In other cases, thrift stores, which operate on paper-thin budgets, cannot meet standards for building appearance or similar requirements and are forced to close their doors. Local officials could actually make a significant step toward solid waste solutions by considering thrift stores in their proper context and working to enable them.

Other tools. Recent research indicates that solid waste behavior varies widely across socioeconomic groups. Higher incomes are associated with higher recycling and lower waste generation. Further, economic incentives have been shown to be effective at reducing waste generation volumes. These suggest that campaigns featuring incentives and rebranding waste as a misplaced resource can increase diversion and reduce waste generation.[23]

Building or Expanding Capital Facilities

As part of the larger planning process discussed above, the district or municipality might have made the decision to build or substantially renovate or expand the landfill, the transfer station, or other facility. The following goes into more detail about some of the issues involved.

Siting and Designing Landfills

The most important issues for new facilities are the site and the design. Landfills fall into the category of locally undesirable land uses, or LULUs. They have bad odors and give off various gases. They attract rodents and other undesirable creatures. They also generate significant amounts of truck traffic. Therefore, they are the last land use that a community wants in its backyard. However, they are also critical for the healthy functioning of the community.

Careful advance planning is required to make the waste disposal facility siting decision as beneficial to the community and as palatable as possible to the neighbors. Decision-makers must consider cost, capacity, and the waste stream. Waste disposal facilities need good transportation access, for they generate large amounts of traffic. For example, to deliver the 1.5 million tons of solid waste generated in Alameda County (in the Bay Area, California) would require over 600 truckloads a day. In a 12-hour day, that's one truck every minute. Transportation is one of the largest cost components of solid waste collection and disposal, so the disposal site cannot be too far away. Minimizing distance also helps reduce carbon emissions from the journey. The following goes into further details about the siting and design aspects of landfills, incinerators, and hazardous waste facilities.

Siting and designing landfills. A landfill is in effect a very large construction site, and so can be a significant source of non-point source water pollution. In addition to normal sediment, the stormwater might also carry off uncovered solid waste. A landfill is often steeper than the surrounding topography, and so accelerated runoff may result. This impact might be exacerbated if the landfill is in a relatively pristine area. In general, since waste disposal facilities are often isolated, it would be easy to overlook environmental factors, but these areas may also be the most environmentally sensitive. For example, it would not make sense to site a landfill in a floodplain, in a canyon, or on seismic fault, but it is surprising how many have been so located.

Wind-blown paper and dust are the most frequently cited concern of the neighbors of many landfills. Prevailing wind conditions should be considered when siting landfills. Wind effects can be mitigated through landfill management practices such as installing wind fences, curtailing operations on windy days, or using dust palliatives on the daily cover.

Landfill gases. Municipal solid waste has a large proportion of organic materials that naturally decompose. This decomposition process has many byproducts, mainly carbon dioxide and methane gas. These gases are known as landfill gas and they must be managed. Methane gas explosions at landfills are not unknown. Both methane and carbon dioxide are greenhouse gases. Landfills therefore have elaborate gas collection and disposal systems that need to be adequately budgeted for. In many cases the gases are collected and sold for industrial uses, but often the methane is burned off or "flared."

Sometimes landfill gas, or LFG, is used to generate electricity. The electricity is generated in the same way as a coal- or natural gas-fired system. An LFG-to-electricity system has three basic components: the gas collection system, which gathers the gas being produced by the landfill; the gas processing and conversion system, which cleans the gas and converts it into electricity; and the equipment which connects to the electrical grid.

An LFG-to-electricity system at the Central Landfill in Johnston, Rhode Island, that receives about 90 percent of the state's solid waste, supplies as much as 12.3 megawatts of electrical power, enough capacity to serve roughly 17,000 households. The landfill sells this electricity to a local subsidiary of New England Power.

Leachate. Waste entering the landfill contains a considerable amount of water that, along with rainwater that enters the landfill, gradually gets squeezed out of the waste and collects at the bottom. This liquid is known as *leachate* and is highly contaminated. If it escapes from the landfill, it can contaminate groundwater resources. The contaminant concentrations in leachate are extremely high relative to other types of liquid waste. Moreover, landfill gas can sometimes enter groundwater instead of escaping through the landfill, further contaminating water.

To prevent leachate from escaping, landfills are lined. Until the mid-1980s, the most common liner material was compacted clay. Clay has been replaced with synthetic fabrics known as geotextiles or geomembranes which are more impermeable than clay and easier to install. Many landfills also have groundwater monitoring wells—small wells around the periphery of the landfill that allow careful monitoring of the groundwater for the presence of contaminants.

There have been some efforts to daylight leachate in order to let natural forces decompose the noxious or toxic elements. Although the concept is appealing, it is expensive and still unproven.

Eventual reuse of the landfill site. It is worthwhile taking into account that the landfill will one day be full, and with proper precautions will be available for other uses. That day may not come for 20 to 30 years, but these are the time horizons often contemplated in comprehensive land use plans, and by then the urban fringe may have extended that far.

To reuse the land (and to comply with current regulations), certain measures must be taken. Contrary to popular belief, wastes in landfills do not biodegrade or decompose quickly. Available oxygen is quickly used up beneath the soil covers, and so the decomposition process becomes anaerobic and proceeds slowly. Although it is not known long how this process lasts, some experts estimate that it will take between 300 and 1,000 years for a landfill to "stabilize," or for the wastes to be completely decomposed. Even after a landfill is closed, all the environmental control measures, such as landfill gas collection and groundwater monitoring, must continue. Very few modern landfills have reached the stage where they no longer need long-term care and management. The idea of a park or even a housing development on top of an old landfill may sound distasteful, but with the right measures just described, it is not an unreasonable notion. In fact, many communities site parks on older landfills. For example, Fresh Kill, New York City's largest municipal landfill (as well as the world's), and the site of much controversy over the years, is being turned into a park. This site was long complained about for odor problems by neighbors across the waters. Although closed in March 2001, it was re-opened temporarily in Spring 2002 to process the debris from the World Trade Center attack. The future park will be three times the size of Central Park, and construction is expected to take 30 years.

Bioreactor landfills. A new approach to landfills may provide more space in existing landfills by speeding up the decomposition process. This approach is called a bioreactor landfill. Presently there are 13 demonstration projects under construction across the U.S. or in early stages of start-up. There are three processes that can be used. The first is the aerobic process where air or oxygen is added to speed up the waste stabilization and is the same process that occurs in a composting system. The second is anaerobic, where liquid is added to the waste, such as from recirculated leachate, or spray or percolation ponds. This process produces landfill gas, which is used for energy. The third approach is a hybrid process where organics are decomposed rapidly in the upper section of the landfill with an aerobic process, while the speeded-up anaerobic process in the lower portion produces landfill gas.[24]

All three processes make the landfill compact more quickly and provide for more capacity. Some estimates indicate that 15–30 percent of airspace within the landfill can be reused. The resulting landfill mixture is also supposed to be less toxic, and to produce less long-term pollution. The accelerated anaerobic processes also are said to be a more efficient way of producing landfill gas. This is no substitute for composting, however.

Siting and designing incinerators. Significant air pollution and control measures are required for any incinerator. In addition, solid waste has water in it that is generally contaminated in some way. Before incineration, an incinerator must collect this water and treat it. These issues need to be considered in the planning and design of the facility.

One byproduct of incineration is ash, and the EPA classifies this as a hazardous material. The ashes are placed in specially designed landfills called monofills. Leachate from ash is also particularly hazardous, so these landfills have multiple liners and sophisticated leachate monitoring systems. The location of the monofill must also be considered when siting the incinerator.

The EPA recommends that all other types of waste reduction and disposal be fully exhausted prior to considering incineration, whether or not energy recovery is contemplated. Options for fully closed, smoke-free incineration have been discussed by some, but have yet to be fully developed.

Conclusion

There are three areas that local officials should be aware of in planning for solid waste that will assist in transforming local waste programs into sustainable infrastructure programs. The first is that today the level of acceptance for diversion programs is quite high. Local officials can easily incorporate these

Table 19.8
Example of carbon emissions reductions from waste management practices

Waste management practice	Metric	CO_2e reduction
Duplex copying and printing (office paper)	2,000 reams	1.9 tons
Recycle plastic film (LDPE)	1 ton	1.9 tons
Recycle paper (mixed general)	1 ton	4.3 tons
Reusable transportation package	Each reusable pallet instead a wood pallet	800 lbs
Recycled/reuse cardboard boxes	1 tons of cardboard	3.87 tons
Compost food scraps	10 tons	10.9 tons
Bay-Friendly landscaping on civic grounds	100 acres	250 tons
Implement a construction & demolition recycling ordinance (50% diversion)	200 new homes	120 tons
Build a LEED-certified civic building	8,500 sf fire stations	33.8 tons
Build GreenPoint rated homes with BFL practices	200 new homes	500 tons

Source: Alameda County Waste Management Authority Assessment, 2008.

programs into their localities. The second is about the changing attitudes toward waste producer responsibility and greenhouse gases. The population is looking for new ways to address waste management that may involve radically new ways of addressing the problem. Finally, even if it is difficult politically for the local general purpose government to move towards a zero waste program with local commercial and residential interests, they can still control what happens in their own agency. Many innovative sustainable programs have been tried out in government jurisdictions before being required for the private sector. Sustainable infrastructure practices for solid and hazardous waste are no exception. See Table 19.8 for examples of possible emissions reductions from enlightened waste management practices.

Additional Resources

Publications

American Public Works Association. *Beneficial Landfill Reuse*. Kansas City, MO: American Public Works Association, 2003.

Franchetti, Matthew J. *Solid Waste Analysis and Minimization: A Systems Approach*. New York: McGraw-Hill Publishing, 2009.

Kreith, Frank and George Tchobanoglous. *Handbook of Solid Waste Management*. New York: McGraw-Hill Professional, 2002.

Porter, Richard C. *The Economics of Waste*. Washington, DC: Research for the Future, 2002.

Rogoff, Marc J., David L. Davis, Roger W. Flint, and Bob Wallace. *Solid Waste Rate Setting and Financing Guide: Analyzing Cost of Services and Designing Rates for Solid Waste Agencies*. Kansas City, MO: American Public Works Association, 2007.

Websites

American Planning Association: www.planning.org/policy/guides/adopted/wastemgmt.htm. The policy statement of the American Planning Association in 2002 supporting waste reduction and other local progressive waste management policies.

U.S. Environmental Protection Agency: www.epa.gov/garbage/recycle.htm. Entry point to a series of practical websites that provide guidance on how to set up a recycling program.

Key Federal Regulations

The following is a list of federal regulations that deal with solid waste and solid waste management. With the exception of the 1990 Pollution Prevention Act, federal policy has tended to focus on better, cleaner disposal methods.

The Rivers and Harbors Act of 1899

Sometimes known as the Refuse Act, this prohibits the disposal of anything into waterways that will disrupt or inhibit navigation.

The Solid Waste Disposal Act of 1965

By the 1960s, more and more cities and towns across the country were practicing open-air burning of trash. In response, Congress passed the Solid Waste Disposal Act in 1965 as an amendment to the Clean Air Act. This was the first federal law that required environmentally sound methods for disposal of household, municipal, commercial, and industrial waste.

The Act gave money to states to develop and execute new waste management programs. The bill made the Secretary of the Interior responsible for solid wastes generated by the mining industry and made the Secretary of Health, Education, and Welfare (HEW) responsible for all other solid wastes.

The Resource Recovery Act of 1970

In 1970, Congress amended the Solid Waste Disposal Act as the Resource Recovery Act, the first nationwide recycling initiative. This new solid waste bill directed the Secretary of HEW to conduct a two-year study to determine a national system for the disposal of radioactive and other hazardous wastes. In addition, the bill called for the Secretary to promote and conduct research into financing waste disposal programs, ways to reduce the amount of waste nationally, methods of waste disposal, proposals for recovering materials and energy from waste, and the health effects of exposure to waste. The EPA gained jurisdiction over these functions upon its creation later that year.

The Resource Conservation and Recovery Act of 1976

With the possible exception of the Clean Air Act Amendments of 1990, there is probably no other piece of environmental legislation as far-reaching, or as troubling to industry, as the Resource Conservation and Recovery Act (RCRA). Passed as a reauthorization of the Resource Recovery Act, it primarily addresses hazardous waste management. It does have a significant nonhazardous waste management component in Subtitle D, which authorized money for solid waste programs—including funding for research, development, and demonstration projects—and created an Office of Solid Waste under the EPA. It also established minimum federal technical standards and guidelines for state plans to promote environmentally sound disposal methods, maximize the reuse of recoverable resources, and foster resource conservation. Non-hazardous waste under RCRA includes garbage, refuse, sludge from water and wastewater treatment plants, non-hazardous industrial wastes, and other discarded materials.

The Comprehensive Environmental Response, Compensation, and Liability Act of 1980

The Comprehensive Environmental Response, Compensation, and Liability Act (CERCLA; also known as Superfund) was passed in 1980 in response to several high-profile environmental disasters, such as Love Canal in New York. It is meant to address sites where hazardous wastes either have caused or will cause damage to the environment. Specifically, CERCLA addresses situations where hazardous waste has been released. The legislation gives a very detailed definition of a release: spilling, leaking, pumping, pouring, emitting, emptying, discharging, injecting, escaping, leaching, dumping, or disposing all count as releasing. CERCLA requires the EPA to identify and list all locations where some sort of hazardous waste release has occurred. The law also established the Hazardous Substance Response Trust Fund, a fund from which clean-up funds could be obtained—the famous Superfund. Finally, the law imposed several compliance requirements on facilities that handle hazardous wastes.

The Pollution Prevention Act of 1990

The Pollution Prevention Act focused industry, government, and public attention on reducing the amount of pollution through cost-effective changes in production, operation, and raw materials use. Opportunities for source reduction are often not realized because of existing regulations, and the industrial resources required for compliance focus on treatment and disposal. Source reduction is fundamentally different and more desirable than waste management or pollution control. Pollution prevention also includes other practices that increase efficiency in the use of energy, water, or other natural resources, and protect the resource base through conservation. Practices include recycling, source reduction, and sustainable agriculture.

PART FIVE

Transportation

Best practices for local and regional transportation infrastructure solutions have been undergoing a radical transformation in the past 15 years. Changes in this area are expected to continue as volatile gasoline prices, concern about greenhouse gas emissions, and aging infrastructure serve as springboards for innovative transportation infrastructure solutions that are not only safe and efficient, but also meet environmental and community goals for sustainability. Since WWII, the United States has relied upon the automobile and the airplane to transport people and goods within the country, and the airplane and ships to move internationally. High-speed rail, used in Japan and Europe, has never played an important role. This may change as national priorities in the United States shift towards greater reliance on mass transit. Demand management solutions for transportation will continue to expand. At the local level, transportation infrastructure is being expanded from an autocentric version to include pedestrian and bicycle travel. Waterfront ports are challenged to view their function in a broader context, addressing real estate as well as shipping concerns.

Transportation infrastructure usually forms a major portion of an area's capital improvement program and budget, whether it is under the purview of the local general-purpose government or a special district responsible for one kind of transportation facility. Local transportation policies are also at the heart of what the practitioner can do to reduce carbon emissions.

All the chapters in this part of the book begin with a history of the system and a discussion of current issues. The elements of the system and its institutional context are presented. Each chapter then goes on to outline how the traditional system provider plans for the system before addressing local planning and sustainability issues.

Chapter 20 (Streets and Streetscapes), looks at the foundational element of local transportation infrastructure—the local street. The three major roles of local government are discussed: (1) planning for streets and their use; (2) maintaining city streets; and (3) managing the right-of-way, including key elements in a right-of-way ordinance.

Chapter 21 (Automobiles and Mass Transit) deals with these modes of transport at the metropolitan level. Key aspects of transportation planning are addressed, including the famous four-step process. Conservation and demand management strategies are presented along with transit-oriented development and other sustainability issues.

Chapter 22 (Airports), the third chapter in this part of the book, describes the airport master planning process in detail. The chapter also addresses land-use compatibility issues of noise and safety that are becoming more urgent these days. Regional airports and the economic development aspects of airports are outlined.

Chapter 23 (Ports and Waterfronts) addresses the impact of containerization and globalization on traditional cargo ports and the rise of the diversified waterfront where concerns about tourism and other real estate developments are part of the infrastructure planning and finance process. Port planning, as practiced by a port authority, is outlined before moving to the local general-purpose government's perspective on the port—both from a financial and a land use viewpoint.

chapter 20

Streets and Streetscapes

In this chapter ...

Importance to Local Practitioners. . . . 385
History and Current Issues. . . . 386
Roman Engineering, Medieval Streets, and the Renaissance 386
Streets in America 387
Streets, Planning, and the FHA 387
Traffic Engineering and Street Standards 388
Elements of the Street and Streetscape 388
The Roadway. . . . 389
Other Elements of the "Traveled Way" 389
Bicycle Facilities 391
The Roadside 391
Poles and signs 392
Street Furniture, Buildings, and Art. . . . 393
Context-Based Framework for Street Design. . . . 395
The Institutions of Streets Today 396
Who Uses the Streets and Pedestrian By-Ways? 396
Who Builds and Maintains the Streets? 397
Local Street Planning 398
Context-Sensitive Solutions and the PRP. . . . 398
Planning for Nonmotorized Transportation. . . . 399
Needs of Bicycles 400
Pedestrian Needs 400
Traffic Calming 401
Parking. . . . 401
Street Maintenance as a Sexy Topic 402
Street Resurfacing and Preventive Maintenance. . . . 402
Street Cleaning. . . . 404
Other Kinds of Street Maintenance 405
Managing the Right-of-Way (ROW) 405
Conclusion 407
Additional Resources 408
Publications. . . . 408
Website 408
Key Federal Regulations 408

Why Is This Important to Local Practitioners?

Streets are the circulatory system of the city. As conduits for the flow of people and goods, they are key to the economic health of a region. But, as Allan Jacobs notes, streets are more than "linear spaces that permit people and goods to get from here to there." They also shape the urban community and have symbolic, ceremonial, social, and political roles.[1] For some people, streets are their only access to the outdoors, and for many the street and streetscape function as a civic living room. A more prosaic reason to think about streets is that they are very often the home to other infrastructure systems. Water, sewer, power, and telephone lines, whether above or below ground, follow the street. Streets are a significant contributor to non-point source pollution, and their surfaces are also a principal source of the urban heat island effect.

This chapter begins with the history of streets. The elements and technology of the street and streetscape system are defined, and the institutions involved in planning and financing them are discussed. The chapter then turns to the three major roles that local governments play with respect to the street and streetscape system: planning, maintaining, and managing. This chapter does not address street design—numerous books and resources are available on that topic—although some of the current changes afoot are referenced. Instead, this chapter sheds light on some aspects of street management for the general practitioner so that capital investment decisions regarding streets can be subject to broader policy objectives, such as sustainability and health concerns.

History and Current Issues

Streets are the oldest form of infrastructure. As soon as humans left their dwellings to visit their neighbors, the concept of a street was born. The earliest cities had streets lined with buildings. Streets not only facilitated the movement of goods, but also permitted access to private parcels of land, be they homes or businesses. The Royal Road, rebuilt in 456 BC and used for millennia to connect the Persian Gulf and the Mediterranean Sea, provides one of the earliest examples of a project sponsored by a government that required land and rights-of-way. India and Egypt had similar road building projects in the fifth century BC.[2]

Roman Engineering, Medieval Streets, and the Renaissance

Modern road building ideas can be traced directly to the Roman Empire. Whereas the earliest streets were simply dirt paths that meandered through cities in an organic fashion, the Romans developed standardized street sections of uniform width with separated pedestrian ways. They also developed the rectilinear grid system with straight streets and uniform block sizes. The Roman *viae militares*—or military roads—formed a vast, 53,000-mile network that connected Rome with its frontier, and were the first truly engineered roads. Roman ideas set the standard for European road construction for the next 1,800 years[3] (Figures 20.1 and 20.2).

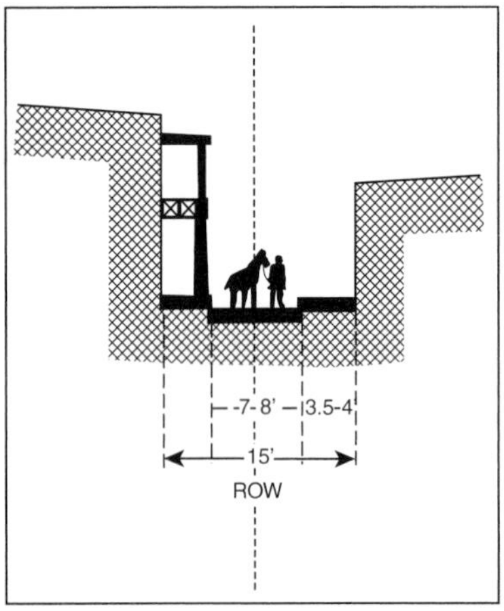

Emperor Augustus set standards for street widths in 15 B.C. The width of *vicinae*—side roads—was about 15 feet. (© *Eran Ben-Joseph*)

Figure 20.1
Cross-section of a Roman standard for construction of a side road in an urban area showing relationships to buildings

After the Roman Empire collapsed, its roads deteriorated into poorly drained dirt paths. Central planning disappeared and frequently the old roads within medieval villages were used as the guidelines for new buildings. The results were narrow, dark, winding streets that were dank, hazardous, and unhealthy. These medieval streets, which we find so charming today, were therefore prime candidates for change when Roman ideals came back into vogue during the Renaissance. In the late 1500s, Italian architect Andres Palladio recommended the construction of straight,

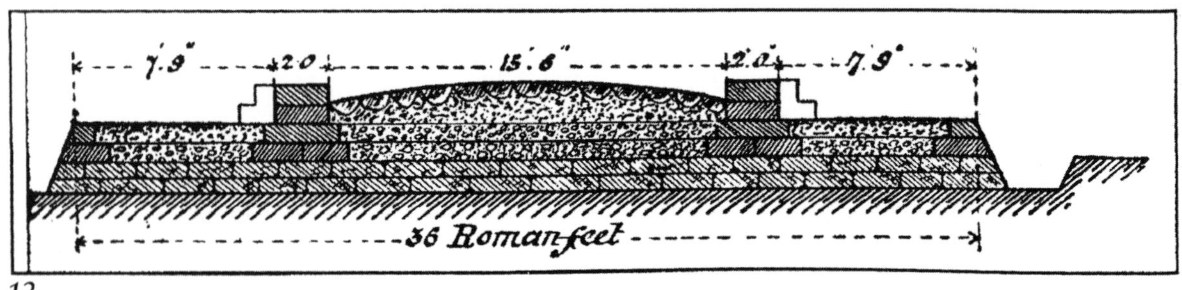

Figure 20.2
Construction drawings of a cross-section of a Roman military road showing measurements, materials, and slope

paved roads, 8 feet wide with separate lanes for pedestrians and carriages. Streets came to be seen as important social instruments, and indeed, streets became the primary focus of city building at that time.[4]

Street construction technology progressed slowly but surely. In 1716, the French king, Louis XIV formed an engineering corps along with a school for civil engineers to improve roads and bridges. This led to the professionalization of civil engineering and a more systematic approach to the road problems of the day. Better methods of construction and better materials for paving made their way onto the scene. In 1765, London implemented a street improvement program that resulted in the first "modern" street section: level paved streets of uniform width, with elevated paved sidewalks. In 1816, John Loudon McAdam developed a unique way of paving roads that consisted of mixing crushed gravel with tar or asphalt, which is the most common paving method today.[5]

Streets in America

European technology slowly made its way to America. The first engineered road in the United States was a private toll road from Philadelphia to Lancaster, Pennsylvania, opened in 1795. In 1816, Virginia created a State Board of Public Works to provide engineering expertise for public works projects. Other states soon followed suit.

However, the invention of the railroad stopped road construction and development during most of the nineteenth century. It was not until 1877 that roads received renewed attention, and from an unlikely source. The "safety bicycle," ancestor to today's bicycle, was invented that year. Cheap, safe, convenient, and mobile, the bicycle captured the public imagination and soon thousands of "wheelmen" were plying the dirt roads. They organized and began lobbying for road improvements. In 1891, New Jersey passed the first road-aid law, and in 1892, the Office of Road Inquiry was created within the Department of Agriculture.

These road improvements came just in the nick of time for the appearance of the next big invention: the motorcar. First developed in the mid-1890s, by 1914, more cars were built than wagons. Motorists formed a more potent lobby than bicyclists, and in 1916, the Federal Aid Road Act was passed to integrate the country's road system and to improve the quality of the roads.

Streets, Planning, and the FHA

Planning theory and its treatment of streets were likewise evolving. In 1927, the garden suburb of Radburn was designed using the idea of a street hierarchy. The architect, Clarence Stein felt that concentrating traffic onto a few main streets would promote a more residential feel to the smaller streets. He also argued that this would be cheaper to build: since only a few streets would carry the bulk of the traffic, only those few would need to be designed for high strength.

In 1929, Clarence Perry and Thomas Adams issued a planning concept known as "The Neighborhood Unit." This concept built on the idea used in Radburn that the residential neighborhood was a special entity that needed to be deliberately planned for and protected from that new urban scourge, automobile traffic.

In 1934, the Federal Housing Administration (FHA) was created to help restructure the private home financing system, which had collapsed during the Depression. The FHA reduced risk for lenders by providing government mortgage insurance. To do so, the FHA required potential lenders and builders to submit detailed plans and specifications for their projects, which the FHA would then evaluate for credit-worthiness. Since, for the home building industry, the FHA was the only game in town, and because the FHA requirements were so rigid, they became the *de facto* standard. The FHA found itself in a position more powerful than that of any planning agency.

Soon, the FHA began issuing technical recommendations concerning street widths, block sizes, and intersection design. In 1936, the FHA embraced Perry and Thomas's idea of the neighborhood unit, and the modern subdivision was born. At this same time, as noted in previous chapters, local municipalities were empowered to adopt rules and regulations for subdivisions within their jurisdiction, and they soon adopted the prevailing set of national standards as put forward by the FHA.

Traffic Engineering and Street Standards

In 1942, the Institute of Transportation Engineers (ITE) published its first traffic engineering handbook. The Institute had been formed 12 years earlier to achieve efficient, free, and rapid flow of traffic using scientific and engineering principles. The 1942 handbook recommended lane widths and street cross-sections that focused on driver comfort and safety at high speeds. This handbook is now in its sixth edition, most recently updated in 2009. Although the ITE standards were merely professionally recommended practices, they soon formed a basis for subdivision regulation by local planning and public works agencies. These standards were responsible for a marked increase in traffic safety, and for the streetscape we know today.

Elements of the Street and Streetscape

A decade ago it was possible to divide up the streetscape into a rigid hierarchy of streets according to their function in a suburban or urban location.

The hierarchy was developed to provide guidance to local practitioners in designing and building safe and efficient streets for automobile traffic where residential land uses were rigidly separated from commercial and retail developments. Today, best practices look at the urban streetscape and its components as part of a larger continuum that varies from natural to highly urbanized. The design of the streets will need to be adapted to where the roadway fits in this hierarchy. Yet, there are common elements to these streetscapes. This section begins by describing the nuts and bolts of streets and streetscapes before outlining new framework for defining the streetscape and the hierarchy of streets.

The Roadway

The design. Roadway design has not changed much since the Roman *viae militares*. The typical street cross-section includes the original soil, called the subgrade. On top of this is the subbase—layers of compacted gravel. On top of this is another layer of a different mix of gravel, called the base course. The base course is topped by asphalt, concrete, or some other paving material. Regardless of the paving material, traditionally the street is designed of impervious material so stormwater will not be absorbed. Such a street will crown in the middle, with a 2–5 percent slope to the gutter. The stormwater will then run off to the gutter, where it will run to the storm drain. As mentioned in Chapter 18 on wastewater, many cities are promoting pervious pavement schemes to reduce stormwater runoff, as well as to lower urban temperatures.

Paving materials. Streets can be paved with many different materials, including concrete, asphalt, cobblestones, bricks, or flagstones. Roslyn Place in Pittsburgh is paved with wooden blocks (which have proven quite long-lived).[7] Each has its advantages and disadvantages, and, even though the design of the roadway changes only slightly with these different materials, each can significantly change the character of the street. In addition, since the paved areas of some cities can range up to 45 percent of total surface area,[8] concerns about increased energy use as a result of urban heat islands, as well as higher air pollution levels and increased carbon emissions, have focused attention on high-reflective pavement materials, pavements that reflect sunlight instead of absorbing it and heating up (Figure 20.3). For the latest in pavement technology, see www.fhwa.dot.gov.

Other Elements of the "Traveled Way"

Curbs. Streets are usually edged by curbs. Curbs separate vehicles and pedestrians, and they form the gutters—that is, a conveyance for stormwater—as well. Curbs are usually continuous, with breaks at the corners for the storm drains. However, a perforated curb, with openings every so often, can allow

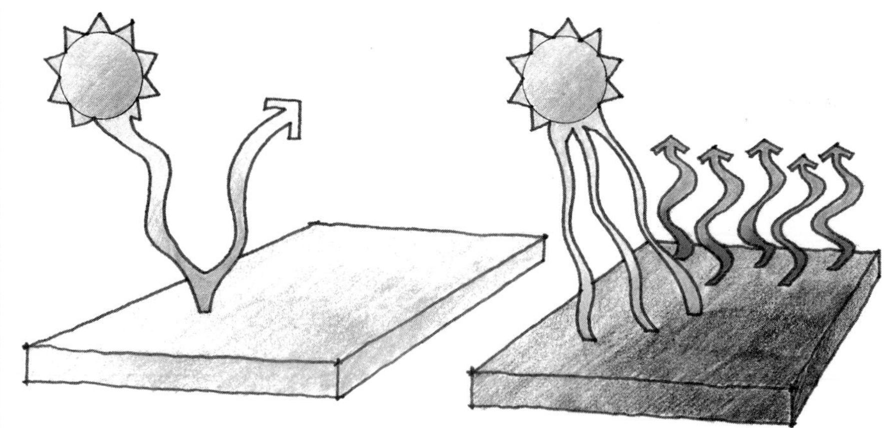

Figure 20.3
Sunlight reflecting off high-albedo pavement but being absorbed and heating up conventional pavement

stormwater to be absorbed more slowly. Effective strategies for stormwater runoff to reduce non-point source pollution start with design of the street, curb, and gutters. Curbs can also have breaks at corners for wheelchair ramps. Curbs are not appropriate for every road.

Intersections. Intersections are always a key element in the street network. A well-designed intersection represents a balance of the needs of each of the intersecting traffic streams, but also of the pedestrians and bicyclists who must negotiate it as well. Different types of intersections have different characteristics with respect to vehicle speed, vehicle safety, and pedestrian safety. For example, a 90-degree turn presents the minimum distance for a pedestrian to walk, but it slows vehicle traffic down and provides less visibility of cross-traffic. Furthermore, the turn radius might be too small for large vehicles like buses or tractor-trailers.

Crosswalks. Crosswalks are familiar features of the streetscape—the painted white ladders or "zebra stripes" that stretch across the street, usually at corners. Most crosswalks are bordered by two white lines. They are usually defined by traffic engineers as safety devices designed to channel the flow of pedestrians and to help them find their way across intersections, especially complex and confusing ones. Most jurisdictions give pedestrians the right of way when they are in the crosswalk.

Crosswalks can be controversial. For example, a City of San Diego study found that nearly six pedestrian accidents occurred in marked crosswalks for every one accident that occurred at an unmarked street crossing. The study concluded that marked crosswalks gave pedestrians a false sense of security.[9] Another analyst has said that jaywalking is actually a good thing, because it ensures pedestrians and motorists are aware of each other.[10] Another more extensive study done for the Federal Highway Administration compared accidents with and without crosswalks and found no difference.[11] On the basis of this idea, the State of California adopted a crosswalk removal policy. However, another report found that the high accident rate is actually due to poorly designed crosswalks and intersections combined with traffic policies that favor high traffic speeds and efficient vehicle movements.[12] Whatever the truth of the matter, it *is* true that pedestrian safety measures only garner less than 1 percent of road-related spending.[13] (See the T.A. Bulletin online at: www.transalt.org for more crosswalk alternatives aimed at increasing safety.)

Bike boulevard

Bicycle Facilities

While bicyclists, except where expressly prohibited, are free to use any road they choose, many cyclists prefer to used dedicated bicycle facilities. These facilities are generally classed into three types: paths, lanes, and routes. Each has its advantages and disadvantages, and a well-balanced mix of them is usually most appropriate for a local jurisdiction. Some cities, such as Berkeley, California, and Portland, Oregon, have designated low-traffic streets where cyclists are encouraged to ride as *bicycle boulevards*. They have special street signs and traffic calming devices, and often the right-of-way is given to a cyclist at an intersection.

The *bicycle box* is a recent addition to the safety toolkit for cyclists. This is a section of pavement at an intersection marked with wide stripes that permit cyclists to wait for stop lights in front of the queue of motor vehicles. It is designed to prevent bicycle/car collisions, especially those between drivers turning right and bicyclists going straight. Portland, Oregon, installed 14 bike boxes in 2008; they are painted green, with green bicycle lanes to and from the bike box. Eugene, Oregon, followed suit in 2011.

Bicycle parking facilities are an important element of road use. In many cities, the odd parking meter or signpost serves this function, but as anyone who has had a front wheel stolen or bent by a parking car will tell you, these are not always completely satisfactory. Some cities provide for secure bike parking at transit stops and at biking destinations. Seattle has a bicycle station downtown, as do Santa Barbara and Chicago. Showers for commuters are a nice touch.

Street scene with people gathering

The Roadside

The sidewalk is a "roadway for pedestrians," but it is also the interface between the street and the land uses that border the street—the heart of the roadway. People use sidewalks not just as means of access but also as gathering places. Sidewalks also serve as extensions of the private spaces they abut—witness the sidewalk cafés that are a sign of a vital urban area. The sidewalk is also the home to many other elements of the streetscape.

Urban avenue

Traffic control signs

Sidewalks can range from the 3-foot-wide concrete strip that graces most residential neighborhoods, to the wide "virtual plaza" of a prime urban space. San Francisco's Market Street has 35-foot-wide sidewalks and hosts over 200,000 pedestrians each weekday.

As the volume of auto traffic increases on a street, so does the necessity for special pedestrian facilities, culminating in grade-separated pedestrian bridges or tunnels. Such bridges are common sights over freeways, but need not be limited to that context. The City of Las Vegas, Nevada, has a system of moving sidewalks and pedestrian overpasses over its famous Las Vegas Boulevard.

Sidewalk design and signalization are subject to the ADA and the ADAAG (ADA Architectural Guidelines) requirements.

Poles and Signs

Street Lighting. Before the advent of natural gas street lighting in the late 1800s, urban streets were dark, dangerous places at night. Today, most streets are well lit with specially designed street lights. Numerous lighting technologies exist, each with differing characteristics such as cost to operate and quality of light, and will be used for different applications. In many communities, street lighting is generally provided at intersections. Pedestrian-scale lighting, which is lower and below the street canopy, is a recent trend in modern streetscape design.

Some communities have light ordinances that are meant to reduce unnecessary glare and light, thereby improving nighttime visibility for safety and security. They also help reduce unnecessary sky glow. These ordinances typically address all exterior lights, not just street lights. Not only do such restrictions cut back on light pollution, but they save energy as well (Figure 20.4).

Traffic Control Devices. Traffic signals, signs, and pavement markings are called traffic control devices. These devices serve three purposes—regulating, warning, and informing. Traffic lights, stop signs, and road centerlines are examples of traffic regulating devices. The familiar yellow diamond-shaped signs warn of curves or falling rock. One-way signs and directional signs are informational. The Institute of Transportation Engineers publishes a manual called the *Manual of Uniform Traffic Control Devices* that has standards for street markings and signs.

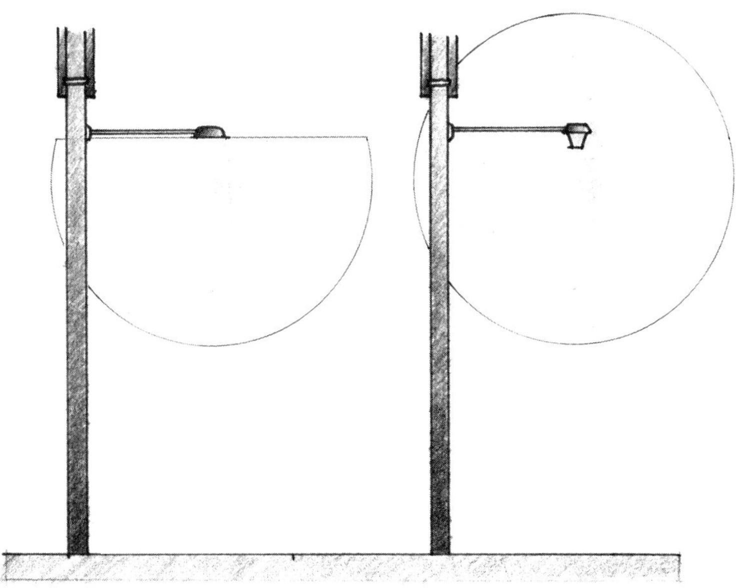

Figure 20.4
Diagram of two alley light fixtures: one promotes light pollution and the other reduces it

Since the presence or absence of a traffic control device involves inherent trade-offs, it should always be the result of careful analysis. Traffic signals, for instance, improve the safety of intersections, improve the flow of traffic, and increase the capacity of side streets. They do so, though, at the cost of increased delay, a decrease in capacity of the main street, and an increase in noise and air pollution.

Street Furniture, Buildings, and Art

Street furniture. Street furniture is the name given to such things as benches, bollards, mailboxes, and utility boxes. Some street furniture has very practical uses. Often, transformers that step down the strength of the electric current are placed in large boxes on the sidewalk. Many utility meters, electrical junctions, and valves are in metal boxes set into the sidewalk itself. Bicycle racks are a necessary piece of street furniture. If we consider streets as social spaces where people meet and interact, then benches and other places to sit are also indispensable.

Bus stop/transit shelters. Like crosswalks, bus stop shelters can be controversial. On the one hand, they provide shelter for waiting patrons and make it easy to locate the bus stop. They also provide space for route maps, transit information, and advertising. On the other hand, many cities consider them maintenance headaches since they are often targets of vandalism. Some cities, such as Seattle, have designed bus and transit shelters with public art that discourages graffiti.

A bus shelter, Eugene, OR

Public art. Public art, in the words of one city, "is a museum with no walls, no fees and no hours posted, and it is accessible to everyone every day." It enriches the city's quality of life, providing a sense of place and adding interest and vitality to the landscape. Public art ranges from monumental sculptures to fountains to paintings. Public art is often funded by private donations or through subscriptions.

Other, more quotidian, elements of infrastructure can also be used to add to the city's stock of public art. The metal plates and covers which serve as access ports to sewers, storm drains, and gas and water mains can be redesigned to add a touch of whimsy or neighborhood flavor to the streetscape. The City of Santa Cruz, California, for instance, embarked on a renovation project for the public areas in the downtown which included camouflaging traffic light control boxes (which are normally unsightly green or beige boxes on the sidewalk) inside classic white marble containers. Some cities have programs for school children to paint fire hydrants and utility boxes. Some architects have also suggested that the utilities themselves, which for years have been placed underground, should be brought back into view as a form of public art so that people can be aware of the structural underpinnings of our urban society.

Street trees in Portland, OR

Street trees. Street trees are important for a number of reasons. First, they reduce energy usage. Properly selected and maintained trees can reduce summer cooling costs by 10–50 percent.

Likewise, in the winter, properly placed trees can shield buildings from cold winds, lowering heating costs. Trees also help reduce greenhouse gas emissions and clean the air by absorbing and storing carbon and by trapping air pollution particles. They intercept and absorb stormwater, reducing runoff and soil erosion and help to mitigate non-point source pollution. Trees can screen out noise pollution. In some areas, trees improve property values by as much as 20 percent. Trees are also important in creating pleasant, walkable streets.

However, all these benefits can only be realized if the appropriately selected trees are properly planted and maintained. The International Society of Arboriculture, for example, has standards on which trees are proper for the urban environment and how these trees should be maintained.

An urban tree must be resistant to road salts and car exhaust, among other urban pollutants. Tree roots can heave sidewalks, damage building foundations, and infiltrate sewer drains. Therefore, trees with relatively shallow root structures are generally unsuitable. See also the City of Portland's *Trees for Green Streets: An Illustrated Guide* at www.oregonmetro.org.

Tree trimming with a cherry picker truck

Context-Based Framework for Street Design[6]

Components of the streetscape. The streetscape is seen as not just the area in the formal right-of-way (ROW), but also as the properties and activities in the private buildings abutting the roadway and the roadway itself. The *context*, which will vary depending upon which part of a city or town the streetscape is located in, consists of the buildings, landscaping, land-use mix, site access, and the public and semi-public open spaces. The *roadside* is the public right-of-way from the curb to the property line. This will include the sidewalk and medians (see Figure. 20.5) The *traveled way* is the public right-of-way bounded by the curbs and includes the roadway, parking lanes, bicycle lanes, medians, turn lanes, transit stops, and gutters.

Traditional street hierarchy. The traditional hierarchy of streets was established to provide guidance to engineers

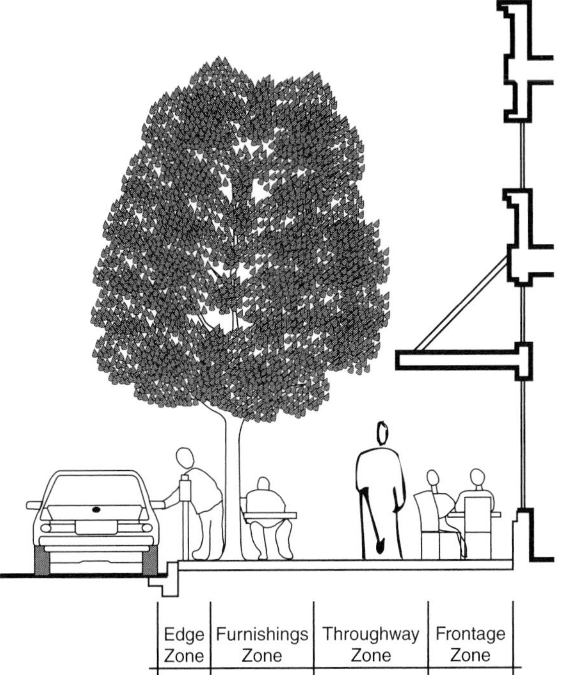

Figure 20.5
Roadside zone schematic from the New ITE Standards

and street designers on capacity and construction requirements for the post-WWII suburban development. Streets are designed to provide both mobility and accessibility. But since a street cannot simultaneously provide both high mobility and high accessibility (defined as trip origins and destinations, not access by disabled), streets were designed and classified according to a tiered system to provide differing degrees of both.

At the bottom are local neighborhood streets, which provide the highest accessibility. At the other end are highways, which provide the highest mobility but the lowest accessibility. Indeed, the highest level of highway is the limited-access freeway, which by its very nature restricts accessibility. (These will be discussed in greater detail in Chapter 21.) In between local streets and freeways lie collectors, which are less accessible than local streets, and arterials, which are less accessible than collector streets. The different classes of streets are distinguished by such characteristics as length, average travel speed, width, average daily traffic, and nature of intersections and traffic control.

Newer approach to street classification. The new ITE system classifies streets and roads by function and by *thoroughfare* type. It was developed to encourage technical roadway designs that would be sensitive to the activities and functions of surrounding land uses and to encourage walkable communities. It uses the older functional classes (such as arterial and collector designations) to develop things like speed and sight distances, but cross-references them to three design-oriented classifications. *Boulevards* are moderate-speed divided arterials that serve many types of users. *Avenues* are moderate-speed urban arterial or collector streets, but shorter than boulevards—these are mainly pedestrian and bicycle routes, but can serve local transit as well. *Streets*, using this classification system, are low-speed (25 mph) and serve local traffic. (See Figure 20.6).

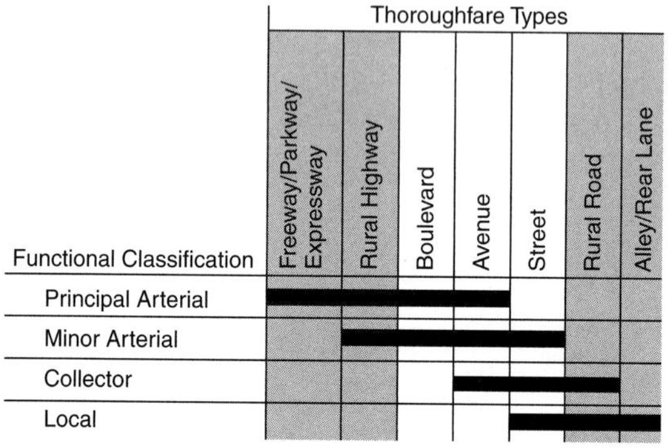

Figure 20.6
New street classification system (context-sensitive solutions) from the Institute of Transportation Engineers

The Institutions of Streets Today

Who Uses the Streets and Pedestrian By-Ways?

In 2009, there were 1,094,946 miles of urban streets in the United States, which comprised 27 percent of total public road, street, and highway miles in the country. On those same streets, Americans traveled 2.0 trillion vehicle-miles, or 67 percent of the total. In other words, urban streets get quite a work-out.

In fact, urban streets get such a work-out that nationally they are in rather poor condition. In 1997, over 70 percent of urban streets were rated "Unacceptable" by the Federal Highway Administration. To just maintain this condition over the next 20 years is estimated to cost $175 billion, while to improve the system to meet future travel and safety goals is estimated to cost $430 billion.

In 1995, the Nationwide Personal Travel Survey reported that people take almost 24,000 walking trips a year, walking a total of 11 billion miles on urban sidewalks, and that was just for work. This does not include the countless miles that children walked to school, or families strolled to the park, or commuters bicycled to work. Unfortunately, streets are often designed and built without pedestrians in mind. The residential subdivision without sidewalks is a common occurrence, as is the sidewalk that inexplicably ends in a muddy footpath.

Who Builds and Maintains the Streets?

Although the federal and state governments have some impact upon local streets, by far the most important institution for streets is the local general purpose government, whether it is the city or the county.

Most local streets are the responsibility of the municipal or county government. Operation and maintenance of streets usually falls under the aegis of the city or county public works department. Functions include traffic light maintenance, marking repainting, pothole repair, crack sealing, and repaving. In addition, street maintenance departments are often also responsible for street lights, street trees, sidewalks, and street furniture. In most jurisdictions, authority over all streets and ROW is the responsibility of the city engineer. The traffic engineer has the authority for the control of traffic. Ideally, these two officials will work closely together. However, wherever located, the traffic engineer has the local responsibility for determining traffic standards and ensuring that the capital needs of streets are cared for, although others may design and build them. Landscape architects are involved with street and sidewalk design, street trees, lighting, and street art.

Local governments have three major functions regarding streets and streetscapes: (1) they *plan* for street improvements in built-up areas and *regulate* the planning of streets for new developments; (2) they *maintain* streets in built-up areas; and (3) they also are responsible for the *management of the ROW*. The following addresses each of these functions from the perspective of the local practitioner.

Context-Sensitive Solutions (CSS)

FHWA defines CSS as "a collaborative, interdisciplinary approach that involves all stakeholders in developing a transportation facility that complements its physical setting and preserves scenic, aesthetic, historic, and environment resources, while maintaining safety and mobility."

Local Street Planning

Planning for streets and streetscapes can occur in the context of a general plan for a new area, or for a plan for an already-built area. It can involve design work as well as the development of regulations that mandate the layout and design of streets. Local practitioners can also be involved in planning the streetscape for individual projects or in regulating what happens in the streetscape through the permit process for private projects.

In undertaking these activities, the local practitioner should be aware that there has been a significant change in philosophy about best practices for street planning in the past 15 years. Traditionally, transportation services have been focused on the safety and mobility of the driving public. However, concern about global warming, decreased funding, and community activism have resulted in a new philosophy for the transportation right of way—that it must serve not only cars but also pedestrians, bicycles, transit facilities, and other multi-modal transfer locations. Furthermore, the transportation needs of the very young, the elderly, and the disabled also need to be part of the plan.

Context-Sensitive Solutions and the PRP

The new Proposed Recommended Practices (PRP) by the ITE and the Congress for New Urbanism typify the changing philosophy for street design. Some of

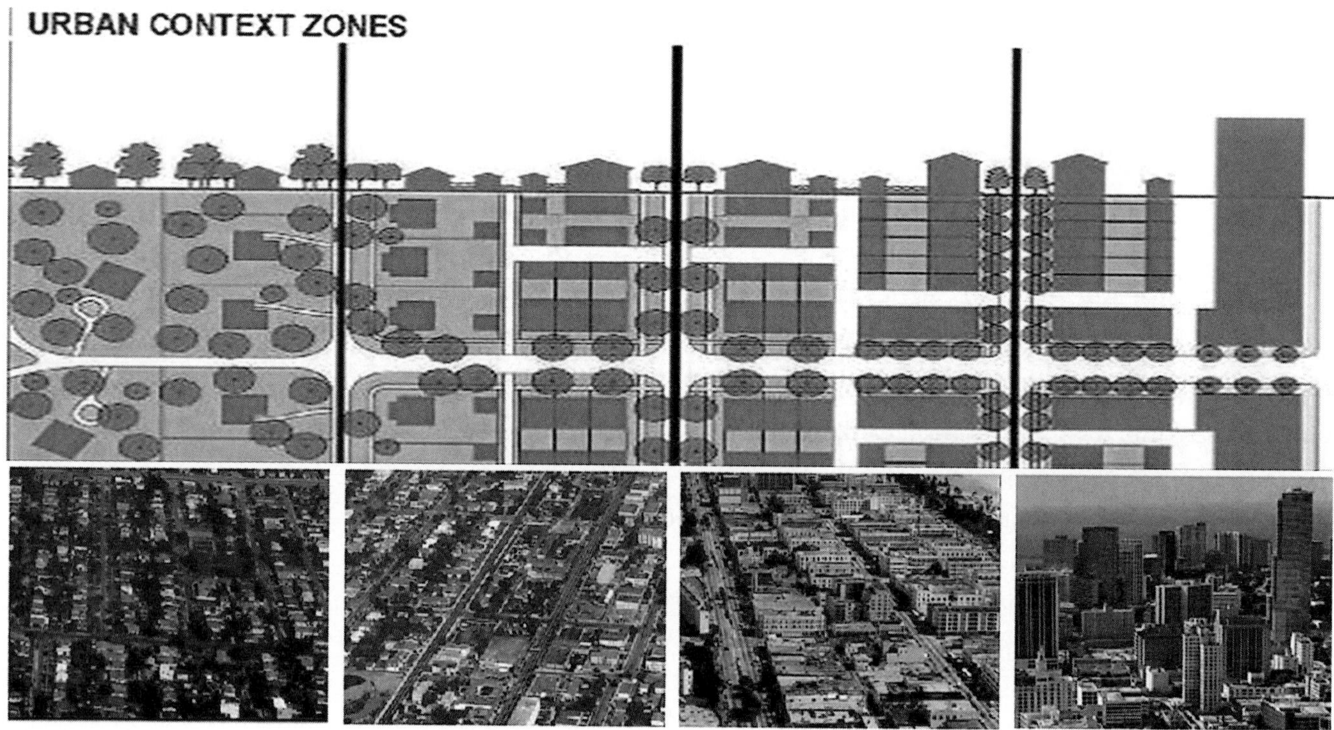

Figure 20.7
Schematic and photos of four urban context zones ranging from suburban to dense urban

the impetus for the PRP came from criticisms of the cul-de-sac street pattern used by subdivisions in the post-WWII era. The cul-de-sac started out as a visionary break from tradition that required less pavement since local streets did not need to be engineered to the standard required for those more heavily traveled, thereby making it cheaper for the developer to build. The traditional street grid pattern has 50 percent more street miles, 50 percent more lane miles of capacity, and 70 percent more acres of pavement than the cul-de-sac alternative. The cul-de-sac was also characterized by low density, segregated land uses, wide streets (to permit fire trucks to turn around), extensive off-street parking, and large setback requirements.

Criticisms of this model focused on the use of the automobile, the lack of connectivity, and the characterization of cookie cutter suburbs as place-less places. Out of these and other concerns arose context-sensitive solutions (CSS), the subject of the PRP document. Instead of beginning with design that is driven by traffic demand and level of service (LOS) objectives, the CSS emphasizes identifying critical issues and community values before establishing design criteria. This means that the roadway project planning process involves a full range of stakeholders so that both transportation objectives and community objectives will be met.

The PRP document turns the CSS concept into a series of tools for transportation design that vary according to the urban context. Instead of one size fits all, local planners and engineers are being encouraged to look to the particular context of the roadway project. The PRP identifies a series of design parameters for each of four context zones that are a subset of seven zones in a continuum of environments from urban to natural (Figure 20.7). The goal of the PRP, incorporating the new philosophy, is to promote compact development with a highly connected, multi-modal circulation network that has buildings, landscaping, and transportation improvements designed at a pedestrian scale. This includes provisions for non-motorized transportation. (For more on this topic, see www.ite.org/css/.)

Planning for Nonmotorized Transportation

Walking and bicycling are gaining favor and attention across the country as viable travel options, for a number of reasons. As concern over air quality and global warming grows, non-polluting modes become more attractive. Bicycling and walking are the cheapest forms of travel, requiring nothing more than a bike or a pair of shoes. Even more recently, the public health community has seized on biking and walking as ways to fight mounting rates of obesity. The Centers for Disease Control have identified transportation infrastructure as a cause of this problem. Cash-strapped and congested cities are rediscovering that whereas a car requires over 24 square feet of roadway

Complete Streets

In 2006, the National Complete Streets Coalition was formed to push for local transportation policies to require the design or re-design of streets and roads to accommodate people using all modes of transportation, not just automobiles. As of 2009, almost 100 cities and states have adopted Complete Streets policies.[1] Complete streets are those that enable safe access for all users. Pedestrians, bicyclists, motorists, and bus riders of all ages should be able to safely move along and across a street. For further information, see www.completethestreets.org.

Note

1 Philip Langdon, "Not Just for Cars Anymore," available at: www.courant.com (accessed July, 2008).

Assessing Pedestrian Needs for the Capital Budget[1]

Several far-seeing communities are preparing needs assessments for walkers that include demand measures, accident rates, volume and speed of automobile traffic, as well as the condition of the sidewalks themselves. These studies are used to prioritize capital expenditures for sidewalks, curb cuts and other pedestrian-friendly improvements. The City of Alexandria, Virginia, used GIS to score walkability by street segment for sidewalks and intersections. Miami-Dade County used data on pedestrian injuries and fatalities to target high-risk locations for specific safety improvements. Alameda County used pedestrian counts at 50 intersections to estimate pedestrian crossings at 7,500 intersections to use as the basis for a capital improvements plan.

Note

1 This is based on Dan Goodman, Robert Schneider, and Trevor Griffiths, "Put Your Money Where the People Go," in *Planning* (Chicago, IL: American Planning Association, 2009).

space to move one person, a bicycle only needs about 5 square feet, and a pedestrian about 2 square feet. Finally, every trip, no matter what the transportation mode, has a walking trip on each end of it. The day has not yet arrived where the commuter can roll out of bed into their automobile, so until that day comes, everyone is a pedestrian at some point. Bicycling and walking, then, are serious alternatives to the car and bus, and not merely for children or the last resort of a former motorist.

Probably the biggest disadvantage to non-motorized modes is that they are slow. Average walking speeds range from 2–4 miles per hours. Bicycle speeds range from 8–20 miles per hours. In our sprawling urban areas, then, these modes are impractical. For this reason, transit must be well integrated with non-motorized modes.

Needs of Bicycles

Bicyclists share some needs with motorists and pedestrians, but also have unique needs for safety and commuting. Establishing a bicycle network should be considered whenever there is a street construction project, although a stand-alone bicycle project is also possible. Best practices for bicycle planning include a network for a city or area that consists of bicycle lanes, boulevards, streets shared with automobiles, and off-streets paths. Bicyclists will benefit from traffic calming, as well as treatments that maximize intersection LOS standards. Perhaps the only truly unique need for bicyclists in urban areas is adequate parking.

Pedestrian Needs

Pedestrians use streets for mobility and access, the same as motor vehicles, but also for social and recreational reasons. Therefore, the ideal pedestrian circumstance will not only emphasize short trip lengths and other similar travel parameters, but also a visually pleasing and safe environment. Places to sit, places from which to watch, things to look at—these are all present in a pedestrian-friendly environment. These things can be found in street furniture, street trees, and compatible land use patterns. Clearly, not every street in the city can be like Main Street, but the fewer streets that present the pedestrian with a blank sound wall on one side and speeding traffic on the other, the better.

Pedestrian access should be an integral part of street design. Studies have shown that people will make pedestrian trips in lieu of auto trips if destinations are accessible by foot. However, other studies have shown that Americans are not willing to walk very far for these trips, on the order of 400 to 1,200 feet, which can make it difficult to accommodate pedestrian destinations within existing land uses. One way to overcome these opposing attitudes is to link pedestrians to transit. This has the effect of enlarging the pedestrian's area within reach.

Traffic Calming

One way to make streets more pedestrian- and bicycle-friendly is to "calm" the traffic. Traffic calming is defined as "changes in street alignment, installation of barriers, and other physical measures to reduce traffic speeds and/or cut-through volumes, in the interest of street safety, livability, and other public purposes." Traffic calming measures can be divided into two types. *Volume control* measures address cut-through traffic problems by prohibiting certain vehicle movements, which diverts traffic to streets better able to handle it. *Speed control* measures address speed problems by changing the vertical or horizontal alignment of the street, or narrowing the street. In practice, many measures actually do a little of both. The physical treatments that are used to achieve these measures are many and varied, and range from simple speed bumps to blocking off the road.

Traffic calming has been used for decades in Europe, but has only caught on in the United States more recently. Traffic calming is not without its opponents. Some motorists, for example, find speed humps annoying at best and argue that they actually do more harm than good. Americans Against Traffic Calming, an Austin, Texas-based group, claims that traffic calming measures "have caused the following: death, injury, pain, discrimination, denied street access for the disabled, increased air pollution and have drastically slowed emergency response times putting you and your family at risk!" Behind this hyperbole lies a real concern for disabled access.

Emergency responders don't like traffic calming either. Fire chiefs, for example, often say that speed tables create additional wear and tear on their cities' emergency vehicles. In fact, some jurisdictions have actually outlawed speed bumps and other traffic calming measures as illegal traffic control devices. (For more on traffic calming, see www.ite.org/traffic/tcdevices.asp.)

Parking

Streets also serve as storage areas for motor vehicles that are not in use, a function otherwise known as parking. On-street parking is a disputed topic. Merchants like it because it makes it easier for patrons to see and visit their shops. Some pedestrians like it because it forms a buffer between them and the moving traffic. On the other hand, bicyclists are at greater risk of getting hit by a car or an opening car door. Some neighborhood associations are successful in prohibiting on-street parking for aesthetic reasons.

When faced with the demand for more parking, decision-makers should not automatically provide it. Parking management and parking control measures are frequently overlooked solutions to perceived parking shortfalls. Parking management can be as simple as brokering shared parking spaces among users with different need patterns such as a church and a movie theater. Parking control measures, which include implementing a pricing structure and limiting the amount of parking, can be used to either eliminate low-value trips, or to encourage the use of alternate modes of transportation.

Building the subgrade

Street Maintenance as a Sexy Topic

It is likely that most local practitioners involved with planning and financing infrastructure think that the topic of street maintenance is deadly dull. However, this is one of the few places where a few policy changes as part of preparing the general plan, the capital improvements program, or a climate change plan can have stunning results for neighborhoods and commercial areas, and can also positively impact environmental concerns. Street maintenance, which we define as street resurfacing and preventive maintenance, as well as street cleaning, is an area often overlooked by planners and local policy-makers, leaving these matters to the engineering community, who may have more limited functional goals.

Laying the asphalt

Street Resurfacing and Preventive Maintenance

Good quality pavements provide a higher quality ride, which increases both the capacity and safety of the road. Poor quality pavements can damage vehicles, goods, and even passengers. Environmental damage to the roadway grows exponentially worse as pavement quality deteriorates. Keeping track of the quality of streets to guide street resurfacing and preventive maintenance efforts throughout the

Smoothing out the asphalt

locality can be a daunting task. One tool that can assist is the pavement management system (PMS) (See sidebar).

The American Association of State Highway and Transportation Officials (AASHTO) defines a pavement management system as "a set of tools or methods that assist decision-makers in finding optimum strategies for providing, evaluating and maintaining pavements in a serviceable condition over a period of time." There were about two dozen different computer software programs known as pavement management systems in 1997, and there

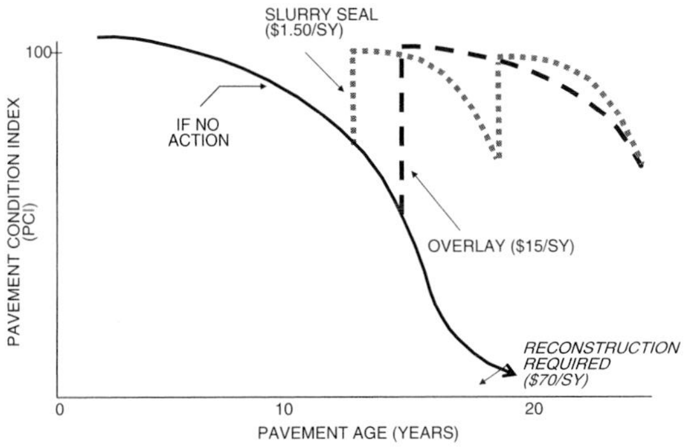

Figure 20.8
Pavement deterioration curves

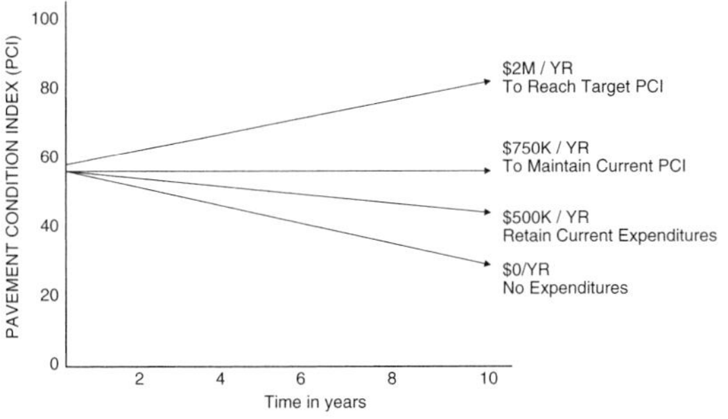

Figure 20.9
Network budget analysis

Pavement Management Systems as a Model

Pavement management systems (PMS) represent one of the earliest applications of computerized asset management. In the 1980s, when personal computers first started appearing on desktops in agency offices, public and private domain management software was soon developed to take advantage of this powerful, affordable tool. The resulting process of creating system segments, evaluating segment condition against a uniform standard, entering the data into a PC, and manipulating the software to create useful system-wide data, has since been expanded to other infrastructure assets.

Figure 20.8 represents—for a single segment—how a new street pavement will deteriorate over its design life. In the first 75 percent of a new pavement's design life, its condition deteriorates by only 40 percent. The next 40 percent of deterioration takes place rapidly in the next 12 percent of its design life. This reality is reflected in the rapidly decreasing slope of the deterioration curve. The underlying premise of pavement management is to treat pavements before they deteriorate to the point of needing expensive reconstruction (at $70/SY). By rejuvenating and sealing pavements at 5- to 7-year intervals (at $1.50/SY), the need for overlaying (at $15–30/SY) is deferred. By performing a combination of these regular treatments and overlays, the major cost and inconvenience of pavement reconstruction is even more significantly deferred.

When an agency's pavement segment data is entered into the PMS, the network (system) condition can be determined. This powerful tool tells an agency the current network-level pavement condition index (PCI), which then permits applying "what if" scenarios to predict future system performance given different maintenance budgets. (See Figure 20.9) For example, if no money is spent on annual maintenance of the street system (line #1), the PCI would drop every year as predicted by the PMS. Current annual expenditures are represented by line #2, showing that the PCI would drop less rapidly. The system can calculate the annual expenditure required to maintain the current PCI (line #3), or the expenditure to achieve a desired target PCI in 10 years (line #4). With this information, an agency can adopt an optimal budget knowing the consequences of their action.

These management and budgetary benefits of a PMS system are also very easy to explain to agency administrators and lay politicians, especially when linked to GIS mapping. In regions where all the localities use the same software, it is possible to aggregate the data from dozens of agencies to determine the regional state of pavement management.

Written by Bob Guletz, PE; of Harris & Associates.

are likely more now. MicroPAVER software, developed in the 1970s by the U.S. Army's Construction Engineering Research Lab (CERL) and provided by the American Public Works Association (APWA), is widely used. One local public works director estimates that about half or more of local governments use a computerized pavement management system. In 1991, the Intermodal Surface Transportation Efficiency Act (ISTEA) required each locality receiving federal funds to have a PMS in place that is updated every two years.

A pavement management system has an information management system that collects and manages the inventory and condition data. It also consists of decision support software that utilizes formulas to analyze the data and provide recommendations on which street projects or segments should be repaired. It should also interface with a maintenance management system or work order system that is used to manage the general workload of an in-house staff in addressing routine maintenance such as potholes.

Collecting pavement quality data for the PMS can be a Herculean task. Sometimes a locality will contract out the initial development of the database and update it with in-house staff. Orange County, California, collected standard pavement condition data for 94 centerline miles of residential streets and over 200 miles of residential right-of-way that included sidewalks, curbs, and gutters using hand-held computers. It is likely that this function will continue to be automated.

Street Cleaning

Street cleaning is an essential although unglamorous part of street maintenance. The public works department of a city or county usually performs this function. Street cleaning is important for safety and public health, for waste management, and also for keeping the sewers clean, which in turn prevents sewer overflows. It is also important for stormwater management as it prevents debris from going into sewers. Particularly important has been the discovery that heavy metals are deposited from certain types of automobiles on the streets. Street sweepers have evolved to the point that some are able to sweep and vacuum up these particles as well as the leaves, paper, and other debris they were originally deployed for.

Street sweeping

Recent advances make it possible for the vacuum sweeper to operate all year long, even in freezing winter areas.

Technology is also continuing to evolve with respect to the use of computerized management of street cleaning. For example, in 2000, the city of Oakland, CA, began a program called Grime Busters that resulted in GPS receivers being installed in some of the street sweepers. The location, speed, and even whether the brushes are up or down can be tracked on a real-time map for the supervisor.

A city worker fills one of many potholes

Other Kinds of Street Maintenance

Other maintenance for the streetscape is also important. Pothole patching is a never ending task, along with re-striping streets and curbs. Replacing graffiti-marked signs is an essential task in some cities, along with removing graffiti from street furniture. Some cities have hired young people for their graffiti buster teams. Others have used non-profit agencies that employ the homeless, while still others contract with specialty painters whose trucks can match any color to paint over the latest application of what some call "street art."

Tree maintenance in the public right-of-way is also a key task and can be labor-intensive process. Trees require watering, pest management, and pruning. The care of street trees does not always fall on those who maintain the streets. In Los Angeles, the Bureau of Street Services manages street trees, but in Chicago, it is the Department of Environment, while in New York, it is the Department of Parks and Recreation.

Keeping street lights burning brightly is also a key task, and finding some efficient system to replace those that are burnt out challenges many cash-strapped local jurisdictions. Similarly, traffic lights must also be maintained.

Managing the Right-of-Way (ROW)

A right-of-way (ROW) is the right to pass across the land of another. The public ROW is, logically enough, open to the use of the public. The public ROW includes the surface, but also the area above and below the ground. Below the surface of local streets many infrastructure providers place permanent facilities such as water mains, electrical conduit, natural gas pipes, sewage systems, power lines, high-pressure steam ducts,

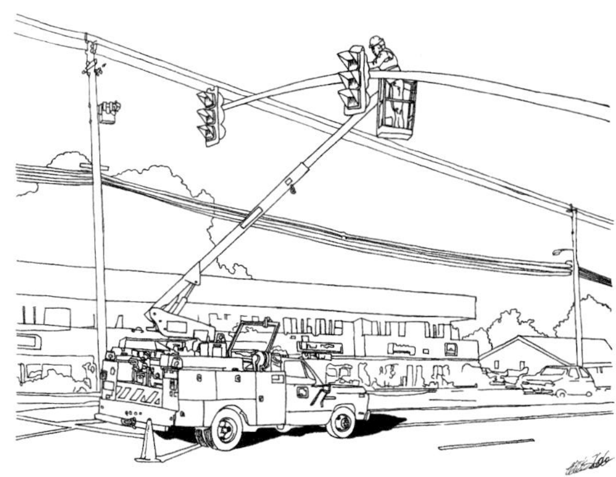

Repairing street signals

telephone and cable wires, and fiber optics. It is imperative that contractors from all agencies are aware of what is underground before they start to dig (see sidebar). Some ROWs for utility corridors are overhead as well as underground. Poles, even if not owned by the locality, are subject to regulation by most local governments.[14]

The ROW is usually dedicated to the local jurisdiction when the subdivisions are formed. In addition to local ROW, there can be state and federal rights-of-way.

The size or width of the ROW will depend on many factors, such as the amount of traffic, the nature of the neighborhood, and the policy goals of the government, and should also consider the other agencies that will want access to it, such as the power company, the water and sewer agencies, and the various telecommunications service providers. (See Figure 20.10 for a schematic of a present-day ROW from Seattle, Washington.)

APWA Color Codes for Call Before You Dig

Red	Electric power lines, cables or conduits, and lighting cables
Yellow	Gas, oil, steam, petroleum, or other hazardous liquid or gaseous materials
Orange	Communications, cable TV, alarm, and signal lines, cables, or conduits
Blue	Water, irrigation, and slurry lines
Green	Sewers, storm sewer facilities, or other drains lines
White	Proposed excavation
Pink	Temporary survey marking

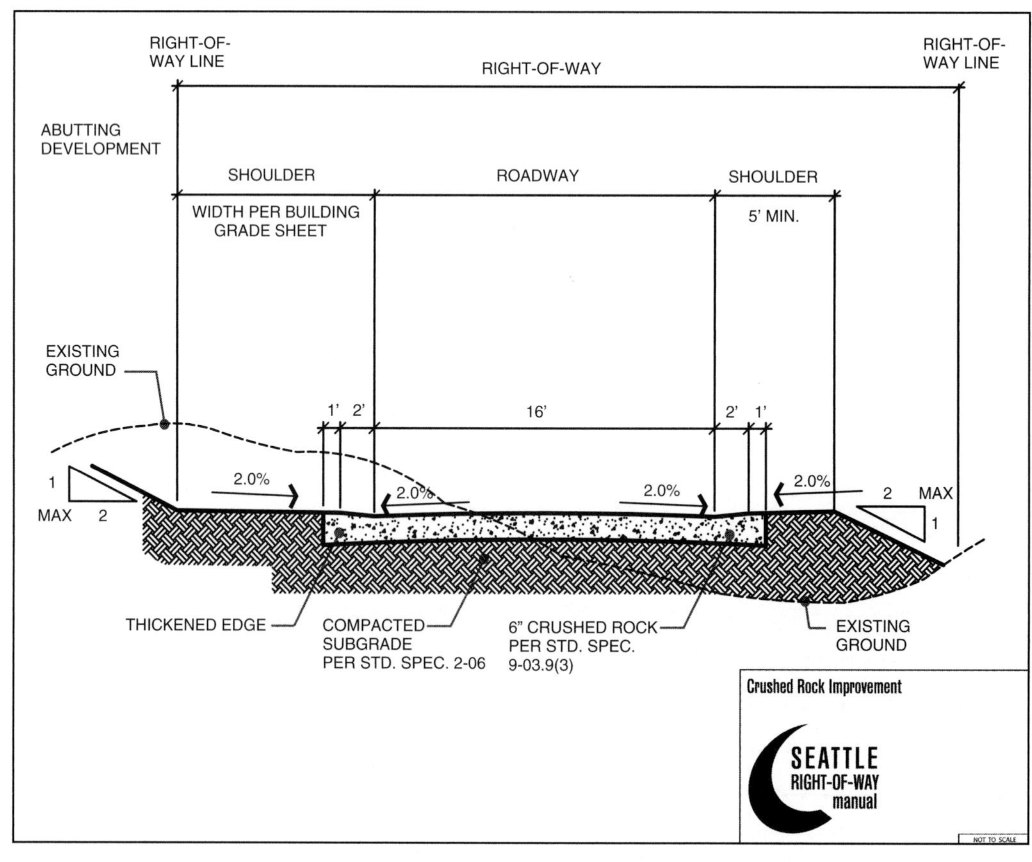

Figure 20.10
Schematic showing placement of water main, sanitary sewer, storm drain, and other right-of-way requirements for streets in Seattle, WA

Local governments (usually the public works department) are responsible for managing the right-of-way on behalf of the public and for obtaining compensation for its use. This is done in order to preserve the public's right to use the ROW and to ensure that the users do not reduce the value of the public's asset. This is done through the use of permits, franchises, and fees for use. The local government and the users of the public ROW have a relationship similar to that between a landlord and tenant.

For many years, policies and procedures governing the ROW at the local level were not codified. The economic boom of the 1990s and deregulation of the telecommunications and energy utilities have caused many local governments to take another look at their ROW practices, since many streets were repeatedly torn up and not repaved properly. (See Chapter 28 for a more thorough discussion of telecommunications infrastructure.) Today it is important for local jurisdictions to have a comprehensive set of procedures and systems to manage the ROW.

Conclusion

Streets and the public ROW require considerable amounts of land. The average city in the U.S. has from 25–35 percent of its area paved over for streets. This is a significant portion, and although streets are the subject of many publications and professional disciplines, until the advent of CSS, the ROW had often received much less critical attention than other land uses. The current suburban street standard is usually accepted without comment. Furthermore, with the exception of some street calming schemes and the infrequent street widening, it is unusual to see any substantial changes made to an existing street network in already urbanized areas. The local practitioner needs to ensure that routine decisions that are taken for granted or embedded within the so-called "technical" standards are exposed to the light of the day. The local practitioner needs to be an advocate for more livable streetscapes designed with uses other than the automobile in mind. In addition, having a wider audience for routine repaving investment decisions may give the locality more bang for its buck when it can be used as a mini-redevelopment tool.

Finally, the local practitioner can be an advocate for more sustainable streetscape solutions, such as pavement surfaces that help to reduce the urban heat island effect, full-cost pricing for parking, and pedestrian- and bicycle-friendly street furniture. The local practitioner can also be an advocate for transforming the ordinary "appliances" of the streetscape into more interesting and artistic public facilities.

Additional Resources

Publications

American Public Works Association. *How to Initiate and Implement a Right-of-Way Program.* Kansas City, MO: American Public Works Association, 2009.

Association of American State Highway and Transportation Officials. *A Policy on the Geometric Design of Highways and Streets.* Washington, DC: Association of American State Highway and Transportation Officials, 2001. This work details the practical requirements for street planning.

Beery, Pamela. J. "Current Right of Way Issues for Local Governments," in *APWA Spring Conference, 2003.* Portland, OR: Beery, Elsner & Hammond, LLP, 2003. This offers legal explanations and information on right-of-way issues.

Dowling, Richard G., David Reinke, National Cooperative Highway Research Program, National Research Council, U.S. Transportation Research Board, et al. *Multimodal Level of Service Analysis for Urban Streets.* National Cooperative Highway Research Program Report 616. Washington, DC: Transportation Research Board, 2008. This document describes a method to plan street corridors that takes into account pedestrians and cyclists as well as automobiles and buses.

Handy, Susan, Robert G. Paterson, and Kent Butler. *Planning for Street Connectivity: Getting from Here to There.* Planning Advisory Service Report 515. Chicago, IL: American Planning Association, 2003.

Jacobs, Allan B. *Great Streets.* Cambridge, MA: MIT Press, 1993. A wonderful book about street design. Very well respected.

McCann, Barbara and Suzanne Rynne. *Complete Streets.* Planning Advisory Service Report Number 559. Chicago, IL: American Planning Association, 2010.

New York City, Department of Transportation. *Street Design Manual.* New York: Department of Transportation, 2009. A comprehensive green approach to street design standards that includes traffic calming, green materials, street and pedestrian lighting, and bicycle-friendly options.

Shoup, Donald. *The High Cost of Free Parking.* Chicago, IL: American Planning Association, 2011.

Southworth, Michael, and Eran Ben-Joseph. *Streets and the Shaping of Towns and Cities.* New York: McGraw-Hill, 1997. A "must-have" that explores street design and its relationship to urban form.

Website

American Public Works Association: www.apwa.org. This has a wealth of materials on right-of-way issues, in addition to the publication listed above.

Key Federal Regulations

Unlike the other infrastructure systems we have looked at, street design has never been subject to explicit federal regulation, although there have been default standards followed by engineers for years. As we saw earlier, the Federal Housing Administration, through its practice of recommendations, has done as much to shape streets as any other organization.[15]

Federal Aid Road Act of 1916

This Act was the first time the federal government concerned itself with roads, and was the culmination of almost 20 years of lobbying by farmers, motorists, and bicyclists—the so-called "good-roads movement"—to improve roads. The Act did not result in many miles of paved road, but it did create in each state a highway agency staffed with engineers. See Chapter 21 for more details.

National Housing Act of 1934

This Act created the Federal Housing Administration, who, through its lending practices, would come to have a profound influence on the shape of suburban streets.

Circular Number Five (Subdivision Development), 1935

This technical standard, the first promulgated by the Federal Housing Administration, stated the FHA's general standards for successful housing developments. It included requirements for such things as street widths, corner radii, and block lengths.

Bulletin on Planning Neighborhoods for Small Houses, 1936

This was another FHA publication, in which the FHA rejected the traditional grid pattern of neighborhoods. Instead, a hierarchical system of curvilinear roads, cul-de-sacs, and courts was the preferred street layout. It also included more refined street width standards.

Federal-Aid Highway Act of 1956

This Act created the Eisenhower Interstate and Defense Highway System, and is considered by some to be one of the most important pieces of legislation ever passed in the USA. It also created the Highway Trust Fund, a dedicated funding source for highways that is supplied by the gasoline tax. Many states have adopted similar trust funds. This is examined more in Chapter 21.

Americans with Disabilities Act of 1990

The 1990 Americans with Disabilities Act says that public programs may not discriminate and must offer all services, programs, and activities in ways that allow equal participation. This includes pedestrian access at street corners. The City of Berkeley, California Disability Compliance Program, for example, says that by ramping the street corners, "The City is working to make each street corner accessible for wheelchair users, blind people and all pedestrians as well as to ensure that sidewalks are not blocked by signs, parking meters and other obstacles."

chapter 21

Automobiles and Mass Transit

In this chapter ...

Importance to Local Practitioners . . . 411
History of Metropolitan Transportation . . . 412
Walking–Horse Car Era . . . 412
Electric Streetcar Era . . . 413
Recreational Automobile Era . . . 414
Freeway Era . . . 415
Federal Involvement in Transit . . . 416
Transportation Modes . . . 417
Automobiles . . . 417
Transit . . . 418
Non-motorized Transportation . . . 420
Who Provides Transportation Services? . . . 422
Federal Level . . . 422
State Level . . . 422
Local level . . . 423
Transportation Planning . . . 423
The Four-Step Process . . . 424
Identifying Needs . . . 428
Level of Service Standards . . . 428
Condition Assessment . . . 430
Facility Inventory . . . 430
Needs and Wants . . . 430
Other Capacity Issues . . . 431
Identify Solutions . . . 432
Evaluate Alternatives . . . 432
Alternative Transportation Approaches . . . 432
Land Use Policies . . . 432
Demand Management . . . 434
Envirnomental and Social Impacts . . . 437
Energy Greenhouse Gas Emissions . . . 437
Transportation and Air Pollution . . . 439
Social Equity . . . 441
Local Transportation Finance . . . 442
Traditional Local Government Sources: Property Taxes . . . 443
User Fee Model . . . 443
Special Purpose Local Option Taxes . . . 444
Choosing a Financing Mechanism . . . 445
Conclusion . . . 446
Additional Resources . . . 447
Publications . . . 447
Websites . . . 448
Key Federal Regulations . . . 448

Why Is This Important to Local Practitioners?

Transportation—the movement of goods and people—is linked to metropolitan physical and economic growth, quality of life, air and water pollution, carbon emissions, and social equity. In recent years, the state of metropolitan transportation planning has received growing scrutiny. Although volatile oil prices and the challenge of global warming demand concerted transportation policy responses at the national level, there is much that local and regional policy-makers and practitioners can do, since over three-quarters of the roadways and 100 percent of transit systems are owned by units of local government (city, county, or region). In addition, transportation infrastructure is a vital component of local and regional land use plans

Transportation has many faces. When examining transportation issues, one must distinguish between passenger and freight; inter- and intra-urban; surface, air, and maritime; motorized and non-motorized; and public and private, to name just a few categories. For the purposes of this chapter, we will look at the issues most relevant to the local practitioner in general purpose government, focusing on passenger transportation within metropolitan regions. As recently as 20 years ago, any metropolitan transportation problems were generally solved by building more highways, using mostly federal funds. Today the subject is much more complex, and transportation issues are at the center of climate change policy and public health discussions.

This chapter begins with the history of metropolitan transportation and then moves to a description of the major transportation modes. Next, the agencies providing transportation services are described. The chapter then provides a primer for local-level transportation planning before discussing transportation finance. The chapter concludes with a restatement of smart and sustainable transportation planning principles.

History of Metropolitan Transportation

The current picture of metropolitan transportation in this country did not emerge overnight. Rather, like that of every other infrastructure system in this book, it is the consequence of various decisions and policies adopted over the course of the past century. In this chapter we will see how the development of metropolitan transportation for passengers has impacted larger urban spatial patterns. In general, land use and transportation have been dancing in an outward spiral that has given us the low-density metro areas we have today.

The evolution of metropolitan transportation can be divided into four eras:[1]

- Walking–Horse Car Era (1800–1890)
- Electric Streetcar Era (1890–1920)
- Recreational Automobile Era (1920–1945)
- Freeway Era (1945–present).

Each of these eras is characterized by a particular technology and by a distinctive pattern of spatial organization. Time budget theory, a common theory of travel behavior, says that people do not like to travel more than 30–45 minutes to make a given, frequent trip. This means that land use is partly shaped by the speed of the prevailing transport technology and that, over time, urban form has shaped itself to transport technologies. As speeds increase, people live farther and farther from where they work, shop, and recreate. This section explores this concept.

Walking–Horse Car Era

Before the middle of the 1800s, the primary mode of transportation was walking. Only the rich could afford the only real alternative, the horse-drawn carriage. Consequently, cities were compact affairs, shaped such that everything was a 30–45-minute walk from the center. The cities were also quite dense: overcrowded, noisy, filthy, and unhealthy. This, combined with the rapid industrialization and its attendant ills that also appeared in the mid-1800s, produced an exodus of the rich to the salubrious nearby countryside.

The arrival of the railroad in the 1830s provided opportunities for easy daily travel to and from the city from newly-built upscale trackside suburban settlements. Soon, businessmen by the scores from New York, Boston, and Philadelphia were making this very trip; over 100 trains a day ran between Boston and its suburbs in 1850. Indeed, the ability of these businessmen to "commute" the price of their fares to lower ones if they bought in bulk produced that term for the journey to work.

Meanwhile, the rest of the population remained in the squalid city, clamoring to get out. Various transport technologies appeared in various cities, but most proved impractical. In 1852, the horse-drawn streetcar was introduced in New York and was an instant hit. The streetcar ran on light rails installed on grade, easy to install and overcoming the problem of rough, muddy, unpaved streets. These "horse cars" traveled at about 5 miles per hour, slightly faster than pedestrians, and so opened up new land on the urban fringe.

Electric Streetcar Era

The horse car, though an improvement, only whet the transport appetites of city dwellers. The speeds were not much greater than walking, and horses were smelly, messy, and prone to infectious epidemics. Disposal of both horse manure and horse carcasses was a serious municipal exercise at this time. So when Frank Sprague invented the electric traction motor in the 1880s, allowing the development of the electric streetcar or trolley, it was a breath of fresh air to the traveling public. The first electric trolley was introduced in Richmond, Virginia, in 1888, and within six years became the dominant form of urban transport across the United States. It spread quickly in part because horse cars were easy to retrofit with the motors, the speeds were much higher (15–20 mph), and electricity was seen as clean and modern. (See Chapter 27, Energy and Power.)

The much higher speeds opened up huge areas of undeveloped lands outside the city. Cities became star-shaped as they grew along the radial lines of the streetcar. Since land was so cheap, there was little incentive to lay track laterally. The typical suburb, then, was centered on the streetcar corridor that led to the city center. Stretching a few blocks on either side of the streetcar rail were residential streets and neighborhoods. The ease and affordability of this new travel mode also allowed the development of specialized land uses, in particular, concentrated retail and commercial districts, located at the hub of transit corridors. This hub became the downtown as we know it today.

By 1920, streetcars had transformed the city into a well-tracked, well-functioning metropolis. "It was at this point in time, many geographers and planners would agree, that

Table 21.1
U.S. public transit ridership

Year	Trips (in millions)
1926	17,234
1935	9,782
1945	18,982
1960	7,521
1970	5,932
1992	724
2002	801

Source: Terry L. Bronson and Christie R. Dawson, *2002 Transit Fare Summary* (Washington, DC: American Public Transportation Association, 2002).

intra-metropolitan transportation achieved its greatest level of efficiency—the burgeoning city truly 'worked.'"[2] It was also at this time that national public transit ridership peaked (excluding the anomalous wartime years) (Table 21.1). (Interestingly, it was also during this time that automobile ownership reach a level it did not attain again until 1949.)

Two things must be noted. First, public transit agencies were for the most part privately operated, owned under city franchise by hotels, real estate developers, or others who had a commercial interest in good transportation. Individual companies might only serve one major street or corridor; any city, then, might have numerous different companies plying different streets. (This is why many train stations are called "Union Station;" it was there that different companies' lines met in "union.")

Second, even with high ridership during this era, transit lines were subsidized. Transit companies were operated as "loss leaders"—by absorbing the costs of an unprofitable transit operation, a businessman could realize larger profits from increased land speculation, home sales, or other business opportunities. Transit subsidies are not new.

Recreational Automobile Era

Meanwhile, the automobile had entered the picture. Invented in the late 1800s, automobiles were initially the playthings of the rich. When Henry Ford revolutionized the automobile manufacturing process in 1913, he made autos accessible to everyone. Auto ownership exploded. By 1916, there were over 2 million autos on the road, up from almost zero, by 1920, 8 million, and by 1929, 28 million. Since then, with the exception of the Great Depression and war years, the increase in car ownership has continued.

Auto ownership really took off in rural areas, as farmers quickly realized the benefits of better access to local markets and service centers. Among city dwellers, cars were used mainly for weekend and recreational outings. Accordingly, most road paving efforts were concentrated in rural areas and among scenic landscapes.

At the time (and still today), the auto represented the "long-hoped-for attainment of private transportation that offered users almost total freedom to travel whenever and wherever they chose."[3] It was also in the mid-1920s that the growth rate of suburbs finally surpassed that of central cities. Couple this with the perceived freedom of the automobile and the famous prosperity of the 1920s, and the resultant dominance of the auto seems inevitable. Indeed, one study found that "as early as 1922, 135,000 suburban dwellings in 60 metropolises were completely dependent on motor vehicles."[4]

It is said that the automobile killed transit, but in one sense, the automobile only finished what the streetcar had started. The streetcar had produced a star-shaped urban form, with a streetcar line running down each arm of the star. High-density development lined each streetcar line. The interstices between the streetcar lines, though, remained inaccessible. The automobile opened those areas up, and suddenly, real estate developers had no more interest in subsidizing transit lines. Faced with decreasing ridership but unable to raise fares or discontinue service due to their city franchises, transit companies entered a death spiral of declining revenues and service.

Transit companies still served large constituencies, and so cities were reluctant to let the private companies fail completely. Instead, local governments bought up floundering companies and kept them afloat, the bare minimum to serve the transit-dependent.

Freeway Era

Meanwhile, auto ownership increased, and so did suburban decentralization. Automobiles finally came of age in the post-WWII era, spurred by many factors: the end of gas rationing and restrictions on automobile production; new-found prosperity after 15 years of Depression and war; new suburbs springing up in the urban hinterland to house returning veterans; and national demographic shifts (from east to west and from south to north) wrought by wartime industrial requirements. These added fuel to a trend that had been in place since the 1920s.

In 1956, President Eisenhower signed the Interstate Highway Act, creating the Interstate Highways. These new, completely auto-oriented facilities soon reshaped "every corner of urban America as the new suburbs they engendered represented nothing less than the turning inside out of the historic metropolitan city."[5] The Act also created the Highway Trust Fund, a special-purpose fund filled by motorists (through the gas tax) for motorists (in the form of highway construction).

Auto-oriented development reshaped the monocentric metropolitan urban area into a polycentric one. The traditional downtown became an anachronism, because now almost any location was within reach of a freeway. Soon, commercial and industrial densities began to mirror residential ones in decline.

Federal Involvement in Transit

As the suburbs waxed, central cities waned. Urban transit was considered a significant development factor, but urban decay and urban transportation were initially considered local problems, and so urban transit systems limped along for 20 years. The first glint of national help appeared in the 1961 Omnibus Housing Bill. As part of a larger effort to improve housing and spur building activity, the Bill provided for loans and grants for pilot projects in mass transportation. This initiated federal involvement in transit and led to the 1964 Urban Mass Transit Act, which provided $375 million in capital assistance over three years.

Transit tunnel, Bay Area, CA

Transit had a renaissance of sorts. It had attention and new infusions of sorely-needed capital. However, urban transit was being handled as a tool of urban development rather than as a mode of transportation. The Urban Mass Transit Administration was not in the Department of Transportation, but in the Department of Housing and Urban Development. Furthermore, federal transit money went straight to cities, bypassing state government (whereas federal highway money went to state departments of transportation). The result was that transit was never really integrated with other modes of transportation.

BART Extension to Milpitas, San Jose and Santa Clara, CA

It was in the 1960s that urban rail began to catch on again, most famously with the construction of the Bay Area Rapid Transit District in the San Francisco Bay area. Washington, Cleveland, Miami, and Los Angeles have also built heavy rail systems. Light-rail trolley lines have been built in many other cities.

However, it was not really until 1991, with the passage of the landmark Inter-modal Surface Transportation Efficiency

Act, that local policy-makers began to take a holistic look at regional transportation issues. This Act and its successor, the Transportation Equity Act, required local policy-makers to look at the whole metropolitan region when making transportation investments. They also increased the importance of transit, bicycling, and walking, modes that had traditionally been overlooked.

Transportation Modes

As stated earlier, transportation is often described in terms of its methods of conveyance, or *mode*. The most common mode is the private automobile. Other modes include buses, airplanes, trains, ferries, and (for freight) pipelines. Non-motorized modes include walking and bicycling. For intra-metropolitan travel, the important modes are the automobile, bus, and certain types of trains. Urban transportation planners usually refer to bus, light rail trains, and heavy rail trains into one mode called "transit." However, systems like the Bay Area Rapid Transit (BART) often call themselves rapid rail or some other term to distinguish themselves from regular intercity or commuter trains. The following addresses the specifics of these modes.

Automobiles

All automobiles share important characteristics. They have pneumatic rubber tires. They are guided by the driver and in most cases can carry at least one other person. Up until recently, all automobiles were powered by internal combustion gasoline or diesel engines; now compressed natural gas, electricity, and hydrogen are appearing as fuel sources in an attempt to address concerns about air pollution, volatile gas prices, and carbon emissions.

Automobiles require space—relative to the size of their cargo, they are quite large. Their operation calls for smooth, engineered surfaces and expansive rights-of-way. They also need to be stored when not in use, which is most of the time. It is these surfaces, rights-of-way, and storage requirements that characterize the automobile experience for planners.

Roadway extent and type and automobile ownership. In 2009, there were 2.24 vehicles per household, or about 823 cars per 1,000 inhabitants. In 2004, 89 percent of the driving age population was licensed to drive a motor vehicle. In 1975, the number of registered vehicles overtook the number of licensed drivers. In fact, in 1996 the number of registered vehicles surpassed even the number of those legally eligible to drive, although commercial fleets were counted in these numbers.[6, 7]

Automobile travel grew 47 percent during the 1960s, 38 percent in the 1970s, 37 percent in the 1980s, 26 percent in the 1990s, and 16 percent in the 2000s. The alternative to *private* automobile traffic is *public* transportation, or transit, discussed below.

Transit

Transit comes in many forms. Since each transit type has different performance characteristics, different physical requirements, and different costs, each type will be ideally suited for a different situation or market. Therefore, decision-makers should be familiar with the general features of each type.

The American Public Transportation Association defines the traditional public transit modes as follows.[8]

Buses. A bus is an internal combustion-powered rubber-tired vehicle. Buses usually have two doors, one in front for loading and unloading, and one in back for unloading only. Fares are usually paid upon boarding, on the vehicle itself. As a result of the passage of the Americans with Disabilities Act, almost all public transit buses have ramps or hydraulic wheelchair lifts to allow a wheelchair user to board. Many transit districts have bicycle racks on their buses. Buses typically carry 45–50 passengers, though recently many localities have been successfully using *articulated* buses, which can carry double that number. They normally share the road with private automobiles, though in some cities, most famously Ottawa, Canada, and Bogotá, Colombia, buses have their own "bus ways."

The *trolley bus* is an electric bus that draws its power from a pair of parallel overhead wires. A trolley bus operates similarly to a regular bus, but the electric motors make it quieter, generate less emissions and a better hill-climber. The overhead wires that trolley buses require are expensive to install and maintain, and many find them unpleasant to look at. This makes trolley buses more expensive and less flexible than regular buses.

Bus rapid transit. Bus rapid transit, or BRT, is a relatively new mode that combines the speed and capacity of a light rail system with the flexibility and low cost of buses. The exact definition of BRT is still open, but in general, a BRT system will have dedicated rights-of-way or other traffic management techniques (such as preferential treatment for buses at signalized intersections) and limited stops to increase bus speed. Further, BRT systems will have faster boarding achieved through vehicle design, station design, or fare collection practices. BRT is controversial (opponents insist that a "bus is still a bus"), but it is receiving a great deal of fanfare and support from the federal government and many transportation experts.

Light rail transit. Light rail transit is a metropolitan electric railway system characterized by its ability to operate single cars or short trains along exclusive rights-of-way at ground level, on aerial structures, in subways or, occasionally, in streets, and to board and discharge passengers at track or car-floor level. Light rail vehicles draw their power from overhead wires. Streetcars, which all but disappeared in the 1950s, are enjoying a renaissance and are considered light rail. Payment of the fare is usually done on the vehicle, but some agencies have a proof-of-payment system which requires the rider to purchase a ticket before boarding and produce it if required by a transit system authority.

Heavy rail transit. Heavy rail refers to traditional high platform subway and elevated rapid transit lines. Heavy rail right-of-way is completely segregated from other uses, through elevated track structures, subterranean tunnels, or fenced areas. Trains consist of anywhere from 2–12 cars, each with its own motor, and drawing power from a third rail (or in some cases from overhead wire). Boarding is from high platforms that are even with the floor level of the car. The cars have many doors, allowing large numbers of people to enter and leave rapidly. Fare payment is usually done in the station or on the platform before boarding.

Commuter rail transit. Commuter rail refers to passenger trains operated on main line railroad track to carry riders to and from work in city centers. The trains are normally made up of a locomotive and a number of passenger cars. Commuter rail lines normally extend an average of 10–50 miles from their downtown terminus. In some cases service is only offered in rush hours. In other cities, service is operated throughout the day and evening and on weekends, though service is rarely offered more frequently than one train every half hour. Station spacing is typically measured in miles.

Comparison of public transit modes. Each transit type has its advantages and disadvantages. Buses, since they use existing roads and streets, are the cheapest and easiest to implement, and are the most flexible in terms of routing. On the other hand, since they mix with traffic, they are subject to the same congestion that motorists are. Buses also have a serious image problem, generally seen as dirty, uncomfortable, and slow. In general, rail transit modes provide a smoother ride than rubber tires. Light rail systems don't mix with traffic as much, so they can have better schedule adherence. Heavy rail systems are extremely expensive to build due to the need for tunnels, elevated structures, or other fully segregated rights-of-way and to accommodate more gentle curves and grades than are needed for light rail or streetcars. Table 21.2 compares various operating characteristics of the different public transit modes.

The Geneva Yard in San Francisco houses light rail transit cars from different decades and cities. In the foreground is one of San Francisco MUNI's modern LRT cars. The background shows historic LRT cars which are used today to carry passengers up and down Market Street

Peachtree Station Concourse in the MARTA Heavy Rail System in Atlanta is the deepest subterranean station in the system

Public art is often found in transit stations since it affects the individual's perception of wait times

Inbound and outbound commuter rail cars on the METRA system in Chicago

Table 21.2
Operating characteristics of public transit modes

	Bus	Light rail	Heavy rail	Commuter rail
Station or stop spacing (in miles)	1/8	1/2–3/4	1–3	2–6
Average speed (mph)	6–12	20–25	20–25	30–40
Capacity (passengers)	50–75	260–520	800–1,600	1,000–2,200
Maximum hourly passengers	3,500	11,000	22,000	48,000

Source: Robert Cervero, *The Transit Metropolis: A Global Inquiry* (Washington, DC: Island Press, 1998).

The size of a public transit system can be measured in terms of directional route-miles, number of stations, and number of riders. One mile of road served by a bus going in each direction equals two directional route-miles. In 2002, nationally, buses had the vast proportion of route miles, and, combined with heavy rail, accounted for the bulk of passengers. See Table 21.3 and Figure 21.1.

Table 21.3
Usage of public transit modes in 2002

Mode	Directional route-miles	Number of stations	Number of passenger trips (in millions)
Bus	158,247	N/A	5,800
Light rail	676	555	320
Heavy rail	1,527	987	2,632
Commuter rail	5,172	972	413

Source: U.S. Department of Transportation, *Transportation Indicators Report, July 2002*, 2002.

Non-motorized Transportation

Walking and bicycling are important transportation modes, but since they were discussed in Chapter 20, they are mentioned only briefly here for purposes of comparison. In 2001, there were approximately 97 million daily walk trips and 9.6 million bicycle trips in the United States. The breakdown of these trips can be seen in Table 21.4.

Table 21.4
Non-motorized transportation by trip purpose, % of total trips by purpose

	Trip purpose			
Mode	Family personal business	Social and recreational	Church and school	Work
Walk	42	34		
Bike	22	60	10	8

Source: Authors, 2013.

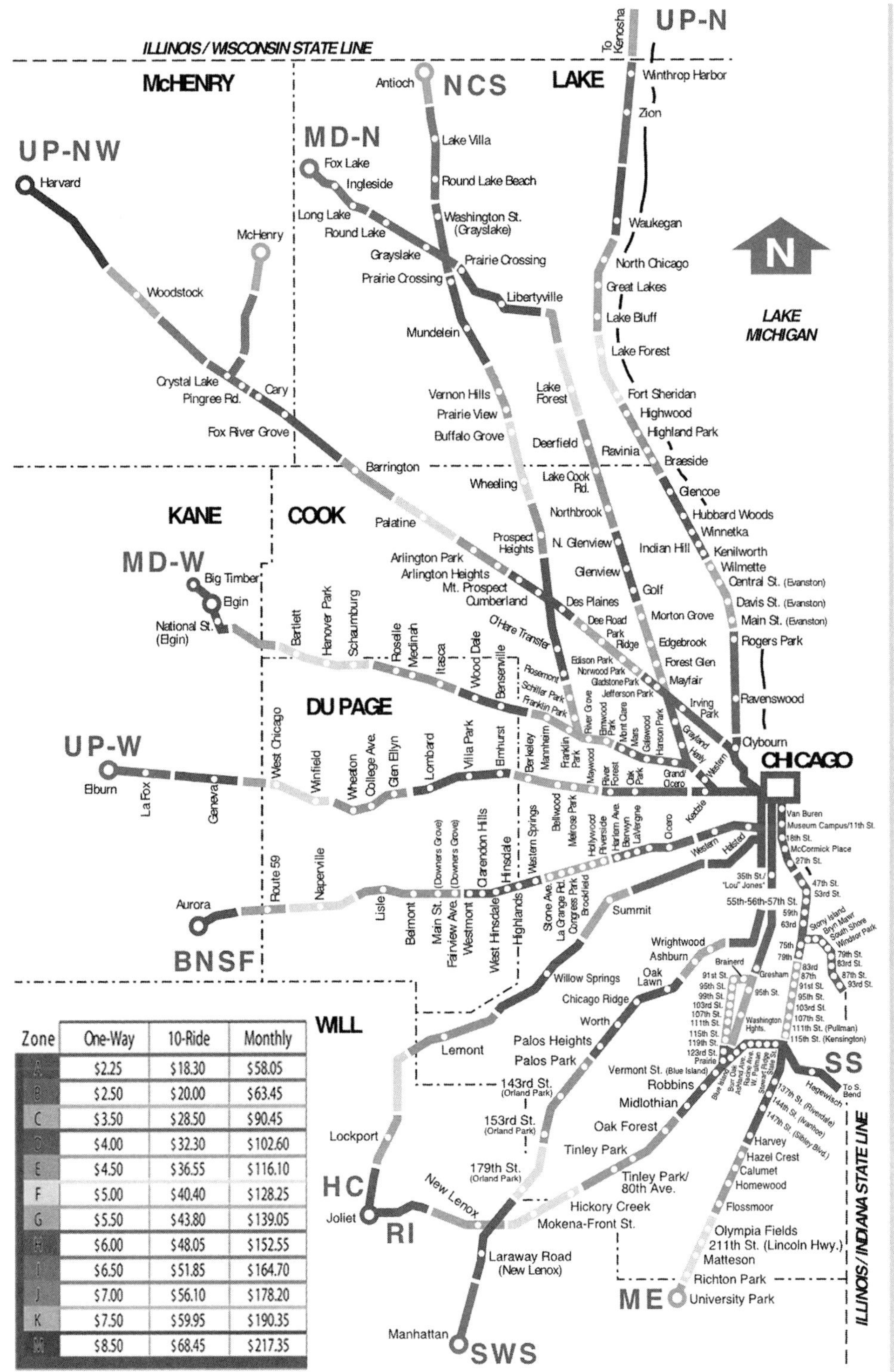

Zone	One-Way	10-Ride	Monthly
A	$2.25	$18.30	$58.05
B	$2.50	$20.00	$63.45
C	$3.50	$28.50	$90.45
D	$4.00	$32.30	$102.60
E	$4.50	$36.55	$116.10
F	$5.00	$40.40	$128.25
G	$5.50	$43.80	$139.05
H	$6.00	$48.05	$152.55
I	$6.50	$51.85	$164.70
J	$7.00	$56.10	$178.20
K	$7.50	$59.95	$190.35
M	$8.50	$68.45	$217.35

Figure 21.1
Chicago commuter rail map

Who Provides Transportation Services?

Transportation is handled in many ways at many different levels. Transportation facility ownership is primarily local. However, many other agencies are involved, sometimes through outright ownership but usually through funding, regulation, or technical consultation.

Federal Level

Federal Transit Administration. The Federal Transit Administration (FTA) administers grants for mass transportation systems for cities and communities nationwide. The FTA also provides technical assistance, information exchange, and leadership for metropolitan transportation agencies. FTA funds are especially important for transit agencies with large capital programs, as the majority of federal transit funding is now confined to capital projects. The New York Metropolitan Transportation Authority receives 28 percent of its capital budget from federal sources.

Federal Highway Administration. The Federal Highway Administration (FHWA) coordinates highway transportation programs in cooperation with the states. The major program is the Federal-Aid Highway Program, which provides federal financial assistance to state departments of transportation to construct and improve the National Highway System, urban and rural roads, and bridges. The FHWA also has a bicycling and walking component.

State Level

Department of Transportation. State Departments of Transportation (DOTs) are the real yeomen of urban transportation. In addition to setting state transportation policy, DOTs are responsible for maintaining all state-owned roads and highways. While state-owned facilities comprise only 13 percent of urban road mileage, they carry about 45 percent of all urban traffic. States also provide a share of transit funding, though it is usually smaller than either the local or federal share. Federal funds are for capital expenses for the most part, while the state provides more operating funds.

State-level transportation offices were first established in the 1930s to spend federal highway monies, but their mandate for the most part has expanded to include all transportation modes. In addition to allocating federal funds, State DOTs manage state-generated funds. Finally, DOTs provide specialized expertise and technical assistance. The DOTs of some states—such as Oregon, Florida, and Minnesota—have emerged as national leaders in various sub-fields of transportation.

Public Utilities Commission. In many states, transit agencies are regulated by the state Public Utilities Commission (PUC). The PUCs enforce safety standards and regulate some bus companies. Charter carriers, such as limousines and charter buses, are also regulated by the PUC.

Local level

Transit agencies. In 2001, there were 6,000 transit agencies in the U.S. Most transit agencies, sometimes called transit districts or transit authorities, are local and publicly owned, and are usually operated by an independent special district. Depending on the size of the service area, the district boundaries may or may not coincide with other political boundaries. Transit agencies receive funds from different sources. Some agencies have dedicated funding sources, such as a sales tax or an assessment on property. Other agencies are funded out of general funds of the cities or counties served. Transit agencies operate and maintain the transit vehicles, set fares and fare policies, and determine routes and schedules. Depending on the type of transit, the agency may also own and maintain its right-of-way.

Metropolitan planning organizations (MPO). At once the most and least important local entity is the Metropolitan Planning Organization (MPO). MPOs are regional planning bodies, created to develop comprehensive transportation blueprints at the local and regional level. MPOs have existed for some time, but the 1991 Intermodal Surface Transportation Efficiency Act gave them increased power in regional planning. MPOs are important because they control the purse strings. They often screen requests from local agencies for state and federal grants for transportation projects to determine their compatibility with the regional plan. On the other hand, MPOs are not very important, because, despite control of federal and state monies, they have very little power over local-level transportation and land use decisions. (Although, if the EPA begins to regulate carbon emissions as part of its conformity requirements for other transportation pollutants, this may change. See sidebar.)

Cities and counties. Local governments own, operate, and maintain local streets and roads. As noted in Chapter 20, city or county departments of public works are often responsible for street repaving, street lights, striping, and traffic signals. Transportation planning may be the responsibility of the planning department or public works. Some cities, recognizing the unique characteristics of transportation, have created separate transportation agencies. San Francisco's Department of Parking and Traffic, for example, handles transportation planning, traffic signal operation and maintenance, street signs, traffic engineering, and parking enforcement.

Transportation Planning

Transportation planning is performed at all levels of government, but since most trips are local and most transportation infrastructure is locally owned,

MPOs, and Carbon Emissions

A lawsuit in the late 2000's confirmed that the EPA had the authority to regulate GHG emissions under the U.S. Clean Air Act. "Conformity" is an existing procedure where the MPO's transportation Improvement Program of capital projects must demonstrate that it does not increase conventional air pollutants. If it does not, the MPO is not authorized to continue funding the project until the problem is fixed. Although there have been 63 U.S. areas that have had a conformity lapse, none have lost funding as a result. Several states, including California and Oregon, have begun to require regional land use and transportation plans that take into account carbon emission as a condition of receiving federal funds.[1]

Note

1. Reid Ewing, Keith Bartholomew, Steve Winkelman, Jerry Walters, and Don Chen, *Growing Cooler: The Evidence on Urban Development and Climate Change* (Washington, DC: The Urban Land Institute, 2008).

the bulk of transportation planning occurs at the local level. Transportation elements are usually required for all comprehensive land use plans since transportation is one of the major determinants of where growth will go.

Transportation planning is arguably the most mature area of infrastructure planning. Indeed, transportation planning is recognized as its own profession. The field has a rich literature, see the Additional Resources section at the end of this chapter for more on the subject.

The goal of transportation planning is to identify transportation needs, and to develop and evaluate ways to meet those needs, taking into consideration issues of cost, equity, environment including climate change issues, and land use. The following sections describe the major steps in putting together a plan for transportation infrastructure or a policy for local transportation. The first section begins with the famous four-step process, then moves on to look more broadly at how transportation infrastructure needs can be determined. A brief discussion of transportation infrastructure and land use follows before a section on the identification of alternatives is presented. The chapter then describes how to evaluate the transportation alternatives before concluding with a discussion of transportation infrastructure financing.

The Four-Step Process

In the 1950s, in two landmark transportation planning studies in Detroit and Chicago, a formalized tool for transportation planning was developed—the so-called four-step process. This process is based on the principles of the rational planning method, and, with certain refinements, is still very widely used today. Since this method is so widespread, and since it is based on many (potentially unquestioned) assumptions, we chose to explicitly present it, although it is not appropriate for all uses.

The basis of the four-step process is the *trip*. A trip is any journey, any movement from Place A to Place B. Every trip has an origin, a destination, a purpose, a route, and a mode. Each of these trip characteristics is important in determining the overall travel patterns, and therefore the transportation needs, of a region.

The four steps are described below; the method as presented is the method at its most basic. The four steps are:

1. Trip generation
2. Trip distribution
3. Mode choice
4. Route selection.

The results of the four-step process are not in and of themselves "planning" results, but are used to help planners identify needs. It must be stressed that the four-step process is a model, and should be used with all the usual cautions associated with using any model. The basis of each of the four steps is mathematical formulae. The process is iterative, and is meant to be run on a computer. Depending on the complexity of the region being studied, it is not unheard of for a four-step model run to take several days on a desktop computer. Before this method can be implemented, the geographic area under consideration must be broken into small, discrete units of analysis. These units, often known as transportation or traffic analysis zones (TAZ), are akin to census tracts and are normally about the same size. Every TAZ can serve as both origin and destination of a trip. Each TAZ is assigned certain parameters depending on the nature of the land use and the demographic factors within the TAZ. For instance, a TAZ that contains mostly low-density residential development will have different parameters than a TAZ that contains a shopping mall, an office park, or factories.

Trip generation. Step one is to determine the number of daily trips that take place in the region being studied. Trip generation estimates the number of trip ends that are produced in or attracted to each TAZ. A land use that creates trips is known as a trip producer. A land use that attracts trips is known as a trip attractor. In general, it is assumed that trips are produced by households, and that trips are attracted to uses like shopping, work, or personal business.

The number of trip ends produced or attracted is estimated using certain assumptions about the number of trips typically made by different types of households (for trip production), or the number of trips typically made to different types of destinations (for trip attraction). Household factors include income, number and age of members, and availability of an automobile. Table 21.5 illustrates an example of household trip production rates according to household income.[9]

Destination factors include type of land use (shopping mall, factory, office park) and the size of the land use. Table 21.6 above illustrates some trip attraction rates by various land use characteristics in one region.

Table 21.5
Household trip production

Household income	No. of trips	% transit
Less than $25,000	5.5	12.5
$25,000 to $44,999	7.5	5.8
$45,000 to $75,000	9.4	4.6
Greater than $75,000	10.5	3.7
All households	7.6	6.6

Source: Metropolitan Transportation Authority, San Francisco Bay Area, 1990.

Table 21.6
Trip attraction rates for various land uses

Land use	Number of trips attracted per:			
	Employee	1000 sq ft of building area	Acre of development	Other
Office	1.5	4.3	15.6	Parking space: 1.4
Restaurant	4.9	14.0	24.5	Seating: 0.4
Hotel	NA	1.0	7.7	Room: 0.5
Bank	17.5	45.4	106.6	Parking space: 8.4
School	9.6	NA	19.2	Student: 1.1
Gas station	110.5	368.3	176.8	NA
Medical clinic	4.3	11.3	13.6	Physician: 17.0; patient: 2.10

Source: Arkoma Regional Planning Commission, 2001.

Table 21.7
Purpose of trips

Trip purpose	% of trips	% of miles
Family/personal business	46	35
Social, recreation	25	31
Work	20	28

Source: U.S. Department of Transportation, *Transportation Statistics Annual Review*, 2000.

Trip distribution. In step two, the production trip ends and the attraction trip ends are linked geographically into complete trips. The assumption behind trip distribution is that time spent traveling is perceived negatively—the more distant the destination, the more burdensome the trip. Therefore, a trip origin is always linked to the closest appropriate trip destination.

Trips have different purposes, and trips of different purposes have different characteristics. For instance, Table 21.7 shows that work trips tend to be longer than trips for personal business. (The average person–trip length, encompassing all trip purposes, is 9.1 miles, and the average commute to work is 11.6 miles.) These factors are considered in the trip distribution step.[10]

Mode choice. If we were to graphically display the information found in the first two steps, we would have a good idea of the community's travel patterns. We would know where people are going, and where they started. But we wouldn't know how they are traveling. Are they driving, walking, taking the bus? The next step is to determine which mode people are using to make their trips.

Mode choice is calculated based on the relative desirability of the different modes for a given trip, where desirability includes issues of cost, passenger comfort and convenience, and travel speed. Table 21.8 illustrates mode choices at the national level for all trips including family and personal business, social and recreational, and the journey to work. Table 21.9 shows mode choices for the journey to work. These two tables show, for instance, that work travelers are more likely to use transit and walk than most travelers. Therefore, when calculating mode choice, the trip purpose must be considered.

Route selection or trip assignment. The final step is to determine the routes travelers choose to reach their destination. Routes are selected based on both travel time and distance. Route selection looks at the average speeds, capacities, and proximity to destinations of all the road segments in the area being studied. As in trip distribution, the assumption is that time spent traveling is perceived negatively, and that a higher-speed, higher-capacity road such as a highway will more likely be chosen by a traveler than a residential street. (Table 21.10, which shows road usage by road type, suggests that this is a valid assumption.)

Performed thoughtfully, the four-step process reveals (as accurately as a model can) the travel patterns of a community and then overlays those patterns onto the existing transportation system. Thus, the four-step process can show where service standards (discussed below) are not met. It can show where or when congestion occurs. It can show where possibilities for transit expansion exist. In sum, the information generated by the four-step process is used to identify needs.

Table 21.8
Trips by mode for all purposes in the United States

Mode	% of trips
Private automobile	88.3
Public transit	2.2
Bicycling	0.1
Walking	0.3
Other (motorcycle, school bus, etc.)	8.1

Source: American Society of Civil Engineers, 2009.

Table 21.9
Journey to work trips by mode in the United States

Mode	% of trips
Drives self	78.2
Carpool	9.4
Public transit	4.9
Walking	3.1
Bicycle or motorcycle	0.6
Taxi	0.1
Other	0.8
Works at home	2.8

Source: U.S. Department of Transportation, *Transportation Statistics Annual Review,* 2000.

Table 21.10
Road usage by classification

Road classification	% of total US road mileage	% of rural/urban travel (VMT*)	% of total U.S. travel (VMT*)
Rural			
Highway	3	46.8	18.3
Arterial	4	16.5	6.4
Major collector	11	20	8
Minor collector	7	5	2
Local	54	12	5
Urban			
Highway	1.9	58	36
Arterial	2.3	20	12
Collector	2.2	8	5
Local	14.9	14	9

Note: *Vehicle miles traveled.
Source: North Carolina Department of Transportation, 2003.

Evaluating Road Capacity Is Complicated and Should Reflect Local Values

Many factors affect the capacity of a roadway: lane and shoulder widths; the location and number of on- and off-ramps; the presence of trucks and other large vehicles; and hills or curves. Ironically, the more congested a freeway is, the higher capacity it has. In free-flow conditions (LOS A), one lane of an urban freeway can carry about 700 passenger cars per hour (pcph) in each direction. At rush hour volumes (LOS E or F), where speeds are low and tempers are high, the same lane can carry up to 2,000 pcph. However, at higher traffic volumes, traffic collisions are more likely, and traffic flow is much more susceptible to interruptions caused by collisions or other incidents.[1]

There are two things to learn from this. First, roadway capacity is not a cut and dried number, but a highly variable one. Second, and perhaps more importantly, congestion is not necessarily a bad thing. In one sense, the roadway is at its most efficient during congestion, since it is moving the highest amount of cars on the same amount of roadway. Urban congestion indices like those developed by the Texas Transportation Institute compare travel times under free flow and congestion conditions. Clearly, individual cars go slower in congestion. The slower travel times are compared to travel times achieved in free flow conditions, and the difference is called delay due to congestion. But in LOS E, more cars are moved in total, so one could say that social benefit is maximized at that point. The point is that local decision-makers must not be slaves to LOS A.

Note

1. Theodore Petritsch, Bruce W. Landis, et al., Florida Department of Transportation, (2006), www.dot.state.fl.us.

Identifying Needs

The four-step process outlined above is usually the first step in identifying need. Another step is to establish the level of service standards the community wishes to provide. An inventory of the transportation facilities can be compiled at any time, and when standards are in place and an inventory exists, the physical condition of transportation facilities can be evaluated to see where there are gaps in service, or where the facility needs replacement. These comprise the needs. The following goes into greater detail about level of service standards, and then briefly describes a condition assessment and facilities inventory.

Level of Service Standards

The concept of level of service, or LOS, is the most common yardstick by which transportation service is measured. LOS is an attempt to quantify the transportation facility user's perceptions of the quality of service provided by the facility. LOS is usually given in terms of a letter, A being the best and F being the worst (see Table 21.11). The determination of the LOS of every travel mode is different (Figure 21.2), and it also depends on the values of the community. LOS criteria are often set forth in a city's general plan.

There have been some critiques of traditional LOS. One reason that automobile-oriented transportation solutions always seem to win out is that the very performance measures we use to evaluate transportation alternatives are skewed in favor of the automobile in the first place. Automobile-based LOS criteria have emerged as the prime criteria for evaluating streets, but unfortunately, levels of service are really proxies for speed (and vehicle, not person, speed at that). This means the transportation systems that focus on speed and efficiency, or *mobility*, will always trump other systems. Furthermore, this attitude has become institutionalized. Indeed, for example, the published mission of Caltrans, the California Department of Transportation, is "Caltrans Improves Mobility Across California."

One noted analyst argues that there should be less emphasis on how fast vehicles move, and more emphasis on how well people's travel needs are met. He proposed that in addition to using *mobility* as the standard, *accessibility*,

Table 21.11
Florida State Department of Transportation level of service definitions

Facility	Level of service					
	A	B	C	D	E	F
4-lane freeway	23,800	39,600	55,200	67,100	74,760	>74,760
2-lane highway	2,000	7,000	13,800	19,600	27,000	>27,000
Suburban arterial	—	4,200	13,800	16,400	16,900	>16,900
Urban arterial	—	—	5,300	12,600	15,500	>15,500
Collector	—	—	4,800	10,000	12,600	>12,600

Source: Florida Department of Transportation, Systems Planning Office, 2002.

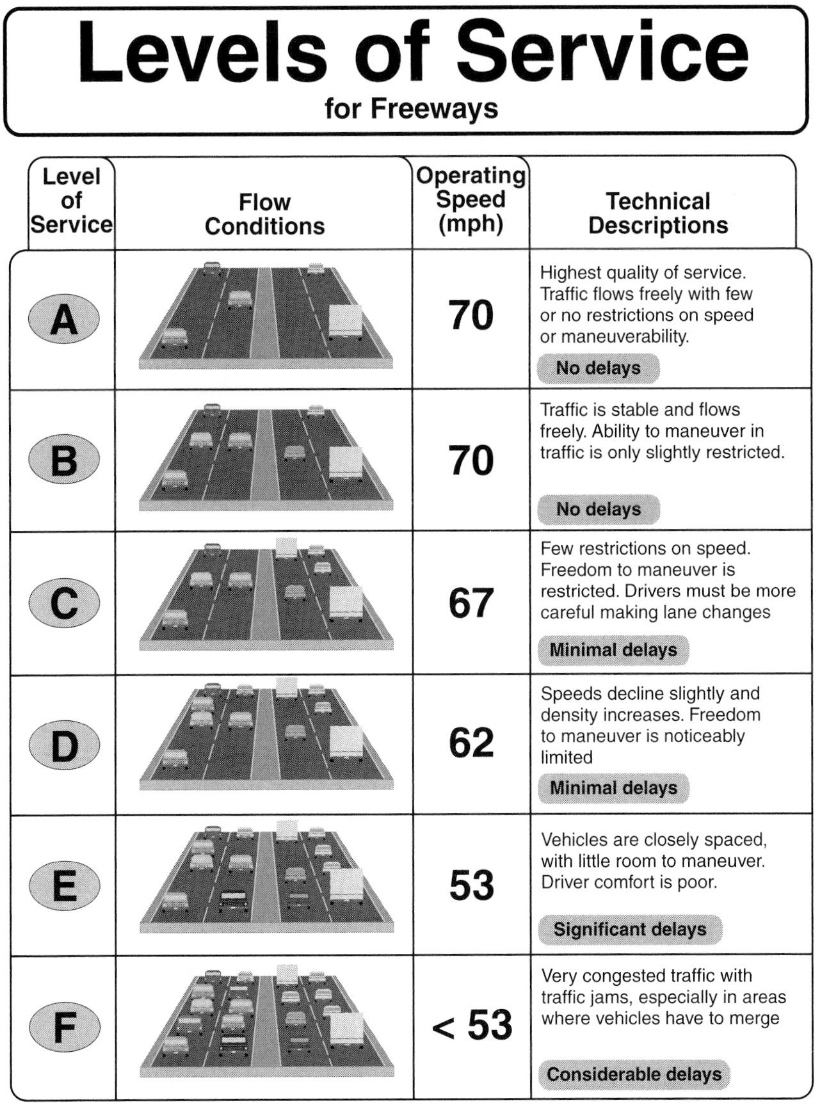

Levels of Service
for Freeways

Level of Service	Flow Conditions	Operating Speed (mph)	Technical Descriptions
A		70	Highest quality of service. Traffic flows freely with few or no restrictions on speed or maneuverability. **No delays**
B		70	Traffic is stable and flows freely. Ability to maneuver in traffic is only slightly restricted. **No delays**
C		67	Few restrictions on speed. Freedom to maneuver is restricted. Drivers must be more careful making lane changes **Minimal delays**
D		62	Speeds decline slightly and density increases. Freedom to maneuver is noticeably limited **Minimal delays**
E		53	Vehicles are closely spaced, with little room to maneuver. Driver comfort is poor. **Significant delays**
F		< 53	Very congested traffic with traffic jams, especially in areas where vehicles have to merge **Considerable delays**

Figure 21.2
Levels of service for automobiles on freeways

livability, and *sustainability* should also be included in the LOS for a jurisdiction. (See Sidebar).

Condition Assessment

The condition of a transportation facility is a function of both its physical condition and the level of service it is providing. The physical condition of a roadway includes the state of the pavement as well as the state of striping, signs, and traffic signals. Roadway pavement has measures like roughness or number of potholes. When one examines transit, one must consider the condition of the buses or vehicles as well as that of the tracks, stations, switching gear, and other operating equipment. Level of service can be measured using the service standards mentioned above.

Facility Inventory

The facility inventory should include a listing of the facility, who owns and/or operates it, and its current usage and current capacity. The inventory should also include current levels of service and current condition (obtained from the condition assessment). The results of the four-step model run are a good place to start for information on current levels of service. Environmental or traffic impact reports are also good sources for particular road segments or intersections. Facility inventories are normally mandated by state and federal funding sources.

Needs and Wants

It is worth noting that of all the infrastructure systems, it is perhaps most difficult to differentiate "needs" from "wants" in metropolitan transportation. Transportation is the business of moving goods and people, and yet decision-makers often equate "transportation service" with a particular transportation mode. Instead of looking at how best to move people from Point A to Point B, a transportation debate may center on how best to accommodate a particular *way* of moving people between the points, a light rail system, for instance. Similarly, decision-makers often overlook prospects of changing travel behavior. Further, there are many instances of valid transportation need that a four-step model can obscure. The needs of the elderly or the disabled will not necessarily show up

Alternate Level of Service Criteria for Streets

- *Mobility*. The ease with which people can move about. Time and cost of travel are low, and many travel options exist. Includes speed but also includes auto ownership, parking availability, transit route density, and sidewalk connectivity.
- *Accessibility*. The ease with which desired activities can be reached from any location. Accessibility is a function of both land use and the transportation systems.
- *Livability*. Autos are dangerous and polluting and consume many raw materials. Therefore a livable transportation system recognizes this and puts the auto in its rightful place, which is to say one among many travel options. This means promoting transit and nonmotorized travel.
- *Sustainability*. A sustainable system meets the needs of the present without compromising the ability of future generations to meet their own needs. This includes issues of air and water quality, consumption of resources, use of land, and carbon emissions.

Adapted from Reid Ewing, "Measuring Transportation Performance," *Transportation Quarterly* 49, no. 1 (1995).

in a travel model, nor will the needs of the transit-dependent. (These are discussed further below.)

Other Capacity Issues

Some shortfalls in capacity are clear-cut. Obviously, if a greenfield site is being developed, streets and roads will need to be developed from the ground up. The shortfall in a built-up area is not so easy to calculate because of the phenomena of *latent demand* and *induced demand.* In addition, *reducing capacity* may actually reduce demand in some instances.

Latent demand. Many planners are familiar with the phenomenon of brand new roadways, built to ease congestion, filling up with traffic almost overnight, and congestion appearing once again. Indeed, it's a truism among transportation professionals that "you can't build your way out of congestion." When new roadway capacity is built, motorists might shift their routes or their time of travel to take advantage of the new congestion-free road. These factors all conspire to cause the new roadway to grow congested. These factors are called "latent demand," demand for roadway space that was simply waiting for an outlet.

Induced demand. There is an even more insidious (and controversial) factor involved called "induced demand." People might make trips that they would not otherwise have, simply because a new roadway makes the trips easier. Transportation professionals have long argued over whether additional transportation capacity actually decreases traffic congestion or whether new roads generate new traffic by the very fact of their construction. A 1995 study showed that in metropolitan regions, a 1.0 percent increase in lane miles induces a 0.9 percent increase in vehicle-miles traveled in 5 years. When a new highway is built, businesses, stores, and other trip attractors move to new locations to take advantage of the easier travel the new highway offers. This in turn induces more people to use the highway than would have otherwise. The result is "[w]ith so much induced traffic, adding road capacity does little to reduce congestion." A study released in 2000 by the Transportation Research Board had similar results. This report, which looked at data for 70 urbanized areas from 1982 to 1996, concluded that increased capacity increases traffic congestion, on average by about 20–30 percent, and in the case of Louisville, Kentucky, by 77 percent. The authors of this report indicated that "induced travel effects strongly imply that pursuit of congestion reduction by building more capacity will have short-lived benefits." Therefore, decision-makers must carefully calculate (or re-calculate as the case may be) the costs and benefits of highway expansion when developing transportation solutions for their communities.

Reducing capacity. What about the opposite effect? Does reducing capacity somehow evaporate trips? Intuition might say no, and that reducing capacity is a sure recipe for pandemonium; if road capacity is suddenly, inexplicably (to the motorist) taken away, without alternatives in place, then traffic chaos might result. However, observers have found that reduced capacity does not create traffic chaos. A reduction in capacity coupled with a well-designed alternative can reduce traffic and can increase social benefit. The key is that transportation policy in such a case be integrated and balanced.

Identify Solutions

After the needs have been identified, alternatives are developed to meet those needs. The most common way to meet transportation service shortfalls is by building additional capacity. This normally means either building new roads or widening existing ones. Increasing transit capacity can mean increasing the frequency of vehicle arrivals, lengthening the schedule, adding cars (in the case of a train), or introducing new routes.

Evaluate Alternatives

After alternative solutions have been identified, they must be evaluated and chosen. There are numerous ways to evaluate transportation projects, and numerous criteria to use. The most obvious criterion is cost. But cost should not be the only or even necessarily the primary one. The Victoria Transport Policy Institute has identified several other criteria:

- *Solving the problem*—does the solution actually address the identified need? This may sound illogical, but it is surprising how often this criterion is ignored or overlooked.
- *Consumer benefit*—will the solution allow system users, i.e. consumers, to save money, avoid stress, and reduce their need to chauffeur non-drivers?
- *Efficiency*—is the solution economically efficient?
- *Livability*—does the solution make a community more "livable," more pleasant, healthy, and economically viable?
- *Security and reliance*—does the option result in a more diverse and flexible transportation system that can accommodate variable and unpredictable conditions?

Alternative Transportation Approaches

Land Use Policies

Many books and articles have been written about the relationship of transportation and land use. This section will highlight only a few of the issues.

Transportation infrastructure affects land use in many ways. First, transportation facilities take up space. Freeways, for example, are voracious consumers of land. Current design principles mean that one high-speed cloverleaf interchange can take up almost 640 acres. A four-lane highway requires 70 feet of right-of-way at an absolute minimum. Rail systems require dedicated right-of-way as well, though not as much. The point is that care must be taken when planning transportation facilities. Seventy feet of concrete might not sound like a lot, but many a neighborhood has been split when an urban freeway cut through its middle. Even widening an arterial may require 10 to 40 feet of new right-of-way, which will most likely come from adjacent private landowners.

Second, transportation technologies influence urban structure. Third, as noted earlier, transportation infrastructure investments are often promoted as keys to economic development in central cities. (Some researchers claim that the connection is weak, inconsistent, and unpredictable. Others suggest that transportation facilities are strong growth shapers, but not necessarily growth generators.)

Finally, land use and transportation infrastructure decisions have a powerful effect on carbon emissions. Smart growth (also called compact development, transit-oriented development, complete streets, and the New Urbanism) has the potential to dramatically reduce vehicle miles traveled. The Center for Neighborhood Technology mapped carbon emissions per acre and per household for major U.S. metropolitan areas and found that although densely populated areas produce more carbon emissions per acre, per capita emissions are far less than in the suburban areas.

Jobs–housing balance. Attempts to square the location of jobs with that of housing within a region, known as achieving a jobs–housing balance, have been suggested as a strategy to reduce traffic congestion, carbon emissions, and air pollution in metropolitan regions. In other words, when the places that people live are near the places that people work, they will make shorter trips and the overall cost of the trip to work will be lower. Indeed, Robert Cervero found that in cities that were "job-surplus"—they had more employment than housing—commuters did indeed travel more, longer, and more often alone.[11] But he also concluded that failures to achieve a jobs–housing balance were attributable to failures in planning and not to the market. In particular, restrictive zoning and NIMBYism were to blame. Of course, it is impossible to control the home and job location of every employed person, but it is not unreasonable to expect a city or region to have an equal amount of housing and employment.

Transit-oriented development. Concentrating development around transit nodes, a concept known as transit-oriented development or TOD, is often suggested as another method of reducing car trips. The theory is that if high-quality public transportation is close to where people live, and that if other trip destinations like shops, day-care centers, and entertainment complexes are near enough to walk to, then automobile use will go down. The idea is not new; in many ways it hearkens back to the streetcar suburbs of the early 1900s. Many cities such as San Diego and Oakland, California, and Portland, Oregon have undertaken real efforts to promote TOD. Barriers to TOD projects include restrictive zoning and building codes, conservative lending practices, and NIMBY resistance to high density developments. These are issues local practitioners can tackle.

Demand Management

The mantra in business during times of economic doldrums is "to do more with less." As local transportation budgets are stretched further and as community opposition to proposed transportation capacity expansion project grows, the same idea can apply to current transportation facilities. In other words, various tools can be used that squeeze more room out of existing transportation facilities without actually building more highways or adding more cars to the trains. This is known as transportation demand management, or TDM. Brief introductions to three current "demand management strategies" follow; for a detailed list of others, see the Victoria Transport Policy Institute's *Online TDM Encyclopedia* (listed in the "Additional Resources" section at the end of the chapter).

Congestion pricing. Congestion continued to worsen in American cities up through 2005. Analysts estimate traffic congestion causes the average traveler in peak periods to spend an extra 38 hours of travel time and 26 gallons of fuel, with a cost to the individual of $710 per year. Together, this is a $78 billion drain on the U.S. economy, with 4.2 billion lost hours and almost 3 billion gallons of wasted fuel.[12] One solution to this problem is congestion pricing. Congestion pricing in transportation is similar to paying more for a long-distance telephone call during business hours—it is based on the assumption that a service should cost more during periods of high demand but limited supply. The idea has been trumpeted by transportation economists for years, but has never really been politically viable for surface transportation (airlines already use a form of congestion pricing).

The idea has two parts. The first one is that transportation facilities or transportation access are priced (Figure 21.3). This could be a toll road, a toll bridge, or a tolled area like a downtown. This is often politically difficult, since motorists are unwilling to pay for something that they currently receive for free, access to a highway or geographic area. The second part is similar to the way airlines price their tickets: during periods of higher demand, prices go up. Everyone knows that airline tickets are more expensive during the holidays. This is because demand for air travel is especially high, and high prices allow the airlines to meter the demand. Similarly, during rush hours, demand for

Congestion and GHG Emissions

Reducing congestion can lower GHG emissions by decreasing travel time spent idling, and thereby lowering fuel consumption. Shifting to other modes of transportation and using electronic signaling can reduce congestion.[1]

Note

1. Pew Center on Global Climate Change 2009. *Climate Tech Book.* Available at: www.pewclimate.org.

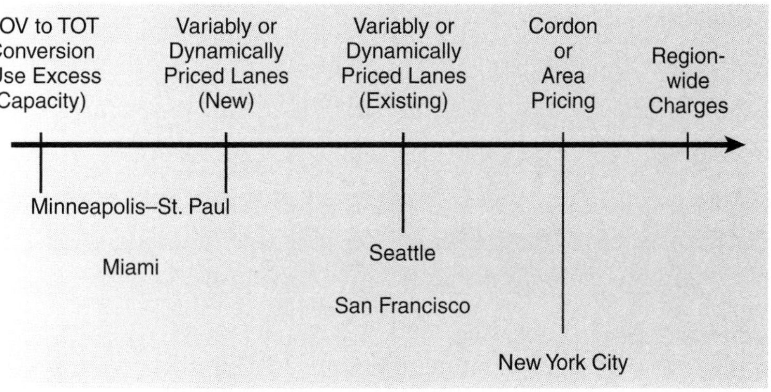

HOV to TOT Conversion (Use Excess Capacity)
Variably or Dynamically Priced Lanes (New)
Variably or Dynamically Priced Lanes (Existing)
Cordon or Area Pricing
Region-wide Charges
Minneapolis–St. Paul
Miami
Seattle
San Francisco
New York City

Figure 21.3
Range of congestion pricing schemes

highway space is high, and a way to meter this demand is to charge for this highway space. If the highway is more crowded, it costs more to use.

Two kinds of pricing strategies are linked to the amount of congestion on highways: variable tolls on single lanes (HOT—High Occupancy Toll lanes) or for entire roadways (Figure 21.4). The second two kinds are the ones that have received more publicity. These can be "cordon" charges for drivers into a designated city area or "area-wide" charges which would be per-mile charges on all roads in an area. Generally, congestion pricing relies on technology such as cameras, software, and algorithms that can read auto license plates and charge the driver or check whether he or she has paid. The cities of London and Stockholm have used congestion pricing (Stockholm estimates that it has reduced congestion by 20 percent), while the cities of San Francisco and New York periodically consider such measures.

Rising gas prices can also contribute to reduced automobile traffic and increased transit use. Gasoline prices in 2007 and 2008, adjusted for inflation, rose above the previous all-time highs from 1981. Consumers adjusted their

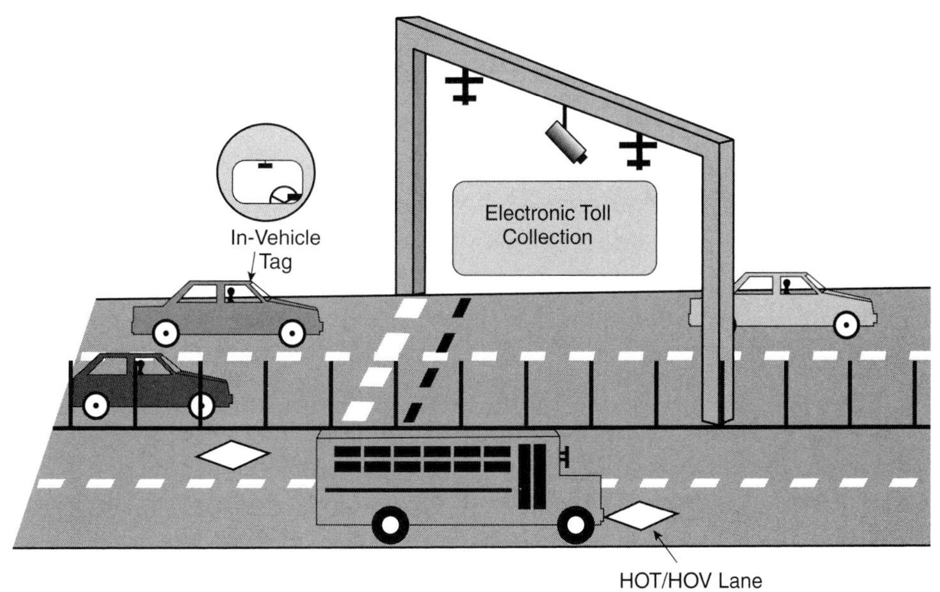

Figure 21.4
Electronic toll collection

behavior and drove less. Nationally, during the period from March 2007 to March 2008, travel on urban arterials (measured in VMT) decreased almost 4 percent nationally, with the North Central and Southern Gulf states the most impacted.[13] Transit usage is also up. For example, during the period from April 2007 to April 2008, high gas prices and increased tolls decreased auto traffic into New York City by 4.2 percent and increased transit ridership by 6.5 percent.[14] One 2008 report estimated that if gasoline prices rose to $7.00 per gallon, VMT would drop by 15 percent and there would be 10 million fewer vehicles on the road.[15]

Parking control. When people drive, they need a place to store their car at their destination. If parking is free and readily available, people might be more willing to drive a car for the trip. In this country, parking is free for over 99 percent of all automobile trips. As one author stated "[f]ree gasoline for employees who drive to work would seem like a reckless offer, yet employer-paid parking is a much stronger incentive to drive to work alone." On the other hand, if parking is difficult or expensive, people might be inclined to look at other ways of making the trip, such as taking the bus. Parking control measures, which include either implementing a pricing structure or just limiting the amount of parking, can be used to either eliminate low-value trips, or to encourage the use of alternate modes of transportation.

As a nation, we pay more for storing our cars the average 23 hours a day they sit immobile than we do for gasoline, but those costs are very rarely directly paid by the individuals actually incurring them. Free parking isn't really free according to a recent publication. Instead the average parking space costs more than the average car. Although the developer pays for it originally, the costs are bundled into the price of housing and goods purchased at the shopping mall. The total subsidy for parking is about the size of the Medicare or national defense budgets.

Intelligent transportation systems. The term "intelligent transportation systems" (ITS) refers to a broad range of various information and electronic technologies that are used to increase the capacity, efficiency, or convenience of transportation systems. Applications of ITS include "smart" traffic signals that anticipate traffic and change green times accordingly, sensors in the roadway that detect ice and automatically notify snowplows, and global positioning systems that notify waiting bus patrons when the next bus will arrive, among many others. The actual list of ITS applications in operation across the country is quite long and diverse.

ITS holds great promise to improve the efficiency of our transportation systems and to stretch transportation dollars by increasing the capacity of the existing system, thereby reducing carbon emissions. It can also be used to implement congestion pricing schemes. Unfortunately, numerous institutional barriers stand in the way of seamless deployment. Since ITS applications are primarily information technologies, they don't fit in the normal framework of road and transit agencies. Therefore, ITS applications often lack dedicated funding sources. Furthermore, agencies often lack the staff to plan, design, or maintain ITS applications. See the Additional Resources at the end of this chapter for more resources on intelligent transportation systems.

Environmental and Social Impacts

Transportation policies have far-reaching environmental and social impacts, as shown in Figure 21.5. Three other factors to consider in evaluating alternatives are transportation energy use and its effect on greenhouse gas emissions, its relationship with air pollution, and social equity.

Energy Greenhouse Gas Emissions

The transportation sector accounts for 28 percent of total U.S. energy consumption, 67 percent of petroleum production consumption, and a third of U.S. greenhouse gas emissions (GHG). Transportation is one of the largest reasons for U.S. dependence on foreign oil, and energy use in this sector was increasing by about 2.1 percent per year prior to 2008, when the rate of increase was impacted by rising oil prices and the weak economy. Within the transportation sector, passenger cars (called light duty vehicles) account for 61 percent of GHG emissions (Figure 21.6). Policy-makers at the state and local levels have been concerned about the energy–GHG connection for passenger cars for several decades, and although support at the federal level has

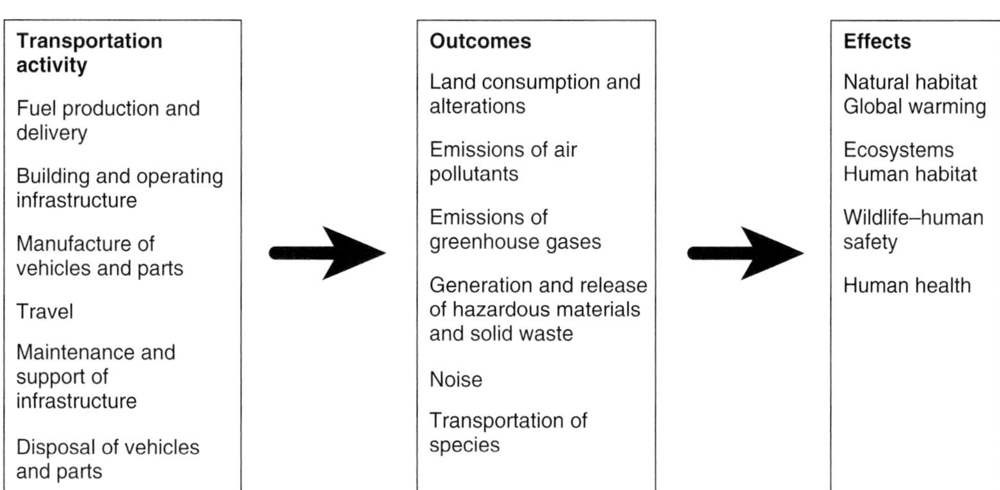

Figure 21.5
General transportation environmental impacts

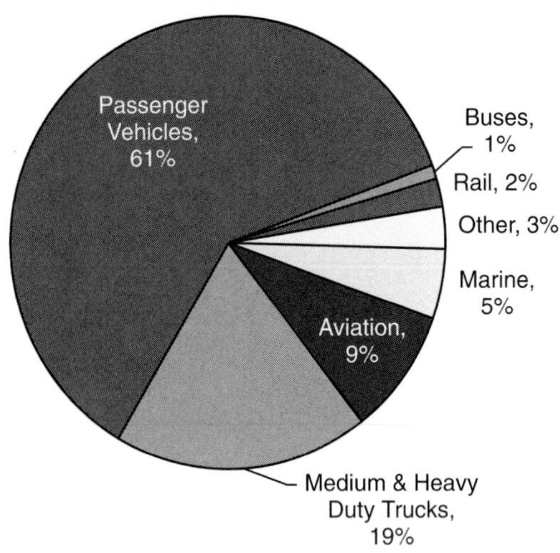

Figure 21.6
Transportation energy use by mode in 2006

Three-Legged Transportation Stool and CO_2 Emissions

There are three major strategies for reducing CO_2 emissions from transportation. The first is to improve *vehicles* so that they use less fuel. The second is to address the carbon content and footprint of the *fuel*, while the third is to reduce VMT through compact growth. The first two strategies are most appropriate at the state and federal levels, but the third is the province of the local practitioner.[1]

Note

1. Reid Ewing, Keith Bartholomew, Steve Winkelman, Jerry Walters, and Don Chen, *Growing Cooler: The Evidence on Urban Development and Climate Change* (Washington, DC: The Urban Land Institute, 2008).

been intermittent, five strategies have emerged to reduce petroleum use and GHG by automobiles to the 80 percent-by-2050 target which is widely believed to be required for climate stabilization, including: vehicle efficiency improvement; alternate fuels; reduction in VMT; system efficiency; and demand management.

Vehicle efficiency improvements. Increases in fuel economy and improvements to car air conditioning systems can reduce emissions. From 1974 to 2004, the efficiency of a typical car in the U.S. doubled—going from 14 to 29 miles per gallon (mpg). These gains leveled off in the 1990s, and in 2000 the average fuel efficiency for cars was 28 mpg, not including trucks, vans, and sport utility vehicles—their fuel efficiency remained at 18 mpg. Efforts at the federal level to mandate increased efficiency have been slow. Although the U.S. passed fuel economy standards for the industry in 1975, known as CAFE (corporate average fuel economy), efforts to raise the bar from 27.5 mpg to 40 mpg in the 1990s failed. States, however, stepped into the breach. In 2004, California required vehicle-makers to reduce GHG emissions (including carbon dioxide, methane, nitrous oxide, and hydrocarbons) by 30 percent by 2016 and to achieve a fuel economy of 35.5 mpg for new light vehicles. Since then, 16 other states representing over 30 percent of GHG emissions, 40 percent of car petroleum use, and over 50 percent of light vehicle sales followed suit. Under the Clean Air Act, California required a waiver from the EPA to implement its own standards. In 2009, the EPA issued the waiver, thus paving the way for these states to move forward. In 2009, the federal government also pledged to set national standards for 2016. In Europe, average fuel efficiency was 44 mpg with future efficiencies likely.

Alternate and low-carbon-footprint fuels. Thirty-one states, representing 72 percent of the GHG emissions and 68 percent of motor gas use, have mandates and incentives for blending bio-fuels into petroleum-based fuels. Minnesota requires 20 percent ethanol by 2013, and Hawaii requires 20 percent renewable by 2020. In 2006, California adopted a "low carbon fuel standard" that was scheduled to take effect in 2010. This calls for the carbon fuel content of on-road vehicle fuels to be reduced by 10 percent by 2020. Regulations to implement these requirements were issued in

2009. One of the controversial aspects of the 10 percent carbon footprint reduction is that the calculations include the entire life cycle of the fuel, from extraction (and cultivation for biofuels) to combustion. The indirect effect of replacing cropland used for energy will also be included. The California standard also permits plug-in hybrids to be counted as well as both types of ethanol.[16]

Reduction in vehicle miles traveled. If all U.S. states adopted the California efficiency and fuel standards, emissions reductions would be about half of what is needed to reach the 1990 baseline, but still considerably less than what is needed to meet the 80-percent-by-2050 target, the goal that scientists indicate is necessary to avoid catastrophe. To meet the latter, an additional 4 percent reduction in VMT is required. Land use planning is therefore a critical tool in reducing GHG emissions and gasoline use. Reduced distances from work, increased residential densities, and the use of transit and other alternate modes of transportation can be affected through local land use policies. As the households-per-acre increase, the annual vehicle miles drop dramatically, with a corresponding decrease in fuel consumption (Table 21.12).

System efficiency and demand management tools. Congestion pricing was discussed above. Consumer choices for slower, lighter vehicles will also lower GHG emissions. The capital costs of these tools are considerably less than the fuel efficiency and alternate fuel strategies. Electronic signaling can also level out traffic flows, which reduces fuel consumption and thereby carbon emissions as well.

Transportation and Air Pollution

Transportation policies also have air-quality impacts aside from carbon emissions. Many of these are not subject to the terms of the Kyoto Protocol, but are still of concern for public health. (Their effect on global warming has not yet been definitively quantified.) "On-road mobile sources," the EPA title for air pollution caused by transportation, make up one of the largest sources of

Table 21.12
Effect of housing density on gasoline consumption

Households per residential acre	Annual VMT per capita	Average annual fuel consumption per capita (gallons)
117.0	2,670	133
32.0	5,090	254
14.0	6,944	347
6.8	7,566	378
3.8	10,216	511

Source: California Energy Resources Conservation and Development Commission, 1993.

Table 21.13
Transportation air pollutants

Air pollutant	% responsible
Volatile organic compounds (VOCs)	29
Oxides of nitrogen (NOX)	34
Carbon monoxide (CO)	51
Particulate matter (PM10)	10

Source: U.S. Department of Transportation. Federal Highway Administration, 2002.

air pollution. Four types of pollutants are of concern. *Volatile organic compounds* (VOCs) are precursors to ground-level ozone, which is an eye, nose, and throat irritant. *Oxides of nitrogen* cause smog. *Carbon monoxide* is a poison that interferes with the body's ability to absorb oxygen and accounts for over half of the pollutants of concern. *Particulate matter* lodges in the lungs (Table 21.13).

Cars contribute to air pollution in four ways: (1) the cold start; (2) normal driving or running emissions; (3) the hot soak; and (4) diurnal emissions:

- *Cold start*. Much if not most of carbon monoxide and VOC emissions of a typical driving cycle occur in the first minute or two of engine operation, when the engine and the catalytic converter are still cold. This is known as the "cold start." In fact, according to the EPA, for a 7.5 mile "model" trip, starting the car cold generates about 16 percent more NOX, 40 percent more CO, and 70 percent more VOC than starting the car when it is warm.
- *Running emissions*. During normal engine operation, combustion byproducts are emitted—the so-called running emissions. In modern cars, this cause of air pollution is largely under control.
- *Hot soak*. After the car has stopped, temperatures under the hood increase since no air is being forced into the engine compartment. This causes the various fluids in the engine, such as gasoline, oil, and coolants, to evaporate. This is called the "hot soak." The hot soak period usually lasts for about an hour after the car has stopped.
- *Diurnal emissions*. This same evaporative phenomenon happens on a slower scale just as a car sits idle. These are known as the diurnal emissions. Fully 50 percent of VOC emissions from a car may occur during these idle periods.

It is important that planners know about these different components of air pollution. Since most air pollution occurs just in starting the car, transportation planning measures that shorten an auto trip but do not actually eliminate it are not improving air quality that much. For example, many people hail the air quality improvements of rapid transit systems such as the San Francisco Bay Area Rapid Transit (BART) system lines that reach far into the suburbs and carry people to employment centers. The reasoning is that people are taking the train to work instead of driving, so there are less auto emissions. Unfortunately, the suburban termini of these systems serve large commuter catchment areas that lie even farther in the suburbs. Commuters drive from home to the station. Thus, though the auto trip is shorter, it is not eliminated, and so there are still two cold starts and two hot soaks for each trip.

Social Equity

Questions of equity must always be at the forefront when one considers any planning issue, and transportation planning is no exception. In general, the issue of equity in transportation involves basic mobility. There are certain segments of the population that, due to our nation's transportation choices, are often denied full participation in society. This includes people who are too old or young to drive, many people with disabilities, and low-income people.

The elderly. As people age, many develop physical limitations that restrict their ability to drive, walk, or use public transportation. Illnesses, medications, or impairments leave them without the transportation they desire, and without mobility, they may decrease their involvement in society. Furthermore, older people are at higher risk when they do travel. Since they are more fragile, they are more likely to be injured in a crash.

In our automobile-dependent society, many older people are forced into immobility as their ability to drive diminishes. A series of forums conducted by the National Highway Traffic Safety Administration identified numerous steps that are needed to equitably accommodate the maturing society, including:

- safer, easier-to-use roads
- safer, easier-to-use pedestrian facilities
- safer, easier-to-use automobiles
- better assessment, rehabilitation, and regulation of drivers
- improved public transportation services
- land use supportive of safe mobility
- intercity transportation.

Ironically, many of the steps needed to accommodate the elderly can also benefit the young. As any teenager too young to drive can attest, being stuck without access to both a car and other transportation alternatives can be very frustrating. The proliferation of "Mom's Taxi" bumper stickers on suburban cars is a symptom of the unsatisfying solution to this problem.

The low-income families. Many low-income families cannot afford automobiles. Owning and operating a car are not cheap. By some estimates, it can cost over $10,000 a year to own and operate a car, factoring in loan payments, registration, insurance, gasoline, and maintenance. Families who cannot afford this are either forced to cut some corners in the cost of car ownership, such as foregoing insurance or driving an unsafe, unreliable car, or become "transit dependent." Unfortunately, most transit systems are focused on the commute to work, ignoring the many other trip types. Furthermore, many low-income workers have nonstandard work hours, which are not supported by transit. Therefore, since they do not have reasonable access to social, recreational, and commercial activities and may not even have access to employment, all for lack of transportation, the transit dependent are often socially marginalized.

People with disabilities. The disabled community has mobility needs just as do non-disabled people. Many of the provisions of the Americans with Disabilities Act (ADA) of 1990 were designed to make sure that these needs were met in all communities and by all public transit agencies. Most people are aware of the set-aside of disabled parking marked by special blue parking spots. But the requirements of the ADA as it pertains to mobility are much more far-reaching. Many travel improvements such as curb cuts at crosswalks and ramps at doorways, Braille markings on elevator controls, and audible stop announcements on buses are mandated by the ADA.

Most members of the disabled community can use regular bus and train service transportation, as long as the stations and vehicles are properly designed. However, some people with disabilities need a special transportation service, usually called *paratransit*. Paratransit is generally door-to-door, on-call transportation service that is meant to replicate the service of regular fixed-route transportation service like a bus or light rail line. The ADA requires all public transit agencies to provide a paratransit service for those riders with disabilities who cannot use the fixed route. Paratransit is a large topic, beyond the scope of this chapter. For more on the subject, see the recommended readings.

Local Transportation Finance

Just as transportation planning is the most mature sector of infrastructure planning, transportation finance is one of the most robust but byzantine of the infrastructure finance regimens. Roads, transit, and bicycle and pedestrian facilities receive funding from a myriad of sources, from all levels of government. Local property tax, sales tax, state and federal gas tax, and local bonds are all common sources of funds. Fares and tolls are also sources. The exact source depends on the nature of the expense (capital or operating) and the nature of the facility (local or state-owned). Originally funded by local governments through property taxes during the nineteenth and early part of the twentieth centuries, after World War II, major transportation facilities were funded by a user fee model, although the upkeep of local roads is still a local matter. More recently, local option transportation taxes are expanding the role of local governments for major facilities.

What follows is a brief summary of various transportation finance tools, since transportation finance is a profession in its own right. Much of this material is explored in greater detail in earlier chapters of this book. The interested reader can find more in the recommended readings at the end of this chapter.

Traditional Local Government Sources: Property Taxes

Local property tax is a major source of the local general fund, which in turn is the source for local transportation funds. The general fund normally pays for street sweeping, street resurfacing, striping and pavement markings, street lights, and other operational elements of local streets. This is one of the oldest forms of financing for local streets and roads. The concept behind using property taxes has been that roads increase property values and this was deemed a fair way of assessing the cost. Bonds can be issued to be repaid by these taxes.

User Fee Model

The user fee model for major facilities arose with the dominance of the automobile from the 1920s onward. This model saw the increasing importance of state and federal governments in funding for major facilities through the creation of trust funds funded by the gasoline tax. Other kinds of user fees include transit fares, tolls, vehicle registration fees, and truck weight fees. These funds are usually planned and allocated at the metropolitan level.

Gasoline tax. The gasoline tax, at both the state and local level, is the primary source of funding for highways and other transportation funds. Today, every state has a gas tax, ranging from $0.08 per gallon in Alaska to $0.324 per gallon in New York. The federal tax of $0.184 per gallon goes on top of that.

The federal gas tax goes into the Highway Trust Fund (HTF). The HTF was created in 1956 by the Interstate Highway Act specifically to finance the construction of the Interstate Highway System. Over its lifetime, it has succeeded in distributing over $215 billion to finance 43,600 miles of the Interstate Highway System in what's been described as a "fair, stable, and proportional way."[17] Today the HTF is more flexible; 16 percent of its funds go to the decidedly non-highway Mass Transit Account and the Leaking Underground Storage Tank Fund.

At the state level, the picture is similar. Generally, gas tax receipts go into a motor vehicle fund, with some small amounts for public transportation. In Wisconsin, for example, 95 percent of gas taxes go to highways, urban arterials, or rural roads, while 5 percent goes to ferries. In California, gas tax funds are constitutionally limited to use on "public streets and highways" and "exclusive public mass transit guideways."

Fares. Many commuters assume that their fares pay the full price of the service they are receiving, but transit fares only cover a fraction of total transit agency expenses. The "Farebox Recovery Ratio" is the term used to describe the ratio of fare revenues to operating expenses. For most transit agencies, the farebox recovery ratio is about 20 to 30 percent. In other words, fares cover only one-fifth to one-third of operating costs. This does not include capital costs at all.

Tolls. Tolls have been used for centuries as a way to pay for transportation improvements and services. The Lancaster Turnpike, from Philadelphia to Lancaster, Pennsylvania, was chartered in 1792 and financed through a series of nine toll gates. However, over the years, tolls have lost favor as a way to pay for highways. Toll highways still exist, mainly in the eastern part of the country, but most people view paying tolls as a nuisance at best, government "double-dipping" at worst.

Tolls are most successful at paying for fixed capital-intensive transportation links like bridges and tunnels. However, it is common for tolls to remain in effect long after the investment is paid off, and then to be used for other transportation projects. For example, in the San Francisco Bay Area, the Golden Gate Bridge, Highway and Transportation District uses toll revenues from the Golden Gate Bridge to subsidize bus and ferry operations.

Special Purpose Local Option Taxes

As responsibilities for transportation infrastructure have devolved to state and local governments or municipalities, the past 35 years have seen a rapid increase in the use of local authorities to raise funds for this purpose. A variety of special purpose financing mechanisms has arisen to supplement the local property tax. Unlike federal and state gas taxes, however, in many jurisdictions expenditures of these funds are not subject to the now traditional metropolitan planning processes for transportation:

Sales tax. Thirty-three states have passed authorizing legislation that permits municipalities to enact sales taxes for transportation projects. Some states require that the sales tax be spent for a specific project proposal, while others merely require that the funds be dedicated to transportation in general. A few also permit the locality to enact sales taxes for other general fund purposes. These taxes must generally be approved by voters. A sales tax can produce a great deal of revenue with only a small increment in a populous area. It is however, regressive, despite the fact that it taxes nonresidents who commute to the jurisdiction. The most well-known of projects funded by sales tax increases have been the new rail transit projects passed by voters in Atlanta, Phoenix, and San Francisco among many others, although ballot measures have failed in some other areas.

Local vehicle and fuel taxes. Thirty-three states authorize localities to assess a local vehicle tax such as a registration fee or license. These are generally used as a general fund source, although a few states earmark these taxes for specific projects, among them safety programs. Local fuel taxes are authorized by 15 states, but are only widespread in about five states, most of them in the South. Generally, these taxes do not require voter approval, have no sunset clause, and are not dedicated for specific projects. The rate must be set very high to generate enough funds to pay for major infrastructure projects.

Income, payroll, and employer taxes. Fifteen states authorize the use of income taxes for general fund purposes, but only four states make connections between these taxes and transportation facilities.

Choosing a Financing Mechanism

In choosing a financing mechanism at the local level, the practitioner needs to weigh the importance of local objectives. The ability of the mechanism to raise sufficient revenues, the costs of administering the collection program, fairness and equity issues—all of these are important policy issues. The following are some questions that the local practitioner should ask:

- *Revenue potential*—What is the amount that can be raised?
- *Stability*—Will the revenue source continue to produce over time?
- *Legal constraints*—Is there state and local enabling legislation in place?
- *Political feasibility*—What do voters and politicians think?
- *Administrative feasibility*—How much does it cost to collect the funds?
- *Equity by use*—Will those who benefit pay? Is there an implicit benefit to only *some groups*?
- *Economic equity and negative externalities*—Will using this funding source have negative effects on the local economy and environment?

Let us now consider these in turn.

Revenue potentials. Table 21.14 shows the revenue potential of several different funding mechanisms in a typical metropolitan area. To raise $15 million, the registration fee must be raised by $22, and the toll by

Table 21.14
Revenue potential of various funding options for transportation facilities in a metropolitan area

You need:	To raise $5 million	To raise $15 million	To raise $30 million
Registration fee ($)	7	22	44
Gas tax	1 cent	3 cents	7 cents
VMT (per mile)	1/14 cent	3/14 cent	3/7 cent
Sales tax (%)	0.07	0.20	0.40
Income tax (%)	0.22	0.65	1.31
Toll	10 cents	30 cents	60 cents

Source: Calculated by authors based on paper by Tamar Henkin, "Overview of Innovative Transportation Revenue Sources," presented at Transportation Research Board's Conference # 33 held in Chicago, 2002, entitled "Transportation Finance: Meeting the Funding Challenge Today, Shaping Policies for Tomorrow." Proceedings available at: www.trb.org/publications/conf/cp33transportationfinance.pdf.

30 cents—both politically not the easiest alternatives. Conversely, to realize $15 million, a sales tax would need to be increased by only 0.2 percent, an income tax by 0.65 percent, and a VMT tax by 3/14ths of a cent. A gas tax would need to be increased by 3 cents per gallon to realize that amount.

Issue of subsidies and revenue sources. Transit critics often point out that whereas public transportation systems are heavily subsidized, motorists pay their own way with gas taxes and other user fees. This is false: motorists only pay about 60 percent of the costs of road construction, maintenance, administration, and law enforcement. In 1998, federal gas tax receipts only met 82 percent of highway construction costs. Furthermore, driving is implicitly encouraged with cheap gasoline and free parking (which motorists enjoy for 99 percent of their trips). In other words, motorists are subsidized just as much as transit is, just not as visibly.

The truth is that every transportation mode is subsidized in one form or another. In 1998, the Federal Aviation Administration granted over $1.5 billion in grants to airports all over the country. Amtrak received over $150 million in state and federal grants in fiscal year 2000. No transportation system can be fully self-sufficient, and so in many ways it comes down to a policy decision as to what transportation system to support.

Conclusion

Transportation is clearly vital to a community's economic vitality, livability, and desirability. The ability to perform daily activities of working, shopping, attending appointments, and seeing friends and families requires that we all can "get around." Our current automobile-centered system, which promotes the consumption of "gas guzzler" vehicles, does not seem well suited to the challenges of volatile oil prices and global warming caused by fossil fuel emissions from automobiles. Therefore, if one word is used to describe a high-quality transportation system, it must be choice. People need choice in the travel options in terms of mode, destination, and vehicle type. A land use/transportation system that is based on dependency, whether that is auto-dependence or transit-dependence, is not meeting people's needs. Although local practitioners cannot directly influence the types of vehicles manufactured, local governments can and have piloted alternative fuel vehicles and transit options for their own employees. Local governments also play a key role in deciding on land use density and design options that can reduce transportation energy use and thereby carbon emissions as well.

Planners, engineers, and policy-makers also must realize that they themselves have choices when it comes to transportation decisions. They need to look beyond traditional forms of transportation. Transit, non-motorized modes, and innovative programs of casual carpooling and other ride sharing opportunities deserve greater attention. The automobile has a place in the urban transportation mix, but perhaps not in the sizes and fuel types presently available. In the end, it is about balance and choice. As Robert Cervero says in his survey of transit-friendly cities around the world:

> It bears noting that a functional and sustainable transit metropolis is not equated with a region whereby transit largely replaces the private automobile or even captures the majority of motorized trips. Rather, the transit metropolis represents a built form and a mobility environment where transit is a far more respectable alternative.[18]

Additional Resources

Publications

Adams, Matthew, et al. *Financing Transportation in California: Strategies for Change (Final Draft).* Berkeley, California: Institute of Transportation Studies, University of California, Berkeley, 2001. Available at: www.repositories.cdlib.org/its/reports/UCB-ITS-RR-2001-22. Although written for California, this is a useful guide to the general world of transportation finance as well.

American Planning Association. *SafeScape, Creating Safer, More Livable Communities.* Chicago, IL: American Planning Association, 2001.

Boarnet, Marlon, ed., *Transportation Infrastructure.* Planning Advisory Service Report Number 557. Chicago, IL: American Planning Association, 2009.

Cervero, Robert, *The Transit Metropolis: A Global Inquiry.* Washington, DC: Island Press, 1998.

Ewing, Reid, Keith Bartholomew, Steve Winkelman, Jerry Walters and Don Chen. *Growing Cooler: the Evidence on Urban Development and Climate Change.* Washington, DC: The Urban Land Institute, 2008. Great synthesis of climate change issues and transportation with particular emphasis on the land use connection. Good recommendations.

Florida Department of Transportation. *Florida's Transportation Tax Sources: A Primer.* Florida: Department of Transportation, 2003. Available at: www.dot.state.fl.us. This was written for

Florida transportation practitioners but it provides a useful overview of the major types of financing mechanisms that can be used in other states.

Handy, Susan, ed., *The Geography of Urban Transportation*. 2nd ed. New York: The Guildford Press, 1995.

Meyer, Michael J., and Eric Miller. *Urban Transportation Planning*. 2nd ed. New York: McGraw-Hill, 2000.

Moore, Terry, Paul Thorsnes, and Bruce Appleyard. *The Transportation/Land Use Connection*. Planning Advisory Service Report Number 546/547. Chicago, IL: American Planning Association, 2007.

Websites

Federal Transit Administration, Office of Grants Management. *Americans with Disabilities Act (ADA) Paratransit Eligibility Manual*. Washington, DC, 1993: www.fta.dot.gov/library/policy/ADA/ada.htm.

www.itsa.orgITS America www.itsa.org and U.S. Department of Transportation's Intelligent Transportation Systems website: www.its.dot.gov. Both are good websites for intelligent systems.

Pew Climate Change Project: www.pewclimate.org. This has an up-to-date website of state and local initiatives along with specialized publications about transportation and climate change.

Victoria Transport Policy Institute. *Online Transportation Demand Management Encyclopedia*: www.vtpi.org/tdm/tdm51.htm.

Key Federal Regulations

Since the mid-nineteenth century, travel, whether on roads or on a public transit system, had been universally recognized as a state and local responsibility. Railroads dominated intercity and interstate travel, and that was felt to be the only place that federal involvement was necessary. However, this attitude has gradually changed, as we see below.

Post Office Department Appropriations Bill of 1913

Long-distance roads first gained federal attention through the delivery of mail. The Post Office Department Appropriations Bill allocated $500,000 for an experimental program to improve post roads—roads on which U.S. mail was carried. The funds would be made available to state or local governments that agreed to pay two-thirds of the cost of the projects.

Federal Aid Road Act of 1916

This Act required each state to establish a highway agency with engineering professionals to carry out any projects that received federal aid. Most states complied, and these highway agencies formed the foundation for today's state departments of transportation, and firmly entrenched highway design and planning as an engineering field. This Act fell in line with Progressive ideals of the day by focusing on rural post roads, thereby enhancing rural life, rather than the long-distance intercity roads wanted by the young American Automobile Association and other urban advocates.

Federal-Aid Highway Act of 1938

This Act directed the Bureau of Public Roads (an ancestor of the U.S. Department of Transportation) to study the feasibility of a national toll road network, to be comprised of six routes. The resultant report concluded that toll roads would never support themselves, so instead recommended a 27,000-mile network of free highways. The report also recommended many of the design features we see in urban freeways today, such as sunken or elevated roadways, access control through on- and off-ramps, and beltways to direct traffic around urban areas.

Federal-Aid to Highway/Interstate Highway Act of 1956

This was a monumentally important Act. Federal aid to highways was rather desultory up to this point, and the state governors were actually calling for Congress to get out of the highway business altogether. However, the 1956 Act specified a national system of highways, and it asked the states to help determine the nature of that system. Furthermore, the federal government pledged to pay up to 90 percent of construction costs. Finally, perhaps most importantly, the Act created a dedicated funding source in the form of the Highway Trust Fund, a pot of money filled by user fees (taxes on gasoline, tires, and truck manufacturers). The highway program was administered by the Department of Commerce, and when Secretary of Commerce Sinclair Weeks announced the allocation of the first $1.1 billion to the states, he called the new program "the greatest public works program in the history of the world."

Omnibus Housing Act of 1961

The Housing Act of 1961 was the most comprehensive housing legislation since 1949. Its objectives were to improve housing for low- and moderate-income families, reduce urban blight and congestion, and stimulate building activity to counter the 1960–1961 recession. President Kennedy, on signing the bill, said that mass transportation is "a distinctly urban problem and one of the key factors in shaping community development." Despite the rhetoric, the 1961 Act did not initiate a wide-scale federal assistance program for mass transportation, but it did provide $50 million in loans and $25 million in grants (taken out of urban renewal funds) for demonstration pilot projects in mass transportation, and it set the stage for increased attention to urban transit.

Urban Mass Transportation Act of 1964

Full federal involvement in public transit had to wait until 1964, with the passage of the Urban Mass Transportation Act (UMTA). The Act authorized $375 million in matching funds for large-scale public or private rail projects in urban areas.

Department of Transportation Established, 1966

In 1966, the federal government established the cabinet level Department of Transportation (DOT). Congress directed the new department to develop a national transportation policy and determine a proper strategy for allocating federal transportation funds. The new agency assumed duties from the Department of Commerce and the Department of Housing and Urban Development.

Rail Passenger Service Act of 1970

This Bill created a semi-public for-profit corporation called the National Passenger Railroad Corporation, known to most people as Amtrak.

Federal Aid Highway Act of 1973

The Interstate Highway System as envisioned in the 1956 act was nearing completion in 1973. Therefore, shifting transportation needs prompted Congress to pass the Federal Aid Highway Act, which permitted states to use a portion of their shares of the Highway Trust Fund for urban mass transportation projects and, for the first time, Congress approved as much money for urban highway projects as it granted for rural highway construction.

Intermodal Surface Transportation Efficiency Act of 1991

The Intermodal Surface Transportation Efficiency Act (ISTEA) was an important, comprehensive transportation package that recognized that metropolitan transportation needs to be approached in an integrated fashion. The purpose of ISTEA was "to develop a National Intermodal Transportation System that is economically efficient, environmentally sound, provides the foundation for the Nation to compete in the global economy and will move people and goods in an energy efficient manner." Beyond the $119.5 billion approved for building and repairing highways, state and local governments were also given more flexibility in determining transportation solutions, whether transit or highways. Furthermore, they were provided the tools, such as enhanced planning and management systems, to help them make their choices. Finally, highway funds were made available for activities such as wetland banking, mitigation of damage to wildlife habitat, air quality projects, wide range of bicycle and pedestrian projects, and highway beautification.

chapter 22

Airports

In this chapter ...

Importance to Local Practitioners. 451
History of Airports and Air Travel 451
Early Flight 452
Airmail and the Beginnings of Commercial Aviation 452
The Great Depression: Federal Entry into the Airport Business 453
Postwar Concern for a Comprehensive National Airport System 453
1960s to Present: Wide-Body Jets, Trust Funds, and Deregulation 454
Current Trends and Issues 455
Trends in Demand and Supply 455
Shortfall in Funding for Airports 455
Low-Cost and Regional Carriers 456
Environmentally Friendly Aircraft. 456
Regional Jets and Personal Travel Aircraft 457
Security 457
Fragmented Aviation System Planning 457
Two Categories of Civil Aviation 458
Description of the Airport 459
Airport Facilities 459
Runways. 459
The Terminal 461
Airways and Air Traffic Control. 461
Institutions of Aviation 463
Who Uses Airports? 463
Who Finances Airports?. 463
Who Owns the Airports? 464
Who Regulates Airports?. . . . 465
Airport Master Planning. 466
The Airport Master Plan. 466
Identifying Needs. 467
Alternatives to Expansion: Demand Management. 468
Airport Location and Siting . . 469
Airport Layout Plan. 470
Environmental and Security Considerations 470
Airports and Land Use 472
Safety Issues 473
Noise 474

(continued)

Why Is This Important to Local Practitioners?

Airports are powerful generators of economic activity in a community. Some experts believe that an airport is the single most important piece of infrastructure a city has to sustain its economic base, providing an important transportation link to the outside world. Airport development projects may also be the largest infrastructure project a city or region ever undertakes. From initial conception to final completion, the process of planning, designing, and building, or expanding an airport can take decades and cost hundreds of millions of dollars. While most of this planning will be done by highly specialized consultants or staff, safety, noise, and sustainability issues make involvement of the local practitioner essential, particularly given that the number of air passenger miles is expected to triple between 2009 and 2030 and that urban areas are encroaching into the environs of many airports. In addition, in many areas the lack of an effective state and regional system of aviation planning makes these issues even more urgent at the local level.

This chapter starts with the history of aviation in this country. It then takes a brief look at emerging trends and current issues in the air travel sector, and continues with a description of the physical elements and the institutional players of the aviation system. Airport master planning is described before land use compatibility issues are addressed. The chapter concludes with a discussion of airports and economic development.

History of Airports and Air Travel

Today, even though all public airports (except one in New Jersey) are owned and operated by local or state governments, the history of airports is very much tied to federal policy. Airports are hugely capital-intensive, well beyond

Local Economic Development, Marketing, and Financing 476
Marketing an Airport 477
Conclusion 478
Additional Resources 479
Publications............... 479
Websites 479
Key Federal Regulations 479

the ability of most localities to finance on their own, and air travel itself implicitly involves at least two airports generally owned by differing jurisdictions. For these reasons, the federal government has been involved in local airport planning since the beginning.[1]

Early Flight

Every schoolchild knows the story of the Wright brothers and their first flight at Kitty Hawk, South Carolina, on December 17, 1903. Although their flight was merely the culmination of a series of "firsts" (the first sustained, powered, controlled, heavier-than-air flight), it marked the beginning of an age. Kitty Hawk beach bears little resemblance to today's modern commercial airport, but they both serve the same purpose, a place for an airplane to take off and land, to be loaded and unloaded, and to be repaired and fueled. Nine years after that historic flight, there were 20 recognized airports in the U.S., all privately owned and operated. World War I fueled a period of airport growth, adding 30 more airports by 1918.

Airmail and the Beginnings of Commercial Aviation

Airmail service opened the public's eyes to the utility of air travel, initially seen as either a plaything or a military tool. The first airmail route opened between New York City and Washington, DC, in May 1918, sparking an increase in publicly owned airfields from 50 to 145 in two years. In 1925, the Postal Service was allowed to enter into contracts with private companies to transport the mail by air, ushering in an era of private sector aviation growth.

In 1926, Congress charged the Department of Commerce with encouraging and stabilizing air commerce by establishing civil airways and other navigational facilities. The Department of Commerce created a separate bureau, the Bureau of Air Commerce, and began regulating airway engineering, airway operation, safety, planning, certification, and inspection. It also began a program of aircraft, mechanic, and pilot licensing.

Up until then, most airports were simply grass fields, drained of water and perhaps covered with gravel or cinders. The fields allowed take-off and landing in any direction: planes were so light that they were affected by the slightest cross-wind and thus needed to take off or land directly into the wind, whichever direction the wind was blowing. However, in the mid-1920s, aircraft became heavier, which meant they were less affected by wind but also more likely to sink into the mud. Paved runways became imperative, and the first one in the United States appeared in Newark in 1928.

The Great Depression: Federal Entry into the Airport Business

Over the years, commercial aviation grew rapidly, and so did the Bureau of Air Commerce. However, the federal government was still prohibited from directly establishing, operating, or maintaining airports. That changed during the Great Depression, when the Civil Works Administration and later the Works Progress Administration entered the airport business. Between 1933 and 1935, 640 new airports, mostly in small communities, were established by these federal agencies, while 943 existing airports received federal aid.

In 1938, President Roosevelt signed the Civil Aeronautics Act, which combined all federal regulations pertaining to aviation into one overarching statute. It also created the Civil Aeronautics Board, housed in the Department of Commerce and concerned with the economic regulation of air carriers; and the Civil Aeronautics Administration, concerned with safety and the technical aspects of civil aviation.

Terminal design and layout received most of the attention in early airports. However, in 1936, the DC-3, the first modern airliner, appeared, and runway design became important to accommodate this larger, heavier plane. In 1941, LaGuardia Airport in New York City was built, the first airport in which the runway and terminal design were treated equally.

The outbreak of World War II sparked another wave of airport building and improvement. Many of the new airfields were built for national defense purposes, but with the understanding that after the war, they would be turned over to local authorities for civil use. Indeed, when the war ended, over 500 new airports were declared surplus to the military and became the responsibilities of cities, counties, or states.

Postwar Concern for a Comprehensive National Airport System

In the postwar period, the federal government turned its attention to developing a national, comprehensive system of airports. The Federal Airport Act was passed in 1946, establishing a federal grant program meant to support both airport construction and airport standardization.

In the ten years after World War II, commercial aviation grew rapidly, and the airport system was straining at the seams, barely keeping pace. In 1956, two airliners collided in the sky over the Grand Canyon, killing 128 people. Suddenly, the sky was a crowded place, and cries for reform rose.

The response was the Federal Aviation Agency (FAA) along with a modernized national system of navigation and air traffic control. This new agency took over the responsibilities of the Civil Aeronautics Administration and that of other agencies that had tangential control of airports or air travel. The Civil

Aeronautics Board was left in place, to handle economic regulation and air safety.

The FAA reported directly to the president until 1966, when the Department of Transportation was created. The Federal Aviation Agency was placed within the new department and renamed as the present Federal Aviation Administration (FAA). At the same time, the National Transportation Safety Board was created, which investigated all transportation-related accidents. This left the Civil Aeronautics Board with only its economic role.

1960s to Present: Wide-Body Jets, Trust Funds, and Deregulation

The late 1960s saw both the advent of the wide-body jet, capable of carrying hundreds of passengers, and an unprecedented boom in commercial air travel. Once again, the national airport system was on the brink of collapse, and once again, the traveling public was indignant. The federal government responded in 1970 with the Airport and Airway Trust Fund, a pot of money to be filled with user fees and to be used only for airport and airway construction and modernization. The user fees came from taxes on passenger fares and aviation fuel, aircraft registration fees, and airfreight.

The new trust infused vast sums of money into the airport system—$1.2 billion over five years. Eighty-five new airports were built and thousands more were improved. When the trust fund came up for reauthorization in 1976, it was expanded. Between 1976 and 1980, over $2.7 billion was spent.

Despite the large sums spent, federal policy had not really changed substantially since the end of World War II. The goal remained to meet demand through airport expansion and technological improvement. Meanwhile, the Civil Aeronautics Board strictly regulated routes and fares. In the 1970s, national momentum gathered to reduce government regulation of private industry, and finally in 1976, following the deregulation of the railroads and the trucking industry, air cargo was deregulated. In 1978, air passenger travel followed suit.

Deregulation represented a major shift in the air travel industry, on many fronts. First, airlines were now free to choose which routes to serve, and many immediately dropped

Loading luggage at Boston Airport during snowstorm

Entrance to Lansing, Michigan Airport

unprofitable ones, invariably those to smaller communities. Second, the small commuter airlines appeared, filling in the gaps left by the retreat of the larger airlines. Third, airports themselves were suddenly forced to be competitive. They could no longer count on a stable base of regulated airlines for revenue. They also had to accommodate any new market entrants. The repercussions of air deregulation continue to reverberate to this day.

Current Trends and Issues

Trends in Demand and Supply

If one word is used to describe air travel in this country and indeed the world, it would be *growth*. The size of airports, number of airports, size of aircraft, performance of aircraft, number of passengers, number of flights, amount of freight—all have grown, and in some cases at quite a rapid rate, recent economic events notwithstanding. Perhaps the only thing shrinking in air travel is the size of the fares; real fares in the U.S. dropped by about 50 percent from 1950 to 2006, and dropped 16 percent between 1995 and 2011.

The FAA projects continued growth of the airline industry. Passenger boardings are expected to increase at 3 percent a year through 2025, with regional carriers outpacing mainline carriers. They estimate that planes will remain crowded, but that point-to-point service will increase, resulting in longer passenger trip lengths. Air cargo activity is highly correlated with gross domestic product, and it is also influenced by security regulations and the use of all-cargo carriers. Cargo activity is expected to grow at an average rate of 5 percent a year through 2025.[2] Over the next 20 years, the FAA estimates that the average seating capacity of each airliner will slowly increase, as will the aircraft utilization rate (the number of times an individual aircraft can be used in a day). Load factors (the ratio of passengers to seats) are also expected to remain high.[3]

These trends suggest that capital investments to expand terminals and runways will be sorely needed in the long run to accommodate more passengers and more cargo and larger aircraft. The FAA projects that, without major improvements in demand management or air traffic control, 27 major airports around the country will need additional capacity by 2025, and that as many as four more new major commercial service airports will need to built in that same timeframe.[4]

Shortfall in Funding for Airports

The federal funding dedicated to the aviation system that characterized the 1970s is not available today. Yet the costs remain substantial. In 1997, the U.S. General Accounting Office estimated that about $10 billion in 1996 dollars would be needed every year to improve and expand airport infrastructure as well as to meet safety, security, and environmental needs.[5] The federal government appears to be concentrating on capacity at the commercial service airports and on upgrading the national airspace system—both important issues. Development of airport infrastructure construction and management technology has not been a priority, and consequently FAA pavement design technology is 20 years behind the highway section.[6] Airport operating and maintenance costs are anticipated to increase substantially over present levels due to more complex operations at larger airports and the need for more skilled workers to develop and use the new technologies.

Low-Cost and Regional Carriers

Low-cost carriers, which offer low-frills point-to-point service, and regional air carriers, which transport passengers over shorter distances (less than 750 miles), are important forces in the airline industry. In fact, while the traditional airlines are facing severe financial pressures and have cut back service in many areas, low-cost carriers grew approximately 20 percent between 1996 and 2001 (compared to 5 percent growth of the traditional carriers in the same period) and now command 30 percent of the air travel market. These airlines typically use smaller aircraft and use a point-to-point routing structure that focuses on delivering the lowest cost travel between two points and does not attempt to provide integrated flight service. Point-to-point service also often utilizes smaller airports.

The bulk of additional runway capacity planned or needed at many airports is to accommodate these airlines' aircraft. Low-cost and regional carriers typically fly smaller aircraft than the traditional airlines. Smaller planes are more vulnerable to turbulence, which means that the separation between consecutive landing planes has to be larger, decreasing the capacity of the runway. In other words, regional carriers actually reduce the capacity of commercial airports because of the planes they fly. This traffic is expected to grow as regional carriers acquire jet aircraft to supplement their propeller-driven fleets. Parallel runways are planned at some connecting hubs to accommodate operations by regional airlines.

Many low-cost and regional carriers operate from a major city's "second airport" to keep costs down. For example, Southwest Airlines chose to serve the Boston metropolitan area not from Logan Airport but from Manchester, New Hampshire, a smaller airport with lower costs.

Environmentally Friendly Aircraft

The airline industry accounts for 13 percent of the transportation sector's greenhouse gas emission, but the industry is responding. For instance,

Boeing's 787 Dreamliner, first delivered to the market in September 2011, is 20 percent more fuel-efficient than its peer aircraft. Given the slow turnover of aircraft and their lifetimes of 25–35 years, introduction of the more fuel-efficient aircraft will not improve aviation emissions to their 1990 levels until the year 2030, but it should be noted that from 1970 to 2000, there have been reductions of about 60 percent in energy use per passenger mile.

Regional Jets and Personal Travel Aircraft

Smaller jets are replacing the commuter prop planes on some of the well-traveled mid-city routes. In the long run, 100-passenger jets are expected to enable the airlines to link many mid-city pairs, over-flying competitor's hubs. Regional jets will improve service in markets that are currently marginal and make more city pairs economically feasible. This might result in less traffic at some hub airports, but will increase traffic at mid-sized and smaller airports. However, regional jets might also add to airside congestion at larger airports. The trend towards regional jets will have a substantial impact on airport infrastructure needs and on the airline industry itself in the next five to ten years.[7]

Security

For several years after the terrorist attacks of September 11, 2001, aviation security crowded out other hot-button issues such as airport capacity shortfalls, airspace congestion, and market liberalization. Heightened security created a new environment of travel time, forcing passengers not only to count the total duration of delays and cancellations as possible inconveniences to their travel plans, but also to reserve ample time to clear security before each flight. This increased time and effort to consumers have added a "hassle factor" that has influenced the fly-or-not-fly decision. For instance, some businesses now choose teleconferencing instead of travel for face-to-face meetings. The decision point for flying as opposed to driving increased from about 200 miles to about 400 miles. Low-cost carriers benefited from this, since smaller planes and smaller airports mean shorter queues, fewer people, and less difficulty getting to the airplane.

Fragmented Aviation System Planning

There is increasing concern in the aviation community about the adequacy of the existing state and regional aviation planning system to cope with both existing demand and the growth that is anticipated in the next 20 years. Many large commercial airports are constrained in their ability to expand, and therefore other, smaller airports within the region may take up the slack. However, their planning staff may lack the technical skills and funds to plan in this increasingly complex environment. In addition, individual jurisdictions and airports may be more concerned about their own particular development agenda than the overall picture. Capital investment decisions at many airports may not be coordinated with land use planning and may not be integrated with other transportation system plans. Planning for the "national aviation system" (the primary commercial airports) is separate from the "state and regional system" (general aviation and the other airports). Private airports are also not part of the official aviation system, yet they meet some of general aviation's operational needs.

The federal government's role in capital investment decisions revolves around the National Plan of Integrated Airport Systems (NPIAS), which identifies about 3,700 airports (all commercial service, reliever and selected general aviation airports) that are significant to national air transportation and thus eligible to receive federal grants under the Airport Improvement Program (AIP). The NPIAS also includes estimates of the amount of AIP money needed to fund infrastructure development projects that will bring these airports up to current design standards and add capacity to congested airports. Rather than being a national aviation system plan, it is more of a funding "program."

Some states are moving to address this problem. Washington and the Puget Sound Regional Council are developing a cooperative planning program to address both state and regional airport system needs. California conducts continuous aviation system planning through the California Aviation System Plan, which includes the participation of MPOs. Massachusetts recently adopted an aviation system plan that explicitly recognized the importance of the regional network of airports. However, states and regions need to do more in this area, since it is unlikely that the federal government will take the lead in developing a national aviation system plan that is inclusive of all local airports.

Two Categories of Civil Aviation

The civil aviation sector covers a surprisingly broad range of aircraft-related activities, much broader than that of motor vehicles, for example. Small, propeller-driven recreational aircraft share the sky with government helicopters, airliners, executive jets, and freight transporters. In general, though, civil aviation is broken down into two categories: general aviation and commercial aviation:

General aviation. General aviation accounts for the bulk of all civil aircraft operations. It includes everything from crop dusting to recreational flying, search and rescue to medical transport, local television traffic reports to special tourist charter flights to the tropics. Basically, any air travel service that is not regularly scheduled is considered general aviation. As Table 22.1 shows, the vast majority of airports in the country are general aviation airports.

Table 22.1
Number of U.S. airports in 2010

Public vs. private	Number	General vs. commercial	Number
Public use	5,175	General aviation	19,251
Private use	14,353	Commercial service (certificated)	551
Military	274		
Total	19,802	Total	19,302

Source: U.S. Department of Transportation, Bureau of Transportation Statistics, 2010.

Commercial aviation. Commercial aviation is what most people first think of when they think of airports. This term describes the scheduled service that airlines provide. It also includes freight and cargo carriers like Federal Express. Airports must be certificated by the Federal Aviation Administration in order to handle commercial service. Table 22.1 also shows the number of certificated airports.

Description of the Airport

An airport is the connection of two transportation environments—ground and air—with facilities to serve each. Airports are marked by incredible diversity in size and complexity. A simple grass strip with only a windsock can be all that an airport, while Denver International Airport covers over 50 square miles in area and has 6 runways, 4 terminals, and 40,000 parking spots. But they all have features in common: a runway, the need for access to ground transportation, and incompatibility with certain adjacent land uses. The following highlights the major elements of an airport.

Airport Facilities

On one side are the *airside facilities*, those on which aircraft operations take place. These include the runways, the taxiways, the apron where planes park, the gate areas where passengers board and alight, and hangar areas. *Groundside facilities* include the terminal, parking, ground transportation, freight handling and storage, and the like. Included in the terminal are often restaurants, car rental agencies, shops, and even hotels. A well-designed airport provides a seamless transition between these two sides.

Runways

The runway is the surface from which the airplane takes off and on which it lands. The runway is the reason the airport exists where it does at all. There are many different runways configurations; the FAA recognizes 22 different types. In general, though, there are four basic types: the single runway, parallel

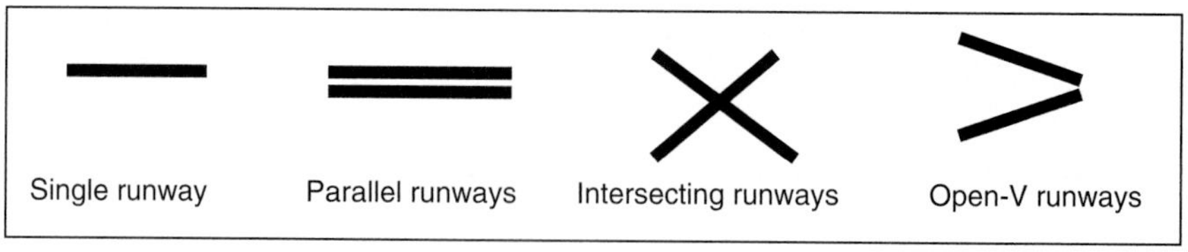

Figure 22.1
Basic runway configurations

Single runway | Parallel runways | Intersecting runways | Open-V runways

Hubbing and Its Downsides

Deregulation and intense competition between airlines, along with increasing urbanization, stimulated the phenomenon of the hub airport during the 1980s and 1990s. Much like the hub of a wheel from which radiate the spokes, a hub airport serves as the center of a radial flight structure. Hubbing maximizes the number of transfer options open to a passenger, and also opens up many more destinations that would not be profitable with direct point-to-point service. For example, in a system with 10 destinations, the spoke-hub system requires only 9 routes to connect all destinations, while a true point-to-point system would require 45 routes.

However, hubbing has some infrastructural downsides. First, it is capital-intensive. Hubbing works best when a large number of planes from a large number of origins all converge on the airport at about the same time. Passengers can then readily transfer to a large number of destinations on an equally large number of departing flights, all leaving at about the same time. Serving this large number of flights at the same time requires many runways, gates, baggage handling equipment, and service personnel, which are all heavily used during the hub rotation but which sit idle the rest of the time. Airports, not airlines, bear these infrastructure costs.

Second, hubbing concentrates air traffic at a few airports. The 30 largest airports in the U.S. (or .05 percent of total commercial airports) together handle over 70 percent of the total national passenger load. This has implications not only for those 30 airports but for the entire system. Delays at one of the major hubs quickly propagate out to every airport serving that hub.

runways, intersecting runways, and open-v runways (Figure 22.1.) The exact configuration has implications for runway capacity, and is chosen based on physical space available, prevailing wind conditions, and the aircraft expected to use it.

The runway at a small private airfield might be a grassy strip about 1,000 feet long, cleared of debris. The runway at a major commercial airport can be over 2 miles long and constructed of carefully engineered concrete.

The length of the runway depends on numerous things. The type of aircraft using the runway is the most important characteristic. Generally speaking, the larger and heavier the plane, the longer the runway needed. Propeller planes generally need shorter runways than similarly-sized jet aircraft. Atmospheric and environmental features matter, too. Since warm air is less dense than cold air, a longer runway is needed in warmer climates. Similarly, runways at higher elevations need to be longer than those lower down. Headwinds allow shorter runways, while tailwinds mean longer ones. An upward sloping runway needs to be longer than a downward sloping one. Water, snow, or slushy conditions necessitate longer runways.

Like any infrastructure system, a runway has a capacity; the capacity of a runway is given as the number of operations (takeoffs or landings) per hour. A rough figure used by airport planners for estimating runway capacity is 60 operations/hour, but a runway's exact capacity is determined by its physical characteristics and by the aircraft types, or more precisely, the particular mix of types, using it. As aircraft move through the air, they create large disturbances in the air stream behind them, called a wake vortex, and this turbulence is dangerous for a plane near the ground. For this reason, consecutive landing airplanes must be separated by enough time to let the turbulence dissipate. The bigger the plane, the larger the turbulence; the smaller the trailing airplane, the greater the separation time needs to be.

The Terminal

The airport terminal is meant to interchange crew, passengers, luggage, freight, and mail between surface transportation and air transportation in a comfortable, convenient, expeditious, and economical manner. That's a tall order, and the design of terminals for large, commercial airports is a specialty unto itself. The local practitioner or policy-maker need only be aware of the distinctions.

Designing a terminal building requires consideration of a number of factors, including runway capacity, growth potential, aircraft types to be served, ground transportation access, layout of the runway and taxiways, commuter and general aviation needs, cargo and maintenance facilities, kitchens, parking garages, car rental agencies, and the numerous retail establishments.

Airways and Air Traffic Control

With something like 10,000 airplanes in the skies above the United States at any one time, it is important for reasons of safety and economy to impose order on the chaos. To achieve this, the skies are divided into established corridors, known as airways and often called "highways in the sky." Airplanes stay in these corridors by means of radio beacons on the ground that the airplanes follow. The corridors are wide and broad.

Coordinating the entire aerial ballet is a national tiered system of air traffic control. Air traffic control towers control aircraft movement on the ground at the airport, as well as the airplanes immediately before landing or after taking off. The second tier, called approach control, controls the

New Large Aircraft

When Airbus sold its first A380, a double-decked wide-bodied airliner capable of carrying up to 853 passengers, in October 2007, it ushered in an era of New Large Aircraft (NLA). NLA will impact airports and airline operations substantially in the next 15 years. Increasing numbers of NLA may complicate air traffic control by requiring greater separation between aircraft, though they are also anticipated to ease airside capacity conditions. They will produce greater traffic peaks for groundside facilities. Major airports will need to increase the distance between gates because of the increased NLA wingspan. NLAs may require longer runways, and they will generate stronger winds as they take off and land. Flight schedules will need to accommodate the longer times to board and unload the increased number of passengers. Baggage systems will need to be upgraded to handle the increased traffic, and terminal waiting areas will need to be expanded.

Airport control tower in Cherry Pointe, NC

Inside the airport

How Planes Fly

Heavier-than-air flight is a balancing act of four forces: weight, thrust, drag, and lift. Weight is obvious. Thrust is the force pushing the airplane forward. Drag is the friction produced by moving an object through the air. Lift is the force that actually keeps the airplane aloft and is perhaps the most "mysterious" (Figure 22.2).

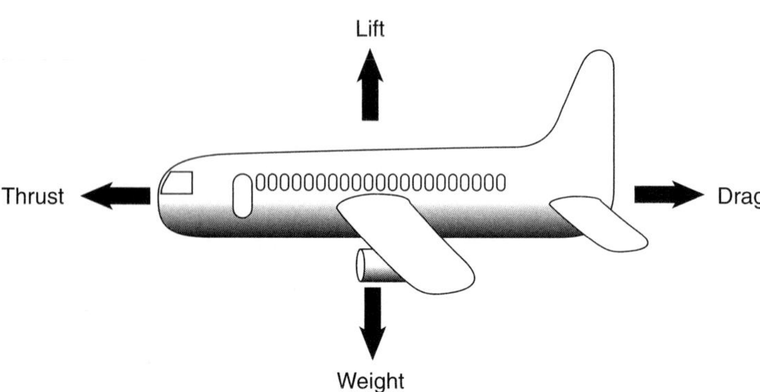

Figure 22.2
The four forces balanced in flight

An airplane wing generates lift by deflecting air downwards, thus pushing the wing upwards. Any surface that is angled against the flow of air will produce lift (anyone who has held their flat hand out of the window of a moving car has experienced this effect). An airplane wing in cross-section is a teardrop shape called an airfoil, the most efficient shape for deflecting air and generating lift (Figure 22.3). Lift is a function of the aircraft's speed relative to the surrounding air, not its speed relative to the ground (the difference between groundspeed and airspeed). During take-off, the plane accelerates to the speed at which the differential air speeds are large enough to lift the plane off the ground. The amount of lift generated is also related to the surface area of the wings, since more area deflects more air.

All moving things, including walking people, produce drag. However, since the drag produced is proportional to the square of the velocity of the object, it becomes a serious concern at higher speeds. In fact, at the cruising speed of most aircraft, the dominant force is drag. Airliners fly at such high altitudes because less drag is produced in thinner air. However, less lift is also produced in thinner air. Airliners fly at an altitude where low drag is balanced against sufficient lift.

At take-off and landing, an aircraft has just enough speed to produce just enough lift to keep itself in the air. If the airspeed changes only slightly, the lift could drop disastrously. Moreover, the plane is quite close to the ground at this point. If something goes wrong, such as an engine failure, the pilots have little room to maneuver or recover. These factors govern much of airfield design and operation.

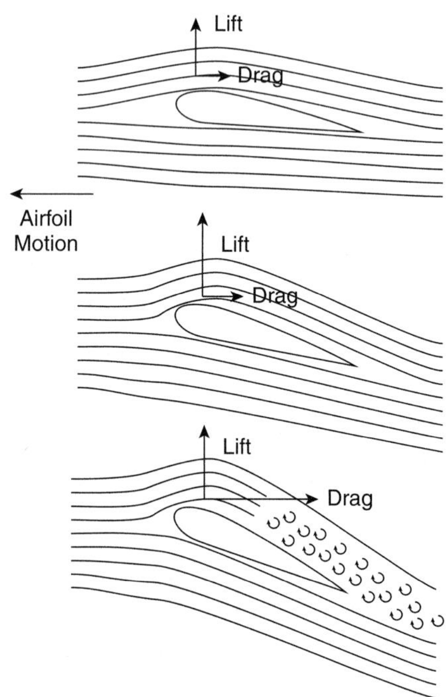

Figure 22.3
Airfoils and flight

approaches to airports out to a distance of 20 to 50 miles, while the highest tier controls the en route traffic, or those planes at cruising altitude traveling between airports.

At many airports, particularly small, low-traffic or isolated ones, this process is relatively easy. In areas with many airports clustered together, managing this process is much harder. The San Francisco Bay Area, metropolitan New York City, and Los Angeles are areas with particularly challenging airspace issues.

Institutions of Aviation

Who Uses Airports?

Airports facilitate the transport of both passengers and cargo. Passenger traffic has been discussed earlier. Cargo, however, is an increasingly important component of aircraft operations that is frequently overlooked in airport planning. Air transportation is the preferred mode for the shipment of high-value, lightweight, and/or perishable goods. In 2009, air transportation moved 2.2 percent of total U.S. freight by value, but only 0.03 percent by weight. The principal need for airport development is related to the cargo sorting and transfer facilities developed by small-package express carriers.

Air cargo is concentrated at busy commercial airports, and surprisingly, the majority of it is carried in the baggage compartments of scheduled passenger aircraft. Less than 5 percent of scheduled flights are by all-cargo aircraft, and these are usually derivatives of passenger aircraft. Cargo-only flights usually occur during off-peak periods and do not substantially contribute to airport congestion and delay problems. However, as passenger aircraft became smaller (reflecting the growth in low-cost and regional carriers), the amount of cargo space available on them has dropped commensurately. This is matched by a steady increase in the payload capacity of all-cargo aircraft (from about 35 tons per airplane in 1996 to an estimated 60 tons per airplane in 2021). The net result should be a shift in the share of cargo that goes by cargo-only airplanes.

Who Finances Airports?

On average, airports draw approximately two-thirds of their total revenue from nonairline sources such as parking fees; rental payments from retail concessionaires; car rental surcharges; and per-passenger facility charges (included in the ticket price), and the remainder from airline rates and charges such as landing fees, terminal fees, aircraft parking fees, gate and hanger rental charges, ground handling service charges, air traffic control charges, and fuel taxes. Both revenue sources are very sensitive to changes in passenger traffic.

Most federal grants are for small commercial airports and general aviation. For example, in 2002, the 71 largest U.S. airports had about 90 percent of the air traffic but received only 40 percent of federal airport grant dollars. In 1990, the Congress passed legislation permitting the airlines to charge a passenger facility charge that was redistributed more equitably to the larger hub airports. In 1996, passenger facility charges plus airport revenues were funding almost 90 percent of the capital improvement programs in the large and medium hub airports, while funding only 38 percent of the same programs in smaller hub and non-hub airports. These smaller airports received 51 percent of federal grants and accounted for 22 percent of all airport spending on capital projects.

Who Owns the Airports?

Most public airports are owned by a city or county government. As Table 22.2 shows, almost all public airports are owned and operated at the local level. In general, the airport owner is responsible for the operation of airports and with implementation of airport master plans.

However, the organizational template for airport ownership and operation varies widely from airport to airport. In some cases, the airport is a department of the city or county government in much the same way as the planning or public works departments. Miami International Airport in Florida and Logan International Airport in Billings, Montana, represent a very large and very small example of this model. This structure is the most common for small, municipally-owned general aviation airports.

Table 22.2
Ownership of public airports, 1992

Level of government	% of total
City or county	61
Special district or port authority	34
State	5
Federal	(1 airport)

Source: Lawrence Gesell, *The Administration of Public Airports*, 3rd ed. (Chandler, AZ, Coast Aire Publications, 1998). See also Lawrence Gesell, and Robin R. Sobotta, *The Administration of Public Airports*, 5th ed. (Chandler, AZ, Coast Aire Publications, 2007).

Sometimes the public airport owned by the city or county can have an airport commission as its governing body. In this case, the members of the commission are usually appointed by the local legislative body. They may be responsible for policy, hiring the chief administrator, and approving the budget. Usually the airport follows the general municipal procedures for personnel, accounting, budgeting, and purchasing.

The airport authority. Although there are many municipalities that operate airports effectively, the trend today is towards independent authorities. Unlike water and sewer services where special districts were created in response to new development, many airport authorities were converted from municipal ownership to respond to specific management or financial issues. A municipal airport is often used by passengers from an area larger than the city's boundaries. As airport use expands along with requirements for new capital facilities, the city may be reluctant to use its bonding capacity due to other local needs. In these situations the creation of an airport authority is useful. An authority can issue bonds for capital improvements paid for with its own income and can do long-range capital planning without the uncertainties of local politics. In addition, analysts evaluating an airport's financial position are more positive when the "loop of funds" is closed—that is, excess airport revenues do not go back into supporting the general government's operations.

In metropolises with seaports as well as airports, it is not unusual to see an autonomous or semi-autonomous port authority that manages all maritime and aviation activities. The Port of Portland, Oregon, the Port of Oakland, California, and the Port Authority of New York and New Jersey are some examples. (See Chapter 23 for a more thorough discussion of ports.)

Privatization. Privatization has been put forward as another institutional arrangement for airports. The 1996 FAA Reauthorization Act approved four privatization initiatives. In 1990, only four primary commercial airports were privately owned, but partial privatization of the airport facilities occurs in some locations. This includes the lease of a publicly owned airport to a private management firm, or private development of key airport facilities, such as a terminal, hangar, parking lots, or hotels. Private operation of concessions in the airport is common. The huge capital needs of the future and the declining role of the federal government in that area may prompt some innovative partnerships.

Who Regulates Airports?

Federal agencies. The Federal Aviation Administration sets civil aviation policy and implements it at the national level. The FAA manages, sets standards for, or regulates air traffic control, pilot training, aircraft registration, airport design and operation, safety and accident investigation, and a host of other aviation-related activities.

State Departments of Transportation. States have varying roles and responsibilities in airport planning and management. The usual seat of responsibility is an office or division of aviation or aeronautics within the state department of transportation. In general, these offices serve as agents for federal airport improvement grants. Some states, such as Washington and California, conduct their own system planning. The extent of regulation varies as well. Ohio requires that all airports, regardless of ownership, be registered with the state. New York, on the other hand, requires only that airports meet federal guidelines.

The offices have various other responsibilities. In Montana, the Aeronautics Division is responsible for air search and rescue operations. In California, the state Division of Aeronautics has a large local land use role—it administers noise regulation and land use planning laws around airports. It also makes recommendations regarding proposed school sites within 2 miles of an airport runway and authorizes helicopter landing sites at or near schools.

Local general-purpose governments. The local government regulates land use and development around airports. It may also be involved in regulating the development process for a new airport or expansion of an existing airport.

Airport Master Planning

The following sections discuss three areas that are important for the local practitioner: airport master planning, land use compatibility concerns, and airports and economic development. From initial conception to final completion, the process of planning, designing, and building or expanding an airport can take decades and cost hundreds of millions of dollars. Airport planning is highly technical and episodic. Because the industry will continue to grow and therefore require increased capacity that will impact adjacent land uses and may require capital financing assistance, local practitioners need to understand how airports are planned. In many instances, local jurisdictions are required to be involved in the preparation of the airport plan.

At the heart of the process is the airport master plan. Preparing or updating the airport master plan is a complex exercise, and a detailed discussion of all the issues is beyond the scope of this chapter. Indeed, whole volumes have been written on the subject. This section will give an overview of the process and then describe some of the issues that the local policy-maker or practitioner will face when contemplating building a new airport or expanding an existing one. For more information, see the Additional Resources at the end of this chapter.

The Airport Master Plan

Both the International Civil Aviation Organization and the Federal Aviation Administration recommend the development of an airport master plan as the basis of orderly capital planning for individual airports.[8]

The magnitude and sophistication of the plan depend on the size and nature of the airport. At major airports, the airport might have its own in-house staff that produce its own forecasts and technical studies. For these airports, the planning process is formal and complex. On the other hand, at small municipally owned airports, the city planning staff, with the help of consultants, might prepare the plan, which will be a simple document or part of the general plan. Moreover, in today's aviation environment, an airport must be considered not in isolation but strategically, as part of a regional and national system.

The master plan should be a guide for development of both the aviation and non-aviation physical facilities of the airport and the development of adjacent land areas. In general, the airport master plan is not too dissimilar to a city's general plan (indeed, some major airports can be considered small cities). Airport master plans should have the following elements:[9]

- an inventory of airport facilities and nearby airport-related uses;
- short-, medium-, and long-term traffic demand forecasts;
- an assessment of current airport capacity;
- site selection if airport expansion or new construction is planned;
- consideration of environmental impacts, including mitigation measures;
- cost-effectiveness and feasibility studies;
- drawings of the airport layout plan, land use plan, terminal area plan, and ground access plan;
- plan implementation, including a capital investment program and potential revenue sources.

Note that the airport master plan should take into account adjacent land uses. Very often, these land uses are not under the control of the airport itself. If the airport is not part of the local government surrounding or close to it, a special effort needs to be made to involve that government in both the preparation and implementation of any airport master plan. The airport planning process follows the basic planning process. (The Additional Resources at the end of the chapter identify other readings with more detail about airport planning.)

Identifying Needs

Before determining future needs, the practitioner needs to identify what role the airport in question plays or will play in the future. Is it an origin-destination airport, a connecting hub, or a regional end point? Is it for passengers or for freight, for commercial flights or for general aviation? Each of these airport types has different requirements and different future demand potential. A similar process should be followed for the expansion of an existing facility.

The facility inventory will detail the current capacities of the various components of the airport. The next step is demand forecasting. Forecasting aviation demand is a complex task—a demand forecast must consider such factors as historical trends in aircraft movements, passenger and/or freight volumes, population, and economic growth in the region; characteristics of national and international air traffic, geography, and airspace issues; and aircraft industry dynamics. In general, air travel demand is cyclical and often seems capricious, so any demand forecasts must be viewed with a healthy skepticism. Nonetheless, such demand forecasting is important.

The capacity of an airport is determined by the weakest link in the chain of airway capacity, runway capacity, apron capacity, terminal capacity, and groundside access capacity.

NextGen Highways in the Sky

NextGen is a satellite-based air traffic control system that is estimated to cost $40 billion and be completed sometime between 2018 and 2025. This system would create an interstate highway system in the sky. It works by using GPS receivers in airplanes and new central systems to replace the current beacon system. NextGen will allow more precise navigation and hence will save fuel and time. It will also permit better routing around weather systems. American Airlines and Delta are already using this system for takeoff and landing into Dallas/FW and Atlanta, where a partial version of this system has been installed. Reductions in cost, energy, and carbon emissions have been noticeable.[1]

Note

1. Eric Torbenson, "Satellite-Based Air Traffic Control System Slow to Get Off the Ground," *The Dallas Morning News* (2009), www.dallasnews.com.

Alternatives to Expansion: Demand Management

As with any infrastructure system, new supply is not the only way to meet future demand. Alternatives to expanding an airport include enhancing the use of existing infrastructure through demand management, technological changes in navigation (such as the FAA's NextGen Plan) and aircraft that can increase runway capacity, and encouraging other modes of transportation.

There are those who argue that it is not the deficit of poured concrete that leads to airport congestion, it is the decision of airlines to schedule take-offs and landings in groups at the same time called banking. If some sort of peak period pricing were established, in which airlines had to pay higher landing fees during peak periods, they would have an incentive to change their schedules and thereby reduce congestion.

Runway capacity is not an immutable feature of the runway, but depends on the mix and types of aircraft using it, as well as the navigational systems that support it. Technological changes can also increase runway capacity. New aircraft designs can make airplanes less susceptible to wake vortices. Enhanced navigation systems can make an airport less vulnerable to poor weather. An individual airport has little control over airplane design, but it can implement changes in navigational and air traffic control systems that will increase its capacity without adding a foot of runway.

Finally, a powerful way to reduce airport demand is to encourage the use of other transportation modes. For short trips, surface transportation can be just as fast, especially in light of increased airport security. High-speed rail is often touted as the best mode for trips of 100 to 300 miles. Airports designed with seamless inter-modal connections that support these surface modes can reduce airport congestion.

Airport Location and Siting

Airport site selection must balance the physical characteristics of the site, the nature of surrounding land use development, any flight path obstructions, atmospheric and climatic conditions, land availability and cost, ground access, and the compatibility of surrounding airspace with the socioeconomic factors creating the need for an airport. These factors are also important for the expansion of existing facilities.

Airports should be relatively close to the economic activities they are supporting, such as population centers and transportation corridors. However, they should not be too close for reasons of safety and noise (discussed below).

Physically, a site should be reasonably flat and clear of impediments like hills or mountains. Drainage is important. An airport should be located away from skyscrapers, smokestacks, television towers, and other obstacles that can obstruct an aircraft's flight path. (See the discussion later in the chapter on land use compatibility issues.)

Prevailing winds should be taken into consideration, both for the reasons mentioned earlier and also because wind can blow in smoke, fog, or water spray, thereby limiting visibility. A cross-wind can make both airspeed and a straight flight path difficult to maintain—numerous airplane crashes have been due to high cross-winds.

Land availability, for both current and future needs, is a prime consideration. Commercial airports are voracious consumers of land. Table 22.3 shows the size of some major U.S. airports.[10] Other considerations include the following.

Transportation access. An airport is a transit point between air and ground. If the airport is to function effectively, local officials must pay particular attention to groundside access.

Automobiles and parking. Airports, particularly large commercial airports, generate significant amounts of traffic. This traffic arises from passengers and those picking up or dropping off passengers, from the transport of freight that is either delivered by air or needed to operate the airport itself, and from the large number of people who actually work at the airport. In 1994, a survey of airport operators revealed that passengers experience more delays on access and circulation

Table 22.3
Area of selected airports

Airport	Area (sq. mi.)
Boston Logan	3.8
New York Kennedy	4.6
Salt Lake City International	11.2
Dallas/Fort Worth	27.8
Denver International	53.1

Source: Paul Stephen Dempsey, *Airport Planning and Development Handbook: A Global Survey* (New York: McGraw-Hill, 2000).

roads than they do on the airfield. Off-airport access roadway congestion, on-airport roadway congestion, and curbside congestion were identified as the three major culprits for these delays.

Intermodal access/transit. Very few modern commercial airports are being built without transit connections, and many existing airports are adding transit service. The transit in most cases is bus, but many communities—such as Cleveland, San Francisco, and Chicago—have invested in rail connections. The airport door-to-door shuttle is fast becoming a major player as well.

Unfortunately, there is a huge economic incentive at many airports to favor auto access over other modes because of the revenues generated by parking fees. An airport can get $15 a day or more from a car parked in long-term parking lots, but will get nothing from passengers brought by taxi, train, or bus. Addressing the ground access issues can be handicapped by the fact that FAA rules prevent the airport from making off-site transportation investments.

Airport Layout Plan

An essential element in the airport master plan is the airport layout plan. This is a set of drawings that lays out the existing conditions and future developments graphically. It is the only element that must be reviewed and approved by the FAA. These drawings and maps may include existing and proposed airport layout plans, airspace drawings, drawings of the terminal area, existing and proposed land use maps, and the airport property map. Figure 22.4 shows the airport layout plan drawing for the Minneapolis Airport.

Environmental and Security Considerations

Airports, by virtue of their size and operating characteristics, place unique, and in some cases extreme, burdens on the environment.

Air quality. Airports are significant sources of air pollution. Aircraft produce emissions during start-up, taxiing, and take-off. While parked, many aircraft are supplied electrical power by small auxiliary gasoline-powered generators. Airport ground vehicle traffic is low-speed and includes many starts and stops and long periods of idling—conditions that lead to high air pollution emissions.

Airport operators have no control over aircraft engine technology, but they can implement other steps to reduce air pollution. Efficient air traffic control can minimize taxiing times. The installation of fixed electrical ground power stations can eliminate the need for auxiliary generators. Airports can utilize zero- or low-emission vehicle fleets.

Non-point source pollution. Airports, especially large commercial airports, have thousands of acres of impervious surfaces. This presents a serious

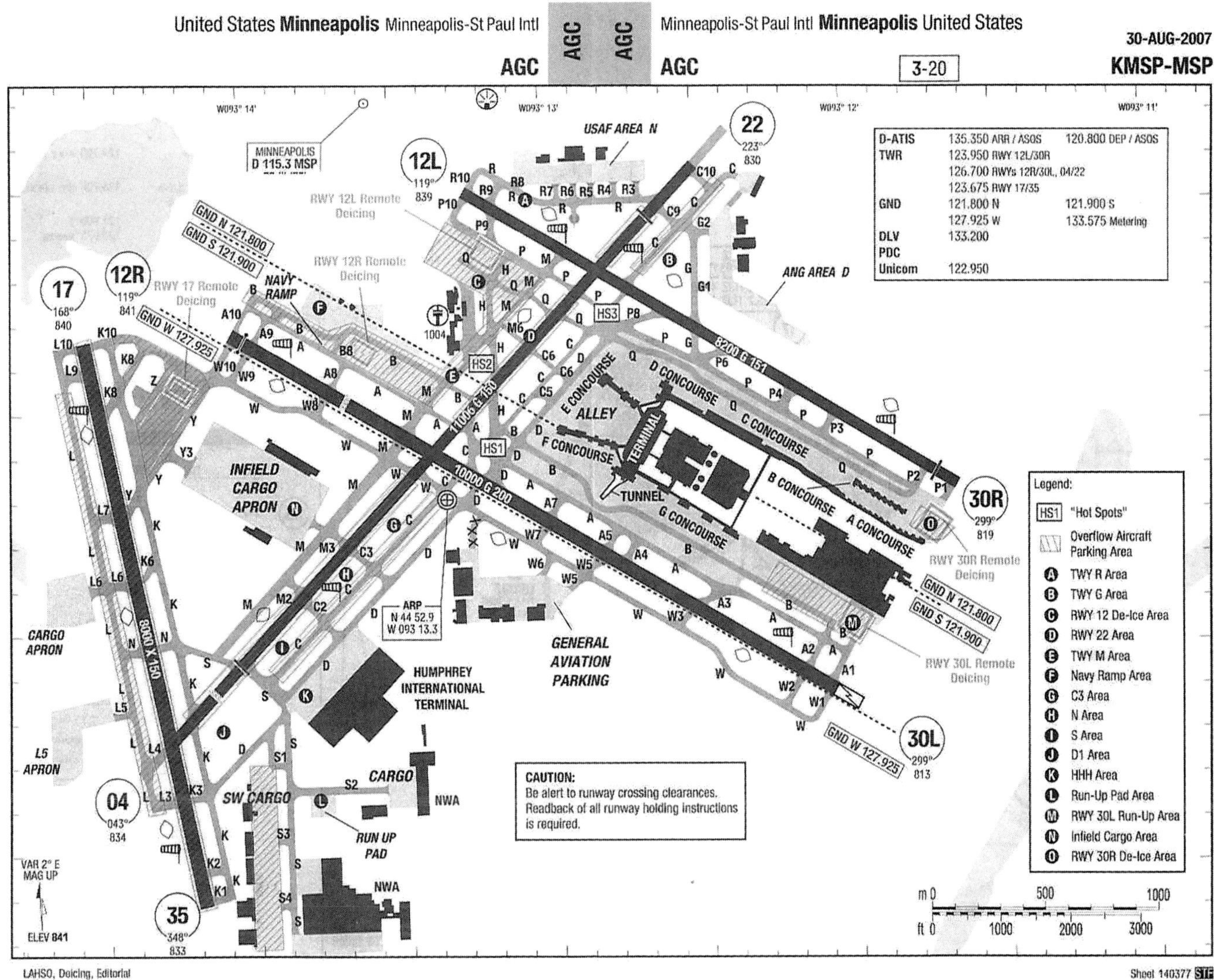

Figure 22.4
Airport layout plan for Minneapolis Airport

problem for stormwater and non-point source pollution management. (See Chapter 18, Stormwater and Flooding.) Making matters worse is that aircraft operations involve many hazardous materials, such as aviation fuel, de-icing fluid, and fire suppression chemicals. These can all make their way into the surface and ground waters unless control measures are taken. (See also the Airports Council International North American sustainability website at www.aci-na.org.)

Security issues. Terrorism, hijackings, and airplane bombings in the 1970s sparked new attention on airport and airline security. The events of September 11, 2001, reinforced the need for security measures. Most security measures

Airport Sustainability Survey

A survey funded by the Transportation Research Board in 2008 found that state, regional, and federal regulations were key drivers for adoption of sustainable best practices. Large and medium airports are focusing on energy reduction, recycling, and carbon emissions. Smaller airports are more concerned with economic viability issues.[1]

Note

1. Fiona Berry, et al., *Airport Sustainability Practices*, Transportation Research Board, Airport Cooperative Research Program Synthesis 10, 2008, www.trb.org.

involve terminal design or operational procedures and are beyond the immediate concern of local officials. Some, however, such as parking restrictions, involve land use issues and affect adjacent land uses. The local law enforcement agencies should be and generally are intimately involved in these security concerns. The Transportation Security Administration is a good source of information for the latest requirements (www.tsa.gov/public).

Airports and Land Use

Many airports that were once in exurban areas are now in suburban or even urban areas due to community growth. Meanwhile, use of the airports has increased dramatically over the past decades and will continue to do so. This has produced encroachment of incompatible land uses onto existing airports, and safety and noise problems. This is a serious issue for both local practitioners and airports and the airline industry. A 2004 survey of state aviation officials about land use issues reported that residential development poses the biggest problem at the local level, especially when developers do not follow noise contours and when disclosure documents are not provided to home buyers. The encroachment of tall structures and bird-attractors also poses problems for airport safety.[11] The officials noted that municipalities seemed reluctant to adopt and enforce zoning ordinances that protect airports, and noted the preference given to land developers, cell towers, and wind farm companies. Smaller areas also seem to feel that the FAA will step in and protect their airport. See Table 22.4 for a listing of incompatible airport land uses.

Table 22.4
Incompatible land use types and issues

Land use type	Potential issue
Residential development	Sound levels Aircraft emergencies
Landfills/stormwater management sites	Bird strikes Aircraft emergencies
Telecommunications or cell towers	Airspace obstruction Obtrusive lighting Electronic interference
Smoke or steam	Sight obstruction for pilot
Recreational use	Aircraft emergencies Obtrusive lighting
Auditorium or outdoor theaters	Sound levels Aircraft emergencies
Power lines	Airspace obstruction Electronic interference
Wetland areas/storm water management sites	Bird and wildlife strikes
Agricultural (crops, orchards)	Bird and wildlife strikes

Source: National Association of State Aviation Officials, 2005.

Safety Issues

Land uses beyond the boundaries of the airport raise safety issues both for those in the airplane and for the occupants of buildings in areas adjacent to the airport in the event of an accident.

Airport safety zones. Over two-thirds of both general aviation and commercial aircraft accidents take place at an airport. Recent analyses have laid the groundwork for more precise spatial designations to ensure safe airport facilities. Accident analyses have helped to generate detailed spatial requirements for safety zones for the airport and its immediate vicinity. Each zone has a relatively uniform risk level and consequently different mitigations and rules. Figure 22.5 shows the sizes of the six safety zones for a long general aviation runway.

Building height. Perhaps the most obvious safety issue is that of the height of buildings and other structures around the airport. Airplanes need clear airspace and smooth controlled paths as they approach or leave the runway. This airspace can be interrupted by tall buildings, antennae, or even trees.

To address this, the FAA has issued guidelines on structure heights around airports. However, land use development is still the responsibility of the local government; the FAA cannot block development even if it affects air safety. If a building that will impact the airspace of the nearby airport is permitted, the FAA will react by changing the procedures regulating the use of the affected runway, such as increasing climb gradients or visibility minimums. These mitigations may impact airline operations and, therefore, airport capacity or efficiency.

Many states have responded to FAA safety guidelines by passing statutes that codify them. Note, though, that this statute only relates to public airports. For private airports, local governments are left to their discretion. For instance, Wisconsin has authorized local governments to pass "Height Limitation Zoning Ordinances" (HLZO) that restrict the use, location,

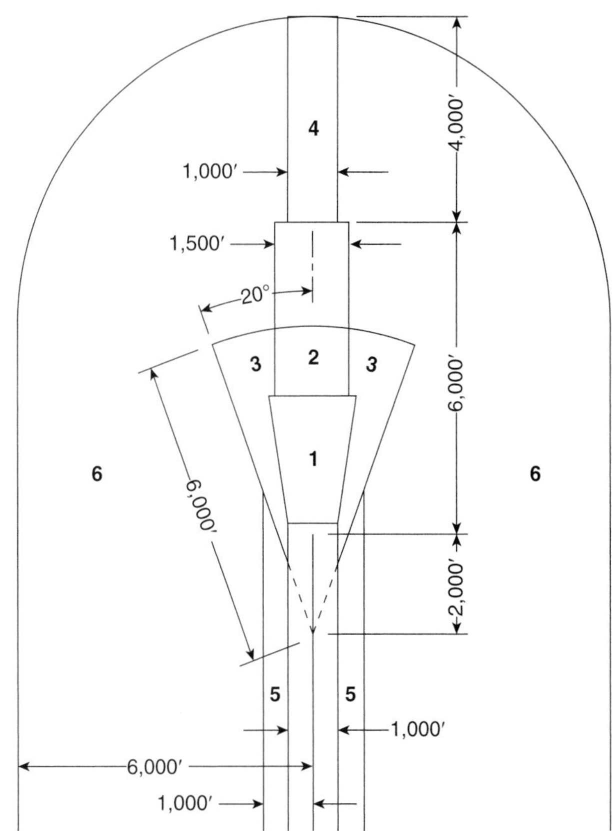

Figure 22.5
Safety compatibility zones

Airport approach lights assists safe landings

height, and size of buildings within 3 miles of the airport. A map is adopted as part of the ordinance.[12]

Bird/aircraft strikes. While they may sound comical, collisions between birds and aircraft are a serious matter, and such collisions have caused numerous air crashes and loss of life. A 10-pound gull striking a plane at 500 miles per hour is enough to crack a cockpit window or destroy a radar antenna. The most common problem is the so-called ingestion of a bird by a jet engine, disabling the engine.[13]

The "Miracle on the Hudson" in 2009 was triggered by the loss of power of US Airways Flight 1549 caused by a flock of Canadian Geese striking the aircraft during climb out. The flight crew successfully landed the plane in the Hudson river with no loss of life.

Unfortunately, airports are an ideal habitat for many birds. There are large amounts of open grassland that is free from most predators. There are often wetlands or areas of standing water between runways. As a sound attenuation measure, some airports plant huge banks of trees, which then serve as homes for birds. Many species of birds like to rest on the runway itself. Crows are known to drop their prey onto runways in order to kill it. Avoiding this safety problem involves manipulating the habitat in such a way to make the airport undesirable for birds. Some airports use decoys and scarecrow-like devices to keep the birds away. Site planning can also be sensitive to this issue so that an airport is not built on any migratory paths. The grass areas can also be kept closely mown or planted with vegetation that birds don't like. Sometimes, however, it is not possible to completely eliminate the "bird problem."

Noise

The Environmental Protection Agency has long recognized that noise is an environmental pollutant, and it has the power to identify levels of noise protective of public health and welfare.[14] Moreover, the U.S. Supreme Court has ruled that if aircraft noise and vibration significantly affect a property owner, then an aircraft overflight constitutes a "taking" under the Fifth Amendment. See *Griggs V. Allegheny County, 1962*. Some states have also held that flights from airports violate nuisance law. In these cases, the financial liability lies with the airport owner.

The FAA has responsibility for regulating aircraft noise, but it has not taken on the task of directly setting the noise level at a given airport. This is considered the responsibility of the local airport operator. The FAA has, however, established policies regarding how airport noise studies should be conducted to be eligible for federal funding, and how the noise portion of any environmental document, prepared for a proposed federally funded aviation related project, should be done.

Aircraft noise is produced in the engines as the fuel ignites and the turbine blades strike the surrounding air. Aerodynamic noise is produced as the airplane moves through the air. In 1969, Congress gave the FAA the responsibility to regulate aircraft design to reduce noise, and in that same year the FAA put forth regulations requiring noise abatement technology on aircraft engines. The FAA established maximum sound levels that could be produced by new aircraft under specified test conditions. To date, three maximum permitted levels have been promulgated, and aircraft that comply with the three levels are called Stage 2, Stage 3, or Stage 4 aircraft, Stage 4 being the quietest. Stage 1 aircraft are those manufactured prior to establishment of the first maximum level. Stage 1 aircraft were phased out of service by law in the U.S. in 1988. Stage 2 aircraft were phased out at the end of 1999. Stage 3 and Stage 4 aircraft are the current standard.

Solutions to the noise problem. Whichever noise level is decided upon, it is up to the airport operator and to local land use officials to enforce it. There are many tools available to control noise. The simplest one is to isolate the airport from any other land uses. This is untenable, however, since airports serve their purpose best if they are reasonably close to activity centers. Zoning or other local land use controls, soundproofing of homes and schools in the flight path, altering flight paths, imposing flight curfews or otherwise curtailing flight operations, and mandating quieter engines are all possible tools for controlling the conflicts between land use and airport noise. In the U.S., however, any proposed limits on use of an airport by any aircraft are subject to extensive study and scrutiny. Sixty-five DNL is the EPA standard used, but enforcement is difficult.

Acquisition and avigation easements. If local governments permit development that encroaches into the airspace of an airport, the airport still has options. The first is the outright *acquisition* of the land and buildings surrounding

Noise Metrics

Noise can be quantified in several ways but airport noise is generally specified as total accumulation or "equivalent levels" over a period of either one hour, symbolized as LAeq,H, or of 24 hours, called "day–night average sound level," and usually denoted as DNL. DNL includes a weighting or penalty of 10 decibels (dB) for sound occurring between 10 p.m. and 7 a.m.

Several organizations have identified levels of cumulative sound exposure below which there should be little risk of adverse effects on human health, and little annoyance. The U.S. EPA identified a value of 55 dB DNL as such a level, while 65 DNL is the standard used by FAA. Figure 22.6 shows typical values of DNL for various locations.

Written by Nicholas Miller, Senior Vice President, Harris, Miller, Miller & Hanson, Inc. www.hmmh.com.

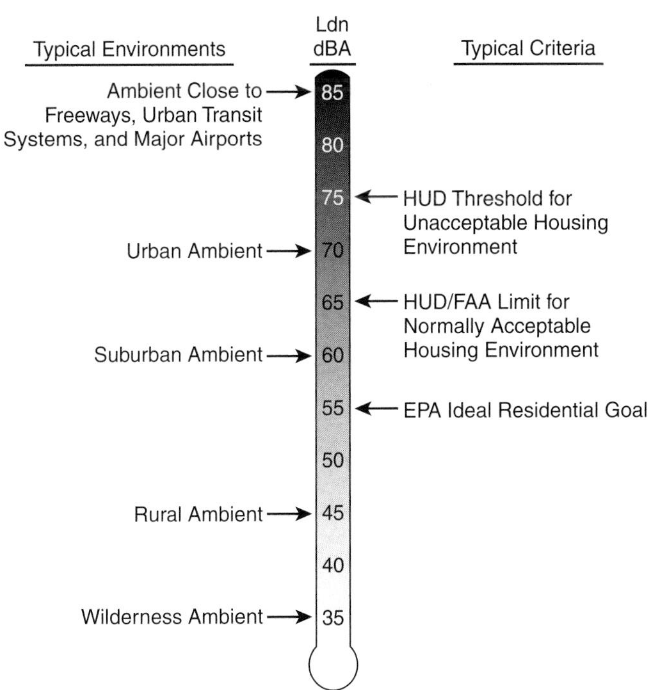

Figure 22.6
Typical DNL (day–night average sound level) values

Selected Model Comprehensive Plan Policies for Airports from State of Washington

The State of Washington's Department of Aviation provides a model list of policies for land use around airports to "Protect the viability of the airport as a significant economic resource to the community by encouraging compatible land uses, densities, and reducing hazards that may endanger the lives and property of the public and aviation users." These policies include:

- Protect the airport from adjacent incompatible land uses and/or activities through standardized land use control measures.
- Encourage open space/clear areas and utilize zoning criteria within key safety areas adjacent to the airport.
- Require a notice to title/disclosure statement for new or substantial redevelopment of lots, buildings, structures, and activities within the airport influence area.
- Encourage the adoption of height restriction regulations that protect the airport from height hazards.
- Identify, preserve, and enhance inter- and multi-jurisdictional planning, goals, policies, and development regulations.
- Encourage economic development opportunities and aviation-related uses adjacent to airports and promote the efficient mobility of goods and services region-wide consistent with the economic development element and the regional transportation strategy.

Adapted from Washington State Department of Aviation, Model Comprehensive Plan Policies.

the airport. This strategy is being used more and more by the Department of Defense around military airfields.

The second is known as the *avigation easement.* An avigation easement is the right, granted by the owner of land adjacent to an airport, to the use of the air space above a specific height for the flight of aircraft. This typically prohibits the property owner from using the land for structures, trees, signs, or stacks higher than a specified altitude. Airports can purchase avigation easements, receive them as gifts, or acquire them through condemnation.

Zoning overlay district. Another tool that can be used to regulate development so that it is not negatively impacted by noise issues is a zoning overlay district. See Figure 22.7 for an example of what such a zoning map might look like. This is developed by using noise contours.

Local Economic Development, Marketing, and Financing

Airports are significant tools for local economic development, and therefore their vitality is inextricably linked to local politics. They are large employment centers. They attract visitors. They require the support and service of other industries. They generate sales and property tax. The City of Houston estimated in 2010 that its system of three airports has a $25.7 billion positive impact on the economy annually.[15] The positive economic impacts are by no means confined to large commercial airports. A study of Las Cruces International Airport, a small, primarily general aviation airport in Las Cruces, New Mexico, estimated that the total value added associated with the airport was $43.1 million in 2008, and it added 567 jobs to the Las Cruces area, a non-trivial sum in a community of 97,618 people.[16]

Airports generate both direct and indirect economic benefits. Direct benefits include the cost of locally purchased goods and services, payroll and benefits, building rental, property tax, gross receipts tax on revenues, landing fees, and aircraft parking fees. Other direct impacts include ground transportation companies that serve the airport and the retail activities within the airport. Indirect economic benefits result from the services provided by travel agencies, hotels, restaurants, entertainment, retail establishments, and recreation areas.

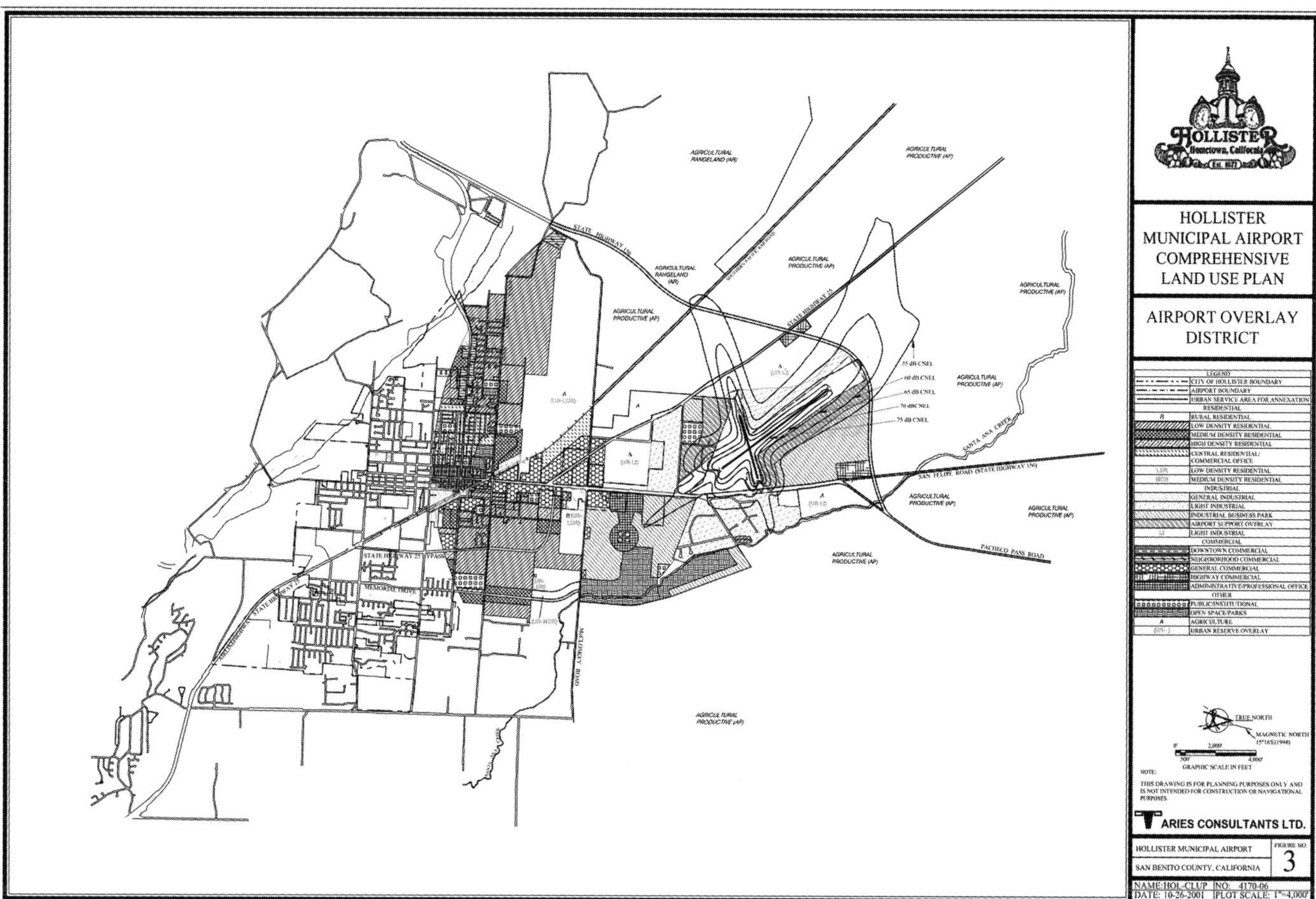

Figure 22.7
Map of airport zoning overlay district

The local general purpose government has a large stake in developing, promoting, and utilizing the economic development potential of an airport. A well-managed airport will always be more attractive to travelers, businesses, and service providers. Many of the benefits, both direct and indirect, derive from off-airport sites, so the land uses and transportation connections to the airport should be carefully managed. In addition, many of the economic benefits require the presence of quality commercial air service. Local governments can attract, recruit, and support commercial air service providers.

Marketing an Airport

Communities have long been in the business of offering large financial incentives to firms to induce them to relocate or expand in the area, a practice known as "chasing smokestacks." Airlines have not been left out, and may even be the subject of more aggressive efforts. The interplay between airlines and airports is complex. Sometimes, expanding the airport and attracting a new airline to the community are linked in a chicken-and-egg situation.

Denver International Airport Opens in 1995

Efforts to expand Denver's existing airport began in the 1960s. However, local environmentalists and neighborhood activists were joined by airlines that feared higher landing and terminal fees and more competitors to block the expansion for many years. In the 1990s, increased congestion began to concern the airlines, and the logjam was broken. The compromise with the community was a new airport—Denver International—located on a flat and uninhabited site 30 miles from downtown Denver. When it opened in 1995, it became the first passenger airport built in more than two decades serving a major metropolitan area in the United States. Experts do not expect other metropolitan areas to follow suit, since most do not have large parcels of uninhabited land so close to the population center.[1]

Note

1. Adapted from A. Altshuler, and D. Luberoff, *Mega-Projects: The Changing Politics of Urban Public Investment* (Cambridge, MA: Brookings Institution Press and Washington, DC: Lincoln Institute of Land Policy, 2003), pp. 123–124.

Denver International Airport, which opened in 1995 after a 30-year process (see sidebar), shared off-airport real estate development benefits with local jurisdictions. In order to obtain a new runway for Hartsfield Airport, Atlanta agreed to pay the City of College Park $82 million over 10 years to pay for the loss of property taxes from land that it needed in that jurisdiction. Cleveland's runway expansion project which required land in adjacent Brook Park resulted in a land swap with Cleveland and 50 percent of all property tax revenue collected from an airport industrial park inside Cleveland's boundaries, among other items.[17]

More often than not, however, a community offers incentives to an airline to start or expand service in the area. Small- and medium-sized cities seem to face greater, and more cut-throat, competition. Airports are known to give money upfront to an airline to guarantee the airline will make a profit in its first year of operation. The Gainesville-Alachua County (Florida) Airport Authority used a U.S. DOT grant award of $650,000 along with a local match to entice Continental to the area. Bloomington, Illinois, offered an airline $1.5 million to initiate service at the Central Illinois Regional Airport.

Conclusion

In today's increasingly connected world, distances are shrinking and borders are opening up. For most communities, an airport is the door onto this new world. It is a powerful tool for economic development. On the other hand, an airport located in the middle of a field will not by itself generate economic activity, nor will expansions that are not sensitive to land use impacts and the regional context.

For these reasons, it is important that local general purpose government include airports in any community planning effort. The complexity, expense, and scope of airports and airport planning can be intimidating and daunting. But local policy-makers already have the basic expertise at hand. Planners can include the airport in the comprehensive plan or create a separate document. Either way, they should apply the same principles and desires to it as to any other aspect of the community. Planners should also encourage low-emission and energy-efficient facilities both when considering a new structure and when re-negotiating a contract with the airlines.

Additional Resources

Publications

Butler, Stewart E. and Lawrence J. Kiernan. *Estimating the Regional Economic Significance of Airports*. Washington, DC: FAA, 1992.

Caves, Robert E. and Geoffrey D. Gosling. *Strategic Airport Planning*. Oxford: Pergamon, 1999.

Dempsey, Paul Stephen. *Airport Planning and Development Handbook: A Global Survey*. New York: McGraw-Hill, 2000.

De Neufville, Richard and Amedeo Odoni. *Airport Systems: Planning, Design and Management*. New York: McGraw-Hill, 2003.

Office of Economic Adjustment. *The Joint Land Use Study Program Guidance Manual*. Washington, DC: Department of Defense, 2006.

Schalk, Susan M. and Stephanie A. D. Ward. *Planners and Planes: Airports and Land-Use Compatibility*. Planning Advisory Service Report Number 562. Chicago, IL: American Planning Association, 2010.

Wells, Alexander T. and Seth Young. *Airport Planning and Management*, 3rd edn, New York: McGraw-Hill, 2004.

Websites

American Association of Airport Executives: www.airport.net. This is the professional organization for airport managers.

Assistant Secretary of Aviation and International Affairs: ostpxweb.dot.gov/aviation/rural/ruralair.htm. This contains information on Essential Air Service.

Federal Aviation Administration Airport Office: www2.faa.gov/arp/index.cfm. This has information on airport funding, airport planning and design, and land use issues.

NASA: virtualskies.arc.nasa.gov/design/tutorial/tutorial1.html. This has a good introductory tutorial on airport design online.

Transportation Security Administration: www.tsa.gov/public.

Key Federal Legislation

As noted earlier, local airport operations have been heavily influenced by federal airport legislation.

Contract Air Mail Act of 1925

This Act allowed the Post Office to contract out the transport of air mail. This energized the development of the private sector. Many of today's commercial airlines started out as airmail companies.

Air Commerce Act of 1926

This Act was passed to promote the stability and modernization of the nascent commercial aviation sector. It charged the Department of Commerce to encourage air travel by establishing civil airways and navigational facilities

Civil Aeronautics Act of 1938

This Act created the Civil Aeronautics Authority and the Air Safety Board. It was the first federal law that attempted to establish a single, coherent federal policy toward civil aviation. These new agencies were independent of the Department of Commerce. This law also allowed the expenditure of funds to build airfields for national defense.

Federal Airport Act of 1946

This Act was passed to give the United States a comprehensive system of airports. It appropriated funds, part to be spent by the Civil Aeronautics Authority and part to be apportioned to the states based on population and land area. The Act was meant to benefit smaller communities who could not afford to build airports on their own.

Airways Modernization Act of 1957

Passed after a series of high-profile air accidents, this Act was passed to develop and modernize the national air navigation and air traffic control system.

Federal Aviation Act of 1958

This Act repealed the Air Commerce Act, the Civil Aeronautics Act, and the Airways Modernization Act, and replaced them with a more cogent set of regulations. It also created the Federal Aviation Agency (later to become the Federal Aviation Administration).

Airport and Airway Development Act of 1970

Tremendous growth in air travel in the 1960s spurred the passage of this Act, which created the Airport and Airway Trust Fund, a special pot of money filled by aviation user fees and spent only on airport expansion and modernization.

Airport and Airway Development Act Amendments of 1976

These amendments reauthorized the Airport and Airway Trust Fund, and sharply increased the expenditure limits. It also increased the number and type of projects available for trust fund monies.

Air Cargo Deregulation Act of 1976 and Airline Deregulation Act of 1978

These two Acts deregulated the commercial aviation industry.

Aviation Safety and Noise Abatement Act of 1979

This Act was passed with the intention of helping airport operators prepare and implement noise abatement programs. It created a section within the FAA that dealt with airport noise, and authorized a noise abatement grant program.Airport approach lights assists safe landings

chapter 23

Ports and Waterfronts

Peter H. Brown and Peter V. Hall

In this chapter ...

Importance to Local Practitioners 481
A Short History of Ports in the United States 481
Description of Ports and Waterfronts 483
Cargo Ports 483
Ports as Destinations 485
Institutional Context of Ports 488
Purposes, Duties, and Powers of Port Agencies 488
Types of Port Agencies 489
Port Finances: Revenues, Expenditures, and Debt 492
Cargo Port Finances 492
Diversified Waterfront Finances 494
Port Planning 496
Types of Port Plans 497
Plans Required by Superior Governments 498
Planning Function within the Port 499
Capital Planning of Cargo Ports 499
Planning on the Diversified Waterfront 501
Conclusion 503
Additional Resources 504
Publications 504
Website 504

Why Is This Important to Local Practitioners?

Ports and waterfronts are zones of great opportunity and diversity, but they also present practitioners with complex and changing infrastructure challenges. Cargo ports play a vital role in today's economy, yet they are also sites of congestion and pollution. While some ports have captured disproportionate shares of containerized trade and others have specialized in niche cargoes, many have been largely deserted by cargo. Most other contemporary ports are trying to manage uses that span from traditional cargo handling to a range of non-maritime commercial, entertainment, and residential activities. Indeed, for former industrial and trading cities, waterfront tourism and other real estate developments are often among the few significant opportunities for revitalizing the rundown urban core although infrastructure planning in these ports is fraught with conflict between competing users and interest groups.

This chapter begins with a short history of ports in the United States, concentrating on the changes associated with containerization. Having described the evolution and current situation of ports, the chapter then turns to the topics of port governance, finance, and planning.

A Short History of Ports in the United States

In the half-century since Malcolm MacLean's converted oil-tanker the *Ideal X* first carried containers between Newark and Houston in 1956, the nature of marine cargo transportation has been fundamentally transformed. How did containerization unleash these changes? What has happened to those ports that have not successfully attracted container cargo? And what do local practitioners need to know about these processes to plan and manage coastal infrastructure successfully?

Cities have grown up around safe harbors since ancient times and until the first half of the twentieth century every major coastal, lake, and river city had some sort of port facility that was central to the trading life of the place. This relationship between city and port was fundamentally disrupted by containerization. Today's specialized vessels are far more selective about which ports they visit and which they bypass. In the past, most cargo ports had a close relationship to their immediate hinterlands—they imported the region's supplies and exported its produce. Large numbers of workers found physically demanding work handling cargo, while the port city had a distinctive trading economy of warehouses and manufacturing plants. Today's container ports offer relatively few jobs, while increased transportation efficiency has allowed many cargo-dependent industries to migrate far inland. The result is the death of the old-style, distinctive port-city economy.

The container eliminates the need for on-dock warehouses. Before there were containers, cargo in bags, nets, boxes, and on pallets would be moved from ship to storage warehouse as quickly as possible. Hence ports were constructed as long, narrow finger-piers with warehouses extending along their full length. In contrast, a modern container-port requires a wide, square terminal area. To get the most return out of costly container ships and cranes, tractor-trailers must be able to efficiently move containers between the dockside, storage stacks, and points of access to land transportation.

Containerization was a continuation of trends in port development that started in the first half of the twentieth century. For example, industries of the automobile age, such as crude oil extraction and refining, resulted in the need for larger sites and access to deeper water. The development of container technology, which also required large sites, further reinforced these trends in the shape, scale, and location of port facilities. The scale of land needed for marine terminals has increased dramatically as well—from finger-piers that were yards long and feet wide to the most recent U.S. mega-terminal, the square 315-acre Pier 400 at the Port of Los Angeles.

Port developers world-wide found that they could not meet these land needs within the old urban core. As a result, today the greatest volume of shipping traffic is handled at points far from the location of the original waterfront. Other forces also contributed to this trend, especially the public's sensitivity to environmental issues and greater concern for the appropriateness of industry on the waterfront and close to inhabited urban areas. These forces have culminated in the near-complete separation of the cargo port and city.

And what of the old waterfront that was left behind? As ports moved to locations outside population centers, and as big manufacturing and industrial businesses moved to the suburbs, they left behind abandoned strips of land that separated cities from their waterfronts.[1] Successive rounds of navy base closures also added to the stock of vacant, developable urban waterfront land.

These lands were covered with decayed, obsolete port and industrial facilities, and they lacked the same density of infrastructure systems found elsewhere in the city, including power, water, sewer, and streets. However, they were a source of potentially valuable developable land within the city's urban core.

The redevelopment of the waterfront has been ongoing in cities across the United States since the 1970s. In many cities, there is a great deal of competition for those waterfront sites that are most attractive and advantageous for non-maritime uses including housing, offices, and restaurants, as well as for recreational uses such as parks and marinas. Continued concern with environmental issues results in the creation of pollution controls that seek to harmonize the old sites with the new uses.

Taken together, these trends have created both challenges and opportunities for port authorities and infrastructure planners. Increased market demand means that cargo ports are now confronted with the difficulties of land acquisition, planning, and development of new port areas and facilities on a much larger scale than ever before. At the same time the abandonment of large land areas has brought both challenges and unprecedented opportunity for many ports—and port cities—as they seek to re-envision their urban waterfronts.

Oakland's Container Port, looking west towards San Francisco.

Description of Ports and Waterfronts

To help practitioners understand the challenges facing infrastructure planning and financing in and around ports today, in this section we describe the various kinds of ports and the economic functions they perform. Ports today act as nodes in international distribution networks, as desirable and attractive sites for recreation and tourism, and as opportunities for new investment.

Cargo Ports

Today's cargo ports are a diverse group; a relatively small number of container ports compete intensely for this valuable cargo, while others have specialized in niche cargoes.

Container ports. In 2009, only 16 U.S. ports could be described as major container ports, each handling more than 500,000 containers that year. Table 23.1 ranks U.S. ports according to the number of containers handled in 2009. All of these ports are located within or near major metropolitan areas, although the cargo itself may have an origin or destination far inland. Indeed, during the 1970s many observers were predicting that the container trade would become concentrated in a very small number of load-center or hub ports, with perhaps as few as one dominant port each on the west and east/Gulf coastlines. It was predicted that these hub ports would capture most inter-continental trade, distributing it via a series of feeder ports in a hub-and-spoke system, much like that used by major passenger airlines.

Table 23.1
U.S. port rankings, 2009

Rank	By containers handled	By total tonnage of cargo handled
1	Los Angeles, CA	South Louisiana, LA
2	Long Beach, CA	Houston, TX
3	New York/New Jersey, NY-NJ	New York/New Jersey, NY-NJ
4	Savannah, GA	Long Beach, CA
5	Oakland, CA	Corpus Christi, TX
6	Houston, TX	New Orleans, LA
7	Hampton Roads, VA	Beaumont, TX
8	San Juan, PR	Huntington Tristate, OH-KY-WV
9	Seattle, WA	Los Angeles, CA
10	Tacoma, WA	Hampton Roads, VA

Source: Authors.

While for the most part these predictions have not come to pass in North America, the trade in containers today is nevertheless highly concentrated. The three largest container ports in the United States today (the ports of Los Angeles, Long Beach, and New York City) account for about half of all the containers imported and exported, and the top 10 container ports account for over three-quarters of all containers handled. Further concentration seems unlikely in the foreseeable future—increased congestion at some major ports, competition between ports, the shortage of land in the major ports, the Jones Act of 1920 that restricts the shipment of goods between U.S. ports to U.S. carriers, and the desire of the largest steamship lines not to place all their eggs in one basket should keep secondary container ports such as Oakland, Seattle, Norfolk, and Charleston in the rankings for the foreseeable future.

Niche cargo ports. Not all cargo fits easily or efficiently into containers. In fact, some of the largest ports today, as measured by total volume of throughput, handle almost no containerized cargo. These cargoes encompass a range of products, typically classified into the following categories that relate to the mode of transportation: (1) dry bulk (e.g., grains, coal, ores); (2) liquid bulk (e.g., crude oil, chemicals); (3) roll-on roll-off or ro-ro cargo (e.g., cars, trucks and other vehicles); (4) break-bulk (e.g., palletized, boxed, or bagged products); (5) neo-bulk (e.g., steel, granite, paper) cargo; and (6) contract cargos including construction materials, equipment, and assemblies, such as wind turbines.

Ports specializing in dry bulk, liquid bulk, and neo-bulk cargos are often located close to the source of a raw material export such as a mine, oilfield, forest, or agricultural region, or a facility to process an imported raw material. For example, the grain- and ore-handling facilities of the Port of South Louisiana which stretches for 54 miles on the Mississippi River made this the eighth largest port (by tonnage) in the United States.

Another consequence of containerization has been the displacement of lower value and non-containerized cargos from the major cargo ports. Many of these cargoes, which include automobiles and fruit, still need to be shipped in bulk, and so they have found a home in ro-ro and break-bulk ports in secondary or peripheral metropolitan areas such as San Diego and Port Hueneme near the Los Angeles metropolitan area, and Wilmington, Delaware, in the heart of the mid-Atlantic population centers.

It should be noted that many container ports also handle a range of non-containerized commodities. For example, in addition to handling almost 4 million containers per year, the port of New York is also a major import location for products ranging from oil to automobiles to cocoa. Likewise, while the port of Baltimore has slipped in the container rankings since 1984, it does command a significant share of the East Coast trade in automobiles and steel.[2]

Rotting finger piers at the urban waterfront in San Francisco, with the port of Oaklands container cranes visible in the distance

Ports as Destinations

As indicated above, the container revolution led to a major and fundamental restructuring of the shipping industry. Many of the country's ports, however, missed this revolution as a result of poor location, a lack of land necessary for expansion, poor infrastructure connections, poor capital investment decisions, or some combination of these factors. Those ports that experienced declines in cargo and related declines in revenues quickly slipped from a period of prosperity into one of deferred maintenance, managed decline, and ultimately decay and abandonment of facilities and waterfront property.

While the container revolution was underway, an enormous shift in the public's perception of the waterfront was also occurring. Until only recently, the port was seen by many as a dirty and dangerous industrial wasteland but this view changed as citizens and developers began to see the waterfront—with its broad vistas and natural beauty—as an increasingly desirable place for housing, offices, recreation, and tourism. By making the most of these two major trends—the port's retreat from the urban core and the new

The Port of San Francisco and the Impacts of Globalization, Technological Innovation, and Geographic Location on Port Fortunes

In 1863, state legislators created the Port of California at San Francisco. The port served as the thriving commercial gateway that fueled the development of northern California for over a century. Then, during the 1950s, a powerful United States Senator steered major federal grants towards Oakland, which used the funds to build a modern new container port on its mud flats. Oakland's new facility opened in 1962, and by 1965 its cargo tonnage equaled San Francisco's. By 1986, Oakland was handling over 90 percent of all containerized cargos on the bay while San Francisco's cranes were operating at only 5 percent capacity.

The Port of San Francisco was located on a congested peninsula, had no large available land areas, and had limited rail competition through a tunnel that was shared with commuter trains. Further, inland-bound cargos shipped by rail from San Francisco pass through Oakland anyhow, so there was little reason not to ship these cargos directly to and from Oakland. Finally, in choosing not to invest in containerization when Oakland did, the Port of San Francisco underestimated the major impact of this revolutionary new technology.

In 1969, the Port of San Francisco gambled on a new technology called "lighter-aboard-ship" (LASH) that was an alternative to containerization, financing a costly new facility on the southern waterfront. However, this technology never proved to be a viable competitor to containers, and the LASH facility went largely unused. In 1971, the Port finally invested in new container facilities but it was too late—Oakland had supplanted San Francisco as the primary port in Northern California. San Francisco's geographic location, lack of land, infrastructure disadvantages, and poor investment decisions together ensured the city's future obsolescence as a major cargo port.

value of waterfront land—politicians sought to transform the waterfront from one type of economic engine into another, leading to the rise of the diversified waterfront.[3]

Non-traditional maritime businesses. Most ports were originally created to centralize and rationalize local maritime operations and to increase seaborne commerce and related employment in a region but in the wake of containerization, many have sought to expand the meaning of the word "maritime." Today's ports engage in a much broader range of water-based activities, including recreational boating, sport fishing, marinas, yacht repair and "mega-yacht" services, heritage ships, harbor tours, dinner cruises, and ferry services.

Perhaps the most important new non-traditional maritime trend has been the rise of the cruise industry. According to the Cruise Line International Association, between 1980 and 2010, the number of cruise passengers increased at an average annual rate of 7.4 percent. The numbers of cruise ships has grown dramatically as well—118 new ships from 2007 through 2010—and like cargo ships, passenger ships have also grown in size as a result of improved designs and shipbuilding technologies.[4] The cruise business is attractive to ports because it offers one of the only ways to retain high-paying union longshoreman jobs while also adding a host of other lower-paying service jobs ranging from food service and catering workers to florists and piano tuners. And the cruise business—with its shiny white ships—is perceived to be visually and environmentally "clean" relative to other waterfront industries although environmentalists argue that cruise ships are major polluters as their emissions go unregulated.

As with cargo, geography limits the ability of many ports to enter the cruise business. Ports located in the middle of the two coasts cannot offer short, round-trip itineraries to foreign locales like Mexico, Canada, and the Caribbean because sailing times are too long. These ports are further constrained by the U.S. Passenger Services Act of 1886, a federal law that prohibits foreign-flagged passenger vessels from calling at two U.S. ports consecutively. For example, it is illegal for these vessels to offer a cruise itinerary that calls at the port of New York and then Boston, or similarly San Francisco followed by Seattle. Ports that do benefit from good locations require modern terminal facilities, good channels and marine infrastructure, and local airports with large airlift capacities.

Finally, climate influences a port's potential as a homeport, a seasonal homeport, or merely a port-of-call. For these reasons, southern ports, and particularly those on the South Atlantic and Gulf coasts that can offer a variety of Caribbean and Latin American itineraries, are best positioned to transform cruising into a major, year-round business and a dependable revenue producer.

Real estate and other new, non-maritime port businesses. Many ports also entered the real estate development business, seeking to re-deploy underutilized maritime property for non-maritime economic development projects. Some port agencies took a more opportunistic approach accepting whatever projects developers proposed while others created comprehensive plans for their waterfronts that incorporated a range of projects including commercial office buildings, housing, hotels, convention centers, parking garages, restaurants, aquariums, museums, retail and entertainment centers, and arenas and stadiums for major league sports teams.

Finally, some state legislatures have required their ports to address other new duties, including water and air quality, environmental preservation, habitat restoration, public access and parks, public art, and regional economic development generally. Cities and counties have also increased the pressure on ports to partially finance or subsidize projects such as waterfront light-rail lines, aquariums, and convention centers. As a result of the port's entry into these many and varied lines of business, the waterfront has become a virtual necklace of attractions connected by river walks, ferries, and light rail lines. And the port authority has become a diversified enterprise that bears little resemblance to its origins as a single-purpose maritime agency (Table 23.2).

Baseball fans in San Francisco can take the ferry or trolley to the waterfront and then promenade along the Embarcadero past parks and restaurants before taking in a ball game at AT&T Park. Visitors to San Diego's waterfront Convention Center can walk to lunch at seaport village or eat dinner in the historic Gaslamp Quarter. In Tampa, cruisers can spend time before and after their cruise visiting the waterfront aquarium and an adjacent retail and entertainment complex, taking the trolley into historic Ybor City for dinner and nightlife, or driving to Disney World, just an hour away.

At the same time, ongoing cargo, ship repair, and commercial fishing operations near or adjacent to these new attractions lend an element of gritty maritime authenticity to what might otherwise be seen as carefully constructed but sterile tourist environments. Finally, as a result of the diversification of businesses on the waterfront, many abandoned zones left behind by the retreat of the port have been transformed into vibrant new places to live, work, and play. Perhaps more important, the urban fabric of many cities, which often ended inland, at the edge of the port district, has been repaired and extended to the water's edge.

Table 23.2
Port activities on the diversified waterfront

Activity	Examples
Traditional maritime	Cargo (bulk, break-bulk, niche, container, etc.); ship building/ship repair; ship services; commercial fishing
Traditional non-maritime	Toll crossings; airports; commuter rail lines; other transportation services; related public safety functions
Non-traditional maritime	Cruise ships; harbor tours and dinner cruises; marinas and recreational boating; yacht and mega-yacht repair and services; water quality; habitat restoration
New non-maritime	Commercial real estate development (new and adaptive re-use) including residential, office, retail, and mixed-use projects; tourism attractions including convention centers, ballparks, arenas, stadiums, hotels, aquariums, museums, entertainment complexes, heritage ships, and waterfront light-rail lines; public access including parks and public art

Source: Authors.

In summary, today's diversified waterfronts play host to a complex mix of activities (summarized in Table 23.2). They include traditional maritime activities related to cargo-handling and fishing; traditional non-maritime activities related to public transportation; non-traditional maritime recreation and tourism-related activities; and new non-maritime activities that encompass the full range of commercial and residential real estate development.

Institutional Context of Ports

Purposes, Duties, and Powers of Port Agencies

Enabling legislation creates and empowers a port agency and describes the port's purpose, duties, powers, governance structure, and jurisdiction. When many legislatures around the country created new ports in the first half of the twentieth century, they sought to build powerful regional economic development engines based on maritime commerce. Therefore, the basic duties of most ports still include the planning and development of maritime facilities for the purpose of increasing sea-borne commerce and trade and the number of maritime jobs in the region. The port's enabling act grants the agency the specific powers it needs to carry out these duties, including the powers to incorporate, hire staff, enter into contracts, incur debt, purchase and dispose of property, and develop, own, and operate cargo and other related facilities. The act empowers the port to collect rents, fees, and other charges from shipping companies using port facilities and to use these revenues to finance operations. The act may grant the port the power to sell revenue bonds, and in some cases the act also grants the port the power to levy taxes within the port district. The act describes how the agency shall be governed and administered, including the size and composition of the board or commission and procedures for hiring the executive director and staff.

The act also describes the physical boundaries of the "port district," or the jurisdiction within which the port is granted the power to build and operate facilities and—in some cases—the power to tax. The district often includes submerged lands, and some ports have the power to develop these lands and to permit and regulate maritime and other activities on the surface of the water above them. In some cases a port's jurisdiction is roughly coterminous with the property it owns or controls. More often, however, a port owns or directly controls a relatively small amount of land within a much larger jurisdiction, although it has the power to acquire new lands and facilities anywhere within this jurisdiction. A port may also have the power to levy taxes in a larger jurisdiction. For example, the Tampa Port Authority only owns approximately 1,400 acres of waterfront property, although it levies an ad valorem property tax within a much larger port district that is roughly coterminous with the surrounding 1,051 square-mile Hillsborough County.[5] As in the case of Tampa, port districts often overlap or are coterminous with the jurisdictions of local general purpose governments (GPG) and other units of government.

At the time of their creation, most ports were confined by their enabling legislation to traditional maritime operations. Legislators have constantly modified their expectations of ports over the decades, however, amending enabling acts regularly and generally expanding the duties, powers, and jurisdictions of ports, sometimes quite dramatically. For example, in the wake of declining cargo volumes, legislators granted many ports the additional duty of engaging in economic development activities that are very broadly defined. These may range from non-traditional maritime facilities such as cruise ship terminals to non-maritime tourism facilities including aquariums, convention centers, and retail and entertainment complexes.[6]

Types of Port Agencies

One oft-cited complaint about port agencies is that they ignore the interests of the citizens of the cities that host them, and are instead focused on what appear to be singular and business-oriented goals. To understand the source of this concern, we need to delve into the question of autonomy and political insulation.

Most ports are structured as "public authorities," as "special district governments," or as line or enterprise departments of state or local GPG. Each of these different structural types offers a port a unique set of powers and opportunities, while at the same time presenting a similarly unique set of constraints and challenges for local practitioners. Most importantly for local practitioners, each type of port structure offers a different balance between autonomy and openness. Autonomy provides the insulation from local political pressures that is needed to execute major infrastructure projects. On the other hand, a degree of representation and involvement in the port's decision-making process has become increasingly important to the public and other stakeholders.

Ports structured as public authorities. Many of the country's oldest port agencies, particularly those created during the first half of the twentieth century are "public authorities." A product of the Progressive Era, the public authority was conceived of as a new form of government that would combine private sector business practices with a public purpose to ensure the efficient and business-like operation of a revenue-generating public enterprise. The public authority derives its unique form of power from two important and interrelated characteristics. First, because appointed boards govern public authorities for fixed terms, they are somewhat insulated from the direct influence of politicians who operate on electoral cycles and must respond to short-term political pressures. Second, because public authorities typically finance infrastructure with revenue bonds backed by a future stream of rents, tolls, fees, and other charges, they are financially self-sustaining—operating like a business and balancing revenues with expenditures. These two characteristics make the authority a powerful form of government for the development of very large and costly infrastructure projects that may take years or even decades to plan and complete, and that may be politically unpopular in the short run.[7]

In practice, many port agencies fall short of the ideal of political and financial autonomy. Port commissions are indeed subject to political influence, grants and taxes have become a larger share of port authority revenues, and some ports have built projects that clearly have not been in the best interests of society. Perhaps most importantly, however, the authority model is often criticized as being non-representative and therefore non-responsive to political pressures and the demands of the public. So while highly autonomous port authorities are effective as providers of maritime infrastructure, they are also inclined to develop projects that better serve their own interests rather than those of the local community.[8]

Variations in structure and governance are based in part upon the period during which an area's port grew, and are therefore largely regional. The geographic location of a port will often explain much about the corporate structure, duties, powers, and overall character of its port agency. In 2000, 29 percent of all U.S. port agencies were non-taxing public authorities, typically found in the North Atlantic region, but also on the South Atlantic and Gulf Coasts and on the Great Lakes.[9]

Ports structured as Special District Governments. Another common port structure that proliferated during the second half of the twentieth century is the "special district government." Special districts represent 43 percent of all U.S. ports, and they predominate on the Gulf and the North Pacific Coasts, where elected boards are also prevalent.[10] In addition to generating revenues from fees and charges, special districts are often granted the power to levy an ad valorem property tax within the port district as well. For this reason they are also often, but not always, governed by an elected commission so as to avoid claims of "taxation without representation."

Therefore the special district lacks the two key characteristics that give the public authority its unique power and autonomy—political insulation and financial independence. On the other hand, because its commissioners must run in contested elections and must represent the interests of the electorate, the special district is often more responsive to the demands of the public. For this reason the special district is sometimes seen as an antidote to the public authority, as it surrenders a high degree of autonomy and independence in the development and operation of infrastructure in exchange for greater responsiveness to public pressure.

While special districts may have greater difficulty in completing large, costly, long-term projects that are unpopular in the short run, they are at the same time less likely to attempt projects that are of questionable value in the eyes of the general public. However, for practitioners in general purpose local government, this does not necessarily mean that a special district port will be responsive to a particular city. The geographic boundaries, let alone the interests of cities and special districts, seldom coincide.

Ports structured as departments within general purpose governments. A smaller number of port agencies are structured as line or enterprise departments of state or local general purpose governments. For example, since 1970, port facilities in Baltimore have been governed by the Maryland Port Authority, which is a department of the Maryland Department of Transportation.[11] Several of the major Californian ports are departments of local governments. In most cases, these city port departments are governed by appointed commissions, with provisions to minimize revenue transfers between the port and the general fund. In many respects, then, local practitioners will find these commission-run ports to be very much like the public authorities described above, with all the concerns about independence and responsiveness already discussed.

However, ports structured as city departments can also be subject to greater influence from elected politicians and other pressures from within the local general purpose government. For example, despite separate finances, the port department may be exposed to claims from the general revenue fund against port surpluses and reserves to cover for the costs of fire, policing, and other city services that the port receives. These can be used to influence port actions and, in general, the more it is subject to political incursions and influence, the more difficult it will be for the port to plan for and develop major infrastructure projects.

Private ports. Finally, contrary to the global trend towards port privatization, there are only a few private ports in the United States. Most of these consist of no more than a single-purpose terminal linked to a bulk distribution facility or extraction industry. They may exist independently, or within the larger jurisdiction of a public port. For example, the Sun Oil Company (Sunoco) operates a large private oil terminal in Philadelphia that accounts for most of the gross tonnage that passes through the combined ports on the Delaware River.

In this section we have discussed the major types of port agency structure with an emphasis on the conflict between autonomy and representativeness and how the formal structure of a port agency influences how it provides infrastructure. This is especially important with regards to the financing of port facilities, to which we now turn.

Port Finances: Revenues, Expenditures, and Debt

As with other infrastructure systems, the financing of ports is changing rapidly as public facilities increasingly face commercial pressures. This section will consider port revenues, expenditures, and debt, beginning with a discussion of finances for cargo ports and then moving to the more complicated topic of financial arrangements for diversified ports.

Cargo Port Finances

Current changes in cargo port revenues, expenditures, and debt are closely tied to containerization and changes in the organization of the global shipping industry. In many parts of the world, cargo ports are being privatized. While this has not happened in the United States, increasingly we have seen terminal-by-terminal "privatization" as the largest, newest, and most advanced container terminals are leased to terminal operators. Port agencies have become landlords, with important consequences for revenue sources and budgeting. To understand these trends we start with a description of the traditional sources of port revenue, and then we discuss the impact of containerization on cargo port finances.

Traditional port revenue sources. When a ship calls at a port, it requires certain services. As the ship approaches the harbor, it may require piloting and tug services. Once docked, the ship itself requires fuel and water, while the off-loading of the cargo requires stevedoring services and specialized equipment. Traditionally, the various charges for port services were listed in a lengthy and complicated list of prices, known as the "port tariff." Port agencies, stevedoring firms, tug operators, and other service-providers have historically charged ship- and cargo-owners a series of user charges to cover the cost of meeting these services. With the exception of some specialized handling equipment (most notably cranes), containerization has not significantly changed the way these service costs are recovered. The same thing cannot be said of the pricing of port infrastructure.

In order to pay for the costs of harbor infrastructure, port agencies collect two main kinds of revenue. "Dockage" is the charge levied on the ship for the berthing space they occupy. Typically this would be priced according to the length of the vessel, since the longer the ship, the more berthing space it occupies. "Wharfage" is the charge levied on the cargo for passing across the wharf. Typically this would be priced according to tonnage, volume, or number of units, since the greater the throughput, the more wharf facilities are required to store and handle the cargo. A third charge, known as "demurrage" is levied if a container or some other cargo stays on the terminal beyond an allowed period. This is typically used in congested ports to encourage higher levels of terminal throughput.

Using this variable price structure, public port agencies typically relied on balance-sheet financing, public subsidies from federal and state allocations, and local taxes to fund infrastructure.[12] The most important effect of this financing system was that it placed both the risks and the rewards of port infrastructure investment on the shoulders of the port agency. Some ports became rich in this way; many made what have turned out to be disastrous investments in wharfs and cranes that now stand idle.

Impact of containerization on revenue structures. Containerization raised the stakes in the port investment game because it changed the scale of port development. As already noted, terminals became bigger, required deeper and wider channels, and required new and improved land transport connections. Equipment needs also changed as the entire business of handling cargo became more capital- and technology-intensive. The old ways of financing and pricing ports have come under pressure and have given way to new arrangements. Although the new arrangements vary from port to port, we can identify some common trends.

During the early years of the container revolution, many public ports received subsidies from federal, state, and local governments in support of economic development strategies. In some cases, ports could rely on a share of general obligation bonds. They also received funding from the federal government in the form of U.S. Army Corp of Engineers appropriations for dredging and coastal engineering, and through connection to the federal highway system. However, with the general cutback in infrastructure spending by the public sector, ports have found that they had to become more self-sufficient. Since the mid-1980s, ports have had to rely more on their own revenues to fund infrastructure through bond finance.[13]

In response to these revenue pressures, ports have changed the way in which they finance infrastructure. Many of the largest ports now lease whole terminals for the exclusive use of a single tenant, typically large multinational terminal operating companies. Many of these terminal operating companies are closely linked to the largest container steamship lines. These leases may range upwards of 25 years in duration, and are typically structured so as to transfer the risk associated with terminal development to the private sector to ensure that the port agency will be able to service the debt incurred in developing the facility. Because it is a government, the port still issues the debt in the form of tax-exempt revenue bonds that have the effect of lowering the cost of money. Port leases also often include participation clauses; the terminal operator may be expected to guarantee a minimum annual rental, with some form of revenue-sharing arrangement if cargo throughput exceeds the minimum amount.

This change in port pricing has occurred in tandem with some profound changes in the shipping industry that have important consequences for port planning and management. The port agency now acts much more as a landlord, negotiating the details of infrastructure provision with their clients, the private port terminal operators. Even those ports that have not switched to long-term leases, and that maintain common-user facilities for use by a number of steamship lines, have incorporated aspects of ground leasing and revenue-sharing in their tariffs. Ports have also simplified their tariffs in order to reduce administrative costs for their customers. For example, whereas in the past they may have priced dockage, wharfage, and storage charges separately, many ports now offer a single charge per container.

All these trends in port revenue, expenditure, and debt contributed to changes in the planning of port infrastructure—towards more project-based planning that has a high degree of tenant involvement, supported by a framework of strategic and long-range capital planning. We take up these planning themes in the discussion below.

Diversified Waterfront Finances

Traditional cargo ports earn revenues from fees, charges, and rents, and they sell revenue bonds backed by anticipated future streams of income in order to finance improvements to new and existing facilities. On the diversified waterfront the infrastructure needs are similar—the maintenance and periodic replacement of seawalls, piers, pilings, bulkheads, and other engineering works are required in order to provide developable land parcels. However, the revenue picture becomes more complex as the port engages in a broader array of business lines.

Cruise revenues. The cruise business is lucrative for the port agency itself as cruise ships help to replace lost cargo ship dockage and wharfage fees, while revenues are further boosted by substantial passenger charges. For example, in 2010, the Tampa Port Authority charged each one of the

802,775 passengers who crossed its docks an average of $11.78 in fees, dockage, wharfage, and parking. Altogether, the port earned $9.5 million dollars in passenger revenue, or 23 percent of the Tampa Port Authority's total operating revenues for the fiscal year.[14] As in the case of modern cargo facilities, cruise terminals are financed with revenue bonds backed by this anticipated stream of future revenues. Also like modern cargo facilities, savvy ports will only construct a new cruise terminal when a cruise line has entered into a long-term lease, guaranteeing rent and a minimum throughput of passengers.

Revenues from ground leases and rents. When ports undertake real estate development projects, they act more like landlords, citing their public trust obligation and a lack of development expertise. Under such arrangements, the port determines the direction of waterfront development and then solicits developers to assemble teams, make detailed proposals, and invest private capital, assuming the lion's share of both the risk and the upside reward. The port subsidizes and stimulates private sector investment by financing the infrastructure improvements—seawalls, bulkheads, pilings, power, sewer, water, and roadways—necessary to create a developable parcel. And like the rent structures for modern cargo facilities described above, the port may also earn additional revenues from participation. Similar to the financial model of a retail shopping mall, in these cases the port receives a share of the profits from food, liquor, hotel rooms, and retail sales when they exceed minimums specified in the lease.

Hotels, marinas, parks, a convention center, and a ballpark combine to create San Diego's tourism driven diversified waterfront.

Subsidizing non-revenue-generating activities. As noted in the previous section, state legislatures have also required many ports to address other new non-revenue-generating duties, including water and air quality, environmental preservation, habitat restoration, public access and parks, public art, and regional economic development. In some cities, other local units of government have pressured ports into partially financing or subsidizing projects such as waterfront light-rail lines, aquariums, and convention centers through loans, loan guarantees, capital and operating grants, and grants of land. Importantly, these types of projects are often loss-makers and may not necessarily fit the port's duties and powers, but the cost in political capital of not contributing is often perceived by port leadership as being greater than the real cost of capital.

Perhaps the greatest challenge facing port agencies responsible for a diversified waterfront is that of matching revenues from profitable lines of business with the costs of building and operating projects that do not generate revenue. Different ports approach this challenge in different ways. For example, the Port of San Francisco faced public demands for new parks that it could not afford to build. At the same time it hoped to increase revenues by renovating several blighted historic piers it owned into office space. The two state agencies that regulate waterfront land use typically prohibit commercial uses on the waterfront, however, so the port and the two agencies

Table 23.3
Potential sources of revenue on the diversified waterfront

Function	Revenue source
Traditional maritime	
Cargo	Dockage, wharfage; ground rents and facilities rents
Ship repair/ship services	Dockage, wharfage; ground rents and facilities rents
Commercial fishing	Slip fees, rents, and charges on fish processing facilities
Traditional nonmaritime	
Transportation (airport, toll crossing, commuter rail)	Passenger fees, landing fees; tolls; fares and parking
Non-traditional maritime	
Cruise	Dockage, wharfage; ground rents and facilities rents; parking; passenger fees
Harbor tour/dinner cruise	Dockage, wharfage, ground rents, and facilities rents
Marina/recreational	
Boating	Slip fees, mooring fees
Yacht repair/services	Ground rent, slip fees, mooring fees
Non-maritime	
Real estate development (commercial office, retail, mixed use, residential, tourist attractions, etc.)	Ground rents; participation (food, beverage, hotel rooms, retail sales); parking

Source: Association of American Port Authorities.

struck a three-way deal. The resulting agreement allowed the port to redevelop its historic piers into new commercial office space and then use the surplus revenues from rents to fund the construction and operation of several new waterfront parks.

Table 23.3 summarizes the potential sources of revenue on the diversified waterfront, which include but extend well beyond the traditional revenue sources of cargo ports. In the final section we consider how to spend this money, through a discussion of port planning.

Port Planning

Planning ports and waterfronts is both technically and politically demanding; while only a small number of highly specialized planners and engineers need to develop the technical skills, all local planners in port cities must develop a keen understanding of the institutional and political issues if they are to positively influence port planning. As already noted, from the perspective of a general purpose government, the challenge of dealing with ports is that for the most part they are governed by independent and autonomous agencies that do not necessarily have the same interests and goals as the city. Before discussing these challenges, this section briefly discusses the range of plans that local practitioners may encounter in their interactions with port agencies, and introduces some of the technicalities of capital planning for cargo ports. Here too, local officials are increasingly called upon to understand the complex infrastructure requirements of modern ports, especially as regards landside transportation access.

Types of Port Plans

While original port enabling acts were often silent on the subject of planning, in recent years legislators have added more specific requirements. Today, enabling acts often require ports to develop and periodically update strategic business plans, capital improvement plans, and in some cases land-use and environmental plans. Other ports choose to engage in these types of planning merely as a good business practice, even when their enabling acts do not require them to do so. The following is a summary of the types of planning activities in which today's ports typically are engaged.

Strategic business plans. Strategic business plans seek to direct resources towards the support and development of lines of business that align with the port's duties and powers, but that will also ensure that revenues are sustained or increased. Business plans are often coordinated with annual operating and capital budgets, with long-range capital improvement plans, and sometimes even with agency reorganization plans. Business plans typically describe goals and objectives as well as performance measurements, including anticipated increases in cargo and/or passenger handling, revenues, and reductions in costs flowing from strategic and operational changes and improvements. Plans may be developed with input from a variety of stakeholders, and are approved and adopted by the port commission.

Capital improvement plans. Enabling acts require virtually all ports to develop and adopt annual operating and capital budgets at a minimum, but some ports also develop longer-range (5–20 years) capital improvements plans (CIPs). Like those developed by general purpose governments, port CIPs frame a long-range strategy and form the basis of the annual capital budget. The port's finance department, with the input of key staff from other operating departments and the port commission, develops the CIP. The CIP is typically developed and updated in coordination with the strategic business plan and the annual operating and capital budgets.

Land use plans. Ports may develop their own comprehensive land use plans, although they often keep them as general as possible in order to preserve flexibility for future development and business opportunities that may not yet be apparent. In the past the public had little input in port land use plans but the rise of public participation has forced many ports to become more inclusive and to work more cooperatively with the GPG and other local governments, particularly when planning for land that abuts growing residential and commercial areas.

Environmental plans. In addition to the regulatory and permitting requirements contained in laws such as the Rivers and Harbors Act, the Clean Water Act, and the National Environmental Policy Act, state and local legislators have also required many ports to engage directly in environmental planning for both uplands and submerged lands within the port district. Environmental surveys of upland property help to influence land-use decisions, and some ports also plan for the remediation and reuse of former industrial sites. Legislation may also require ports to develop plans for the improvement of water quality, including bilge-water regulation, and the protection and restoration of wildlife habitats. Some ports even replace lost habitats with new islands created from dredge spoils. Major cargo ports are also becoming increasingly involved in air quality planning, since both ships and port-bound trucks are major emission sources. Environmental plans can be separate or a part of a larger land-use plan, and they are approved and adopted by the port commission.

Plans Required by Superior Governments

Port infrastructure planners also find that they have to respond to a complex web of federal and state planning requirements. For example, the California Coastal Act creates a powerful incentive for ports to prepare master (or land use) plans and submit them to the California Coastal Commission for certification. The port authority will then be granted permitting authority for coastal developments consistent with the master plan within the certified area. The federal Coastal Zone Management Act is likewise designed to promote proactive and sound planning.

Integration with the transportation plans of MPOs. The creation of metropolitan planning organizations (MPOs) under the Integrated Surface Transportation Efficiency Act (and its successors) provided a powerful incentive for port authorities to become involved in regional planning efforts with the goal of securing adequate landside transportation access to their facilities. In some cities, ports are working with the regional metropolitan planning organization and other local units of government to coordinate plans that affect infrastructure development.

However, in the final analysis, it is important for practitioners and public officials to understand that because they are often not required to do so by law, many public ports, particularly those that see themselves as business organizations, do not always recognize or accept the benefits of coordinating their plans with other units of government, let alone with the demands of the public. Nonetheless, these other interests continue to play an increasingly important role in planning for ports and waterfronts.

Planning Function within the Port

Until the 1980s, the planning function at most ports was carried out informally, typically under the leadership of the engineering department. Today, most ports have dedicated planning and development departments. Ports on diversified waterfronts sometimes maintain quite large planning departments and their staffs manage community planning processes for large and often controversial redevelopment projects on the urban waterfront.

More typically, however, ports rely upon a combination of in-house staff, private consultants, and private shipping companies, terminal operators, cruise ship companies, and commercial real estate developers for their planning. Because major projects are often highly specialized and non-recurring, ports obtain specific expertise and capacity for individual projects from outside sources, avoiding the need to maintain large and costly planning departments. Likewise, with the rise of long-term dedicated terminal leases, many of the planning and design functions at cargo ports are assumed by private terminal operators. In these cases, port planners and engineers work closely with the tenant's planners or hired private consultants.

Capital Planning of Cargo Ports

The trend in container ports towards larger and larger terminals (sometimes called mega-terminals) that are leased as single entities to terminal operating firms has changed the role of the port agency in the planning process. Tenant terminal operators now have more say in the physical planning process because they bear much of the development risk and because they want to tailor terminal layouts, berthing configurations, and so on to suit their specific needs. Private engineering consultants play an important role in this site-level planning and design work.

However, public port agencies still require a minimum capacity to develop long-range capital plans. This kind of planning typically involves what is known as a gap analysis, in which the demand for cargo handling is compared to the supply of handling capacity. The results of this exercise will typically be presented in the form of a series of port-wide development scenarios that will then be used in negotiations with actual and potential tenants. Under this approach, specific investments can be timed to meet anticipated handling needs.

Estimating the demand for handling capacity involves a long-range cargo forecast typically derived from econometric models of trade flows and the apportionment of expected cargo growth between competing ports. The expected demands for cargo handling are then compared to existing and planned facilities in a capacity analysis. Capacity analysis is a complicated task involving advanced operations research and engineering expertise, but in essence it proceeds on the understanding that a modern cargo port is a pipeline through which cargo moves as it is transferred between land- and ocean-based transportation modes.[15] Each step or process has certain infrastructure requirements: for example, the approach of the ship to mooring requires a channel dredged to an appropriate depth, cranes that have enough reach to serve the vessels that are likely to call, and so on. The attributes of the given infrastructure element in the pipeline determine the handling capacity of that process. However, because each step or process is connected to the entire pipeline, the efficiency of any one process depends on the efficiency and integration of the entire system, and conversely, the entire system can be no more efficient than the least efficient element in the system. The processes in cargo handling, the infrastructure requirements associated with each, and the key interest groups are summarized in Table 23.4. From the local practitioner's perspective, it is especially important to understand that the cargo pipeline is

Table 23.4
Processes in cargo handling, infrastructure requirements, and key interest groups

Step/process	Infrastructure requirements	Key interest groups
Ship approach	Channel dredging Pilotage	Army Corp of Engineers Port Authority
Vessel docking	Sufficient berth space Berth depth/breadth	Port Authority Terminal Operator
Cargo unloading	Cranes Specialized lift equipment	Shipping line Terminal Operator Stevedore Longshoremen
On terminal transportation	Tractors and trailers Terminal handling equipment	Terminal Operator Stevedore Longshoremen
Transit storage	Sufficient storage space Refrigerated container plugs Bulk storage tanks/silos Storage sheds	Terminal Operator Port Authority Longshoremen Cargo owner
Load to inland transport	Terminal handling equipment On-dock rail Intermodal transfer facilities	Railroads Short- and long-haul truckers Terminal operator
Inland departure	Terminal gate Inland access routes	Railroads Short- and Long-Haul Truckers Local/State/Federal roads and highways, Local communities
Customs and other services	Inspection facilities Cargo processing facilities Services to ships (victuals, fuel) Container services (repair, stuffing)	Homeland Security Port Authority Marine services firms

Source: Authors.

not simply a set of technical requirements—it is also a complex network of actors each pursuing their own goals. Hence the task of the practitioner trying to improve cargo throughput efficiency is not only technically demanding—it also requires the co-operation of a range of actors.

The need for co-operative planning is becoming more and more important since applications of information technology are allowing port operators to experiment with logistics systems that integrate port terminal operations more closely with activities in warehouses and distribution centers that may be many miles inland. The quality and congestion of land transport connections—from local roads and highways to rail links and crossings—are thus increasingly important to the overall efficiency of the cargo handling pipeline.

For these reasons, port planners are finding themselves increasingly involved in planning processes that involve state highway authorities and local roads and planning departments. The prime example of this kind of activity is the Alameda Corridor that extends inland from the ports of Los Angeles and Long Beach, but similar hinterland access planning activities have been undertaken by ports such as Oakland, Seattle and Tacoma, Portland, and New York. For their part, local officials in GPGs need to anticipate and act proactively to address these challenges.

Planning on the Diversified Waterfront

On the diversified waterfront, local practitioners will encounter port agencies engaged in a broad range of activities that are reshaping the interface between the city and the waterfront. However, as discussed above, the diversified waterfront creates something of a mismatch between the original goal of powerful, independent port agencies with a singular focus on maritime commerce, and the new requirement for ports to provide a variety of services. Indeed, many port agencies are now required to act more like general purpose governments. For these reasons, port plans are often not coordinated with the plans of other units of government. This mismatch is often the source of tension between ports and cities, but it also presents local practitioners in GPGs with new opportunities to achieve greater coordination in planning efforts along the waterfront.

Competing interests. One of the biggest planning challenges facing practitioners and administrators is that of reconciling the interests of the port with those of the local GPG and other units of government that share borders or overlapping jurisdictions. It is not uncommon for a port agency, a local general purpose government, a redevelopment authority, an economic development authority, a local waterfront redevelopment agency, and a state regulatory agency all to be wrestling for control of land use on exactly the same portion of waterfront land. Uncoordinated and conflicting goals may also exist within different departments of the GPG, including city planning, public works, licenses and inspections, and economic development. Political pressures from the executive and legislative branches of the GPG will further influence the goals and objectives of these individual departments. In such cases the city or local GPG may seek to control land use decisions in the port district through land use plans and zoning maps that dictate allowable uses and height and bulk requirements. Some cities also develop plans and design guidelines for public "river walks" along the edge of waterfront property that might be owned or controlled by the port. In either case, the port may choose to either adopt or ignore these types of plans at its discretion when developing its own plans and projects.

Conflicts over land use and development. The GPG's economic development department, or an independent redevelopment/economic development agency, may also express interest in specific parcels of property within the port district. In many cases, these types of agencies seek to attract investment and jobs to the region by providing land, infrastructure, and financing for new commercial or industrial facilities. Underutilized port property is often targeted for such non-maritime uses, and economic development advocates argue that the region would be better off with a new, tax-revenue-generating use than with underutilized property that may generate revenues at some unknown future date. However, the port, and representatives of maritime industries, may argue that the loss of underutilized port property to non-port uses hinders the future growth of the port. Perhaps more importantly, port agencies also argue that changes in use send the wrong signals to shipping companies by suggesting that the city is no longer committed to supporting port and cargo operations.

Some ports may be inclined to engage in activities that will increase port revenues at the cost of jobs in other less profitable lines of business. For example, while the cruise business can be quite lucrative for the port, some representatives of the cargo industry believe that it hurts the regional economy by sacrificing high-wage union stevedoring jobs for new, lower-wage service jobs.

Finally, one important new influence in port and waterfront planning is the ever-increasing demand from the public, and particularly residents who live near the waterfront, for a role in the decision-making process. For while citizens had little say in the past, since the 1970s increased concerns over abandoned land, environmental issues, and changed perceptions of the value of waterfront property have together led to heightened public interest. Citizens have since lobbied politicians and organized through non-profit neighborhood and environmental groups, all in an attempt to pressure the port into including the public in its planning process. The outcomes of these efforts have varied widely, but in all cities with waterfronts the port is under greater pressure than ever before to take into account the views of the public when planning for new projects.

Conclusion

This chapter began with a history of ports in the United States, and traced the implications of containerization for infrastructure planning and provision, both in the context of cargo ports, and also along the nation's increasingly diversified waterfronts. In both cargo ports and diversified waterfronts, infrastructure planners face the complex challenges that arise from many competing interests seeking access to scarce, valuable, and ecologically sensitive land. Effective planning must address the legitimate interests of a complex and varied set of users of waterfront property, from maritime interests to economic development advocates to residents and environmentalists.

In the most successful container ports, public officials now find themselves working closely with major tenants and clients to plan ever-larger and more costly facilities. At the same time, the organizational context for planning and infrastructure delivery is also often quite challenging. Today's port planners work in agencies with confined jurisdictions. In order to ensure that cargo circulates from the waterfront to the hinterland as efficiently as possible, these planners find themselves working with transportation planners and officials in inland jurisdictions to improve landside access to the port. In less successful cargo ports, officials face tough choices about which investments are most likely to pay for themselves by attracting more cargo, and whether the available resources may be better invested in non-traditional or non-maritime uses.

Much has changed on the urban waterfront too, and as a result the planning challenges are more daunting than ever. The retreat of maritime cargo operations has led to new opportunities for a vibrant and diversified waterfront comprised of a variety of attractions and infrastructure systems. In addressing these opportunities, old port agencies have sought to remake themselves as entrepreneurial organizations generating revenues from a wide array of business lines, maritime and otherwise. Maritime interests have struggled to ensure that underutilized waterfront property is preserved for traditional port uses, while others have pushed for a wide variety of real estate development approaches, particularly during the building boom of the 2000s. More recently those maritime interests have been proven to be right as population growth has driven increased consumer demand, shipping volume, congestion at major ports, and thus the resurgence of formerly marginalized ports. At the same time, some cities have started to make plans for the preservation of industrial land. Finally, politicians and residents have come to view the waterfront as a different kind of asset that should be developed as a place for the public, rather than purely for its economic value.

Added to the challenges presented by diverse interest groups are those that stem from multiple, overlapping, and sometimes competing units of government, all of whom seek to exert their powers and planning prerogatives on the waterfront. Cities find themselves redeveloping their waterfront plans through a combination of formal and informal processes, involving repeated rounds of planning and complicated financial structures that mix public and private capital. As a result, virtually every project is the product of a unique set of conditions, actors, and negotiations, rather than a grand master plan. So for the local practitioner in the GPG, the first step is to develop an understanding of the unique powers and limits of their own port, and then find ways to work with its leadership and staff to ensure that planning is consistent with the goals and aspirations of the citizens of the region.

Additional Resources

Publications

Breen, A. and D. Rigby. *Waterfronts: Cities Reclaim Their Edge*, 2nd ed. Washington, DC: Waterfront Press, 1997.

Brown, P. H. *America's Waterfront Revival: Port Authorities and Urban Redevelopment*, Philadelphia: The University of Pennsylvania Press, 2009.

Desfor, G., J. Laidley, D. Schubert, and Q. Stevens, eds. *Transforming Urban Waterfronts: Fixity and Flow*. New York: Routledge, 2010.

Hall, P.V. and M. Hesse, eds. *Cities, Regions and Flows*. Abingdon: Routledge, 2013.

Thoresen, Carl A. Port Designers Handbook. 2nd ed. London: Thomas Telford Publishing, 2010.

Website

The American Association of Port Authorities (AAPA): www.aapa-ports.org.

PART SIX

Community Facilities

Community facilities are usually provided by local general purpose governments. In growing areas their location often follows the decisions about transportation and water/sewer infrastructure. In already urbanized areas, community facilities can be upgraded to attract business, tourists, or for a general sense of civic pride. Community facilities include schools, civic buildings, parks, and recreational and open space facilities, as well as buildings and facilities operated by a private or non-profit organization. The first exposure a local practitioner man have to infrastructure is frequently through the development or renovation of a civic facility.

Part VI consists of three chapters. Chapter 24 (Public and Quasi-Public Buildings) contains an overview of different types of public buildings such as city halls, libraries, public safety buildings, criminal justice facilities, and hospital buildings, as well as a short section on quasi-public buildings such as sports arenas, convention centers, and cultural facilities. The chapter describes the basic elements in a building, including green building innovations.

Chapter 25 (Public Schools as Public Infrastructure) discusses the capital needs of schools before going on to summarize the major planning issues in education today, including how to plan for the capital needs of schools in the inner city and rapidly growing areas. The impact of capital needs for schools on local general purpose governments is also described.

Chapter 26 (Parks, Recreation, and Open Space) presents the historical evolution of parks in the United States before turning to financing and planning options. Level of service standards for different park elements are outlined, as well as how to consider their positive economic impacts during the planning and budgeting process and their role in the sustainable community.

chapter 24

Public and Quasi-Public Buildings

In this chapter ...

Importance to Local Practitioners. 507
Public Buildings Today: From Civic Buildings to Hospitals . . . 507
Municipal Government Buildings. 508
Civic Centers and City Hall 508
Public Libraries, Museums, and Cultural Buildings 508
Warehouses and Corporation Yards. 510
Municipal Public Safety Buildings 510
Criminal Justice Facilities 512
Court houses. 512
Prisons and Jails 512
Public Hospitals and Community Health Centers. . . 513
Quasi-Public Buildings 514
The Lure of Tourism 514
Convention Centers and Hotels. 515
Sports Facilities. 516
Cultural and Historic Districts, Festival Malls, and Entertainment Centers. 517
Sustainability and Public Buildings. 518
Climate Change Issues. 518
Green Building Codes 519
Leadership in Energy and Environmental Design (LEED) 519
Elements of a Building 520
The Building Envelope. 520
Glazing Systems and Lighting . 521
Water and Wastewater. 521
Heating and Cooling 522
Energy Systems and Electricity. 523
Land Use Planning and Public Facilities 524
Comprehensive and Area Plans 524
Performance Indicators 525
Other Building Design Considerations 526
Seniors and Americans with Disabilities Act of 1990. 526
Emergency Management Considerations 527
Building Appearance and Design Competitions. 527
Conclusion 528
Additional Resources 528
Publications. 528
Websites 529

Why Is This Important to Local Practitioners?

Public and quasi-public buildings include a wide range of facilities, from civic centers to football stadiums; from libraries to hospitals. They are important for local practitioners for a number of reasons. First, these facilities are a significant part of the local capital improvement budget—and many existing facilities are facing significant replacement or renovation costs. Second, the design and operation of these facilities can be a critical component of a local climate change plan. Greening new buildings and retrofitting existing facilities with more energy- and water-efficient solutions is one of the first steps a local general purpose government can take to reduce carbon emissions. Third, public buildings have land use implications. They can be a significant element in an urban design plan and can have important social, cultural, and equity impacts. Public and quasi-public buildings can and do play a substantial role in economic development as well. Finally, some local practitioners may end up being the project manager for one of these facilities as it is conceptualized, financed, and constructed.

Public Buildings Today: From Civic Buildings to Hospitals

The use of one relatively simple term—public building—to describe buildings owned by the public sector belies their vast range in size and function. Local public and quasi-public buildings range in size and function from modest civic centers to large-scale convention centers and sports stadiums.

Public buildings normally have more complex requirements than many private counterparts, such as commercial office buildings. They also have different life cycles, timelines, and levels of risk acceptance. While private

What Is a Public Building?

Is it a building paid for and managed with public tax money? Is it a building that is open and accessible to general populace? Is it a building in which public business is done, even if the building itself is not publicly owned? In a sense, the answer to all these questions is yes. Generally, local practitioners are concerned with civic buildings such as city halls, court buildings, libraries, schools, and public safety buildings, including jails. Those in large jurisdictions may also be involved with convention centers, sports facilities, and other quasi-public facilities, although these facilities are beginning to be issues for smaller suburban governments as well.

Seattle Public Library, Ballard Branch. This library uses the latest in environmentally-friendly design including a sod roof and air intakes that exploit prevailing breezes.

Denver Art Museum Expansion doubled the size of the existing facility when it opened in 2006.

buildings must make a profit, public buildings are built to meet a broad set of societal goals, only one of which is the "bottom line."[1]

Municipal Government Buildings

Civic Centers and City Hall

The most ubiquitous public building at the local level is the city hall or the county's administrative headquarters. These buildings and their counterparts for state and federal administration house elected officials and the staff that administer government functions. Often the administrative building includes the police (or sheriff) and fire department, but these can also be in separate buildings.

Although it is tempting to think that government administrative buildings are most like commercial office buildings, there are some important differences associated with the ceremonial and public meeting function of government. The city hall and its accompanying park or civic square are often the basic ingredient in a civic center plan (Figure 24.1), with other government buildings being located in close proximity. County administrative buildings are frequently located in urban areas and can be an important component of a government district.[2]

Public Libraries, Museums, and Cultural Buildings

Public libraries are an important piece of the urban social fabric. Americans make 3.5 billion library visits each year, borrowing more than 1.6 billion books. According to the Institute of Museum and Library Services, higher average student reading scores correlate with higher funding for libraries, regardless of a community's economics or adult education levels.[3] In addition, public libraries are the largest source of Internet access for those without computers at home or work, a critical factor in bridging the digital divide. A 2007 study found that 73 percent of public libraries (the majority of which were in rural areas) were still the only source of free public access to the Internet.[4] In 2005, there were over 9,000 public libraries in the United States, and almost 17,000 including local branches.[5]

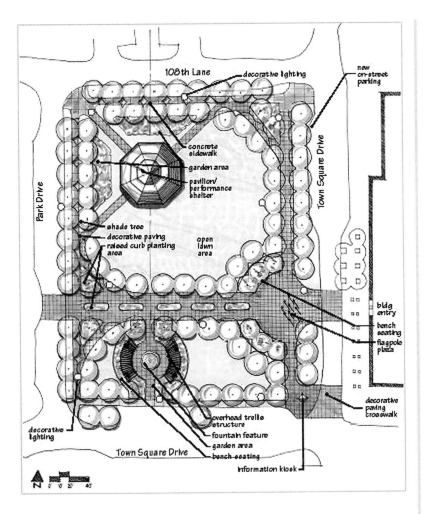

Figure 24.1
Blaine, MN, town square park plan

Principles of Great Civic Squares

- A great civic square has an image and identity.
- Streets and sidewalks around the square (the outer square) should be composed of active and welcoming uses, as should the square itself (the inner square).
- Civic squares should have a variety of smaller places within them to attract people all day.
- A square should be easy to get to by foot and should not be surrounded by fast-moving traffic.
- Uses should change as do the seasons so that people are attracted year round.
- Civic squares should be clean and well maintained.

Adapted from Project for Public Spaces, Inc., E. Moulder and S. Clark, *Parks and Recreation, Cultural and Arts Programs: Alternative Service Delivery Choices* (Washington, DC: International City/County Management Association, 1999).

Operating expenses for libraries often suffer during economic contractions, but capital bond referenda for renovations and new libraries have done well over the past 20 years. Indeed, during the prosperous 1990s, and from 2003 to 2006, library construction enjoyed a renaissance.[6] Even in 2007, a period of economic slowdown, most library ballot measures passed for both operating expenses (69 percent) and capital construction (74 percent).[7]

Art museums and other cultural buildings, including public auditoriums, can also be municipally based. Museums are usually run by non-profit organizations independent of the general purpose government, but often receive hotel taxes and other public subsidies. Regional and local art centers are another form of civic infrastructure. All these facilities are an important part of the central city's cultural life, and planning

Merced Multicultural Arts Center, Merced, CA

for new facilities or the upkeep of their physical condition should be of concern to the local practitioner.

Warehouses and Corporation Yards

In addition to administrative buildings, local governments also own and operate facilities that support the operations of the city and county. These include what are frequently called "corporation yards" (Figure 24.2). These are facilities that house the operations and maintenance functions for streets, sidewalks, sewers, facilities, vehicles, and equipment. These can include office space for administrative functions, as well as warehouses for storing old lampposts, furniture, and other items and equipment. They are usually located in an industrial area or on the outskirts of town. In some older cities, corporation yards are now in the middle of the urbanized area and expansion or upgrade poses difficult issues for the jurisdiction.

Adeline Maintenance Facility interior

Municipal Public Safety Buildings

Municipal public safety buildings include police stations, fire stations, ambulance headquarters, police, fire, and rescue training centers, and emergency operations centers. Public safety buildings have very specific purposes, and therefore their design must include some specialized facilities. Police stations need firing ranges, photography labs, internal segregation of circulation (for the public, staff, and prisoners), and parking for both the public, staff, and police cars. In these volatile times, police stations also require special security considerations, such as structural reinforcement, shatterproof glass, and controlled access. (See Chapter 28 on telecommunications for innovative public safety telecommunications systems.[8])

The technical requirements of a public safety building must also be balanced with specific siting requirements. For instance, many cities have time criteria for fire response. This means that fire stations have to be located in particular locations to provide proper coverage for the city.[9]

Interior of the Alameda County EOC, Oakland, CA

More recently there has been interest in specialized facilities for emergency operations centers (EOCs). In the wake of earthquakes in California, many local governments and utilities have built these centers and have trained all staff in effective

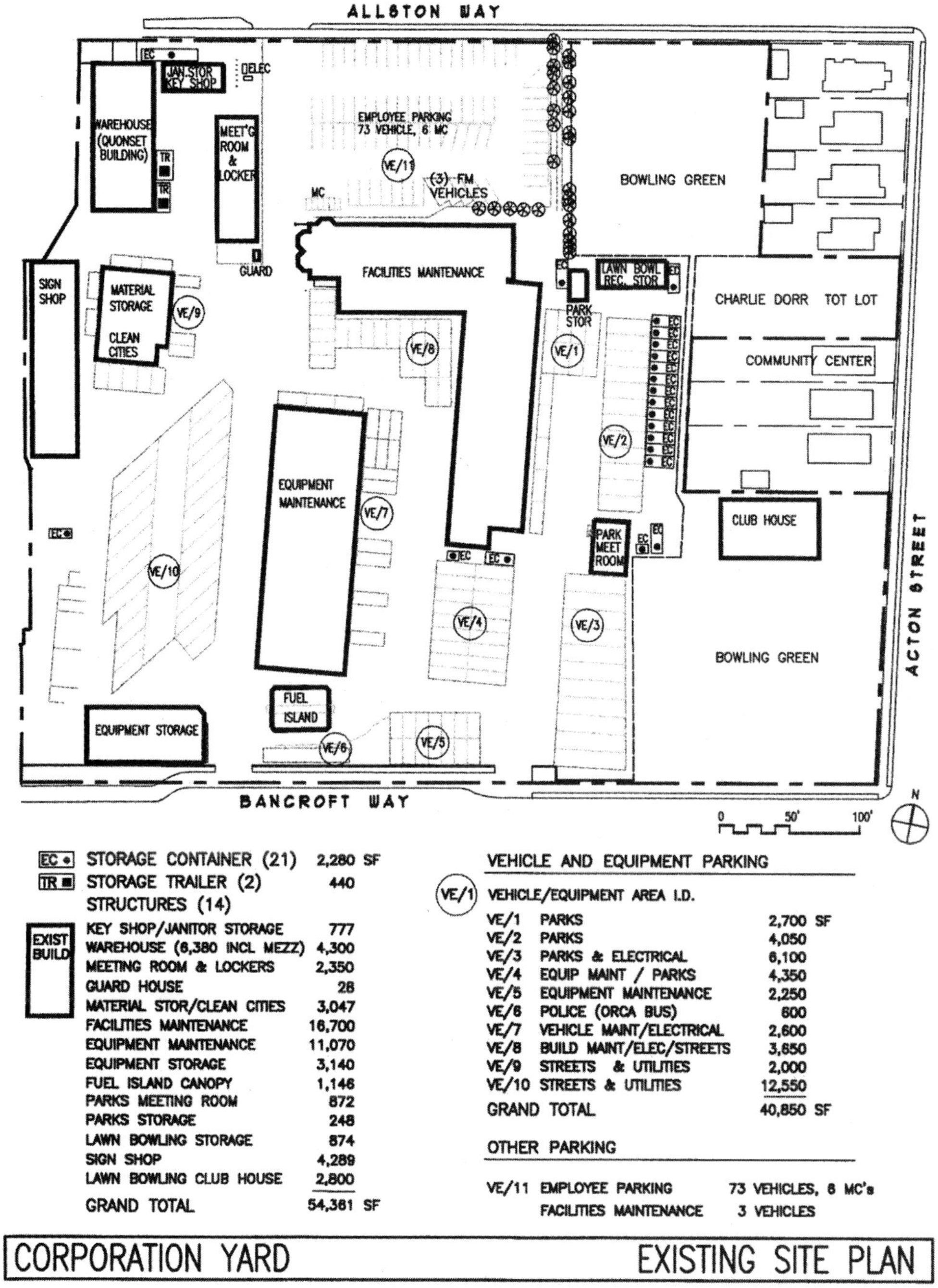

ALLSTON WAY

BANCROFT WAY

ACTON STREET

EC	STORAGE CONTAINER (21)	2,280 SF
TR	STORAGE TRAILER (2)	440
	STRUCTURES (14)	
EXIST BUILD	KEY SHOP/JANITOR STORAGE	777
	WAREHOUSE (6,380 INCL MEZZ)	4,300
	MEETING ROOM & LOCKERS	2,350
	GUARD HOUSE	28
	MATERIAL STOR/CLEAN CITIES	3,047
	FACILITIES MAINTENANCE	16,700
	EQUIPMENT MAINTENANCE	11,070
	EQUIPMENT STORAGE	3,140
	FUEL ISLAND CANOPY	1,146
	PARKS MEETING ROOM	872
	PARKS STORAGE	248
	LAWN BOWLING STORAGE	874
	SIGN SHOP	4,289
	LAWN BOWLING CLUB HOUSE	2,800
	GRAND TOTAL	54,361 SF

VEHICLE AND EQUIPMENT PARKING

VE/1 VEHICLE/EQUIPMENT AREA I.D.

VE/1	PARKS	2,700 SF
VE/2	PARKS	4,050
VE/3	PARKS & ELECTRICAL	6,100
VE/4	EQUIP MAINT / PARKS	4,350
VE/5	EQUIPMENT MAINTENANCE	2,250
VE/6	POLICE (ORCA BUS)	600
VE/7	VEHICLE MAINT/ELECTRICAL	2,600
VE/8	BUILD MAINT/ELEC/STREETS	3,650
VE/9	STREETS & UTILITIES	2,000
VE/10	STREETS & UTILITIES	12,550
	GRAND TOTAL	40,850 SF

OTHER PARKING

VE/11	EMPLOYEE PARKING	73 VEHICLES, 6 MC's
	FACILITIES MAINTENANCE	3 VEHICLES

CORPORATION YARD EXISTING SITE PLAN

Figure 24.2
Site plan of a corporation yard prior to renovation

Modesto Police Headquarters, Modesto, CA

Police Station in Southwick, MA

disaster management procedures. The events of 9/11 and Hurricane Katrina also made the effective and coordinated functioning of public safety and administrative staff essential. Federal funding made possible many new EOCs. Some double for other functions as well. The City of Berkeley's EOC, built in 1998, is also used as a training center. Alameda County's EOC doubles as the fire dispatch center.

The majority of public safety buildings are owned by the government. However, buildings that house volunteer fire departments, approximately one-third of the total in the United States, tend not to be publicly owned.

Historic Portland Hotel Gate at Pioneer Courthouse Square in Portland, Oregon

Criminal Justice Facilities

Court houses

Counties are usually responsible for the administration of the state's criminal justice system. Offices need to be found for district attorneys, court facilities, and frequently, public defenders. Municipal courts need similar facilities. Planning and designing a new courthouse or even an upgrade of an existing facility are often the largest single financial decision that a court administrator or a judge will face. Court functions also have specialized needs and are usually designed by architects or consultants with experience in this area.[10]

The planning process begins with a condition assessment and needs analysis, as it does for most public buildings. However, a critical component is the analysis of court security and security design. Courts may also need special facilities for higher profile court cases. Financing issues are similar to other public safety functions—they rely upon general funds and usually a dedicated bond issue repaid by state taxes or local property taxes.[11]

Cherokee County, NC, Courthouse in 1927

For additional information on planning for courthouse facilities, the National Center for State Courts has a wealth of publications on their website under "Courthouse Design and Finance." (www.ncsconline.org). One report of interest, written in 2008 for Kitsap County, Washington, contains a lengthy literature review that evaluates the available guidance as well as the capital plan for that county's court system.[12]

Prisons and Jails

Nationally, the number of people behind bars is steadily increasing, at a rate 25 percent higher than general population growth. In 1999, there were 3,365 jails nationwide, holding over 605,000 inmates, up from 3,338 jails and 460,000 inmates in 1993. Meanwhile, the municipal police service, which includes operation and maintenance of jails, can consume as much as 30 percent of a city's general budget.[13]

Prisons and jails are both detention facilities. However, prisons are used for incarceration arising from criminal conviction, whereas jails are used to hold people as they wend their way through the justice system (awaiting sentencing, for instance) or those incarcerated for minor offenses. In general, jails hold people for periods under one year. Furthermore, prisons are operated by state and federal governments, while jails are operated by local governments.

The trend in jails is towards facility centralization, with jail inmates being held in a relatively few large jails as opposed to many smaller ones. Gone are the days of the two-cell city lock-up, although a small number of local police departments still have their own jails.

Funding for jails comes out of the General Fund. However, a significant source of revenue for local jails comes from housing state and federal inmates for a fee. Approximately 70 percent of all jail jurisdictions charged fees to house inmates for other correctional authorities in 1999.

Contracting for jail services became more common between 1993 and 1999, when the number of private jails operating under contract to local government authorities increased from 17 to 47. These 47 private jails held 14,000 inmates.[14]

Public Hospitals and Community Health Centers

Another type of specialized public building is the public hospital and the health center. Psychiatric hospitals may also be public. These can also be called safety net hospitals, which include both public and not-for-profit private hospitals whose mission is to provide medical services, regardless of an

Santa Barbara County Courthouse

Cutaway of design for San Francisco Jail No. 3

Rendering of exterior of San Francisco Jail No. 3, San Bruno, CA

The Birthing Center of the Cleveland Regional Medical Center in Shelby, NC

Community Health Center Palm Springs, FL

individual's ability to pay. Public health facilities are usually operated by counties and large cities or by special hospital districts. The special hospital districts can be independent from the local jurisdiction or can be a district that depends upon the local jurisdiction for bond authority. According to the U.S. Census from 2002, which did not report on public hospitals alone, there were approximately 1,521 special hospital or health and welfare districts.[15]

In addition to public hospitals, local governments may be involved in federally supported health centers, called community health centers (CHCs). These are community-based organizations that serve populations with limited access to health care, such as the low income and uninsured, the homeless, those living in public housing, and seasonal farmworkers. CHCs provide primary care services, while the public hospitals—the safety net hospitals—provide inpatient or emergency care for the poor.

The future of capital investment in public hospitals promises to be tighter. Those with limited access to capital markets are public hospitals in urban areas.[16]

Public Financing of a County Health System

In 2003, voters in Maricopa County, Arizona, authorized a ballot initiative to create a special health district and to increase the property tax by about $22 a year per home in order to subsidize the county-run health system. The initiative will raise $40 million a year to pay operating costs and renovate older hospital facilities for the Maricopa Integrated Health System (MIHS). MIHS runs several medical and mental hospitals in Phoenix, along with 11 clinics that serve the county's poor and uninsured.

Quasi-Public Buildings

The Lure of Tourism

Another major class of public buildings that a local practitioner should know about is often referred to as "quasi-public" buildings. Sometimes these are also thought of as "mega-projects" because of their size relative to the more ordinary government building. These include facilities such as football stadiums, sports arenas, convention centers, and the complexes of cultural buildings, private restaurants, and recreational facilities that cities have been building or subsidizing in great numbers in the past decade to attract tourism.

Although the first generation of sports stadiums and arenas were built in the 1920s and the modern-era convention centers in the 1950s, it was not until the 1980s that serious competition for tourism began between cities. The competition was led by "messiah" mayors who brought together the public sector with the business and civic community to invest in the "infrastructure of play" to revitalize the urban core.[17] By 1992, bonds issued by state and local governments for convention centers ($1.4 billion in 1992), stadiums ($560 million), and other economic development activities ($2.6 billion) came close to the amount issued for streets, highways, and bridges.[18]

The four most important public financing methods for quasi-public buildings were local infrastructure support, local bond financing, local land/facilities donation, and tax increment financing.[19] Sixty percent of convention centers and auditoriums are operated under contract to a profit-making firm.

Convention Centers and Hotels

The meeting center business is profitable. In 1998, it was 2 percent of the U.S. Gross Domestic Product, producing $81 billion a year.[20] Either as a matter of local pride, or for economic returns, cities try to be meeting centers. A key ingredient of a successful "meeting city" is the convention center and hotel complex.

By 1990, the square footage of convention facilities in U.S. cities had almost doubled from 1970. When Chicago's McCormick Place Convention Center opened in 1960, it had 320,000 square feet of space, but by 2007 it had grown to 2.7 million square feet. Atlanta had 1.4 million square feet available in 2002. In 1998, 409 convention centers were open for business with 70 percent of them built since 1970.[21] From 1994 to 2004, public capital spending on convention centers has doubled, and convention center space has increased by 50 percent.[22]

Exterior of McCormick Place, Chicago

Convention centers (and sports facilities, discussed below) need downtown hotels to house the temporary residents, and restaurants to feed them. Originally, city officials and planners thought that building the convention center itself would be enough to stimulate the private sector to build the supporting facilities. However, many cities found that public subsidies were necessary for the hotels. This practice is widespread, particularly given the increasing competition between cities for conventions and conferences. Another type of subsidy used by cities is deep discounts for tradeshows. This practice has resulted in some convention centers, such as in Washington, DC, and St. Louis, operating at a loss.

Austin Convention Center

The overall market for conventions had been declining even before the economic downturn of 2008. In 2004, attendance

Portland, Oregon, Convention Center

Coors Field Baseball Stadium, Denver, CO

Gillette Stadium

Cowboy Stadium under construction

at the 200 largest tradeshows was at 1993 levels. Advances in communications technology have likely added to this decline. Even at the height of the boom of the 2000s, larger convention centers saw a decline in attendance, because of the increased competition by smaller cities, must offer deep discounts to attract conferences.

Sports Facilities

City leaders attach great symbolic value to having a major league team in residence. Sports teams are cited as essential to attract tourism, and cities compete intensely for the teams, many of which have moved frequently in the past several years. To keep teams, cities finance the construction of stadiums and guarantee attendance or revenues. However, despite the fervor with which they are pursued, countless research projects done by highly respected economists and planners have shown that sports facilities do not make money for a city.[23]

Beginning in the 1980s, big league teams became more sophisticated in their ability to play off cities against each other. All four major sports leagues (football, baseball, basketball, and hockey) were able to negotiate new and subsidized facilities as conditions of staying in an existing city or moving to a new one. For example, in the 1980s and early 1990s, 11 teams threatened to leave their existing city and six actually did. During this same period, 20 cities actively recruited baseball teams, while 24 sought football teams.[24]

The average cost of a new sports facilities (in 2001 dollars) went from $51 million in the 1950s to $226 million in the 1990s. The average cost for facilities scheduled to be built in 2000–2005 was $314 million. Part of the increased cost is due to amenities, but it is also due to downtown locations and more expensive land costs. Table 24.1 shows the number of facilities built from 1950 through 2005, along with the average cost and public subsidies received.

Municipal government participates in many ways to ensure that these facilities are built and operated. Usually a public–private partnership is involved, with the public sector contribution outweighing the private. The sports projects can also be combined with an area-wide redevelopment effort, and therefore have the potential to negatively impact residents in the immediate neighborhood.

Therefore, in recent years, community groups have been successful in negotiating community benefits agreements (CBA) that provide for mitigations in the form of affordable housing, living wage jobs, green design, and open space requirements. The Landmark Stables Center Agreement in Los Angeles was followed by CBAs for San Diego's Petco Park and Ballpark Village, and Brooklyn's Atlantic Yard.[25]

Cultural and Historic Districts, Festival Malls, and Entertainment Centers

In addition to convention centers and sports facilities, many cities have pursued redevelopment of the central city to attract tourism. One strategy is to create a special identifiable district, such as a historic district, which may include enclosed festival and shopping malls, or cobblestone streets and historic street lights. Complementing this may be a cultural district with subsidies for art museums, opera halls, and other performance facilities. These facilities serve not only tourists, but improve the quality of life for local and regional residents as well.

The National Endowment for the Arts indicates that the nation's 50 largest cities are providing some public support for the arts. The 1988 Indian Gaming Regulatory Act, which permitted tribes to negotiate with states to operate gaming facilities, has resulted in quite a few of these facilities being built, and in substantial sums of money for other local infrastructure.[26]

Not all observers find these new tourist facilities benign. One of the criticisms about

Table 24.1 Sports facility construction by decade from 1950 through 2005

Total stadia and arenas	1950s	1960s	1970s	1980s	1990s	2000–2005	Total
Number built	6	25	28	14	52	20	145
Average cost ($ millions)	51	126	132	157	226	314	1,006
Public subsidy ($ millions)	51	99	117	104	129	211	711
Public share (%)	100	78	89	66	57	67	71
Percent built by decade	4	17	19	10	36	14	100

Source: Adapted by authors from A. Altshuler, and D. Luberoff, *Mega-Projects: The Changing Politics of Urban Public Investment* (Washington, DC, and Cambridge, MA: Brookings Institution Press and Lincoln Institute of Land Policy, 2003).

Denver Lower Downtown Historic District. An historic streetlamp is visible in the foreground and the historic renovated Union Station in the background

Denver Lower Downtown Historic District

the central city revitalization campaigns for tourism is that they create exclusionary playgrounds for suburbanites, tourists, and "day-trippers," and that they are isolated from the surrounding neighborhoods of the city.[27] One urban sociologist, Michael Sorkin, describes the tourist complex as:

> [A] theme park which presents its happy regulated vision of pleasure—all those artfully hoodwinking forms—as a substitute for the democratic public realm, and it does so appealingly by stripping troubled urbanity of its sting, of the presence of the poor, of crime, of dirt, of work. In the "public" spaces of the theme park or the shopping mall, speech itself is restricted: there are no demonstrations in Disneyland. The effort to reclaim the city is the struggle of democracy itself.[28]

Sustainability and Public Buildings

Climate Change Issues

Today's buildings consume more energy than any other sector of the U.S. economy, including transportation and industry. They consume 40 percent of the energy and account for 76 percent of all power plant-produced electricity.[29] Buildings also produce 43 percent of the carbon emissions from fossil fuel combustion in the United States. Buildings also account for 14 percent of all freshwater use, 45–65 percent of landfill waste, and 31 percent of the mercury in solid waste.[30] Buildings produce wastewater, and their sites and roofs often contribute to stormwater runoff. Although buildings have become more efficient in recent years, the use of equipment that runs on electricity is expected to increase. The building stock of the U.S. is also expected to double in the next 30 years,[31] and almost 75 percent of existing buildings (including residential) were built prior to 1980 and will soon need upgrading. All of this provides both the incentive and opportunity to employ more sustainable features in buildings.[32]

Public and quasi-public buildings are a natural venue to jumpstart the development and use of green building techniques. Indeed, many local sustainability or climate change plans produced to date include goals for new construction and retrofit of public facilities. Although there have been several studies that indicate that green building costs more (from 1–2 percent of construction costs for large buildings, and 5 percent for small buildings), advocates indicate that by involving the engineers during the conceptualization of the project, and by considering life-cycle costs and total social costs, green costs may be negligible.[33] Indeed, a study by McKinsey and Company found that existing technologies could significantly reduce carbon emissions in buildings and save money at the same time.[34]

Green Building Codes

During the past ten years "Green Building Codes" have been adopted by a variety of jurisdictions. These codes are usually designed to improve energy efficiency, reduce water consumption and carbon emissions. Aside from LEED, the major certification systems are Green Globes, EarthCraft House, and the National Association of Home Builders Green Building Guidelines. Most of the green building codes make LEED rating standards mandatory. Although paying lip service to water and site efficiencies, most of the rating systems target energy efficiency.[35]

The federal government provides incentives for the use of LEED, and many state and local governments mandate a certain level of LEED certification for public construction, as well as private construction benefiting from state subsidies or tax advantages.

The American Institute of Architects has also established a set of goals as part of the "The 2030 Challenge" that seeks to make all U.S. buildings carbon-neutral by 2030. Federal legislation in 2009 also contained programs to make U.S. buildings net zero energy. Landscape architects, the authors of the "Blue-Green" movement, are active in promoting more sustainable water use.

Leadership in Energy and Environmental Design (LEED)[36]

The United States Green Building Council (USGBC) is a non-profit organization which developed and administers the LEED certification system. Today USGBC includes 9,000 organizations, including companies, governments, schools, non-profits, and trade associations which promote sustainability and green design.[37]

The LEED Green Building Rating System evaluates a building's performance against a variety of environmental factors (Table 24.2) to come up with an overall rating of Certified (26–32 points), Silver (33–38 points), Gold (39–51 points), or Platinum (52–69 points). The first system was for commercial buildings, but as of 2007 there were also systems for commercial interiors, existing buildings, core and shell buildings, residential, and neighborhoods. The ratings are done by inspectors certified by USGBC and their fees are paid for by the developer.

LEED certification mark

Table 24.2
LEED certification points by category

Category	Points
Sustainable sites	14
Water efficiency	5
Energy and atmosphere	17
Materials and resources	13
Indoor environmental quality	15
Innovation and design processes	5
Total	69

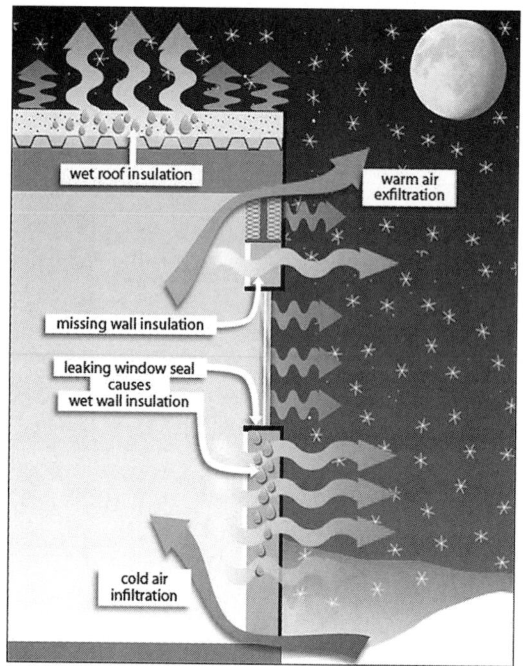

Diagram of heat loss in conventional building

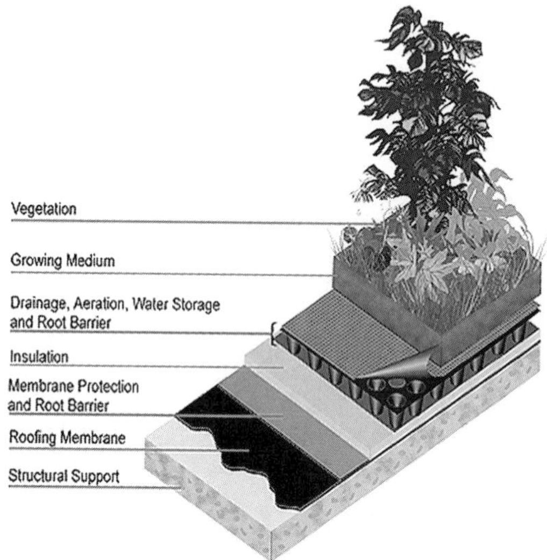

Green roof diagram

Minneapolis Central Library green roof

The following describes the major elements or technologies of the building and also suggests some current "green" building improvements that can be used for public facilities.

Elements of a Building

All buildings are much more than four walls and a roof. They are complex systems of systems. Structural systems; electrical, lighting, and communication systems; heating, cooling, and ventilation systems; plumbing and draining systems; fire protection and other safety systems; acoustic mediation systems; and finally, the specific elements of the function for which the building was built in the first place—all must find a place within the finished building.

The Building Envelope

The building envelope is the external shell of the building, the "four walls and the roof." As with windows, the envelope serves many purposes, including energy efficiency, environmental protection for the occupants, and aesthetics. A well-designed building envelope will be well insulated, will look good, and will be constructed of easy-to-maintain and long-lasting materials. It will also reduce heat loss during cold seasons and retain heat during hot seasons. Many municipal facilities have structural steel "bones."

Roofs are constantly exposed to sun and rain, and play an important role in the overall energy efficiency of the building and the urban heat island effect. They are best used for heat management in parts of the ountry with large swings in daily temperatures since their mass will mute the highs and lows. They are better at cooling than in protecting the building from the cold and therefore greater energy-savings benefits will be seen in warmer states.[38] Green roofs can also be part of an on-site water management program to reduce stormwater runoff. For more on green roofs, see www.greenroofs.org.

Glazing Systems and Lighting

The collection of windows and other transparent surfaces in a building is called the glazing system that provides daylight, solar heat, ventilation, views, and emergency access. Windows can have different qualities of solar heat gain, thermal efficiency, air-leakage rates, transmission of visible light, and materials of construction. In addition, lighting accounts for 25 percent of the energy use in a commercial building.[39] Energy for lighting can be reduced by using more natural light from large windows, skylights, and a smart lighting system that uses photo sensors to open or close blinds or sunshades (Figure 24.3)

Water and Wastewater

Public buildings have restrooms and drinking fountains and some might have kitchens and laundry rooms. Still others might have landscaping irrigation needs. Finally, almost all public buildings have overhead sprinklers for fire protection. All of these require the delivery of water and the removal of wastewater. A typical plumbing system will include water pipes, drain pipes, ventilation pipes, and natural gas lines for the water heater (Figure 24.4). In buildings that use recycled water, a separate piping system from the centralized or satellite treatment plant or an on-site treatment facility is required.

Gray water systems are of interest in drought-prone areas because of their potential to reduce potable water use by 50–80 percent. Many public buildings and recreational facilities throughout the USA have used gray water for irrigation for many years. If treated, gray water can be used for toilet flushing, showers, and laundry.

Blue-green features in a public building can also prevent the harmful effects of stormwater runoff. Such features include green roofs, living walls, green façades and vine walls, rainfall harvesting with water reuse and courtyard landscapes with biofiltration. The

Natural Resources Defense Council's Green Building

The Natural Resources Defense Council's headquarters in Santa Monica, California, employs a number of energy- and water-saving devices that result in using 60 percent less energy than a conventional office building of the same size, and 60 percent less water. They state that if all commercial and public buildings in the United States were "as efficient as ours, the country could cut energy production enough to meet 70% of the requirement of the Kyoto Protocol."

Energy-efficient innovations include a solar electric array that produces 20 percent of the building's power, and they also buy wind certificates. The design of the building reduces artificial lighting demand by having sunlight reach every storey through three light wells. Sensors dim hallway lights when daylight is sufficient and sensors turn off lights when offices are vacant. The windows directly facing the sun have a reflective coating to prevent excess heat from coming through. Natural ventilation is used where possible and in other areas, peak load cooling is provided by a high efficiency, non-ozone-depleting air conditioner. Each office has a heating "convector" with its own thermostat which turns off if a window is opened.

The building has an integrated, gray water recycling system that captures gray water and rainwater, and treats it so that it can be used for showers, toilet flushing and irrigation. All rainwater is captured on site through a porous paving system and landscape planters. The building uses dual flush toilets, and waterless urinals. Landscaping includes both a drip system and a high efficiency subterranean system.

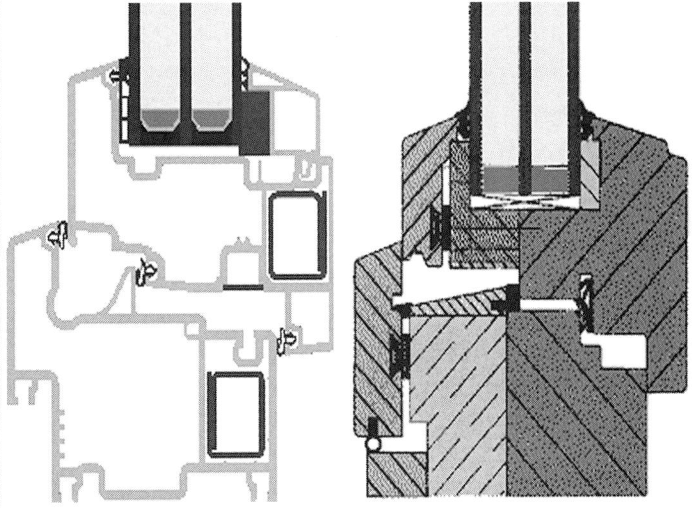

Figure 24.3
Energy-efficient windows

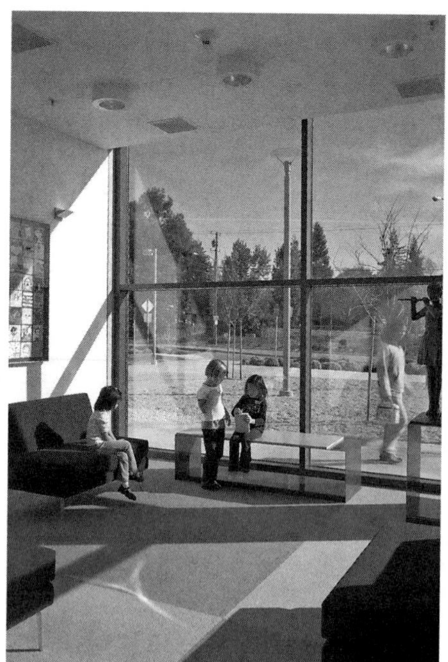

Interior view of window wall system of the Community School of Music and Arts in Mountain View, CA

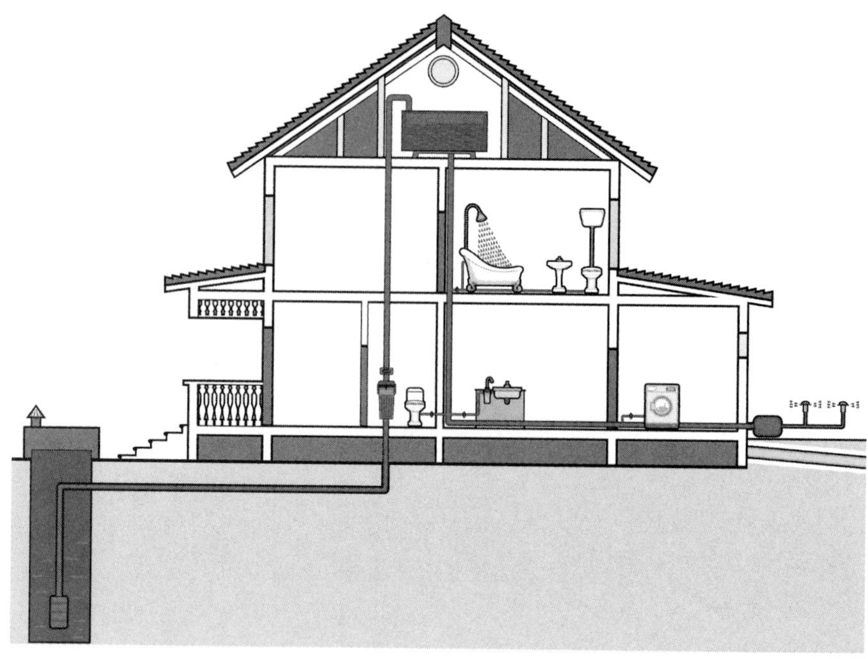

Figure 24.4
Schematic of an interior plumbing system

Cities of Portland and Philadelphia have stormwater ordinances that treat green roofs and courtyards as undeveloped open space. Philadelphia requires that all new or redevelopment proposals increase the pervious area by 20 percent.[40] The City of San Francisco has embarked on a Green Infrastructure program, looking at buildings and water systems in the neighborhood as a whole (Figure 24.5).

Dramatic gains in water efficiency can be also obtained by retrofitting a public building with water-efficient toilets, low flow faucets, and by using a more efficient irrigation system. Wastewater can also be used for heating. A Salt Lake City, Utah, law firm uses a heat pump in the winter to transfer heat from the sewers back into the building with a series of glycol and water pipes, making use of the fact that the contents of city sewers are warmer than the surrounding air in the winter. [41]

Heating and Cooling

Public buildings have systems for heating and cooling. Together, these systems are usually known as heating, ventilation, and air conditioning, or HVAC, systems. HVAC systems, depending on age and application, consist of furnaces, boilers, ducting, air handlers (units that move the

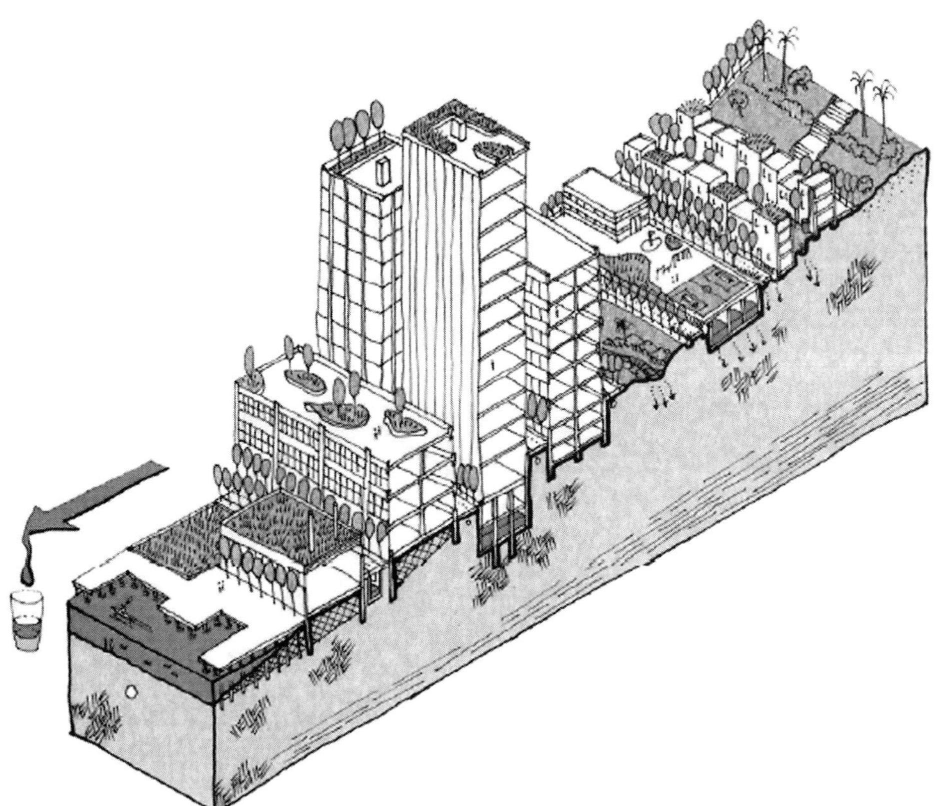

Figure 24.5
Green Infrastructure schematic

heated or cooled air throughout the building), and air conditioners. HVAC systems also include the thermostatic controls.

The heating and cooling systems in a commercial building account for 32 percent of its energy use. Energy use and carbon emissions can be reduced in this area by using natural ventilation or integrating new efficient technologies such as ductless systems and evaporative coolers. If the building envelope's insulation is improved along with window performance, less energy will be needed to cool or heat the building. Additionally reduced heat from appliances and lights will assist energy conservation and reduction of carbon emissions.[42]

A crane lifting a heating, ventilation, and air conditioning unit

Energy Systems and Electricity

In the early days of electrification, the electrical power requirements of any building were simple: to provide for lighting (though proper lighting design is itself a complex procedure). In today's public building, the situation is more complex. Emergency lighting, for example, must be on a separate circuit from other lighting, and must be relatively immune to power outages. Additionally, today's office equipment has special electrical power needs, such as a higher voltage or conversion to direct current.

On-site or distributed energy generation technologies using renewable sources can supplement grid-produced electricity for buildings. Photovoltaic cells or wind turbines are increasingly being designed into state-of-the-art buildings. Currently the U.S. Department of Energy's Building America program is partnering with private industry to develop more carbon-neutral renewable energy sources so that net-zero energy buildings can become a reality. In Europe, there are several instances of buildings generating biogas on-site from food scraps and blackwater. (See also the Database of State Incentives for Renewable Energy (DSIRE) at www.dsireusa.org for more information on incentives to promote renewable energy.)

Land Use Planning and Public Facilities

Local practitioners can influence how other parts of the organization plan for public buildings by changes to local development regulations regarding these facilities that are then used to process the permit for the public or quasi-public facility. Local practitioners can also be involved in developing a capital needs study and performance indicators for public facilities that are then used as back-up for the capital improvement plan and budget. The local practitioner can also be a project manager for a public building or quasi-public building.

Comprehensive and Area Plans

Local officials should ensure that both the regulations about green building standards and the location of public buildings are considered during the development of a comprehensive infrastructure plan or the community facilities plan. There are some local controls available over public buildings (Table 24.3).

If such a plan does not exist when the locality is contemplating a major renovation or the development of a new public building, this is an excellent opportunity to revisit the land use for the area surrounding the building. The development or major rehabilitation of a public building, whether it

Retrofitting Existing Buildings

The Sears Tower in Chicago, Illinois, has been retrofitted with an in-building wireless system for cell phones and WiFi computer access in all offices, on all floors, and in the elevators. The system supports supports fire, life, and safety communications for emergency first responders.

In addition, in 2009, the Sears Tower embarked on a $350 million renovation to use 80 percent less energy. This include replacing its 16,000 single pane windows, adding more efficient water fixtures, sensor lighting systems, and updating the building's 104 high-speed elevators. On-site energy will be generated from wind turbines, photovoltaics, and solar hot water heating. Green roofs will be added both to help stormwater runoff and as insulation.

Table 24.3
Local controls over public buildings

Feature	Local regulation
Building location	Land use element of general plan; subdivision ordinance; zoning ordinance
Green building standards	Ordinances; building codes
Seniors, ADA	Building codes
Emergency preparedness building performance	Performance indicators in capital budget

Source: Compiled by the authors.

is city, country, state, or federal, can be a catalyst to rethink the area plan for the civic center. The preparation of a detailed design plan can be used to mobilize public support for the bond issue to pay for the public building.

Performance Indicators

Level of service standards are used in assessing the need or demand for a particular type of public building. They can be referred to during the condition assessment stage while preparing a capital needs study or during the development of an individual building. They can also be used while preparing a long-range comprehensive plan or as part of the capital budget process. They are often hard to determine, and oftentimes have nothing to do with the building itself. For example, service standards for city hall and the county administrative buildings might be certain opening hours or a certain minimum waiting time to see a clerk. Many communities choose not to use service standards at all when evaluating some public facilities, instead relying on a method of citizen "demand and response."[43]

However, many public facilities do have quantifiable service standards, and many of these standards affect or are affected by the physical and land use context of the building. Here we discuss some service standard issues for some common public facilities.

- *City hall and other specialized municipal buildings.* Although it is difficult to set an output-oriented service standard for city halls and corporation yards, every year the International City/County Manager's Association Center on Performance Measurement publishes statistics about square footage and costs for these facilities by different types of cities.[44]
- *Public libraries.* Public libraries often measure service quality by the size of their collection, or by the number of books per person in the jurisdiction. Similarly, libraries measure their patrons' activity.

Nationwide, the average is 4.2 visits per capita. Libraries also might measure community accessibility, through such things as opening hours, proximity to residents, number of branches, or parking lot size.

- *Police stations.* Police departments measure their performance by looking at response times, crime rates, and officer "free time." Community policing or proactive crime prevention is important, and can be measured with the concept of officer free time. Free time is time that officers are not responding to service calls. Thirty percent free time means officers are able to respond to urgent service calls within five to eight minutes (a commonly accepted average for crimes in progress) and still have time to fill out necessary reports, attend department meetings, and miss work due to sickness or vacation. At an average of 40 percent free time, officers can take the time to investigate reoccurring problems in their beat area and develop mutually-supportive relationships with the community to solve problems proactively. Police standards are not linked to land use *per se*, but key ways of reducing response time for incidents and increasing free time certainly include building new police stations and expanding existing ones.
- *Fire stations.* Fire departments measure their service quality with a seemingly simple measure—response time. Response time is defined as elapsed time (in minutes) from when the first fire company is dispatched to when the first company arrives at the emergency scene. A response time of five minutes is a common standard. It is a function of many things: distance from the fire station to the emergency, condition of the fire trucks, and condition of the streets and roads on the way.
- *Jails.* Standards for jail normally are determined by the state, and include physical as well as operational proscriptions. The State of Nebraska, for example, requires 25 square feet of floor space for every inmate in holding areas, and 50 square feet of area in cells. Similar standards govern the size of guardrooms, number of showers, size of inmate processing areas, and size of dining facilities. These standards will determine the size, location, and layout of jails.

Other Building Design Considerations

Seniors and Americans with Disabilities Act of 1990

An issue of escalating importance is the need to design for the aging population. In 1960, there were 12.7 million people over the age of 65. In 1991, the figure was 31.5 million, and by 2020 it is expected to be 52.1 million, over 5 million of whom will be over the age of 85.[45]

The Americans with Disabilities Act gives civil rights protections to individuals with disabilities similar to those provided to individuals on the basis of race, color, sex, national origin, age, and religion. It guarantees equal opportunity for individuals with disabilities in public accommodations, employment, transportation, and state and local government services. Accommodations such as wheelchair access, ramps and slopes, parking and loading, elevators, doors, handrails, fire escapes, assembly, signage, drinking fountains, and bathrooms are all building elements that must be considered during the design of a public building.[46]

Wheelchair access to balcony in San Antonio Municipal Building

Emergency Management Considerations

Some public buildings also need to be physically "hardened" to withstand the rigors of earthquakes, floods, hurricanes, and terrorist attack. Public safety buildings, which will be the center of the emergency response effort, need special structural improvements and communications equipment. Other public buildings that represent a significant public investment should also be designed to be able to survive a disaster and to protect the lives of the public employees.[47]

Building Appearance and Design Competitions

Design competitions are an oft-overlooked method of achieving attractive buildings. Noted civil engineer and historian David Billington has said that design competitions, especially those with juries composed of engineering and architectural laypeople, can

Figure 24.6a
Design competition winner for Berkeley Public Safety Building, 1996

Figure 24.6b
Exterior of new Berkeley Police Dept building as built, 2005

produce great results.[48] Others disagree: Witold Rybczynski wrote: "Public competitions for architectural commissions don't necessarily produce the best buildings" when design is promoted over a sense of place for the community. See Figures 24.6a and 24.6b for the competition winner for the Berkeley Public Safety Building and a photo of what was actually built.

Conclusion

Public buildings are often the first contact the local practitioner will have with infrastructure, yet they are often the most difficult to plan and finance because they have no independent revenue source. Sizing, location, and timing of construction are all dependent on the local political process. Quasi-public buildings and facilities such as convention centers also must be subsidized, but they depend on generating market demand as well. They are usually more popular with elected officials since they are often tied to the national or regional prestige of a city. Although the technical aspects of actually designing and constructing a building are quite developed, planning and financing are usually specific to the particular facility and the local policy context. Regardless, the local practitioner can use the development or planning opportunity to promote a more sustainable approach to the facility.

Additional Resources

Publications

Noll, Roger and Andrew Zimbalist. *Sports, Jobs, and Taxes.* Washington, DC: Brookings Institution, 1997. This book looks at the economic impact of new stadiums and sports franchises on the local economy.

Sanders, Haywood. *Space Available: The Realities of Convention Centers as Economic Development.* Washington, DC: Brookings Institution, 2005. A sober analysis of what convention centers actually do for local economies and what the future market holds.

Tobin, Robert, et al. *The Courthouse: A Planning and Design Guide for Court Facilities.* Denver, CO: National Center for State Courts, 1999. A useful guide for the court administrator facing his or her first development project.

Wilson, Alex, U.S. Federal Energy Management Program. *Greening Federal Facilities: An Energy, Environmental, and Economic Resource Guide for Federal Facility Managers and Designers.* Washington, DC: U.S. Department of Energy, 2001. A comprehensive guide to green facilities useful at the local level as well.

Yaudelson, Jerry. *The Green Building Revolution.* Washington, DC: Island Press, 2007. A good introduction to the LEED system, with many case studies of best practices for green public buildings, public health, and green education facilities.

Websites

Center for Climate and Energy Solutions: www.c2es.org/.

Database of State Incentives for Renewables and Efficiencies: www.dsire-usa.org.

U.S. Green Building Council: www.usgbc.org.

chapter 25

Public Schools as Public Infrastructure: Schools, Community, and Land Use Planning

Jeffrey Vincent

In this chapter ...

Importance to Local Practitioners . . . 531
A Brief History of Public Schools in the United States . . 532
Schools in the Early Nineteenth Century . . . 532
The Establishment of Public Schools . . . 532
Progressivism in the Late Nineteenth Century . . . 533
The Early Twentieth Century . . . 533
Post WWII: The 1950s . . . 534
The 1960s–1970s . . . 534
The Excellence Era: The 1980s–1990s . . . 535
The Twenty-First Century . . . 535
The Institutional Context of Public Schools . . . 536
Types of School Districts . . . 536
School District Consolidation . . . 537
Who Uses the Schools? . . . 537
How Are Schools Financed? . . . 537
What Is the Current Condition of School Infrastructure? . . . 539
Public Schools in the Context of Cities and Metropolitan Areas . . . 539
Public Schools and Neighborhood Demographics . . . 539
Schools and Housing Nexus . . . 540
Planning Public Schools Today . . . 541
Small Schools Emphasis . . . 541
Small Schools and Urban Planning . . . 542
Types of Small Schools Being Developed . . . 542
Impediments to Small Schools . . . 543
Innovative Design Practices . . 544
Schools and Land Use Planning . . . 547
Providing Schools in High Growth Areas . . . 548
Smart Growth and Schools . . 548
Schools in Redevelopment . . 550
Conclusion . . . 551
Additional Resources . . . 552
Publications . . . 552
Websites . . . 552

Why Is This Important to Local Practitioners?

Public schools exist in nearly every community. As elements of public infrastructure, public schools present unique opportunities and challenges for local planning practitioners. Their quality, both real and perceived, is directly related to surrounding land uses, yet they are run not by the city or county, but typically by a local school district separate from the other coordinated public infrastructure bodies. Prior to the 1970s, local planning and school planning were fairly in sync, and generally development was sequenced to match public school capacity. Today, the norm is poorly aligned planning systems that do little to systematically link public school capacities and operating considerations with new development or redevelopment. This is beginning to change as many cities and states are more concretely addressing the need to coordinate land use and school planning. In many locales, innovative school planning is moving away from the traditional "one-size-fits-all" approach to school facilities design.

This chapter highlights the relationships between cities and public schools. It describes ways in which local planning practitioners can better consider public schools as important, and unique, elements of public infrastructure that should be a part of comprehensive city planning. First, the chapter provides a brief history of school planning in the United States. Next, we describe the institutional context of public school districts, including school district types, financing issues, and the conditions of school infrastructure. From there we discuss the relationship between cities, metropolitan areas, and public schools by considering demographic changes, the tax base, and the housing–schools nexus. The remainder of the chapter focuses on the major land use considerations involving public schools as infrastructure, current issues, trends and innovations, followed by some recommendations and a presentation of best practices.

A Brief History of Public Schools in the United States[1]

Historically, U.S. public schools have been rooted in neighborhoods under the control of local leaders. Although this changed during the Progressive Era, when schools came under the purview of elected school boards that were separate from city hall, today's schools are moving back towards stronger connections with the community. The evolution from the historic one-room school houses to current facility models enables greater understanding of the contemporary, complex set of laws governing the siting, design, and construction of public schools as public infrastructure.

Schools in the Early Nineteenth Century

Schools emerged in the nineteenth century largely as independent, community-based methods for educating the children in small towns and neighborhoods. The school and the community were intricately intertwined—the school was frequently the physical and social focal point within the community, taking on roles in providing space for educational, social, dramatic, political, and religious activities. One-room school houses came to dot the landscape. In many instances, school houses and churches were one and the same. In the predominantly rural context of the United States at this time, school gave children the chance to come together and socialize with their peers. These schools, in what is now described as a "village model," were a largely voluntary institution with attendance varying by weather conditions and seasonal agricultural labor necessities at home. In the early 1800s, the Lancasterian model of schooling, an English model that sought to educate great numbers of children to advanced skills at low cost, emerged and helped pave the way for the current system of free mass education supported by public taxes. During this time, little consideration was given to how the design and placement of school buildings accommodated learning.

The Establishment of Public Schools

Advocates such as Horace Mann, who would go on to become the first U.S. Secretary of Education, promoted the importance of free public education as an integral component in a free and democratic society. They argued that schools should become the "great equalizer" to eliminate poverty and inequality. As the notion of free public education began to take hold across the nation, the first free public common school was established in 1839 in

Massachusetts, which was also the first state to mandate a six-month minimum school year. In the late 1800s, advocates also pressed issues of comfort, efficiency, and design—including location, size, construction methods, and external considerations—into school facility planning.

Progressivism in the Late Nineteenth Century

Led by John Dewey, the late nineteenth-century Progressive movement in education pushed centralizing schools together under districts and standardized the curriculum between sites. As a result, there was a general increase in administrative staff, creating a more formal system of schooling. Education was regarded as the means of transforming pre-industrial society—with its perceived lackadaisical attitudes and work habits—to a fully modernized society ready to prosper in the industrial age. Schools were seen as the mechanism to change society. Public schools came to be regarded as the only institution that could counteract or compensate for neglectful families. Progressives argued for greater governmental involvement in addressing social ills and supported public schools as a mechanism for Americanizing new immigrants. Progressivism marked a shift from the Lancasterian model of mass learning to more individualized learning, which demanded smaller class sizes and uniquely designed buildings. However, the dominant school facility design of the time was the "Quincy Box," consisting of four floors of identical rooms, a basement, and an attic. Radical at the time, Progressive educational reforms were based on supporting the industrializing economy.

The Early Twentieth Century

The early twentieth century continued to see changes in education and school facilities planning. Two major changes included the addition of vocational and physical education programs, which were supported for secondary school levels by the federal government. In addition to the normal college preparatory courses, new classes in science and commerce were added. There was also a new interest in the school's role in supporting children's physical health. Partly in response to the large number of young people rejected by the armed forces in World War I, schools began to build sports fields, swimming pools, gymnasiums, and playgrounds. Some schools also began including space for nurses offices and health clinics. As a result, there were new physical design requirements to accommodate the changing and growing curriculum structure. During this time, the needed space for schools grew, and new school facility sites increased to sometimes more than 10 acres. To accommodate the need for more space, most new schools were built on the urban fringe where land was plentiful.

A significant school facility design change came in the form of the "finger plan" in the 1930s. Used primarily in elementary schools and seeking to accommodate the curriculum changes occurring at the time, the finger plan was characterized by buildings with wings typically 30–40 feet apart, containing four to five classrooms on one side in line with an open corridor and an "outdoor classroom" on the other side. Usually built on 10-acre sites, this design utilized aspects of daylighting and cross-ventilation. Although a poor energy user due to the great expanses of outside walls, the finger design continued through the 1960s, supported by a plentiful supply of inexpensive energy. For many years, school planners were guided by states' specific site size recommendations based on projected school capacity. These recommendations differed for elementary, middle, and high schools, and worked well during the early- to mid-1900s when land was relatively cheap and readily available.

Post WWII: The 1950s

The second half of the twentieth century saw continued educational changes that brought changing school facility design needs. Along with a national population boom following World War II, there was a great increase in school facility construction, especially in the growing suburbs. Innovations in plastics, glass, and concrete led to single storey school buildings as the norm in the 1950s. As new schools sought to expand educational programs to include health and food services, physical education programs, specialized administrative space, large auditoriums, and libraries—all of which often occupied 50–80 percent of the building—there was a need for increasingly larger school sites. In line with the construction innovations of rapid suburban development during this time, there was also a growing prevalence of prefabricated modular building plans and components to support the boom in construction.

The 1960s–1970s

The 1960s and 1970s brought two new school designs as educators and architects questioned the traditional conception of the classroom as a self-contained unit. Various examples of "cluster plans" and "open space plans" emerged that pursued greater interior school flexibility such as new concepts of open spaces with flex walls, pods, and team teaching designs. As a result, the facility designs of the 1940s and 1950s became inadequate. The 1960s also brought about more formalized school facility specifications in many states. Labeled a "functional approach" to school design, school planners began to study predicted enrollments, the spatial relationships of different internal uses, building design, needed parking, outdoor recreation needs, and other factors to guide the total facilities planning process. In addition, the

1970s brought about federal legislation that affected school design, including mandating gender equity in physical education and equal access for low-income students and the disabled, which further required additional acreage on-site. However, the national energy crisis of 1974 forced many states to require school buildings to be designed to conserve more energy, ushering in a new era of energy-efficient school design.

The Excellence Era: The 1980s–1990s

The 1980s and 1990s, dubbed the "excellence era," marked a generation of education reforms aimed at enhancing student learning and at addressing community context. During this time, there were continued changes in school facility design that reflect the demands of our educational and social culture. Today, the trend is towards new arrangements of decentralization and site-based management.

The 1980s saw an increase in school-age children across the country, and many districts found they did not have enough space at the elementary school level. During this time, additions were made to existing schools, modular classrooms were purchased, and many new schools were planned. The cycle for needed school infrastructure fluctuates with the changing population of school-age children. For example, the post-World War II baby boom spawned a focus on building elementary schools in the 1950s. As those Baby Boomers aged, the 1960s saw an increased need for additional high schools. Then, in the 1970s, there was a nation-wide drop in enrollments and many districts suddenly had surplus schools. School planners wrestled with what to do with school buildings that were no longer needed for traditional education purposes.

During the 1980s and 1990s, many former school buildings were sold and/or converted to shopping centers, corporate offices, apartments, municipal offices, and community centers. In many instances, schools sat vacant and suffered severe deterioration as a result of deferred maintenance. Older schools that became empty, often located in prime locations within neighborhoods, aided in neighborhood blight. School facilities planning involves considering the effects of "booms and busts" in the population of school-age children over time to avoid having too few, or too many, schools.

The Twenty-First Century

Today, student populations are on the rise. However, these trends are not homogeneous across the country; there are unique regional and city-level differences. Likewise, the number of immigrant students, especially in central cities, continues to increase, adding specific challenges to schools. Amidst these population characteristics, public policy changes in education have ushered in a new era of student choice, school accountability, and public schools run by other groups rather than school districts. Charter schools, magnet schools, and small schools are increasingly common phenomena.

The twenty-first century finds educators and school designers questioning the outcomes of learning based on the new reality of the postindustrial, information age. People need to be creative, critical thinkers and have the ability to work in teams, which is in contrast to the needs for manufacturing productivity in the industrial age. Innovators in school design are re-envisioning schools in the context of a rapidly changing economy and trying to create schools and built environments that support these outcomes. In addition, schools are increasingly being scrutinized in terms of how they fit into redevelopment and Smart Growth strategies. Likewise, local planning practitioners are called to think about how redevelopment and growth decisions impact schools and learning. Above all, our schools need to be designed and redesigned to foster innovative teaching and learning methods so that students are best prepared for the changing nature of work and citizenry in our dynamic society.

The Institutional Context of Public Schools

School districts are the most well-known special districts. By 2007, there were close to 99,000 public schools in nearly 14,000 school districts.[2] Most school districts are governed by locally elected boards and conduct business separately from the general purpose government, although some districts are run by municipalities or counties. School districts' primary sources of funding come from state funds and local property taxes.

Types of School Districts

School district organization and geographic structure vary by state.[3] Most U.S. schools offer a comprehensive curriculum, and often provide other programs and services as well. School districts, also known as local education agencies, are classified as unified (primarily serving children of all grade levels), secondary (primarily serving children in grades 7–12), or elementary (primarily serving children in grades K–6). At the federal level, the U.S. Department of Education establishes, administers, and coordinates federal assistance in education. The Department's mission is to promote excellence in the nation's schools by ensuring that all students have equal access to education.

Public school districts provide or support elementary, secondary, and sometimes postsecondary education services. Most school districts are classified as independent, but about 10 percent are classified as agencies of other governments. These districts are typically dependent on either state, county, municipal, or township governments. The way in which this is set up comes from the varied state legislative provisions. In 2007, 30 states had only independent school districts. While being the dominant form of school governmental organization in the country, it is near universal in the western states. Fifteen states have both independent and dependent school districts. State-dependent school districts tend to be located in sparsely populated areas. The District of Columbia, Alaska, Hawaii, Maryland, and North Carolina have no independent school districts—they are all administered by either county, municipal, or state governments. Thirteen states have some public school districts administered by municipal governments.

School District Consolidation

While the number of other special districts tripled from 1952 to 2007, the number of school districts decreased nearly 80 percent from 67,355 to 13,051 during the same time; about one-fifth of the total. This trend was highest between 1942 and 1972 due to consolidation and reorganization of school districts, particularly rural districts. Since then, school district consolidation has continued, but at a much slower rate. About 75 percent of the school district declines since 1997 can be attributed to four states: California, Minnesota, Nebraska, and Oregon. In 22 states, there was either no change or an increase in the number of school districts from 1997 to 2007. Whereas 25 states had at least 1,000 school districts in 1942, only two (California and Texas) have at least 1,000 in 2002. The consolidation statistics reveal how school districts have become large public service and infrastructure institutions.

Who Uses the Schools?

Today, school-age children account for almost 20 percent of the nation's population. The U.S. Department of Education projects 7 percent overall national enrollment growth between 2010 and 2011.[5] Certain areas, primarily those that attract younger families will show a greater increase in school age population. Federal projections to 2012 show increases of 2 percent in the Northest, 2 percent in the Midwest, 9 percent in the South and 13 percent in the West.

How Are Schools Financed?

Public school districts are financed by a variety of revenue sources, from all levels of government. School financing predominantly comes from three sources: state funds, local funds, and developer fees. Locally, revenue comes from such sources as: local taxes, primarily property tax; investments; and revenues from student activities, textbook sales, transportation and tuition fees, and food service. State sources include funds for transportation and pupil-targeted programs (special, gifted, vocational, and adult education), textbook funds, capital outlay, debt service payments on local school debt, property-tax-relief payments, child nutrition matching payments, employee benefit payments, and loans to local education agencies. Finally, the federal government provides direct grants-in-aid to schools or agencies, funds distributed through a state or intermediate agency, and revenues in lieu of taxes to compensate a school district for nontaxable federal institutions within a district's boundary (known as impact aid). However, it is important to remember that education in the USA remains primarily the responsibility of state and local governments. The federal government fills gaps in state and local support for education when critical needs arise.

As independent special districts schools typically have the authority to enter into bonded indebtedness. Long-term borrowing in the form of bonds is a mechanism utilized by the majority of public school districts for funding school construction and modernization. A bond is a security whereby the issuer agrees to pay a fixed principal sum on a specified date and at a specified rate of interest. School bonds typically go before the voters before being issued, which can be difficult for cash-strapped school districts when voters reject a proposed bond. Most states require a simple majority, as opposed to a two-thirds majority, approval to go forward with a bond.

Other features of school finance sources are:

Revenue mix. In 2007, the national average for revenue source mix was 44 percent local, 48 percent state, and 8 percent federal, but that mix varies greatly from state to state and district to district. Districts in poorer states rely much more heavily on higher levels of government. For example, revenues from local sources as a share of total budget ranged from 7 percent in Vermont to 66 percent in Nevada. An average of $9,683 was spent on each student nationally in 2005. Again, this varies by state.

Financing facility maintenance and increased capacity. Although there is an elaborate funding structure for public education as described above, there is still much debate about how to pay for needed maintenance and additional capacity. Providing and financing schools to accommodate new residential development has been labeled "school mitigation." Typically, developers pay fees to support building the new schools that will be needed to serve their developments. A key question is whether it is more equitable to finance schools from a metropolitan-wide, statewide, or even federal base than to tax new development. But for reasons of tradition and local control, Americans have preferred local finance. School mitigation is discussed in more detail later in this chapter.

What Is the Current Condition of School Infrastructure?

Today, many argue that the United States is in a national school-facilities crisis characterized by aging infrastructure, overcrowding, and inadequate spending compared to need. In 1995, the U.S. General Accounting Office (GAO) estimated that U.S. schools had to make $112 billion worth of repairs.[6] As of 2000, only about $12 billion of those had been completed.[7] The GAO estimated that 14 million children attended schools in serious disrepair. Low-income and minority children, who are often already disadvantaged, were most likely to attend these schools.

A 2013 report by the U.S. Green Building Council found that many schools across the country continue to be in states of disrepair that threaten the health, safety and education of students.[8] The report estimated that $542 billion was needed nationally to comply with modern laws and building codes and to modernize all schools. This is an average of $5,450 per student. Most states and local school districts do not have an adequate infrastructure plan for remedying these needs.[9] Still about $20 to $30 billion is spent annually on K-12 capital investments by school districts across the country.[10] However, severe inequities have been found in how these funds are spent.[11]

This section has provided the institutional context of public schools by describing different district types, trends in consolidation, the population they serve, school infrastructure conditions, as well as investigating issues of finance. The next section will discuss public schools in the context of the changing nature of cities and metropolitan areas.

Public Schools in the Context of Cities and Metropolitan Areas

Understanding public schools as unique elements of public infrastructure entails understanding how the broader trends within metropolitan areas have impacted our educational system.

Public Schools and Neighborhood Demographics

As neighborhood demographics change, so do schools. Public schools have been shown to be important indicators of community health, both currently and for the future, for two main reasons: (1) when schools reach a certain threshold of poverty, middle- and upper-class families of all races choose not to live nearby; and (2) a community's school children are likely to become its next generation of adults.[12] Schools enrolling increasing numbers of poor students tends to lead to a softening of middle-class demand for housing in the vicinity, because these families do not perceive the school to be high quality.

This transition accelerates as the racial diversity (i.e., non-white) percentage of students in the school increases, leading to neighborhoods and schools that are economically and racially segregated. Non-white enrollment in U.S. public schools was about 12 percent between 1940 and 1960, rose to 36 percent in 1996, and is projected to be nearly 60 percent by 2050.

Understanding this process of neighborhood and school change requires a better understanding of middle-class family mobility choices. While little is known about when and where middle-income families choose to live, the impact of poverty concentration on public school quality is significant.[13] Various recent public policies by municipalities (such as tax abatements, below market-rate mortgages, and other financial incentives) have sought to attract and retain middle-class families. However, few have addressed public schools specifically within these policies. Cities need to attract and retain middle-income families in order to remain viable economically, socially, and politically. Both real and perceived problems of public schools remain a major obstacle to middle-class retention within, and attraction to, inner cities.

Schools and Housing Nexus

Until court-ordered desegregation policies came into effect in the 1950s, classrooms mirrored the demographic compositions of local neighborhoods. However, desegregation policies have been met with mixed success. While many schools have been successfully integrated along race and/or class lines, these policies have created vast logistical issues including transportation and school assignment, because most neighborhoods remain segregated. Increased attention is being given to policies to promote integration in neighborhoods through housing policies that may, in turn, promote school integration.[14] The trend in building mixed-income housing developments is one example. However, systematic policy linkages between housing integration and educational integration in order to attract and retain middle-class families in urban neighborhoods have rarely been created.

The quality of schools impacts the residential location choices families make. The 2003 Public Policy Institute of California (PPIC) Statewide Survey on Government[15] found education to be the third most important issue for citizens, and the similar 2002 PPIC California Statewide Survey on Land Use[16] found schools to be the third most important factor people considered in choosing neighborhoods to live in. Also, the American Planning Association/American Institute of Certified Planners (APA/AICP) 2000 Millennium Survey found the highest concern of voters (76 percent) to be having adequate schools and educational facilities.[17] Moreover, when voters in suburbs and small-to-medium cities were asked what might lead them to live in an urban setting, better schools ranked first.

Both urban and educational researchers note the close link between housing segregation, by both class and race, and school segregation. The result in many metropolitan areas is a bifurcated public educational system: a suburban one of adequate resources and student achievement, and an urban one of inadequate resources (though not necessarily less) and student failure.

This section has described the city and metropolitan contexts in which public schools find themselves. The remainder of the chapter focuses on planning for schools with a particular emphasis on the local planning practitioner's role.

Planning Public Schools Today

Generally, school facility planning changes occur in accordance with the prevailing theories and practices of learning and education. Because of this, planners need to understand current educational issues to assist in developing and redeveloping metropolitan areas to support schools. At the same time, this must be balanced with the reality of increasingly expensive and scarce real estate on which to site new schools. The rising cost and scarcity of land have led school designers to rethink the one-size-fits-all approach with its rigid site requirements and to adopt a more "functional approach."

Small Schools Emphasis

The small school movement has been the biggest trend in school reform and design since the 1990s. Educational researchers have shown that smaller schools tend to improve student academic performance, as well as enhancing school safety. The movement towards smaller schools is counter to recent trends. Since 1950, the number of schools has declined by about 70 percent, while the average size of schools grew fivefold.

The intimate settings of smaller environments tend to encourage more effective learning because more personal relationships between students and teachers can develop. Evidence has shown that small schools seem to benefit children in poverty the most, which makes a strong case for considering small schools in community revitalization efforts in urban neighborhoods. Educators disagree on the recommended sizes of schools, but generally agree to limits of 300 to 900 students, with class size limited to about 20 students. Compared to larger schools, students in small schools have been shown to fight less, feel safer, have more regular attendance, and feel more attached to their school.[18] In addition, research has found that small schools tend to produce greater teacher satisfaction and enable greater parent and community involvement. However, some more recent study findings on small schools do not all find as many positive benefits as the earlier studies did.

There are numerous ways schools are incorporating small school concepts. Some schools are simply built for lower enrollments and others can have multiple small learning communities within one larger school. Such designs are very different from the large complexes of the mid-twentieth century. A great challenge for cities is how to retrofit existing schools to create spaces that optimize learning and embrace new educational theories. Equally challenging can be paying for smaller schools; the debate persists on whether or not smaller schools lose the economic efficiencies that helped originally drive the move towards larger schools.

Small Schools and Urban Planning

The notion of small schools has been met with great enthusiasm from several urban planning philosophies such as Smart Growth and the New Urbanism, as well as agreeing with the desired goals in "livable communities" and "walkable neighborhoods." These movements promote the use of smaller, community-based, and pedestrian-oriented urban design. Planners, architects, and developers have been touting the role that civic buildings, and especially public schools, can play in the life of a neighborhood, both as social and physical centerpieces. Thus, among Smart Growth advocates and New Urbanists, small schools, often referred to as "neighborhood schools," can be a cornerstone around which older neighborhoods are redeveloped and new neighborhoods are established.

However, school decisions are made by educators, not community planners, whose main goals are student performance and learning outcomes. Therefore, it is important for planners to address the core issues of student experience and educational achievement when attempting to work with school districts to site, design, or retrofit schools. Some planning practitioners have embraced the small schools concept as a mechanism for doing this, and they stress the importance of both size and location of schools in order to foster community, educational achievement, and neighborhood improvement.

Types of Small Schools Being Developed

The following section describes two types of schools that commonly incorporate aspects of small schools: charter schools and magnet schools:

Charter schools. As a relatively recent creation, the charter school concept has been one of the fastest growing public education innovations of the last decades. Unlike traditional public schools, charters operate largely independently of school district bureaucracies. Schools are issued a "charter" from their sponsor, typically a state or local school board, that serves as a contract between the board and the school that outlines operational details and performance criteria. While laws differ in each state, charters are typically issued for a period of 3–5 years, after which the school is evaluated and a decision is made to renew or revoke the charter—an arrangement that allows charter schools to exercise increased autonomy in return for accountability.

Minnesota passed the first charter law in 1991, and was soon followed by California in 1992. By 2005, 41 states, Puerto Rico, and Washington, DC, had charter schools. Charter schools have received wide support from governors, state legislators, school districts, and the federal government. The U.S. Department of Education has provided grants to support states' charter school efforts since 1995. As charter schools become more prevalent, the traditional models of large school design are decreasing in prominence, thus changing the issues of siting and space requirements. The Congress for the New Urbanism has promoted charter schools as important elements in their higher-density, mixed-use developments.[19]

Magnet schools. Magnet schools are a special type of public school meant to encourage voluntary integration, and tend to be located in older urban areas. Magnet schools have been created, usually with specialized and unique curriculum (such as fashion, art, or technology), to entice students of varied social and economic backgrounds to attend the school. With their introduction in the 1970s, the goal was to create public schools of high quality that would act as "magnets" for a racial cross-section of students beyond the apparent segregated neighborhood boundaries. It was hoped that they would bypass the political opposition caused by mandatory busing policies. Magnets meet their racial quotas by open enrollment access for students beyond established attendance zones.

Impediments to Small Schools

While small schools are a recent major educational trend, it remains to be seen how long their popularity will last. However, their creation requires resisting some established regulations of school planning. The following describes two current public policies and decision-making procedures around public schools that make school facility planning difficult, contentious, and misaligned: school acreage requirements and state funding biases.

Acreage requirements. Individual state education departments recommend or mandate the minimum number of acres for schools, but most have been influenced by the school facility guidelines set forth by the Council of Educational Facility Planners International (CEFPI).[20] The CEFPI school facility guidelines document released in 2004, however, does not list minimum site size recommendations.[21] School siting requirements are a major point of contention between city planners seeking to create mixed-use environments and school planners seeking to meet the requirements for adequate athletic fields, ample parking, and school buildings themselves.

The old CEFPI recommendations mean that an elementary school with 400 students would need 14 acres, a middle school with 600 students would need 26 acres, and a high school with 2,000 students would need 50 acres. Because sites this large are difficult to find in built-up urban areas, new schools tend to be built on the urban fringe, attracting new housing and commercial development nearby. In this manner, some planners have argued that schools become advance scouts for typical sprawl development. Older schools in existing urban areas tend to occupy much smaller sites and be surrounded by densely developed neighborhoods with little to no room for expansion. Therefore, the system of school site decisions and design is situated in a way that makes it difficult for cities, especially dense cities, to meet the acreage standards for new schools.

Many have argued that states' site size requirements are often overinflated, encourage land consumption, and counter mixed-use and higher density trends found in Smart Growth policies. It is for this reason that critics have pointed to schools as major generators of sprawl.[22] However, an increasing number of states are revising their site-size guidelines; currently, 25 states no longer have site-size requirements. However, current school standards often inadvertently favor the desertion of older schools in built-out neighborhoods for the construction of new schools on the urban fringe. Table 25.1 shows the different sizes and facilities of small and larger schools.

State funding biases toward building new schools. In considering modernizing existing schools within urban areas, state reimbursement policies, which tend to favor new schools rather than upgrading existing schools, are a significant factor. Known as the "60 percent rule," but varied by state, this rule states that if the cost of renovating an older school exceeds 60 percent of the cost of a new school, the school district is encouraged to build a new school instead. The percentages vary between states; for example, in 2003 Massachusetts had a 50 percent rule, Minnesota a 60 percent rule, and Washington State an 80 percent rule. These percentage rules can prevent a full cost analysis by state and local governments to support renovation projects because they rarely factor in the added infrastructure costs of new schools such as purchasing the site, establishing water and sewer connections, and the transportation and road work implications.[23]

Innovative Design Practices

Many unique models of public schools exist that differ from the schools' traditional relationship with its surrounding land uses. The following section

Table 25.1
Examples of small schools compared to larger, conventional school sizes

Former CEFPI Guidelines*(per 100 students) **plus 1 acre for each additional 100 students		**Examples of small public school facilities**	
Elementary School	10 acres**	Tenderloin Community School, San Francisco, CA (325 Students)	0.73 acres 3 floors Rooftop playground
Middle School	20 acres**	Gonzalo and Felicitas Mendez Fundamental Intermediate School, Santa Ana, CA (1,240 students)	12 acres 2 floors
High School	30 acres**	Minneapolis Interdistrict Downtown School, Minneapolis, MN (450 students)	0.8 acres

Note: **Even though CEFPI no longer lists these site size recommendations in their 2004 guide, these old recommendations continue to define the regulations in many states.

Source: Connie Chung, *Using Public Schools as Community Development Tools: Strategies for Community-Based Developers* (Cambridge, MA: Joint Center for Housing Studies of Harvard University and Neighborhood Reinvestment Corporation, 2002). Available at: http://www.nw.org/network/pubs/studies/documents/schoolsCommDevelop2002.pdf.

describes some recent innovative design practices that are especially relevant to local planning practitioners.

Infill development and adaptive reuse. As many urban school districts experience increased overcrowding, obtaining new sites within the urban landscape for new schools is becoming increasingly difficult. As noted previously, school districts have been critiqued for having obscenely large site requirements, not having to adhere to local urban plans, and being magnets for sprawl when built on the urban fringe. Historic conservation and Smart Growth efforts agree with educators on the need to seek ways of preserving old school buildings or using vacant urban sites for school infill projects rather than building large new schools in outlying areas.[24] Adaptive reuse strategies have utilized old strip malls, abandoned warehouses, and other underutilized properties to site schools. The use of pre-existing properties can prevent negative disruption of the neighborhood fabric and keep schools in central locations. This recycling of land fits within strategies of community and economic development and in the general considerations for Smart Growth.[25]

Community schools. While not an entirely new idea, the community school concept gained increased prominence in the 1990s, and has been used as a tool in community revitalization. Community schools (sometimes called full-service schools) bring educational, recreational, and health services together under one roof, particularly in disadvantaged neighborhoods, to better meet the wide variety of needs for children and their families. The Beacon Schools of New York City are one popular and successful example.

While there are numerous models around the country, they share the idea of making the school a hub to integrate the efforts of educators, families, and community partners. These partners can include social service providers, family support groups, youth development organizations, institutions of higher education, community organizations, and businesses, as well as civic and faith-based institutions. Common partnership efforts produce activities such as family support centers, early childhood and after-school programs, health and mental health services, dental services, job training programs, counseling, adult education, family nights, and community-based learning. The school becomes a place that seeks to integrate and provide the space to co-locate these activities to better meet the needs of children and families.

Joint-use facilities. Some schools, communities, agencies, and other organizations have constructed joint-use facilities or negotiated sharing physical space.[26] Examples include building one shared library or gym instead of separate city and school ones to maximize the use of financial capital and space. Alternatively, schools choose not to build these amenities and work out arrangements with cities and other agencies to use those already present in the community. This practice enables schools to site on smaller infill spaces within cities rather than looking to the urban fringe for adequate acreage. An underlying theme in this strategy is creating unity between schools and communities through shared space.

Green school design. Many states and school districts are increasingly looking at ways to build schools that reflect "green building" techniques in order to create better teaching and learning environments.[27] Additional benefits include increasing energy savings and improving indoor air quality. Research has shown that natural daylighting and better air circulation can improve academic performance. While some schools have been known to utilize solar or wind power, less obvious green techniques are far more common, such as orientation of the building to the sun to maximize natural light and moderate temperature, and innovative ways of reducing wastewater on site. These strategies will yield the most significant savings over time. Money saved in terms of heating, ventilating, and cooling (HVAC), electricity, and water can be redirected toward instruction.

Supported by the U.S. Green Building Council in Washington, DC, which established the LEED Green Building Rating System, many states have begun green building programs of their own, some geared specifically towards schools, others geared toward all public buildings. Massachusetts has joined with the Renewable Energy Trust to initiate a "green schools" program that awards design and construction grants to school districts in the state. The program uses financial incentives to encourage districts to build green, with $150 million funded by ratepayer surcharges. California has established the Collaborative for High Performance Schools (CHPS), which provides information, services, and incentive programs for California school districts to design schools that are healthy and energy- and resource-efficient.

Green schoolyards. In conjunction with green building techniques, schools are also rethinking the use of the surrounding school grounds. Numerous schools are creating or recreating more natural environments instead of the traditional playground equipment, asphalt courts, and grass fields, in order to use them as outdoor classrooms.[28] Examples include daylighting creeks, constructing wetlands, and planting gardens and native plants, all of which can be incorporated into various school curriculums.

For example, South Elementary School in Waltham, Massachusetts, includes a solar thermal system on the roof that provides hot water, while water runoff from the building will be reintegrated into a constructed wetland area on the school site. The solar and water cycles, in conjunction with the wetland, have been integrated into the school's curriculum. The solar power produces enough electricity to power the library/media center, the art room, and the cafeteria. The school plans to be able to give power back to the grid for city credit when not in use.

Schools and Land Use Planning

While city planners use zoning and other growth management laws to ensure coordinated and adequate location and provision of the various elements of development, in many states, school districts are exempt from local planning and zoning laws. In these instances, it is not at all uncommon for a school district to be planning a new school in a location in which city planners are envisioning something entirely different. When these circumstances occur, school district plans can supersede the city's general plan, which can lead to school districts ignoring the community's broader plans for sensible, well-managed growth. For example, in 1994, the courts ruled that South Carolina school districts were not obligated to work with local planners or other government officials when selecting a new site.[29]

Some states have begun to address this problem by proposing legislation that would require school districts to defer to local zoning laws, and many of these proposals also require that general plans contain provisions to accommodate the need to build or renovate the public schools. While this may seem straightforward, many general plans do not address planning for new schools or the modernizing of older school buildings. Thus, it is not just that school districts need to align their planning with a city's general plan and zoning, but also that cities need to systematically consider the public school element within their general plans.

Across the country, the school mitigation issue has been problematic. Many states have seen the need for schools grow faster than anticipated and have found conventional funding mechanisms inadequate. Largely, the persistence of the school mitigation issue has been the result of the tension between local school districts, the local government, and developers. States differ in their guidelines for when school bonds can be passed and how development impact fees may be assessed to pay for public schools. In addition, the needs of different locales require different strategies. For example, urban school districts cannot rely on conventional school mitigation strategies such as development fees, which are designed around a suburban growth model, because the school-age population may be increasing even though there is little-to-no new housing development occurring.

Providing Schools in High Growth Areas

Siting new schools within urban and suburban areas is fraught with complexities, especially in attempting to bring urban growth management and school planning together. As noted, there is typically a major disconnect in intergovernmental coordination between local general planners and school facilities planners. This disconnect manifests itself in the mismatch between local growth plans and public school capacity limits and, in particular, the inadequacies of many APFOs (adequate public facilities ordinances) to mediate this relationship.[30]

In Florida, the legislature debated whether or not to add schools to its concurrency requirement. The legislature decided not to mandate this, but have instead opted to require school boards to collaborate with local governments on school siting through inter-local agreements. The Florida legislature has recently taken an unprecedented move by requiring that if local governments want to use APFOs, then they must work with the local school district to create common growth management plans, population projections, developmental review bodies, and funding strategies.

Smart Growth and Schools

Increasingly, Smart Growth conversations are engaging in issues around the role of public schools.[31] According to the North Carolina State Department of Public Instruction, planning new schools following Smart Growth principles means:[32]

- Involving community stakeholders early and continuously in the planning process for new schools, additions, and renovations to improve relations, enhance facility improvements, and potentially improve funding.
- Locating schools with and within the urban or community fabric. Avoid developing larger sites with their own self-contained parking lots, drives, and extensive, stand-alone playfields, which contribute to

urban sprawl. Make use of existing infrastructure: water, sewer, pedestrian ways, transit systems, and parking, as well as nearby businesses (food service, office support, etc.) that can provide outside or contracted services and support what would normally be a part of the school. Note that this can be a substantial construction savings also. On-site water and sewer (wells and septic systems) costs have escalated dramatically.

- Designing buildings that relate to the existing neighborhood fabric: as close to the street as adjacent buildings for friendliness and urban context.
- Using two or three-stories where possible to promote density and reduce sprawl, and develop façades and aesthetics that relate to its surroundings, yet still say school.
- Sharing or make use of other joint amenities: parks, libraries, restaurants, civic facilities, etc. rather than constructing duplicate ones.
- Opening the school for other community uses, and work out joint-use arrangements (including funding) to promote the school as a community center rather than just a school.

The American Planning Association's Smart Growth Policy Guide specifically notes the importance of public schools and educational infrastructure within urban planning.

Traffic impacts. When considering locations for siting new schools, consolidating schools, or closing older schools, a careful examination of the transportation impacts is necessary. Typically, school planners mitigate potential traffic congestion caused by the new school, but do not consider how far students will have to travel to get to the school. Siting new schools on large sites on the urban fringe requires that existing residents often have to travel great distances to get there, especially if the new school means that one or more smaller, neighborhood schools will be closing. A study of the costs of sprawl in Maine found that between 1970 and 1995 the number of students in the state declined by 27,000, but the school busing costs rose from \$8.7 million to \$54 million.[33] In addition, large site requirements often mean that school distances from even nearby neighborhoods are prohibitive to walking or biking, resulting in increased automobile traffic during the drop-off and pick-up times.

Walkable and bikeable schools. Increasing attention is being given to the relationship between the built environment and children's health, especially in relation to schools. School location and the built environment play a major role in determining whether or not students walk or bike to school.[34] The Centers for Disease Control and Prevention report that one in five children and one in three teens are overweight or at risk of being overweight —a 50 percent increase in just 10 years.

The 2001 National Household Travel Survey (NHTS) found that 15 percent of students aged 5–15 walked to or from school and only 1 percent biked.[35] In contrast, the first NHTS survey in 1969 found that nearly 50 percent of students walked or biked to school. The decline in walking and biking to school has largely been a result of long distances between home and school, as a result of the historical trend in building larger, more consolidated schools.

Walking or biking to school has been found to be more prevalent when a home is within one mile of a school, a strong argument in favor of smaller, neighborhood-based schools.[36] A poor walking environment consisting of low densities, little mixing of land uses, long blocks, and incomplete sidewalks discourages walking and cycling. The federally-backed Safe Routes to School Program financially supports walking and biking enhancements around schools such as sidewalks and bike lanes. In addition, the U.S. Department of Health and Human Services and the Centers for Disease Control have started a Kids-Walk-to-School Campaign, citing rising rates of childhood obesity, diabetes, and asthma.

The main critique of small, walkable "neighborhood" schools is their resulting lack of student diversity, because, by definition, they draw students from the immediate school surroundings. Thus, this arrangement may work to recreate the social and economic isolation of the neighborhood, as well as resegregating some schools. Some success has been achieved in locating small walkable schools on the border between diverse neighborhoods.

Schools in Redevelopment

Many of the innovative school plans described in this chapter have been developed in response to the critical needs of declining urban areas and land pressures. Research suggests a good school is an important factor in the future economic development of a community, especially in smaller communities.[37] In addition, school closure (often through consolidation) has been shown to have hard impacts on local retail and business establishments, as well as being a lost source of employment in itself.[38] For these reasons, practitioners working on general planning issues need to consider the public school system as an integral element of public infrastructure, not just in providing education, but also impacting adjacent and nearby land uses.

Increasingly, public schools are being considered as magnets for urban development designed to encourage new inner-city housing and employment opportunities, improve mobility, and reduce suburban migration. The trend in these initiatives is coordinating projects with joint planning, including the school. Using the need for new schools in urban areas as a mechanism for wider community improvement is gaining steam, particularly in California.

In this state, schools are increasingly being considered in urban redevelopment projects, often in conjunction with various mixed-use ideas.

The task for local general purpose government practitioners is to figure out ways to enable redevelopment efforts to benefit local public schools rather than working against them. Doing so means coordinating across various agencies that may have little history of direct communication. Not surprisingly, funding is a major issue for school districts. Local practitioners can work to assist financing through coordinating development that supports school mitigation and concurrency.

Conclusion

This chapter began with a brief history of public schools in the United States and described their relationship to the planning work of general purpose governments. From there we explored the unique institutional status of school districts as autonomous bureaucracies, often disconnected from the general purpose government—a disconnect that results in the minimal inclusion of the public school element within urban general plans and underlies the lack of understanding of how public schools fit into urban redevelopment strategies. The historic "silo planning" separation between public school districts and cities makes coordination between the two challenging, but not impossible. At the same time, there is increased evidence of coordination between cities and school districts as an emerging point of integration between schooling and city planning. A school facility planner was recently quoted as saying, "It doesn't make sense to go on building the usual stand-alone schools … the future lies in mixed use, collaboration, and urbanity."[39]

Many of the current trends in school facility planning are happening at the nexus of school planning and Smart Growth policies. These ideas include mixed use developments that include schools, considering options for public transit, walking and bicycle commuting, and increasing density. There is a general critique of the entrenched standards of the post-WWII educational model of large, factory-like school buildings, standardized curriculum and community detachment, and rigid school district control. However, the historic institutional separation between school policy-making and larger urban policy-making serves as a major structural context in which local planning practitioners must navigate.

Public schools exist as a unique aspect of city infrastructure—they are viewed as assets when they are "good" and as liabilities when they are "bad." In the creation of policy around public school quality and local planning, we must continue to come back to the dynamic questions of understanding how children learn and the promotion of desirable learning outcomes as the driving forces in school planning and design. City quality and school quality are intricately linked.

Additional Resources

Publications

ICMA. *Local Governments and Schools: A Community-Oriented Approach.* IQ Report, Volume 40, Special Edition. Washington, DC: International City/County Management Association, 2008. This guide provides strategies for how local governments and schools can bring their respective planning efforts together to take a more community-oriented approach to schools and reach multiple community goals—educational, environmental, economic, social, and fiscal.

Kuhlman, Renee. 2010. *Helping Johnny Walk to School: Policy Recommendations for Removing Barriers to Community-Centered Schools.* Washington, DC: National Trust for Historic Preservation. Report identifies strategies, policies, and best practices for the retention and development of community-centered schools.

U.S. Environmental Protection Agency. *Travel and Environmental Implications of School Siting.* Washington, DC: U.S. Environmental Protection Agency, 2003.This study examines the relationship between school locations, the built environment around schools, how students get to school, and the impact on air emissions of those travel choices.

U.S. Environmental Protection Agency. 2011. *Voluntary School Siting Guidelines.* Washington, DC: EPA. The guidelines present recommendations for evaluating the environmental and public health risks and benefits of potential school locations during the school siting process.

Vincent, Jeffrey M. 2012. California's K-12 Educational Infrastructure Investments: Leveraging the State's Role for Quality School Facilities in Sustainable Communities. Available on-line at: http://citiesandschools.berkeley.edu/reports/CCS2012CAK12facilities.pdf. This report takes a comprehensive look at the state of K-12 school facilities in California and presents a detailed policy framework for the state's role in K-12 infrastructure to appropriately support educational quality and contribute to healthy, sustainable communities goals.

Weiss, Jonathan D. *Public Schools and Economic Development: What the Research Shows.* Cincinnati, OH: KnowledgeWorks Foundation, 2004. Summarizing the research linking public schools and economic development, this report details the ways in which public schools impact business attraction, real estate values, and economic growth.

Websites

Building Educational Success Together: www.bestschoolfacilities.org. BEST works towards a vision where all children learn in school buildings that are safe, educationally adequate, and able to serve as community anchors in vibrant, healthy neighborhoods.

Center for Cities & Schools: citiesandschools.berkeley.edu. CC&S, at the University of California, Berkeley, positions high-quality education as an essential component of urban and metropolitan vitality to create equitable, healthy, and sustainable cities and schools for all.

Environmental Protection Agency, Smart Growth and Schools: www.epa.gov/smartgrowth/schools.htm. The EPA maintains a website with articles and links to information on smart growth and school planning.

chapter 26

Parks, Recreation, and Open Space

In this chapter ...

Importance to Local Practitioners . . . 553
The History of Parks in the United States . . . 554
The Park as Nature in the City . . . 554
The Reform Park: 1900–1930 . . . 556
Meeting the Demand for Park Recreation Facilities . . . 556
Parks and the Cultural Revolution . . . 557
Private Parks, Tourist Cities, and a Broader Role for Parks and Open Space . . . 558
Institutional Context of Parks and Recreation Facilities . . . 559
In-House Organization . . . 559
The Parks Department's Relationship with School Districts . . . 560
Privatization and Contracting Out . . . 561
Financing for Parks and Open Space . . . 562
Property Taxes and Bonds . . . 562
Sales Taxes . . . 562
Open Space Assessment Districts . . . 563
Development Impact Fees and Mitigations . . . 563
User Fees and Other Non-tax Revenues . . . 564
Planning Considerations . . . 565
Assessing the Needs for Parks and Open Space . . . 565
Highlighting the Economic Impact of Parks and Open Space . . . 571
Sustainability and Green Programs . . . 572
Conclusion . . . 575
Additional Resources . . . 575
Publications . . . 575
Key Federal Legislation . . . 576

Why Is This Important to Local Practitioners?

Sustained increases in discretionary income over the past few decades, environmental concern about the loss of green space (nationally and locally), and the persistence of inner city park problems make park and recreation infrastructure an important local topic. Some practitioners will need to know about issues in the park and recreation area in order to work with a developer or landscape architect to ensure that a project contains enough open space and provides for sufficient funds to maintain it. Others might be working with the parks and recreation staff to develop a short-term strategic plan for a neighborhood where the park is in disrepair. Parks and recreational facilities are also important to the physical health of the populace, particularly given the concerns about obesity. Parks and open space can be key elements in a local climate change plan. Local practitioners also need to be aware of the capital needs associated with parks and recreational facilities as they prepare capital improvement plans and budgets.

This chapter begins with a history of parks and open space in the United States, before describing the institutional context for park planning and operations at the local level. Elements of the park and open space system are outlined. Different options for financing parks are presented before describing a process for developing a park and recreation plan that includes developing local level of service (LOS) standards. An example of how parks and tree canopies are used in a local sustainability plan is presented.

Park design is not covered, since this is a profession in its own right and there is a large volume of literature and training available elsewhere in this topic.

The History of Parks in the United States

Two themes have wound their way through the history of park infrastructure in the United States—parks as a natural preserve and refuge from the urban crush, and parks as a location for organized recreational activities. At some periods, first one and then the other theme is emphasized. More recently, a more holistic view has emerged that seeks to integrate parks and open space preservation into the overall social and environmental vision of the city.

In some respects, the origin of park infrastructure in the United States is no different from the arenas, circuses, gardens, baths, and spas that were available for communal use in ancient civilizations. Many of these traditions persisted in post-feudal Europe, where the parks and gardens associated with the grand castles of the aristocracy were turned into public places. When the earliest settlers arrived in America, they carried these traditions with them. However, the park movement in the United States did not formally begin until the middle of the nineteenth century.

The Park as Nature in the City

Rapid industrialization in the late nineteenth century turned American cities into places that were unhealthy and almost uninhabitable. Many lacked paved streets, city squares, sewers, and other amenities. Public funds were targeted primarily to water and police protections. In addition, although only about 15 percent of the population lived in cities in 1850, the conditions were crowded and unhealthy. In the face of this, a demand arose for parks to reproduce the natural aspects of the countryside in the city.

Most of the great urban parks in the United States were built between 1850 and 1900, among them Central Park in New York City and Golden Gate Park in San Francisco, while in Chicago, a variety of smaller parks such as Jackson and Humboldt were developed.

Golden Gate Park playground

Central Park owes its origin to the campaign of Mayor Ambrose Kingsland in 1850 and legislation submitted by him to the New York Council in 1851. In 1857, when Frederick Law Olmsted became the first superintendent of the site destined to be Central Park, almost a century of planning for public parks was initiated that has influenced the shape and design of our cities today. Although Olmstead was originally a "failed farmer and sometime journalist," he is regarded today as the father of landscape architecture in the United States.[1] Not only did he and his colleague Calvert Vaux design Central Park, their landscape architecture firm was later responsible for the design of numerous parks, subdivisions, and playgrounds throughout the United States up until 1950.

Olmstead's philosophy for parks, open space, and conservation captured the values of the time, and endures to this day. Olmstead felt that parks and open space were necessary for both the physical and mental health of urban families. He believed that park vegetation and street trees screen and purify the air of the city, and that they were the "lungs of the city." In addition, he believed that parks and open space were essential to the mental health of city dwellers as a place for leisure activities outside the crowded tenements and unpaved streets.

The design of the parks of this era was informal and natural. Plantings were to be irregular and natural. Active recreational activities were conceived as an integral part of the park, but they were unstructured: "rowing, romping and playing." Playing fields were put at the edges of parks to guarantee access rather than to be an integral part of the park. Buildings were "necessary evils required to make parks usable," and for a decade at least, the design was almost exclusively rustic.

The city leaders who brought these parks into being were "moral entrepreneurs" and part of the social and economic elite. Although the parks of this era were designed to promote the mingling of the classes, they functioned primarily as a middle- and upper-class equivalent to the aristocratic preserves of the wealthy in Europe for America's growing middle class. The fees for a working family to use transportation to get to the park could amount to a day's wage. The practice of racial segregation in the parks or this era was common. Many who used the park continued the European tradition of wearing rich clothing and driving expensive carriages in the parks. The concerts, design, and sculpture in these parks reflected the tastes of the upper class. Activities which enjoyed a more popular appeal, such as gambling, animal fights, and roller skating, were forbidden.

During this same period, 1850–1900, the national parks also came into being. In the 1870s, the photographs of William Henry Jackson and Carlton Watkins helped to convince Congress to designate some of the most beautiful and unsettled areas of the country as national parks. Today, the assets belonging to the national park system are valued at over $35 billion.[2]

The Reform Park: 1900–1930

If the "natural park" epitomized the pre-eminence of the bucolic, natural Emersonian ideal, then the "reform park" in the late 1890s and early decades of the twentieth century added another aspect to the urban park—that of social reform. As Jacob Riis, Lincoln Steffans, and others organized to end slum housing and dirty food and water, they also shifted the focus of the park movement to neighborhood parks oriented to the betterment of the working class. Reformers, seeking better living conditions for urban dwellers, banded with property owners, who saw parks as a way of increasing property values, and with political bosses, who saw construction of public parks as a means of patronage.

The reform park featured organized recreational activities to provide structure for the many unattended children whose parents worked. Some park reformers used the death rates of children to advocate for more parks and organized recreational activities. It was also at this time that the playground movement arose, spreading through the settlement houses before becoming institutionalized in municipal government. Many of these programs were run with "a reforming zeal, and at times their advocates spoke of them as if they were municipally run settlement houses."[3]

The model "reform" park in the 1920s had both an indoor and outdoor plant. It had game fields, running tracks, swimming and wading pools, and playgrounds. The field house consisted of an assembly hall, indoor gymnasiums, lockers, and equipment. The design and placement of the fields and buildings were symmetrical—grounds were flat.

Meeting the Demand for Park Recreation Facilities

The appointment of Robert Moses as Park Commissioner in 1930 signaled the end of efforts to use parks as a mechanism of social reform. Tight budgets because of the Depression coincided with the rise of city planning and its emphasis upon rational analysis and efficiency. Park officials echoed these sentiments, focusing on recreational standards and efficiency. During World War II, parks were used for civil defense programming. Park construction was suspended, and did not begin again until the 1950s.

This period did not have the same guiding vision about the purpose of parks as the two earlier eras. Instead, the emphasis was on meeting the demand for recreational services. The literature of the time described the numbers and kinds of parks, playgrounds, and buildings constructed, instead of outlining the underlying purpose for which they were built. The program of the park and playground had only a loose relationship to the design.

This change in approach to park maintenance and development was related to several factors. First, cities themselves had changed. They were no longer the commercial center for an agriculture hinterland, or the industrial center for the region. Instead, both industry and households were abandoning them for rapidly growing suburbs, which of course had their own parks, parkways, and open space. Second, the people living in the cities and suburbs had changed as well. Families were smaller and had more money, and people worked shorter hours than their counterparts 100 years previously. In addition, many suburban families had their own open space—in their suburban backyards. "Starting in the 1920s, and intensifying after World War II, the middle class took care of itself by moving to the suburbs, and the city park became the almost exclusive domain of the poor."[4]

Sign for César Chávez Park

Wild kites casting a shadow in César Chávez Park

Parks and the Cultural Revolution

The inner city riots in the 1960s, the beginning of the environmental movement, and the affluence of the time coincided with, or perhaps were responsible for, a change in the political agenda for urban parks. The inner city riots shocked public officials into concern about the deterioration of the inner city and the urban parks, which were perceived as crime-ridden and unsafe. The anti-war movement, the Flower Children, and the Baby Boomer generation changed the dominant culture of America from the Brooks Brothers' suited organization man to one that was far more eclectic and open-minded.

Kite festival in César Chávez Park for kids

Welcome sign at Adventure Playground

Outside Adventure Playground

Inside Adventure Playground

Park officials responded with a proactive and stunningly different approach to the provision of park services in order to get the middle class back into the parks and to provide safe experiences for inner city residents. "If the pleasure ground had been a pious patriarch, the reform park a social worker, and the recreation facility a waitress or car mechanic, the new park was something of a performance artist."[5]

The design and parks programming put in place by New York City's Commissioner of Parks in the 1960s, Thomas Hoving, typifies the kind of programming that eventually became commonplace. "Hoving's Happenings," such as rock music, blue grass festivals, mulled wine, and kite flying, updated the kinds of cultural activities permitted in parks. Athletic activities were also updated—a trend that has continued to this day with the provision even of skate board parks. The adventure playground was initiated—where construction materials could be used by children to change the park from day to day. New freeform models of playground equipment were designed and built.

Private Parks, Tourist Cities, and a Broader Role for Parks and Open Space

As the leisure time and household income of the average American increased, the private sector became more active in the provision of recreational activities and services. The proliferation of theme parks, convention centers, baseball stadiums, and football arenas began to be big business. This coincided with the rise of tourism as one of the most dynamic economic sectors in the world economy.[6] The parks and recreation story of the post-industrial, global city is part of the tourist industry and its impact upon recreation. Public parks have been forced to justify themselves in economic terms, both as an element in the "Destination City" and as a component of the city that is attractive to new economy industries.

There has also been a resurgence of interest in the natural park, with "eco-parks" and "eco-tourism" on the rise. Finally, there is interest today in integrating the site-specific park design with the larger vision of the community for parks, open space, and green communities. Current thinking looks at parks and open spaces as places to help achieve public health and civic engagement goals or even to play a role in reducing the urban heat island effect along with more well-known recreational goals. Park planners are being encouraged to move beyond traditional park classification schemes to think of parks and open space as a system.[7] From the opposite direction, community revitalization projects in older cities with large amounts of vacant and unmanaged parcels incorporate open space as an essential effort in rebuilding a community. Non-profit community groups in Boston, the South Bronx, and Philadelphia, for example, have turned urban vacant land into play spaces and community gardens, either on a temporary or permanent basis.[8]

Institutional Context of Parks and Recreation Facilities

Planning and financing parks and recreation programs is usually the responsibility of the city or county government, although many jurisdictions have a special park district with its own elected board. Early twentieth-century parks and recreation programs were run by commissions that operated independently of the municipal government. Around World War II, the parks and recreation commissions became part of city government, often as separate departments. This resulted in a more secure stream of funds and acceptance of these functions as necessary municipal services. Being located within the municipal or county structure also meant that these departments could successfully apply for federal and state funds. Finally, this gave park systems the police powers necessary for planning and operating parks.[9]

In-House Organization

The parks and recreation functions may be combined in one department reporting directly to the city manager or the county administrator. Alternatively, the parks department may be located in the public works department. Recreational activities taking place in the parks can be a part of the parks division, or can be located in a separate leisure services department, or human services department.

Frequently, cities and counties alternate between the two models. As economic times are better or where there is strong local interest in development of new park facilities or recreational programs, localities have separate departments. As times get tighter, and overhead costs must be cut, the parks responsibility moves back to the public works department. Despite the fact that parks and recreation facilities are among the first to be put on the chopping block in the event of an economic downturn, a 1992 study looking at staffing and revenues for local parks and recreation departments found that the anti-tax movement of the 1980s had not caused substantial reductions. Instead, nationally, park revenues increased 240 percent from 1964 through 1992, total employment had increased 20 percent since 1976, and per capita expenditures on parks increased steadily and consistently from the mid-1960s and early 1970s.[10] In Fiscal Year 2007, non-seasonal employees in major city park agencies ranged from a high of 1.93 employees per 1,000 residents in Tampa and 1.72 in Seattle, to lows of 0.35 in Philadelphia and 0.33 in Toledo.[11]

The focus of the parks and recreation agencies, regardless of where they are located administratively, is on the operations and maintenance of the parks and their facilities, and the recreational programs. Sometimes there are separate plans for each that need to be integrated into the long-range plan for the community and into the capital improvements program. The parks department may take the lead on capital improvement planning, or the planning may be managed by the local department with responsibility for all capital construction projects. (See Figure 26.1 for a planning map that integrates parks, natural areas, and recreational facilities.)

The Parks Department's Relationship with School Districts

Park and recreation agencies play an important role in after-school programs, either as a principal service provider or as partner. Local recreation and park

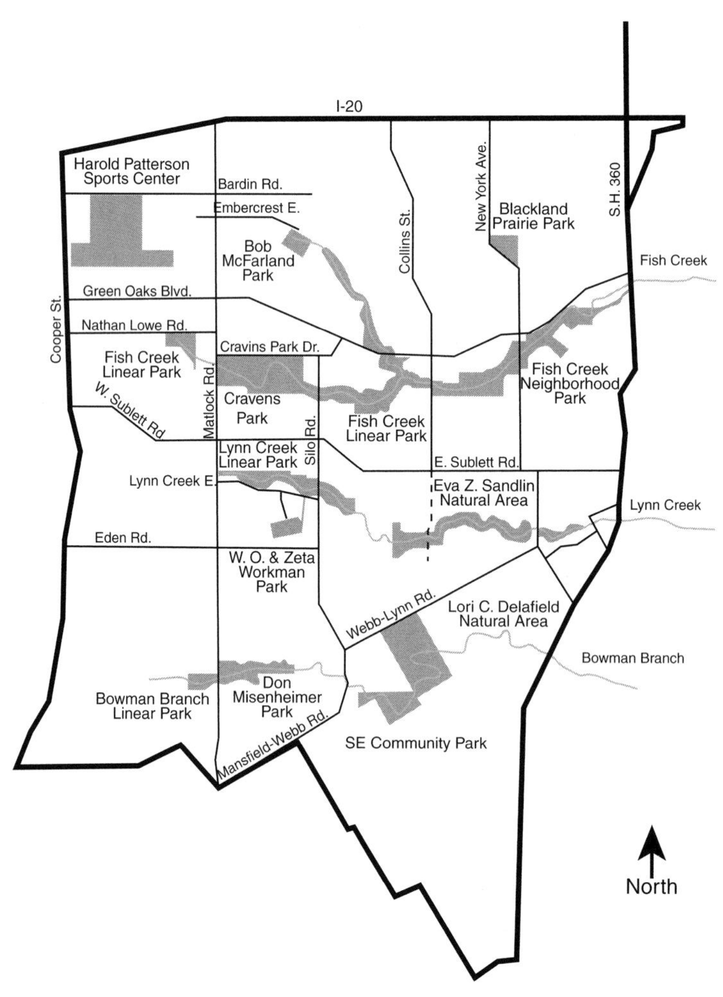

Figure 26.1
Planning map of District 5 parks in Arlington, TX

systems manage an estimated 80,000 sites, according to the National Recreation and Park Association. Local public recreation agencies also provide most of the playing fields and centers used by other entities. In 1998, according to the U.S. Department of Education, 87 percent of successful 21st Century Community Learning Center grantees included a recreation component.

Another very popular relationship between school districts and local governments is joint park development. For example, Henrico County, Virginia, is planning to fund with a $4 million bond issue on the ballot an expansion of an existing 11-acre park originally planned with the Henrico County Public School District to 38 acres. The funding will pay for a floodlit, sodded, and irrigated football field, expanded parking, and other amenities that will benefit the school district as well as serving the surrounding community.

Privatization and Contracting Out

Parks departments have not been immune to privatization pressures. In the 1980s, some cities began to "outsource" park activities to the private or non-profit sector. Indianapolis was one of the first cities to do so, followed by many of the less unionized cities in the southwestern United States. Parks maintenance functions were contracted out in many southern California localities during the economic downturn of the early 1990s. Two major cities, New York and Chicago, also outsourced some of their non-park operations. An ICMA survey on this topic found that most local governments operate and maintain recreation facilities with local government employees. About 70 percent of the functions are performed by in-house employees. Larger jurisdictions also make use of volunteers—about 14 percent of the work was done by them. Many contract with another government authority (11 percent) or the private sector, both profit and non-profit (15 percent).[12]

One prominent parks specialist, Alexander Garvin, describes the advantages and disadvantages of three different institutional relationships that can be used by parks and recreation departments to finance park development and operations:[13]

- The *public model* has broad-based funding from taxes and is open to citizen and business involvement in

Parks, Children and Distance

Although New York City has only 4.6 park acres per 1,000 residents, more than 91 percent of its children live within walking distance of a park, compared to only one-third of all children in Los Angeles, which has 9.1 acres of park per 1,000 residents. Boston also has a higher proportion of children living within walking distance of a park (Table 26.1).

Table 26.1
Children's park access in seven major cities

City	Percentage of children within one-quarter mile of a park	Number of children without access to a park
Boston	97	2,900
New York	91	178,500
San Francisco	85	16,700
Seattle	79	18,600
San Diego	65	102,300
Dallas	42	182,800
Los Angeles	33	657,700

Source: Center for Park Excellence, *No Place to Play.* Trust for Public Land, 2004, available at: www.tpl.org/ccpe (accessed July 24, 2008).

design. However, park budgets are generally cut first in time of fiscal crisis, and many urban parks are not well maintained. Volunteers are often used for clean-up.

- The *public–private co-venture* allows for user fees to raise capital. This type of venture can be open to citizen involvement. However, few private corporations are able to subsidize improvements for an extended period of time.
- The *market-oriented civic model* relies heavily on local citizen help and can use self-perpetuating innovative fees. In addition, management motivation will continuously improve the level of service. However, the volunteer management structure may not work in the long term.

Financing for Parks and Open Space

The local practitioner will be involved in parks and open space financing as part of the yearly budget process, or when a park or recreational facility is being developed or upgraded. The practitioner may be involved in assessing parks and open space needs as part of processing the permit application for a new development. The following illustrates how some jurisdictions have made use of the major sources of infrastructure funding for park purposes.

Property Taxes and Bonds

The primary source of funding for parks from WWII until recently has been property taxes and other local general funds. In the past decade, many states authorized local governments to issue special-purpose bonds for park and recreational facilities, as well as including funding for parks and open space by the traditional general purpose bond funded through property taxes or other local funds. States also issue bonds that are used for state, local, and regional projects. In the period from 1988 to 2011, voters have approved almost $58 billion dollars to fund open space, parks, wildlife habitats, watersheds, farms, and ranchlands. As with other infrastructure funding, parks and open space protection follows the economic cycle[14] (Figure 26.2).

Bonds can be a safer method of securing capital funds for parks and open space than a direct allocation from the operating fund. In economically tough times, these funds can be more easily cut. For example, in 2003, a difficult budget year, Pennsylvania froze $100 million from its "Growing Greener" initiative for open-space preservation and watershed protection, while Utah cut funds for open space land purchases by $2 million.[15]

Sales Taxes

Another fund source for parks and open space programs is local sales tax increases. This method was popular in the 1990s and is attracting

Per Capita Expenditures on Parks

Figures for the amount spent on parks and recreation services are not uniformly available for all jurisdictions. However, for the 84 largest city park agencies in the United States, the per capita dollar amount of expenditure was $76 in FY2006, with a range from $31 for Houston and $48 for Los Angeles, up to $242 in Seattle, $151 in Minneapolis, $131 in Chicago, and $108 in Denver. These figures include grounds and facilities maintenance and repair, recreational programs and activities, and funds spent for capital improvements during that year.[1]

Note

1. Center for Park Excellence, *Fact Sheets*. Trust for Public Land, 2008, available at: www.tpl.org/ccpe (accessed July 24, 2008).

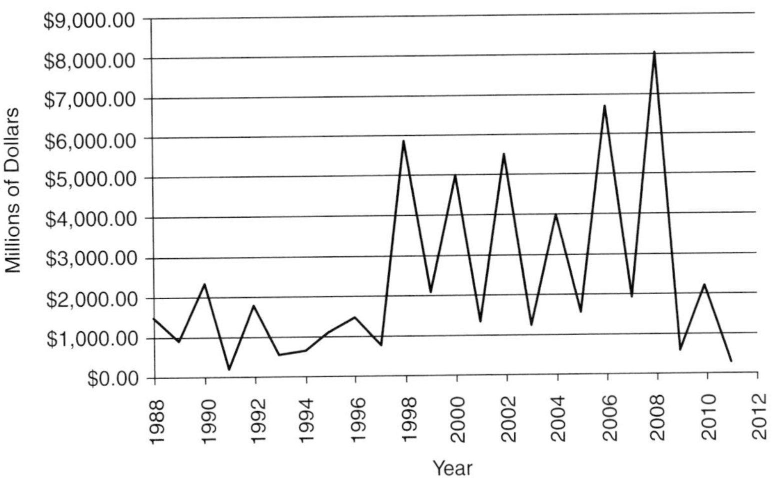

Figure 26.2
Proceeds of local government bond measures for parks

renewed interest. Sacramento County, California, will consider a 0.1 percent sales tax increase to fund the regional park system at its November 2012 election. In Missouri, St. Louis, St. Louis County, and St. Charles County are considering raising the sales tax by three-sixteenths of a cent (0.1875) to pay for the Arch project and other area parks.

Open Space Assessment Districts

Special assessment districts can also be used to preserve open space. Usually these have to be authorized by voters since they are repaid by taxes. A 2001 National Association of Realtors survey found that voter support for preserving open space was dependent upon the cost and use. Voters would support open space initiatives for neighborhood parks, playgrounds, playing fields, and walking trails, but not golf courses. Over 80 percent of the voters would support the preservation of farmland, natural areas, stream corridors, and true wilderness areas, but not areas no longer used for farming. Those surveyed would support taxes of up to $50 a year to pay for these bonds.[16]

Development Impact Fees and Mitigations

Park or open space land can be required through the local development permit process. Dedications of land for park areas in newly developing areas are common and should be insisted upon. Lands over a certain slope, wetlands, and lands in the floodplain are candidates for dedication.[17]

Development impact fees can also be used to fund parks and recreation needs caused by new development. Lancaster, California, has an Urban Structure Program surcharge for parks that is incorporated into the calculation of impact fees. Both Nantucket and Martha's Vineyard have a real estate transfer tax dedicated to park and recreational needs.[18] In Florida, municipalities are authorized to set up mitigation land banks for parks and open space purposes. Impact fees for supporting vest pocket parks can also be levied as appropriate mitigation for large-scale developments in urban areas.

User Fees and Other Non-tax Revenues

The levying of user fees for parks rose dramatically in the 1990s, according to a 1999 report by the Trust for Public Lands. Although most park districts do not approach the experience of Wheeling, West Virginia, which brought in 99 percent of its revenues from non-tax sources at that time, charging fees for some part of parks and recreational activities is increasingly common. The San Francisco Parks and Recreation Department received $30.3 million annually from fees, almost 40 percent of its budget at that time (Figure 26.3).

PUBLIC OPEN SPACE SERVICE AREAS **Map 2**

 EXISTING PUBLIC OPEN SPACE

 OPEN SPACE SERVICE AREA
Areas within acceptable walking distance

Note:
Because of the scale of this map it is not possible to show precise boundaries or exceptionally small open spaces.

Open Space Category	Size in Acres	Service Area Radius in Miles
Citywide	varies 1-1000	1/2
District	over 10	3/8
Neighborhood	1-10	1/4
Subneighborhood	less than 1	1/8

Figure 26.3
Map of open space in San Francisco with areas within walking distances shaded

In 2005, this figure was considerably higher, with golf fees, camps, naming rights, and admissions from the zoo and the Japanese tea garden being among the largest components. Portland raised 43 percent of its $50 million annual budget from fees in 1999, while the rest came from the city. New York City reported 22 percent of its parks and recreation budget from non-tax sources, Chicago 30 percent, and Seattle only a modest amount. Municipal marinas usually pay for themselves, and may actually generate income for the city.[19]

Municipalities enacting user fees have found that four factors are responsible for success: (1) good service; (2) explaining what the fee is for; (3) providing alternatives; and (4) instituting them at the end of the season for the activity rather than during the season.

Planning Considerations

Previous chapters have outlined the steps in preparing a comprehensive infrastructure plan, which include setting up a process to involve stakeholders, and a capital improvements program and budget. They have also addressed how to develop an infrastructure project and described the factors involved in assessing infrastructure needs of a new development project. These general processes will not be repeated here. Nor will design issues be addressed, as noted previously. However, parks and open space planning have three special considerations that the local practitioner should be aware of.

The first area is the needs assessment process for parks and recreation facilities, which can be used as part of the preparation of the comprehensive infrastructure plan, or during the development of an impact fee schedule for parks for new development, or as part of the preparation of the CIP and budget. The second area touches on how to evaluate park projects for their economic benefit, either during the budget process or during the planning for an individual park project. The third area contains some suggested linkages between parks and recreation and a more comprehensive local strategy for sustainability and "green" programs. The next three sections of this chapter address each in turn.

Assessing the Needs for Parks and Open Space

From 1950 to the 1980s, park professionals used a national per capita standard to estimate need. Originally a single standard was used for all kinds of parks: 10 acres of park land per 1,000 persons. As parks became more differentiated with respect to use, a variety of individual measures were promoted. These standards, however, were based on professional judgments and were not tailored to the needs of a specific community. Consequently they were often ignored.

The Recreational Fee Demonstration Program

The National Park Service began collecting fees for hiking, fishing, hunting, and skiing on public lands that were previously accessible at no cost in 1996 with the Recreational Fee Demonstration Program. This demo program was made permanent in 2004 by the Recreation Enhancement Act. These fees are returned to the site of collection and are used to provide, maintain, and improve visitor centers, trails, campgrounds, and other facilities. This program was actively opposed by the Sierra Club, which was concerned that the fee would provide incentives to replace hiking and camping with motorized sports and more elaborate capital facilities. "It's the snowmobilers (and the recreation industry) against the high-country hikers (and the environmentalists) in the battle for the future of America's parkland (and soul)," according to the Sierra Club.[1]

Note

1 J. Margolis, "Park Wars," *The American Prospect* 13(16) (2004).

Amount of Park and Open Spaces in U.S. Cities

Today, the amount of park and open space in the largest 25 cities in the United States varies considerably. Predictably, the number of acres per 1,000 residents is the lowest in high density cities such as New York, Los Angeles, Chicago, and Boston, with an average of 7.2 acres. Medium-density cities have an average of 13 acres per 1,000 residents, while lower-density cities such as Dallas, San Diego, and Indianapolis have an average of 19 acres per 1,000 residents. The average for all cities is 13 acres.

However, when the amount of park and open space is looked at as a percentage of the land area, the pattern is reversed. Cities with a high population density have a much higher percentage of land devoted to parks and open space than do their less-dense cousins. For example, 27 percent of the land area in New York City is devoted to parks and open space, followed by San Francisco with 25 percent. The average for all 25 cities is 13 percent of their land.

In the 1980s, when exactions for park land dedication and other recreational facilities began to be used for new development, the "rational nexus" test required an approach that could be justified by specific community needs caused by the development. Accordingly, the quantitative level of service (LOS) approach used in other functional areas such as transportation, water, and sanitation was developed by the National Recreation and Park Association (NRPA) and the American Academy for Park and Recreation Administration to estimate parks and recreation needs.

Today, the concept of the needs assessment for parks and open space has been expanded to include qualitative information obtained from site visits, interviews, and focus groups with park users, community residents, and other stakeholders, including staff and elected officials. Site visits and observations of parks and gathering places can also be done. Quantitative techniques such as interviews, surveys, and the use of the GIS to develop baseline inventory information are now common for capital planning for parks and other infrastructure (Figure 26.4). Local governments are usually interested in seeing how their locality compares to others in terms of indicators such as acres of park per capita, or cost of service per capita, or acres of park land. Using a GIS-based program, resident access to existing facilities can also be mapped and evaluated.[20] Level of service standards can be developed for different classes of facilities in order to quantify the amount of land and facilities needed by the locality. This technique is well suited to be the basis of impact fees and indicators of service that can be used in the capital budget and its reporting systems. The latter, as well as service demand estimates, is discussed further below:[21]

1 Determine which elements of the parks, recreation, and open space system will be part of the locality's plan. The first step that the local practitioner should take is to determine the types of parks and open space elements that will be part of the plan. This should be done as part of an open citizen

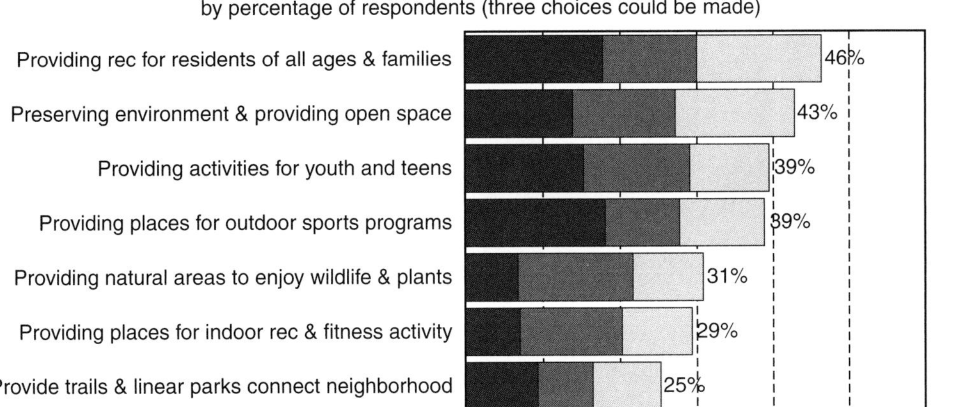

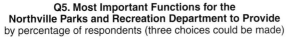

Figure 26.4
Importance of various parks and recreation functions to local citizens, Northville, MI

participation process. The classification system outlined in Table 26.2 can also be used as a framework to inventory and assess the condition of existing facilities. Some special-use facilities that are becoming increasingly common are skateboard parks and dog parks. Competing uses for scarce space often make decisions about land use within and adjacent to parks contentious.

The connecting greenways and pathways are as important as the parks themselves as both recreational venues and open space. The NRPA classification (see Table 26.3) can be used by the local practitioner either to inventory existing systems or as a framework to develop needs for developing areas.

Other open space and natural resource areas can also be identified. These may be lands that are unsuitable for development but would benefit the public if preserved, such as parcels with steep slopes and natural vegetation, drainage ways, ravines, manmade ponds, and utility easements. The objective of preserving these lands is to maintain the livability of the community with as many natural amenities as possible. Functioning ecosystems, forests and woodlands, and wildlife habitat also fall into this category. See Chapter 7 for a discussion of an innovative subdivision ordinance that could be used to preserve this kind of open space.

2 Determine the Recreation Activity Menu. The Recreation Activity Menu (RAM) is the list of all recreational activities and their facilities, such as tot lots and picnic units, that will be part of each park classification. This list begins with all the activities, programs, and facilities that are currently being used and/or are needed to satisfy the needs of the present population. Market surveys, informal interviews, or hearings can be used to elicit this information.

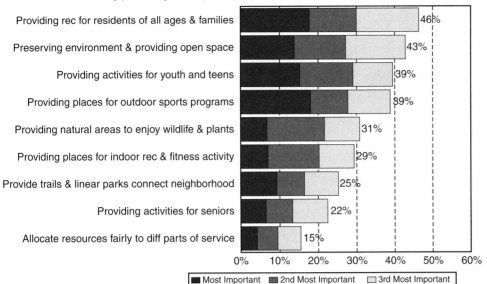

Table 26.2
Parks and open space classifications and their location and size criteria

Classification	General description	Location criteria	Size criteria
Mini-park	Used to address limited, isolated, or unique recreational needs	Less than a ¼-mile distance in residential setting	Between 2500 sq. ft. and one acre in size
Neighborhood park	Neighborhood park remains the basic unit of the park system and serves as the recreational and social focus of the neighborhood. Informal active and passive recreation	¼ to ½-mile distance and uninterrupted by nonresidential roads and other physical barriers	5 acres is considered minimum size, while 5 to 10 acres is optimal
School park	Depending on circumstances, combining parks with school sites can fulfill the space requirements for other classes of parks, such as neighborhood, community, sports complex, and special use	Determined by location of school district property	Variable—depends on function
Community park	Serves broader purpose than neighborhood park. Focus is on meeting community-based recreation needs, as well as preserving unique landscapes and open spaces	Determined by quality and suitability of the site. Usually serves two or more neighborhoods and ½ to 3-mile distance.	As needed to accommodate desired uses. Usually between 30 and 50 acres
Large urban park	Serve broader purpose than community parks and used when community and neighborhood parks are not adequate. Focus is on meeting community-based recreational needs and preserving unique landscapes and open spaces	Determined by the quality and suitability of the site. Usually serves the entire community	As needed. Usually a minimum of 50 acres, with 75 or more being optimal
Natural resource areas	Lands set aside for preservation of significant natural resources, remnant landscapes, open space, and visual aesthetics/buffering	Resource availability and opportunity	Variable
Special use	Covers a broad range of parks and recreation facilities oriented toward single-purpose use	Variable—dependent on specific use	Variable
Sports complex	Consolidates heavily programmed athletic fields and recreation facilities to larger and fewer sites strategically located throughout the community	Strategically located community-wide facilities	Determined by project demand. Usually a minimum of 25 acres, with 40 to 80 acres being optimal
Private park/recreation facility	Parks and recreation facilities that are privately owned yet contribute to the public park and recreation system	Variable, dependent on use	Variable

Source: James D. Mertes and James R. Hall, *Park, Recreation, Open Space and Greenway Guidelines* (Alexandria, VA: National Recreation and Park Association and the American Academy for Park and Recreation Administration, 1995).

The National Parks and Recreation Association has a set of standards for determining the size of the buildings and other constructed facilities, and hence the amount of land needed. The standards cover ice hockey rinks, field hockey, and soccer fields, among others. See the National Parks and Recreation Association publication under the resource section at the end for further specific standards.

Table 26.3
Greenways and parkways classification and description

Classification	General description	Description of each type
Greenways	Effectively tie park system components together to form a continuous park environment	N.A.
Park trail	Multipurpose trails located within greenways, parks, and natural resource areas. Focus is on recreational value and harmony with natural environment	Type I: Separate/single-purpose hard-surfaced trails for pedestrians or bicyclists/in-line skaters Type II: Multipurpose hard-surfaced trails for pedestrians and bicyclists/in-line skaters Type III: Nature trails for pedestrians. Hard or soft surface
Connector trails	Multipurpose trails that emphasize safe travel for pedestrians to and from parks and around the community. Focus is as much on transportation as it is on recreation	Type I: Separate/single-purpose hard-surfaced trails for pedestrians or bicyclists/in-line skaters located in independent ROW Type II: Separate/single-purpose hard-surfaced trails for pedestrians or bicyclists/in-line skaters. Typically located within road ROW
On-street bikeways	Paved segments of roadways that serve as a means to safely separate bicyclists from vehicular traffic	Bike route: Designated portions of the roadway for the preferential or exclusive use of bicyclists. Bike lane: Shared portions of the roadway that provide separation between motor vehicles and bicyclists, such as paved shoulders
All-terrain bike trail	Off-road trail for all-terrain (mountain) bikes	Single-purpose loop trails usually located in larger parks and natural resource areas
Cross-country ski trail	Trails developed for traditional and skate-style cross-county skiing	Loop trails usually located in larger parks and natural resource areas
Equestrian trail	Trails developed for horseback riding	Loop trails usually located in larger parks and natural resource areas. Sometimes developed as multipurpose with hiking and all-terrain biking where conflicts can be controlled

Source: James D. Mertes and James R. Hall, *Park, Recreation, Open Space and Greenway Guidelines* (Alexandria, VA: National Recreation and Park Association and the American Academy for Park and Recreation Administration, 1995).

3 Determine the open space size standards. Open space size standards are the minimum acreage needed for facilities that support the activity menu for each park classification. Using the activity menu for each, the amount of acreage needed for facilities to support those activities is calculated. Passive versus active areas should be identified using a community participation process since these are local standards reflecting the values of the community. The resulting standards are community-specific.

4 Inventory the supply, condition, and current use of parks, recreation, and open space. As the level of service standards are being developed for the community's park and open space menu, the current conditions and use of existing facilities should be inventoried. These should be translated into measurements of how many visits per year each park unit or classification gets, and how many it could accommodate. Expected use is a

Ways to Meet ADA Requirements

1. *Alterations to existing buildings.* Functions in the building such as telephones, restrooms, or drinking fountains must be made accessible when rehabilitating an existing facility.
2. *Removal of barriers.* Curb cuts, the installation of grab bars in restrooms, widening entrances, accessible parking spaces, removing high pile carpet, and installing a paper cup dispenser at existing drinking fountains are ways of removing barriers in park and recreation facilities.
3. *Outdoor recreation facilities.* The National Park Service has a publication with excellent guidelines that shows how changes to promote accessibility can be consistent with the degree of development of the facility. For example, highly developed sites with parking lots, restrooms, and swimming pools should have highly developed routes for accessibility, while fragile natural areas with limited development should have the minimum accessible routes.
4. *Services and programs.* Parks offering programs in non-accessible buildings should remove the barrier, shift the program to another site, provide the service in an alternative method, or construct a new facility.

James D. Mertes, and James R. Hall, *Park, Recreation, Open Space and Greenway Guidelines* (Alexandria, VA: National Recreation and Park Association and the American Academy for Park and Recreation Administration, 1995).

combination of average daily use and peak use combined with the availability of the facility.

The condition of the facilities in terms of health, safety, and ADA requirements should be evaluated, including a risk assessment performed by the entity's risk officer or a contractor. Frequently, liability concerns rather than demand or other planning issues drive infrastructure funding and decisions about parks.

National standards and guidelines exist for public playgrounds: the U.S. Consumer Product Safety Commission's (CPSC) *Handbook for Public Playground Safety* and the American Society for Testing and Materials' (ASTM) Standard Consumer Safety Performance Specification for Playground Equipment for Public Use. The CPSC contains voluntary safety recommendations for the design and maintenance of playground equipment, while the ASTM document is a technical one for manufacturers. Some states also have adopted mandatory safety requirements for play equipment.

5 Assess the demand for parks, recreation, and open space. The local practitioner needs to determine the per-capita demand for these services. Actual recreation participation is "expressed demand," which can be measured by keeping records of facility use. "Latent demand" is more difficult to determine and usually requires surveys through workshops, questionnaires in utility bills, and focus groups. Future population estimates used in developing the land use plan for the jurisdiction should be used for park and recreation planning as well. Projecting future demand on the basis of past use must be handled carefully since demographic preferences for activities are changing. For example, the growth in girls soccer during the past 25 years has added a new level of demand for playing fields to many localities, and the recreational and open space needs of seniors need to be taken into account for the next 25 years.

Once the demand is calculated, the minimum necessary for each facility should be assessed and summed. This analysis is done for the city as a whole, as well as for individual neighborhoods and communities. The calculations can also be performed for each park classification, and used as the basis for development impact fees once costs are assigned.

Highlighting the Economic Impact of Parks and Open Space

The positive effects that park and recreation infrastructure can have on health and the environment are well known. However, the impact of parks on the economy is frequently overlooked. The following discusses approaches the local practitioner can take during the preparation of the comprehensive infrastructure plan, as well as during budget discussions.[22]

Position the park as part of the tourist industry. Many park officials argue that the municipal parks and recreation department needs to be seen as an important part of the local tourism industry. These officials argue that parks have not only the social and health benefits commonly associated with them, but they also have economic benefits that are greater than their costs. Parks can attract both tourists and high-tech businesses to a locality. Parks also enhance real estate values, and they attract retirees who can be an asset to the economic health of a city. The new businesses, tourism, and increased population can also provide jobs and increased income for local residents.

Emphasize the impact of parks on the tax base. The positive impact of parks and open space on real estate values and hence on the tax base is also important. This occurs because people are willing to pay more to live near parks and open space. This results in higher property taxes, which can be used to pay for the cost of acquiring and developing the parkland. The value of the park is capitalized into the real estate values. This is called "proximate capitalization" or "the proximate principle," and can amount to an increase in value of 20 percent. Frederick Law Olmsted pioneered this argument to successfully persuade New York City officials to invest $14 million in Central Park.

Look at the bigger picture during the budget process. In analyzing the role that parks and recreational facilities play during the annual budget process, it is important to look at the economic impact outside of the organization itself. In one city, when both the parks and recreation department and the local visitors' center were asked to analyze the impact of hosting a national softball championship, they came up with completely different numbers. The parks and recreation department showed a loss of $9,375, while the visitors' center calculated the economic benefits to the community of $525,000. This is because typically municipal departments are interested in the financial impact on their budget, while commercial enterprises are interested in the wider economic impact. Both kinds of data, however, are needed to make a decision about the benefit of the program to the locality.

Guidelines for the economic analysis or business plan. A new park or recreational facility or event can be evaluated in terms of revenues and expenditures just like other infrastructure. For example, Reston, Virginia, analyzed two locations for a skateboard park in terms of projected use, revenues, and operating expenses. The city used the experience of comparable skate parks as well as a demographic analysis of skateboarders, in-line skaters, and freestyle bikers living near the proposed park to determine use and revenue schemes.

A comprehensive analysis of the costs and benefits of an entire system was commissioned in 2000 by the East Bay Regional Park District, which provides park and recreational services to several counties in California. This analysis looked at user costs and benefits, as well as direct and indirect economic benefits of the park use. The replacement value of the park assets were also calculated, as were fee structures of the various services. The report noted that direct expenditures by the park district were about $80 million annually, with about $9.1 million from non-local sources or user fees that resulted in $18 million in positive net new economic impacts to the area. The remainder of the district's revenues is from property and other local taxes. Park users expend about $254 million in park-related activities, with about one-quarter of that being net new expenditures, that is attributable to the presence of the park, and because of multiplier effects, worth $148 million.

Other kinds of economic analyses have been done at the local level for special events. For special events, care must be taken in doing this analysis. One expert notes the following common pitfalls that should be avoided:

- Expenditures by local residents should be excluded.
- Expenditures of those who would have visited the municipality anyway, or who are already there, but are "time switching" to attend the event should be excluded.
- Income, rather than sales, should be used to measure economic impact.
- Jobs going to outside residents should not be counted.
- The present unused capacity of many organizations should be taken into account.

Sustainability and Green Programs

The development of an open space program or a parks plan can also be used as a catalyst to develop a more comprehensive sustainable or "green" program at the local level—parks help to address climate change since they can sequester carbon and other pollutants. Open space and parks are often viewed as the responsibility of only one particular department. However, with an overall program that involves participation by all functions in a jurisdiction, the impact on the livability of the locality can be vastly improved.

One example is the Sustainable City program in Santa Monica, California, that has goals for all city departments. As of 2011, the city was in the process

Table 26.4
City of Santa Monica's open space and land use goals for sustainability

Goals

1. Develop and maintain a sufficient open space system so that it is diverse in uses and opportunities and includes natural function/wildlife habitat as well as passive and active recreation with an equitable distribution of parks, trees, and pathways throughout the community.
2. Implement land use and transportation planning and policies to create compact, mixed-use projects, forming urban villages designed to maximize affordable housing and encourage walking, bicycling, and the use of existing and future public transit systems.
3. Residents recognize that they share the local ecosystem with other living things that warrant respect and responsible stewardship.

Indicators—System Level	Targets
Open Space	
Number of acres of public open space by type (including beaches, parks, public gathering places, gardens, and other public lands utilized as open space)	Upward trend
Percent of open space that is permeable	Upward trend
Trees	
Percent of tree canopy coverage by neighborhood	Upward trend
Percent of newly planted and total trees that meet defined sustainability criteria to be developed by 2007	Target to be developed by 2007
Parks—Accessibility	
Percent of households within ¼ and ½ mile of a park by neighborhood	Upward trend in park accessibility for Santa Monica residents
Land Use and Development	
Percent of residential, mixed-use projects that are within 1/4 mile of transit nodes and are otherwise consistent with Sustainable City Program goals	Upward trend
Regionally Appropriate Vegetation	
Percent of new or replaced, nonturf, public landscaped area and nonrecreational turf area planted with regionally appropriate plants	Target to be developed in 2007

Source: City of Santa Monica, CA, *Santa Monica Sustainable City Plan*, 2003.

of developing numerical performance indicators that are used as part of the on-going management program for the city. See Table 26.4 for a statement of the city's goals for open space and land use, as well as performance measure indicators. Figure 26.5 shows the tree canopy coverage for the city as a whole.

There are specific techniques that can be used by a city or county to "green" the urban environment (this section is adapted from the Urban Land Institute's project in the Vermont Corridor Station, in Los Angeles, California):

- Installing climate-appropriate plants that provide shade and food to local flora and fauna.
- Upgrading open spaces on private land.
- Removing asphalt from alleys and replace it with rockcrete, grass, or the best available technology for permeable surfaces.
- Training city landscaping crews and vendors in the proper trimming and care of mature trees.

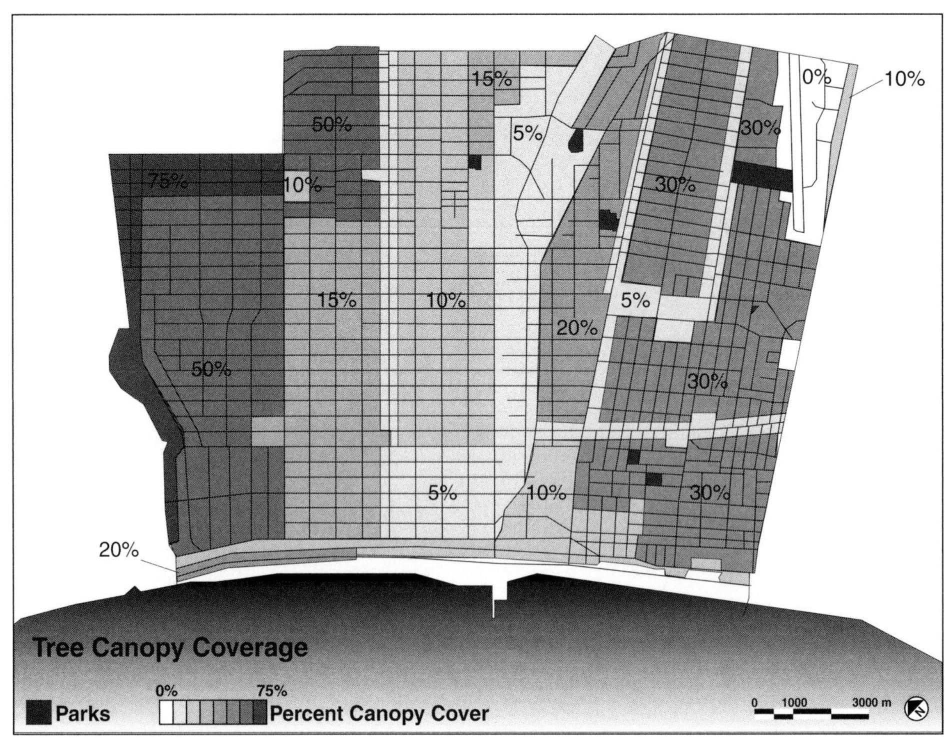

Figure 26.5
Tree canopy coverage in the City of Santa Monica, CA

- Supporting neighborhood gardens.
- Refurbishing freeway median areas and embankments and medians along major boulevards.
- Installing vertical surface landscaping along prominent walls.
- Maintaining and upgrading areas beneath billboards.
- Public spaces can be "greened" by:
  - installing street furniture for the community that depends on public transit;
  - creating a balance between street/site lighting levels necessary for pedestrian and vehicle safety and the need to reduce light pollution.
- Removing security fences and razor wire.
- Providing access to and use of open areas owned and operated by local community colleges and school districts.
- Installing rooftop photovoltaic systems and rooftop gardens on public buildings.

Conclusion

Planning and financing parks and open space projects in a locality require more ingenuity from local practitioners than do some other infrastructure areas where there is a well established history about level of service standards. It is important to develop park and open space needs that are tailored to the needs and desires of the individual locality instead of relying upon national standards or those of an outside expert. An open participatory process should be used for planning and designing the parks, and to develop the standards that will be used for new development within the jurisdiction. An open process will also mobilize community support and allow the locality to take advantage of some of the new institutional arrangements for developing or redeveloping local parks, and then to maintain them. The economic benefits that parks, recreation, and open space bring to a locality should not be overlooked during the budget process. Finally, the catalytic effect that the parks and recreation program can have on developing a comprehensive local sustainability program should be remembered.

Additional Resources

Publications

Bonham Jr., Blaine, Gerri Spilka, and Darl Rastorfer. *Old Cities/Green Cities: Communities Transform Unmanaged Land.* PAS Report 506/507. Chicago, IL: American Planning Association, 2002. Lots of great case examples on how to use vacant and unmanaged land in urban areas for community gardens and other community development activities.

Cranz, G. *The Politics of Park Design: A History of Urban Parks in America.* Cambridge, MA: MIT Press, 1982. A classic history of parks for the aficionado.

Crompton, J. L. *Parks and Economic Development.* Planning Advisory Service Report 502. Chicago, IL: American Planning Association, 2001. A thorough approach for evaluating and promoting the positive economic impacts of parks and recreation programs. Designed for the park professional.

Forsyth, Ann and Laura R. Musacchio. *Designing Small Parks: A Manual for Addressing Social and Ecological Concerns.* Hoboken, NJ: Wiley and Sons, 2005.

Garvin, Alexander, *Parks, Recreation, and Open Space: A Twenty-First Century Agenda.* Planning Advisory Service Report No. 497. Chicago, IL: American Planning Association, 2000. An overview of issues and a program that can be used for parks and recreation programs at the local level.

Gavin, Alexander, *Public Parks: The Key to Livable Communities.* New York: W.W. Norton & Co., 2011.

Harnik, P., Paying for Urban Parks without Raising Taxes, in *Local Parks, Local Financing.* Vol. 2. San Francisco: Trust for Public Land, 1999.

Hopper, K., Increasing Public Investment in Parks and Open Space, in *Local Parks, Local Financing.* Vol. 1. San Francisco: Trust for Public Land, 1999. These two are excellent publications that provide a comprehensive overview of innovative funding sources for parks and open space. Top notch work.

Lewis, Megan, ed., *From Recreation to Re-Creation: New Directions in Parks and Open Space System Planning.* PAS Report 551. Chicago, IL: American Planning Association, 2008.

A comprehensive look at park planning that integrates planning for specific park sites and planning for the parks and open space element with the larger social and environmental vision of the city. Has chapters on needs assessments and parks funding as well as a CD with additional brief papers on subjects as diverse as climate change and safe neighborhoods.

Key Federal Legislation

Urban Park and Recreation Recovery of 1978

The Urban Park and Recreation Recovery (UPARR) program was established in November 1978 to provide direct federal assistance to urban localities for rehabilitation of critically needed recreation facilities. The program does this by providing matching grants and technical assistance to economically distressed urban communities. The law also encourages systematic local planning and commitment to continuing operation and maintenance of recreation programs, sites, and facilities. To be eligible for rehabilitation grants, a jurisdiction is required to maintain a current Recovery Action Program plan approved by the National Park Service, the federal agency which administers the program. The Recovery Action Program plan serves both as a guide for local action planning and as a statement of a community's commitment to the revitalization goals of the UPARR program.

The UPARR provides three types of grants: Rehabilitation, Innovation, and Planning. Rehabilitation grants provide capital funding to renovate or redesign existing close-to-home recreation facilities. Innovation grants usually involve more modest amounts of funding aimed at supporting specific activities that either increase recreation programs or improve the efficiency of the local government to operate existing programs. Planning grants provided funds for the development of a Recovery Action Program plan. In FY2002, about $30 million was awarded to 71 cities and counties.

Americans with Disabilities Act of 1990

See the description in Chapter 20 on developing the infrastructure project, and the sidebar in this chapter for specific ways park development can meet ADA requirements.

Land and Water Conservation Fund Act of

The Land and Water Conservation Fund (LWCF) was established for the acquisition and management of land at the federal level, and to provide matching grants to states for recreation planning, land acquisition, and facility development, but not for operations and maintenance. The LWCF is a trust fund that accumulates revenues from federal outdoor user fees, federal motorboat fuel tax, surplus property sales, and oil and gas leases on the Outer Continental shelf. The latter are the largest single source of LWCF revenues.

Expenditures from the LWCF have fluctuated widely over the past 30 years due to political differences between Congress and the Administration. Annual appropriations peaked in FY1978, at $806 million, after dropping to $76 million in FY1974. Appropriations are usually around $200 to $300 million. In 1995, the accumulated balance in the LWCF was over $10 billion. Most of this fund has been spent on land acquisition for the Forest Service, the National Park Service, the U.S. Fish and Wildlife Service, and the Bureau of Land Management.

Youth Conservation and Service Corps

This program provides full-time, paid training, education, and community service opportunities for youth and young adults living in cities and urban counties to rehabilitate existing recreation facilities. This program is administered in connection with the UPARR program.

PART SEVEN

Energy and Telecommunications

This part of the book turns to the infrastructure systems that are mostly provided by private regulated monopolies: energy and telecommunications. However, a small but substantial number of local governments also own and operate these systems. The chapters in Part VII describe the history of the technology before outlining the major elements in the system for each area. The institutions involved in providing and regulating the infrastructure are also addressed before going into the role of local government in dealing with public and private providers. Land use and policy concerns for each are discussed.

Chapter 27 (Energy and Power) looks at the changing nature of providing energy to the community, and of the technologies available today. It also covers energy conservation, including a discussion of local policy and land-use tools that can be used to promote energy conservation and reduce greenhouse gas emissions.

Chapter 28 (Telecommunications) provides advice for the local practitioner on the four roles that local governments can play in this area: the provider; the regulator; the facilitator or promoter; and the user. How to develop a telecommunications plan is addressed, along with zoning and other kinds of regulations that protect a locality's interest. Strategies that local governments can use to attract private telecommunications providers are discussed. The chapter concludes with a description of how local governments are using telecommunications today for more effective and efficient service delivery.

chapter 27

Energy and Power

In this chapter ...

Importance to Local Practitioners . . . 579
The History of Power . . . 579
The Origins of Electricity . . . 580
Early Electrical Systems in the United States . . . 580
The Rise of the Energy Trusts . . . 580
The Origins of Regulation . . . 581
The Deregulation of Electricity . . . 582
Petroleum and Natural Gas in the United States . . . 582
Flows of Energy and Climate Change in the United States . . 583
Energy Uses and Sources . . . 583
Energy, Greenhouse Gases, and Climate Change . . . 584
Mainstream Energy Technology and Facilities . . . 585
Electrical Generation Facilities . . . 585
Types of Fuels Used in the United States . . . 587
Electrical Transmission and Distribution . . . 589
Natural Gas Facilities . . . 590
Low Carbon Energy and Decentralized Solutions . . . 591
Low Carbon or Renewable Energy Sources . . . 591
Decentralized Energy Production . . . 594
Energy Institutions . . . 594
Who Produces Power? . . . 594
Who Regulates Power? . . . 595
Power and the Local General Purpose Government . . . 596
Control of Municipally-Owned Systems . . . 597
Land Use Regulation . . . 598
Regulations for Local Development . . . 599
Specific Regulations . . . 600
Decreasing Energy Costs in Public Facilities . . . 601
The Energy Element . . . 602
Background . . . 602
Creating the Energy Element . . . 603
Conclusion . . . 607
Additional Resources . . . 607
Publications . . . 607
Websites . . . 608
Key Federal Regulations . . . 608

Why Is This Important to Local Practitioners?

Energy is in the midst of far-reaching changes due to rising energy costs, geopolitical shifts, and concern about climate change. Energy supply and use are linked to local economic development, jobs, and quality of life. Residents, industry, and transportation systems all rely on energy. Any local planning process should therefore consider energy. Adequate energy supply must keep pace with community growth, or the local economy will suffer (the California energy crisis of 2000/2001 illustrated this; the Silicon Valley Manufacturing Group reported that each rolling blackout, which averaged 90 minutes in length, cost local industry upwards of $50 million). Energy conservation is also important, not just to save money and reduce our dependence on foreign oil, but to reduce our carbon footprint.

The chapter opens with a brief history of the harnessing of the two largest sources of municipal energy: electricity and natural gas. It then moves onto the technology of energy production and distribution facilities. The chapter concludes with a discussion of state and local actions to reduce energy consumption and an outline of the major steps involved in preparing a sustainable energy plan for the locality. This section also addresses energy flows between energy sources and end users. Carbon emissions are described and low carbon energy sources evaluated.

The History of Power

Even though sources of energy include natural gas, coal, petroleum products, wind, and nuclear fission, most of these energy sources are primarily used as fuel to generate electricity. Furthermore, electricity and natural gas account for 90 percent of the energy used in residences and businesses.

The Origins of Electricity

In 1877, Charles Brush designed the first lighting system, consisting of a generator and a single arc lamp. Two years later, in 1879, he patented a system of electrical distribution. Later that year, he connected 12 of his lamps to his new distribution system in the Public Square of Cleveland and ushered in an era of electrical power when, at 8:05 p.m. on April 29, he created the first illuminated street in the United States. Two months later, the California Electric Light Company formed in San Francisco. Using Brush's technology, this company became the first in the world to enter into the business of generating and selling electricity. This inaugurated the role of the private sector in the electrical power industry.

Early Electrical Systems in the United States

Brush's system soon lost out to those of other innovators, in particular Thomas Edison and William Stanley. Edison designed a system of electrical generation and lighting, and soon received a franchise to operate a central power system in New York City. Edison's system used direct current, or DC, which was inefficient over long distances and was primarily useful for electric lights, not electric motors. Stanley, working for the Westinghouse family, developed a transmission system based on alternating current, or AC, which was more efficient over long distances. At first, both AC and DC systems existed side by side, but in 1893, the Westinghouse Company unveiled a "universal" system that joined AC and DC in the same network.

At the municipal level, the normal procedure was to grant a franchise to a supplier, who would then install a system to light a house, a block, or a neighborhood. The conventional wisdom was that competition would keep rates low, so cities granted franchises with impunity, and often these franchises overlapped or conflicted. It was truly a free market, which is a polite way of saying it was "frontier-style chaos." Hundreds of entrepreneurs received a franchise, only to quickly go under when they lacked capital or expertise. The small service areas of the franchises also created an atmosphere ripe for corruption. Politicians often granted electrical contracts to personal favorites. Rates could be extortionary.

The Rise of the Energy Trusts

Soon, large trust companies began to vigorously acquire the smaller franchises. Perhaps the most famous of these trusts was that formed by Thomas Edison and J. P. Morgan, the notorious Robber Baron banker. Morgan, soon imitated by John Rockefeller, James Hill, the Armour family, and others, embarked on a path of consolidation that by 1929 gave him control over 33 percent of the nation's electricity supply.

These powerful trusts did nothing to stop the corruption or high rates; if anything, they exacerbated the problems. In response, a grassroots push for municipal ownership of power companies emerged in the late 1880s. Almost invariably, these municipally owned systems, or "munis," produced power more cheaply and reliably. Not surprisingly, the trusts did everything in their capacity to suppress there municipal systems. As a result, publicly owned systems have never held more than a fraction of the power market. Even today, public systems only supply about 30 percent of the nation's total electricity.

The Origins of Regulation

While the public and private sector battled it out, others debated over what the structure of the industry should look like. In 1898, Samuel Insull, former assistant to Thomas Edison who was then an electricity mogul himself, proposed a compromise to the public–private dispute: public regulation of private monopolies.

In 1907, both the National Electric Light Association, the premier industry organization, and the National Civic Federation recommended state regulation of private utilities. By 1916, 33 states had regulatory agencies in place to regulate private electricity providers.

The private utilities soon controlled the state regulatory agencies, however, leading to what New York Governor Franklin Roosevelt called the "Insull Monstrosity." The electric utilities had built up "what has justly been called a system of private socialism which is inimical to the welfare of a free people," Roosevelt stated. Once he became president, his New Deal Congress quickly passed the Public Utilities Holding Company Act in 1935, which attempted to break up the private utility trusts. This Act set in place the regulated power industry we know today.

In 1954, Congress allowed the private development of nuclear power with the Atomic Energy Act. Although initially reluctant, both public and private electricity companies soon began building nuclear reactors to generate electricity, pushed along both by significant boosterism at the federal level and by state regulatory policies that allowed a private utility to garner profits based on the amount of capital investment.

Electricity Explained

Electricity is the flow of electric current through wires. In many ways, it is analogous to the flow of water through a pipe. The basic unit of charge is the coulomb, which is analogous to a gallon of water. Both water and electricity flow. While the flow rate of water might be measured in gallons per minute, the flow of electricity is measured in amperes, which is one coulomb per second. In a water system the pressure is expressed in pounds per square inch. With electricity, the pressure is expressed in volts. Materials have differing abilities to conduct electricity. The ability to conduct flow is a measure of the resistance of the material, which has the unit of ohms. A fundamental law in electricity is Ohm's Law, which says that:

Current in amperes = Voltage in volts/ Resistance in ohms

In other words, the higher the voltage (pressure), the greater the current (flow), which makes sense intuitively. Alternatively, if you increase the resistance, you decrease the current. Ohm's Law is important in the transmission of electricity. Of course, electricity by itself is not worth much. We want to make it do work. In fact, work has a specific definition, which is to apply a force through a distance. The unit of work is the joule. We also care about the time it takes for electricity to do work, which is to say we care about the power of electricity. Work divided by time is power, and the unit of power is the watt, which is one joule per second. One thousand watts, of course, is one kilowatt.

However, a string of high-profile nuclear accidents, such as the one at Three Mile Island, and the rise of environmentalism sparked an anti-nuclear movement.

The Deregulation of Electricity

The 1990s saw a push for deregulation. The primary goal of deregulation is to allow consumers the ability to choose their electricity provider. Advocates say that reducing government control of electrical utilities would result in lower rates and expanded services.

The consensus among those involved in deregulation seems to be that deregulation can work, but that no one has hit on the ideal formula for making it do so. The issue is a complex one; the Center for Responsive Politics provides a cogent discussion of deregulation at its website: www.opensecrets.org/news/electricty.htm.

Petroleum and Natural Gas in the United States

It is widely accepted that August 27, 1859, is the birthdate of the American petroleum industry. On that day, outside Titusville, Pennsylvania, Edwin Drake struck oil 69 feet below the surface of the ground. Although the properties of oil and gas were well known before this, these materials were rare and hard to obtain. Drake's discovery, and subsequent construction of a 5½-mile pipeline from the well to Titusville, opened up the possibilities of petroleum products, of which natural gas was initially the most promising.

Since natural gas distribution requires capital-intensive pipe networks, the natural gas industry began to parallel the electric industry in the development of oligopolistic firms. In response, in 1938, the transmission of natural gas was federally regulated under the Natural Gas Act.

Since a viable interstate pipeline system was not operational until after World War II, industrial and electric utility markets for natural gas were comparatively small at first. These markets continued to rely on oil and coal. However, once the transmission and distribution system was in place after the war, the use of natural gas exploded. Soon, demand began to outstrip supply, and costs began to rise. In 1954, the federal government stepped in with price caps at the well, and the sale and transmission of natural gas became regulated from the source all the way to the end user.

The oil crisis of the 1970s changed this. Slowly, the federal government rolled back regulation, first at the well and later over the entire transmission system. There never was a push for municipal gas supply ownership, and so in many ways the natural gas industry developed without controversy. Today, while the electricity industry remains largely regulated, the natural gas industry is largely unregulated and is entirely private.

Flows of Energy and Climate Change in the United States

Energy Uses and Sources

The United States uses one-quarter of the world's energy, with less than 5 percent of the population.[1] Within the United States, the Department of Energy divides end use into four major categories: (1) transportation (28 percent); (2) residential (20 percent); (3) industrial (33 percent); (4) and commercial (17 percent) (Figure 27.1). Petroleum is the major energy source for transportation, while in the other three end use areas, energy is delivered directly with natural gas or indirectly through the use of electricity, which is generated from

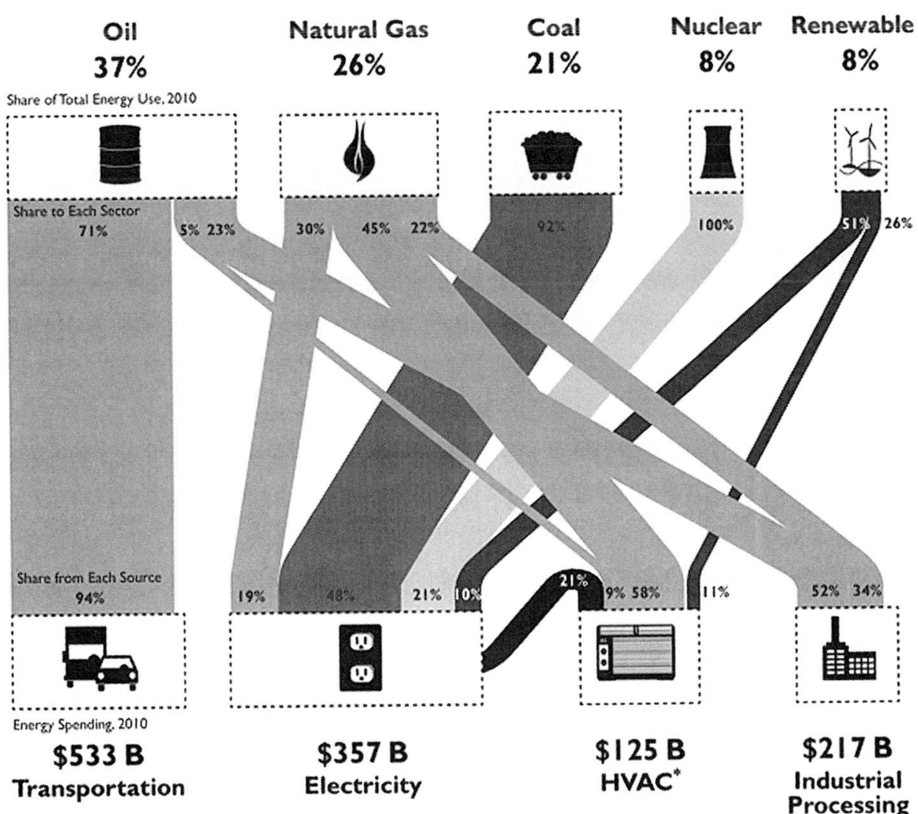

Figure 27.1
Sources and uses of energy in the United States in 2010

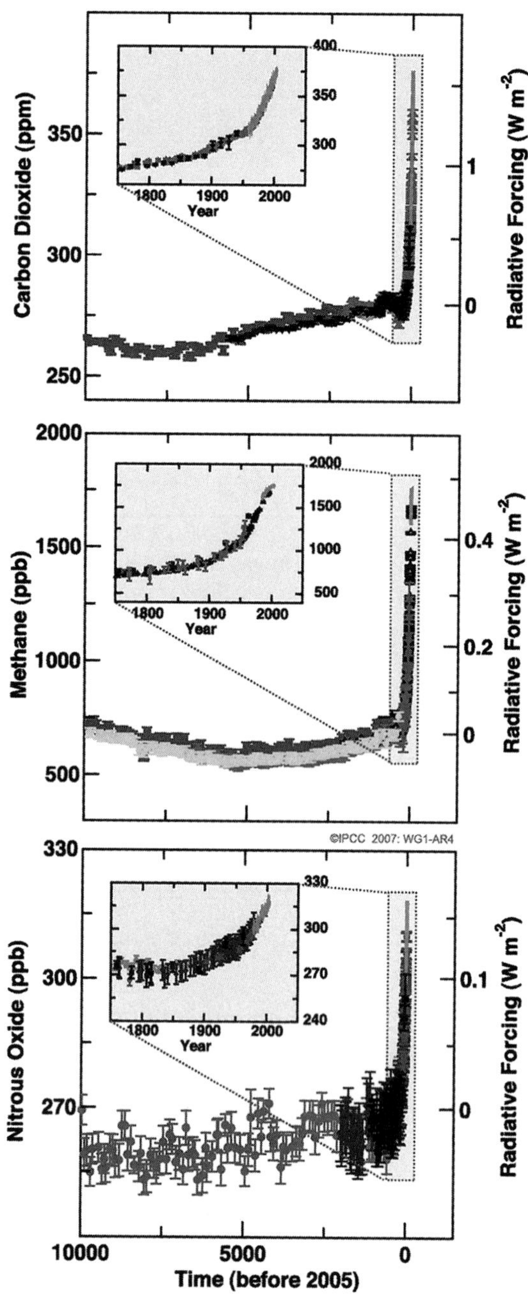

Figure 27.2
Changes in the concentration of green house gases in the past ten thousand years and changes from 1970 to 2005 in the inset

primary energy sources such as coal, natural gas, oil, nuclear power, and other natural sources.[2]

Most electricity is generated by turbines that are driven by water, wind, or steam—the latter of which is produced by burning fossil fuels, nuclear fission, or biomass products. Fossil fuels (coal, natural gas, or petroleum) account for 68 percent of U.S. electricity generation, while 21 percent is generated by nuclear plants. Energy sources for electricity can be renewable or non-renewable, but electricity itself is neither.[3]

Energy, Greenhouse Gases, and Climate Change

Greenhouse gases (GHGs) are those gases present in the Earth's atmosphere that reduce the loss of heat into space. They include carbon dioxide, methane, nitrous oxide, and fluorinated gases (see Figure 27.2). Ozone depletion, caused by CFCs, has only a minor role in greenhouse warming, though the media has in the past confuses the two. GHGs are emitted by many sources, but the production of energy is the single largest source, accounting for 88 percent of total emissions in 2010.[4]

Increasing levels of greenhouse gases result in increased global temperatures. The average global air temperature has increased almost 1 degree Centigrade in the 100 years prior to 2005. Eleven of the past 12 years have been among the warmest since 1850. Climate model projections indicate that during the twenty-first century, the average global surface temperature will rise from 1 to 6 degrees Centigrade Global warming has resulted in rising sea levels, an increase in extreme weather events, and changes to the amount and pattern of precipitation, including the shrinking of the Arotic ice pack, the prolonged drought in Africa, and the increased intensity of hurricanes in the North Atlantic. Additionally, the large wildfires in the Western United States and extreme weather events in other parts of the United States are other manifestations of climate change.

A rich portfolio of actions has been taken at the state and local level to reduce greenhouse gas emissions. At their core are efforts to reduce: (1) transportation emissions with the "triple-legged stool" of increased vehicle fuel economy, reductions in the carbon content of fuel, and a reduction in vehicle miles traveled through more compact development and better transit alternatives; (2) industrial use through market-based solutions; and (3) residential and commercial energy use through "green building" and energy conservation measures. Noting the important role that electricity plays as an intermediate user of energy that is then used by residential, commercial, and industry end uses, there are also efforts targeted specifically to increase the efficiency of its production.

Mainstream Energy Technology and Facilities

The mainstream model of energy provision consists of large, centralized, supply-side solutions that have relied for many years on increasing returns to scale to lower the price of energy. Conventional energy facilities consist of the power plant that produces electricity and the transmission and distribution system (Figure 27.3). Similarly, although natural gas is piped from wells to the user, its system also benefits from centralized provision. The following describes the systems for electricity and natural gas.

Electrical Generation Facilities

Electricity is generated at a power plant. The heart of a power plant is the generator, a device that converts mechanical energy into electrical energy. There are several different types of power plants, each using different fuels and requiring different infrastructure support (Figure 27.4). Power plants need access to water (a combined-cycle plant can use five million gallons a day), access to the electrical grid, and access to their fuel. Power plants also need to be accessible to those who work there. Moreover, the power industry contends that the closer power generation is to power consumption, the better.

In addition to the building housing the actual generation equipment, power plants may also require land area for storing fuel and waste products like ash. Most combined-cycle power plants (see below) require about 10–20 acres. However, a 300-megawatt coal power plant can take 300 acres. The proposed Toquop Power Plant in Nevada is exceptionally large, taking up 640 acres.

GHG Goals

Many experts indicate that the United States needs to reduce greenhouse gas emissions by 60–90 percent below 1990 levels by 2050 and that we need to be significantly below 1990 levels by 2030. However, as of 2013, some knowledgable scientists state that it is not longer possible to roll back the clock to the emissions of 2 decades ago. Irreversible change in the oceans and icepacks have triggered feedback loops that are accelerating the rate of global warming despite anthropogenic reductions. This is not to say that carbon emission reductions are unnecessary, but their short term impact on warming will not be as great as hoped and therefore planning to adapt to warmer conditions is necessary.[1]

Note

1. Paltzev, S., J. Reilly, and A. Sokolov. "What GMG Concentration Targets Are Reachable in this Century?" MIT Joint Program on Science and Policy of Global Climate Change, 2013.

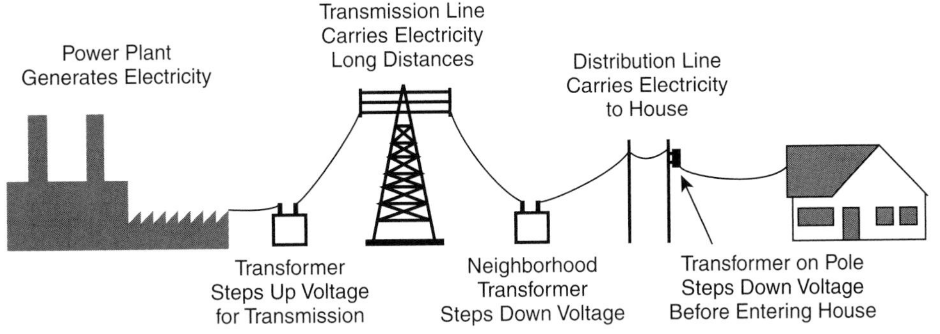

Figure 27.3
Transporting electricity

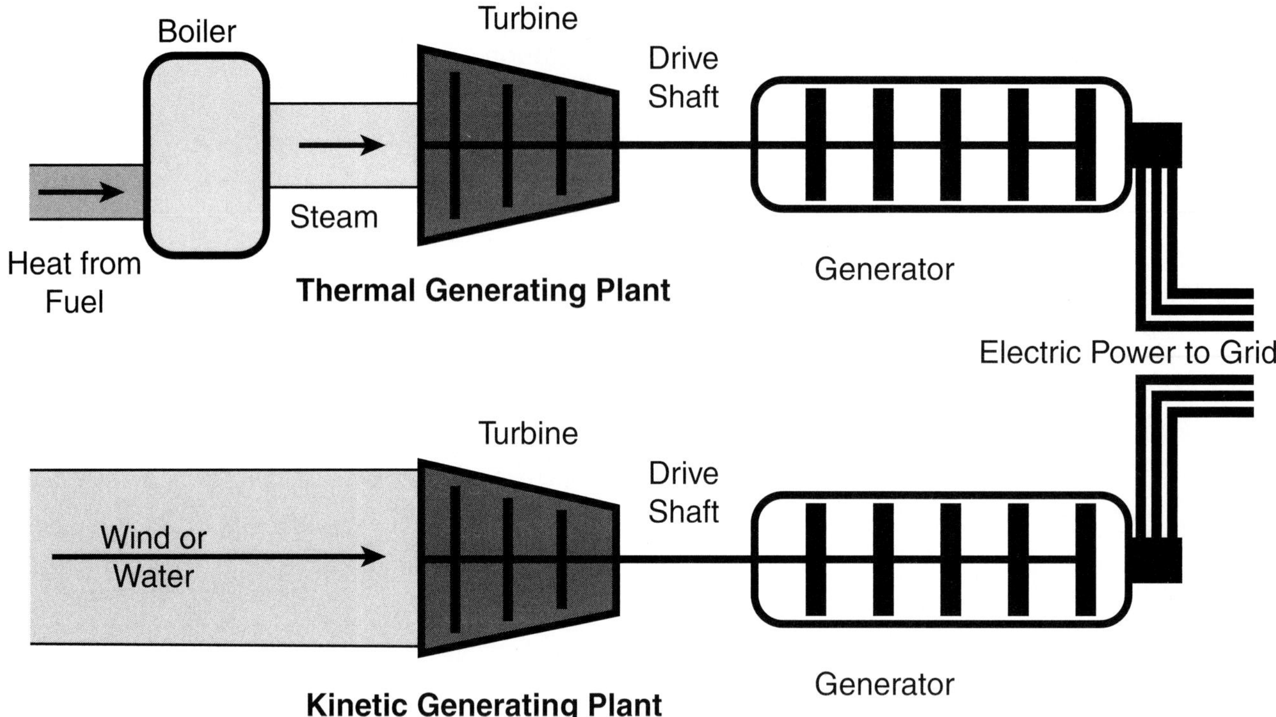

Figure 27.4
How electricity is generated from fuel, wind, and water sources

AC versus DC

Electrical current can come in two varieties, direct current or alternating current. With direct current, the current only flows in one direction. A battery produces a DC current. Direct current is not efficient to transmit over long distances. But DC is better on a micro scale. Many electrical appliances, including computers, use DC power.

With alternating current, on the other hand, the current changes direction with a certain frequency. The frequency with which the current changes direction is measured in cycles per second, or hertz. In much of the Western hemisphere the electrical power standard is 60 hertz, with most household appliances needing 120 volts. In Europe, the standard is 50 hertz and most appliances take 240 volts. This is why electrical devices made for one region do not work in the other.

Peaker plants. In addition to the main large power plant, the electric companies have also been building peaker plants. Electricity companies are in the business of supplying electricity whenever the customer demands it. Electricity cannot be stored, and so the companies must have the equipment on hand to meet maximum demand. For this reason, electricity companies build peaking plants, power plants that only become activated during periods of high or peak demand.

Cogeneration. The process of burning a fuel, creating steam, and then using the steam to turn a generator is very inefficient. For example, a coal-fired plant might only convert 35 percent of the energy in the coal into electricity (35 percent efficiency). The rest of the energy goes into the steam, where it is wasted. In the 1970s, cogeneration began to be used. In a cogeneration process, the electricity is produced in the normal fashion, but the steam is captured and sold to some other industrial user, who would have to burn fuel to make steam otherwise.

Types of Fuels Used in the United States

The preponderance of fuels used to generate power in the United States are either fossil fuels—coal (48 percent), natural gas (19 percent), and petroleum (1 percent)—or nuclear (21 percent). Renewable sources, including hydroelectric, wind, solar, and geothermal, accounted for only 10 percent in 2010. (See Figures 27.5 and 27.6.) The following describes the traditional fuel sources used in the United States.

Coal. Historically, coal has been an important source of energy, supplanting wood as the most used source in the U.S. in 1885. Today, coal plants are the most common method for generating electricity, despite the significant pollution they cause. A report by the Environmental Integrity Project in 2008 found that in the United States the 50 worst coal-fired power plants produce 40 percent of the sulfur dioxide emissions per megawatt-hour, while providing just 14 percent of the electricity.

The normal method of transporting coal to the power plant is by train, and therefore these plants are near rail lines. Coal slurries are also used, which are composed of pulverized coal mixed with water that can be transported by pipelines.

Electricity generated from coal also results in substantial amounts of carbon-dioxide emissions, and one solution is to "sequester" the carbon dioxide emissions from coal-fired electrical plants. This is also called carbon

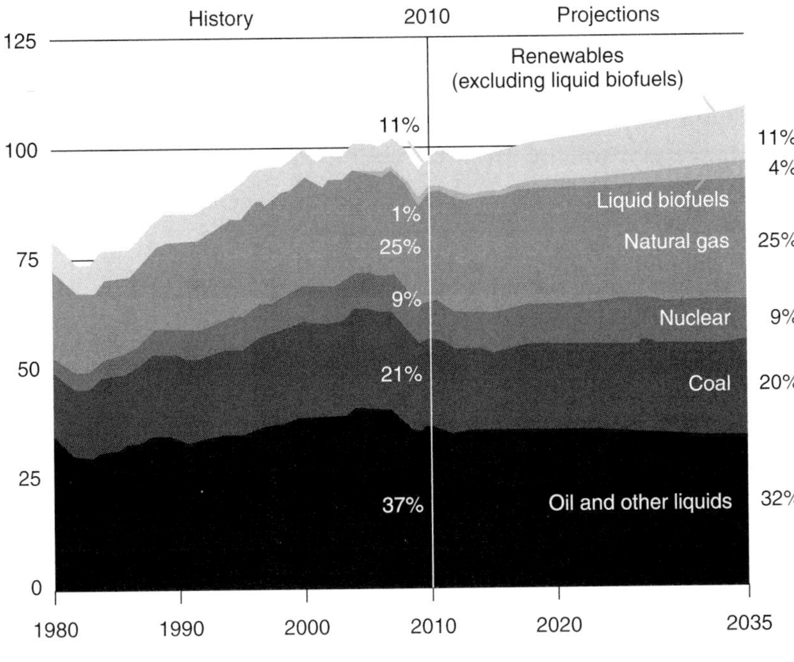

Figure 27.5
Energy consumption history in the United States by source from 1980 and projections to 2035 in quadrillion BTU per year

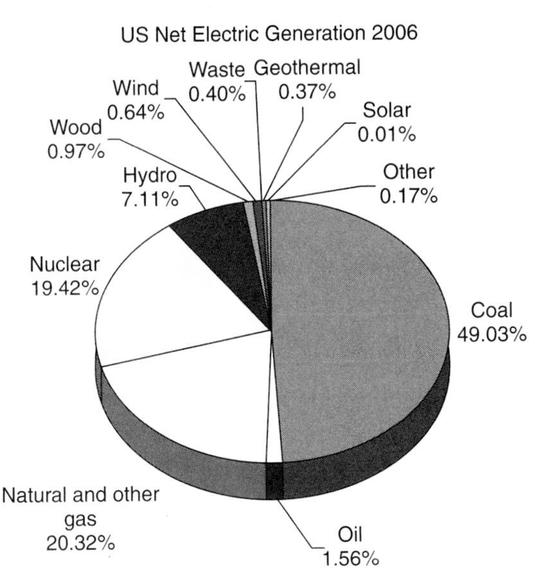

Figure 27.6
U.S. net electric generation, 2006

capture and storage (CCS). Terrestrial sequestration occurs naturally, when plants absorb CO_2 from the air during photosynthesis and ultimately store it as biomass or transfer it to the soil. Geological sequestration is a manmade solution for CO_2 that involves capturing the CO_2 where it is emitted, and storing it in deep underground geologic formations such as older oil and natural gas reservoirs, unmineable coal seams, deep saline formations, and others.

Natural gas. In 2010, 19 percent of the nation's electricity was fueled by natural gas. Natural gas is also a major source of space heating. Natural gas is the cleanest burning of the fossil fuels. It is odorless, colorless, and tasteless. (For safety, natural gas companies add a chemical called mercaptan to give it that rotten egg smell.) It is mostly methane. Natural gas is distributed through a nation-wide network of pipes, and the plants need access to these lines. Natural gas can be burned to heat water for steam and can also be burned to produce hot gases that pass directly through a turbine to generate electricity. These are the cleanest type of combustion-based plant, and with the addition of a cogeneration or combined-cycle element are the most efficient.

In the 1970s, combined-cycle gas and steam plants appeared, and they were more widely adopted in the 1990s. The gas-fired plants were a technological breakthrough in terms of efficiency after the scale economies of the large plants peaked in the 1960s. This type of a plant first burns natural gas in a turbine similar to a jet engine, which then powers a generator that produces energy. The exhaust from this turbine, which is at about 1,000 degrees F, is used to generate steam, which in turn powers a second steam generator. In this way, a combined-cycle plant captures more of the energy of natural gas, and can be up to 50 percent efficient.

Petroleum. Petroleum can also be used to make steam to turn a turbine, and in 2006 this source generated 1.56 percent of the nation's electricity. During the early 1970s, electric utilities used petroleum extensively to generate electricity, but this was curtailed in the 1970s and 1980s as petroleum prices rose and instability occurred in the oil-producing countries in the Middle East.

Nuclear power. Nuclear power is an important generator of electricity. Since 1984, nuclear plants have provided almost as much of total U.S. electricity as gas fired plants.

Nuclear power plants use heat generated from the radioactive decay of uranium to turn water into steam, which then turns the generator. From an air pollution and carbon emission point of view, nuclear power is also benign.

Nuclear power has several downsides. First, it produces thermal pollution. Second is the issue of what to do with the spent uranium fuel rods. They are considered high-level radioactive waste that must be stored at the plant that produced it. Any discarded equipment, clothing, or machinery that has been exposed to uranium is considered low level waste. This waste is transported off-site to various storage areas around the country. Oftentimes, community opposition to a nuclear plant centers on the transportation of radioactive waste through the area.

Nuclear power plants can be controversial because they emit radiation and pose a threat of catastrophe should they fail. Studies have shown that U.S. nuclear plants actually emit a negligible amount of radiation. According to the National Cancer Institute, those living next door to a nuclear power plant, receive less radiation each year than received in one round-trip flight from New York to Los Angeles. More serious are events that disable emergency mechanisms or damage the reactor itself.

Hydropower. Water is currently the leading renewable energy source used by electric utilities to generate electric power. It accounted for 7 percent of U.S. electricity generation in 2006. Seventy percent of the hydroelectric power in the United States is generated in the Pacific and Rocky Mountain States. Generating electricity using water costs less than the fossil fuel plants, and because there is no fuel combustion, there is little air pollution or carbon emissions.

Hydroelectric plants generate electricity by using the power of falling water to spin a turbine connected to a generator. The greater the distance that the water falls, the greater the amount of energy it can transmit to the generator. There are two basic types of hydroelectric systems. In the first system, flowing water accumulates in reservoirs created by the use of dams. The water falls through a pipe to apply pressure on turbine blades to drive the generator. In the second system, called run-of-river, the force of the river current drives the turbine blades without the need for a dam.

Hydroelectric plants are normally located at naturally occurring waterfalls or at river narrows where the water level can be artificially raised through a dam.

Electrical Transmission and Distribution

Once electricity is generated, it must be transmitted to the end user. This is the responsibility of the electrical transmission and distribution systems. Transmission is the transport of electricity from its point of generation to the various areas of consumption, and can be considered analogous to an interstate highway. We are all familiar with the immense towers that carry power lines across the rural landscape, which are transmission lines.

If electrical transmission is the interstate highway, then electrical distribution is the local street system. Once the transmission lines arrive at their destination, the electricity must be distributed to the various end users. The distribution system can be overhead on poles, or, as is growing increasingly the case, in underground channels. The transmission and distribution systems form a network or grid.

Ohm's Law states that for a given resistance, the higher the voltage, the greater the amount of current. For this reason, the voltage in long-distance transmission lines is very high, on the order of 10,000 volts (by comparison, the voltage in the average household outlet is 120 volts). The voltage leaving the generator is not this high, though, so the voltage must be transformed. A step-up transformer raises, or steps up, the voltage of the electricity. One reason why AC is better than DC is that AC is easy to transform in this fashion. The transformers are located at substations.

Multiple transmission lines

Electrical users do not normally need 10,000 volts, however. Therefore, when the electricity enters the distribution system, the voltage is stepped down at another substation. Different users need different voltages. Industrial users or transit systems might need high voltages, while residential customers need low voltages. In fact, the distribution system operates at around 1,000 volts, and the voltage is stepped down one final time before it enters an individual house or business. The step-down transformers that accomplish this final step can often be seen as small canisters mounted on telephone poles or as green boxes at the side of the road.

Utility pole with transformer, power and telephone lines

Natural Gas Facilities

The natural gas production, transmission, and distribution system parallels that of electricity, but it is much less complex. Natural gas is found in underground pockets, usually accompanying other petroleum product deposits. These pockets are tapped by drilling wells, and the gas is then pumped to the surface. Once the gas reaches the surface, it is refined to remove impurities like water, sand, and other gases. Much as electricity is transmitted at high voltage, natural gas is transmitted through large pipes at high pressure, on the order of 1,500 pounds per square inch. The gas is transmitted to a local distribution company, which can be the same company that distributes \electricity. Here, the pressure is reduced and the gas enters the distribution system through which it reaches homes and businesses. Almost all natural gas distribution lines are buried underground.

Low Carbon Energy and Decentralized Solutions

Scale economies in electrical generation through the 1960s resulted in large plants, top-down planning, and supply-side solutions to energy needs. Although this model is still being used today to develop new energy sources, a more decentralized approach is also guiding many research efforts. The energy technology is in a state of extreme flux and innovation, spurred by the rising price of petroleum, concern about carbon emissions, and the potential of information technology to provide real-time information to consumers and producers about energy use.

Low Carbon or Renewable Energy Sources

Solar power. Solar power is the most popular of the renewable energy sources. Sunlight is converted into electricity using photovoltaics (PV), or solar thermal panels, which capture the heat in the sunlight. Photovoltaic cells are modules covered with a chemical coating, usually silicon, that, when exposed to sunlight, gives off electrons. The electrical current is DC, which requires conversion to AC. Photovoltaic cells have efficiencies of about 8–11 percent.[5] Solar thermal power is generated using mirrors or lenses in a parabolic dish or trough to concentrate the light to be used as a heat source for a conventional power plant.

Solar thermal has been used for large-scale generation, but recently multi-megawatt PV plants (grid-connected photovoltaics) have also been built. The latter was the fastest growing energy source internationally in 2008. Germany is the largest consumer of PV electricity, followed by Japan, while large-scale solar plants exist also in Spain and one is currently under construction in Portugal. In the United States, the Alamosa Photovoltaic Solar Plant is being built by SunEdison and is expected to produce 8.22 megawatts. PV can also be used at the individual building level for decentralized electrical generation.

The attractiveness of solar power is based on the fact that 40 minutes of sunlight contains enough energy to supply global energy needs for a year. Converting 2.5 percent of the sunlight reaching the sunny southwest would result in as much energy as the United States consumes in a year.[6] If only 0.02 percent of the available power in sunlight were tapped, it could entirely replace fossil fuels and nuclear power as an energy source. See Figure 27.7.

Currently solar energy is used to generate less than 0.1 percent of the electricity used in the USA, and government projections for 2030 show that

Changes to the Grid

Most power grids are based on 1950s technology and have outmoded control systems. Communication systems between the grids are not adequate. Most energy infrastructure is built to close down when trouble occurs, which spreads blackouts. However, today's technology makes it possible to have a self-healing "smart" grid. Utilities are experimenting with ways to monitor the grid in real time in order to use that information to control the flow of power to avoid blackouts. Utilities are also upgrading their networks with higher capacity infrastructure, and are trying to develop ways to produce and store power closer to consumers. Tomorrow's electrical grid may have many small generating facilities that use alternative energy sources such as solar and wind power and are coordinated with real-time control systems. This would reduce transmission losses, operating costs, and negative environmental impacts.[1]

Note

1. Department of Energy, Office of Energy Delivery and Reliability. www.energy.gov

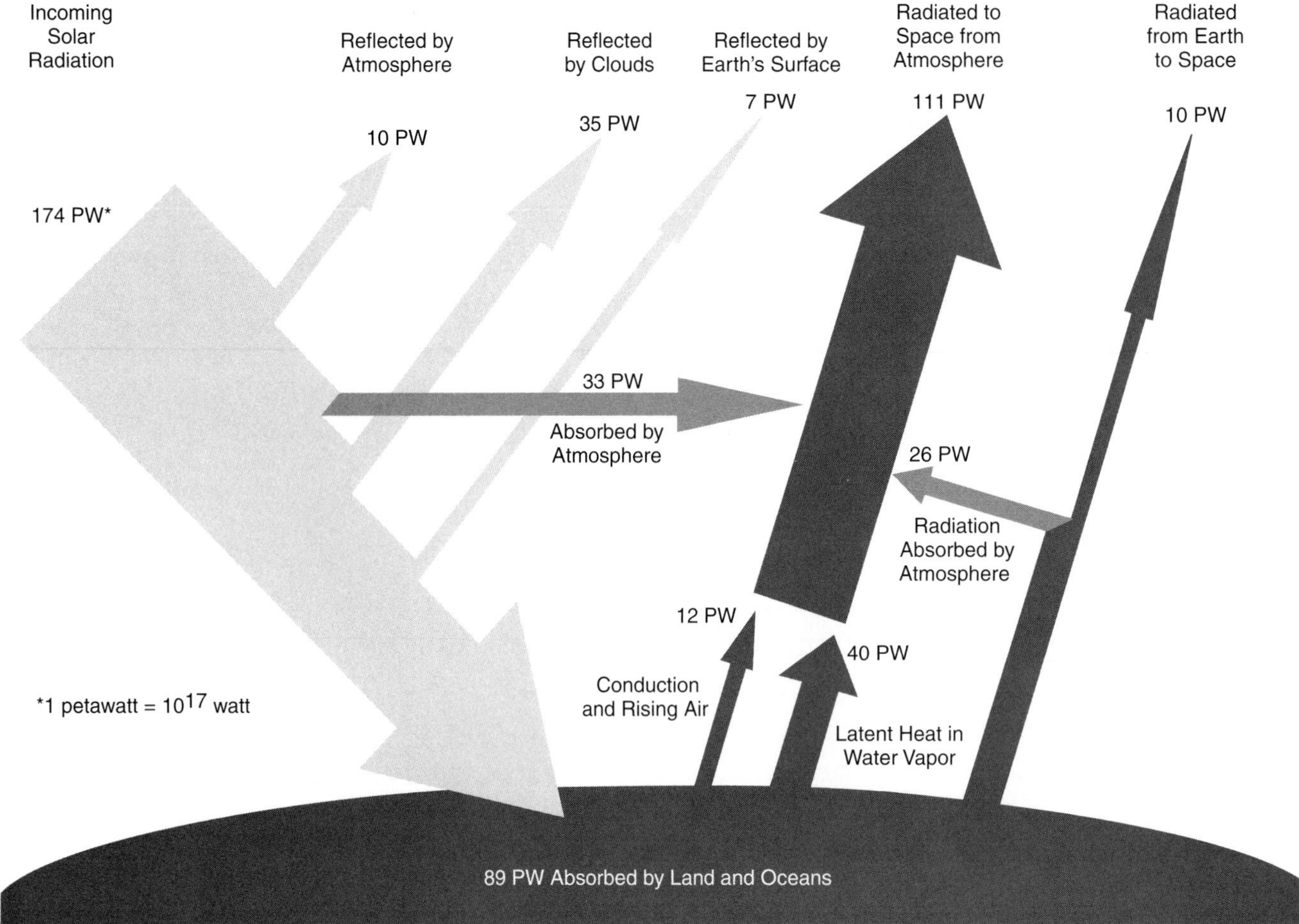

Figure 27.7
Diagram of flow of incoming solar energy to the Earth and its disposition

although the kWh of solar energy will increase from 0.9 billion to 7 billion kWh, it will still generate only 0.1 percent of the nation's electricity.[7] A "Grand Plan" proposed in *Scientific American* in 2008 explored the costs and benefits of developing enough large-scale PV plants to eliminate the need to import foreign oil. For an investment of $420 billion in plants and transmission lines (the equivalent of the yearly amount of farm subsidies), 70 percent of the nation's electricity needs could be provided by 2050. The downside is the requirement to cover 46,000 square miles of land in the southwest with PVs.[8]

Wind power. Wind power is also a rapidly growing source of electricity (wind generation increased from about 6 billion kilowatt hours in 2000 to about 120 billion kilowatt hours in 2011). Windmills have been widely used in rural areas for years. Today's wind farms consist of a battery of metal windmills that convert the energy contained in wind into electricity. The Department of

Energy estimates that wind-powered electricity could rise to 20 percent by 2030 if transmission lines are built, among other things.

In 2013, a $7 billion wind-power transmission project that will carry electricity from the wind generators in the remote western parts of the state to Dallas, Houston, and San Antonio is nearing completion. However, the project has been plagued by cost overruns. In addition, low wholesale prices for electricity have put a damper on wind generator construction.

There are three issues with wind. First, good wind sites are not near cities, and transport costs for a trip of 1,000 miles, for instance, would double the cost per kilowatt hour. Second, wind does not blow all the time, and generally not when demand is high. Finally, winds are variable, and a storage mechanism is needed. Scientists are currently working on improvements to capacitors, a substitute for batteries that does not degrade over time or produce heat. The goal is to produce a "box" to hold the wind's energy.

Generation costs, however, are much lower than for most conventional sources. Wind-powered electricity can be generated for a cost of 8–8.5 cents/kWh. Power costs in Texas range from 4–7 cents, Pennsylvania 6–8 cents, New England 8–10 cents, and 7–9 cents in California.

Geothermal power. Geothermal power comes from heat energy buried beneath the surface of the earth. In some areas of the country, enough heat rises close to the surface of the earth to heat underground water into steam, which can be tapped for use at steam-turbine plants. The steam that billows out from an underground hot spring is used to power generators. Hot bedrock actually can be found throughout the country, which, if it produced steam, could be used to produce energy without high transmission costs. Scientists at MIT are working to tap this source. Geothermal power generated less than 1 percent of the electricity in the country in 2006.

Biomass and other biofuels. Biomass fuels account for about 1 percent of the nation's electrical generation. Biomass includes wood, municipal solid waste (garbage), and agricultural waste such as corn cobs and wheat straw—all of which can be used to replace fossil fuels in utility boilers. To date, just a few electric utility generating units have been built that use wood or waste products as a primary fuel. Food scraps and feces can generate biogas to generate electricity and to power vehicles but most of these efforts are in Europe and Asia. This area has the potential for a decentralized approach.

Other experimental efforts. Some of the research efforts designed to augment transportation fuels may also be useful for electrical production. The U.S. Congress has provided tax subsidies for ethanol and biodiesel, while researchers at UC Berkeley, along with some private companies, are currently developing microbes to turn cellulosic biomass into a fuel molecule similar to gasoline. This molecule is expected to have higher energy content than ethanol and will be easier to extract and distribute. In addition, one large-scale company is exploring bio-butanol, while another is seeking to refine diesel from leftover poultry, pork, and beef fat.[9]

A 240-megawatt tidal power generating plant in southeastern France uses the ebb and flow of tides to turn turbines. Two exotic technologies are ocean thermal energy conversion, which exploits the temperature differential between deep ocean waters and the surface waters warmed by the sun, and wave power plants, which use the up-and-down motion of waves on the open ocean to turn turbines. These latter technologies are proven, but at this point are prohibitively expensive for widespread implementation.

Decentralized Energy Production

Centralized power production achieves economies of scale with the generation equipment, but experiences power losses in transmission and distribution. Peaker plants are also needed for high demand periods, which are highly polluting and energy-inefficient. Decentralized power production could avoid these problems by placing the energy source near the demand, and by allowing fine-grain customization of power production. Today technologies exist that would permit decentralized power production.

Perhaps the most exciting is the micro-turbine. Micro-turbines are essentially miniature jet engines that are connected to small electric generators. They are relatively small—about the size of two filing cabinets standing back-to-back—and can produce up to 30 kilowatts of electricity. They are attractive because they produce very little emissions (up to 600 times less per kilowatt compared to coal-fired plants). Micro-turbines can be particularly cost effective when the waste heat from the exhaust is captured and used to offset the energy needed to heat or cool a building or preheat a water boiler. They can be designed to either augment power supply during periods of high demand or eliminate the need to be connected to the local grid.

Energy Institutions

Who Produces Power?

In this section, we will look at the institutions that generate, regulate, and consume electricity and natural gas. As we saw in the history section, the electric power industry is a mix of public and private. In 2000, there were 2009 publicly owned electrical utilities, 240 were privately owned (often called investor-owned), 894 were cooperatives, and nine were federal power agencies. Publicly owned electric utilities, although large in number, are small in terms of megawatt hour sales. Investor-owned utilities produced about two-thirds of the megawatt-hours sold in 2000. Publicly owned systems served 15 percent of total customers, cooperatives 12 percent, federal agencies 0.03 percent, and privately owned served 73 percent. Unlike the electricity industry, the natural gas industry is entirely privately owned. In 2000, there were over 8,000 natural gas producers in the U.S., who domestically produced 85 percent of the natural gas the nation consumed (Canada supplied the remaining 15 percent).

Note that for both the electrical and natural gas industries, production or generation, transmission, and distribution should all be considered separate and distinct processes. In other words, the company or agency that owns the power plant need not be the same one that owns and operates the transmission lines, and a third entity may own the distribution system. Indeed, much of the deregulation process has involved the explicit de-coupling of these three processes.

Who Regulates Power?

The Federal Energy Regulatory Commission (FERC) is an independent regulatory agency within the Department of Energy. FERC regulates the interstate transmission and sale of natural gas, the interstate transmission of oil by pipeline, and the interstate transmission and wholesale sales of electricity. It also is responsible for licensing and inspecting all non-federal hydroelectric projects. Specifically, this means, among other things, that FERC approves all wholesale electricity and natural gas rates. It also means that FERC has a say in who gets to sit in top positions at top power companies.

States have separate agencies to regulate electric and natural gas services and prices. These agencies are often called Public Utilities Commissions or Public Service Commissions. Each state is unique, of course, but all these bodies serve the same general purpose of regulating the intrastate transmission and sale of electricity and natural gas.

Power companies must balance electricity supply and demand over the entire network, ensure the proper voltages and currents, and maintain reliability. For this reason, and also to facilitate planning, avoid duplication, and increase the availability of power to purchase if need be, power companies have formed various voluntary organizations. The most powerful of these organizations is the North American Electric Reliability Council (NERC). NERC is divided into ten regional councils that cover not only the United States, but also Canada and parts of Mexico. (See the NERC Regions map, Figure 27.8.) In 2006, NERC became the electric reliability organization (ERO), certified by the FERC to establish and enforce reliability standards for the national bulk-power system.

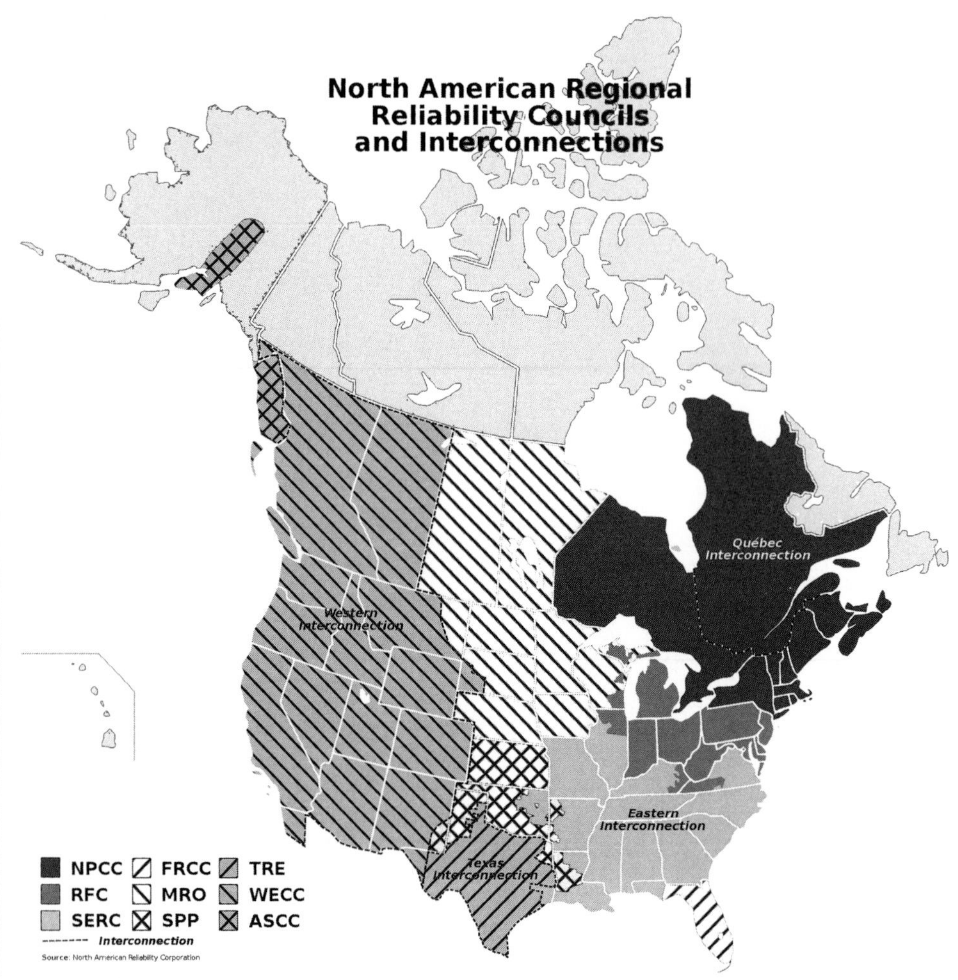

Figure 27.8
North American Electrical Reliability Corporation (NERC) Regions

Power and the Local General Purpose Government

Local general purpose government (GPG) has many roles in relation to the production and provision of power, as well as with the carbon emissions of such power. First, it may own and operate the power production and distribution facilities. It is rare, however, for a local government to own transmission facilities. Fifteen percent of power is provided by such municipally owned systems. Second, the GPG is a key player in the location of power production facilities. Third, the GPG (working through the state) must ensure the adequate provision of power, to both existing and new development. The GPG also has an obligation to reduce the energy use and costs of not only its own facilities (such as schools, court houses, and libraries), but also all other facilities that it controls through land use regulations, subdivision ordinances, or building codes. The GPG also purchases energy and can try to influence how its provider produces that energy. All of these roles are best served if the

Table 27.1
U.S. electric utility statistics, 2000

	Number	Sales (Megawatt-hours)
Municipal (public) utilities	2,009	516,681
Privately owned utilities	240	2,437,982
Cooperatives	894	305,792
Federal power agencies	9	49,094
Total	3,152	3,309,549
% Municipal (Public)	64	16

Source: American Public Power Association, 2002.

local government has an energy element as part of its general plan. This can also be part of a climate action plan.

Control of Municipally-Owned Systems

Some 2009 cities and counties own and operate their own power plants and distribution systems (Table 27.1). Some of the largest cities include Seattle (which has had municipal power since 1902), Cleveland (since 1906), and San Antonio. The Los Angeles Department of Water and Power is the largest publicly owned electric utility in the country, with 1.5 million customers. By comparison, the privately owned Southern California Edison Company, which serves most of the rest of southern California, has 4.3 million electric customers.

The control of these power systems resides at various locations in different city governments. In Cleveland, the power system is located within the Department of Public Utilities. In Seattle and San Antonio, the utility is a separate city department. These departments must perform functions normally performed by the private sector, such as developing demand figures, determining the service area, and establishing rate structures. Energy provision service standards is set by the state regulatory body such as the Public Utilities Commission, although local governments can set internal efficiency standards for their own power systems.

Public power costs less. In 2003, rates for publicly owned systems were 17 percent less than rates of privately owned systems.

Running a municipal utility is a specialized topic that is beyond the scope of this book. See the recommended readings in the Additional Resources section for more information, in particular the resources offered by the American Public Power Association.

Sly Creek power plant, Easter, CA

British Thermal Units

Since energy use can come in many forms, the Department of Energy tracks energy use in British Thermal Units (BTU). One BTU is the amount of energy needed to raise one pound of water one degree Fahrenheit. One candy bar has about 1,000 BTU. On average, in one month a microwave oven uses 50,000 BTUs, while a refrigerator uses 350,000 BTUs. Table 27.2 shows the conversion of common energy providers into BTUs.

Table 27.2 BTU conversion

Energy source	Converted to BTU
1 kilowatt of electricity	3,413
1,000 cubic feet of natural gas	1,027,000
1 gallon of gasoline	125,000
1 cord of white oak	27,000,000
1 gallon of propane	91,000

Land Use Regulation

Local government must regulate the location of energy facilities just as it does any other land use. Facilities for the energy industry include power plants, dams, power lines, gas wells, and coal mines. One study estimated that the amount of land taken up by energy-related activities approaches that taken up by the Interstate Highway System. These land uses can all be regulated within the standard framework.

Power plant location. Power generation and processing facilities are of concern for local planners. As we saw earlier, power plants are normally located in or near populated areas. Unfortunately, power plants make undesirable neighbors. They may be noisy, produce smoke and steam, and are visually unattractive. Public concern about nuclear power plants remains high. Power plants are probably best viewed as necessary evils, and siting them requires outreach and community involvement.

Rights-of-way and utility corridors. Transmission and distribution lines and natural gas pipelines require rights-of-way. It is not unusual for transportation rights-of-way such as railway or highways to double as electricity or gas corridors as well, since many electrical power lines and gas lines are too small to need their own dedicated right-of-way. Gas lines are usually underground, so they are not visually intrusive, but land use above them is limited for reasons of safety and access. Power lines, especially high-voltage transmission lines, are normally above ground. The land underneath them is available for agriculture, recreation, and perhaps even development. The effects of electromagnetic radiation on human health (high-voltage power lines give off this radiation) are still disputed. The City of Alameda, California, owns its electrical distribution system and uses the rights-of-way to provide a high-quality telecommunication service.

Local planners and engineers, when reviewing plans for any energy facility, should seek opportunities to share right-of-way, for "undergrounding" currently aboveground power lines, and for sharing trenches and street cuts (see more on this in Chapter 20).

Regulations for Local Development

Overall, spaceheating accounts for over half the total energy use by households, with 27 percent being used for lighting and appliances (see Table 27.3). These figures vary by sub population, however. For instance, Table 27.4 shows average annual household energy consumption by age of house. Not surprisingly, the older the house, the more energy it uses. The use of energy changes as well. In older houses, a greater share of energy is used for space heating, while in newer houses, approximately equal shares are used for appliances and space heating. Armed with this knowledge, officials could potentially see the largest gains in energy efficiency by better insulating older houses. At any rate, a community should be aware of the age of its housing stock when considering energy efficiency measures.

Table 27.5 shows household energy consumption by household income. The highest income group uses almost twice as much energy as the lowest income group. This is for several reasons. Higher income households normally own and use more electric and electronic devices. Higher income households often have larger homes, which require more energy to heat and cool. Higher income households are less sensitive to price. An energy-efficiency strategy in this case might be to implement an increasing block tariff structure for electricity. Another might be the

Table 27.3
Percentage use of energy by U.S. households, 2001

Use	Total energy use (%)
Space heating	51
Appliances, lighting	27
Water heating	19
Air conditioning	4

Source: U.S. Department of Energy, *Annual Energy Review*, 2002.

Table 27.4
Residential energy consumption by age of house, 2005

Year of construction	Average annual household energy consumption (BTU x 10^6)	Average annual energy consumption per square foot (BTU x 10^3)	Average annual energy expenditure per household ($)
Before 1940	120	52	2,047
1940–1949	104	51	1,810
1950–1959	98	48	1,770
1960–1969	95	48	1,764
1970–1979	83	45	1,654
1980–1989	81	41	1,685
1990–1999	94	38	1,907
2000–2005	94	33	1,936

Source: Department of Energy, *Residential Energy Consumption Survey*, 2005.

Table 27.5
Residential energy consumption by household income, 2005

Household income ($)	Average annual household energy consumption (BTU x 10^6)
Less than 10,000	73.7
10,000–14,999	76.2
15,000–19,999	78.8
20,000–29,999	84.9
30,000–39,999	86.2
40,000–49,999	95.0
50,000–74,999	99.2
75,000–99,999	112.4
100,000 or more	130.5

Source: Department of Energy, *Residential Energy Consumption Survey*, 2005.

vigorous promotion of energy-efficient electric and electronic devices (those given the "Energy Star" appellation by the Department of Energy).

Specific Regulations

Development density. Low density development means higher per capita energy consumption. This is so for a variety of reasons. Some are obvious: in a low-density area, there are more streets, which in turn require more street lights. Others are less so: electrical distribution lines must be longer, which means they lose more power from electrical resistance. Detached dwellings also use more energy simply because they have more surface area exposed to the outside, which means higher energy loss. Buildings that share a wall are more efficient, as are apartments that share a floor/ceiling. Detached dwellings also tend to be larger—there is more space to heat and cool, which again means more energy consumption.

A study of a planned new community in Colorado showed that increasing development densities from 2.24 dwelling units per acre to 4.4 dwelling units per acre would decrease energy consumption by 16 percent. Housing developments with a mix of residence types—apartments, duplexes, and town houses, as well as various sizes of single-family units—can reduce heating and cooling costs by 40 to 56 percent.

Zoning ordinances, subdivision regulations, and the development review process are all tools that local officials have to promote energy-efficient development patterns. The comprehensive or neighborhood plan is also another opportunity to promote energy efficiency and reduce carbon emissions. Opportunities for energy-efficient development include mixed-use development, clustered development, and infill or redevelopment areas. (See Table 27.6.

Table 27.6
Selected energy conservation land use policies at the local level

Policy	Mechanism
Mixed-use development	Allow mixed-use development through zoning
Infill and redevelopment	Revise zoning to allow infill; craft special brownfield policies
Compact development and clustering	Revise zoning to allow compact development
Impact fees	Tie impact fees to energy consumption to encourage conservation measures
Urban forestry	Encourage or require street trees in subdivision regulations
Housing	Encourage or require energy-efficient housing designs in subdivision regulations
Site design and building efficiency	Encourage or require energy-efficient building and site designs in permit process
Urban services boundary/urban growth boundary	Encourage higher density development
Decentralization	Encourage decentralized electricity production in development review
Centralization	Examine centralized heating/cooling in master planned communities or campuses

Source: South Carolina Energy Office, 2000.

Also see www.smartcommunities.ncat.org for the Place^{3}S manual and software.)

Building design. Buildings can be designed to be energy-efficient. Energy-efficiency requirements can be incorporated into local building codes. The local government can also offer local incentives for energy-efficient measures such as fluorescent instead of incandescent bulbs in lights, energy-efficient windows, and insulation standards. Green building is a rich topic, and is addressed in more detail in Chapter 24.

Solar access ordinances. The concept of solar access is that buildings are sited and designed to take advantage of active and passive solar heating. Zoning ordinances and building codes can create both problems and opportunities for solar access, depending on how they are written. Most ordinances and codes pertain to building height, setback from the property line, exterior design restrictions, yard projection, lot orientation, and lot coverage requirements. These can be designed to guarantee access to the sun. The most important solar access regulation for subdivision development governs building orientation. An east–west orientation is the best for solar access. Boulder, Colorado, and Port Arthur, Texas, have solar access ordinances, while Multnomah County (Portland) in Oregon also has a written ordinance that applies to all new development. It is supplemented with a series of graphics to indicate how the regulations are to be interpreted.

Tree ordinances. Urban forests, in addition to their many other benefits, provide energy savings by moderating temperatures for an entire neighborhood or community and thereby protecting the area from climatic extremes. Strategically-placed shade trees can reduce air conditioning costs by up to 30 percent. Trees used as windbreaks can save from 10–50 percent in energy used for heating. The City of Eugene, Oregon, requires that street trees be provided by all new development that includes the creation of a new street.

Use of decentralized energy sources. Opportunities for efficiencies achieved through decentralization can be identified during the project review process. The use of microturbines, photovoltaic cells, or similar systems can be specified or mentioned in subdivision regulations, development agreements, or the development permit process. Photovoltaic cells can be located anywhere, and are probably best used to eliminate the need for peak production.

Decreasing Energy Costs in Public Facilities

Local governments have a fiscal responsibility to save energy. Saving energy cuts energy expenses, which frees up money for other uses (Table 27.7). According to the National Science Foundation, cities can often reduce energy costs by 15 percent without affecting the services they provide to their citizens. Berkeley, California, estimates that its public building energy retrofit

Table 27.7
Cost savings and payback periods for selected energy-efficient building improvements

Description	Cost ($)	Savings ($)	Payback (years)
Upgrade the lighting	2,861	683	0.3
Improve programmable thermostat	230	722	4.2
Install high efficiency motors	8,738	1,378	6.3
Install LED exit signs	400	103	3.9
Totals	12,229	2,886	4.2

Source: New York State Energy Research and Development Authority.

work saves $370,000 a year. South Carolina saved $12.7 million over four years as a result of more efficient public buildings. Economists for the State of Nebraska estimated that 80 percent of every dollar spent on energy bills leaves the state economy without generating further economic activity.

Methods to reduce energy use in public facilities include efficient building design as discussed in Chapter 24. Energy audits of existing public facilities can also be done. An energy audit looks at consumption and cost figures by building for the past year to establish a baseline against which conservation measures can be taken. Often the local energy utility or a state agency will do such an audit at no-cost or at cost and suggest some changes in local practices to reduce consumption.

Even simple changes to office procedures can save energy. Powering down a computer at night and over weekends can save $35 per computer per year. Similarly, one overhead fluorescent light fixture, the ones most common in public buildings, saves almost $2 for every hour that it's turned off. Adjusting the office thermostat up or down by a few degrees can reduce heating and cooling costs by 5 percent.

The Energy Element

Background

From 2004 to the present, local efforts to conserve energy and reduce green house gones (GHG) have gained a great deal of momentum. The U.S. Mayors Climate Protection Agreement, launched in 2005, encourages cities and counties to pledge to be a "cool city" and to sign an agreement to reduce citywide greenhouse gas levels to 7 percent below 1990 levels by 2012. The agreement also calls for participating members to develop a carbon emissions inventory, to set reduction targets, and to develop a series of actions to reduce the GHGs. The International Council for Local Environmental Initiatives (ICLEI) has a "Cities for Climate Protection" program that can be used to fulfill the Cool Cities pledge. Their website has information on how to conduct an emissions inventory, along with suggested programs for the action plan (www.iclei.org).

This section of the chapter concentrates on the energy "element," which brings together the energy conservation and carbon emission strategies discussed thus far. It can provide a focal point for the complex local issues of energy use and land use planning and development, affordable housing, economic growth, and carbon emissions. An energy element or plan may, in some localities be more politically palatable than a carbon emissions reduction plan.

Portland, Oregon, provides a good example of how energy planning can become an integral part of comprehensive urban planning. Under its energy policy, the city improved energy efficiency in municipal buildings, residential buildings, commercial and industrial facilities, transportation, and energy supply.

Creating the Energy Element

The energy element of the comprehensive plan is no different than any other element. It should contain goals, objectives, and strategies. It should be complementary to the other elements, and should serve as a guide to local policymakers as they consider issues related to land use and the built environment. It should look 10–20 years into the future—the goals and objectives of the energy element will aid decision-making for issues directly and indirectly related to energy use. In this way, energy conservation becomes a consideration in all aspects of land use planning and regulation in a community. Table 27.8 illustrates some of the linkages between an energy element and other elements of the comprehensive land use plan, along with the opportunities and factors for decreasing energy use.

Likewise, preparing the energy element is similar to the preparation of the other elements. However, the initial process, known as the energy assessment, may be new to planners. The energy assessment is a picture of energy consumption in the community, and it can provide most of the information that local governments need to develop their goals, objectives, and strategies.

Inventory of local energy sources and costs. The sources of the community's energy supplies are an important baseline for energy planning. The inventory should include an assessment of the diversity, availability, reliability, and affordability of energy sources, local dependence on these sources, and the economic and environmental consequences of dependence. In particular, three questions should be answered:

- How diverse is the mix of supplies?
- Are the supplies from renewable or nonrenewable sources?
- What is the breakdown of energy source by sector?

Table 27.8
Linkages between energy and other general plan elements

General plan element	Opportunities to include energy issues	Factors affecting energy conservation
Population	Population distribution	Population growth impact energy use Population distribution impacts energy use
Economic	Recruitment of clean industries Industrial and business recycling Brownfield redevelopment Green industrial parks Siting and design issues Telecommuting Infill development Adaptive reuse of existing facilities	As the largest user of energy, industry presents the potential for significant energy conservation impacts Proximity of employers to residential areas impacts transportation energy use
Natural resource	Air and water quality Forest management Open space and greenway preservation Urban forestry programs	Trees and other vegetation impact air temperatures
Cultural resources	Energy-efficient facilities Alternative transportation access Reuse and preservation of historic facilities	Proximity of entertainment and religious facilities to residential areas impacts energy used in transportation
Community facilities	Energy-efficient facilities Community and institutional recycling programs Site selection and design Telecommunications infrastructure	Proximity of recreational, educational, and other community facilities to residential areas impacts transportation energy use Building types impact energy use
Transportation	Multi-modalism Street and parking design Travel alternatives Traffic signal optimization Energy-efficient street lights and traffic signals	Existing road system and planned road additions or improvements impact energy use Street and parking design requirements impact energy use Existing and future transit plans impact energy use
Housing	Housing density Energy-efficient construction Housing types Affordability Infill development	Existing and planned housing densities impact energy use Existing and planned housing types impact energy use Higher densities reduce energy use
Land use	Compact development Mixed-use development Multi-modalism Redevelopment Zoning and land development regulations	Redevelopment of existing sites and facilities reduces energy consumption Proximity of trip origins and trip destinations impacts transportation energy use

Inventory of current usage and projected future needs. Once energy sources and related costs have been identified, the next step is to determine how that energy is being used. In this step, the community is classified by end-use sector, and consumption characteristics are surveyed within each sector. (The U.S. Department of Energy tracks energy according to four end-use sectors—residential, commercial, industrial, and transportation.) Again, five questions should be answered as part of the planning process:

- Which sectors are the largest consumers of energy now?
- In the future?
- How do local consumption patterns compare with other local, state, and national patterns?
- What are the costs of energy use?
- What are the carbon emissions of such use?

Each end-use sector should be surveyed to determine special characteristics and needs that influence energy consumption and carbon emissions. The transportation sector usually is the largest end-use sector in a community, using about 40–50 percent of the total energy. Residential/commercial/industrial buildings is another sector that needs to be inventoried. The residential sector normally accounts for about 20–30 percent of a community's energy use, while the commercial and industrial account for another 20–25 percent. Public building energy use should be quantified. Use of electricity for street lights, traffic signals, and water and sewer systems needs to be included—this is typically 5–10 percent of the total.

Corbon emission inventory A carbon emission inventory measures the total amount of carbon emissions or greenhouse gas emissions produced by a community, municipal operations, organization, supply chain, project, product, individual, or other entity. The sources of greenhouse gas emissions may be a result of the use of natural gas, fuels, electricity, and refrigerants or associated with business air travel, employee commute, purchased goods, materials management, product distribution, land use management, and waste. A project-based carbon inventory measures the sources of emissions over the full life-cycle of a project from design concept through operation and decommissioning. Various computer programs exist to translate these into carbon emissions and costs. See, for example, The Energy Yardstick at www.sustainable.doe.gov. As an example, the Portland, Oregon, metropolitan government found in 2010 that transportation generated 25 percent of emissions; the production, manufacture, and disposal of materials, goods, and food, 48 percent; and energy, 27 percent.[10]

For standardized reporting purposes, GHG emissions are broken down into three categories, or scopes:

1. All direct GHG emissions.
2. Indirect GHG emissions from consumption of purchased electricity, heat, or steam.
3. Other indirect emissions, such as the extraction and production of purchased materials and fuels, transport-related activities in vehicles

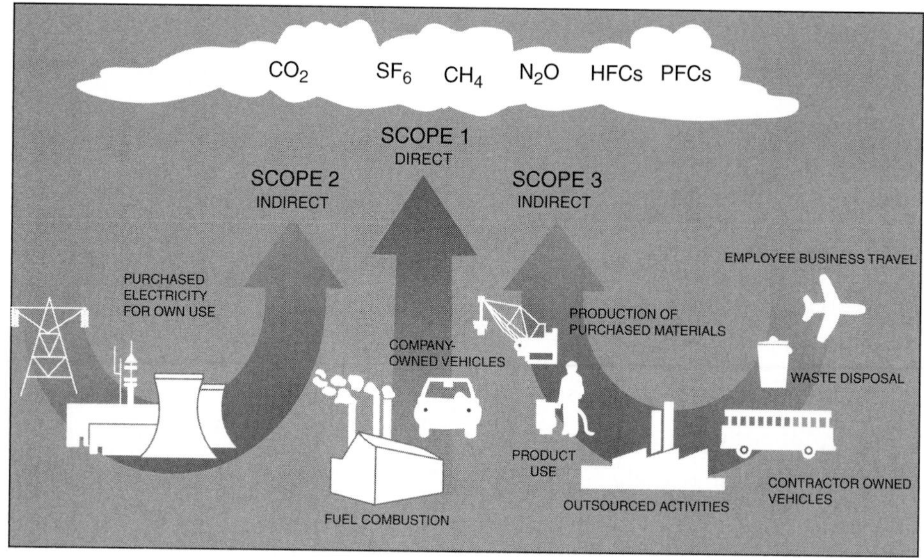

Figure 27.9
Sources of Scopes 1, 2, and 3 emissions

Source: World Resource institute and World Business Council for sustainable development

not owned or controlled by the reporting entity, electricity-related activities (e.g., T&D losses) not covered in Scope 2, outsourced activities, waste disposal, etc. (Figure 27.9).

Assess energy conservation and carbon emission reduction opportunities. Once the energy sources, costs and usage, and future use projections have been identified, the next step is to assess specific opportunities for saving energy. An evaluation of energy use sectors can reveal opportunities for cost-effective conservation measures. This is perhaps the most important step. This chapter has offered many instances of energy conservation opportunities.

Assess local renewable resource potential. After the plans are developed to maximize energy conservation efforts within the community, the next step is to assess the community's ability to develop local renewable resources. Developing local energy resources will reduce the need to import energy, thus increasing the diversity of energy supplies. Developing renewable energy sources, such as wind, solar, or hydroelectric, furthers goals of sustainability and improved air quality. New York City is experimenting with using tidal power to produce electricity, while the City of Chicago has an active cogeneration project.

Summarize findings. The final step in the energy assessment process is the summation of findings from the previous steps. This will provide local officials with the information they need to establish realistic goals and objectives and make informed policy choices. The summary will feed directly into the final energy element.

Conclusion

Energy use is one of the most significant aspects of providing smart and sustainable infrastructure at the local level. Energy is also important to the health, quality, and economic vitality of a community. For this reason, local governments must ensure that the energy needs of a community are met. However, local governments also have the responsibility to consider energy conservation and efficiency as part of local planning. Conserving energy and reducing carbon emissions saves money and preserves environmental quality and resources for future generations.

Local governments have many tools at their disposal to do this. Land use and other local regulations can be implemented to conserve energy. Local governments that operate municipal power supplies can apply pricing schemes that encourage conservation and the reduction of GHG emissions. Local governments can encourage the development of renewable energy sources. All of these tools can best be compiled, presented, and employed if the community creates a local energy policy or "sustainability" plan, which can be a stand-alone plan or included in the general plan as its own element or form the heart of a local climate change or sustainability plan.

Additional Resources

Publications

The many manuals and publications for energy conservation in land use, facility design, and regulation: www.energy.ca.gov

Sustainable Systems, Inc., Federal Energy Management Program, Greening America. *Greening Federal Facilities: An Energy, Environmental, and Economic Resource Guide for Federal Facility Managers*. Washington, DC: U.S. Department of Energy, 2001.

Lovins, Amory and the Rocky Mountain Institute. *Reinventing Fire: Bold Business Solutions for the New Energy Era*. White River Junction, Vermont: Chelsea Green Publishing, 2011.

Matheny-Burns Group, South Carolina Energy Office, and Office of Regional Development. *Energy: Preparing an Energy Element for the Comprehensive Plan*. South Carolina Energy Office and Office of Regional Development. Columbia, South Carolina: South Carolina Energy Office, 2000. In addition to being an excellent resource in itself, this document contains a comprehensive bibliography and a detailed glossary: www.state.sc.us/energy/PDFs/Planning_Guide.pdf.

Rynne, Suzanne, Larry Flowers, Eric Lantz, and Erica Heller, Eds. *Planning for Wind Energy*. Planning Advisory Service Report Number 566. Chicago, IL: American Planning Association, 2011.

Shuford, Scott, Suzanne Rynne, and Jan Mueller. *Planning for a New Energy and Climate Future*. Planning Advisory Service Report Number 558. Chicago, IL: American Planning Association, 2010.

Websites

American Public Power Association: www.appanet.org/publications/index.cfm. This maintains a detailed publication list.

Center for Climate and Energy Solutions: www.c2es.org/. Another great source of ideas to slow down climate change for the local practitioner.

Good Company: www.goodcompany.com. A sustainability consulting firm with source documents from many cities on the west coast that address sustainability issues.

Local Governments for Sustainability (ICLEI): www.iclei.org. Their mission is to support local governments to make deep reductions in carbon emissions. Good resources.

National Center for Appropriate Technology: www.smartcommunities.ncat.org. An excellent website that contains energy-efficiency tools including Place^{3}S.

Rocky Mountain Institute: www.rmi.org/Home. A think tank whose mission "is to drive the efficient and restorative use of resources."

U.S. Department of Energy: www.doe.gov. This has both information and links to other sites.

U.S. Department of Energy, Energy Efficiency and Renewable Energy program: www.eere.energy.gov. This is a very informative website.

Key Federal Regulations

The following is a list of key federal regulations that illustrates the gradual change in national energy policy over the years, and is based on material from the website of Kanner & Associates, a law firm specializing in energy matters.

Right-of-Way Act of 1901

This law permitted the use of rights-of-way through forest reserves (later national forests) and national parks for electrical power transmission and telephone and telegraph communication.

Withdrawal Act of 1910

This law had many parts, not all related to power, but it authorized the president to withdraw public land from public use and reserve it for water-power sites. This was the first step toward a national electrical power policy.

Federal Water Power Act of 1920

This Act was an attempt to establish a coherent national hydroelectric program. It created the Federal Power Commission, which had authority to permit hydroelectric facilities and to regulate interstate sales and transmissions of electricity. The Federal Power Commission was to address navigation, irrigation, and flood control, as well as hydroelectric power in its water planning.

Tennessee Valley Authority Act of 1933

This Act created the Tennessee Valley Authority (TVA). The TVA was one of President Roosevelt's New Deal programs. It was designed to address the extreme poverty of the Tennessee River Valley by implementing a host of modernization measures. Most of these measures involved building dams for flood control and power generation. The TVA was also authorized to transmit and sell electric power, to build transmission systems, and to acquire other electrical generation facilities in the Tennessee River Valley. The TVA is now the largest public utility in the country in terms of capacity, with 29,500 MW.

Public Utilities Holding Company Act of 1935

The Public Utilities Holding Company Act (PUHCA) was passed to break up the large and powerful trusts that controlled the electricity and natural gas distribution networks. PUHCA gave the Securities and Exchange Commission the authority to break up the trusts and to regulate the reorganized industry in order to prevent their return. PUHCA also gave states the power to set up public utility commissions (PUCs), which would regulate intrastate utility activities, including setting wholesale and retail rates. This Act established the regulated power industry that we know today.

Federal Power Act of 1935

This law was passed to regulate the transmission of electricity across state lines, under the interstate commerce clause of the Constitution. The Federal Power Commission's powers were extended to handle this new task, including authority over the electricity utility industry, wholesale rates, interconnections, and wheeling of wholesale electricity. The FPC was also charged with ensuring adequate and reliable service.

Rural Electrification Act of 1936

This law created the Rural Electrification Administration to furnish financial and technical assistance to organizations for providing electricity to rural areas and small towns. In 1935, only 17 percent of farms had electricity, but by 1953, that figure had reached 90 percent. This law was amended in 1947 to include telephone service.

Bonneville Project Act of 1937

This Act created the Bonneville Power Administration (BPA), a federal agency responsible for transmitting and marketing the energy produced by federally-owned dams in the Northwest, particularly those along the Columbia River.

The National Gas Act of 1938

The National Gas Act gave the Federal Power Commission (FPC) the power to regulate natural gas pipelines, though not wellhead prices. (The wellhead price is the value of crude natural gas at the mouth of the well. It is essentially the cost of extracting and processing the gas for shipment.) The purpose of the law was the "protection of consumers against exploitation at the hands of natural gas companies."

Flood Control Act of 1944

This Act formed the basis for the creation of the federal Power Marketing Administrations (PMA) based on the Bonneville Power Administration (BPA) model. It also made the BPA permanent. These PMAs are federal agencies that market and generate electric power primarily from dams operated by the Bureau of Reclamation and U.S. Army Corps of Engineers.

First Deficiency Appropriation Act of 1949

The Act allowed the Tennessee Valley Authority to build fossil fuel-fired power plants. Up to this point, the TVA had only hydroelectric plants.

Atomic Energy Act of 1954

The Atomic Energy Act was passed to promote the peaceful uses of nuclear energy through private enterprise and to implement President Eisenhower's Atoms for Peace Program. This Act allowed the Atomic Energy Commission to license private companies to use nuclear materials and build and operate nuclear power plants. It amended the Atomic Energy Act of 1946, which had placed complete power over atomic energy development with the Atomic Energy Commission.

Phillip's Decision of 1954

Natural gas demand in the 1940s and 1950s grew faster than the infrastructure to supply it, which led to price fluctuations and supply shortages. Natural gas producers asked the Federal Power Commission for wellhead price caps to remedy the situation, but the FPC said it did not have that authority. In 1954, in what became known as the Phillip's Decision, the Supreme Court disagreed, saying the Federal Power Commission did in fact have authority over the wellhead price. This decision created an industry structure that consisted of federally regulated gas producers, who sold to federally regulated pipelines, who in turn sold gas to state-regulated local distribution companies (LCDs). LCDs then sold the gas to end users.

Price-Anderson Act of 1957

This Act limits the liability of the nuclear industry in the event of a nuclear accident. It covers incidents that occur through operation of nuclear plants as well as transportation and storage of nuclear fuel and radioactive wastes. This is essentially a huge subsidy to the nuclear industry.

Energy Supply and Environmental Coordination Act of 1974

The Energy Supply and Environmental Coordination Act (ESECA) gives federal government the authority to prohibit electric utilities from burning natural gas or other petroleum products.

Department of Energy Organization Act of 1977

In addition to forming the Department of Energy, this Act created a fifth federal power marketing administration, the Western Area Power Administration (WAPA). The Act also transferred the PMAs from the Department of the Interior to the Department of Energy, and abolished the Federal Power Commission, creating in its stead the Federal Energy Regulatory Commission (FERC).

National Energy Act of 1978

This Act was actually an omnibus bill that incorporated five separate laws: the Public Utility Regulatory Policies Act, the Energy Tax Act, the National Energy Conservation Policy Act, the Power Plant and Industrial Fuel Use Act, and the Natural Gas Policy Act. Passed in response to the 1970s oil embargo, its general purpose was to ensure sustained economic growth, while also permitting the economy time to make an orderly transition from inexpensive to more costly energy.

Public Utility Regulatory Policies Act of 1978

The Public Utility Regulatory Policies Act (PURPA) is a landmark piece of legislation. PURPA was passed in direct response to the energy crisis of the 1970s, specifically to address the perceived shortage of natural gas. Its primary purpose was to promote more efficient energy generation, through both cogeneration technology and through greater use of alternative sources of power generation. To do this, PURPA established a class of non-utility energy suppliers comprised of small power producers and cogenerators (known as qualifying facilities or QFs). Furthermore, Section 210(a) of PURPA required utilities to buy electricity from these QFs at low rates. Importantly, PURPA was intended to accomplish its objectives while protecting consumers from having to pay more for power from QFs than they would pay for power produced or purchased by the utility.

Energy Tax Act of 1978

The Energy Tax Act encouraged investment in cogeneration equipment and solar and wind technologies by allowing a special tax credit. The credit was later expanded to include other renewable technologies. However, this credit was ended in tax reform legislation of the 1980s.

National Energy Conservation Policy Act of 1978

This Act required utilities to provide energy conservation services to residential consumers in order to encourage slower growth of electricity demand.

Natural Gas Policy Act of 1978

This law ended FPC control over the wellhead price of "new" gas as of 1985, but kept in place the wellhead price controls for older vintages of gas. This was the first step in the deregulation of the energy industry.

Energy Policy Act of 1992

The Energy Policy Act (EPAct) took the first step toward electricity deregulation by creating a new category of electricity producer, called exempt wholesale generators, who were exempt from PUHCA restrictions. The law mandated that FERC open up the national electricity transmission system to wholesale suppliers on a case-by-case basis. EPAct was meant to decrease wholesale rates, thereby lowering retail rates.

FERC Order 636 of 1992

Order 636, the Restructuring Rule, required the "unbundling" of natural gas supply from natural gas transportation. It made it possible for customers to select supply and transportation services from any competitor in any quantity and combination.

Energy Independence and Security Act of 2007

This Act contains funding to increase the production of clean renewable fuels, and to promote research on greenhouse gas capture and storage options.

chapter 28

Telecommunications

In this chapter ...

Importance to Local Practitioners . . . 613
Industrial Era Telecommunications . . . 614
Point-to-Point Communications . . . 614
Wireless Telephony . . . 615
Content Transmission: Radio and Video . . . 616
Information Age Telecommunications . . . 618
The Rise of the Internet . . . 618
Bandwidth and Speed . . . 619
The Institutions of Telecommunications . . . 621
Who Provides Telecommunication Service? . . . 621
Who Regulates Telecommunication? . . . 622
The Role of Local Government . . . 623
Local Government as a Telecommunications Planner . . . 624
Planning for Telecommunications . . . 624
Elements of a Telecommunications Plan . . . 624
Local Government as Regulator . . . 625
Overview . . . 625
Zoning Ordinances . . . 626
Thinking about Zoning for Telecoms . . . 627
Right-of-way . . . 628
Franchiser of Cable Telecommunications Providers . . . 630
Local Government Options . . . 630
The Decision to Intervene in the Market . . . 630
Local Government as Provider . . . 631
Local Government as Promoter . . . 633
Local Governments and Broadband Supply . . . 633
Local Government as User of Telecommunications . . . 634
E-Government Service Delivery . . . 634
Internal Communications . . . 635
Public Internet Access . . . 635
Conclusion . . . 636

(continued)

Why Is This Important for Local Practitioners?

Telecommunication services are in the midst of far-reaching changes that are likely to continue and indeed accelerate as the Internet evolves. All local governments are deeply involved in telecommunications as users and as regulators of land use and the right-of-way (ROW). Some governments have also become telecommunications providers or facilitators as the infrastructure of telecommunications is emerging as the "fourth utility" along with water, sewers, and energy. For many jurisdictions, advanced telecommunication systems are a critical piece of the state and local economic development strategy while for others adequate access to high-speed broadband is an equity question.

The impact of the broadband Internet on telephony along with the digitization of video, voice, and data has resulted in fundamental changes in the way telecommunication services are delivered. The increasing power and range of broadband, both wired and wireless, will result in many opportunities for local governments. At the same time, local governments must stay vigilant to ensure that public interests are protected.

This chapter begins with the history of industrial age telecommunications and their technologies. The history of the Internet and its impact on telecommunications and the entertainment industry are outlined. Issues of broadband and economic development as well as access are touched on. The institutions of telecommunications are described. The final sections of the chapter present advice on "how to do it" for the four roles that local government plays with respect to telecommunications infrastructure: planners, regulators, provider/promoters, and users.

Additional Resources 636
Publications. 636
Websites 637
Key Federal Regulations 638

Industrial Era Telecommunications[1]

Telecommunications is the electronic transmission of information. The infrastructure of the twentieth-century telecommunications industry is a product of the industrial era and scientific advances made possible by taming electricity. This system owes its rapid growth from 1837 onward to the invention of devices that convert information into electrical signals that are then broadcast through the air, or sent through copper wires (landlines) to a receiver, which reconverts the signal into the original message.[2] The following two sections trace the history and technologies, first, of point-to-point telecommunications, and then content transmissions such as radio and television.

Point-to-Point Communications

For over a century the most important two-way communication systems have been connected by a series of wires, beginning with the telegraph and then the telephone system, which recently has taken advantage of wireless transmission.

Telegraph. The first practical suggestion for the transmission of information by electricity came in 1753, when Scotsman Charles Morrison published an article called "An Expeditious Method of Conveying Intelligence." In this article, he described a system of 26 wires, one for each letter of the alphabet, which could spell out words by energizing the respective circuit. In 1837, Samuel Morse exhibited the first modern telegraph machine in New York. It had only one wire so Morse developed his now-famous code of dots and dashes to spell out words. While eventually overtaken, the technology set the standard for instantaneous communication over long distances through wires and was instrumental in the early history of this country.

Telephone. In 1821, Sir Charles Wheatstone coined the word "telephone" after he developed methods for transmitting sounds over wires. In 1876, Alexander Graham Bell uttered the first sentence transmitted over wires, to his assistant—"Mr. Watson, come here, I want you"—and in the process developed the first telephone system and the American Bell Telephone Company.

In 1882, the American Bell Telephone Company began buying up local telephone licenses and soon Bell telephones were the only ones legally allowed in the country. The American Telegraph and Telephone Company (AT&T) was formed in 1885 to develop American Bell's long-distance telephone network and to connect the local Bell companies. By 1899, AT&T had absorbed American Bell and had built an impressive coast-to-coast system of high-quality telephone lines and local exchanges.

The expiration of Bell's patents in 1894 opened the door for independent telephone companies. Between 1894 and 1904, over 6,000 such companies were born, and the number of telephones nationwide exploded from 285,000 to over 3,300,000. Competition was fierce as these companies fought each other or, more usually, the Bell System.

In 1913, AT&T agreed to become a regulated monopoly in return for connecting competing local companies to its long-distance network and for agreeing to be subject to Federal Communication Commission decisions regarding prices and policies. This monopoly was successful in providing phone service to 50 percent of American households by 1945, 70 percent in 1955, and 90 percent in 1969. In 2009, 95.7 percent of U.S. households had a land-line telephone.[3]

The AT&T monopoly went virtually unchallenged until the early 1980s. In 1984, however, in response to increasing criticism about the lack of innovation and high prices of the system (and as part of the wave of deregulation that swept through the U.S. and other countries at that time), the courts forced AT&T to give up its 22 local Bell companies which were then reorganized into seven regional holding companies or "Baby Bells." The Baby Bells were to provide regulated local service while the parent body of AT&T retained long-distance service. The Telecommunications Act of 1996 continued the path towards greater competition for the telephone service by allowing the Baby Bells to enter the long-distance market if they gave up their local monopoly. Long-distance rates have dropped from 1982 to the present, and in California, for example, some consumers can choose from 150 different long-distance companies.

Wireless Telephony

The first cellular telephone trials were authorized in 1977 and by 2011 there were 328 million cell phones in operation, or 104 percent of the U.S. population (Figure 28.1). Cell phone use for talk and, increasingly, data, changes so rapidly that figures are obsolete as soon as they are recorded, but in the 12 months ending in June 2011, Americans spent 2.3 billion minutes talking and sent 2.2 billion text messages.[4]

Cell phones use radio waves to send and receive information. They derive their name from the fact that the

Analog and Digital

Telecommunications are based on turning information, be it visual, written, or spoken, into electrical energy. This electrical energy is then transmitted either through wires or through the air, and is turned back into information at the other end. Information can be turned into electrical energy in two ways. The first case, known as *analog*, occurs when the electrical energy varies continuously and directly as the information. The other case is known as *digital*, where the information is coded into a series of zeros and ones.

In systems that rely on wires, the signal is sent in the form of changes in voltage. In an analog system, the voltage varies continuously; this "voltage wave" is what is seen on an oscilloscope. In a digital system, the signal is a series of on–offs: there is either voltage or there is not. A device that transforms a signal from digital to analog and back is called a *modem*, short for *mo*dulator–*dem*odulator. In systems that transmit through the air, the signal is sent in the form of radio waves.

Information in the original telephone system, radio and television, and early cell phones was transmitted via analog. Digitization of information transmission is a relatively recent phenomenon. Digitization makes it possible to transmit voice, video, and data via the Internet.

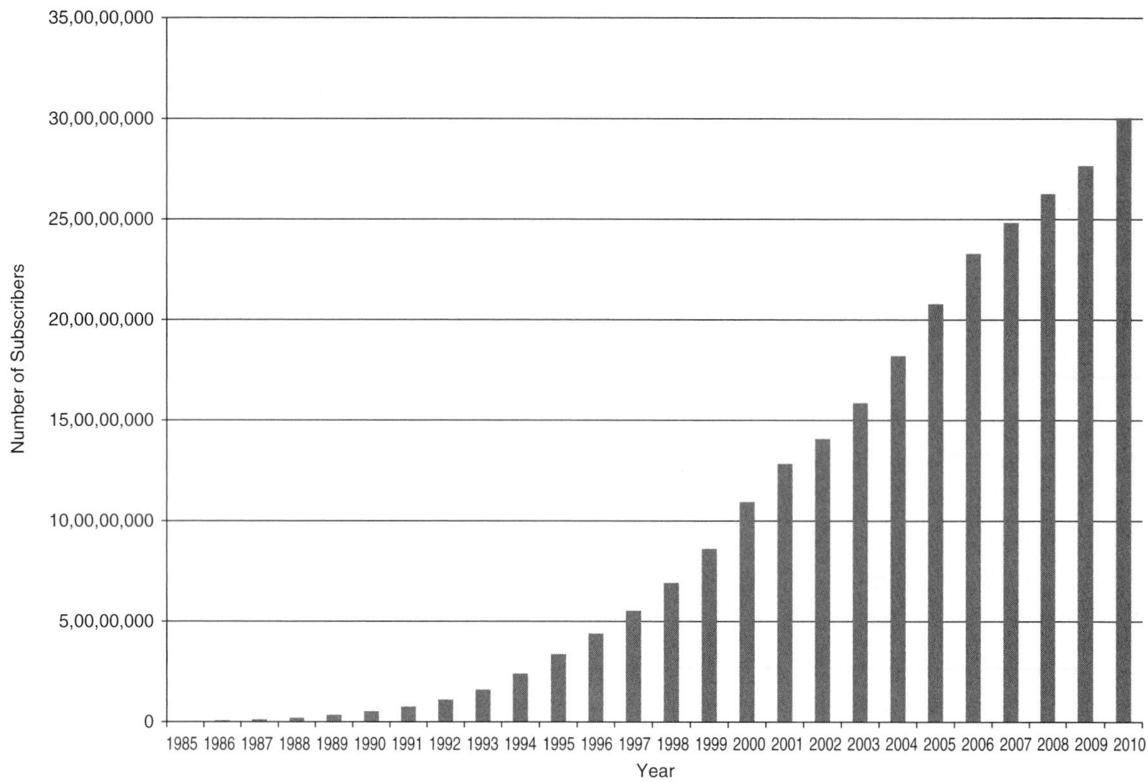

Figure 28.1
Cell phone subscribers in the United States from 1995 to 2010 (millions)

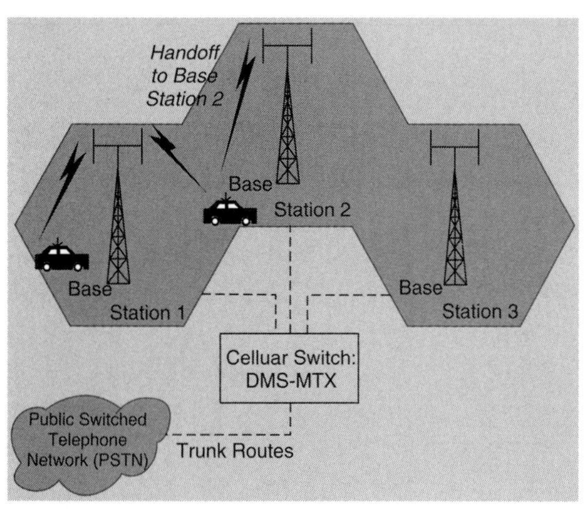

Figure 28.2
Cell phone system

reception area for a particular system is broken into small cells (Figure 28.2). Each cell has its own complete transmission and broadcasting system which includes an antenna. The cells are relatively small; during a call, a user may move in or out of a number of cells. Since the cells are so small, the cellular telephone does not need to send a very powerful signal. Fourth-generation, or 4G, technology became the standard in 2011.

Content Transmission: Radio and Video

Electrical signals can also transmit information by broadcasting from a central location. Traditional radio and television programs took advantage of this form of transmission. This is suitable for the widespread dissemination of "content" from one point to many listeners or viewers.

Radio. In 1896, Guglielmo Marconi received the first patent for "wireless telegraphy"—radio. At first, his system was based on a tightly focused beam aimed at the receiving antenna, but he soon abandoned this in favor of a system that sent signals out in all directions—that is, the signals were *broadcast*. Marconi built the first broadcast station in 1898 and transmitted signals over a distance of 14 miles.[5]

Initially wireless radios were employed for maritime use since telephone technology was good and getting better while the quality of wireless point-to-point communication was poor and subject to atmospheric conditions. The development of vacuum tube equipment in the 1910s made higher quality radio transmission possible but private radios did not become commonplace until after WWI, when the wartime ban on private use was lifted. In 1921, General Electric bought a struggling wireless telegraph company called American Marconi to form the Radio Corporation of America, or RCA. Suddenly, radio broadcasts were everywhere.[6]

Broadcast TV. In 1884, German Paul Nipkow patented the first electromechanical television system. John Logie Baird of London took Nipkow's idea, improved it, and in 1925 successfully transmitted his first picture: the head of a dummy.[7] RCA had a working electronic television system by 1933. In 1939, President Franklin Roosevelt was televised at the opening of the World's Fair in Flushing, New York, and the next day, modern electronic sets went on sale. WWII stopped television development but the postwar economic boom

Electromagnetic Spectrum (EM)

Wireless transmission of information uses what is known as the electromagnetic (EM) spectrum. The EM spectrum refers to the range of different types of radiation—energy that is emitted in the form of a wave (Figure 28.3). Radiation can be characterized based on wavelength, which is the distance between consecutive peaks on the energy wave, or by the frequency, which measures the number of waves per second. The common unit of measurement is the hertz. One hertz is equal to one wavelength per second. Wavelength and frequency are inversely proportional, i.e., the larger the wavelength, the smaller its frequency.

The Federal Communication Commission (FCC) controls the EM spectrum from 9 kilohertz (KHz) to 400 gigahertz (GHz). This corresponds to most of the radio range and the bottom end of the microwave range. The FCC has divided its portion of the spectrum into small chunks, and each chunk goes to a different application; this information is maintained in what is called the spectrum inventory. This system was put in place after the *Titanic* disaster, when multiple home operators obscured emergency calls for help. Maritime navigation, for example, is allocated to the 50 KHz range, while satellites are allocated to the 20 GHz range.

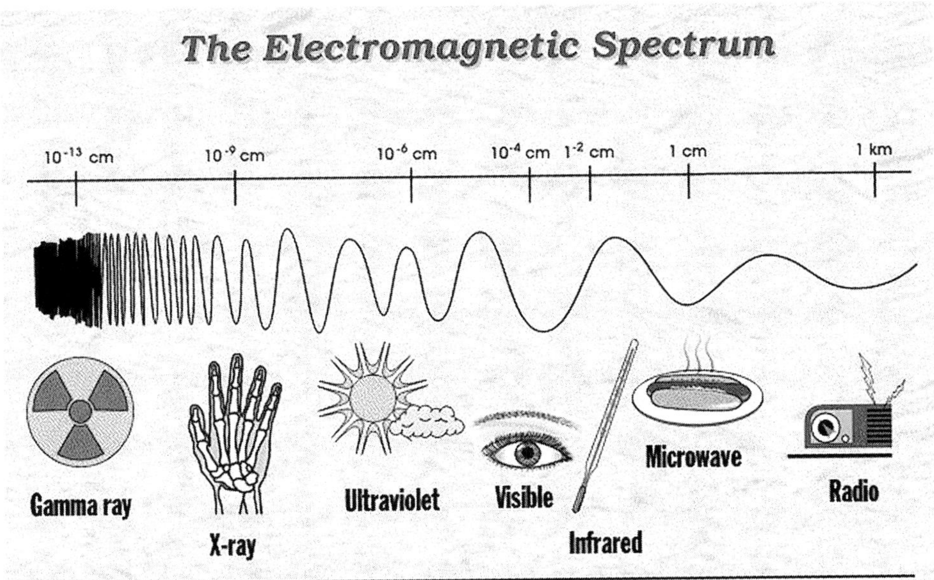

Figure 28.3
The electromagnetic spectrum

resulted in the number of TV sets in the United States growing from 7,000 in 1946 to 10,000,000 in 1950.[8]

In 1954, the National Television System Committee (NTSC) established the technical standards for color broadcast television and by 1965 all of NBC's programs were in color. By 1974, 97 percent of American households owned at least one TV set; this reached 98.9 percent by 2010. Of that group, 83.9 percent owned more than one TV, and 99.9 percent of those TVs were color. In 2010, the average person watched over 34 hours of TV a week.

Cable television. Cable was initially known as Community Antenna Television (giving us the acronym CATV). At first, it was simply a way to bring quality broadcast TV to areas that otherwise could not receive the regular TV signal. System operators, though, started using the system to import new channels from neighboring cities, and so an area that originally received only three channels—ABC, NBC, and CBS—could have six, seven, or even more different channels. The first pay cable TV company, HBO (Home Box Office), was launched in Wilkes-Barre, Pennsylvania, in 1972.

Information Age Telecommunications

The Rise of the Internet

Many refer to the rise of the Internet as the Internet Revolution because of the changes it has caused, and is still causing, in the communications industry, on industrial business practices and technologies, and our culture. "If information technology is the present-day equivalent of electricity in the industrial era, in our age the Internet could be likened to both the electrical grid and the electric engine,"[9] because it carries both the information and the engine and its very existence increases the power of information.

The information age was inaugurated with the invention of computers in the middle of the last century. The first generation of computers consisted of large mainframes that needed specialized technicians to operate. Mini-computers were developed in the 1970s and then in the 1980s personal computers (PCs) hit the market. Early computer users began linking PCs together to share data in what was known as "Local Area Networks" (LANs). LANs themselves were tied together within large organizations and the phenomenon of e-mail arose. Out of these early systems grew the Internet.

The Internet is a world-wide computer network (or networks of networks—Internet is short for *Internetworking*) that lets different computers, each with different hardware, communicate with each other. Its development was a consequence of the American response to the 1957 launch of Sputnik I into space by the Russians. At that time the Department of Defense formed the Advanced Research Projects Agency (ARPA), a research funding agency designed to regain the U.S. lead in defense-related science and technology.

Table 28.1
Internet usage in the U.S. and Worldwide

	2001	2004	2011
United States			
Internet users (millions)	149	193	236
Wireless Internet user share (%)	4.5	27.9	46.3
Worldwide			
Internet users (millions)	533	945	1,460
Wireless Internet user share (%)	16.0	41.5	56.8

Source: www.akamain.com. state of the internet, 2012.

ARPA created a network in 1969 called ARPANET to link computers owned by the military, defense contractors, and universities.[10]

In 1990, ARPANET was declared obsolete for military purposes and was turned over to the National Science Foundation for civilian use. By this time, computer networking technology was in the public domain and most home computers in the country had the capability to network. In 1995, the Internet became privately operated and browsers were released that permitted all networked computers to talk to one another—for most users, this was the real start of the Internet.[11]

Today, there are literally billions of websites, and thousands of commercial applications.[12] In 2010, over 72 percent of U.S. households had access to a high-speed Internet connection.[13] Wireless is also ubiquitus with 46 percent of internet users in the U.S., and 57 percent worlwide using this medium (Table 28.1). See also a topo map of Salt Lake city in 2004 showing intensity of wi-fi signals (Figure 28.4). In addition, high-speed Internet connections have proliferated around the world in the European Union, Japan, and Korea, so much so that analysts currently are concerned that the United States is lagging behind (Table 28.2).[14]

Bandwidth and Speed

Information, whether transmitted with or without wires, "takes up room" in a way analogous to water in pipes. A pipe of a certain size can only carry so much water. In the same way, a wire of a certain size can only carry so much information. The information-carrying capacity of a wire is called "bandwidth."

Packet Switching

The Internet sends information using a technology called packet switching. Packet switching works by breaking down an e-mail or piece of a voice message into small packets. Each packet contains its piece of the message, plus its ultimate destination. The packets are then sent out by the host computer. Since each packet knows the destination, it can take its own route to get there, depending on what lines are available. Along the way, the packets are passed from one computer to the next. Since each computer in the chain only needs to know its next step, the transaction costs are minimized.

Table 28.2
Average broadband speeds (in Mbps) of OECD countries in 2007

	Average upload speed	Average download speed
Wireless	.736	1.840
Cable	.722	8.619
DSL	1.603	8.993
FTTx	58.591	77.120
Total	5.920	13.707

Source: David Chaffee and Mitchell Shapiro, *The Municipal and Utility Guidebook to Bringing Broadband Fiber Optics to Your Community*. American Public Power Association, 2008, available at: http://www.appanet.org (accessed October 18, 2008).

Bandwidth is the number of bits of data per second that can be transmitted, either when downloading or uploading.

Common bandwidth designations are: thousands of bits or kilobits per second (Kbps); millions of bits or megabits per second (Mbps); or even Gigabits (billions of bits) and Terabits (trillions of bits per second) (Table 28.3). The term "broadband" refers to having a high or large bandwidth: the FCC defines broadband as 4 Mbps download, 1 Mbps upload.[15] Broadband speeds vary by the transmission medium, both in terms of theoretical and actual capacity. Fiber directly to the home (FTTH) is the fastest, followed by cable, DSL, and wireless.

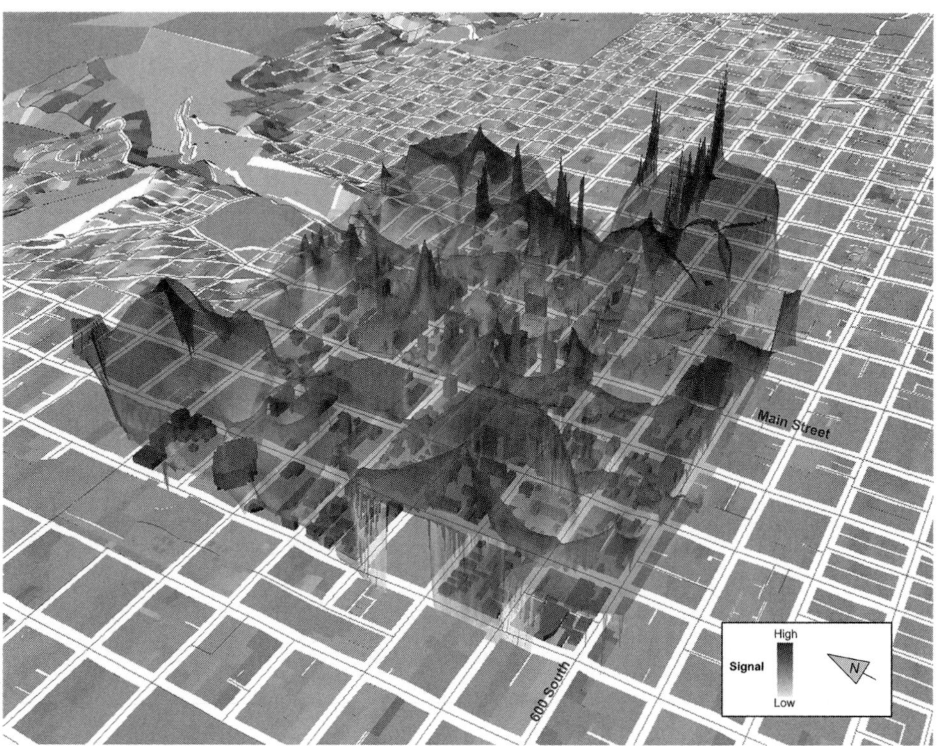

Figure 28.4
Wi-Fi signals in downtown Salt Lake City in 2004

Table 28.3
Measures of data speed

Speed measure	Term	Magnitude	Example
Bits per second	bps	One	56,000 dial up modem
Kilobits per second	kbps	Thousand	56 kbps, dial up modem
Megabits per second	Mbps	Million	Most local area networks using T-1 lines run at 10 Mbps
Gigabits per second	Gbps	Billion	One Gbps is the fastest speed of Ethernet
Terabits per second	Tbps	Trillion	In development

Source: Thomas Asp, Harvey L. Reiter, Jerry Schulz, and Ronald L. Vaden, "Electrical Wires for Data Transmission," IQ Report 36, no. 2 (Washington, DC: ICMA, 2004).

Speed matters because of the increasing demand for consumer and business services to download more and more complicated information, such as X-ray photos, instructional videos, and DVD movies. Table 28.4 compares the times it takes to download these items on a traditional dial-up modem, DSL (both slow and fast), and several fiber service levels.

The Institutions of Telecommunications

Who Provides Telecommunication Service?

The legacy providers. It's been some time since the telecommunication field could be easily divided up into: *telephone service* provided by AT&T; *television service* provided

FCC Broadband Definitions

The FCC developed a seven-category designation for bandwidth, reflecting the increased speeds now possible.

- First generation broadband: 200Kbps to 768 Kbps;
- Basic Broadband, tier 1: 768 Kbps to 1.5 Mbps;
- Basic Broadband, tier 2: 1.5 Mbps to 3 Mbps;
- Broadband tier 3: 3 Mbps to 6 Mbps;
- Broadband tier 4: 6 Mbps to 10 Mbps;
- Broadband tier 5: 10 Mbps to 25 Mbps;
- Broadband tier 6: 25 Mbps to 100 Mbps;
- Broadband tier 7: more than 100 Mbps.

Robert D. Atkinson, Daniel K. Correa, and Julie A. Hedlund, *Explaining International Broadband Leadership*. The Information Technology & Innovation Foundation, 2008. available at: www.itif.org.

Table 28.4
Speed of service for different content packages by broadband technology type

Broadband technology and download speed		Content packages by download size			
		E-mail w/2 MB Att	X-Ray photo (8 MB)	Video (600 MB)	DVD movie (5 GB)
Dial-up	56 Kbps	7.11 min	28.43 min	1.48 days	11.60 days
DSL Light	416 Kbps	50 sec	3.33 min	4.17 hrs	1.63 days
DSL	2 Mbps	9.50 sec	38.01 sec	47.51 min	6.20 hours
Fiber	10 Mbps	2.13 sec	8.53 sec	10.67 min	1.39 hours
Fiber	100 Mbps	.21 sec	.85 sec	1.07 min	8.36 min
Fiber	1 Gbps	.02 sec	.09 sec	6.40 sec	50.10 sec

Source: David Chaffee and Mitchell Shapiro, *The Municipal and Utility Guidebook to Bringing Broadband Fiber Optics to Your Community*. American Public Power Association, 2008, available at: http://www.appanet.org (accessed October 18, 2008).

by NBC, ABC, and CBS; and *radio* provided by independent radio stations. Today, the field is a complex, rich stew of cellular telephone service providers, paging and messaging service providers, specialized mobile radio service providers, wireless data service providers, payphone providers, prepaid calling card providers, satellite TV service providers, cable TV service providers, Internet service providers, long-distance telephone providers, and local telephone service providers. The vast majority are private firms.[16] In the six years following the passage of the 1996 Telecommunications Act, there was substantial consolidation of media ownership (cable, satellite, newspaper, radio, and television). That consolidation appears to have leveled off since then.

Who Regulates Telecommunication?

Federal, state, and local agencies have varying degrees of authority to regulate different aspects of the telecommunications industry. The 1996 Telecommunications Act amended the Communications Act of 1934 to result in the current federal law that guides state and local governments. Presently there are three categories of telecommunication—voice, video, and data—that are subject to different regulations, but these categories are becoming increasingly irrelevant as the advent of Internet Protocol Enabled Services (IPes) allows any information to travel on any medium, telephone, cable networks, or wireless systems.[17]

Federal Communications Commission (FCC). The principal agency charged with implementing the 1996 Act is the Federal Communications Commission (FCC). It is an independent agency, appointed by the President, that has responsibility for radio, television, wire, satellite, and cable (but not the Internet, which remains largely unregulated). The FCC has a consumer protection role and antitrust oversight functions. For example, the FCC reviewed the massive merger of America Online and Time-Warner in 2000. The FCC also manages the electromagnetic spectrum and handles issues of international communications. The FCC also has some limited control over broadcast content. The FCC also oversees the quest toward universal service.[18]

State and local roles. At the state level, usually the state public utilities commission is responsible for regulating telecommunications service with respect to customer service, service standards, installation and repair of telephone lines, DSL lines, and standards for reporting major service outages. Local governments have the authority to regulate cable and open video services, but not telephone. They have the authority to manage the right-of-way for all types of telecommunications, and to set and enforce zoning and building regulations regarding telecommunications.

Local governments have the ability to tax local cable TV and telephone services but not the Internet. Local governments have the authority to impose utility user taxes on consumers of cable TV and telephone services. While theoretically they could tax the cable TV company directly, virtually all local governments charge the cable TV company a franchise fee (which is not a tax) of 5 percent of gross revenues, the maximum allowed under federal law with any taxes offset against the fee. Federal law does not preempt taxes on telephone companies but some state laws do. Federal law preempts taxes on the Internet, but it remains an open question whether local agencies can charge franchise fees if Internet lines use public rights-of-way.

Internet protocol enabled services (IPes). The Internet is not regulated by the federal, state, or local governments, although the placement and design of physical structures are subject to local zoning and building codes. The Internet Tax Freedom Act (ITFA) placed a moratorium, most recently extended in 2007 to November 2014, on local taxes on Internet purchases. Since IPes can travel on any of the existing telecomm transmission platforms, they are currently moving to the least regulated. Various advocacy groups, such as the National League of Cities, are trying to ensure that IPes services regardless of the platform, be subject to normal taxation, franchise fees, and ROW management rules and that IPes platforms include a set-aside for non-commercial users.

The Role of Local Government

High-speed broadband access is increasingly being seen as critical to the economic competitiveness of the United States, and to economic development for individual regions within the U.S. A 2007 Brookings Institution study found that for every percentage point increase in broadband penetration in a state, employment increased from 0.2 to 0.3 percent a year, with the most impact on finance, education, and health care industries.[19] This study found that although 67 percent of Americans households had broadband in 2007, substantial divisions in broadband access exist. For instance, only 59 percent of African-American and 49 percent of Hispanic households had access to broadband.[20] These access equity issues are important public policy concerns.

Meanwhile, e-governance, or the use of electronic communication tools to support public engagement with local government, is on the rise. A 2010 study found that 82 percent of Internet users had looked for information or completed a transaction on a government website in the past year.[21]

State and local governments are not waiting for federal action. Instead many of them have established their own broadband task forces, recognizing the link with economic development and local and regional competitiveness.

For example, in 2006, California established a California Broadband Task Force to "remove barriers to broadband access, identify opportunities for increased broadband adoption and enable the creation and deployment of new advanced communication technologies."[22] The City of Seattle established its own Broadband Task Force in 2005, while the city of San Francisco and the State of North Carolina among others commissioned a broadband feasibility report at the same time.[23]

Thus, local governments have four roles with respect to telecommunications: they plan; they act as a regulator and franchiser; they may be a promoter, facilitator, or provider; and they are one of the largest users of telecommunications. The following sections describe these roles further.

Local Government as a Telecommunications Planner

Planning for Telecommunications

Given the sweeping and accelerating changes in technology that have been underway since the 1990s, the increasing importance of telecommunications as the fourth utility, the economic and access issues described above, increased demands for public safety and disaster response that make a city or county's internal telecommunications needs a higher priority, it is time for most local jurisdictions to prepare, expand, or update a local telecommunications plan.

Local governments should plan for telecommunications in the context of all four possible roles. Telecommunications plans provide the overall policies that are then translated into local zoning amendments, right-of-way management activities, and local telecommunications investment decisions.

Elements of a Telecommunications Plan

A telecommunications plan should address the locality's responsibility for regulating the location of telecommunications facilities, including franchising responsibilities for cable and addressing right-of-way management. Economic development that includes attracting high-technology firms must have adequate broadband, and this issue should be addressed if appropriate. The locality's role as a user of telecommunications also should be a part of the plan. Capital projects or improvements to telecommunications facilities and computer networks of the local government should be included as well.

The telecommunications plan rests on an assessment of the current situation and local goals and objectives. A variety of background studies to assist in this process are outlined below, and the issues section of the plan should summarize them. The plan should include a statement of policies, guidelines, and actions to be incorporated into the long-range implementation plan. It can also include a summary map that shows existing telecom facilities, public rights-of-way, and public structures that can be used as locations for new telecom facilities, along with preferred locations in other parts of the jurisdiction. Agreements between telecom firms and local government for use of telecom facilities by police, fire, and emergency service personnel as needed should be part of the plan.

The Telecommunications Plan should contain a capital needs study for telecommunications infrastructure that will be publicly provided in the jurisdiction. This will provide a bridge to the capital improvements program (CIP) and budget.

Local Government as Regulator

Overview

The telecommunications boom in the late 1990s challenged the capacity of local government as cable and telephone companies made substantial investments in underground wires, cellular towers, and other physical equipment. Internet providers built server farms and ran cable across the continent and under the seas. Right-of-way, siting, compensation, and competitive access to local networks dominated the telecommunications agendas of many local governments, often giving way to courtroom battles. Public works officials lamented the impact of repeated trenching on the integrity of local streets. Neighbors complained about intrusive antennas and transmission towers while at the same time buying cell phones. The focus of all this activity became the responsibility of the local government to manage and regulate.

Prior to the 1996 Telecommunications Act, local governments had little regulatory authority over telephony, although they managed access to the public right-of-way. The 1996 Act placed limitations on local government authority to obtain compensation from telephone companies for use of the public rights-of-way, but it preserved local franchising authority for cable television. While it also placed some limitations on local zoning authority depending upon the type of facility, the Act also reaffirmed the authority of state and local governments to manage the public rights-of-way.

In the late 1990s, some states adopted policies to facilitate the new telecommunications grid rollout by setting limits on fees and requirements for in-kind compensation from

Steps in Preparing the Telecommunications Plan

The City of San Francisco used the following process to develop its 2002 telecommunications plan:

1. Conducted preliminary assessment of market conditions.
2. Convened working groups of industry, citizens, and informed individuals.
3. Prepared telecommunications needs assessment based on:
   a. survey of telecommunications companies doing business in San Francisco;
   b. surveyed residents on telecommunications issues as part of citizen survey;
   c. conducted small business (telecom users) focus groups;
   d. interviewed major telecommunications institutional users.
4. Synthesis and outline of telecom strategy.
5. Held targeted public meetings for local businesses, economically challenged and homeless, youth, seniors, people with disabilities, and general public.
6. Prepared initial plan draft.
7. Workshops with telecommunications commission to refine the plan.
8. Second draft of telecom plan.
9. Adoption.
10. Prepared implementation plan.

Adapted from the City and County of San Francisco Telecommunications Plan, 2001, City of San Francisco Telecommunication Commission.

Sunnyvale's Telecommunications Goals

The City as Regulator

- To retain control of public property within the confines of state and federal legislation to regulate telecommunications services provided to Sunnyvale citizens.
- To promote universal access to telecommunications services for all Sunnyvale citizens.

The City as Service Provider

To use telecommunications to maintain and enhance information resources and services provided to Sunnyvale citizens.

The City as Facilitator

- To promote use of telecommunications technology, where appropriate and within the scope of available resources, to enhance the economic vitality of Sunnyvale.
- To facilitate the creation of an advanced telecommunications infrastructure, within given resources, for Sunnyvale citizens, businesses, industry, and schools.

City of Sunnyvale, CA, Telecommunications Policy (1996), Policy 7.2.16, Council Policy Manual, December 2012.

The Death of Distance?

Futurologists have long claimed that improved telecommunications technology and infrastructure will be the "death of distance and the end of cities."[1] It seems, however, that as it has become easier to communicate across the globe, the role of metropolitan areas in the global and national economy have been strengthened. Local policy-makers have significant powers to influence how telecommunications affects the physical and social form of urban areas. The new technologies should be looked at to help shape land use and equity issues in a purposeful way.[2]

Notes

1 Frances Cairncross, *The Death of Distance: How the Communications Revolution Is Changing our Lives* (Boston: Harvard Business Review Press, revised ed. 2001).
2 Ibid.

local governments. Some also standardized permit application requirements and established permit processing time limits for local agencies. At the same time, almost half of local governments had a zoning ordinance in place regulating the siting of cellular and wireless communications towers, or had an ordinance regulating use of local ROW.

The following addresses zoning and then right-of-way as issues for telecommunications infrastructure. Cable franchises, the third local regulatory arena, is treated only briefly. A complex and litigious area, it is beyond the scope of this book.

Zoning Ordinances

Local governments have traditionally dealt with the impact of telecommunications facilities on land use through zoning ordinances focused on aesthetics and safety. Utility poles and wires that were built in the 1930s caused many communities to require that new developments underground their wires or to otherwise regulate the size, location, and height of the poles. The 1950s saw the construction of freestanding telecommunications transmissions towers on private property. Some could be 400 feet tall and might occupy 10,000 square feet of land. Although initially these were located away from residential uses, as communities grew, land use conflicts arose and new zoning restrictions were put into place. Sometimes existing towers were changed from a permitted to a non-conforming use, while new towers were required to go through a conditional use permit process.

Technological advances in the 1990s intensified these conflicts. Roof top antennas and satellite dishes made for irate neighbors. A 30-foot diameter satellite dish, typical for a cable TV company, might need up to 2,000 square feet of space. Internet companies needed room for servers and had unique power requirements, including poles, wires, and fences. Towers and relay stations were needed to support the increased use of cell phones, pagers, and PDAs. Cellphone monopoles in dense urban settings were viewed as eyesores.

While some communities tried to regulate the appearance and location of towers, dishes, and antennas, others were concerned about adequate reception. The City of Seattle in 2001 amended its zoning ordinance to permit minor communication utilities at any location for a commercial service if it filled a gap in coverage and was the least intrusive location and facility. In 2004, the City of Pasadena proposed an amendment to their zoning ordinance to allow flexibility for "stealth" cellular antennas in their Open Space Zone to improve cell coverage in the hills that abut the city. During the next several years, the advent of new wireless devices may make these changes less necessary as coverage became greater while their size became smaller. In addition, some of the newer technologies do not require line of sight because they use the lower radio frequencies and can be hidden inside buildings.

Section 704 of the 1996 Telecommunications Act reaffirmed local governments' right to control the siting, construction, and modification of cellular and other wireless telecommunications facilities unless their decisions both "prohibit or have the effect of prohibiting the provision of personal wireless services," and "unreasonably discriminate among providers of functionally equivalent services." This was a negotiated compromise between local governments and the cellular industry for "personal wireless communications." It is likely that new legislation will see the cellular industry push for restrictions on local governments in this area. However, depending upon the timing of any new federal legislation, new wider range fixed base wireless systems combined with the Internet-based telephony may render these concerns moot on both sides.

Section 207 of the 1996 Act restricted the control of local government on the placement of satellite dishes and television antennas. Local pole attachment policies must comply with federal and state regulations. See the sidebar for some creative ways of mitigating visual impacts.

Thinking about Zoning for Telecoms

These are the major objectives a locality might have for telecommunications facilities to the local mechanism that might be employed.

Conditions zoning ordinances should meet. The following five conditions should guide local decisions about

Mitigating Visual Impacts of Antennas and Satellite Dishes

- Antennas and satellite dishes should be integrated with the design of the building to provide an appearance compatible with the structure.
- Satellite dishes shall be screened to the top of the dish on at least three sides and shall be enclosed in the direction of the signal to the elevation allowed by the azimuth of the antenna. If screening on the remaining side is not to the top of the antenna, the antenna and the inside and outside of the screen shall be painted the same color to minimize visibility and mask the contrasting shape of the dish with building or landscape elements.
- New antennas shall be consolidated with existing antennas and mechanical equipment unless the new antennas can be better obscured or integrated with the design of other parts of the building.
- Antennas mounted on permitted accessory structures, such as a free standing sign, shall be integrated with the design, material, shape, and color and shall not be visibly distinctive from the structure.

Adapted from City of Seattle's Proposed Zoning Ordinance, Section 23.57.016, 2001.

the content of local zoning ordinances to regulate cellular antennas and towers:

- Cannot result in prohibiting local service.
- Cannot discriminate among providers.
- Reasonable processing time for permits.
- Written findings for denial.
- Cannot deny based on emissions hazards. (See sidebar.)

Electromagnetic Radiation

Section 704 of the 1996 Act pre-empts the right of local governments to regulate the placement of cell towers "on the basis of the environmental effects of radio frequency emissions to the extent that such facilities comply with the Commission's regulations concerning such emissions." In recent years, there have been questions raised about electromagnetic radiation increasing the risk of cancer. Cell phones have caused particular concern since the broadcasting and receiving antenna is directly next to the cell phone user's brain, and since early studies had weakly suggested that there is a correlation between leukemia and EM exposure. Many experts have concluded that no cancer risk has been shown for exposure to electromagnetic radiation or for use of cell phones.[1] Indeed, some researchers claim that there exists no biologically plausible mechanism by which the low level of electromagnetic radiation emitted by towers or cell phones could even cause cancer.[2] The issue, however, is still hotly debated in some jurisdictions.

OSHA has a website devoted to the health impacts of electromagnetic radiation, with links to articles on both sides of the debate. (See www.osha.gov/SLTC/radiofrequencyradiation/.)

Notes

1 National Cancer Institute, "Cell Phones and Cancer Risk," June 28, 2012. Available at: http://www.cancer.gov/cancertopics/factsheet/Risk/cellphones.

2 Michael Shermer, "Can You Hear Me Now? The Truth about Cell Phones and Cancer," *Scientific American*, Oct. 4, 2010.

Right-of-way

Perhaps the most challenging aspect for local governments has been the voracious appetite that telecoms have for public rights-of-way for wire, towers, and other equipment. The right-of-way may account for 20 percent of the land area of a city. Although antennas for TV, radio, and even cell phone monopoles may be located on private property, Wi-Fi and cell phone providers may prefer that their antennas be located on public telephone poles and street lights. If this is not possible, they too can locate on private property. The 1996 Telecommunications Act affirmed right of the local governments to manage the use of the right-of-way by all types of telecommunications services and to obtain compensation for the use of the right-of-way.

The locality should have a general strategy for all use of rights-of-way including a right-of-way ordinance that sets out permitted uses, fees, and compensation matters. However, the following highlights ROW management and compensation issues for telecommunication infrastructure.

Use of the right-of-way. The 1996 Act has been supplemented in many states with additional legislation that further defines the authority of the locality over the ROW. In California, for example, telephone companies have been given specific authority to put poles and lines on any public road or highway. This law applies to city streets as well. However, the local governments have authority over the "time, place and manner." Local agencies in California can designate which streets are used, the precise location of the line within the street, whether it is underground or not. They can also determine the location of utility boxes. Telecom companies do not have the right to install facilities on administrative buildings, parks, or light poles if the local government does not permit it.

Compensation and telecommunications. A Master ROW ordinance should set out a compensation schedule for all users. It should address permit processing fees for digging or locating equipment in the public right-of-way, fees for the use of the right-of-way, and franchise fees. Issues about the fees for processing the permit for the use of the public right-of-way do not differ from those of other right-of-way users. Fees or "rent" for use of the public's right-of-way are often negotiated as part of the franchise agreement with the provider. Provisions for fees must be consistent, however, with state requirements, which vary from state to state.

Fees or "rent" for the use of the right-of-way have been bitterly contested by many telecommunications providers. There are two reasons for this. The first is the competitive nature of many portions of the telecommunications industry. The single water or sewer provider in a community using the ROW can pass the cost of using it to the customer without fear of losing that customer. However, many telecommunications providers are seeking cost and time advantages over their competitors. Their interests are best served by the lowest payment to the locality.

The second reason why the fees for ROW use have been contested is that federal legislation has different franchise rules depending on whether the service is "voice," "video," or "data." The 1996 Act allows the locality to obtain 5 percent of gross revenue from cable providers. Although the locality is pre-empted from regulating quality of service for telephony, federal legislation still permits the "franchise" or taxing ability. Some states however, have pre-empted local governments from this as well. In addition, information services are not subject to local regulation or taxation. Therefore, much of conflict with the industry has concerned the classification of the electrical impulses being transmitted over the line. With increasing convergence of voice, video, and data over the same line, or from the same transmitter through the air, it is difficult to quantify the exact amount of revenue due from each type. Some of the litigation seeks to prevent localities from basing compensation on revenues and instead to use actual costs of administering the ROW trenching process.

It is likely that any legislative update will attempt to resolve some of these issues. In the meantime, one attorney

Local Control over ROW

The right-of-way provisions in the 1996 Act are not only under fire in the courts, but the telecommunications industry has long sought to pre-empt some provisions. NATOA (National Association of Telecommunications Officers and Advisors) noted that the telecomm industry does "not accept the validity of the local government argument that public health and safety regulations relating to rights of way access imposed upon other utilities (including municipal utilities) should also apply to entities that provide broadband services." This organization has adopted as a "core value" the need to maintain local government control over the public right-of-way. It remains to be seen how the trickledown from the wireless revolution will affect the outcome.

Source: NATOA website, www.natoa.org

notes that "there is still good authority for using gross revenues as a reasonable measure of the value of the right-of-way" even though this is disputed by the industry. Localities need to carefully specify the revenues that serve as the basis for the fee. If they are not earned locally, such as with wireless and wireless Internet, other formulas such as a per pole or per linear foot should be used.

Finally, although setting the fee is complex, collecting it should not be forgotten. From 2000 to 2002, the City of Portland audited the revenues from non-cable telecommunications providers for ROW use and recovered over $5 million in one-time revenues and $700,000 in on-going annual revenues, for an investment of only $115,000 in special audit contracts plus in-house staff time. Their "audits" range included re-computing the fee for those companies that pay by the linear foot, and "financial audits" for revenue-based fees. During these audits, each line of revenue earned in the city was examined to ensure that it was in the franchise fee base. This is an important revenue source for Portland since franchise fees are the second largest source of discretionary general fund revenues, and 94 percent of them are non-cable. Table 28.5 shows the land use mechanisms that a local government can use to advise local telecom objectives.

Franchiser of Cable Telecommunications Providers

Cable services are regulated by the federal, state, and local governments. The 1996 Telecommunications Act continued to permit local governments to "franchise" local cable companies—that is, to set service standards and to enter into a contract with them to use the ROW in return for a percentage fee and other forms of in-kind compensation. A survey in 1999 by ICMA found that 93 percent of local governments had franchise agreements with cable companies, with an average contract length of 12 years. Many of these contracts will need to be renewed, and often they are transferred to other providers as the industry consolidates. Local governments need to be prepared for lengthy, complex, and often confrontational negotiations during this process.

Local Government Options

The Decision to Intervene in the Market

If a local government feels that the private sector is not providing the desired level of telecommunications service within its jurisdiction, it may choose to intervene in the market. There is a spectrum of actions ranging from complete provision to easing of policy that make it easier or cheaper to provide adequate broadband capacity for residents and businesses. These include aggregating demand; providing access to local facilities for commercial

Table 28.5
Land use mechanisms to achieve local telecom objectives

Local objective	Mechanism
Ensure adequate number of sites for future telecomm facilities	General Plan; Zoning ordinance and map
Allow use of public rights-of-way and public buildings for telecom facilities; Set fees for use	Zoning ordinance, Master ROW ordinance
Ensure coordination of construction in rights-of-way to minimize public inconvenience and disruption	Master Right-of-way Ordinance
Establish design criteria to promote public safety, maintain community character, and minimize impact on adjacent lands	Zoning ordinance, general ordinance, subdivision regulations
Provide for removal of obsolete or unused facilities	Zoning ordinance; Master Right-of-way Ordinance
Agreement between private telecom providers and local government for use of telecomm by police, fire, and emergency services	Development agreements, Memoranda of agreement
Ensure investment in telecom infrastructure for internal city uses as well so that all citizens can have access	Zoning ordinance, permit process, subdivision regulations, development agreements, general ordinance, capital improvements program, and Budget

Source: Stuart Meck, "Executive Summary: Status of State Planning Reform," in *Planning Communities in the 21st Century* (Chicago, IL: American Planning Association, 2002).

providers; and subsidizing all or part of a commercial venture. Table 28.6 lists questions and options for local government that are interested in becoming involved.

Local Government as Provider

Municipal telecommunication service providers. In the 1980s and 1990s, many local governments chose to become telecommunications providers, in almost all cases building upon existing municipal electrical utilities. The ideas seemed to make sense. Electric utilities and telecommunications share complementarities, and public utility tax-free status makes the cost of capital, the main barrier to entry for telecom, artificially low. But expectations were rarely met, and public telecom providers could not compete with private providers. For example, in 1999, the City of Alameda, California, took its Bureau of Electricity, changed the name to Alameda Power and Telecom, and began offering cable TV and high-speed Internet service over its city-wide fiber optic cable network. By 2008, however, after years of trouble and debt, the city council sold its telecommunications business line to a private company for a loss. Indeed, a 2004 examination of eight municipal telecom providers across the country found that all of them were generating negative returns. The report concluded, "The municipal

The Last Mile Issue

To fully capitalize on the power of the Internet, a household will need a high bandwidth communications line that can handle all the information. However, up to now the cost of extending high-speed broadband wiring to every single house has been too high. This problem is known colloquially as the problem of the "last hundred feet" or the "last mile" issue.

The new wireless technologies will change the economics of the last mile issue since they are low cost. They will permit both local governments and commercial providers to enter the field with a modest outlay of funds. Alternatively, if broadband over power lines proves feasible on a widespread basis, this would also solve the last mile problem.

Table 28.6
Should the locality become involved in telecom provision?

Decision Factor	Options
What is legally permitted?	• Precluded from being provider? • Rules about financing private provision?
Who is the desired user? What is the purpose of providing the service?	• Government: Public Safety • Government: General Purpose • Businesses • Residents • Local Universities/Library
What is the geographic scope?	• Neighborhood • Commercial Area • City-wide • Metropolitan/County (MAN)
Which aspect of telecom infrastructure needs government provision or facilitation?	• Ducts/Conduit (including "dark" fiber) • First Mile connections to customers premises • Interconnection points • Middle Mile connections (links smaller networks to Internet backbone: backhaul)
What kinds of technology will be used?	• Wireless Mesh (grassroots, or Metropolitan Area Network sponsored by governments) • Fixed Wireless (Broadband Fixed Wireless with large cells) • Wired (usually fiber) • Combination of above
What types of telecommunication services will be provided?	• Broadband (Internet) • Video (cable TV) • Voice (telephony)
What types of government action are feasible, and politically possible?	• Financing through bonds etc. • Subsidizing grassroots organizations • Build or contract to be built • Operate or contract for operations
What kind of a business model does the agency want to use?	• Wholesale (local government sells capacity to private providers) • Retail (local government sells service to end-users) • Public–Private Partnership where locality provides physical facility and private party operates (franchise model)

Source: Gillette, Lehr, and Osorio, 2003.

governments that are using their taxpayers' money to enter the telecom business are not investing that money wisely."[24] Similarly, in 2004 and 2005, many high-profile U.S. cities, such as San Francisco, Milwaukee, Chicago, Houston, and Philadelphia entered into public/private partnerships with telecom providers to provide a municipal Wi-Fi network. Telecoms such as EarthLink and MetroFi entered into agreements to pay for the network build-out, operations, maintenance, and upgrades. By 2007, however, it became apparent that the initial forecasts for Wi-Fi subscriptions to pay for the networks were unrealistic.[25] In cities embarking on Wi-Fi networks, the phone and cable companies would lower their broadband prices. The slower speeds and lower reliability of Wi-Fi also make it less competitive for a private party thinking of paying for such service. The cities which have been successful in deploying a municipal Wi-Fi system, such as Portland, Oregon, are those which have treated it as a public utility and have paid for it publicly.[26] This suggests that local governments should meet telecom goals in less direct ways. These economic failures lead some to question whether the locality should become involved in telecom provision.

Local Government as Promoter

Better than becoming a direct provider of broadband is becoming a promoter. Demand-side initiatives are popular with local governments because they involve little risk and they are low cost. Communities can measure demand, stimulate it, or aggregate it. Communities can conduct or contract for a market study to measure potential demand for the needed telecommunications service to encourage private sector investment. They can also stimulate demand by sponsoring web pages for local businesses, as is done by the Blacksburg, Virginia, Electronic Village, establishing pilot telecommunications projects and supporting local education projects. Educating citizens and businesses about broadband also seems to result in increased demand.

The local government can take the lead in cooperating with other governments or business users to make the economies of scale worthwhile for private investment. They can do this by entering into joint purchasing arrangements or partnerships, and/or by sharing bandwidth. This type of effort requires many users to be successful and is often done at the regional or even state level. For example, Sunnyvale, California, formed a partnership with nine other cities to aggregate their purchasing power to induce a telecommunications provider to rebuild the county's voice infrastructure for voice, video, and data. The Berkshire Connect project in New England is another example of aggregating demand.

Local Governments and Broadband Supply

Local governments can also change local policies or provide funds directly to induce the private sector to provide needed broadband infrastructure. In most areas of the United States today, the incumbent cable and phone companies are not planning to provide fiber to the home (FTTH) and therefore there are active efforts by cities and states to look at the feasibility of publicly provided FTTH or public–private partnerships in this area. Forward-thinking cities and states have concluded that FTTH is the best choice for next-generation applications because it has more capacity than other existing or emerging systems (perhaps coupled with a city-wide wireless system for mobile applications and those requiring less bandwidth). The City of Seattle notes that "to support next-generation applications, broadband networks should have bandwidth capable of simultaneously delivering voice, switched video and data" which would require at least 20–25 Mbps downstream and that 25 Mbps should be considered a minimum.[27]

Policy changes to make it easier for telecom providers to build infrastructure. Just as some communities wish to restrict the use of local buildings and ROW for telecommunications devices, others wish to promote them. Local governments can make their communities more attractive to private sector providers by re-visiting local permit procedures so that they reduce the cost or processing time for telecommunications providers.

Localities can also make available locations in the ROW such as streets, parks, and public buildings for telecommunications. Rules for adding wires and equipment to utility poles might be eased, and zoning requirements for wireless on private property relaxed.

Financial subsidies. Subsidies can be provided to commercial providers to enter the community as a one-time or on-going grant, tax incentives, or low cost loans. ROW fees may be reduced for telecommunications (although the local city or county attorney needs to be consulted here). Subsidies can also be provided to non-profit organizations and grassroots organizations for more informal deployments. An MIT study on broadband noted that subsidies appear to be more common at the state level, and they cite Michigan's tax credits and low-cost financing for telecommunications providers that invest in high bandwidth infrastructure.

Local Government as User of Telecommunications

Local governments are users of telecommunications infrastructure for the delivery of e-government services to the public, for internal communications, including public safety, and for the provision of public Internet access. All three roles rely on mastering a complex of technical information. Only the highlights of this interesting topic will be described here, but additional resources will be noted.

E-Government Service Delivery

Local governments offer e-government in two categories: information dissemination and interactive tools. Information dissemination, by far the most common service, includes the publication of meeting agendas and minutes, ordinances, maps, and plans. Interactive tools are somewhat less common, but communities offer on-line applications for voter registration, zoning and building permits, and business license application and renewals, inspection status and scheduling, and work order requests (Table 28.7). On-line payment of taxes, utility bills, fines, and fees are also offered by an increasing number of local governments. It is important to remember that there are two faces to e-government service delivery: the first is the ability of the government to provide the service, while the other is the ability of the constituent to access the service. In 2010, the FCC found that minority, low-income, lower-educated, and disabled constituents had up to 60 percent less access to Internet service than national averages.[28]

More detailed information on how to set up on-line services can be found in three ICMA publications.[29]

Internal Communications

Local governments are often the largest users of telecommunications in a community. They are responsible for administrative buildings, schools, libraries, public works facilities, utility monitoring, police and fire stations, courthouses, and other buildings that need basic telephone service and often more sophisticated telecommunications devices. Police and fire departments operate PSAPs (public safety access point) for 911 and reverse 911 services and police and fire communications. Schools, libraries, and the e-government websites noted above need access to the Internet. They sometimes operate their own government access television channels to provide information to citizens.

Public Internet Access

Local governments have also taken the lead on providing no-cost access to the Internet for the public, either with a series of local "hot-spots" or through a public mesh system.

Wireless e-911

Communities are beginning to deploy wireless 911 systems for emergency calls from cell phones that identify the caller's location, much like the current wired 911 system. The first phase calls for public safety access points (PSAPs) to be able to identify the cell phone tower that processed the call, but unfortunately, this can sometimes include an area as large as 100 miles. The second stage calls for the ability of a wireless phone to be located within 50 to 100 meters. As of 2004, 65 percent of the PSAPs nationally have implemented the requirements of the first phase, but only 19 percent have implemented the second phase. Funding constraints are primarily responsible for the lack of progress. See the ICMA IQ Report on Wireless E911 for further discussion, available at www.bookstore.icma.org.

Table 28.7
Percentage of government entities offering public participation on their websites from 2000 to 2008

	2000	2001	2002	2003	2004	2005	2006	2007	2008
E-mail	68	84	81	91	93	92	92	89	88
Search	48	52	43	—	—	—	—	—	—
Comments	15	5	10	24	29	28	46	44	48
E-mail updates	5	9	5	12	24	21	31	39	44
Broadcast	2	7	4	—	—	—	—	—	—
Personalization	0	1	2	2	3	3	6	10	25
PDA access	—	—	—	1	1	1	1	1	3

Source: Brookings West, 2009.

City of San Francisco's Telecommunications Plan

The City of San Francisco updated its Telecommunications Plan in 2004 to ensure emergency communications interoperability, robust wireless networks during an emergency, and strong cyber-security protocols. The plan calls for developing a mobile broadband infrastructure to provide redundancy for wireless public safety networks.

Community "hot spots" or Wi-Fi hot spots can be deployed with very little cost. Wi-Fi equipment is cheap and easily installed, and is robust. These hot-spots can be interconnected to provide wider area coverage.

Some local governments have funded community agencies and encouraged grassroots organizations that use open source software to build mesh networks out of 802.11 end-user PCs. In Austin, Texas, four parks and all rest areas offer free Internet. The project cost of $3,000 was underwritten by a deli sandwich chain. New York City has at least 32 free hot-spots, and San Francisco has 27.

Portland State University and the Oregon Graduate Institute of Science and Technology are building a network that will be deployed on the roof of the university to cover the downtown Portland Area.

Conclusion

Telecommunications technology and its regulatory context at the federal level are still in the midst of rapid change. At the same time, the infrastructure of telecommunications is becoming more necessary for everyday life: civic, personal, and commercial. Telecommunications, particularly access to high-speed broadband, is becoming the fourth utility. Local practitioners should become familiar with the outlines of the various debates at a minimum, and should work to ensure that the jurisdiction has telecommunications plan and implementing ordinances. Public expenditures for telecommunications should be included in the capital improvement plan and budget.

Additional Resources

Publications

Evans-Cowley, Jennifer and Joseph Kitchen. *E-Government*. Revised edition. Planning Advisory Service Report Number 564. Chicago, Illinois: American Planning Association, 2011.

Gillett, Sharon E., William H. Lehr, and Carlos Osorio. "Local Government Broadband Initiatives." *Telecommunications Policy* 28, no. 7 (2004). MIT's Program on Internet Convergence. Good survey of different types of government initiatives.

Goldsmith, Jack and Tim Wu. *Who Controls the Internet?: Illusions of a Borderless World*. New York: Oxford University Press, 2008. For those who are interested in how the original idea of the Internet as a force that would erase national borders and be beyond government control has changed.

Hurley, Deborah and James H. Keller. *The First 100 Feet Options for Internet and Broadband Access*. Cambridge, MA: MIT Press, 1999. A series of articles discussing various government and business strategies to connect households to the Internet via broadband. Dated, but the issues are still with us in many communties.

McMahon, Kathleen, Ronald Thomas, and Charles Kaylor. *Planning and Broadband: Infrastructure, Policy, and Sustainability*. Planning Advisory Service Report Number 569. Chicago, IL: American Planning Association, 2012 Recomended.

Neuchterlein, Jonathan E. and Philip J. Weiser. *Digital Crossroads: American Telecommunications Policy in the Internet Age*. Cambridge, MA: The MIT Press, 2007. A masterful text on how the Internet is changing the economics of the telecommunications world. Has very clear descriptions of the new technologies as well.

Servon, Lisa. *Bridging the Digital Divide: Technology, Community and Public Policy*. Malden, MA: Blackwell Publishing, 2002. Very readable book describing the dimensions of the digital divide in the United States and the role of local efforts, to reduce this, such as Community Technology Centers (CTCs), training efforts and government policies. Good chapter on what Seattle has done.

Valle-Riestra, Paul. *Telecommunications: The Governmental Role in Managing the Connected Community*. Point Arena, CA: Solano Press Books, 2002. A practical guide to government rules and regulations for telecommunications with the emphasis on California law.

Research and Legal Websites

www. akamai.com/state of the internet. Great report updated frequently.

www.baller.com/library-art-practical.html. Website hosted by attorney Jim Baller with a guide to state and federal pole attachment rules. See the websites of private attorneys who specialize in municipal telecommunications law. www.gov-law, www.ballard.com, and www.millervaneaton.com are websites for attorneys from Washington, DC, Oregon, and California, respectively.

Georgia Center for Advanced Telecommunications Technology: gcatt.org. Click on to their Office of Technology Policy and Programs (OTP) for policy papers and studies.

Internet SOCiety (ISOC): www.isco.org. A myriad of Internet resources are on this site from the organization that is the home for the groups responsible for Internet infrastructure standards, including the Internet Engineering Task Force (IETF) and the Internet Architecture Board (IAB).

Massachusetts Institute of Technology Program on Internet and Telecoms Convergence: itc.mit.edu/. The latest in research on business and government Internet technology and local applications. Good links The journal published quarterly by the National Association of Telecommunications Officers (www.natoa.org) is also a good resource. U.S. Department of Commerce. National Tele communications and information agenans. This website at www.ntia.doc.gov/ntiahome/staterow/statelocalrow.html with links to rights-of-way success stories for telecommunications.

U.S. Department of Housing and Urban Development: www.hud.gov/offices/hsg/mgh/nnw/nnwindex.cfm. HUD's website describes the Community Technology Centers they have spearheaded for disadvantaged neighborhoods to promote digital capacity.

Organization Websites

Federal Communications Commission: www.fcc.gov.

International City/County Manager's Association: www.icma.org. Excellent publications for local governments.

National Association of Regulatory Utility Commissioners: www.naruc.org.

National Association of Telecommunications Officers and Advisors: www.natoa.org. First stop to see what's new on the regulatory and local telecommunications scene from the local perspective.

National Telecommunications and Information Administration: www.ntia.gov.

Key Federal Regulations

The following is a list of key federal regulations that illustrates the gradual change in national telecommunications policy over the years.

Radio Act of 1912

The first national telecommunication regulation that gave federal authorities the responsibility to allocate different portions of the EM spectrum to different users was precipitated by the sinking of the *Titanic* in April, 1912. The radio operators on board the *Titanic* had sent out a distress call, which was received by a station in Newfoundland, but amateur radio operators all along the East Coast of the USA filled the airwaves with such radio noise that the distress signal was not relayed properly.

Radio Act of 1927

In 1927, Congress created the Federal Radio Commission, which was responsible for enforcing the "public interest" standard for licensing. In exchange for the privilege of using federal spectrum resources, the licensee was required to "serve the public interest, convenience, and necessity." Many felt that the purpose behind this standard is that Congress feared radio's potential power to prompt radical political or social reform, spread indecent language, and to monopolize opinions. In other words, Congress sought a "voice that would articulate middle class ideology."

Communications Act of 1934

This law consolidated regulation of wired and wireless telecommunications and has been the cornerstone of national telecommunications policy for 60 years. This law also established the FCC and set regulations for rate-of-return on capital investment. It also determined the procedure of averaging rates to achieve Universal Service.

Rural Electrification Act of 1949

This Act amended the original Rural Electrification Act to make long-term, low-interest loans available to rural telephone systems. Rural telephone service had been lagging; this new availability of low-interest loans sparked a period of rapid growth.

Communications Satellite Act of 1962

The Communications Satellite Act of 1962 gave a monopoly on international communications via satellite to a new corporation called Comsat. Ironically, AT&T designed and built the satellite, but lost the right to use it.

Telecommunications Act of 1996

This law was the first major modification of national telecommunication policy since 1934. It attempted to introduce competition into the telephone market. Specific provisions regarding local government's role to manage the right-of-way and local land use are noted within the main body of this chapter. It also repealed the FCC's authority to regulate rates of programming service tiers of cable service, usually referred to as "expanded basic."

Section THREE

Conclusion

PART EIGHT

Conclusion

Chapter 29 (A New Paradigm for Infrastructure) revisits the need for smarter infrastructure planning at the local level. It brings together the themes of the individual chapters into some general strategies that can be followed by innovative and persistent local practitioners to address some of the problems with the current system of infrastructure planning and finance. The chapter then concludes with recommendations for long-term changes that should be taken to improve the system.

chapter 29

A New Paradigm for Infrastructure

In this chapter ...

Importance to Local Practitioners. 643
Time for a New Infrastructure Paradigm 643
The Old Paradigm 643
Towards a New Paradigm for Infrastructure 644
Smart and Sustainable Planning Can Improve Infrastructure Delivery 646
Integrate Existing Infrastructure Planning in a "Virtual Jurisdiction" 646
Changes to the Content and Process of What Is Planned and Built 647
Improve Infrastructure Decision-Making and Management Tools. 648
Long-Term Structural Changes to the Infrastructure Provision System 649
Inefficient Local General Purpose Government. 649
Local Capacity Grants for Interdisciplinary Efforts. 650
State-Level Changes 650
Federal Program Changes. . . 651
University and Professional Organizations Can Help. 651
Conclusion 652

Why Is This Important to Local Practitioners?

The current system of infrastructure provision is not ready for the challenges of the twenty-first century. A patchwork quilt of infrastructure providers work at cross-purposes at the local level, while outdated infrastructure management systems in many cities and counties lead to infrastructure failures and the deterioration of the environment. Growth pressures often result in infrastructure investment decisions that propagate sprawl during boom times, and under-maintenance when the economy is less strong. Central cities invest heavily in infrastructure for global commerce, while deteriorated sidewalks, streets, schools, and parks may co-exist in the adjacent poorer areas. The need to plan for infrastructure that mitigates carbon emissions is urgent. Equally pressing is the need for infrastructure that enables cities to adapt to the changing climate and more frequent and extreme weather events.

At the same time, we can see the outline of a new paradigm for infrastructure that is being explored by forward-thinking cities and made possible by the dawn of the information age. This is an infrastructure that can operate at many scales from neighborhood to subnational; one that is multidisciplinary and integrates its various functions into a systemic whole that respects the metabolism of the city. The following summarizes the challenges of the current system, describes the parameters of the NextGen infrastructure system before moving to an action agenda for today's practitioner, and for policy-makers at other levels of government.

Time for a New Infrastructure Paradigm

The Old Paradigm

Recently, the virtues of the special district, single discipline-based infrastructure planning model have been eclipsed by its shortcomings. The system of multiple single-purpose infrastructure providers worked well to roll out the grid to meet the post-World War II growth requirements. Special districts independent from city hall and county governments were developed to provide transportation, water, sanitation, bridges, hospitals, schools, and the mega projects. Regulated private monopolies built information systems and provided energy and power. Departments within cities and counties well insulated from the winds of politics provided smaller-scale street-level infrastructure, while parks and municipal buildings became the infrastructure focus of local politicians. All of this has provided legacy infrastructure that at its best is the triumph of the vertically integrated, command and control model pioneered by private industry for the large manufacturing corporations.

But as we enter the heart of the twenty-first century, we see that the lack of attention to land use implications of major transportation grants and water and sewer investments has contributed to sprawl and negative ecological impacts. Focus on infrastructure construction instead of maintenance and conservation has resulted in unnecessary public expenditures for rebuilding. Lack of cross-functional coordination between systems has led to piecemeal and unplanned development.

This is exacerbated by serious inefficiencies in local general purpose governments themselves, such as outmoded accounting systems, calcified contracting processes, and personnel processes that seem almost to perish with their own weight. The current infrastructure planning and finance system is also confounded by the rapid pace of change and uncertainty about technology, about growth, and about the availability of capital.

In the 1920s, the excesses of decentralized development—subdivisions with roads that did not connect with neighboring roads for example—resulted in the passage of enabling legislation to give local governments the ability to control these excesses. The equivalent is needed today, either through better planning at the local and regional level, and/or real changes to the structure of the infrastructure service system. When both land and energy were cheap in this country, and when climate change and its impacts had not yet emerged as a serious issue, this was not a problem. Today, however, a new more integrated way of thinking about infrastructure is needed.

Towards a New Paradigm for Infrastructure

Today, we are at the dawn of a new era of infrastructure. Earlier chapters have described the evolution of infrastructure by discipline and how it is shaped by the economic, environmental, and social imperatives of the time. The industrial era of the 1950s is long gone, but its legacy to us remains in the massive infrastructure works designed to capture economies of scale in providing services to a young country where land, water, and optimism seemed endless.

This era is long gone. Today we are more sober about the ability of the environment to absorb the excesses of production. We decry the deterioration of legacy infrastructure, but its very deterioration is an opportunity to rebuild and restructure how vital eco-system services, energy and communication functions, and transportation are conceived. The twenty-first century is one where the economic base and resulting cultural mores have been and will be fundamentally changed by innovation and the products of the information industry. Just as widespread availability of cheap electricity worked its way through every aspect of production and consumer tastes and social structures during the twentieth century, the high tech information revolution will be working a similar transformation in the twenty-first century. This will both require and make possible an infrastructure system that will be quite different than that of the last century. If properly conceived, this century's infrastructure will address the impacts of climate change and, indeed, even mitigate them. At its core it will borrow from the intellectual tradition of sustainability, which is interdisciplinary, intergenerational, and views the built environment as a metabolism where the flows of materials and energy need to be balanced.

In its ideal incarnation the new infrastructure paradigm will be:

- *Multi-disciplinary.* Infrastructure will be conceived of holistically. Major investment decisions in capital projects will be made considering the larger development context and interrelationships with other systems. Land use plans will be integrated with plans for infrastructure that identify the impact of LIFE-CYCLe carbon emissions and other materials for all the systems.
- *Biophilic.* Infrastructure will promote the integration of nature within the built environment. Multi-purpose parks will slow the rainfall, cool the urban heat island, and provide recreational opportunities and indeed, be the "lungs of the city." Greenways will link central business districts to outlying areas; traditional bridges and highways will be eco-friendly.
- *Multi-scalar.* Gone are the days of the monopoly of the regional wastewater treatment plan; auto-centric development; large-scale electrical production plants; the district primary school; one location for City Hall. The infrastructure of the twenty-first century will be: multi-modal transportation; storefront government services; community schools; district energy systems; distributed water and waste infrastructure; water–waste–energy systems at the neighborhood level.
- *Regional and mega-regional.* Not only will infrastructure be multi-scalar, it will also be planned at the appropriate scale. For the major networked systems, this is regional or mega-regional.

Smart and Sustainable Planning Can Improve Infrastructure Delivery

This section moves from the grand concept to specific steps that can be taken by the local practitioner today to begin to move towards the twenty-first-century infrastructure paradigm. There is already a rich tradition of conceptual discussion and exciting pilot projects to draw from to further these aims. The Smart Growth movement focused attention on designating certain areas for urban growth to foster more efficient use of land and minimize infrastructure costs. Sustainability planning also emphasizes a systems and interdisciplinary approach. This concept can be transferred to local infrastructure planning. The following are strategies that can be used by the local practitioner as he or she works within the existing system.

Integrate Existing Infrastructure Planning in a "Virtual Jurisdiction"

Good peer relationships combined with establishing or upgrading the capital budgeting process are the most important changes that can be made at the local level to improve local infrastructure investment decisions. The impact on outcomes that will occur by involving practitioners from another discipline in the CIP process, or another agency with infrastructure responsibility, cannot be over-estimated. A recurring nightmare for any public works director is the newspaper story about one set of engineers paving a street one year, while another set just several feet away in the same office plan to tear them up the following year to install sewers. Establishing a "virtual jurisdiction" for infrastructure planning and budgeting can avoid this dark scenario and can also maximize the impact of public dollars. Specifically, the practitioner can:

- *Establish or participate in the capital budgeting process in order to focus infrastructure investments strategically within the jurisdiction and to ensure LIFE-CYCLe budgeting.* The most important thing that a practitioner in a general purpose government can do to improve infrastructure investments is to focus on the existing budget process. If a routine capital budgeting process does not exist, one should be established. If it does exist, attention should be given to ensuring that the program plans for the individual infrastructure systems are based on accurate condition assessments, and that the program plans for individual systems address land use impacts both individually and cross-functionally. In the best of all possible worlds, the capital program and budget should be linked to long-term program plans for each infrastructure system, which would form a cross-functional infrastructure plan linked to the jurisdiction's comprehensive land use plan. Regardless, a "good enough" strategic plan which focuses all capital investments by neighborhood or geographic area can get more bang for the buck. Ideally the capital budget should provide a forum for LIFE-CYCLe budgeting so that decisions to undertake a project are tempered by the on-going operating and maintenance costs.

- *Establish cross-functional relationships.* It is not common at the local level for individual system practitioners to work closely with other system counterparts. This is true within an agency and truer between local agencies. Although some supervisors, managers, and politicians may discourage this kind of interaction, this can and must change. In the short run, a manager or elected official can require that cross-functional contacts occur during the planning and budgeting process. It is also possible for the local practitioner to reach out to colleagues without such mandates. These relationships may enable the local practitioner to influence the timing and location of infrastructure by other providers. He or she should actively participate in or initiate cross-jurisdictional efforts. At a minimum, the practitioner can take back information on what others are doing to his or her own jurisdiction in order to better inform their own strategic capital plan. Second best, the practitioner should try to influence other infrastructure providers to locate their investments to maximize their own jurisdiction's strategic plan. The best of all possible worlds occurs if joint voluntary cross-functional planning takes place with all infrastructure providers.

Changes to the Content and Process of What Is Planned and Built

Earlier chapters in this volume have documented many innovative practices with respect to demand management. Yet this is not the norm in today's world of infrastructure. Regulating demand for infrastructure can have a great impact on our environment. Similarly, ensuring that effective public participation processes are present for infrastructure decision-making can also have positive results. Specifically, the local practitioner can:

- *Employ demand management and conservation strategies for their own infrastructure and require it for others.* Change comes hard to those who design infrastructure. The dominant paradigm for building roads, water supply facilities, sewers, and even public buildings has been to meet the demand, even if this results in an inefficient use of resources and damage to the environment. Planners in cities and counties can promote local regulations that require "green building," and water and energy conservation plans from all those who require a local use permit, including their own jurisdictions. Within the city or county government, planners can be advocates for the use of fees as a demand regulator for other departments that deliver infrastructure services.

- *Expand meaningful public participation in the hidden infrastructure decisions.* The past several decades have seen the rise of public participation with respect to mega-projects. The public usually is active in decisions about community facilities provided with property taxes. However, a continuing challenge is to expand meaningful public participation to the investment decisions of the other infrastructure projects funded through user fees—many of which fly below the radar in special districts that were set up to avoid some of the time-consuming public debates in the local general purpose government arena.

Improve Infrastructure Decision-Making and Management Tools

Upgrading the practice of infrastructure provision at the local level is also important. Improvements in information technology and advances in thinking about how to function in complex institutional environments can improve local infrastructure investment decisions. Scenario-building capacity is important at the front end, while asset management systems and performance indicators are important once the infrastructure is built. Training in collaborative planning techniques can assist in helping to create the virtual jurisdiction.

- *Invest in scenario-building capacity and decision support tools.* Today's information technology makes it possible to mine census data and internal data systems to model the impact of infrastructure decisions in many complex and exciting ways. The local practitioner should push for budget investments in this area. This may be in the local agency, or it may involve participating in funding for the regional planning organization. The power of maps, three-dimensional computer simulations, and other visual aids cannot be underestimated in assisting local investment decisions. Capacity in this area can also contribute to the effective management of existing facilities.

Local citizens can focus great attention during the budget process on $20,000 worth of funding for a recreation center but ignore a multimillion dollar sewer investment because it's "too technical." The loss of a half-time position is instantly recognized in shorter pool hours or the loss of an after-school program while the land use impact of a street maintenance program or a sanitary sewage treatment facility is difficult to understand. A new generation of decision support tools is making possible sophisticated visual and spatial representations of the impacts of different infrastructure decisions. Agent-based modeling, GIS, and other systems have the potential to make these decisions more real to the average citizen, and to quickly provide analyses to respond to changing local conditions and opportunities.

- *Use performance indicator systems and infrastructure management systems.* Performance indicators and infrastructure management systems can be used for condition assessment as part of the capital budgeting process, and also to help make difficult budgeting decisions about the need for maintenance during the budget process. It can also help to raise awareness within the community, at least by the elected governing body, about the need to make preventative budget decisions to avoid catastrophic failures and expensive fixes. Performance indicators can be linked with practical daily and monthly work order and work management systems used by operating staff as well. These systems are costly, especially if they are linked to the jurisdiction's general accounting system—a feature that will allow timely cost comparisons and the ability to shift staff to high priority functions. Many still do not function as effectively as they should, and they require special staffing, but they are an investment that will repay the jurisdiction tenfold.
- *Invest in training that promotes cross-functional relationships.* A wide variety of methods exist today that can help local practitioners cross the gulf between the differing disciplines of infrastructure. These include collaborative planning practices and negotiation techniques that can be used within the agency and between agencies. These skills are increasingly important given that the jurisdictions for capital investment in an area form a crazy patchwork quilt of boundaries, each with its own set of governing bodies and/or goals.

Long-Term Structural Changes to the Infrastructure Provision System

Although it is difficult to change the formal governance structure at the state and local level, some states and local governments have already made important changes in infrastructure planning, while others are beginning to look critically at their own infrastructure provision systems. A rationalization of service delivery boundaries within a metropolitan area—or regional decision-making for land use and capital investment decisions—would be ideal. Short of that, incentives to plan at the regional level for transportation, water/wastewater/solid waste, and for distributed energy solutions would be a good first step.

Inefficient Local General Purpose Government

A difficult problem in local infrastructure delivery is that many cities and counties are not effectual in dealing with anything but routine infrastructure capital improvements. Local city and county politicians themselves may be

Five Actions to Begin an Infrastructure Program

- Use capital budgeting process as a lever.
- Establish cross-functional relationships.
- Promote demand management and "green building."
- Make technical assumptions accessible to public input.
- Improve infrastructure decision-making and management tools.

instrumental in setting up special districts to meet urgent infrastructure needs, which fractures the system even further. There are, however, notable local governments (albeit in growing areas) that have invested in productivity systems for contracting and for personnel that take the fruits of the high tech and IT revolution to cut through the older systems, and which work even in highly unionized environments.

However, without changes in city and county contracting and procurement procedures, changes in personnel systems, and the widespread adoption of integrated accounting systems and other productivity enhancements, special districts will continue to proliferate. The exciting large-scale projects will be given to public authorities to manage. Similarly, line departments managing infrastructure within the city or county will be reluctant to participate in cross-functional planning. The pressures for fractured infrastructure provision will remain strong without these other changes.

Local Capacity Grants for Interdisciplinary Efforts

Although some cities and counties are able to reinvent themselves, many do not have the funds even to prepare a strategic infrastructure plan while funding work management systems, performance indicators, in-depth condition assessments, and improved contracting and personnel procedures are beyond their imagination. Yet the amounts of money needed to fund a cross-functional strategic plan are small indeed compared to the total amount of infrastructure construction funds it would influence. Planning funds could either come from a national or state program or could be "tithed" from special purpose districts. The ability to be able to respond quickly at the local level with a reasoned analysis to economic and climate changes requires more capacity than presently exists in local general purpose governments. Similar comments can be made about construction capability. Training for local staff in collaborative planning, public participation, and negotiation techniques, as well as in budget and financing options would also assist more effective capital decision-making.

State-Level Changes

States can require special districts to obtain local jurisdiction approvals for capital expenditures as part of their enabling legislation, but without capacity improvements for cities and counties, this will defeat the purpose of forcing a cross-functional look at land use impacts of infrastructure investment. States can streamline contracting requirements for local governments, but again, without similar capacity improvements at the local level to provide for speed and transparency to avoid corruption, nothing will be gained.

States can and should have their own cross-functional capital strategic plans and capital budgets. Some already do, but this is not the norm. States can also adopt concurrency or consistency regulations, as many states have, which forbid local governments from approving new development without adequate infrastructure. States can also require that all government districts address local zoning, and not exempt these districts from local land use controls. Some states have mandated demand management plans from infrastructure providers, but this can be expanded. These changes are important, but should be done in tandem with the local capacity building as noted above.

States can make as a condition of receiving state and federal pass through funds, that local governments collaborate on regional infrastructure plans. They can also set goals for carbon emissions, water and energy use and provide incentives for achieving these goals.

Federal Program Changes

This book is designed for the local practitioner, not for the federal policy-maker. However, two comments are in order. Some of the problems with infrastructure, particularly the need to invest in new technology, need to be jump started at the federal (or state) level. New technology development involves more capital than is normally available at the regional or local level, and sometimes more than is available from the private sector. Second, some of the infrastructure solutions (such as a rail system) are subnational in scope.

The federal government has led the way with requirements for long-term planning in order to receive capital funds. These requirements date back to the 701 Comprehensive Planning Assistance Act of 1954, and were followed by a requirement for what we today would call a single purpose, albeit long-term, master plan for sanitation plants, airport facilities, and transportation as a condition of receiving federal funds. However, the infrastructure planning requirements should require both cross-functional coordination (outside transportation, for example), and the analysis of land use impacts of the federal investment.

University and Professional Organizations Can Help

In many agencies, bread and butter infrastructure is handled by engineers who are not familiar with land use planning issues, while some planners are not comfortable or familiar with the impact of technical infrastructure decisions on land use. Civil engineering schools should require their graduates to take a course in land use planning, capital budgeting, and public participation. Curricula for land use planners should include a class on infrastructure either offered by planners, or by the local civil engineering department in the university. Training in financing infrastructure in both disciplines can be improved.

Professional planning and engineering organizations should offer training for practicing professionals in the areas mentioned above. They should also encourage joint membership and board participation at all levels, as some already do.

Conclusion

The industrial era infrastructure built by the private sector and local government has been effective in providing the United States with some of the highest quality of infrastructure in the world. However, this system has increasingly resulted in negative ecological and social impacts as the world becomes smaller through innovations in transportation and information technology. New patterns of infrastructure planning and provision are needed for our changed world. Central to this new standard for planning is the importance of communication—between individuals responsible for the individual infrastructure systems, developers of projects and subdivisions, and between agencies. Although the local practitioner may wish for additional powers to force recalcitrant infrastructure providers to take coordinated action, persistent and concerted communication between providers is a doable first step.

Similarly, although many may point to the lack of funding for infrastructure, without effective, imaginative, and coordinated planning, even existing infrastructure dollars will not be maximized. "Smart planning" needs to complement "smart growth." The amounts of money spent on infrastructure are large enough to warrant better and more coordinated planning.

Change will not come quickly to an area where capital investment decisions from 50 or 100 years ago impact and guide our decisions today. However, today's planners and engineers have tools at their fingertips that their colleagues of yesteryear only dreamed of. The tools and techniques are here—they need only to be applied in a framework that rewards coordinated infrastructure planning and finance to make a sea change of difference.

Glossary

Activated Sludge Process. Some sewage treatment plants use activated sludge instead of trickling filters to consume the bacteria. This process speeds up the work of the bacteria by bringing air and sludge heavily laden with bacteria into close contact with the wastewater.

Adequate Public Facilities Ordinance (APFO). An adequate public facilities ordinance sets measurable standards for the quality and adequacy of certain infrastructure systems (usually water, sewers, and transportation) that must be in place before new development can be approved.

Adequate Public Facilities Ordinance (APFO) for Schools. Locally adopted requirement that stops new development approvals once a school's enrollment exceeds a certain established percentage. An APFO is meant to sequence development to match existing or planned school capacity levels.

Aircraft Utilization Rate. The number of times an individual aircraft can be used in a day.

All-Terrain Bike Trail. An all-terrain bike trail is an off-road trail for all-terrain or mountain bikes.

Alternating Current. Electric current changes direction with a certain frequency. The frequency with which the current changes direction is measured in cycles per second, or hertz.

Analog. Analog information is carried in voltage waves (electricity) that differ with respect to frequency and amplitude.

Appropriative Water Rights. Rights granted with a permit or license from a state or federal agency that specifies the right to use a certain amount, in a specific place, and at certain times.

Aquifer. Aquifers are underground water sources. They often consist of many interconnected layers of materials with varying degrees of storage capacity, permeability, and connectivity.

ARPANET. A computer network created in 1969 by the Advanced Research Projects Agency (ARPA) to link computers owned by the military, defense contractors, and universities.

Arterial. Any high-capacity road, including highways and freeways.

Artificial Recharge. This is a form of water management where excess surface water is stored in the ground through injection wells or spreading basins.

Asphalt. A mix of gravel and a tar-like substance made from petroleum byproducts. It is cheaper and easier to apply than concrete and it provides a much smoother road surface, but is not as curable as concrete.

Asset Management. Asset management is a systematic approach to managing capital assets that seeks to minimize costs over the useful life of the facilities while keeping service levels adequate. It implies making repairs over time to maintain the facility in good working order instead of suffering a catastrophic and expensive failure.

Authority. See *Public Authorities*.

Average Cost Pricing. Average cost is the total cost to provide the infrastructure service divided by the quantity provided. Under this method, the user charge is set at a level that covers all the costs associated with the provision of the services. Also known as full cost pricing.

Avigation Easement. The right to the use of the air space above a specific height for the flight of aircraft. It may prohibit the property owner from using the land for structures, trees, signs, or stacks higher than the altitude specified.

Bandwidth. The term bandwidth has come to mean the general capacity of any method of information transmission and is expressed in bits per second, usually as mbps (a million bits per second).

Banking. When a large number of planes from a large number of origins (called a *bank* of planes) all converge on the airport at about the same time.

Benchmarking. Benchmarking is the process of comparing some aspect of performance in one jurisdiction to another, or to past performance of the same jurisdiction. Can be used with performance indicators.

Bicycle Lane. A bicycle lane (called a Class 2 facility) is a section of the roadway that has been separated from motor vehicle traffic, usually by a painted stripe but occasionally by a more substantial feature like raised pavement.
Bicycle Path. A bike path (called a Class 1 facility) is a physically separate right of way that may or may not parallel a road. It is often shared by pedestrians.
Bicycle Route. Bike routes (called a Class 3 facility) are routes on streets marked with signs where bicycles share the road with cars.
Biosolid. Biosolid is another name for sewage sludge that can be used to amend local soils and as a fertilizer.
Black Water. Water from plumbing fixtures containing urine and feces.
Bps (Bits Per Second). A measure of information speed on the Internet. One bit is transmitted each second.
Break Bulk Cargo. Loose but differentiated, non-containerized cargo that may be packaged in bales, bags, boxes, or on pallets.
Building Code. The term building code refers to local codes that regulate construction of development. They usually consist of electrical, plumbing, mechanical, and building requirements. Also referred to as the *Uniform Building Code*.
Building Envelope. The building envelope is the external shell of the building, the "four walls and roof."
Build–Own–Operate–Transfer (BOOT). This term refers to an arrangement when the private company builds, owns, and operates the facility. The facility may revert to the government at the end of some time period.
Build–Transfer–Operate (BTO). This is an arrangement where the private company builds the facility and transfers ownership to the government but continues to operate it.
Bulk Cargo. Bulk cargo is loose liquid or undifferentiated solid aggregate cargos, often the product of extraction and agricultural industries that are poured directly into a ship's hold. Examples include oil and petroleum products, coal, sand, gravel, phosphate and other dry bulk cargos that are the products of mining, and unbagged grains.
Bus. An internal combustion-powered rubber-tired vehicle that carries many passengers.
Business Improvement Districts (BID). A type of special assessment district used in commercial areas to pay for capital improvements such as street resurfacing, parking, sidewalk repair. Can also be used for operating expenses.
Bus Rapid Transit (BRT). A new transportation mode whereby buses have dedicated rights-of-way or preferential treatment at signalized intersections to increase their speed.
Bypass and Blending. Some wastewater plants divert stormwater around the secondary treatment units into the primary treatment areas for storage during a storm—"bypass" and when the sewage flows return to normal blend the stored stormwater with processed wastewater—"blending."
Cable Television. Cable TV (Community Antenna Television CATV) is a television system where broadcast TV is received by a regional facility and then distributed to consumers via a cable network instead of antennas.
Capital Budget. The capital budget is the formal budget document for one or two years containing the funds for projects in the capital improvements program (CIP) that will be designed and built in the following years.
Capital Improvements Program (CIP). This is a budget document that lists the infrastructure projects that the agency believes will be funded for the next five years. It is often a rolling document that is updated every year.
Capital Needs Study. The capital needs study is the long-range infrastructure program for that function with plans extending 20 and sometimes 50 years into the future. It consists of an inventory and condition assessment for existing facilities and plans for new facilities. It may also contain a budget and an analysis of fee structure alternatives. It is also called a "capital needs assessment."
Capital Project. A capital project usually includes all long-lived infrastructure, such as water, water and sewers, streets, parks, buildings, along with equipment such as fire trucks, radios, police cars, telecommunications equipment, furniture, and computers. The definition differs from jurisdiction to jurisdiction.
Cellular Telephones. Cell phones are portable phones that use radio waves to send and receive information. They derive their name from the fact that the reception area for a particular system is broken into small cells each with its own complete transmission and broadcasting system.
Certificates of Participation (COP) or Lease-Rental Bonds. Certificates of participation are shares in a debt obligation created with a lease that are sold to investors like bonds. The developer or non-profit agency uses the funds to build a facility that is leased on a long-term basis to the local or state government.
Charter School. A charter school is a nonsectarian public school, which unlike a traditional public school, operates largely independent of school district regulations. Charter schools are issued a "charter" from their sponsor, typically a state or local school board that serves as a contract between the board and the school.
Cluster Zoning. See *Planned Unit Development (PUD)*.
Coaxial Cable or Coax. Coaxial cable (or coax) is composed of two concentric conductors separated by an insulator.
Cogeneration Plant. A cogeneration plan produces electricity as do other plants, the steam arising from excess electricity is produced in the normal fashion, but the steam is captured and sold to other users.
Collaborative Planning. Collaborative planning is a formal process emphasizing shared learning among stakeholders with diverse interests in order to solve a mutual problem.
Combined-Cycle Plant. Natural gas is burned in a turbine which powers a generator. The exhaust from the turbine, which is at about 1,000°F, is then used to generate steam, which in turn powers a second generator.

Combined Sewer System. A combined system carries domestic, commercial, and industrial wastewaters and stormwater runoff through a single pipe system to the treatment plant, where it is treated and discharged into the surface water.
Commercial Service Airports. Airports that have scheduled passenger services and handle 2,500 or more passengers a year.
Community Park. A community park serves a broader purpose than neighborhood park. The focus is on meeting community-based recreation needs, as well as preserving unique landscapes and open spaces.
Community Water Systems. As defined by the EPA, these are systems that serve more than 25 people a day all year round.
Comprehensive Plan. The comprehensive plan is a document that sets out, at a minimum, the long-range physical plan for the entire jurisdiction. It is also called the general plan or in some places, the master plan.
Computerized Maintenance Management Systems (CMMS). Computerized maintenance management systems (also known as Work Order Management Systems, WMS) organize the daily, monthly, or annual work of the public works crews. These systems track sewers, street trees, medians, parks, sidewalks, and often public buildings in order to manage system failures, repairs, citizen complaints, and preventive maintenance.
Concrete. A mixture of gravel, sand, water, and a limestone-derived cement. It must be mixed, poured, and then allowed time to "cure," and so is more expensive than asphalt.
Concurrency. This is a term used to describe regulations for processing development applications that require the timely availability of infrastructure when the new development is occupied.
Condemnation. Condemnation is when a government agency acquires property from a private party through forced acquisition, or eminent domain. The agency must pay the fair market value, also called just compensation.
Condition Assessment. The condition assessment is a description of the current state of the infrastructure system, with an evaluation of the performance and the expected life and capacity of each of the facilities or elements that make up the system.
Congestion Pricing. Congestion pricing charges more for peak period usage when demand is high and supply is limited. It is a way of rationing the use of a fixed infrastructure resource rather than increasing capacity.
Conjunctive Use. This is when the water purveyor withdraws water from the stream or river when it is plentiful, and during lower flows switches over to groundwater and does not use the river. This is also called in lieu recharge.
Connection Fees. Connection fees are the charges that a locality, water or sewage company, or utility company imposes to hook up the individual residence to the larger infrastructure network.
Connector Trails. Connector trails are multipurpose trails that emphasize safe travel for pedestrians to and from parks and around the community. Focus is as much on transportation as it is on recreation.
Container. A metal box or detachable truck trailer body that may be carried by ship, road, or rail. Standard containers may be modified for ventilation, insulation, or refrigeration. Their interiors may also be modified by racks, tanks, and other devices to hold irregular objects and liquids. Containers come in various standard sizes, namely 20′, 40′, 45′, 48′, or 53′ in length, 8′0″ or 8′6″ in width, and 8′6″ or 9′6″ in height.
Contracting Out. The term "contracting out" has come to have a special meaning at the local level in the United States. This usually means contracting out something that is now done by in-house staff whether it is for the entire maintenance and operation of the infrastructure service or for a limited aspect of the service.
Conventional Pollutants. Conventional pollutants are human wastes, food wastes from a garbage disposal, and gray water such as laundry and shower water. They include pathogens, among other substances.
Convergence. Convergence is the integration of different kinds of telecommunications services—voice, video, or data—into a single service or a single means of transmission.
Council of Government (COG). A Council of Government is a formal government body or agency that has been formed to deal with metropolitan or multi-government issues at the regional level. The COG only has the powers granted to it by member jurisdictions.
Coupon Rate. This is the rate of interest for the municipal bond. It is usually set by the market when the bond is issued.
Cryptosporidiosis. Cryptosporidiosis is a disease caused by a parasite in drinking water that causes intestinal upsets and sometimes death.
Debt Affordability Model. An analysis that shows how much debt the jurisdiction can "afford" by comparing levels of per capita debt, debt as a percentage of the assessed value of taxable properties in the jurisdiction, debt as a percentage of personal income, and annual debt service on net debt as a percentage of general fund expenditures.
Debt Limit or Cap. Statutory regulations about the amount of debt, usually general obligation bonds that a jurisdiction can incur. It is most frequently characterized as a percentage of assessed valuation of the property in the jurisdiction but it can also be a percentage of the local budget or in some cases a total amount. Revenue bonds and certificates of participation are exempt from debt limits in most states.
Decreasing Block Rates. Decreasing block rates is an approach to setting fees for a service where the more that is consumed, the lower the price per unit. The theory is that economies of scale from more consumption result in lower costs.
Dedications. Dedications are when a developer constructs certain infrastructure such as streets, sidewalk, sewers, and water lines and then turns them over ("dedicates" them) to the county, city, or homeowners association for maintenance. The receiving government usually sets the standards for construction of these items.
Demand Management. Demand management is an approach to infrastructure provision that looks at ways to decrease or shift demand so that existing, or new facilities that are smaller, can meet the need for the service. Managing demand often

includes behavior modification schemes and price incentives to consume less. Part of the general approach may include physical improvements to the existing system to make it more efficient.

Deregulation. Deregulation refers to the rollback of regulations for telecommunications, energy, and the airline industry among others so that consumers can choose their own providers and so that the providers are more free to determine the kind of service to provide.

Desalination. Desalination is the process of removing salt, other minerals, or chemical compounds from impure water, such as brackish water or seawater, for the purpose of creating potable water.

Design–Build. Design–build is when the same entity designs and builds the infrastructure facility.

Design–Build–Operate (DBO). Design–build–operate is an arrangement where the government has a long-term contract with a private company to design–build and operate the infrastructure facility. The government generally finances the project with the use of public debt.

Development Caps. A growth management technique used by local governments that places a restriction on the amount of development that can take place in a local jurisdiction. Can be tied to the number of building permits allowed.

Development Regulations. Development regulations translate the land use designations of the general plan into more specific ordinances or regulations in order to regulate private market development projects as they go through the local permit process. They can set the requirements for the kind and amounts of infrastructure that the private developer is expected to provide for the project.

Development Standards. Development standards are provisions in the subdivision ordinance (or the PUD) that contain the engineering specifications for the on-site infrastructure improvements.

Digital. A type of electronic transmission that uses a series of on–offs for voltage. Becoming the preferred method of transmission for music, video, and voice.

Direct Current. Electric current only flows in one direction as in a battery.

Distillation. This is a technique used to create potable water from salt water. Seawater is heated to produce steam which is then condensed to produce water with a low salt concentration.

Divestiture. See *Full Privatization*.

Effluent. Effluent is the term used to describe both treated and untreated sewage.

Electromagnetic Spectrum. The electromagnetic (EM) spectrum refers to the range of different types of radiation—energy that is emitted in the form of a wave. Radiation can be characterized based on wavelength, which is the distance between consecutive peaks on the energy wave, or by the frequency, which measures the number of waves per second.

Element. An element in the comprehensive plan is the plan for a particular substantive area such as land use or transportation. Some states mandate certain elements while others are optional.

Eminent Domain. Eminent domain is the power of the government to take private property for a public purpose without the owner's consent. (See *Condemnation* and *Fair Market Value*).

Enterprise Funds. An accounting category or "fund" that tracks revenues generated by a local government "enterprise" such as garbage service, or water or wastewater services.

Environmental Impact Assessments. An analysis of the social, economic, environmental, and often institutional impacts of the effect of a development project. The National Environmental Protection Act (NEPA) requires an EIA for certain kinds of projects and certain states have passed mini-NEPAs with their own environment review requirements. They should ensure that adequate infrastructure will be in place for the development.

Equestrian Trail. An equestrian trail is one developed particularly for horseback riding.

Exactions. Exactions are a contribution of money or property by a developer that the locality uses to provide infrastructure and other services that are assessed during the permit process. Exactions is a broad term that encompasses impact fees, land dedications for parks and roads, and in lieu fees.

Expressed Demand for Parks. Expressed demand is a term used in recreation planning. It is actual recreation participation by activity that is measured by keeping records of facility use.

Facility Design Review. A facility design review is similar to value engineering but this review also looks at the relevance of the design itself. This kind of review can be done at the beginning or end of the design phase or even after construction bids are received. See *Value Engineering*.

Fair Market Value. Fair market value is the highest price that a seller who knows what the property will be used for would accept, and that a buyer would agree to, provided that neither party is compelled to buy or sell.

Fare Box Recovery. The amount of operating expenses in a transit system that is covered by fare revenues.

Fiber Optic Cable. Fiber optic cable is made of glass. Information is transmitted along a fiber optic cable using light, reflecting internally along its distance. It is capable of high speeds of data transmission.

Fiber to the Home (FTTH). An experimental system for cable TV that is entirely fiber instead of fiber and coaxial cable.

Fiscalization of Land Use. Fiscalization of land use refers to the practice of some localities that decide on land uses based on what generates the largest amount of impact fees or taxes with the smallest service requirements.

Flat Rate. A flat rate fee is one where all users pay the same amount regardless of quantity consumed.

Four-Step Method. Used for transportation planning and consists of process that estimates: (1) trip generation; (2) trip distribution; (3) mode choice; and (4) route selection.

Freeway. An arterial highway where access is only through off and on ramps.
Full Cost Pricing. Full cost pricing recovers all costs, including operations, maintenance, and capital costs.
Full Privatization. Full privatization is where the infrastructure service responsibility is transferred to a private or non-profit entity and the capital asset is sold or becomes part of an investor-owned utility. The purchasing company pays a sales price. This is also called divestiture.
Fund Accounting. Local governments create separate funds for major revenue sources to ensure that those that must be used for certain purposes can be easily tracked.
Gbps (gigabits per second). A measure of information speed on the Internet. Gbps is a billion bits per second or one thousand mbps.
General Aviation Airports. Airports which are not classified as commercial or primary airports.
General Fund. This is the main account for many governments. Revenues in this fund are those that can be spent on anything the local politicians decide to use the funds for, and most often they are used to support police, fire, and planning. These funds are also called discretionary funds since their uses are not prescribed for a particular function.
General Obligation (GO). GO bonds are backed by the full faith and credit of the issuing (borrowing) government. This means that the government is obligated to use its unlimited taxing power to repay the debt in case of default. Many states require that voters approve the issuance. See *Revenue Bonds*.
General Purpose Government (GPG). These are governments with multiple service delivery responsibilities, as compared to a special district which usually just has one or two programs. At the local level, these are cities and counties. States and the federal government are also GPGs.
Geosynchronous Orbits. An orbit for communications satellites that results in the satellite always remaining above the same spot on earth.
Geothermal Systems. Naturally-occurring steam is used to power the turbine to produce electricity.
Glazing System. The collection of windows and other transparent surfaces in a building.
Government Accounting Standards Board (GASB). A private non-profit organization formed in 1984 that develops standards for state and local government accounting practices. Although its standards are not legally required, state and local governments and districts follow them because it helps in the bond market. In addition, most accountants regard their standards as authoritative when auditing a local agency.
Government Accounting Standards Board Rule 34 (GASB 34). GASB 34 requires that state and local government financial statements include the value of all infrastructure assets each year. Setting the value involves depreciating the original purchase price two ways, one of which allows the locality to account for maintenance of the facility but only if it has an inventory and performs a condition assessment every three years.
Green Building. "Green building" is the design and construction practices that significantly reduce or eliminate the negative impact of buildings on the environment and occupants.
Greenways. Greenways are vegetation areas that link different parks or open spaces in a city together to form a continuous park environment.
Gray Water. Gray water is domestic wastewater that has not been contaminated by toilet discharge or other potentially infectious bodily wastes. It includes wastewater from bathtubs, showers, washbasins, and washing machines but generally the term does not include water from kitchen sinks or dishwashers.
Groundwater. Groundwater is water that comes from a well connected to an underground aquifer.
Hazardous Waste. Waste that poses an immediate or long-term threat to human, plant, or animal life. The federal government indicates that it is "ignitable, corrosive, reactive or toxic."
Headend. The headend of the cable television system is the satellite-signal reception system where the local cable TV provider transfers television and radio signals to the distribution system.
Heating, Ventilation, and Air Conditioning (HVAC) System. HVAC (Heating, Ventilation, and Air Conditioning) systems are those for heating and cooling.
Horizontal Consistency. The local plan has horizontal consistency when its provisions are consistent with the general plan of a neighboring city or with the infrastructure plan of a special purpose district such as a water or school district.
Hub Airport. An airport which serves as the center of a radial flight structure and maximizes the number of transfer options open to a passenger.
Hub-and-Spoke System. A transportation network in which traffic is directed along "spokes" through one or more dominant nodes called hubs. This is different from a network in which each node maintains direct connections with other nodes.
HVAC. See *Heating, Ventilation, and Air Conditioning (HVAC) System*.
Hybrid Fiber/Coax (HFC). The hybrid fiber/coax (HFC) system for cable TV has fiber carrying the signals from the headend to nodes and coaxial cable carrying the signal from the nodes to the subscriber.
Hydroelectric plants. Electricity is produced using the power of falling water to turn a generator.
Hydromodification. Hydromodification is when the normal flow of a stream is disrupted, either with too much flow or too little. This can cause the ecology of the stream to be out of balance with the loss of plant and aquatic life.
Impact Aid for Schools. This is federal aid that compensates school districts for loss of local tax base caused by federal activity, such as the presence of a military base or other governmental property.

Impact Fees. Impact fees are charges levied during the development process for a new commercial, industrial, or residential project to pay for off-site or community facilities that must be built or expanded because of the new development. The terms "impact fees," "development fees," and "in lieu fees," are used interchangeably.
Incineration. The burning of waste.
Increasing Block Rates. Increasing block rates refers to a fee structure where the consumer pays a higher rate per unit when his or her use passes a designated volume or threshold of use. This pricing scheme can be used for conservation purposes.
Induced Demand. Induced demand is when new infrastructure built for existing or future demand actually increases the demand or use of the system. This is often noticed with new highways or transportation improvements.
Infiltration. Water entering into a sewer system from the ground through holes in the pipe.
Inflow. Inflow is water discharged into the sewer pipes, usually from illegal connections with roof leaders, or illegal cross-connections between sanitary and storm sewers.
Injection Wells. Most commonly used to dispose of hazardous wastes. They are pipes that extend several thousand feet into rocks bounded by impermeable layers and have no contact with aquifers.
Inlets. Inlets are grated openings often found at street corners which collect stormwater runoff from gutters and guide it into storm sewer submains under the street.
In Lieu Fees. In lieu fees are payments by a developer instead of dedicating land within the development or providing necessary off-site infrastructure improvements. See *Impact Fees*.
Integrated Resource Planning. Integrated resource planning is a planning paradigm promoted in the 1990s that required that demand management strategies for water be explicitly considered and that citizens be actively involved in the process.
Intelligent Transportation Systems (ITS). Information or electronic technologies used to increase capacity, efficiency, or convenience of the transportation system.
Interceptors. Interceptors or trunk lines are large lines that terminate at the wastewater treatment plant. They can be from 15 inches to over 60 in diameter.
InterLATA and IntraLATA. The country is divided into local telephone service areas called LATAs (Local Access and Transport Areas). IntraLATA calls are local calls while InterLATA calls are long-distance calls.
Internal Consistency. Planning rules are internally consistent if implementing regulations and budgets for the jurisdiction carry out the intention of the comprehensive plan.
Interoperability. Interoperability is where all electronic devices (computers, phones, or radios) can be used with any kind of network, or at least the network of the organization or jurisdiction.
Jobs–Housing Balance. A comparison of the amount or changes in employment compared to housing units.
Kbps (kilobits per second). A measure of information speed on the Internet. One Kbps is one thousand bits per second.
Large Quantity Generator (LQG). A hazardous waste producer that produces 2,200 pounds of hazardous waste per month.
Large Urban Park. Large urban parks serve a broader purpose than community parks and are used when community and neighborhood parks are not adequate. Focus is on meeting community-based recreational needs, and preserving unique landscapes and open spaces.
Latent Demand for Parks. Latent demand for recreational activities is desire for facilities that would be used if they were provided. It is difficult to determine and usually requires surveys through workshops, questionnaires in utility bills, and focus groups.
Latent Demand for Roads. Latent demand for roadway space is traffic that is not present because the existing roadways are so congested but would be on the road if new facilities were available.
Laterals. Laterals are pipes leading from individual houses or buildings to a larger pipe, called the sub-main, in the public right-of-way. Laterals are usually 6 inches in diameter and made of clay, concrete, or PVC.
Leachate. Water that becomes contaminated by the wastes in the landfill.
Lease-Rental Bonds. See *Certificates of Participation*.
Legionnaires' Disease. Legionnaires' disease is a form of pneumonia spread by contaminated mist or moist air. Outbreaks of the disease are often associated with problems in the cooling towers and evaporative condensers of large air conditioning systems.
Level of Service (LOS) Standards. Level of service (LOS) standards are a performance standard for the infrastructure system that allows the locality to specify how much of the system is needed in response to the size of the development, the number of new households, and the quality of service desired.
Life-cycle Analysis. Life-cycle analysis, also known as life-cycle cost analysis, is the consideration of all agency expenditures and user costs throughout the entire service life of an infrastructure system, not just the initial capital investments.
Lift. The force that keeps the airplane aloft.
Lift Station. The lift station contains a collection basin (called the wet well) into which sewage flows from an area which lies below the grade of the main system. It also contains a pump which pumps the sewage uphill to connect to the rest of the system which then flows by gravity to the treatment plant.
Light Rail Transit. An electric railway system where passenger trains operate along exclusive rights-of-way in order to board and discharge passengers at track or car-floor level.

Linkage Fees. Linkage fees are used to mitigate the secondary effect of commercial and industrial projects, such as the need for housing and day care. See *Exactions*; *Impact Fees*; *In Lieu Fees*.
Load Factors. The ratio of passengers to seats on a commercial airlines flight.
Local Exchange or Central Office. The central office is a facility that houses the switching and transmission systems for the telephone system, and delivers basic dial tone service, voice mail, and caller ID.
Loop Detectors. Loop detectors are devices imbedded in the pavement that register a car's magnetic field and send that information to the traffic signal controller.
Magnet School. A special type of public school, often located in older urban areas, designed to encourage voluntary integration by offering specialized and unique curriculum to attract students of varied social and economic backgrounds.
Managed Competition. Managed competition is when the public agency and the private sector both submit bid packages to operate a particular infrastructure service. This enables the public agency to see if it can compete with the private sector.
Managing for Results. See *Performance Management*.
Manholes. Manholes, also called utility access ports (UAPs), are located at any changes in direction, pipe size, slope, or any time two lines intersect.
Marginal Cost Pricing. Marginal cost pricing is based on the cost of producing an additional unit of the service, or the last unit to be produced. Used by the private market and a tool for demand management for the local agency since it permits the consumer to decide whether the benefit he or she will derive from higher consumption justifies the additional cost.
Market-Oriented Civic Model for Parks. The market-oriented civic model for parks relies heavily on local citizen help and may use self-perpetuating financing, innovative fees. One of several proposed by Alexander Garvin.
Maturity Structure of the Bond. Bonds can be issued as "serial bonds," when each individual certificate has its own maturity date, or as a "term bond" where there is a single final date for all the certificates.
Mbps (Megabits Per Second). A measure of information speed on the internet. One Mbps is a million bits per second.
Mercaptan. A chemical with a rotten egg smell added to natural gas so users can smell if there is a leak.
Mercury-Vapor Lamps. Mercury-vapor lamps have a bluish-white light and are often used on rural roadways.
Metal Halide Lamps. Metal halide lamps have a white light but are expensive to operate. They are used in sports stadiums and in some pedestrian areas.
Metropolitan Planning Organization (MPO). The federal transportation program requires areas with populations of over 50,000 to establish MPOs to plan and allocate federal transportation funds. The composition of the MPO is determined by the state transportation department.
Microturbines. Miniature jet engines that are connected to small electric generators.
Middens. Used by ancient human settlements to dispose of animal bones, plant residues, and other debris. The first landfill.
Mini-Park. A mini-park is a small park located often on a single parcel and used to address limited, isolated, or unique recreational needs.
Mitigation Fees. Mitigation fees are used to pay for environmental impacts identified as part of the NEPA or state and local environmental assessment processes. See *Exactions*; *Impact Fees*; *In Lieu Fees*.
Modem. A modem is the device that transforms a signal from digital to analog and back. Modem is short for *mo*dulator-*dem*odulator.
MTBE. MTBE (methyl tertiary butyl ether) was required to be added to gasoline in the 1980s to improve air emissions. It is a potentially carcinogenic compound and has been found in drinking water wells recently.
Multiple Chemical Sensitivity. Multiple chemical sensitivity is a condition in which a person is sensitive to a number of chemicals, all at low concentration.
Municipal Wastewater. Municipal wastewater is composed of household wastes and industrial wastewater from manufacturing and commercial uses. This is contrasted with agricultural or rural runoff.
Natural Monopoly. A service where economies of scale and the nature of capital investment naturally favor one firm having a monopoly position in the market.
Natural Resource Areas. Natural resource areas are lands set aside for preservation of significant natural resources open space. They may also be used for buffering other land uses or set aside for visual and aesthetic reasons.
Neighborhood Park. The neighborhood park is a small park designed to serve the recreational and social focus of a neighborhood. It is the basic unit of the park system and supports informal active and passive recreation.
Neo-Bulk Cargo. Separately differentiated but uniformly packaged goods which stow as solidly as bulk cargo, such as paper bales or coils of steel.
Nessie Curve. A Nessie curve is a graph that projects aggregate capital replacement needs by year for sewers and water infrastructure. It was developed by the Australians and is named for the Loch Ness monster because it is huge, and much of it lurks beneath the surface.
Niche Cargo. Refers to non-containerized cargo that requires special handling facilities such as refrigeration and fumigation for fruit and meat, cleaning and repairs for automobiles.
Non-Point Source Pollution. Non-point source pollution (NPS) comes from many sources and cannot be easily collected into a single sewer pipe.

NPDES. The National Pollutant Discharge Elimination Program (NPDES) was authorized by the 1972 Clean Water Act to regulate point source pollution. It was amended in 1987 to regulate non-point source pollution as well.
Off Stream Storage. This is when available water is diverted or pumped into a valley or canyon with little or no aquatic ecosystem and recreational benefit and used as a reservoir.
On-Street Bikeways. On-street bikeways are paved segments of roadways that serve as a means to safely separate bicyclists from vehicular traffic. This is the term used by park and recreation planners. For the official transportation classifications, see *Bicycle Paths*, *Bicycle Routes*, and *Bicycle Lanes*.
Open Space Size Standards. Open space size standards are the minimum acreage needed for facilities that support the activity menu for each park classification.
Overdrafting. Overdrafting means that more water is being taken out of the aquifer or the basin than is being replaced and water withdrawal exceeds the recharge rate.
Package Unit. A package unit performs all the functions of a municipal wastewater treatment plant but on a much smaller scale.
Packet Switching. The Internet sends information using a technology called packet switching. This technology breaks down an email into small packets that take their own routes to get to the destination depending on which lines are available.
Park Trail. A park trail is a passageway of any kind in a greenway, park, or natural resource area.
Pathogen. A pathogen is a pollutant that can cause disease in humans, including bacteria and viruses.
Pavement Management System (PMS). A pavement management system is a database that contains information on current pavement conditions and software that evaluates and prioritizes alternative reconstruction, rehabilitation, and maintenance strategies against a predetermined condition level.
Peaker Plants. Power plants that are active only during periods of high or peak demand.
Peak Load Pricing. Peak load pricing is where capacity is increased to provide services at the peak use period and capacity costs are paid for by those who cause the peaks if possible.
Perchlorate. Perchlorate is a by-product of the jet and rocket fuel industry that has shown up in groundwater supplies in some parts of the country.
Performance Indicators. A performance indicator is a measurable data item that shows how well the jurisdiction is doing with regard to its goals and objectives. For infrastructure facilities, indicators usually monitor the services delivered once the facility is built.
Performance Management. Performance management is a process that involves developing a strategic plan for the organization, deploying resources to achieve these goals and objectives, developing performance indicators that reflect the goals and objectives, reporting on progress, and making changes as needed. Also called "Managing for Results."
Personal Air Travel. Tiltrotor aircraft and magnetic levitation (maglev) vehicles are two innovative small aircraft which many predict will be used by individuals to bypass congested airports.
Photovoltaic Cells. Modules covered with a chemical coating, usually silicon, that, when exposed to sunlight, gives off electrons and produces electricity.
Plain Switched Telephone Network (PSTN). The traditional telephone network that consists of customer telephones connected by wires that run directly to a central office of the telephone company. When a customer places a call, the switch creates a circuit that remains open until one party hangs up.
Planned Unit Development (PUD). Planned unit development or cluster zoning ordinances permit the developer to "cluster" development in certain areas of a tract provided that the overall density remains the same as the underlying subdivision and zoning ordinances for the area.
Planning Commission. A planning commission provides advice to elected officials about land use issues and can provide a forum for hearings and public workshops. It usually consists of citizens appointed by the mayor or the elected body of the city or county.
Plat. Plat is another term for map, and is used to designate the map submitted and approved during the subdivision application process. The map or plat shows the layout of the streets, lots, public utilities such as water, sanitary and storm sewers, and private utility placement for cable, gas, electricity, and telecommunications.
Plumbing System. The plumbing system includes water pipes, drain pipes, ventilation pipes, and natural gas lines for the water heater.
Point Source Pollution. "Point" sources of pollution are those produced from identifiable sources such as a home, a factory, or a commercial building.
Pollution Prevention. Changing practices which produce excessive or toxic wastes.
Primary Airports. Commercial service airports that handle more than 10,000 passengers a year.
Primary Treatment. In the primary treatment stage for both sewage and raw water, solids are allowed to settle out of the water in a physical process. For water, substances like alum are added to precipitate out foreign matter.
Private Park/Recreation Facilities. Private park/recreation facilities are those parks and recreation facilities that are privately owned yet contribute to the public park and recreation system.
Private Placement. Private placement is also known as a negotiated agreement. This is when a bond is sold to a specific institution or individual instead of being put on the market.

Privatization. See *Full Privatization.*
Project Management System. A project management system monitors progress in the development of a capital project against time and budget milestones.
Property Taxes. Property taxes are local taxes that are a function of the value of the property. They are part of a larger class of taxes based on value called "ad valorem" taxes.
Proximate Capitalization. This is the increase in value that a property has as a result of being located near a park or open space. This is also called "the proximate principle" and can result in an increase in capitalization of the value of property by 20 percent.
Public Authorities. The U.S. Census does not distinguish between public authorities and special districts, but often the term is used to denote independent agencies formed to take on large and risky infrastructure projects with the potential to be self-sufficient financially. Boards of public authorities are usually appointed rather than elected. See *Special Districts*; *Special Assessment Districts.*
Public Model for Parks. The public model for parks is where there is broad-based funding and it is open to citizen and business involvement in design. One of several proposed by Alexander Garvin.
Public Prices. Public prices are fees charged by local governments for infrastructure and non-infrastructure services that cover the costs of operation and maintenance and sometimes part of the capital costs.
Public–Private Co-Venture for Parks. The public–private co-venture model for parks allows user fees to raise capital and is also open to citizen involvement. One of several proposed by Alexander Garvin.
Public–Private Partnerships. There is no standard recipe for a public–private partnership. The basic idea is that the government contracts in some manner for the private sector to build an infrastructure facility. Alternate arrangements are: *Design–Build–Operate*; *Build–Own–Operate–Transfer*; *Build–Transfer–Operate*; and *Design–Build.*
Rational Formula. The rational formula is the formula that estimates the volume of stormwater runoff in a particular area.
Rational Nexus. Rational nexus is the test used to see if impact fees are legal. This test requires that there must be a linkage between the dollar amount assessed to the development, or the amount and location of land dedicated to the development itself.
Recreation Activity Menu. The recreation activity menu (RAM) is the list of all recreational activities and their facilities, such as tot lots and picnic units that will be part of each park's classification.
Recycling. Re-using materials and objects in their original or changed forms rather than discarding them as wastes.
Re-engineering. This refers to an organization's efforts to voluntarily give up old ways of doing business, also called internal restructuring.
Regulatory Fees. Regulatory fees are defined by the U.S. Census as fees authorized under the police power of the agency rather than its taking authority. Such fees include licenses, franchises, and permit fees, and are generally set to cover the administrative costs of regulating the activity, that is, the costs of issuing the permit.
Reliever Airports. Small general aviation airports that can relieve congestion at nearby commercial service airports. These airports are usually a metropolitan area's "second" airport.
Resource Recover Park. Co-location of reuse, recycling, compost processing, waste disposal facilities, and retail businesses which sell or give away reused and recycled materials in a central facility.
Revenue Bonds. Revenue bonds (also called enterprise bonds) use the fees from a government service such as water or parking to retire the debt and pay the interest. Usually these types of bonds do not require voter approval.
Reverse Osmosis. One technique used for creating potable water from salt water. The process works by forcing seawater through a semi-permeable membrane that restricts salt and other minerals but allows water molecules to pass through.
Right-of-Way (ROW). Right-of-Way is the right to pass across the land of another and the public ROW is open to the use of the public.
Riparian Rights. Rights granted to land owners along natural water courses such as streams, rivers, and lakes.
Roadway Level of Service. A measure of capacity of a roadway segment in vehicles per hour, average vehicle speed, and average per-vehicle delay. Goes from A to F.
Ro/Ro. A shortening of the term, "Roll On/Roll Off." This is cargo that has wheels, allowing it to be loaded and discharged from ships with loading ramps. Also refers to car ferries and to other vessels designed to carry automobiles and/or trains.
Rough Proportionality. Rough proportionality is the requirement for impact fees that they be proportionate to the amount of use. The entire cost of a facility that will be used by others cannot be charged to one development.
Rubberized Asphalt Concrete. Rubberized asphalt concrete (also known as asphalt rubber hot mix) is regular asphalt mixed with crumbs of rubber that are made by shredding old tires.
Sanitary Landfill. A sanitary landfill is a site where wastes collected from individual households and businesses are disposed of using engineering techniques to control the putrefaction of organic wastes. These techniques can include a liner and specified depths of garbage and soil or clay which is subsequently compacted.
Sanitary Sewer Overflow (SSO). Occasional unintentional discharge of raw sewage from municipal sanitary sewers.
School Desegregation. Sometimes called integration, desegregation is a term used to describe the process of breaking up concentrations of racial or ethnic groups in public schools in attempts to promote equity in educational access. Desegregation plans involve the bringing together of students of different races (or other factors, e.g. sex and national origin).

Seasonal Rates. Price structures that charge more during times when supplies are scarcer or when demand is higher. Used for conservation purposes.
Secondary Treatment. In the secondary treatment stage for sewage, biological processes attack the pathogens in the water. Secondary treatment for raw water consists of adding a disinfectant of chlorine, or using ultraviolet light, to kill the bacteria.
Septic Tank System. A septic tank system consists of a tank where solids in sewage settle out and the resulting effluent goes into pipes in a drainfield or leachfield where natural processes attack the pathogens. A properly functioning system is the equivalent of the primary and secondary phases of sewage treatment.
Service Area. The service area is the geographic boundary for the area that the infrastructure system will serve and defines the area to be planned for. Often, however, the ideal geographic area for an infrastructure system crosses political jurisdictions that have the responsibility for planning.
Sick Building Syndrome. This is a situation in which occupants of a building experience acute health effects that seem to be linked to time spent in the building.
Socioeconomic Models. Computer models that use simulations to project jobs and housing growth in order to project the need or demand for infrastructure facilities.
Sodium Lamps. High pressure sodium lamps have a gold-white light and are very efficient and long-lived. They are replacing low-pressure sodium lamps that are more efficient but give off a deep yellow light and have poor color rendition.
Solid Waste. Any solid, semi-solid, liquid, or contained gaseous materials that have reached the end of their useful life and can be discarded.
Source Control. Source control refers to efforts to deal with pollutants where they are generated, rather than down the line.
Source Reduction. Change in design, manufacture, purchase, or use of products to reduce their amount or toxicity before becoming municipal solid waste.
Special Assessment. Special assessments are charges to a property or a business in a specific geographic area based on the cost of public services rendered but distributed to each payer according to an estimate of the benefit received.
Special Assessment Districts. According to the U.S. Census, special assessment districts are taxing districts that are subordinate units of the city or county. They can cover the entire jurisdiction or a part of it and may require a vote of the electorate to be established.
Special Districts. According to the U.S. Census, special districts refer to government agencies with their own elected governing body. They can raise taxes but usually they deliver services that are paid for through fees, such as water and sewer districts.
Sports Complex. A sports complex consolidates heavily programmed athletic fields and recreation facilities to larger and fewer sites strategically located throughout the community.
Step-Up and Step-Down Transformers. These transformers raise or lower the voltage of the electricity from the generator to the transmission lines, and then to the consumer.
Subdivision Ordinance. The subdivision ordinance specifies the requirements for turning raw land into individual parcels or lots and is the primary tool for regulating infrastructure provision in developing areas.
Subgrade. The subgrade is found at the bottom of a typical street cross-section and it is the original soil which may or may not be treated.
Surface Impoundment. A contained pile of waste on the surface of the earth (as opposed to underground impoundment).
Surface Water. Surface water includes lakes, rivers, estuaries, and is an important source of water. Surface water is diverted directly from a stream, river, or canal and stored in natural lakes, reservoirs, and sometimes in aquifers.
Sustainable Development. Sustainable development has been defined by many as that which meets the needs of the present-day population without negatively impacting the ability of future generations to meet their needs.
Tax Increment Financing. A mechanism to repay bonds based on the difference between the original assessed value and the increase over time due to rising property values caused by the improvements funded by the bonds.
Tertiary Treatment. Tertiary treatment for sewage consists of processes that seek to rid the water of heavy metals.
Thermal Treatment. Method of treating hazardous waste involving incineration.
Time-of-Day Pricing. Usually price structures that attempt to shift consumption to time of day when less is normally consumed. Used for conservation. See *Peak Load Pricing*.
Time Switching. Time switching refers to tourists who visit a locality or facility in response to a special event but who would have visited the municipality anyway, or who are already there, but are "time switching" to attend the event.
Total Maximum Daily Load (TMDL). The total maximum daily load (TMDL) represents the maximum amount of all pollutants taken together that a water body can receive and still meet water quality standards.
Total Water Management. Total water management looks at water as a limited rather than unlimited resource and envisions the water utility as the good steward of land and water.
Toxic Pollutants. Some 100 toxic substances have been defined by the EPA as harmful to animal or plant life. These are organics (pesticides, solvents, polychlorinated biphenyls, PCBS, and dioxins) and metals (lead, silver, mercury, copper, chromium, zinc, nickel, and cadmium).
Traffic Calming. Changes in street alignment, installation of barriers, and other physical measures that reduce traffic speeds to make the street safer and more livable.
Transit-Oriented Development. Concentrating development around transit nodes such as a subway stop.

Transmission. Transport of electricity from its point of generation to the consumer.
Transportation Analysis Zones (TAZ). Similar to census tracts, this is the unit of analysis for transportation planning.
Trickling Filter Process. A trickling filter is used in secondary sewage treatment processes to attack the organic matter. Bacteria gather and multiply on the filter as it consumes most of the organic matter.
Trip. Traffic planners define a trip as a journey and note its origin, destination, purpose, route, and mode.
Trip Attractor. A land use that attracts trips.
Trolley Bus. An electrically powered bus that draws its power from a pair of parallel overhead wires by means of two poles.
Uniform Building Code. See *Building Code*.
Uniform Development Standards. These are the design requirements for infrastructure constructed by private developers that are set by the local general purpose government. They generally include detailed engineering specifications for the capacity, location and placement, composition, and dimensions of the infrastructure.
Uniform Rate Structures. Uniform rate structures charge the same price per unit for all customers or rate payers.
United States Army Corps of Engineers (ACE). The U.S. Army Corp of Engineers is responsible for flood protection of coastal and riverside lands, and for dredging navigation channels. The ACE also reviews applications for dredging projects and prioritizes these projects as well as related federal spending.
Urban runoff. Urban runoff occurs during heavy rainfall periods and includes metals from the brakes of cars, sewage, pesticides, and sediment among others.
User Charges and Fees. User charges or fees are prices charged by governments for consumers of publicly provided infrastructure (or other) services. Examples of user fees include charges for each can of garbage or the use of public recreational fields, golf courses, and swimming pools.
Utilidors. Utilidors are conduits that contain several utility systems, such as electricity, water, sewer, gas, and telecommunications. They may be above ground in the attic or below ground.
Utility Access Ports (UAPs). See *Manholes*.
Utility Charges. Utility charges or "rates" are user fees charged for infrastructure services that the community has either chosen to provide publicly or have provided privately but where the federal and state government regulates the enterprise to some extent. See *Public Prices*.
Value Engineering. Value engineering refers to the process of hiring an engineer or architect to review the detailed plans and specifications of a project to eliminate tasks or design features in order to reduce the size of the construction contract. See *Facility Design Review*.
Vertical Consistency. Local plans are vertically consistent when the comprehensive plan is consistent with a regional plan and a regional plan is consistent with a higher-level state plan.
VoIP (Voice over Internet Protocol). VoIP works by transferring small bits of conversation in digital packets over data networks (the Internet) instead of over the regular telephone system.
Voltage. The amount of pressure in a stream of electricity.
Wake Vortex. Large disturbances created in the air as behind an aircraft. Any plane that flies into a wake vortex is in for a rough ride.
Waste Diversion. Reducing the amount of solid waste a community must dispose of, usually into a landfill.
Waste Stream. The mix of waste produced by a community. Many states and local governments have goals for reduction by specific type of waste in the waste stream.
Waste to Energy. The recovery of energy from burning trash and garbage.
Water Cycle. The water cycle is the process where water evaporates into the air, becomes part of a cloud, falls to the earth as precipitation, and evaporates again.
Water Duty Factors. A water duty factor is the amount of current use per household, or square foot area for the land use type. This is also called land use-based unit water demands.
Water Pipes or Lines. The largest pipes in the water network are called feeder mains which can be 18 inches in diameter. Branch mains lead off the feeder mains. Service lines or pipes link the branch mains to the user and can be 2 inches in diameter.
Water Reclamation or Recycling. This is the reuse of wastewater (usually municipal) as groundwater recharge, agricultural irrigation, landscape irrigation for golf courses, parks and use in decorative water bodies.
Wholesaler. A water district that is a wholesaler purchases water from the federal and state government and sells it to local entities such as cities, smaller water districts, or private water companies.
Wi-Fi. Wi-Fi is the term that refers to wireless Internet transmissions. Wi-Fi operates on a frequency band that is unlicensed worldwide and transmits data up to 10 Mbps. It can transmit data about 100 yards from the transmitter. The transmitter is connected to a hard-wired broadband Internet connection. It uses an unlicensed radio frequency.
WiMax. WiMax (Worldwide Interoperability for Microwave Access) is a new computer chip for wireless Internet connection that is expected to be available in future. WiMax products may achieve speeds of 70 Mbps over a distance of 30 miles, although probably not both at the same time.
Work Order Management Systems (WMS). See *Computerized Maintenance Management System*.
Zoning Ordinance. The zoning ordinance is a local regulation that implements the general plan by translating broad land use designations into rules that apply to specific parcels. It also includes rules for processing development applications.

Notes

chapter 1

1 Herbert Muschamp, "New York Times," in *Building the Public City: The Politics, Governance and Finance of Public Infrastructure*, ed. David Perry (Thousand Oaks, CA: Sage Publications, 1995).

2 Joel A. Tarr, "The Evolution of the Urban Infrastructure in the Nineteenth and Twentieth Centuries," in *Perspectives on Urban Infrastructure*, ed. R. Hansen (Washington, DC: National Academies Press, 1984), 4.

3 David C. Perry, "Building the Public City," in *Building the Public City: The Politics, Governance and Finance of Public Infrastructure*, ed. David Perry (Thousand Oaks, CA: Sage Publications, 1995).

4 Claire L. Felbinger, "Conditions of Confusion and Conflict: Rethinking the Infrastructure-Economic Development Linkage," in *Building the Public City: The Politics, Governance and Finance of Public Infrastructure*, ed. David Perry (Thousand Oaks, CA: Sage Publications, 1995).

5 Stephen Graham and Simon Marvin, *Splintering Urbanism: Networked Structures, Technological Mobilities and the Urban Condition* (London: Routledge, 2001), 3.

6 Kazys Vanelis, *The Infrastructural City: Networked Ecologies in Los Angeles* (Los Angeles, CA: Los Angeles Forum for Architecture and Urban Design, 2009).

7 William W. Rostow, *The Stages of Economic Growth: A Non-Communist Manifesto* (Cambridge: Cambridge University Press, 1960).

8 Perry, "Building the Public City," op. cit.

9 Robert W. Burchell, David Listokin, and William R. Dolphin, *Development Assessment Handbook* (Washington, DC: Urban Land Institute, 1994).

10 For another typology, see Brian Hayes, *Infrastructure: A Field Guide to the Industrial Landscape* (New York: W.W. Norton, 2005).

11 Some people include parks and open space in the environmental category. We place it under community facilities since these are infrastructure services usually provided by a local general purpose government, often with local taxes, while water and waste are often, but not always, the responsibility of special districts or enterprise departments within the local general purpose government.

12 Intergovernmental Panel on Climate Change, *Climate Change 2007: Synthesis Report. Contribution of Working Groups I, II, and III to the Fourth Assessment Report on the Intergovernmental Panel on Climate Change* (Geneva, Switzerland: IPCC, 2007), available at: www.ipcc.ch. This report was prepared based on all available research on the topic by a group of 650 scientists from around the world under the auspices of the World Meteorological Organization and the United Nations Environment Programme.

13 Chrisna du Plessis and Raymond J. Cole, "Motivating Change: Shifting the Paradigm," *Building Research & Information* 39, no. 5 (2011): 436–449.

14 *The Penn Manifesto: Educating Urban Designers for Post-Carbon Cities* (Philadelphia, PA: PennDesign and Penn Institute for Urban Research, 2009). This manifesto was produced as a result of a conference, "Re-Imagining Cities: Urban Design After Peak Oil," in November 2008 at the University of Pennsylvania, available at: www.upenn.edu/penniur/afteroil/Educating%20Urban%20Designers.pdf. The original statement referred to "urban design" not "infrastructure."

15 Jennifer Cheeseman Day, "Population Profile of the United States," U.S. Census Bureau, available at: www.census.gov/population/www/pop-profile/natproj.html (accessed November 6, 2011).

16 American Water Works Association and Water Industry Technical Action Fund (U.S), *Dawn of the Replacement Era: Reinvesting in Drinking Water Infrastructure: An Analysis of Twenty Utilities' Needs for Repair and Replacement of Drinking Water Infrastructure* (Denver, CO: American Water Works Association, 2001).

17 "2009 Report Card for America's Infrastructure," American Society of Civil Engineers, available at: www.asce.org/reportcard/ (accessed January 15, 2012).

18 Gregory Inghram and Anthony Flint, "Cities and Infrastructure: A Rough Road Ahead" *Land Lines* (July 2011). Available at: www.lincolninst.edu/pubs/1925-Cities-and-Infrastructure.

19 William Ascher and Corinne Krupp, eds., *Physical Infrastructure Development: Balancing the Growth, Equity, and Environmental Imperatives* (New York: Palgrave Macmillan, 2010).

20 Nathan Musik, *Public Spending on Transportation and Water Infrastructure* (Washington, DC: The Congress of the United States, Congressional Budget Office, 2010).

21 Robert B. Ward, "State Revenues in an Era of Fundamental Change," Nelson A. Rockefeller Institute of Government, available at: www.rockinst.org/pdf/government_finance/2011-12-02-state_revenues_ppt.pdf.

22 Graham and Marvin, 2001, op. cit.

23 Karen Stromme Christensen, *Cities and Complexity: Making Intergovernmental Decisions* (Thousand Oaks, CA: Sage, 1999).

24 Elisa Barbour, *Metropolitan Growth Planning in California, 1900–2000* (San Francisco: Public Policy Institute of California, 2002).

25 Stuart Meck, "Executive Summary: Status of State Planning Reform," in *Planning Communities in the 21st Century* (Chicago, IL: American Planning Association, 2002).

26 David E. Dowall and Jan Whittington, *Making Room for the Future: Rebuilding California's Infrastructure* (San Francisco: Public Policy Institute of California, 2003); and Michael Neuman and Jan Whittington, *How Does California Make Its Infrastructure Decisions?* (San Francisco: Public Policy Institute of California, 2000).

27 Robert McNamara, "Albert Gallatin's Report on Roads, Canals, Harbors, and Rivers: Jefferson's Treasury Secretary Envisioned a Great Transportation System," available at: history1800s.about.com/od/canals/a/gallatinreport.htm (accessed February 22, 2012).

28 See www.asce.org; Jonathan D. Miller, Urban Land Institute, and Ernst & Young, *Infrastructure 2011: A Strategic Priority* (Washington, DC: Urban Land Institute, 2011); and Barry B. LePatner, *Too Big to Fall: America's Failing Infrastructure and the Way Forward* (New York: Foster Publishing, 2010).

29 Vladimir Novotny, John Ahern, and Paul Brown, *Water Centric Sustainable Communities: Planning, Retrofitting, and Building the Next Urban Environment* (Hoboken, NJ: John Wiley & Sons, 2010).

chapter 2

1 Professor Michael Teitz, personal communication with author, May 21, 2003.

2 Joel A. Tarr, "The Evolution of the Urban Infrastructure in the Nineteenth and Twentieth Centuries," in *Perspectives on Urban Infrastructure*, ed. R. Hansen (Washington, DC: National Academies Press, 1984).

3 Ibid.

4 Eric H. Monkkonen, *America Becomes Urban: The Development of U.S. Cities and Towns 1780–1980* (Berkeley, CA: University of California, 1988).

5 Claire L. Felbinger, "Conditions of Confusion and Conflict: Rethinking the Infrastructure-Economic Development Linkage," in *Building the Public City: The Politics, Governance and Finance of Public Infrastructure*, ed. David Perry (Thousand Oaks, CA: Sage, 1995).

6 Charles D. Jacobson and Joel A. Tarr, *Ownership and Financing of Infrastructure: Historical Perspectives*, Policy Research Working Paper 1466 (Washington, DC: The World Bank, 1995).

7 Jerry Mitchell, *The American Experiment with Government Corporations* (Armonk, NY: M.E. Sharpe, 1999).

8 Jacobson and Tarr, 1995, op. cit.

9 Heywood T. Sanders, "Public Works and Public Dollars: Federal Infrastructure Aid and Local Investment Policy," in *Building the Public City: The Politics, Governance and Finance of Public Infrastructure*, ed. David Perry (Thousand Oaks, CA: Sage, 1995).

10 David Perry, "Building the City through the Back Door: The Politics of Debt, Law, and Public Infrastructure," in *Building the Public City: The Politics, Governance and Finance of Public Infrastructure*, ed. David Perry (Thousand Oaks, CA: Sage, 1995), 203–306.

11 American Society of Civil Engineers, *History and Heritage of Civil Engineering*, available at: www.asce.org/About-ASCE/History-of-ASCE/.

12 Tarr, 1984, op. cit.

13 Ibid.

14 Felbinger, 1995, op. cit.

15 Tarr, 1984, op. cit.

16 Alfred D. Chandler, *The Visible Hand: The Managerial Revolution in American Business* (Cambridge, MA: Belknap Press, 1977).

17 Monkkonen, 1988, op. cit.

18 Alberta M. Sbragia, *Debt Wish: Entrepreneurial Cities, U.S. Federalism, and Economic Development*, Pitt Series in Policy and Institutional Studies, ed. Bert A. Rockman (Pittsburgh, PA: University of Pittsburgh Press, 1996).

19 Ibid.

20 Perry, "Building the City through the Back Door," op. cit.

21 Jacobson and Tarr, 1995, op. cit.

22 Ibid.

23 Marc A. Weiss, *The Rise of the Community Builders: American Real Estate Developers, Urban Planners and the Creation of Modern Residential Subdivisions* (Ithaca, NY: Columbia University Press, 1987).

24 Rolf Pendall, "Septic Systems, Rural Roads, and Exurbanization: Lessons about Infrastructure and Sprawl from Rhode Island," in *Rebuilding Nature's Metropolis: Growth and Sustainability in the 21st Century* (Chicago, IL: Association of Collegiate Schools of Planning, 1999).

25 John M. Levy, *Contemporary Urban Planning* (Upper Saddle River, NJ: Prentice Hall, 2000).

26 Peter Hall, *Cities of Tomorrow: An Intellectual History of Urban Planning and Design in the Twentieth Century* (Oxford: Blackwell, 1997).

27 Laurence Conway Gerckens, "The Comprehensive Plan in the 20th Century," American Planning Association, available at: www.asu/edu/caed/proceedings01/GERCKENS/gerckens.htm.

28 Ibid.

29 Ibid.

30 American Society of Civil Engineers, *History and Heritage of Civil Engineering*, op. cit.

31 Ibid.

32 Tarr, 1984, op. cit.

33 Ibid.

34 Jacobson and Tarr, 1995, op. cit.

35 "Top Ten Public Works Projects of the Century," American Public Works Association, available at: www.apwa.net/About/Awards/TopTenCentury (accessed July 12, 2001).

36 Jacobson and Tarr, 1995, op. cit.

37 Perry, "Building the City through the Back Door," op. cit.

38 Hall, 1997, op. cit.

39 Stephen Graham and Simon Marvin, *Splintering Urbanism: Networked Structures, Technological Mobilities and the Urban Condition* (New York: Routledge, 2001).

40 Ibid.

41 R. Capello and A. Gillespie, "Transport, Communication and Spatial Organization: Future Trends and Conceptual Frameworks," in *Transport and Communications in the New Europe*, ed. G. A. Giannopoulos and A. E. Gillespie (New York: Belhaven Press and Halsted Press, 1993).

42 Carl Muehlmann and Keith Mattrick, *Trends in Public Infrastructure Spending* (Washington, DC: The Congress of the United States, Congressional Budget Office, 1999).

43 G. Burtless, "Effects of Growing Wage Disparities and Changing Family Composition on the U.S. Income Distribution," CSED Working Paper (Washington, DC: The Brookings Institution, 1999).

44 Muehlmann and Mattrick, 1999, op. cit.

45 Dennis R. Judd and Susan S. Fainstein, *The Tourist City* (New Haven, CT: Yale University Press, 1999).

46 Gerckens, "The Comprehensive Plan in the 20th Century," op. cit.

47 T. J. Kent, *The Urban General Plan* (San Francisco: Chandler Publishing Company, 1964).

48 William H. Claire, *Urban Planning Guide*, ASCE Manuals and Reports on Engineering Practice (New York: American Society of Civil Engineers, 1969).

49 Norman Williams, "The Three Systems of Land Use Control," *Rutgers Law Review* 25, no.1 (1970): 80–101.

50 See, for instance, Clark Binkley et al., *Interceptor Sewers and Urban Sprawl* (Lexington, MA: Lexington Books, 1975).

51 Frank S. So et al., *The Practice of Local Government Planning* (Washington, DC: International City Management Association, 1979).

52 Charles J. Hoch, Linda C. Dalton, and Frank S. So, *The Practice of Local Government Planning*, 3rd ed. (Washington, DC: International City Management Association, 2000).

53 Professor Emeritus, Frederick Collignon, personal communication with the author, August, 1999.

54 Peter Choate and Walter S. Choate, *America in Ruins: The Decaying Infrastructure* (Durham, NC: Duke Press Paperbacks, 1981); National Council on Public Works Improvements, *Fragile Foundations: A Report on America's Public Works. Final Report to the President and Congress* (Washington, DC: National Council on Public Works Improvements, 1988); and U.S. Department of Transportation, *Financing the Future: Report of the Commission to Promote Investment in America's Infrastructure* (Washington, DC: U.S. Department of Transportation, 1993).

55 Stuart Meck, *Growing Smart Legislative Guidebook 3: Model Statutes for Planning and the Management of Change* (Chicago, IL: APA Planners Press, 2002).

56 See, for example, William Fulton et al., *Who Sprawls Most? How Growth Patterns Differ across the U.S.* (Washington, DC: Brookings Institution Press, 2001); Philip R. Berke and David R. Godschalk, *Urban Land Use Planning*, 5th ed. (Urbana, IL: University of Illinois Press, 2006); Eric Damian Kelly, *Planning, Growth, and Future Facilities: A Primer for Local Officials*, PAS Report No. 447 (Chicago, IL: American Planning Association, 1993); and Eric Damian Kelly, *Community Planning: An Introduction to the Comprehensive Plan*, 2nd ed. (Washington, DC: Island Press, 2010).

57 See Barry B. Lepatner, *Too Big to Fall* (New York: Foster Publishing, 2010); and Felix G. Rohatyn, *Bold Endeavors: How Government Built America and Why It Must Rebuild Now* (New York: Simon & Schuster, 2009).

chapter 3

1 "2008 National Population Projections," U.S. Census Bureau, available at: www.census.gov/population/www/projections/2008projections.html (accessed August, 2008), Table 3, Statistical Abstract of the United States, 2012. See also Jennifer M. Ortman and Christine E. Guarneri, "United States Population Projections: 2000–2050," U.S. Census Bureau, available at: www.census.gov/population/www/projections/analytical-document09.pdf, which models different net international migration assumptions.

2 The following is based on John Pitkin and Dowell Meyers, "U.S. Housing Trends: Generational Trends and the Outlook to 2050," in *Special Report 298: Driving and the Built Environment: The Effects of Compact Development on Motorized Travel, Energy Use, and CO_2 Emissions* (2008), available at: onlinepubs.trb.org/Onlinepubs/sr/sr298pitkin-myers.pdf; Martha Farnsworth Rich, "How Changes in the Nation's Age and Household Structure Will Reshape Housing Demand in the 21st Century," in *Issue Papers on Demographic Trends Important to Housing* (Washington, DC: U.S. Department of Housing and Urban Development, 2003); and Brookings Institution, "State of Metropolitan America," available at: www.brookings.edu/metro/StateOfMetroAmerica.aspx (accessed February 29, 2012).

3 Farnsworth Rich, 2003, op. cit., Table 1.

4 *Demographic Challenges and Opportunities for U.S. Housing Markets* (Washington, DC: Bipartisan Policy Center, 2012), available at: www.urban.org/UploadedPDF/412520-Demographic-Challenges-and-Opportunities-for-US-Housing-Markets.pdf.

5 Grayson K. Vincent and Victoria A. Velkoff, *The Next Four Decades: The Older Population in the United States: 2010 to 2050, Population Estimates and Projections* (Washington, DC: U.S. Department of Commerce, Economics, and Statistics Administration, U.S. Census Bureau, 2010).

6 John Pitkin and Dowell Myers, "U.S. Housing Trends: Generational Changes and the Outlook to 2050," report prepared for the Committee on the Relationships Among Development Patterns, Vehicle Miles Traveled, and Energy Consumption, Transportation Research Board (2008); and Dowell Myers and John Pitkin, "Demographic Forces and Turning Points in the American City, 1950–2040," *The Annals of the American Academy of Political and Social Science* 626 (November 2009): 91–111.

7 Arthur C. Nelson, "The New Urbanity: The Rise of a New America," *The Annals of the American Academy of Political and Social Science* 626, no. 1 (2009): 192–208.

8 Dowell Myers, John Pitkin, and Julie Park, "Estimation of Housing Needs Amid Population Growth and Change," *Housing Policy Debate* 13, no. 3 (2002): 567–596.

9 "Economic Policy Program, Housing Commission," Bipartisan Policy Center, available at: www.bipartisanpolicy.org/category/projects/housing-commission?page=1 (accessed February, 2012).

10 Ibid.

11 Alan Altshuler and Jose A. Gomez-Ibanez, *Regulation for Revenue: The Political Economy of Land Use Exactions* (Washington, DC: Brookings Institution Press, 1993).

12 Ibid.

13 Lawrence Mishel, Jared Bernstein, and John Schmitt, *The State of Working America, 2000–2001* (Ithaca, NY: Cornell University Press, 2001).

14 Economic Policy Institute, "The State of Working America," available at: stateofworkingamerica.org/inequality/poverty/ (accessed February 20, 2012).

15 Robert Cervero, "Jobs–Housing Balance Revisited: Trends and Impacts in the San Francisco Bay Area," *Journal of the American Planning Association* 62, no. 4 (1996): 492–511; and Jerry Weitz and Tom Crawford, "Where the Jobs Are Going: Job Sprawl in U.S. Metropolitan Regions, 2001–2006," *Journal of the American Planning Association* 68, no. 1 (2012): 53–69.

16 Robert W. Burchell et al., *Costs of Sprawl—2000*, Transit Cooperative Research Program Report no. 74 (Washington, DC: National Academy Press, 2002).

17 William H. Frey, "Population Growth in Metro America Since 1980: Putting the Volatile 2000s in Perspective," Brookings Institution, available at: www.brookings.edu/papers/2012/0320_population_frey.aspx.

18 "State of Metropolitan America," Brookings Institution, available at: www.brookings.edu/metro/stateofmetroamerica.aspx (accessed March 24, 2012).

19 Ibid.

20 James Heintz, Robert Pollin, and Heidi Garrett-Peltier, *How Infrastructure Investments Support the U.S. Economy: Employment, Productivity, and Growth* (Amherst, MA: Political Economy Research Institute, University of Massachusetts/Amherst, 2009), available at: www.americanmanufacturing.net/files/peri_aam_finaljan16_new.pdf (accessed March 24, 2012).

21 For a review of economic studies, see Alicia H. Munnell, "How Does Public Infrastructure Affect Regional Economic Performance?," *New England Economic Review* Sep./Oct. (1990): 11–33.

22 David Alan Aschauer, "Is Public Expenditure Productive?" *Journal of Monetary Economics* 23 (1989): 177–200.

23 Congressional Budget Office, *The Economic Effects of Federal Spending on Infrastructure and Other Investments* (Washington, DC: Congressional Budget Office, 1998), available at: www.cbo.gov/sites/default/files/cbofiles/ftpdocs/6xx/doc601/fedspend.pdf (accessed February 20, 2002).

24 U.S. Treasury Department and Council of Economic Advisors, "An Economic Analysis of Infrastructure Investment," available at: www.treasury.gov/resource-center/economic-policy/Documents/infrastructure_investment_report.pdf.

25 "Statement of Douglas W. Elmendorf, Director, Congressional Budget Office, Policies for Increasing Economic Growth and Employment in 2012 and 2013 before the Committee on the Budget, United States Senate" (Washington, DC: Congressional Budget Office, November 15, 2011).

26 Alfred Marshall, *Principles of Economics*, 8th ed. (New York: Macmillan, 1986).

27 William Alonso, *Regional Policy* (Cambridge, MA: MIT Press, 1975).

28 Jay M. Stein, "The Economic Context of the Infrastructure Crisis," in *Infrastructure Planning and Management*, ed. Jay M. Stein (Thousand Oaks, CA: Sage, 1988).

29 Bennett Harrison, "Industrial Districts: Old Wine in New Bottles?" *Regional Studies* 26, no. 5 (1992): 469–483.

30 Thomas M. Power, "The Economic Base: Distracting Vision, Distorting Reality," in *Environmental Protection and Economic Well-Being*, 2nd ed. (Armonk, NY: M.E. Sharpe, 1996).

31 Michael E. Porter, *Clusters and Competition: New Agendas for Companies, Governments, and Institutions* (Boston, MA: Division of Research, Harvard Business School, 1998).

32 Susan E. Clarke and Gary L. Gaile, *The Work of Cities, Globalization and Community* (Minneapolis, MN: University of Minnesota Press, 1998).

33 Samuel Nunn, "Built Investments and Planning Environments in the New Economy: Exploring Regional Differences, 1990–2000,"

in *Proceedings of the Association of Collegiate Schools of Planning 43rd Annual Conference*, Cleveland, OH, 2001.

34 Mitch Moody and Arthur C. Nelson, "Are Development Fees Bad for Local Economic Development?" in *Proceedings of the Association of Collegiate Schools of Planning 43rd Annual Conference*, Cleveland, OH, 2001.

35 Saurav Dev Bhatta and Matthew P. Drennan, "The Economic Benefits of Public Investment in Transportation," *Journal of Planning Education and Research* 22, no. 3 (2003): 288–296.

36 Norman Krumholz and John Forester, *Making Equity Planning Work: Leadership in the Public Sector, Conflicts in Urban and Regional Development* (Philadelphia, PA: Temple University Press, 1990).

37 Samuel Nunn, Drew Klacik, and Carl Schoedel, "Strategic Planning Behavior and Competition for Airport Development," *American Planning Association Journal* 62, no. 4 (1996): 427–441.

38 Eric Damian Kelly, *Planning, Growth, and Future Facilities: A Primer for Local Officials*, PAS Reports no. 447 (Chicago, IL: American Planning Association, 1993).

39 David Neumark, "California's Economic Future and Infrastructure Challenges," in *California 2025: Taking on the Future*, ed. Ellen Hanak and Mark Baldassare (San Francisco: Public Policy Institute of California, 2005).

chapter 4

1 Congressional Budget Office, *Public Spending on Transportation and Water Infrastructure* (Washington, DC: Congressional Budget Office, 2010), available at: www.cbo.gov/sites/default/files/cbofiles/ftpdocs/119xx/doc11940/11-17-infrastructure.pdf (accessed March 25, 2012).

2 U.S. General Accounting Office, *Leading Practices in Capital Decision Making*, GAO/AIMD-99-32 (Washington, DC: U.S. General Accounting Office, 1998).

3 Congressional Budget Office, 2010, op. cit.

4 U.S. Census Bureau, *Federal, State, and Local Governments: Government Finance and Employment Classification Manual* (Washington, DC: U.S. Census Bureau, 1992), available at: www.census.gov/govs/www/class.html.

5 U.S. Census Bureau, "Government Organization," in *Census of Governments*, Vol. 1, No. 1 (Washington, DC: U.S. Census Bureau, 2002).

6 Ibid.

7 U.S. Census Bureau, *2007 Census of Governments* (Washington, DC: U.S. Census Bureau, 2007).

8 Sometimes there is a separate citizen board to review the permits for individual land use decisions. This may be known as the Board of Adjustments, the Zoning Board, or the Development Review Board. There are many organizational configurations and titles for these functions. If there is only one commission or board, it handles policy issues as well as individual land use decisions.

9 David Perry and A. J. Watkins, *The Rise of the Sunbelt Cities* (Thousand Oaks, CA: Sage, 1977).

10 Not including housing authorities.

11 U.S. Census Bureau, *2007 Census of Governments*, op. cit.

12 Kathryn A. Foster, *The Political Economy of Special-Purpose Government* (Washington, DC: Georgetown University Press, 1997).

13 Ibid.

14 Annmarie Hauck Walsh, *The Public's Business: The Politics and Practices of Government Corporations* (Cambridge, MA: MIT Press, 1978).

15 Jerry Mitchell, "Policy Functions and Issues for Public Authorities," in *Public Authorities and Public Policy: The Business of Government*, ed. Jerry Mitchell (New York: Greenwood Press, 1992).

16 Jerry Mitchell, *The American Experiment with Government Corporations* (Armonk, NY: M.E. Sharpe, 1999).

17 Foster, 1997, op. cit.

18 H. V. Savitch and Ronald K. Vogel, "Regional Patterns in a Post City Age," in *Regional Politics: America in a Post-City Age, Urban Affairs Annual Review*, ed. H. V. Savitch and Ronald K. Vogel (Thousand Oaks, CA: Sage Publications, 1996).

19 John M. Levy, *Contemporary Urban Planning* (Upper Saddle River, NJ: Prentice Hall, 2000).

20 Charles Sabel, Archon Fung, Bradley Karkkainen, Joshua Cohen, and Joel Rogers, *Beyond Backyard Environmentalism (New Democracy Forum)* (Boston, MA: Beacon Press, 2000).

21 Stephen Graham and Simon Marvin, *Splintering Urbanism: Networked Structures, Technological Mobilities and the Urban Condition* (London: Routledge, 2001).

22 Richard T. LeGates, *The Region Is the Frontier*, California State University, Sacramento, 2001, available at: www.csus.edu/calst/government_Affairs/Region_is_the_Frontier.html.

23 Ibid.

24 California's SB 375, adopted in 2008, requires largely voluntary regional planning agencies to prepare regional plans to achieve state-mandated carbon emission reductions. Additionally, the environmental review process in that state requires consideration of carbon emissions. See Elisa Barbour and Elizabeth A. Deakin, "Smart Growth Planning for Climate Protection," *Journal of the American Planning Association* 78, no. 1 (Winter 2012): 70–86.

25 Ethan Selzer and Armando Carbonell, eds., *Regional Planning in America: Practice and Prospect* (Cambridge, MA: Lincoln Institute of Land Policy, 2011).

26 Petra Todorovich, "An Infrastructure and Economic Recovery Plan for the United States," *Land Lines* 21, no. 1 (January 2009). Available at: www.lincolninst.edu/pubs/1550-An-infrastructure-and-Economic-Recovery-Plan-for-the United-States.

chapter 5

1 See Stephen M. Wheeler, *Planning for Sustainability* (New York: Routledge, 2004), for details of these earlier reports.

2 Jonathan D. Weiss, "Local Governance and Sustainability: Major Progress, Significant Challenges," in *Agenda for a Sustainable America*, ed. John C. Dernbach (Washington, DC: The Environmental Law Institute Press, 2008).

3 Wheeler, 2004, op. cit. p. 179.

4 Ibid.

5 Patricia Salkin, "Land Use: Blending Smart Growth with Social Equity and Climate Change Mitigation," in *Agenda for a Sustainable America*, ed. John C. Dernbach (Washington, DC: The Environmental Law Institute Press, 2008).

6 T. J. Kent, Jr., *The Urban General Plan* (San Francisco: Chandler Publishing Company, 1964).

7 Jerry Weitz, *Sprawl Busting: State Programs to Guide Growth* (Chicago, IL: APA Planners Press, 1999).

8 Elisa Barbour, *Metropolitan Growth Planning in California, 1900–2000* (San Francisco: Public Policy Institute of California, 2002).

9 American Planning Association, *Planning for Smart Growth: 2002 State of States* (Chicago, IL: American Planning Association, 2002).

10 Paul Lewis, *California's Housing Element Law: The Issue of Local Noncompliance* (San Francisco: Public Policy Institute of California, 2003), available at: www.ppic.org/main/publication.asp?i=350.

11 Stuart Meck, *Growing Smart Legislative Guidebook 3: Model Statutes for Planning and the Management of Change* (Chicago, IL: APA Planners Press, 2002).

12 Stuart Meck, "Executive Summary: Status of State Planning Reform," in *Planning Communities in the 21st Century* (Chicago, IL: American Planning Association, 2002).

13 Ibid.

14 U.S. General Accounting Office, *Local Growth Issues* (Washington, DC: U.S. General Accounting Office, 2000).

15 Meck, *Growing Smart Legislative Guidebook 3*, 2002, op. cit.

16 David E. Dowall and Jan Whittington, *Making Room for the Future: Rebuilding California's Infrastructure* (San Francisco: Public Policy Institute of California, 2003); and Michael Neuman and Jan Whittington, *How Does California Make Its Infrastructure Decisions?* (San Francisco: Public Policy Institute of California, 2000).

17 Jerry Weitz, *Growing Smart User Manual* (Chicago, IL: APA Planners Press, 2002).

18 Charles Sabel, Archon Fung, Bradley Karkkainen, Joshua Cohen, and Joel Rogers, *Beyond Backyard Environmentalism (New Democracy Forum)* (Boston, MA: Beacon Press, 2000).

19 Meck, *Growing Smart Legislative Guidebook*, 2002, op. cit.

20 See www.cnu.org.

21 American Planning Association, *Sustaining Places: The Role of the Comprehensive Plan*, Sustaining Places Task Force Report (Chicago, IL: APA, 2011).

22 The Government Performance Project, *County Grade Report 2002* (New York: Syracuse University Campbell Public

Affairs Institute, 2002), available at: www.maxwell.syr.edu/gpp/grade/county_2002/index.asp?id=1.

23 Ibid.

24 Ibid.

25 U.S. Congress General Accounting Office, *Leading Practices in Capital Decision Making* (Washington, DC: U.S. Congress, General Accounting Office, 1998).

26 Kelly Hoell, *Good Company* (2012), available at: www.goodcompany.com.

27 City of New York, *PlaNYC* (2007), available at: www.newyork.gov.

chapter 6

1 U.S. General Accounting Office, *Environmental Protection: Federal Incentives Could Help Promote Land Use that Protects Air and Water Quality*, Report GAO-02-12 (Washington, DC: U.S. General Accounting Office, 2001).

2 Clark Binkley et al., *Interceptor Sewers and Urban Sprawl* (Lexington, MA: Lexington Books, 1975); Urban Systems Research & Engineering, *The Growth Shapers: The Land Use Impacts of Infrastructure Investment* (Washington, DC: Council on Environmental Quality, 1976); and Richard D. Tabors, Michael H. Shapiro, and Peter P. Rogers, *Land Use and the Pipe: Planning for Sewerage* (Lexington, MA: Lexington Books, 1976).

3 Philip K. Berke, and David R. Godschalk, *Urban Land Use Planning*, 5th ed. (Urbana, IL: University of Illinois Press, 2010); and Eric Damian Kelly and Barbara Becker, *Community Planning: An Introduction to the Comprehensive Plan* (Washington, DC: Island Press, 2000).

4 Richard D. Tabors, "Utility Services," in *The Practice of Local Government Planning*, 1st ed., ed. Frank S. So, et al. (Washington, DC: International City Management Association, 1979).

5 The following is based on Dowell Myers, John Pitkin, and Julie Park, "Estimation of Housing Needs Amid Population Growth and Change," *Housing Policy Debate* 13, no. 3 (2002): 567–96.

6 Binkley et al., 1975, op. cit.; Urban Systems Research & Engineering, 1976, op. cit.; and Tabors, Shapiro, and Rogers, 1976, op. cit.

7 John R. Ottensmann, "LUCI: Land Use in Central Indiana Model and the Relationships of Public Infrastructure to Urban Development," *Public Works Management and Policy* 8, no. 1 (2003); 62–76; see also John R. Ottensmann and Don Reitz, "luci2, Scenarios, and the Hendricks County USA Comprehensive Plan," in *The Future of Cities and Regions* (New York: Springer, 2012), 125–146.

8 U.S. General Accounting Office, 2001, op. cit.

9 S. H. Putman and S-L. Chang, "The Metropolis Planning Support System: Urban Models and GIS," in *Planning Support Systems: Integrating Geographic Information Systems, Models, and Visualization Tools*, ed. Richard K. Brail and Richard Klosterman (Redlands, CA: ESRI Press, 2001); and Paul Waddell, "Between Politics and Planning: Urbanism as a Decision Support System for Metropolitan Planning," in *Planning Support Systems: Integrating Geographic Information Systems, Models, and Visualization Tools*, ed. Richard K. Brail and Richard Klosterman (Redlands, CA: ESRI Press, 2001).

chapter 7

1 John Delafons, *Land-Use Controls in the United States*, 2nd ed. (Cambridge, MA: MIT Press, 1969).

2 Frank S. So, "Public Works Planning," in *Management of Local Public Works*, ed. Sam M. Cristofano and William S. Foster (Washington, DC: International City Management Association, 1986).

3 Eric Damian Kelly and Barbara Becker, *Community Planning: An Introduction to the Comprehensive Plan* (Washington, DC: Island Press, 2000).

4 So, 1986, op. cit.

5 Delafons, 1969, op. cit.

6 Kelly and Becker, 2000, op. cit.

7 Stuart Meck, *Growing Smart Legislative Handbook: Model Statutes for Planning and the Management of Change* (Chicago, IL: American Planning Association, 2002).

8 Ibid.

9 Delafons, 1969, op. cit.

10 Arthur C. Nelson and James B. Duncan, *Growth Management Principles and Practices* (Chicago, IL: American Planning Association, 1995).

11 Ibid.

12 Ruth Steiner, "Florida's Transportation Concurrency Planning," *University of Florida Journal of Law and Public Policy* 12, no. 2 (2001): 269–297.

13 Meck, 2002, op. cit.

14 Steve Dotterrer, personal communication with the author, April, 19, 2005.

15 Douglas R. Porter, *Performance Standards for Growth Management*, PAS Reports no. 461 (Chicago, IL: American Planning Association, 1996).

16 Kelly and Becker, 2000, op. cit.

17 Meck, 2002, op. cit.

18 Delafons, 1969, op. cit.

19 Meck, 2002, op. cit.

20 Chris Duerksen, Rocky Mountain Land Use Institute, available at: www.rmlui.law.du.edu (accessed June, 2009).

21 Christopher Wood, *Environmental Impact Assessment: A Comparative Review* (Harlow: Longman Scientific & Technical, 1995).

22 Kelly and Becker, 2000, op. cit.

23 Mires Rosenthal, "Code Administration," in *Management of Local Public Works*, ed. Sam M. Cristofano and William S. Foster, Municipal Management Series (Washington, DC: International City Management Association, 1986).

24 Stuart Meck, *Subdivision Control: A Primer for Planning Commissioners* (Chicago, IL: APA Planners Press, 1996).

25 Ibid.

26 Ibid.

27 Ibid.

chapter 8

1 Those interested in the development of on-site infrastructure as part of a private development are referred to the capable Donna Hanousek, *Project Infrastructure Development Handbook* (Washington, DC: Urban Land Institute, 1989).

2 Mike E. Miles, Bayle Berens, Marc J. Eppli, and Marc A. Weiss, *Real Estate Development* (Washington, DC: Urban Land Institute, 2007).

3 Richard G. Rypinski, *Eminent Domain: A Step-by-Step Guide to the Acquisition of Real Property* (Point Arena, CA: Solano Press, 2002).

4 Flyuberg, Bert, Nils Bruzelius, and Werner Rothengatter. *Megaprojects and Risk: Anatomy of Ambition* Cambridge, England: Cambridge University Press, 2003.

5 U.S. Department of Transportation, Federal Highway Administration, *Improving Transportation Investment Decisions through Life-Cycle Cost Analysis* (2003), available at: www.fhwa.dot.gov/infrastructure/asstmgmt/lccafact.htm.

6 U.S. General Accounting Office. Role of Facility Design Reviews in Facilities Construction. Washington, DC: U.S. General Accounting Office, 2000.

7 Ortolano, Leonard. *Environmental Regulations and Impact Assessment*. New York: Wiley and Sons, 1997.

8 Rypinski, op. cit.

9 Ibid.

10 Ibid.

11 Pecca, S. P. *Real Estate Development and Investment*. Hoboken, New Jersey: Wiley and Sons, 2009.

12 Knutson, Kraig. Clifford J. Schexnayder, Christine Fiori and Richard Mayo. *Construction Management Fundamentals*. New York: McGraw Hill, 2004.

chapter 9

1 Lawrence W. Hush and Kathleen Peroff, "The Variety of State Capital Budgets: A Survey," *Public Budgeting and Finance* 8, no. 2 (1988): 71.

2 John Forrester, "Municipal Capital Budgeting: An Examination," *Public Budgeting and Finance* 12, no. 2 (1993): 85–103.

3 Robert Kee, Walter Robbins, and Nicholas Apostolou, "Capital Budgeting Practices of U.S. Cities: A Survey," *The Government Accountants Journal* (1991): 16–22.

4 Kurt Svendsen, personal communication with the author, December 6, 2003.

5 Eric Damian Kelly, *Planning, Growth, and Future Facilities: A Primer for Local Officials*, PAS Reports, no. 447 (Chicago, IL: American Planning Association, 1993).

6 Ibid.

7 Svendsen, 2003, op. cit.

8 Anne Arundel County, *Charter of Anne Arundel County*, Maryland (2004).

9 John A. Vogt, *Capital Budgeting and Finance: A Guide for Local Governments* (Washington, DC: International City/County Management Association, 2009).

10 This material was adapted from Fairfax County, VA, *Capital Improvement Program 2006–2010* (2003), available at: www.co.fairfax.va.us.

11 Susan Robinson, "Capital Planning and Budgeting," in *Local Government Finance*, ed. John E. Petersen and Dennis R. Strachota (Chicago, IL: Government Finance Officers Association, 1991).

12 Ibid.

13 Government Finance Officers Association, "Planning and Budgeting for Capital Improvements," in *Budgeting: A Guide for Local Governments* (Chicago, IL: Government Finance Officers Association, 1993), 170–196.

14 GASB was formed in 1984 under the auspices of a non-profit organization. Although it is not a government agency, its rules are the standards against which audits are conducted. Failure to comply with GASB's standards can result in audit findings that affect the ability of the agency to issue bonds.

15 Massachusetts Department of Revenue, *A Practical Guide for Implementation of Government Accounting Standards Board Statement # 34 for Massachusetts Local Governments* (2001).

16 GASB 34 implementation deadlines vary according to the jurisdiction's annual revenues for the first fiscal year after June, 1999. Those with annual revenues of more than $100 million must report in the budget year after June, 2001; those with annual revenues of more than $10 million after June, 2002, and those with revenues under $10 million in the target year must begin reporting after June, 2003.

17 "GASB's Reporting Model Project," Rutgers University (2001), available at: accounting.rutgers.edu/raw/gasb/repmodel/background.html (accessed June 12, 2001).

18 Robert A. Bowyer, *Capital Improvements Programs: Linking Budgeting and Planning* (Chicago, IL: American Planning Association, 1993).

19 Michael Brown, personal communication with author, December 6, 2003.

20 Ibid.

21 Bowyer, 1993, op. cit.

22 Government Finance Officers Association, 1993, op. cit.

chapter 10

1 U.S. Congressional Budget Office, *Public Spending on Transportation and Water Infrastructure* (Washington, DC: W.S. Congressional Budget Office, 2009), available at: www.cbo.gov.

2 The Government Performance Project, *City Grade Report 2000*, Syracuse University Campbell Public Affairs Institute (2000), available at: www.maxwell.syr.edu/gpp/grade/city_2000/press_national.asp.

3 David Perry, "Building the City through the Back Door: The Politics of Debt, Law, and Public Infrastructure," in *Building the Public City: The Politics, Governance and Finance of Public Infrastructure*, ed. David Perry (Thousand Oaks, CA: Sage Publications, 1995), 203–236.

4 The Government Performance Project, 2000, op. cit.

5 Ralph Gakenheimer, "Infrastructure Shortfall: The Institutional Problems," *American Planning Association Journal* Winter (1989): 14–23.

6 Ibid.

7 Joel A. Tarr, "The Evolution of the Urban Infrastructure in the Nineteenth and Twentieth Centuries," in *Perspectives on Urban Infrastructure*, ed. R. Hansen (Washington, DC: National Academies Press, 1984), 4–66.

8 U.S. General Accounting Office, 2004, op. cit.

9 "Asset Management Guidance for Transportation Agencies," Cambridge Systematics Inc. (2003).

10 Ronald W. Hudson, Ralph Haas, and Waheed Uddin, *Infrastructure Management* (New York: McGraw-Hill, 1997).

11 U.S. Environmental Protection Agency, *Fact Sheet: Asset Management for Sewer Collection Systems* (Washington, DC: U.S. Environmental Protection Agency, 2004).

12 U.S. Environmental Protection Agency, 2004, op. cit.

13 H. E. Reid, "Beyond Speed: The Next Generation of Transportation Performance Measures," in *Performance Standards for Growth Management*, PAS Report no. 461, ed. Douglas R. Porter (Chicago, IL: APA Planning Press, 1996).

14 This draws from U.S. Department of Transportation. *Asset Management Primer* (Washington, DC: U.S. Department of Transportation, 1999).

15 David N. Ammons, Erin S. Norfleet, and Brian T. Coble, *Performance Measures and Benchmarks in Local Government Facilities Maintenance* (Washington, DC: International City/County Management Association: ICMA Center for Performance Measurement, 2002).

16 Ibid.

17 J. M. Kelly and W. C. Rivenbark, *Performance Budgeting for State and Local Government* (New York: M. E. Sharpe, 2003).

18 International City/County Managers Association, "Fact Sheet on the ICMA Center," ICMA Center for Performance Measurement (2003), www2.icma.org.

19 Kelly and Rivenbark, 2003, op. cit.

chapter 11

1 U.S. Congressional Budget Office, *Congressional Budget Office Issues and Options in Infrastructure Investment* (Washington, DC: U.S. Congressional Budget Office, May 2008), available at: www.cbo.gov/publications/19633/05-16-infrastructure.pdf.

2 U.S. Congressional Budget Office, *Public Spending on Transportation and Water Infrastructure* (Washington, DC: U.S. Congressional Budget Office, 2010), available at: www.cbo.gov/publication/25116 (accessed March 28, 2012).

3 U.S. Congressional Budget Office, *Congressional Budget Office Issues and Options in Infrastructure Investment*, 2008, op. cit.

4 U.S. Congressional Budget Office, *Trends in Public Spending on Transportation and Water Infrastructure, 1956 to 2004* (Washington, DC: U.S. Congressional Budget Office, 2007).

5 U.S. Department of Commerce, Bureau of Economic Analysis.

6 U.S. Census Bureau, *State and Local Government Finances Summary: 2009* (Washington, DC: U.S. Census Bureau, 2011).

7 Dick Netzer, "Property Taxes: Past, Present and Future," in *Urban Finance under Siege*, ed. Thomas R. Swartz and Frank J. Bonello (Armonk, NY: M.E. Sharpe, Inc., 1993).

8 Michael E. Bell and John H. Bowman, "Property Taxes," in *Local Government Finance: Concepts and Practices*, ed. John E. Petersen and Dennis R. Strachota (Chicago, IL: Government Finance Officers Association, 1991).

9 Netzer, 1993, op. cit.

10 Ibid.

11 Jack Huddleston, *The Property Tax and Planning* (Cambridge, MA: Lincoln Institute of Land Policy, 2005).

12 Mary Edwards, *State and Local Resources Beyond the Property Tax* (Cambridge, MA: Lincoln Institute of Land Policy, 2006).

13 Robert L. Bland and Irene S. Rubin, *Budgeting: A Guide for Local Governments*, Municipal Management Series (Washington, DC: International City/County Management Association, 1997).

14 Gregory K. Ingram, "Infrastructure: Spending More and Spending Well," *Land Lines* 21, no. 1 (2009). Available at: http://www.lincolninst.edu/pubs/1549_Report-From-The-President.

15 Vicki Elmer, "Fragile Foundations Redux: Assessing the Cost and Impact of Future Infrastructure Costs," paper delivered at the American Collegiate Scholars Association (October 2005).

16 The 10 percent benchmark is only for taxes, not fees and charges.

17 In fact rating agencies use this figure as a benchmark only for debt paid by taxes. The U.S. figure includes debt financed by revenues as well, such as water fees, parking fees, and so on.

chapter 12

* An earlier version of this chapter appeared in White, S.B. and Z. Kutual (eds). *Financing Economic Development in the 21st Century* (3rd edition). New York, NY: M.E. Sharpe, 2012.

1 Joe Mysak, *Handbook for Muni-Bond Issuers* (Princeton, NJ: Bloomberg Press, 1998).

2 "Bond Basics," GMS Group, www.gmsgroup.com/?q=Municipal-Bonds-Basics-0.

3 P. S. Maco, "A Country Built by Bonds: A Short History of the U.S. Municipal Bond Market and the Building of America's Infrastructure," PowerPoint presentation to China Environment Forum, Environmental Financing in China Initiative, Vinson & Elkins LLP, Washington, DC (2010), available at: www.wilsoncenter.org/news/docs/MACO.PPT.

4 "Bond Basics," op. cit.

5 J. W. Temel for The Bond Market Association, *The Fundamentals of Municipal Bonds* (New York, NY: Wiley and Sons, 2001).

6 T. Agriss, *Municipal Bond Market Issues: Recent Developments* (Overland Park, KS: Black & Veatch Corporation, 2008), available at: www.bv.com (accessed August 2010).

7 U.S. Census Bureau, *Statistical Abstract of the United States*, Table 502 (Washington, DC: U.S. Census Bureau, 2002).

8 U.S. Census Bureau, *Statistical Abstract of the United States 2012*, Table 440 (Washington, DC: U.S. Census Bureau, 2012).

9 The following is drawn from Temel, 2001, op. cit.; John E. Petersen and Thomas McLoughlin, "Debt Policies and Procedures," in *Local Government Finance: Concept and Practices*, ed. John E. Petersen and Dennis R. Strachota (Chicago: Government Finance Officers Association, 1991); and John A. Vogt, *Capital Budgeting and Finance: A Guide for Local Governments* (Washington, DC: International City/County Management Association, 2009).

10 S. Provus, *The Basics of Industrial Development Bonds*, CDFA Spotlight (2006), available at: www.cdfa.net (accessed on August 27, 2010).

11 Ibid.

12 Vogt, 2009, op. cit.

13 J. L. Leithe and J. Joseph, "Financing Alternatives," in *Financing Growth*, ed. Susan Robinson (Chicago, IL: Government Finance Officers Association, 1990), 91–107.

14 Vogt, 2009, op. cit.

15 This section is based on Temel, 2001, op. cit.; Vogt, 2009, op. cit.; and Mysak, 1998, op. cit.

16 D. Litvak, *Special Report: Default Risk and Recovery Rates on U.S. Municipal Bonds*, Fitch Ratings (2007), available at: www.fitchratings.com (accessed August, 2010).

17 Ibid.

18 Public Bonds (2011), "Municipal Bonds and Defaults," available at: www.public bonds.ord/public_fin/default (accessed April, 2011).

19 Mysak, 1998, op. cit.

20 Temel, 2001, op. cit.

21 Ibid.

22 Vogt, 2009, op. cit.

23 Ibid.

24 Ibid.

25 "Municipal Bond Credit Report Q1 2010," 2010, op. cit.

26 Temel, 2001, op. cit.

27 *Bond Buyer*, 2010, available at: www.bondbuyer.

28 Mysak, 1998, op. cit.

29 Vogt, 2009, op. cit.

30 Temel, 2001, op. cit.

chapter 13

1 Kurt. C. Zorn, "User Charges and Fees," in *Local Government Finance*, ed. John E. Petersen and Dennis R. Strachota (Chicago, IL: Government Finance Officers Association, 1991).

2 Margaret Sohagi, "Defining the Terms," in *Exactions and Impact Fees in California*, ed. William W. Abbott, Peter M. Detweiler, Charles D. Jacobson, Margaret Sohagi, et al. (Point Arena, CA: Solano Press, 2001).

3 Ibid.

4 Zorn, 1991, op. cit.

5 U.S. Census Bureau, *Statistical Abstract of the United States: 1998* (Washington, DC: Government Printing Office, 1998).

6 Frederick Collignon, personal communication with the author, December 15, 2003.

7 Stephen J. Bailey, "User-Charges for Urban Services," *Urban Studies* 31, no. 4/5 (1994): 745–765.

8 Ibid.

9 Ibid.

10 Janice A. Beecher, in Michigan State Utility Conference, 2003.

11 Frederick Collignon, op. cit.

12 Ibid.

13 Daniel Rubenfeld, "Public Pricing," in *Public Economics* (New York, NY: Macmillan, 1996), 300–315.

14 Zorn, 1991, op. cit.

15 W. Michael Hanemann, "Price and Rate Structures," in *Urban Water Demand Management and Planning*, ed. Duane D. Baumann, John J. Boland, and W. Michael Hanemann (New York, NY: McGraw-Hill, 1998), 137–179.

16 Ibid.

17 Collignon, op. cit.

18 Hanemann, 1998, op. cit.

19 Collignon, op. cit.

20 Ibid.

21 Hahnemann, 1998, op. cit.

22 Ibid.

23 Hanemann, 1998, op. cit.

24 Collignon, op. cit.

25 Ibid.

26 U.S. General Accounting Office, *Environmental Protection: Federal Incentives Could Help Promote Land Use that Protects Air and Water Quality* (Washington, DC: U.S. General Accounting Office, 2001).

27 Douglas R. Porter, Ben C. Lin, Susan Jakubiak, and Richard B. Peiser, *Special Districts: A Useful Technique for Financing Infrastructure* (Washington, DC: The Urban Land Institute, 1992).

chapter 14

1 California Office of Planning and Research, "A Planner's Guide to Financing Public Improvements" (1997), available at: ceres.ca.gov/planning/financing/.

2 Arthur C. Nelson, ed., *Development Impact Fees: Policy Rationale, Practice, Theory and Issue* (Chicago, IL: APA Planners Press, 1988).

3 Ibid.

4 Clancy Mullen, "State Impact Fee Enabling Acts," Duncan Associates (2005), available at: www.impactfees.com/publications%20pdf/summary%20of%20state%20acts.pdf.

5 Alan Altshuler and Jose A. Gomez-Ibanez, *Regulation for Revenue: The Political Economy of Land Use Exactions* (Washington, DC: Brookings Institution Press, 1993).

6 Neil S. Mayer and Bill Lambert, "Flexible Linkage in Berkeley: Development Mitigation Fees," in *Development Impact Fees: Policy Rationale, Practice, Theory and Issues*, ed. Arthur C. Nelson (Chicago, IL: APA Planners Press, 1988), 242–256.

7 Linda Hausrath, "Economic Basis for Linking Jobs and Housing in San Francisco," in *Development Impact Fees: Policy Rationale, Practice, Theory and Issues*, ed. Arthur C. Nelson (Chicago, IL: APA Planners Press, 1988).

8 Altshuler and Gomez-Ibanez, 1993, op. cit.

9 Robert W. Burchell, David Listokin, and William R. Dolphin, *Development Assessment Handbook* (Washington, DC: Urban Land Institute, 1994).

10 Margaret Sohagi, "Defining the Terms," in *Exactions and Impact Fees in California*, ed. William W. Abbott, Peter M. Detweiler, Charles D. Jacobson, Margaret Sohagi, et al. (Point Arena, CA: Solano Press, 2001).

11 Stuart Meck, *Growing Smart Legislative Guidebook 3: Model Statutes for Planning and the Management of Change* (Chicago, IL: APA Planners Press, 2002).

12 Sohagi, 2001, op. cit.

13 Brian W. Blaesser and Christene M. Kentopp, "Impact Fees: The Second Generation," *Washington University Journal of Urban and Contemporary Law* 55, no. 64 (Fall 1990): 68.

14 Mullen, 2005, op. cit.

15 Wes Clarke and Jennifer Evans, "Development Impact Fees and the Acquisition of Infrastructure," *Journal of Urban Affairs* 21, no. 3 (1999): 281–288.

16 Mullen, "State Impact Fee Enabling Acts," 2005, op. cit.

17 Ibid.

18 Clancy Mullen, "2007 National Impact Fee Survey," Duncan Associates (2007), www.impactfees.com/pdfs_all/2005%20impact%20fee% 20survey.pdf.

19 John Landis, Michael Larice, Deva Dawson, and Lan Deng, "Pay to Play: Residential Development Fees in California Cities and Counties, 1999," California Department of Housing and Community Development (2001), available at: www.hcd.ca.gov/hpd/pay2play/pay_to_play.html.

20 Marla Dresch and Steven M. Sheffrin, *Who Pays for Development Fees and Exactions?* (San Francisco: Public Policy Institute of California, 1997).

21 William W. Abbott, Peter M. Detweiler, M. Thomas Jacobson, Margaret Sohagi, and Harriet A. Steiner, *Exactions and Impact Fees in California: A Comprehensive Guide to Policy, Practice, and the Law* (Point Arena, CA: Solano Press, 2001).

22 Clarke and Evans, 1999, op. cit.

23 Gene Bunnell, "Analyzing the Fiscal Impacts of Development: Lessons for Building Successful Communities," *Journal of the Community Development Society* 29, no. 1 (1998), 38–57.

24 James E. Frank and Paul B. Downing, "Patterns of Impact Fee Use," in *Development Impact Fees: Policy Rationale,*

Practice, Theory and Issue*, ed. Arthur C. Nelson (Chicago, IL: APA Planners Press, 1988).

25 Clarke and Evans, 1999, op. cit.

26 Paul A. Jargowsky, "Sprawl, Concentration of Poverty, and Urban Inequality," in *Urban Sprawl: Causes, Consequences and Policy Responses*, ed. Gregory Squires (Washington, DC: The Urban Institute, 2002).

27 Douglass B. Lee, "Evaluation of Impact Fees against Public Finance Criteria," in *Development Impact Fees: Policy Rationale, Practice, Theory and Issues*, ed. Arthur C. Nelson (Chicago, IL: APA Planners Press, 1988).

28 Mullen, "State Impact Fee Enabling Acts," 2005, op. cit.

29 Ibid.

30 Charles D. Jacobson, *Exactions: Statutory Authority and Limitations, Exactions and Impact Fees in California* (Point Arena, CA: Solano Press, 2001).

31 M. Thomas. Jacobson, "Authority for and Limitations on Exactions—Constitutional Issues," in *Exactions and Impact Fees in California*, ed. William W. Abbot, Peter Detweiler, Charles D. Jacobson, Margaret Sohagi, et al. (Point Arena, CA: Solano Press, 2001).

32 The landowner and the City were in a hostile relationship throughout most of the planning process—one side wanting very intense development, and the city wishing to preserve the open space. The plan itself was challenged both by voters who wanted no development as well as the landowner who wanted more intense development. The plan sustained both challenges, although the landowner's lawsuit was dismissed by the U.S. Supreme Court on procedural grounds, thereby avoiding the rational nexus test that the City had prepared the plan to address.

33 Ibid.

34 Burchell, Listokin, and Dolphin, 1994, op. cit.

35 Frederick Collignon, personal communication with the author, December 15, 2003.

36 Mitch Moody and Arthur C. Nelson, "Are Development Fees Bad for Local Economic Development?" in *Annual Meeting of the Association of Collegiate Schools of Planning* (Cleveland, OH: 2001).

37 Jargowsky, 2002, op. cit.

38 Robert A. Blewitt and Arthur C. Nelson, "A Public Choice and Efficiency Argument for Development Impact Fees," in *Development Impact Fees: Policy Rationale, Practice, Theory and Issues*, ed. Arthur C. Nelson (Chicago, IL: APA Planners Press, 1988).

39 Gerrit Knapp, "If Sprawl Is the Problem, Impact Fees Are Not the Answer," in *Annual Meeting of the Association of Collegiate Schools of Planning* (Cleveland, OH: 2001).

40 Clarke and Evans, 1999, op. cit.

41 Burchell, Listokin, and Dolphin, 1994, op. cit.

42 Ibid.

43 James C. Nicholas, Arthur C. Nelson, and Julian C. Juergensmeyer, *A Practitioner's Guide to Development Impact Fees* (Chicago, IL: APA Planners Press, 1991).

44 Meck, 2002, op. cit.

chapter 15

1 Werner Z. Hirsch, "Contracting out by Urban Governments: A Review," *Urban Affairs Review* 30 (1995): 458–472.

2 Emanual S. Savas, *Privatization and Public-Private Partnership* (New York, NY: Chatham, 2000).

3 Daniel A. Yergin and Joseph Stanislaw, *The Commanding Heights: The Battle between Government and the Marketplace that Is Remaking the Modern World* (New York, NY: Simon and Schuster, 1998).

4 Peter Osborne and Peter Plastrick, *Banishing Bureaucracy* (New York, NY: Addison-Wesley, 1997).

5 David Perry, "Urban Tourism and the Privatizing Discourses of Public Infrastructure," in *The Infrastructure of Play: Building the Tourist City in North America and Europe*, ed. Dennis R. Judd (Armonk, NY: M.E. Sharpe, 2003), 23–70.

6 Willard Price, "Innovations in Public Finance: Implications for the Nation's Infrastructure," *Public Works Management and Policy* 7, no. 1 (2002): 63–78.

7 Savas, 2000, op. cit.

8 Price, 2002, op. cit.

9 Elliott D. Sclar, *You Don't Always Get What You Pay For: The Economics of Privatization* (Ithaca, NY: Cornell University Press, 2000).

10 Touche Ross & Co, "Privatization of Facilities," in *Capital Projects* (Washington, DC: International City/County Management Association, 1989), 185–190.

11 Ibid.

12 Amir Hefetz and Mildred Warner, "Privatization and Its Reverse: Explaining the Dynamics of the Government Contracting Process," *Journal of Public Administration, Research and Theory* 14, no. 2 (2004): 171–190.

13 Mildred Warner and Amir Hefetz, "Cooperative Competition: Alternative Service Delivery, 2002–2007" in *2009 Municipal Yearbook* (Washington, DC: International City and County Managers Association, 2009).

14 OECD, "The Involvement of Private Capital and Management," in *Urban Infrastructure: Finance and Management* (Paris: OECD, 1991), 65–81.

15 David E. Dowall, "Rethinking Statewide Infrastructure Policies: Lessons from California and Beyond," *Public Works Management and Policy* 6, no. 1 (2001): 5–17.

16 Ibid.

17 Michael Nadal, Paul Seidenstat, and Simon Hakim, *America's Water and Wastewater Industries: Competition and Privatization* (Vienna, VA: Public Utility Reports, 2000).

18 OECD, 1991, op. cit.

19 Nadal et al., 2000, op. cit.

20 U.S. General Accounting Office, *Privatization: Lessons Learned by State and Local Governments* (Washington, DC: U.S. General Accounting Office, 1997).

21 Price, 2002, op. cit.

22 Osborne and Plastrick, 1997, op. cit.

23 Price, 2002, op. cit.

24 Ibid.

25 Ibid.

26 Ibid.

chapter 16

1 "Roman Aqueducts," InfoRoma, available at: www.inforoma.it/aqueduct.htm (accessed February 10, 2002).

2 U.S. Geological Survey, *Estimated Water Use in the United States in 1995* (Washington, DC: U.S. Geological Survey, 1998).

3 Ibid.

4 U.S. Environmental Protection Agency, *Factoids: Drinking Water and Ground Water Statistics for 2000* (Washington, DC: U.S. Environmental Protection Agency, 2001).

5 Keith Christman, "The History of Chlorine," *WaterWorld* (September 1998).

6 U.S. Environmental Protection Agency, *The History of Drinking Water Treatment* (Washington, DC: U.S. Environmental Protection Agency, 2000), available at: www.epa.gov/ogwdw/consumer/pdf/hist.pdf.

7 Brys Sarte, *Sustainable Infrastructure: Design and Engineering Solutions* (Hoboken, NJ: Wiley and Sons, 2011).

8 G. Daigger, "Evolving Urban Water and Residuals Management Paradigms: Water Reclamation and Reuse, Decentralization and Resource Recovery," *Water Environment Research* 81, no. 8 (2009): 809–823; H. Dreiseitl, D. Grau, and K. H. C. Ludwig, eds., *Planning, Building and Designing with Water*, 3rd ed. (Basel, Switzerland: Birkhäuser Verlag AG, 2009); S. Hermanowicz, "Sustainability in Water Resources Management: Changes in Meaning and Perception," *Sustainability Science* 3, no. 2 (2008): 181–188.

9 Vladimir Novotny, Jack Ahern, and Paul Brown, *Water Centric Sustainable Communities* (Hoboken, NJ: Wiley and Sons, 2010); Cynthia Mitchell, Kumi Abeysuriya, and Dena Fam, *Development of Qualitative Decentralised System Concepts for the 2009 Metropolitan Sewerage Strategy for Melbourne Water* (Melbourne: Institute for Sustainable Futures, UTS, 2008).

10 Modis Land Surface Temperature Group, Institute for Computational Earth System Science, University of California, available at: www.icess.ucsb.edu/modis/modis-lst.html.

11 National Resources Defense Council (NRDC) *Water Efficiency Saves Energy: Reducing Global Warming Pollution through Water Use Strategies* (Los Angeles, CA: NRDC, 2009), available at: www.nrdc.org/water/files/energywater.pdf.

12 U.S. EPA, water.epa.gov/infrastructure/sustain/waterefficiency.cfm, (accessed June 21, 2012).

13 International Energy Agency (IEA), *Renewable Energy Essentials: Hydropower* (Paris: IEA, 2010), available at: www.iea.org/papers/2010/Hydropower_Essentials.pdf.

14 James E. McMahon and Sarah K. Price, "Water and Energy Interactions," *Annual Review of Environmental Resources*, 36 (2011):17.1–17.29.

15 American Water Works Association, "Dawn of the Replacement Era: Reinvesting in Drinking Water Infrastructure: Estimated Life Expectancy of Water Pipes by Era of Installation," in *Water Infrastructure: Information on Financing, Capital Planning, and Privatization* (Washington, DC: U.S. Congress, General Accounting Office).

16 Steve Albee, "Future Water Trends," in *Water Conference* (Washington, DC: U.S. Environmental Protection Agency, Office of Water, 2003).

17 King County, Washington, Department of Natural Resources and Parks, Wastewater Treatment Division, *Regional Wastewater Services Plan, 2006 Comprehensive Review and Annual Report*, September 2007. Available at: http://your.kingcounty.gov/dnrp/library/wastewater/wtd/construction/Planning/RWSP/CompReview/06/06CompReviewAR.pdf.

18 Sanjay Jeer et al., "Cross-Section of Ground Water Showing How a Reservoir Is Replenished from Water Tables," in *Nonpoint Source Pollution: A Handbook for Local Government*, PAS Report no. 476 (Chicago, IL: American Planning Association, 1997).

19 McMahon and Price, 2011, op. cit.

20 American Water Works Association, *Desalination (2002–2004)*, available at: www.awwa.org/Advocacy/pressroom/Desalination.cfm.

21 McMahon and Price, 2011, op. cit.

22 Larz T. Anderson, *Planning the Built Environment* (Chicago, IL: APA Planners Press, 2000), Figure 4.1.

23 Daniel A. Okun, "Dual Water Systems Can Save Drinking Water While Improving Its Quality," *Environmental Engineer* December (2005): 10.

24 Anderson, 2000, op. cit.

25 Leonard S. Hyman, *The Water Business: Understanding the Water Supply and Wastewater Industry* (Vienna, VA: Public Utilities Reports, 1998).

26 Ibid.

27 Ibid.

28 Karen Johnson and Jeff Loux, *Water and Land Use: Planning Wisely for California's Future* (Point Arena, CA: Solano Press, 2004).

29 Ibid.

30 Peter Shanaghan, "Remarks," in *The Second National Drinking Water Symposium* (National Regulatory Research Institute, 2003).

31 David Haarmeyer and Debra G. Coy, "An Overview of Private Sector Participation in the Global and U.S. Water and Wastewater Sector," in *Reinventing Water and Wastewater Systems: Global Lessons for Improving Water Management*, ed. David Haarmeyer and Debra G. Coy (New York, NY: Wiley & Sons, 2002).

32 N. L. Barber, "Summary of Estimated Water Use in the United States in 2005," U.S. Geological Survey Fact Sheet 2009–3098 (2009).

33 Gary Wolff and Peter H. Gleick, "The Soft Path for Water," in *The World's Water 2002–2003* (Oakland, CA: The Pacific Institute, 2004).

34 Ibid.

35 Johnson & Loux, 2004, op. cit.

36 American Water Works Association, *White Paper on Integrated Resource Planning in the Water Industry* (Denver, CO: American Water Works Association, 1993).

37 Janice A. Beecher, "Integrating Water Supply and Water Demand Management," in *Urban Water Demand Management and Planning*, ed. Duane D. Baumann, John J. Boland, and W. Michael Hanemann (New York, NY: McGraw-Hill, 1998).

38 American Water Works Association, *Water Resources Planning Manual* (Denver, CO: American Water Works Association, 2001); Vista Consulting Group, *Guidelines for Implementing an Effective Integrated Resource Planning Process* (Denver, CO: American Water Works Association Research Foundation, 1997); Beaudet, Bevin, Bill Bellamy, Mike Matichich, and John Rogers, *A Capital Planning Strategy Manual* (CD-Rom) (Denver, CO: American Water Works Association and American Water Works Association Research Foundation, 2001).

39 Andrew Dzurik, *Water Resources Planning*, 2nd ed. (Totowa, NJ: Rowman & Littlefield, 1996).

40 Ibid.

41 U.S. Environmental Protection Agency, *About Water and Wastewater Pricing*, available at: www.epa.gov/water/infrastructure/pricing/About.htm (accessed January 10, 2005).

42 American Water Works Association, "Residential Water Rate Structures," in *Water Stats* (Denver, CO: American Water Works Association, 1998).

43 Ibid.

44 U.S. Environmental Protection Agency, *Cases in Water Conservation: How Efficiency Programs Help Water Utilities Save Water and Avoid Costs* (2002), available at: www.epa.gov/owm/water-efficiency/utilityconservation.pdf.

45 North Carolina Department of Environment and Natural Resources, *Water Efficiency Manual for Commercial, Industrial and Institutional Facilities* (1998), available at: www.p2pays.org/ref/01/00692.pdf.

46 U.S. Environmental Protection Agency, *Water Conservation Planning Guidelines* (Washington, DC: U.S. Environmental Protection Agency, 1998).

47 Richard Anderson, "Water Infrastructure and Conservation Issues Discussed at Urban Water Summit," *U.S. Mayor Newspaper* (May 10, 2004).

48 William Maddaus and Michelle Maddaus. "Evaluating Water Conservation Cost-Effectiveness with an End Use Model," in *Proceedings of 2004 Water Sources Conference*, AWWA, Austin Texas, January 12–14, 2004.

49 Rutherford H. Platt, "The 2020 Water Supply Study for Metropolitan Boston: The Demise of Diversion," *Journal of the American Planning Association*, 61, no. 2, June (1995): 185–199.

50 U.S. Environmental Protection Agency, *Cases in Water Conservation*, 2002, op. cit.

chapter 17

1 Richard D. Tabors, Michael H. Shapiro, and Peter P. Rogers, *Land Use and the Pipe: Planning for Sewerage* (Lexington, MA: Lexington Books, 1976).

2 Joel A. Tarr, *The Search for the Ultimate Sink: Urban Pollution in Historical Perspective* (Akron, OH: University of Akron Press, 1996).

3 Martin V. Melosi, *The Sanitary City: Urban Infrastructure in America from Colonial Times to the Present* (Baltimore, MD: Johns Hopkins University Press, 2000).

4 Ibid.

5 Ibid.

6 Ibid.

7 Leonard S. Hyman, *The Water Business: Understanding the Water Supply and Wastewater Industry* (Vienna, VA: Public Utilities Reports, Inc., 1998).

8 Ibid.

9 U.S. Environmental Protection Agency, Office of Water, *Clean Water Act History* (Washington, DC: U.S. Environmental Protection Agency, 2002), available at: www.epa.gov/region5/water/cwa.htm (accessed February 26, 2002).

10 U.S. Environmental Protection Agency, *Overview of Current Total Maximum Daily Load – TMDL – Program and Regulations* (Washington, DC: U.S. Environmental Protection Agency, 2004).

11 Department of Urban & Regional Planning, University of Wisconsin-Madison/Extension and Wisconsin Department of Natural Resources, *Planning for Natural Resources: A Guide to Inclusive Natural Resources in Local Comprehensive Planning* (2002), available at: dnr.wi.gov/org/es/science/publications/SS_964_2002.pdf.

12 U.S. Environmental Protection Agency, *Wastewater Primer*, 1998, EPA 833-K-98-001. available at: www.epa.gov/owm/: Office of Water.

13 Ibid.

14 Larz T. Anderson, *Planning the Built Environment* (Chicago, IL: APA Planners Press, 2000).

15 "New Findings on Emerging Contaminants," *Medical News Today* (February 18, 2008), available at: www.medicalnewstoday.com.

16 Lisa Alvarez-Cohen and David L. Sedlak, "Emerging Contaminants in Water," *Environmental Engineering Science* 20, no. 5 (2003): 387–388.

17 U.S. Environmental Protection Agency, *Report to Congress on the Impacts and Control of CSOs and SSOs* (Washington, DC: U.S. Environmental Protection Agency, 2004).

18 Water Infrastructure Network, "Maumee Valley Sewer Needs," *WIN News* (2004).

19 Ty Tagami, "Sewer Fix at Crucial Step: Pollutant Limits Enters Debate," *Atlanta Constitution Journal*, December 18, 2004.

20 U.S. Environmental Protection Agency, *Report to Congress on the Impacts and Control of CSOs and SSOs* (Washington, DC: U.S. Environmental Protection Agency, 2004).

21 Ibid.

22 Robert Pitt, Melinda Lalor, and John Easton, *Potential Human Health Effects Associated with Pathogens in Urban Wet Weather Flows* (Tuscaloo: University of Alabama, 2003), available at: unix.eng.ua.edu/~rpitt/Publications/MonitoringandStormwater/Stormwater%20Pathogens%20JAWRA.pdf, (accessed February 5, 2005).

23 Texas Department of Health, "Cryptosporidia: TDH Issues Final Brushy Creek Study Results," *Disease Prevention News* 58, no. 18 (1998).

24 U.S. Environmental Protection Agency, *Report to Congress on the Impacts and Control of CSOs and SSOs*, 2004, op. cit.

25 As cited in Cynthia Mitchell, Kumi Abeysuriya, and Dena Fam. *Development of Qualitative Decentralised System Concepts for the 2009 Metropolitan Sewerage Strategy for Melbourne Water* (Melbourne: Institute for Sustainable Futures, UTS, 2008).

26 U.S. Congress, General Accounting Office, *Water Utility Asset Management* (Washington, DC: U.S. Congress, General Accounting Office, 2004).

27 Ken Kirk, "Water and Sewer Replacement Needs," *Engineering News Record* (October 14, 2002).

28 The American Society of Civil Engineers estimates that there are about 600,000 miles of publicly owned pipe.

29 U.S. Environmental Protection Agency, *The Clean Water and Drinking Water Infrastructure Gap Analysis* (Washington, DC: U.S. Environmental Protection Agency, 2002).

30 Ibid.

31 U.S. Congress, General Accounting Office, *Water Utility*, 2004, op. cit.

32 Janice A. Beecher, Senate Committee on the Environment, H_2O Coalition, March 27, 2001.

33 National Small Flows Clearinghouse, "Septic System Information," *Septic News* (2004).

34 Juli Beth Hoover, "Decentralized Wastewater Management: Linking Land Use, Planning and Environmental Protection," paper presented at American Planning Association Annual Meeting (2001).

35 D. Cordell, A. Rosemarin, J. J. Schröder, and A. L. Smit, "Towards Global Phosphorus Security: A Systemic Framework for Phosphorus Recovery and Reuse Options," *Chemosphere*, Special Issue on Phosphorus (2011), DOI: 10.1016/j.chemosphere.2011.02.032.

36 D. L. Childers, J. Corman, M. Edwards, and J. J. Elser, "Sustainability Challenges of Phosphorus and Food: Solutions from Closing the Human–Phosphorus Cycle," *BioScience* 61 (2011): 117–124.

37 P. Malmqvist, G. Heinicke, E. Korrman, T. Bergstrom, and G. Svensson, *Strategic Planning of Sustainable Urban Water Management* (London: IWA Publishing, 2006), 161–173.

38 U.S. Environmental Protection Agency, *Evaluation of Energy Conservation Measures for Wastewater Treatment Facilities*, EPA 832-R-10-005 (Washington, DC: U.S. Environmental Protection Agency, 2010).

39 U.S. Environmental Protection Agency, *Inventory of U.S. Greenhouse Gas Emissions and Sinks: 1990-2006*, EPA 430-R-08-005 (Washington, DC: U.S. Environmental Protection Agency, 2008), Chapter 8.

40 Water Environment Research Foundation, *Energy Production and Efficiency Research: The Roadmap to Net-Zero Energy*, Fact Sheet, undated, available at: www.werf.org.

41 Cynthia Mitchell, Kumi Abeysuriya, and Dena Fam. *Development of Qualitative Decentralised System Concepts for the 2009 Metropolitan Sewerage Strategy for Melbourne Water* (Melbourne: Institute for Sustainable Futures, UTS, 2008).

42 This section is based on Takashi Asano, Franklin L. Burton, Harold L. Leverenz, Ryujiro Tsuchihashi, and George Tchobanoglous, *Water Reuse: Issues, Technologies, and Applications* (New York: McGraw-Hill, 2007).

43 Alan Hals, "New Solutions for an Old Problem: Managing Wastewater Sludge," *Water On-Line* (June 19, 2008), available at: www.wateronline.com.

44 Chris Baber, "Tapping into Waste Heat," *Water Environment & Technology*, December 2010, available at: www.wef.org/magazine.

45 G. Daigger, "Evolving Urban Water and Residuals Management Paradigms: Water Reclamation and Reuse, Decentralization and Resource Recovery," *Water Environment Research*, 81, no. 8 (2009): 809–823.

46 Asano et al., 2007, op. cit.

47 S. Hermanowicz, "Sustainability in Water Resources Management: Changes in Meaning and Perception," *Sustainability Science* (New York, NY: Springer, 2008).

48 Ralf Otterpohl, "Design of Highly Efficient Source Control Sanitation and Practical Experiences," in *Decentralised Sanitation and Reuse: Concepts, Systems and Implementation*, ed. P. Lens, G. Zeeman, and G. Lettinga (London: IWA Publishing, 2005).

49 Carol Steinfeld and David Del Porto, *Reusing the Resource: Adventures in Ecological Wastewater Recycling* (Concord, MA: Ecowaters Books, 2007), available at: www.ecowaters.org.

50 Lucy Allen, Juliet Christian-Smith, and Meena Palaniappan, *Overview of Greywater Reuse: The Potential of Greywater Systems to Aid Sustainable Water Management* (Oakland, CA: Pacific Institute, 2010).

51 P. Lens, G. Zeeman, and G. Lettinga, eds., *Decentralised Sanitation and Reuse: Concepts, Systems and Implementation* (London: IWA Publishing, 2005).

52 Brooke Ray Smith, Master's thesis, University of California at Berkeley, Department of Landscape Architecture, 2008.

53 Spokane County, WA, *Comprehensive Wastewater Management Plan Background* (Spokane County, WA, 2002).

54 Council on Environmental Quality, *The Growth Shapers: The Land Use Impacts of Infrastructure Investment* (Washington, DC: Council on Environmental Quality, 1976).

55 Richard D. Tabors, "Utility Services," in *The Practice of Local Government Planning*, 1st ed., ed. Frank S. So, et al. (Washington, DC: International City Management Association, 1979).

56 Eric Damian Kelly and Barbara Becker, *Community Planning: An Introduction to the Comprehensive Plan* (Washington, DC: Island Press, 2000).

57 Warren Viessman, Jr. and Mark J. Hammer, *Water Supply and Pollution Control*, 6th ed. (Menlo Park, CA: Addison-Wesley Longman, 1998).

58 For example, see: "Sewer Model Helps Postpone $2 Million BC Sewer Project for Five Years," Accessed at: http://www.pizer.com/pdf/surrey.pdf, reprinted from *Environmental Science and Engineering* 13, no. 1 (2000).

59 Cameron Speir and Kurt Stephenson, "Does Sprawl Cost Us All? Isolating the Effects of Housing Patterns on Public Water and Sewer Costs," *Journal of the American Planning Association* 69, no. 1 (2002): 56–70.

60 U.S. Environmental Protection Agency, *Wastewater Primer*, 1998, op. cit.

61 Deborah Galardi and Barbara Buus, "Development Fee Trends and Tucson Case Study," in *Joint Management Conference* (Water Environment Federation, 2004).

62 Ibid.

chapter 18

1 Vladimir Novotny, Jack Ahern, and Paul Brown, *Water Centric Sustainable Communities: Planning, Retrofitting and Building the Next Urban Environment* (Hoboken, NJ: John Wiley & Sons, 2010).

2 Ibid.

3 National Research Council, Committee on Reducing Stormwater Contributions to Water Pollution, *Urban Stormwater Management in the United States* (Washington, DC: National Academy Press, 2009).

4 Novotny et al., 2010, op. cit., and National Research Council, 2009, op. cit.

5 U.S. Public Interest Research Group, *Troubled Waters: An Analysis of Clean Water Act Compliance, January 2002–June 2003* (2004).

6 U.S. Environmental Protection Agency, *America's Wetlands: Our Vital Link between Land and Water*, EPA 843-K-95-001 (Washington, DC: Office of Water, Office of Wetlands, Oceans and Watersheds, U.S. EPA, 1995).

7 Memorandum dated October 27, 2011, "Achieving Water Quality Through Municipal Stormwater and Wastewater Plans," available at: cfpub.epa.gov/npdes/integratedplans.cfm.

8 Draft Integrated Planning Approach Framework, January 13, 2012.

9 U.S. Environmental Protection Agency, *Wastewater Primer* (Washington, DC: U.S. Environmental Protection Agency, 1998), available at: www.epa.gov/owm/ (accessed December 21, 2004).

10 Ibid.

11 John F. Damico and Lamont W. Curtis, *Financing Stormwater Utilities* (Kansas City, MO: American Public Works Association, 2003).

12 Janice Kaspersen, "The Stormwater Utility: Will It Work in Your Community?" *Stormwater* (2000).

13 John Randolph, *Environmental Land Use Planning and Management* (Covelo, CA: Island Press, 2004).

14 U.S. Environmental Protection Agency, *Water Quality Conditions in the United States* (Washington, DC: U.S. Environmental Protection Agency, 2000).

15 Keith Lichten, personal communication with the author, 2004.

16 T. R. Schueler, P. A. Kumble, and M. A. Heraty, *A Current Assessment of Urban Best Management Practices: Techniques for Reducing Non-Point Source Pollution in the Coastal Zone* (Washington, DC: Metropolitan Washington Council of Governments, 1992).

17 Environmental Defense Fund, "Final Federal Plan Won't End Pollution from Factory Farms," (1999), available at: www.commondreams.org/pressreleases (accessed December 30, 2004).

18 U.S. Environmental Protection Agency, *Wastewater Primer* Environmental Protection Agency, 1998, op. cit.

19 Lichten, op. cit.

20 U.S. Environmental Protection Agency, *Wastewater Primer*, 1998, op. cit.

21 Intergovernmental Panel on Climate Change, *Climate Change 2007: Summary for Policy Makers* (2007), available at: www.ipcc.org.

22 CH2MHill Inc., *Confronting Climate Change: An Early Analysis of Water and Wastewater Adaptation Costs* (National Association of Metropolitan Water Agencies, 2009), available at: www.amwa.net/cs/climatechange.

23 K. Emanuel, "Increasing Destructiveness of Tropical Cyclones over the Past 30 Years," *Nature* 436 (2005): 686–688.

24 U.S. Global Change Research Program, *Scientific Assessments, Global Climate Change Impacts in the United States* (Washington, DC: U.S. Government Printer, 2009), available at: www.globalchange.gov/publications/reports/scientific-assessments/us-impacts/full-report/national-climate-change.

25 Rocky Mountain Climate Organization (RMCO) and the Natural Resources Defense Council (NRDC), *Doubled Trouble: More Midwestern Extreme Storms* (2012), available at: http://www.rockymountainclimate.org/reports_3.htm.

26 Ibid.

27 Ibid. (see Table 3.1).

28 B. Douglas, "Global Sea Level Rise: A Redetermination," *Surveys in Geophysics*, 18, nos 2–3 (1997), 279–292.

29 Randolph, 2004, op.cit.; Bay Area Association of Stormwater Management Agencies, *Start at the Source: Design Guidance Manual for Stormwater Quality Protection* (1999), available at: www.basmaa.org/documents/index.cfm?fuseaction=documentdetails&documentID=23.

30 Randolph, 2004, op. cit.

31 Larry W. Mays, *Stormwater Collections Systems Design Handbook* (New York: McGraw-Hill, 2001).

32 Randolph, 2004, op. cit.; John Sansalone and Steven G. Buchberger, "Partitioning and First Flush of Metals in Urban Roadway Storm Water," *Journal of Environmental Engineering* 123, no. 2 (1997): 134–143; and Water Environment Federation and American Society of Civil Engineers, *Urban Runoff Quality Management: WEF Manual of Practice* no. 23 (1998).

33 EPA, available at: www.epa.gov/LID.

34 Adapted from EPA's description of Green Infrastructure.

35 Novotny et al., 2010, op. cit. NEW NOTE

36 U.S. Environmental Protection Agency, *Nonpoint Source Pointers (Factsheets)* (Washington, DC: U.S. Environmental Protection Agency, 1997), available at: www.epa.gov/OWOW/NPS/facts/.

37 Mays, 2001, op. cit.

38 Schueler, et al., 1992, op. cit.

39 Johnson Foundation, *Financing Sustainable Water Infrastructure* (Racine, WI: The Johnson Foundation at Wingspread, 2012).

40 Steve Veal and Allen Mullins, "Stormwater Phase I Fees," *Stormwater* (March, 2003).

41 Larry Levine and Alisa Valderama. *Financing Stormwater Retrofits in Philadelphia and Beyond*, NRDC, February 2012, available at: www.nrdc.org/water/files/StormwaterFinancing_report.pdf.

42 Veal and Mullins, 2003, op. cit.

43 Grant Hoag, "Developing Equitable Stormwater Fees," *Stormwater*, no. 1 (2004).

44 Deborah Galardi and Barbara Buus, "Development Fee Trends and Tucson Case Study," in *Joint Management Conference* (Water Environment Federation, 2004).

45 See water.epa.gov/grants_funding/cwsrf/cwsrf_index.cfm.

46 Donna Cooper and Jordan Eizenga, *Financing Water Infrastructure* (August 2, 2011), available at: www.americanprogress.org/.

47 See www.americanprogress.org/issues/2011/08/water_infrastructure_howto.html.

48 Novotny et al., 2010. op. cit.

49 Andrew Dzurik, *Water Resources Planning*, 2nd ed. (Totowa, NJ: Rowman & Littlefield, 1996).

50 U.S. Environmental Protection Agency, Office of Water, *Clean Water Act History* (Washington, DC: U.S. Environmental Protection Agency, 2002), www.epa.gov/region5/water/cwa.htm (accessed February 26, 2002).

chapter 19

1 U.S. Environmental Protection Agency, *Basic Facts about Solid Waste* (Washington, DC: U.S. Environmental Protection Agency, 2002).

2 Gordy Slack, "Emeryville Shell Game," *California Wild* (1999), available at: www.calacademy.org/calwild/sum99/habitat.htm (accessed March 6, 2002).

3 Martin V. Melosi, *The Sanitary City: Urban Infrastructure in America from Colonial Times to the Present* (Baltimore, MD: Johns Hopkins University Press, 2000).

4 Robert Gottlieb, "A Waste Management Crisis," in *Solid Waste Management Planning: Issues and Opportunities* (Chicago, IL: APA Planners Press, 1990).

5 San Francisco Bay Conservation and Development Commission, *Welcome Page* (1999), available at: www.bcdc.ca.gov/allink/allink.htm (accessed March 10, 2002).

6 H. S. Peavey, D. R. Rowe, and G. Tchobanoglous, "Solid Waste: Definitions, Characteristics, and Perspectives," in *Environmental Engineering* (New York: McGraw-Hill, 1985).

7 J. W. Vincoli, *Basic Guide to Environmental Compliance* (New York: Van Nostrand Reinhold, 1993).

8 Jon W. Kindschy, Marilyn Kraft, and Molly Carpenter, *Guide to Hazardous Materials and Waste* (Point Arena, CA: Solano Press, 1997).

9 E. W. Repa, "The U.S. Solid Waste Industry: How Big is It?" *Waste Age* 32, no. 12 (2001): 60.

10 Waste Management, Inc., "About WM," (2002), available at: www.wm.com/about.asp (accessed March 10, 2002).

11 California Integrated Waste Management Board, *Waste Stream Measurement and Analysis* (2003), available at: www.ciwmb.ca.gov/LGCentral/WasteStream/.

12 Zero Waste International Alliance, available at: www.zwia.org.

13 U.S. Environmental Protection Agency, Office of Solid Waste, *Reduce, Reuse, Recycle* (Washington, DC: U.S. Environmental Protection Agency, 2001).

14 U.S. EPA, "Waste," in *Inventory of U.S. Greenhouse Gas Emissions and Sinks, 1990–2010* (2008), available at: http://www.epa.gov/climatechange/Downloads/ghgemissions/US-GHG-Inventory-2012-Main-Text.pdf Page ES-5 and ES-9 (accessed April 15, 2012).

15 U.S. Environmental Protection Agency, *Questions and Answers about Full Cost Accounting* (Washington, DC: U.S. Environmental Protection Agency, 2008).

16 Jon Yates, "Community Group Exults in Thrift Store Rejection," *The Tennessean* (February 18, 2000).

17 U.S. Environmental Protection Agency, *Introduction to Laws and Regulations* (Washington, DC: U.S. Environmental Protection Agency, 2002).

18 New Jersey Housing and Mortgage Finance Agency, *The Homeowner's Resource Guide to Recycling and Disposing of Hazardous Materials* (New Jersey, 2009).

19 California Department of Resources Recycling and Recovery, "Diversion Rate, Local Government Central Glossary of Terms," available at: www.calrecycle.ca.gov/LGCentral/Glossary/.

20 This analysis is patterned after one in the Seattle Zero Waste Report, 2007.

21 This section is based on material from the California Integrated Waste Management Board website, particularly caproductstewardship.org, the Northwest Product Stewardship

Council website, EPA, the Seattle Zero Waste Report, 2007 and the 2008 Five Year Audit and Program Assessment of the Alameda County Waste Management Authority. The latter two documents contained surveys and descriptions of best and emerging practices for diversion from around the country. Material was also obtained from www.grrn.org.

22 U.S. EPA, "Management of Electronic Waste in the United States," Draft, 2007 (accessed July 2, 2008).

23 Ning Ai, "Challenges of Sustainable Urban Planning: The Case of Municipal Solid Waste Management," Ph.D. dissertation at the School of City and Regional Planning, Georgia Institute of Technology, August 2011.

24 See www.epa.gov/osw/nonhaz/municipal/landfill/bioreactors.htm.

chapter 20

1 Allan B. Jacobs, *Great Streets* (Cambridge, MA: The MIT Press, 1993).

2 Michael Southworth and Eran Ben-Joseph, *Regulated Streets* (Berkeley, CA: University of California, Berkeley, 1993).

3 Ibid.

4 Arizona Department of Transportation, *Pedestrian Crosswalks: How Safe Are They?* (2000), available at: www.dot.state.az.us/ROADS/traffic/xwalk.htm.

5 Richard Untermann, "Taming the Automobile: How We Can Make Streets More 'Pedestrian Friendly,'" *Planning Commissioners Journal* 1, no. 1 (November/December 1991).

6 James M. Daisa, *ITE Committee Report Summary: Context-Sensitive Solutions in Designing Major Urban Thoroughfares*, available at: www.ite.org.

7 U.S. Department of Transportation, Bureau of Transportation Statistics, *National Transportation Statistics* (Washington, DC: U.S. Department of Transportation, Bureau of Statistics, 2002), available at: www.bts.gov/publications/national_transportation_statistics/.

8 U.S. EPA and "Life Cycle Costs and Market Barriers of Reflective Pavements," Lawrence Berkeley National Laboratory, available at: enduse.lbl.gov/Projects/pavements.html.

9 Reid Ewing, "Measuring Transportation Performance," *Transportation Quarterly* 49, no. 1 (1995): 91–104.

10 U.S. Department of Transportation, Office of Highway Policy Information, *Nationwide Personal Travel Survey Databook* (Washington, DC: U.S. Department of Transportation, Office of Highway Policy Information, 2001), available at: www-cta.ornl.gov/npts/1995/Doc/databook95/contents.pdf.

11 Joey Ledford, "Legality, Efficiency of Speed Bumps Are Debated," *Journal-Constitution* (July 28, 1998).

12 William Fulton, *The New Urbanism*, PAS Reports (Chicago, IL: APA Planners Press, 1996).

13 Hollie Lund Person, "Local Accessibility, Pedestrian Travel and Neighboring: Testing the Claims of New Urbanism," paper presented at American Planning Association National Planning Conference, New Orleans, LA, 2001, available at: www.asu.edu/caed/proceedings01/PERSON/person.htm#Anchor-Author-49575.

14 Barbara McCann and Bianca DeLille, *Mean Streets 2000*, Surface Transportation Policy Project (2000), available at: www.justtransportation.org/resources.html.

15 See also Edward Weiner, *Urban Transportation Planning in the United States: An Historical Overview*, available at: tmip.fhwa.dot.gov/clearinghouse/docs/utp/.

chapter 21

1 P. O. Muller, "Transportation and Urban Form: Stages in the Spatial Evolution of the American Metropolis," in *The Geography of Urban Transportation*, ed. S. Hanson (New York, NY: Guilford Press: 1986).

2 Rick Hall, "Americans against Traffic Calming" (2002), available at: www.io.com/~bumper/ada.htm (accessed July 14, 2002).

3 Muller, 1986, op. cit.

4 Ibid.

5 Ibid.

6 American Society of Civil Engineers, *Specific Policy Recommendations for America's Infrastructure* (Washington, DC: American Society of Civil Engineers, 2001), available at: www.asce.org/reportcard/index.cfm?reaction=policy.

7 John Landis, "Simulating Highway and Transit Effects," *Access* 12 (Spring, 1998).

8 American Public Transportation Association, Glossary of Transit Terminology (Washington, DC: American Public Transportation Association, July 1994), available at:

http://www.apta.com/resources/reportsandpublications/Documents/Transit_Glossary_1994.pdf.

9 U.S. Department of Energy, Energy Information Administration, *Annual Energy Review 2000* (Washington, DC: U.S. Department of Energy, Energy Information Administration, 2000).

10 Richard F. Weingroff, "Federal Aid Road Act of 1916: Building the Foundation," *Public Roads* (Summer, 1996).

11 R. Cervero, "Jobs-Housing Balance Revisited Trends and Impacts in the San Francisco Bay Area," *Journal of the American Planning Association* 62, no. 4 (1996).

12 Texas Transportation Institute, *2007 Urban Mobility Report* (College Station, TX, 2007).

13 U.S. Department of Transportation, Federal Highway Administration, *Traffic Volume Trends* (April, 2008).

14 William Neuman, "Politics Failed, but Fuel Prices Cut Congestion," *New York Times* (July 3, 2008).

15 CIBE World Markets, 2008.

16 Nicholas Lutsky, *Prioritizing Climate Change Mitigation Alternatives: Comparing Transportation to Options in Other Sectors*, UCD-ITS-RR-08-15 (Davis, CA: University of California at Davis, 2008).

17 Robert W. Poole, Jr. and Adrian T. Moore. *Restoring Trust in the Highway Trust Fund* (Reason Foundation Policy Study 386), August 2010.

18 Robert Cervero, *The Transit Metropolis: A Global Inquiry* (Washington, DC: Island Press, 1998), 4.

chapter 22

1 A. T. Wells, *Airport Planning & Management*, 2nd ed. (Blue Ridge Summit, PA: TAB Books, 1992).

2 Federal Aviation Administration, *National Plan of Integrated Airport Systems (NPIAS) (2001–2005)* (Washington, DC: Federal Aviation Administration, 2002).

3 Federal Aviation Administration, *FAA Aerospace Forecast Fiscal Years, 2012–2032* (Washington, DC: Federal Aviation Administration, 2012).

4 Federal Aviation Administration, *Capacity Needs in the National Airspace System, 2007–2025* (Washington, DC: Federal Aviation Administration, 2007).

5 P. S. Shapiro, et al., *Intermodal Ground Access to Airports: A Planning Guide* (Washington, DC: Federal Highway Administration and Federal Aviation Administration, 1996).

6 H. Blokpoel, *Bird Hazards to Aircraft: Problems and Prevention of Bird/Aircraft Collisions* (Ottawa: Clarke, Irwin & Company, Ltd., 1977).

7 Aleksandra Mozdzanowska and R. John Hansman, *Observations and Potential Impacts of Regional Jet Operating Trends* (Cambridge, MA: Massachusetts Institute of Technology International Center for Air Transportation, 2003).

8 International Civil Aviation Organization, *Infrastructure Management*, available at: www.icao.int/sustainability/Pages/eap-infrastructure-management.aspx.

9 Federal Aviation Administration Office of Airport and Programming, *Airport Master Plans. Advisory Circular 150/5070-6B* (Washington, DC: Federal Highway Administration and Federal Aviation Administration, May 2007).

10 Authors' research, various sources.

11 Airport Cooperative Research Program, *Enhancing Airport Land Use Compatibility*, ACRP Report 27 (Washington, DC: Transportation Research Board, 2010).

12 Wisconsin Statute 114.136, Airport and Spaceport Approach Protection.

13 Federal Aviation Administration Wildlife Hazard Mitigation Program, Wildlife Strike Database, available at: www.faa.gov/airports/airport_safety/wildlife/database/.

14 42 U.S.C. §4901 et seq. *Noise Control Act*.

15 GRA, Incorporated, *Houston Airport System Economic Impact Study* (Jenkintown, PA, June, 2011).

16 New Mexico Department of Transportation, Aviation Division, *New Mexico Airport System Plan Update 2009*, available at: http://www.donaanacounty.org/works/airport/docs/ExecutiveSummary.pdf.

17 Rena A. Koontz, "Council Goes Along with Airport Pact," *Cleveland Plain Dealer*, October 1, 2001.

chapter 23

1 B. S. Hoyle, D. A. Pinder, and M. S. Husain, eds., *Revitalising the Waterfront: International Dimensions of Dockland Redevelopment* (London: Belhaven, 1988).

2 P. V. Hall, "Regional Institutional Convergence? Reflections from the Baltimore Waterfront," *Economic Geography* 79, no. 4 (2003): 347–363.

3 P. H. Brown, *America's Waterfront Revival: Port Authorities and Urban Redevelopment* (Philadelphia, PA: University of Pennsylvania Press, 2009).

4 Cruise Line International Association, *CLIA Cruise Market Overview: Statistical Cruise Industry Data Through 2010* (Fort Lauderdale, FL: CLIA, 2011).

5 Tampa Port Authority, *Tampa Port Authority: Comprehensive Annual Financial Report for Fiscal Year Ended September 30, 2010* (Tampa, FL: Tampa Port Authority, 2011).

6 Brown, 2009, op. cit.

7 M. N. Danielson and J. W. Doig, *New York: The Politics of Urban Regional Development* (Berkeley, CA: The University of California Press, 1982); J. Mitchell, "Policy Functions and Issues for Public Authorities," in *Public Authorities and Public Policy: The Business of Government*, ed. J. Mitchell (New York: Greenwood Press, 1992), 1–14.

8 Mitchell, 1992, op. cit.

9 R. B. Sherman, *Public Seaport Agencies in the United States and Canada* (Alexandria, VA: The American Association of Port Authorities, April 2000).

10 Ibid.

11 Hall, 2003, op. cit.

12 D. Luberoff and J. Walder, *U.S. Ports and the Funding of Intermodal Facilities: An Overview of Key Issues* (Cambridge, MA: New England University Transportation Center, Massachusetts Institute of Technology, 2000).

13 Ibid.; and U.S. General Accounting Office, *Freight Transportation: Strategies Needed to Address Planning and Financial Limitations*, Report by the General Accounting Office to the Committee on Environment and Public Works, U.S. Senate (Report GAO-04-165) (Washington, DC: U.S. General Accounting Office, 2003).

14 Tampa Port Authority, 2011, op. cit.

15 E. Van de Voorde, "Sea Ports, Land Use and Competitiveness: How Important Are Economic and Spatial Structures?" in *Transport and Urban Development*, ed. D. Banister (London: E&FN Spon, 1995).

chapter 24

1 J. A. Lackney, P. Park, and L. P. Witzling, *The Cost of Facility Development: A Comparative Analysis of Public and Private Sector Facility Development Processes and Costs* (Milwaukee: Center for Architecture and Urban Planning Research, 1994).

2 K. Madden, "The Return of the Civic Square," in *Making Places* (Project for Public Spaces, Inc., 2002).

3 Institute of Museum and Library Services, *Museums and Libraries: An Investment in Learning* (Washington, DC: Institute of Museum and Library Services, n.d.).

4 American Library Association, *The State of America's Libraries 2008* (2008), available at: www.ala.org.

5 Natural Resources Defense Council, "State Budgets Pinching Anti-Sprawl Programs," (2003). Available at: http://www.nrdc.org/.

6 American Library Association, *Fact Sheet: Renaissance in Public Libraries* (Chicago, IL: American Library Association, 1999).

7 Christopher Freeman, "Library Referenda 2007: A Mixed Ballot Bag," *Library Journal* (March 15, 2008), available at: www.libraryjournal.com (accessed July 20, 2008).

8 N. Darwick, K. J. Matulia, and M. J. Varner, *Police Facility Design* (Gaithersburg, MD: International Association of Chiefs of Police, 1978).

9 *A Look at Building Activities in the 1999 Commercial Buildings Energy Consumption Survey: Public Order and Safety Buildings* (Washington, DC: Energy Information Administration, 2002).

10 Robert Tobin, et al., *The Courthouse: A Planning and Design Guide for Court Facilities* (Denver, CO: National Center for State Courts, 1999).

11 Robert Tobin, *A Court Manager's Guide to Court Facility Financing* (Denver, CO: National Center for State Courts, 1995).

12 National Center for State Courts, *A Comprehensive Evaluation of the Kitsap County Courthouse* (Washington, DC: Institute for Court Management, Court Executive Development Program, 2007–2008 Phase III Project, May 2008), available at: www.ncsconline.org.

13 E. D. Kelly and B. Becker, *Community Planning: An Introduction to the Comprehensive Plan* (Washington, DC: Island Press, 2000).

14 J. J. Stephan, *Census of Jails: 1999* (Washington, DC: Bureau of Justice Statistics, 2001).

15 Ingrid Singer and Betsy Carrier, National Association of Public Hospitals and Health Systems, *Capital Investment in America's Safety Net: Results of the NAPH Capital Expenditure and Financing Survey for FY 2001* (National Association of Public Hospitals and Health Systems, 2003), available at: www.nphhi.org. See their website for the results of their annual surveys on patients served, sources of funds for operations.

16 Healthcare Financial Management Association, *Financing the Future Report 1* (Healthcare Financial Management Association, 2003), available at: www.hfma.org.

17 D. R. Judd, et al., "Tourism and Entertainment as Local Economic Development: A National Survey," in *The Infrastructure of Play: Building the Tourist City in North America*, ed. D. R. Judd (New York: M.E. Sharpe, 2002).

18 H. T. Sanders, "Public Works and Public Dollars: Federal Infrastructure Aid and Local Investment Policy," in *Building the Public City*, ed. David Perry (Thousand Oaks, CA: Sage, 1995).

19 D. C. Petersen, *Sports, Convention, and Entertainment Facilities* (Washington, DC: Urban Land Institute, 1996).

20 H. T. Sanders, *Space Available: The Realities of Convention Centers as Economic Development Strategy* (Washington, DC: Brookings Institution Press, 2005).

21 R. G. Noll and A. Zimbalist, *Sports, Jobs and Taxes: The Economic Impact of Sports Teams and Stadiums* (Washington, DC: Brookings Institution Press, 1997).

22 David Perry, "Urban Tourism and the Privatizing Discourses of Public Infrastructure," in *The Infrastructure of Play: Building the Tourist City in North America and Europe*, ed. D. R. Judd, pre-publication document, 2002, pp. 23–70.

23 H. T. Sanders, *Flawed Forecasts: A Critical Look at Convention Center Feasibility Studies* (Boston, MA: Pioneer Institute for Public Policy Research, 1999).

24 N. deMause and J. Cagan, *Field of Schemes* (Lincoln, NB: University of Nebraska Press, 2008).

25 A. Altshuler and D. Luberoff, *Mega-Projects: The Changing Politics of Urban Public Investment* (Washington, DC, and Cambridge, MA: Brookings Institution Press and Lincoln Institute of Land Policy, 2003).

26 Perry, "Urban Tourism and the Privatizing Discourses of Public Infrastructure," 2002, op. cit.

27 J. Hannigan, *Fantasy City: Pleasure and Profit in the Postmodern Metropolis* (New York: Routledge, 1998).

28 Michael Sorkin, "Introduction," in *Variations on a Theme Park: The New American City and the End of Public Space*, ed. M. Sorkin (New York: Hill and Wang, 1992).

29 "The 2030 Challenge," *Architecture 2030*, available at: architecture2030.org/2030_challenge/the_2030_challenge.

30 U.S. Green Building Council, available at: www.usgbc.org/.

31 Pew Center for Global Climate Change, *Building Solutions to Climate Change* (2006), available at: www.pewclimate.org (accessed June 30, 2009).

32 See www.eere-energy.gov (accessed June 30, 2009).

33 Jerry Yudelson, *The Green Building Revolution* (Washington, DC: Island Press, 2008), Chapter 4.

34 Jon Creyts, Anton Derkach, Scott Nyquist, Ken Ostrowski, and Jack Stephenson, *Reducing U.S. Greenhouse Gas Emissions: How Much at What Cost?* (New York, NY: McKinsey & Company, 2007).

35 Rebecca C. Retzlaff, "Green Building Assessment Systems," *Journal of American Planning Association* 74, no. 4 (2008): 505–519.

36 This section is based on material from Yudelson, *The Green Building Revolution*, 2008, op. cit.

37 Ibid., Chapter 1.

38 Charlie Miller, "Blue-Green Practices: Why They Work and Why They Have Been So Difficult to Implement through Public Policy," in *Growing Greener Cities*, ed. Eugenie L. Birch and

Susan M. Wachter (Philadelphia, PA: University of Pennsylvania Press, 2008).

39 "Building Overview," Center for Climate and Energy Solutions, available at: www.pewclimate.org/technology/overview/buildings (accessed June 30, 2009).

40 Miller, 2008, op. cit.

41 Francis Johnson, "Local Law Firm Equips Salt Lake City Mansion with Fossil Fuel-Free Power," *The Enterprise*, Salt Lake City, Utah, April 16, 2007.

42 "Building Overview," Center for Climate and Energy Solutions, op. cit.

43 ICMA Center for Performance Measurement, "Facilities Management," in *Comparative Performance Measurement: FY 2002 Data Report* (Washington, DC: International City/County Management Association, 2002).

44 K. Whiteman, "Green Building," in *IQ Report* 35 no. 12 (Washington, DC: International City/County Management Association, 2003).

45 *Design for Aging* (Washington, DC: American Institute of Architecture Aging Design Research Program, 1993).

46 M. E. Miles, B. Berens, and M. A. Weiss, *Real Estate Development* (Washington, DC: Urban Land Institute, 2000).

47 D. E. Dowall and J. Whittington, *Making Room for the Future: Rebuilding California's Infrastructure* (San Francisco, CA: The Public Policy Institute of California, 2003).

48 D. P. Billington, "The Challenges to Engineers: Bridges as Art," in *The Art of Designing Bridges and Highways* (Berkeley, CA: University of California, 2002).

chapter 25

1 For a more detailed history, see David B. Tyack, "The One Best System: A History of American Urban Education," in *Planning and Designing Schools*, ed. C. William Brubaker (New York: McGraw-Hill, 1997).

2 U.S. Department of Education, National Center for Education Statistics, *Digest of Education Statistics* (Washington, DC: U.S. Department of Education, 2010). Unless otherwise noted, data in this chapter are from this source.

3 U.S. Department of Education, *National Public Education Financial Survey* (Washington, DC: U.S. Department of Education, 1998).

4 U.S. Census Bureau, *2007 Governments Integrated Directory (GID)*. Available at http://harvester.census.gov/gid/gid_07/options.html.

5 U.S. Department of Education, National center for education Statistics. *Projections of Education Statistics: 2012, Section I: Elementary and Secondary Enrollment*. Available at http://nces.ed.gov/programs/projections/projections2012/seclb.asp.

6 U.S. Congress, Government Accounting Office, *School Facilities: The Condition of America's Schools* (Washington, DC: U.S. Congress, Government Accounting Office, 1995).

7 The Neighborhood Capital Budget Group, *Rebuilding Our Schools Brick by Brick* (Chicago, IL: The Neighborhood Budget Group, 1999).

8 The Center for Green Schools. *The State of our Schools* (Washington, DC: U.S. Green Building Council, 2013).

9 Filardo, Mary. *Good Buildings Better Schools: An economic Stimulus Opportunity with long-term benefits* (Washington, DC: Economic Policy Institute, 2008).

10 Joe Agron, "Growth Spurt: 30th Annual Official Education Construction Report," *American School & University Magazine* (2009).

11 Mary Filardo, Jeffrey M. Vincent, Ping Sung, and Travis Stein, *Growth and Disparity: A Decade of U.S. Public School Construction* (Washington, DC: Building Educational Success Together, 2006).

12 Myron Orfield, *American Metropolitics: The New Suburban Reality* (Washington, DC: Brookings Institution Press, 1997).

13 David P. Varady and Jeffrey A. Raffel, *Selling Cities: Attracting Homebuyers through Schools and Housing Programs* (New York: State University of New York Press, 1995), Chapter 4.

14 Deborah McKoy and Jeffrey M. Vincent. "Housing and Education: The Inextricable Link," in *Segregation: The Rising Costs for America*, ed. James H. Carr and Nadinee Kutty (London: Routledge, 2008).

15 Mark Baldassare, *Statewide Survey: Californians and Their Government* (San Francisco, CA: Public Policy Institute of California, 2003).

16 Ibid.

17 American Planning Association, *The Millennium Planning Survey* (Washington, DC: American Planning Association, 2000).

18 D. Gottfredson, *School Size and School Disorder* (Baltimore, MD: Center for Social Organization of Schools, Johns Hopkins University, 1985).

19 Michael P. Garber, R. John Anderson, and Thomas G. DiGiovanni, *Scale and Care: Charter Schools and New Urbanism* (1998).

20 Council of Educational Facility Planners International, *Guide for Planning Educational Facilities* (Scottsdale, AZ: Council of Educational Facility Planners International, 1991).

21 Council of Educational Facility Planners International, *Creating Connections: The CEFPI Guide for Educational Facility Planning* (Scottsdale, AZ: Council of Educational Facility Planners International, 2004).

22 Edward T. McMahon, "School Sprawl," *Planning Commissioners Journal* 39 (Summer 2000).

23 State of Massachusetts, Executive Office for Administration and Finance, *Reconstructing the School Building Assistance Program* (2000).

24 Constance E. Beaumont and Elizabeth G. Pianca, *Historic Neighborhood Schools in the Age of Sprawl: Why Johnny Can't Walk to School* (Washington, DC: National Trust for Historic Preservation, 2003), available at: www.nthp.org/issues/schoolsRpt.pdf.

25 Stephen Spector, *Creating Schools and Strengthening Communities through Adaptive Reuse* (Washington, DC: National Clearinghouse for Educational Facilities, 2003).

26 Filardo, Mary, Jeffrey M. Vincent, Marnie Allen, and Jason Franklin. *Joint use of Public Schools: A Framework for a New Social Contract* (Washington, DC: 21st Century School Fund and Center for Cities and Schools, 2010).

27 Pamela Shorr, "It's So Easy Being Green," *American School Board Journal* 19, no. 10 (2004).

28 Cheryl Wagner, *Planning School Grounds for Outdoor Learning* (Washington, DC: National Clearinghouse for Educational Facilities, 2000).

29 Christopher Kouri, *Wait for the Bus: How Lowcountry School Site Selection and Design Deter Walking to School and Contribute to Urban Sprawl* (Charleston, SC: South Carolina Coastal Conservation League, 1999).

30 Steve Donnelly, "A Toolkit for Tomorrow's Schools: New Ways of Bringing Growth Management and School Planning Together," *Planning* (2003).

31 U.S. Environmental Protection Agency, *Schools for Successful Communities: An Element of Smart Growth* (Washington, DC: Council of Educational Facility Planners International, 2004).

32 North Carolina Department of Public Instruction, Division of School Support, *Making Current Trends in School Design Feasible* (North Carolina, 2000).

33 Maine State Planning Office, Executive Department, *The Costs of Sprawl* (Maine, 1997).

34 U.S. Environmental Protection Agency, *Travel and Environmental Implications of School Siting* (Washington, DC: U.S. Environmental Protection Agency, 2003), available at: www.epa.gov/livability/school_travel.htm.

35 U.S. Department of Transportation, Bureau of Transportation Statistics, *National Household Travel Survey* (Washington, DC: U.S. Department of Transportation, 2003).

36 Tracy E. McMillan, "The Influence of Urban Form on a Child's Trip to School," in *The Association of Collegiate Schools of Planning Annual Conference* (Baltimore, MD: ACSP Press, 2002).

37 David L. Barkley, Mark S. Henry, and Shuming Bao, "Good Schools Aid Rural Development in South Carolina," *Issues in Community and Economic Development* 5, no. 1 (1995).

38 Randall S. Sell, F. Larry Leistritz, and JoAnn M. Thompson, *Socio-Economic Impacts of School Consolidation on Host and Vacated Communities*, Report no. 347 (North Dakota State University Department of Agricultural Economics, 1996).

39 Philip Langdon, "Stopping School Sprawl," *Planning* (2000).

chapter 26

1 Alexander Garvin, *Parks, Recreation and Open Space: A Twenty-First Century Agenda*, Planning Advisory Service Report No. 497 (Chicago, IL: American Planning Association, 2000), 9.

2 U.S. Congressional Budget Office, *Investing in Capital and Information: Deferred Maintenance in the National Parks*. Available at: http://www.cbo.gov/showdoc, 1999.

3 G. Cranz, *The Politics of Park Design: A History of Urban Parks in America* (Cambridge, MA: MIT Press, 198), 61.

4 Ibid., p. 186.

5 Ibid., p. 138.

6 D. R. Judd et al., "Tourism and Entertainment as Local Economic Development: A National Survey," in *The Infrastructure of Play: Building the Tourist City in North America* ed. D. R. Judd (New York: M. E. Sharpe, 2002).

7 Megan Lewis, *From Recreation to Re-Creation: New Directions in Parks and Open Space System Planning*, PAS Report 551 (Chicago, IL: American Planning Association, 2008).

8 Blaine Bonham Jr., Gerri Spilka, and Darl Rastorfer, *Old Cities/Green Cities: Communities Transform Unmanaged Land*, PAS Report 506/507 (Chicago, IL: American Planning Association, 2002).

9 L. S. Eplan, "Planning Urban Park Systems," in *Infrastructure Planning and Management*, ed. J. M. Stein (Thousand Oaks, CA: Sage, 1986).

10 B. P. McGregor, and J. L. Crompton, "National Trends in the Financing and Staffing of Local Government Park and Recreation Departments," paper given at 1992 Leisure Research Symposium, Cincinnati, OH, October 15–18, 1992.

11 "Center for Park Excellence, Fact Sheets," Trust for Public Land, available at: www.tpl.org/ccpe (accessed July 24, 2008).

12 E. Moulder and S. Clark, *Parks and Recreation, Cultural and Arts Programs: Alternative Service Delivery Choices.* (Washington, DC: International City/County Management Association, 1999).

13 Garvin, 2000, op. cit.

14 "LandVote 2007: National Overview," Trust for Public Land (2008), available at: www.tpl.org (accessed July 23, 2008).

15 Natural Resources Defense Council, *State Budgets Pinching Anti-Sprawl Programs* (2003), available at: www.nrdc.org/news/news.

16 Ibid.

17 Eplan, 1986, op cit.

18 K. Hopper, "Increasing Public Investment in Parks and Open Space," in *Local Parks, Local Financing*, vol. 1 (San Francisco, CA: Trust for Public Land, 1999).

19 P. Harnik, "Paying for Urban Parks without Raising Taxes," in *Local Parks, Local Financing*, vol. 2 (San Francisco, CA: Trust for Public Land, 1999).

20 Lewis, 2008, op. cit.

21 James D. Mertes and James R. Hall, *Park, Recreation, Open Space and Greenway Guidelines* (Alexandria, VA: National Recreation and Park Association and the American Academy for Park and Recreation Administration, 1995); Lewis, 2008, op. cit.

22 This section is based on J. L. Crompton, *Parks and Economic Development*, Planning Advisory Service Report 502 (Chicago, IL: American Planning Association, 2001).

chapter 27

1 Paul Harrison and Fred Pearce, "Atlas of Population and Environment," American Association for the Advancement of Science (2001), available at: atlas.aaas.org/natres/energy_popups.php?p=consume (accessed July 29, 2008).

2 U.S. Department of Energy, available at: www.doe.gov.

3 Ibid.

4 U.S. Environmental Protection Agency, *Draft Inventory of U.S. Greenhouse Gas Emissions and Sinks, 1990–2010* (U.S. Environmental Protection Agency, February 27, 2012), available at: www.epa.gov/climatechange/emissions/usinventoryreport.html.

5 Solar Electric Power Association, "Photovoltaic Q&A" (2000), available at: www.SolarElectricPower.org/power/pv_q&a.cfm (accessed March 21, 2002).

6 Ken Zweibel James Mason, and Vasilis Fthenakis, "A Solar Grand Plan," *Scientific American* (January 2008), 64–71. Available at: http://www.scientificamerican.com/article.cfm?id=a-solar-grand-plan.

7 *New York Times* (July 15, 2007).

8 Zweibel et al., 2008, op. cit.

9 John Carey, "Ethanol is Not the Only Green in Town," *Business Week* (April 30, 2007).

10 See library.oregonmetro.gov/files//regional_greenhouse_gas_inventory.pdf.

chapter 28

1 National Electrical Manufacturers Association, *A Chronological History of Electrical Development* (New York, NY: National Electrical Manufacturers Association, 1946); Anthony P. Nuciforo, "Telecommunications," in *Understanding Infrastructure: A Guide for Architects and Planners*, ed. George Rainer (New York, NY: Wiley & Sons, 1990); and Thomas H. White, *United States Early Radio History* (March 1, 2002), available at: www.ipass.net/~whitetho/part1.htm (accessed April 2, 2002).

2 "Electrical Engineering," Wikipedia, available at: en.wikipedia.org/wiki/Signal_(electrical_engineering) (accessed September 22, 2008).

3 Alexander Belinfante, *Telephone Subscribership in the United States* (Washington, DC: Industry Analysis Division, Federal Communications Bureau, 2002), available at: www.fcc.gov/Bureaus/Common_Carrier/Reports/FCC-State_Link/IAD/subs0701.pdf.

4 CTIA, *Semi-Annual Wireless Industry Survey* (June 2011), available at: files.ctia.org/pdf/CTIA_Survey_MY_2011_Graphics.pdf.

5 White, 2002, op. cit.

6 Ibid.

7 "The History of TV," About.com (2002), available at: inventors.about.com/library/inventors/bltelevision.htm?terms=history+of+television (accessed April 5, 2002).

8 Ibid.

9 Manuel Castells, *The Internet Galaxy: Reflections on the Internet, Business, and Society* (New York, NY: Oxford University Press, 2001).

10 Ibid.; "Various History Documents," Internet Society (ISOC) (2004), available at: www.isoc.org/internet/history; and David Kristula, *The History of the Internet*, available at: http://www.davesite.com/webstation/net-history.shtml.

11 Ibid.

12 Robert L. Lucky, "The Evolution of the Telecommunication Infrastructure," in *The Changing Nature of the Telecommunications Information Infrastructure* (Washington, DC: National Academy Press, 1995); U.S. Department of Transportation, Federal Highway Administration, *Transportation Air Quality: Selected Facts and Figures* (Washington, DC: U.S. Department of Transportation, 2002), available at: www.fhwa.dot.gov/environment/aqfactbk/index.htm.

13 Robert Crandall, William Lehr, and Robert Litan, "The Effects of Broadband Deployment on Output and Employment: A Cross-sectional Analysis of U.S. Data," *Issues in Economic Policy*, no. 6 (July 2007); John B. Horrigan, "Broadband Adoption and Use in America," OBI Working Paper Series No. 1, FCC, February 2010.

14 Jim Baller and Casey Lide, "America Needs a Fiber-Based National Broadband Policy Now, If Not Sooner," *The FTTH Prism*, 3, no. 2 (October 2006), available at: www.baller.com (accessed on September 20, 2008).

15 Sixth Broadband Deployment Report, FCC 10-129, July 2010, available at: transition.fcc.gov/Daily_Releases/Daily_Business/2010/db0720/FCC-10-129A1.pdf.

16 Investopedia.com, *The Telecommunications Handbook* (2008), available at: www.investopedia.com/features/industryhandbook/ (accessed October 12, 2008).

17 Paul Valle-Riestra, *Telecommunications: The Governmental Role in Managing the Connected Community* (Point Arena, CA: Solano Press, 2005).

18 "Cable Internet Basics," Denver, CO: The Cable Modem Information Network (2002), available at: www.cable-modem.net/gc/basics.html.

19 Crandall et al., 2007, op. cit.

20 John B. Horrigan, *Broadband Adoption and Use in America*, FCC Omnibus Broadband Initiative (OBI) Working Paper Series (February, 2010).

21 *E-Government*, APA PAS 564 (April 2011).

22 California Broadband Task Force, *The Connectivity Report* (2007), available at: www.ca.gov (accessed August 19, 2008).

23 City of Seattle, Department of Information Technology, Office of Cable Communication, available at: seattle.gov/cable among others.

24 Thomas M. Lenard, "Government Entry into the Telecom Business: Are the Benefits Commensurate with the Costs?"

(February 2004), available at: www.pff.org/issues-pubs/pops/pop11.3govtownership.pdf.

25 Olga Karif, "Why Wi-Fi Networks Are Floundering," *Business Week* (August 15, 2007).

26 Tim Wu, "Where's My Free Wi-Fi?" *Slate* (September 27, 2007), available at: www.slate.com (accessed August 19, 2008).

27 City of Seattle, *Report of the Task Force on Telecommunications Innovation* (2005), www.seattle.gov/cable/docs/SeaBTF.pdf (accessed August 18, 2008).

28 Horrigan, 2010, op. cit.

29 See ICMA publications: *E-Government: What Citizens Want; What Local Governments Provide*, 2002. Item number 42834; *E-Government: Online Services and Procurement*, 2001, Item 42695; *E-Government: Planning, Funding and Outsourcing*, 2001. Item 42694, available at: bookstore.icma.org.

Illustration credits

Figure Sources

Fig. 1.1 IPCC. "IPCC, 2007: Summary for Policymakers." In *Climate Change 2007: The Physical Science Basis. Contribution of Working Group I to the Fourth Assessment Report of the Intergovernmental Panel on Climate Change*, ed. Solomon, S., D. Qin, M. Manning, Z. Chen, M. Marquis, K.B. Avery, M. Tignor and H.L. Miller. New York, NY: Cambridge University Press, 2007.

Fig. 1.2 Environmental Protection Agency. *Inventory of U.S. Greenhouse Gas Emissions and Sinks: 1990–2009.* Office of Atmospheric Programs, 2011. Available online at www.epa.gov/climatechange/index.html

Fig. 2.1 U.S. Congressional Budget Office. *Public Spending on Transportation and Water Infrastructure.* Washington, DC: U.S. Congressional Budget Office, 2010.

Fig. 3.1 Frey, William H. "Population Growth in Metro America since 1980: Putting the Volatile 2000s in Perspective." Washington, DC: The Brookings Institution, 2012. Available at www.brookings.edu/~/media/Files/rc/papers/2012/0320_population_frey/0320_population_frey.pdf

Fig. 4.1,2 & 3 Elmer, Vicki, 2009.

Fig. 4.4 U.S. Census Bureau. *Census of Governments 2000–2001.* Washington, DC: U.S. Census Bureau, 2001.

Fig. 4.5 Regional Planning Association. "America 2050 Megaregions Map." Available at www.rpa.org/america2050/images/2050_Map_Megaregions_Influence_150.png

Fig. 5.1 Authors, 2013.

Fig. 5.2 U.S. Environmental Protection Agency, Office of Solid Waste and Emergency Response. *Opportunities to Reduce Greenhouse Gas Emissions through Materials and Land Management Practices.* Washington, DC: U.S. Environmental Protection Agency, 2009. ES-1.

Fig. 6.1 Weitz, Jerry. *Sprawl Busting: State Programs to Guide Growth.* Chicago, IL: Planners Press, 1999. page 199, Figure 8.3.

Fig. 6.2 Avin, Uri. Placemaking Inc. "Images of Queens County Maryland." 2005.

Fig. 8.1 Authors based on material in Miles, et al, Urban Land Institute, 2007

Fig. 8.2 Authors based on material in Miles et al, Urban Land Institute, 2007

Fig. 8.3 Authors adapted from Bass et al, Solano Press, 2001 and Ortolano, Wiley & Sons, 1997.

Fig. 9.1 U.S. General Accounting Office. *Leading Practices in Capital Decision Making.* Washington, DC: U.S. General Accounting Office, 1998.

Fig. 9.2 City of Berkeley, CA. *Proposed CIP.* 1997.

Fig. 9.3 Personal communications, Kurt Svendsen, Dec. 6, 2003.

Fig. 10.1 U.S. Congressional Budget Office. *Public Spending on Transportation and Water Infrastructure.* Washington, DC: Congressional Budget Office, 2010.

Fig. 10.2 U.S. Environmental Protection Agency. *Fact Sheet: Asset Management for Sewer Collection Systems.* Washington, DC: U.S. Environmental Protection Agency, 2002.

Fig. 10.3 Lemer, A.C. "Progress Toward Integrated Infrastructure-Assets-Management Systems: GIS and Beyond." APWA International Public Works Congress 1998.

Fig. 10.4 City of Los Angeles. "L.A. Infrastructure Report Card." 2002.

Fig. 10.5 Drohde, David. "Edmonds Street Condition: Streets Rated from Poor to Severe." City of Edmonds, WA: February 2012.

Fig. 11.1 U.S. Congressional Budget Office. *Public Spending on Transportation and Water Infrastructure.* Washington, DC: U.S. Congressional Budget Office, 2010.

Fig. 11.2 Authors, 2013, from U.S. Congressional Budget Office. *Public Spending on Transportation and Water Infrastructure.* Washington, DC: U.S. Congressional Budget Office, 2010.

Fig. 11.3 Authors, 2013, from U.S. Census Bureau. *State and Local Government Finances by Level of Government and by State: 2008–2009.* Washington, DC: U.S. Congressional Budget Office, October 2011.

Fig. 11.4 U.S. Census, State and Local Government Finances Summary: 2009. Figure 2. Issued October 2011.

Fig. 11.5 DeKalb County, Georgia County Tax Commissioner. Web.ca.dekalb.ga.us/taxcommissioner/tc-home.html. Dunwoody, GA. "2010 Property Tax Facts." Available at dunwoodyga.gov/departments/Finance_Administration/2011-Sample-Property-Tax-Bill.aspx

Fig. 11.6 Fairfax County, Virginia. *FY2013–2017 Advertised Capital Improvement Program.* 2012.

Fig. 12.1 Authors, 2013, from The Bond Buyer Archives. *Municipal Bond Issuances.* 2010.

Fig. 12.3 Brown, P. "Port authorities and urban redevelopment: Politics, organizations, and institutions on a changing waterfront." PhD diss. Pennsylvania: University of Pennsylvania, 2004. Available at repository.upenn.edu/dissertations/AAI3125792

Fig. 13.1 Beecher, Janice. "IPU Trends." Institute of Public Utilities, Michigan State University, 2009.

Fig. 13.2 Beecher, Janice. "IPU Trends." Institute of Public Utilities, Michigan State University, 2009.

Fig. 14.1 Authors, 2013.

Fig. 14.2 Authors, 2013.

Fig. 15.1 Warner, Mildred E. and Amir Hefetz. "Managing Markets for Public Service: The Role of Mixed Public/Private Delivery of City Services," *Public Administration Review* 68, no. 1 (January/February 2008), page 155, figure 1.

Fig. 15.2 Compiled by authors from Hefetz, Amir and Mildred E. Warner. 2007. "Beyond the Market vs. Planning Dichotomy: Understanding Privatisation and its Reverse in US Cities," *Local Government Studies* 33, no. 4: page 18, figure 1.

Fig. 15.3 U.S. General Accounting Office. *Privatization: Lessons Learned by State and Local Governments.* Washington, DC: U.S. General Accounting Office, 1997.

Fig. 16.1 Wong, T.H.F. and R.R. Brown. "The Water Sensitive City: Principles for Practice." *Water Science & Technology* 60, no. 3 (2009): 673–682.

Fig. 16.2 Simmon, Robert. "Land Surface Temperature in the World." In *Global Warming.* Earth Observatory NASA. www.icess.ucsb.edu/modis/modis-lst.html

Fig. 16.3 King County, Washington State. Land Resources Division. www.kingcounty.gov

Fig. 16.4 Buddemeier, R.W. "Water Table Drawdown and Well Pumping." 2000. Available at www.kgs.ku.edu/HighPlains/atlas/apdrdwn.htm

Fig. 16.5 City of New York Department of Environmental Protection. *New York City's Water Supply System.* 2005. Available at www.nyc.gov/html/dep/html/drinking_water/wsmaps_wide.shtml

Fig. 16.6 Anderson, Larz T. *Planning the Built Environment.* Chicago, IL: APA Planners Press, 2000.

Fig. 16.7 U.S. Geological Survey. Freshwater Withdrawals in the United Staes, 2005. Ga.water.usgs.gov/edu/wateruse-total.html.

Fig. 16.8 Wolf, G. and P.H. Gleick. "The Soft Path for Water" in *The World's Water, 2002-2003.* Washington, D.C.: Island Press. Private communication with Pacific Institute to obtain most recent data.

Fig. 16.9 American Water Works Association. "1996 Water:\Stats: The Water Utility Database." 1996. Available at www.awwa.org/Resources/Surveys.cfm?ItemNumber=39303&navItemNumber=39456

Fig. 16.10 American Water Works Association. "1996 Water:\Stats: The Water Utility Database." 1996. Available at www.awwa.org/Resources/Surveys.cfm?ItemNumber=39303&navItemNumber=39456

Fig. 17.1 Anderson, Larz T. *Planning the Built Environment.* Chicago, IL: APA Planners Press, 2000. Reprinted with permission. Copyright 2000 by the American Planning Association.

Fig. 17.2 City of Berkeley, CA. Department of Public Works.

Fig. 17.3 Anderson, Larz T. *Planning the Built Environment.* Chicago, IL: APA Planners Press, 2000. Reprinted with permission. Copyright 2000 by the American Planning Association

Fig. 17.4 U.S. Environmental Protection Agency. Wastewater Division, *Report to Congress: Impacts and Controls of CSO's and SSO's*, 2004.

Fig. 17.5 U.S. Environmental Protection Agency, Region I website.

Fig. 17.6 U.S. Environmental Protection Agency, Wastewater Division, 2004.

Fig. 17.7 American Public Works Association. Clip Art. CD Rom. 1998.

Fig. 17.8 Anderson, Larz T. *Planning the Built Environment.* Chicago, IL: APA Planners Press, 2000. Reprinted with permission. Copyright 2000 by the American Planning Association

Fig. 17.9 Childers, D.L., J. Corman, M. Edwards and J.J. Elser. "Sustainability challenges of phosphorus and food: Solutions from closing the human phosphorus cycle." *BioScience* 61 2011:117–124.

Fig. 17.10 Baber, Chris. "Tapping into Waste Heat." Water Environment & Technology December 2010. Available at www.wef.org/magazine

Fig. 17.11 Asano, Takashi, Franklin Burton, Harold Leverenz, et al. *Water Reuse: Issues, Technology, and Applications.* New York, NY: McGraw-Hill, 2007.

Fig. 17.12 Lange, J., and Ralf Otterpohl. Handbuch zu einer zuKunftsfaehigen Abwasserwirtschaft, Mall-Baton-Verlag, Germany, 2001.

Fig. 17.13 Kujawa-Roeleveld, Katarzyna. "Resource recovery from source separated domestic waste(water) Streams: Full Scale Results." *Water Science & Technology* 64, no. 10 (2011).

Fig. 17.14 Lange, J., and Ralf Otterpohl. Handbuch zu einer zuKunftsfaehigen Abwasserwirtschaft, Mall-Baton-Verlag, Germany, 2001.

Fig. 17.15 King County, WA. "Regional Wastewater Services Plan." Available at www.kingcounty.gov/environment/wtd/Construction/planning/rwsp.aspx

Fig. 18.1 U.S. Environmental Protection Agency. Office of Water. *Clean Water Act History.* 2002.

Fig. 18.2 U.S. General Accounting Office. *Water Infrastructure: Information on Financing, Capital Planning, and Privatization.* Washington, DC: U.S. General Accounting Office, 2002.

Fig. 18.3 Webster, P.J., G.J. Holland, J.A. Curry, and H.-R. Chang, Changes in Tropical Cyclone Number, Duration, and Intensity in a Warming Environment, Science 309 2005: 1844–46.

Fig. 18.4 United States Global Change Research Program. *Global Climate Change Impacts in the United States.* 2009. Available at www.globalchange.gov/publications/reports/scientific-assessments/us-impacts/full-report/national-climate-change

Fig. 18.5 Rohde, Robert, based on data from Bruce C. Douglas. *Global Sea Rise: A Redetermination,* from Surveys in GeoPhysics, 18. P 279-292.

Fig. 18.6 Dreiseitl, Herbert. Atlier Dreiseitl. 2008. Used with Permission.

Fig. 18.7 U.S. Environmental Protection Agency. *Urban Runoff Pollution Prevention and Control Planning.* Cincinnati, OH: U.S. Environmental Protection Agency, Office of Research and Development, Center for Environmental Research Information, 1993.

Fig. 18.8 Guillette, Anne. *Achieving Sustainable Site Design,* in Whole Building Design Guidelines. Figure 1. At National Institute of Building Sciences website.

Fig. 18.9 U.S. Environmental Protection Agency. *Urban Runoff Pollution Prevention and Control Planning.* Cincinnati, OH: U.S. Environmental Protection Agency, Office of Research and Development, Center for Environmental Research Information, 1993.

Fig. 19.1 King County Washington Solid Waste Division. "Current System." Available at your.kingcounty.gov/solidwaste/about/documents/system.pdf

Fig. 19.2 U.S. Environmental Protection Agency. *Characterization of Municipal Solid Waste in the United States: 1998 Update.* Washington, DC: U.S. Environmental Protection Agency, 1998.

Fig. 19.3 Pacific Consultants and Engineers. Available at www.pacificincinerators.com/incinerators.htm

Fig. 19.4 Bournay, Emmanuelle. "Solid Waste Management Cost for Selected Cities." UNEP/GRIS-Arendal. Available at www.grida.no/graphicslib/detail/solid-waste-management-cost-for-selected-cities_1139

Fig. 19.5 U.S. Environmental Protection Agency. *Characterization of Municipal Solid Waste in the United States: 1998 Update.* Washington, DC: U.S. Environmental Protection Agency, 1998.

Fig. 19.6 City of Seattle, Seattle Zero Solid Waste Plan, 2007.

Fig. 19.7 East Bay Municipal Utility District, Used with permission.

Fig. 19.8 Taub, Christian. "Various fluorescent light bulbs." 2009. wikimedia commons. Available at commons. wikimedia.org/

Fig. 20.1 Southworth, Michael and Eran Ben-Joseph. Streets and the Shaping of Towns and Cities. 2nd edition. Washington, D.C.: Island Press, 2003.

Fig. 20.2 Aitken, Thomas. *Road Making and Maintenance.* 2nd edition. London: Charles Griffin and Co., Ltd., 1907.

Fig. 20.3 Chicago Department of Transportation. *The Green Alley Handbook: An Action Guide to Create a Greener, Environmentally Sustainable Chicago.* Chicago, IL: Department of Transportation.

Fig. 20.4 Chicago Department of Transportation. *The Green Alley Handbook: An Action Guide to Create a Greener, Environmentally Sustainable Chicago.* Chicago, IL: Department of Transportation.

Fig. 20.5 Institution of Transportation Engineers. *Designing Walkable Urban Thoroughfares: A Context Sensitive Approach.* An ITE Recommended Practice. Washington, DC: Institute of Transportation Engineers, 2010. Available at www.ite.org/css/online/index.html

Fig. 20.6 Institution of Transportation Engineers. *Designing Walkable Urban Thoroughfares: A Context Sensitive Approach.* An ITE Recommended Practice. Washington, DC: Institute of Transportation Engineers, 2010. Available at www.ite.org/css/online/index.html

Fig. 20.7 Institute of Transportation Engineers. *Fact Sheet 1.* Washington, DC: Institute of Transportation Engineers.

Fig. 20.8 Bob Guletz, PE, of Harris and Associates

Fig. 20.9 Bob Guletz, PE, of Harris and Associates

Fig. 20.10 City of Seattle Department of Transportation. *Seattle Right of Way Improvements Manual.* Seattle, WA: City of Seattle, 2011. Available at www.seattle.gov/transportation/rowmanual/manual/default.asp

Fig. 21.1 aristophanes2000hotmail.com. "The Chicago Metra Commuter Rail System." Wikimedia Commons. Available at commons.wikimedia.org

Fig. 21.2 California Department of Transportation. "Levels of Service for Freeways." Available at www.dot.ca.gov/ser/forms.htm

Fig. 21.3 U.S. Department of Transportation. FHWA, 2007

Fig. 21.4 U.S. Department of Transportation. FHWA, 2007

Fig. 21.5 Compiled by authors.

Fig. 21.6 U. S. Department of Transportation 2008.

Fig. 22.1 Authors, 2013

Fig. 22.2 Nakamura, Mealani. "Reports on How Things Work: Air Foil." Massachusetts Institute of Technology, 1999.

Fig. 22.3 Nave, C.R. "Hyperphysics." Georgia State University, 2005. Available at hyperphysics.phy-astr.gsu.edu/hbase/pber.html#airf

Fig. 22.4 "Minneapolis St. Paul Airport." Global Cargo Virtual Airlines. Available at www.globecargova.org/docs/MSP_AFC2.jpg

Fig. 22.5 California Department of Transportation, Division of Aeronautics. *Airport Land Use Planning Handbook.* Copyright 2012. All rights reserved. Safety Compatibility Zone Example 3. Produced initially by Institute of Transportation Studies, University of California, Berkeley, 2002. Available at www.dot.ca.gov/hq/planning/aeronaut/documents/ALUPHComplete-7-02rev.pdf

Fig. 22.6 Miller, Nicholas P. "Housing and the Sound Environment." *Land Development National Association of Home Builders* 18, no. 3 (Summer 2005). Available at www.hmmh.com/cmsdocuments/NAHBArticle.pdf

Fig. 22.7 Aries Consultants, Ltd. *Comprehensive Land Use Plan: Hollister Municipal Airport.* Prepared for the City of Hollister, California. October 2001). Available at www.sanbenitocog.org/clup/CLUP_Report-HollisterAirport.pdf

Fig. 24.1 City of Blaine, Montana. Proposed Town Square, Retrieved from website, July 2005.
Fig. 24.2 City of Berkeley, Dept of Public Works, Engineering Division. Berkeley, CA, 2005.
Fig. 24.3 iStock International, 2005. Available on-line at www.istockphoto.com
Fig. 24.4 iStock International, 2005. Available on-line at www.istockphoto.com
Fig. 24.5 Jencks, Rosey. San Francisco Public Utilities Commission. 2011.
Fig. 24.6a Hinshaw, Mark, Holt Hinshaw Architects. 1996.
Fig. 24.6b Elmer, Vicki, 2005.
Fig. 26.1 City of Arlington, TX, Parks & Recreation. *Planning, Design and Construction.* 2005.
Fig. 26.2 Trust for Public Land, "LandVote 2007: National Overview," www.tpl.org (accessed July 23, 2008).
Fig. 26.3 City of San Francisco, Planning Department. *Plan Element: Recreation and Open Space.* 1991.
Fig. 26.4 Northville MI Parks and Recreation Department. *Recreation Strategic Plan Survey.* 2005.
Fig. 26.5 "Open Space Locations." City of Santa Monica. Available at www.smgov.net/uploadedFiles/Departments/OSE/Categories/Sustainability/Sustainable_City_Progress_Report/Open_Space_and_Land_Use/OSLU3_Chart1.pdf
Fig. 27.1 Congressional Budget Office. *Energy Security in the United States.* May 2012. Available at go.usa.goc/Vmg
Fig. 27.2 Intergovernmental Panel on Climate Change, 2007.
Fig. 27.3 U. S. Department of Energy. Energy Information Administration, 2004.
Fig. 27.4 U.S. Department of Energy, Energy Information Administration. *Energy Kid's Page: Energy Facts.* 2004. Available at www.eia.doe.gov/kids/energyfacts/index.html
Fig. 27.5 U.S. Department of Energy. *Early Release Overview Full Report.* DOE/EIA-0383ER (2012). 2012.
Fig. 27.6 U. S. Department of Energy, Energy Information Administration. 2006.
Fig. 27.7 Mierto, Frank. Wikimedia Commons. www.wikimedia.commons.org. Accessed July, 2012.
Fig. 27.8 Bouchecl. "NERC Map." 2009. Wikimedia Commons. www.wikimedia.commons.org
Fig. 27.9 Figure in Putt del Pino, Samantha, Ryan Levison, and John Larson. *Hot Climate, Cool Commerce: A Service Sector Guide to Greenhouse Gas Management.* World Resources Institute, 2006. Available at www.wri.org/chart/operational-boundaries-ghg-emissions. WRI and World Business Council for Sustainable Development.
Fig. 28.1 CTIA, Estimated Subscribers, in Semi Annual Wireless Industry Survey, June 2011.
Fig. 28.2 MichNet Backbone 2005 www.merit.edu.mn.resourcs.network.backbone.pdf
Fig. 28.3 National Aeronautics and Space Administration. http://missionscience.nasa.gov/ems/01_intro.html
Fig. 28.4 Torrens, Paul M. University of Maryland. 2005.

Photo Sources

Page No.		Source
6	Photo 1.1	"iStockphoto." iStock International. 2005. Available at www.istockphoto.com/index.php
6	Photo 1.2	EarthTech Inc. Available at www.earthtech.com
6	Photo 1.3	Waliczek, Adam. "iStockphoto." iStock International. 2005. Available online at: www.istockphoto.com/index.php
7	Photo 1.4	City of Portland, OR. *Portland Online: West Side Big Pipe Sewer Project Photos.* 2005. Available online at: www.portlandonline.com/cso/index.cfm?&a=125125&c=42944
7	Photo 1.5	EarthTech Inc. www.earthtech.com
7	Photo 1.6	"MTA Maryland MCI D4500CTH #1746C." Wikipedia Commons. Available at en.wikipedia.org/wiki/File:MTA_Maryland_MCI_D4500CTH_-176C.jpg
9	Photo 1.7	SimonP. "Ashbridges Bay Wastewater Treatment Plant." Wikimedia Commons. Wikimedia.commons.org
10	Photo 1.8	"iStockphoto." iStock International. 2005. Available at www.istockphoto.com/index.php
10	Photo 1.9	Goldstein, Beth, 2005.
10	Photo 1.10	City of Berkeley, CA Marina Division, 2005.
11	Photo 1.11	Gough, Joe. "iStockphoto." iStock International. 2005. Available at www.istockphoto.com/index.php
11	Photo 1.12	Kohr, Aaron. "iStockphoto." iStock International. 2005. Available at www.istockphoto.com/index.php
11	Photo 1.13	Szafranski, Keith. "iStockphoto." iStock International. 2012. Available at www.istockphoto.com/index.php
14	Photo 1.14	Rofidal, Kevin. "I35W Collapse - Day 4 - Operations $ Scene." Wikimedia Commons. 2007. Available at commons.wikimedia.org/wiki/File:I35W_Collapse_-_Day_4_-_Operations_%26_Scene_95.jpg
20	Photo 2.1a	American Public Works Association. *Historical Photos.* Ed. Graeme Hunt. CD Rom. 1998.
20	Photo 2.1b	American Public Works Association. *Historical Photos.* Ed. Graeme Hunt. CD Rom. 1998.
20	Photo 2.1c	American Public Works Association. *Historical Photos.* Ed. Graeme Hunt. CD Rom. 1998.
21	Photo 2.2	American Public Works Association. *Historical Photos.* Ed. Graeme Hunt. CD Rom. 1998.
24	Photo 2.3	American Public Works Association. *Historical Photos.* Ed. Graeme Hunt. CD Rom. 1998.
25	Photo 2.4	Staub, David, K. "Chicago Cultural Center, Chicago, Illinois." Wikimedia Commons. 2007. Available at commons.wikimedia.org/wiki/File:Chicago_Cultural_Center.jpg
25	Photo 2.5	O'Rear, Charles. "Chicago's Union Station is the Heart of the City." Wikimedia Commons. 1974. Available at commons.wikimedia.org
25	Photo 2.6	Kerr, Doug. BayShore New York Firehouse. Wikimedia Commons. 2010. Available at commons.wikimedia.org
26	Photo 2.7	American Public Works Association. *Historical Photos.* Ed. Graeme Hunt. CD Rom. 1998.
26	Photo 2.8	American Public Works Association. *Historical Photos.* Ed. Graeme Hunt. CD Rom. 1998.
28	Photo 2.9	Freytag, Lawrence. "iStockphoto." iStock International. 2012. Available at www.istockphoto.com/index.php
28	Photo 2.10	Gilder, David. "Hoover Dam Hydroelectric Power Plant." Shutterstock.com. Available at www.shutterstock.com/pic-1273574/stock-photo-the-famous-and-historic-hoover-damn-hydroelectric-power-plant-at-the-nevada-arizona-border.html
29	Photo 2.11	Lordkinbote. "Unit No. 863 of the now-defunct Los Angeles Railway LARy makes its rounds." Wikimedia Commons. Available at commons.wikimedia.org
30	Photo 2.12	American Public Works Association. *Historical Photos.* Ed. Graeme Hunt. CD Rom. 1998.
31	Photo 2.13	American Public Works Association. *Historical Photos.* Ed. Graeme Hunt. CD Rom. 1998.
31	Photo 2.14	American Public Works Association. *Historical Photos.* Ed. Graeme Hunt. CD Rom. 1998.
31	Photo 2.15	American Public Works Association. *Historical Photos.* Ed. Graeme Hunt. CD Rom. 1998.
105	Photo 7.1	Dobrowolski, Anya, 2012

Page No.		Source
105	Photo 7.2	Dobrowolski, Anya, 2012
106	Photo 7.3	Dobrowolski, Anya, 2012
115	Photo 7.4	Dobrowolski, Anya, 2012
121	Photo 8.1	Goldstein, Beth 2005
124	Photo 8.2	Dobrowolski, Anya 2005
124	Photo 8.3	City of Berkeley Marina, Division, 2005
129	Photo 8.4	Dobrowolski, Anya, 2012
129	Photo 8.5	Dobrowolski, Anya, 2012
131	Photo 8.6	Dobrowolski, Anya, 2012
133	Photo 8.7	Barnes, Richard. Courtesy of Mark Cavagnero Associates, 2005.
133	Photo 8.8	EarthTech, Inc. 2005.
275	Photo 16.1	Michael Willis Architects. "Projects." 2005. www.mwaarchitects.com/projects/index.html
275	Photo 16.2	EarthTech Inc. Available at www.earthtech.com
278	Photo 16.3	sbjjk. "Ma Wat River water pipe under construction 2001." Wikimedia Commons. 2005. Available at commons.wikimedia.org/
301	Photo 17.1a	Dániel, Csörföly. "Manhole cover in Hungary." Wikimedia Commons. commons.wikimedia.org
301	Photo 17.1b	Azuma,Daiju. "Manhole cover in Osaka Japan." Wikimedia Commons. commons.wikimedia.org
301	Photo 17.2	Michael Willis Architects. "Projects." 2005. Available at www.mwaarchitects.com/projects/index.html
302	Photo 17.3	American Public Works Association. *Clip Art.* CD Rom. 1998.
311	Photo 17.4	Orange County Water District, 2012, used with permission.
311	Photo 17.5	Orange County Water District, 2012, used with permission
312	Photo 17.6	Orange County Water District, 2012, used with permission
312	Photo 17.7	Orange County Water District, 2012, used with permission
312	Photo 17.8	Orange County Water District, 2012, used with permission
314	Photo 17.9	Aguabio, Ltd. "Ultrafiltration Membrane System Used on an Activated Sludge Wastewater Treatment Plant." Wikimedia Commons. Available at commons.wikimedia.org
329	Photo 18.1	U.S. Department of Transportation
329	Photo 18.2	U.S. Department of Transportation
329	Photo 18.3	American Public Works Association. *Clip Art.* CD Rom. 1998.
354	Photo 19.1	Berkeley Historical Society.
357	Photo 19.2	American Public Works Association. *Clip Art.* CD Rom. 1998.
359	Photo 19.3	American Public Works Association. *Clip Art.* CD Rom. 1998.
391	Photo 20.1	Dobrowolski, Anya 2012.
391	Photo 20.2	Dobrowolski, Anya 2012
392	Photo 20.3	Dobrowolski, Anya, 2012
392	Photo 20.4	U.S. Department of Transportation, Federal Highway Administration. Manual of Uniform Traffic Control Devices MUTCD. 2004. Available at mutcd.fhwa.dot.gov/HTM/2003r1
393	Photo 20.5	Dobrowolski, Anya, 2012
394	Photo 20.6	Dobrowolski, Anya, 2012
395	Photo 20.7	American Public Works Association. *Clip Art.* CD Rom. 1998.
402	Photo 20.8	American Public Works Association. *Clip Art.* CD Rom. 1998.
402	Photo 20.9	American Public Works Association. *Clip Art.* CD Rom. 1998.
402	Photo 20.10	American Public Works Association. *Clip Art.* CD Rom. 1998.
404	Photo 20.11	American Public Works Association. *Clip Art.* CD Rom. 1998.
405	Photo 20.12	American Public Works Association. *Clip Art.* CD Rom. 1998.
405	Photo 20.13	American Public Works Association. *Clip Art.* CD Rom. 1998.
416	Photo 21.1	EarthTech Inc. Available at www.earthtech.com

Page No.		Source
416	Photo 21.2	EarthTech Inc. Available at www.earthtech.com
419	Photo 21.3	Bell, Jon. "Mostly Rail Transit Pages." Available at web.presby.edu/~jtbell/transit
419	Photo 21.4	Bell, Jon. "Mostly Rail Transit Pages." Available at web.presby.edu/~jtbell/transit
419	Photo 21.5	Bell, Jon. "Mostly Rail Transit Pages." Available at web.presby.edu/~jtbell/transit
419	Photo 21.6	Bell, Jon. "Mostly Rail Transit Pages." Available at web.presby.edu/~jtbell/transit
454	Photo 22.1	Schwen, Daniel. Wikimedia commons, 2005.
454	Photo 22.2	Dj1997. Wikimedia commons, 2012.
461	Photo 22.3	Horton, John. "iStockphoto." iStock International. 2005. Available at www.istockphoto.com/index.php
461	Photo 22.4	Barskaya, Galina. "iStockphoto." iStock International. 2005. Available at www.istockphoto.com/index.php
473	Photo 22.5	Ingram, Jonas. "iStockphoto." iStock International. 2005. Available at www.istockphoto.com/index.php
483	Photo 23.1	Courtesy Port of Oakland, California
485	Photo 23.2	Brown, Peter H.
495	Photo 23.3	Courtesy San Diego Convention Center
508	Photo 24.1	Courtesy of the Seattle Public Library 2005. Available at www.spl.org/default.asp?pageID=branch_open_features_detail&branchID=3&cid=1118937763511
508	Photo 24.2	Goldstein, Beth. "Photos." Denver, CO and Berkeley, CA.
509	Photo 24.3	Michael Willis Architects www.mwaarchitects.com
510	Photo 24.4	Courtesy of Beverly Prior Architects, Available at www.bparch.com/Pages/designC/design1F.html
510	Photo 24.5	Michael Willis Architects www.mwaarchitects.com
511	Photo 24.6	Courtesy of Beverly Prior Architects www.bparch.com
511	Photo 24.7	Town of Southwick, Massachusetts. http://southwickma.org/Public_Documents/F0000F6B4/S002CC090.-1/station.jpeg
512	Photo 24.8	Courtesy of Pioneer Courthouse Square, "Portland Hotel Gate." ed. Theresa Vetsch. 2004. Available at www.pioneercourthousesquare.org
512	Photo 24.9	North Carolina Court System. http://www.nccourts.org/County/Cherokee/Courthouse/Default.asp
513	Photo 24.10	County of Santa Barbara, http://www.santabarbaracourthouse.org/sbch/
513	Photo 24.11	Courtesy of Beverly Prior Architects, Available at www.bparch.com/Pages/designC/design1F.html
513	Photo 24.12	Courtesy of Beverly Prior Architects, Available at www.bparch.com/Pages/designC/design1F.html
514	Photo 24.13	Courtesy of Wilkerson Associates Architects. Photo by Dennis Nodine, Charlotte, NC. "Featured Projects: Healthcare." 2005. Available at www.wilkersonarch.com/Frametest_Charlotte_Architects_Healthcare.htm
514	Photo 24.14	Florida Public Health Institute. www.flphi.org/fphi-in-the-news
515	Photo 24.15	admairoa.blogs.com
515	Photo 24.16	John Tornow, Wikimedia commons. 2007
516	Photo 24.17	Fourquare.com,"Oregon Convention Center". https://is0.4sqi.net/userpix/APQPCFATVJYJE5C4.jpg
516	Photo 24.18	Goldstein, Beth. "Photos." Denver, CO and Berkeley, CA
516	Photo 24.19	Greenspun, Philip. http://philip.greenspun.com/images/20061019-boston-aerials/gillette-stadium-1.tcl
516	Photo 24.20	Courtesy of stadiumsofnfl.com
517	Photo 24.21	Goldstein, Beth. "Photos." Denver, CO and Berkeley, CA.
517	Photo 24.22	Goldstein, Beth. "Photos." Denver, CO and Berkeley, CA.
519	Photo 24.23	U. S. Green Building Council
520	Photo 24.24	Courtesy of Infra-Red Analyzers, Inc. www.iranalyzers.com/building-envelope.htm
520	Photo 24.25	Low Impact Development Center. www.lowimpactdevelopment.org/
520	Photo 24.26	Courtesy of the City of Minneapolis, MN
522	Photo 24.27	Courtesy of Mark Cavagnero Associates Photo by Tim Griffith
523	Photo 24.28	Courtesy of Big Bend Community College. http://library.bb.cc.wa.us/Library-ATEC/construcgion/images04-03-29/default.htm

Page No.		Source
527	Photo 24.29	City of San Antonio, Texas.
554	Photo 26.1	San Francisco Public Library. "San Francisco Historical Photograph Collection." 2002. Available at sfpl.org/librarylocations/sfhistory/sfphoto.htm
557	Photo 26.2	City of Berkeley. Images of Berkeley. Parks and Marina Division. 2005.
557	Photo 26.3	City of Berkeley. Images of Berkeley. Parks and Marina Division. 2005
557	Photo 26.4	City of Berkeley. Images of Berkeley. Parks and Marina Division. 2005
558	Photo 26.5	City of Berkeley. Images of Berkeley. Parks and Marina Division. 2005
558	Photo 26.6	City of Berkeley. Images of Berkeley. Parks and Marina Division. 2005
558	Photo 26.7	City of Berkeley. Images of Berkeley. Parks and Marina Division. 2005
590	Photo 27.1	Croswhite, Don. "iStockphoto." iStock International. 2005. Available at www.istockphoto.com/index.php
590	Photo 27.2	Plett, Ran. "iStockphoto." iStock International. 2005. Available at www.istockphoto.com/index.php
598	Photo 27.3	EarthTech Inc. Available at www.earthtech.com

Table Sources

Table 1.1 American Society of Civil Engineers. *2009 Report Card for America's Infrastructure*. 2009. Available at www.asce.org/reportcard

Table 3.1 Riche, Martha Farnsworth. "How Changes in the Nation's Age and Household Structure Will Reshape Housing Demand in the 21st Century." In *Issue Papers on Demographic Trends Important to Housing*. Washington, DC: U.S. Department of Housing and Urban Development, 2003. Table 1. Available at www.huduser.org/Publications/PDF/demographic_trends.pdf

Table 3.2 Burchell, Robert W., et al. *Costs of Sprawl—2000*. Transit Cooperative Research Program Report no. 74. Washington, DC: National Academy Press, 2000.

Table 4.1 Dowall, David E. and Jan Whittington. *Making Room for the Future: Rebuilding California's Infrastructure*. San Francisco: Public Policy Institute of California, 2003.

Table 4.2 U.S. Census Bureau. *2002 Census of Governments, Volume 1, Number 1, Government Organization*. GC02(1)-1. Washington, DC: U.S. Government Printing Office, 2002. Available at www.census.gov/prod/2003pubs/gc021x1.pdf

Table 4.3 U.S. Census Bureau. *2002 Census of Governments, Volume 1, Number 1, Government Organization*. GC02(1)-1. Washington, DC: U.S. Government Printing Office, 2002. Available at www.census.gov/prod/2003pubs/gc021x1.pdf

Table 5.1 U.S. General Accounting Office. *Local Growth Issues-Federal Opportunities and Challenges*. Washington, DC: U.S. General Accounting Office, 2000; and LeGates, Richard T. *The Region Is the Frontier: Frameworks, Goals, and Mechanisms for Collaborative Regional Decision-Making in Twenty-First Century California*. Sacramento, CA: Faculty Fellows Program, Center for California Studies, California State University, Sacramento, 2001.

Table 5.2 King County, Washington and ICLEI. *Preparing for Climate Change: A Guidebook for Local, Regional, and State Government*. King County, Washington, 2008.

Table 8.1 U.S. General Accounting Office. *Role of Facility Design Reviews in Facilities Construction*. Washington, DC: U.S. General Accounting Office, 2000.

Table 9.1 Anne Arundel County, MD. *Budget for FY 2004*. 2004.

Table 9.2 Santa Barbara County, CA. *Proposed CIP*. 2010.

Table 9.3 City of Los Angeles, CA. *L.A. Infrastructure Report Card*. 2003.

Table 10.1 National Research Council. *Measuring and Improving Infrastructure Performance*. Washington, DC: National Academy Press, 1995.

Table 10.2 Hudson, W. Ronald, Ralph Haas, and Waheed Uddin. *Infrastructure Management*. New York: McGraw-Hill, 1997.

Table 10.3 National Research Council. *Measuring and Improving Infrastructure Performance*. Washington, DC: National Academy Press, 1995.

Table 10.4 National Research Council. *Measuring and Improving Infrastructure Performance*. Washington, DC: National Academy Press, 1995.

Table 11.1 National Income and Product Account. Available at www.bea.gov (accessed January 15, 2012)

Table 11.2 U.S. Census Bureau. *State and Local Government Finances by Level of Government and by State: 2008–2009*. Washington, DC: U.S. Congressional Budget Office, October 2011.

Table 11.3 U.S. Census Bureau. *U.S. Statistical Handbook*. Washington, DC: Government Printing Office, 2001.

Table 11.4 U.S. Census Bureau. *U.S. Statistical Handbook*. Washington, DC: Government Printing Office, 2001.

Table 12.1 U.S. Census Bureau. *Statistical Abstract of the United States 2012*. Washington, DC: U.S. Census Bureau, 2012.

Table 13.1 U.S. Census Bureau. *Statistical Abstract of the United States: 2001*. Washington, DC: U.S. Census Bureau, 2001. Available at www.census.gov/govs/www/estimate.html accessed July, 9, 2002

Table 14.1 Mullen, Clancy. *2011 National Impact Fee Survey*. Duncan Associates, 2011.

Table 14.2 Mullen, Clancy. *2011 National Impact Fee Survey*. Duncan Associates, 2011.

Table 14.3 Mullen, Clancy. *2011 National Impact Fee Survey*. Duncan Associates, 2011.

Table 16.1	U.S. Department of the Interior. U.S. Geological Survey. Available at ga.water.usgs.gov/edu/2010/gallery/global-water-volume.html
Table 17.1	Tabors, Richard D., Michael H. Shapiro, and Peter P. Rogers. *Land Use and the Pipe: Planning for Sewerage*. Lexington, MA: Lexington Books, 1976.
Table 18.2	Knox County Tennessee. *Stormwater Management Manual*. 2 vols. Available at www.knoxcounty.org/stormwater/proposed_stormwater_ordinance.php
Table 19.1	U.S. Environmental Protection Agency. *Municipal Solid Waste in the United States: 2010 Facts and Figures*. Washington, DC: U.S. Environmental Protection Agency, 2010.
Table 19.2	U.S. Environmental Protection Agency. *Municipal Solid Waste in the United States: 2010 Facts and Figures*. Washington, DC: U.S. Environmental Protection Agency, 2010.
Table 19.3	California Department of Resources Recycling and Recovery. *Resource Recovery Parks a Model for Local Government Recycling and Waste Reduction*. 2002. Available at www.ciwmb.ca.gov/lglibrary/innovations/recoverypark/default.htm
Table 19.4	U.S. Environmental Protection Agency. *Municipal Solid Waste in the United States: 2010 Facts and Figures*. Washington, DC: U.S. Environmental Protection Agency, 2010.
Table 19.5	Alameda County Waste Management Authority. 2008.
Table 19.6	Alameda County Waste Management Authority. 2008.
Table 19.8	Alameda County Waste Management Authority. 2008.
Table 21.1	Bronson, T. and C.R. Dawson. 2002 Transit Fare Summary. Washington, D.C.: American Public Transportation Association.
Table 21.2	Cervero, Robert. *The Transit Metropolis: A Global Inquiry*. Washington, DC: Island Press, 1998.
Table 21.3	U.S. Department of Transportation. Bureau of Transportation Statistics. *Transportation Indicators Report-July 2002*. 2002. Available at www.bts.gov/publications/transportation_indicators/july_2002/index.html
Table 21.5	Metropolitan Transportation Authority. *San Francisco Bay Area 1990 Regional Travel Characteristics: Working Paper #4*. 1990. Available at Ntl.bts.gov/DOCS/SF.html
Table 21.6	Arkoma Regional Planning Commission. *Bi-State Transportation Trip Generation Rate Study*. 2001.
Table 21.7	U.S. Department of Transportation. Bureau of Transportation Statistics. *Transportation Statistics Annual Review*. 2000.
Table 21.8	American Society of Civil Engineers. *Specific Policy Recommendations for America's Infrastructure*. Washington, DC: 2001. Available at www.asce.org/reportcard/index.cfm?reaction=policy
Table 21.9	U.S. Department of Transportation. Bureau of Transportation Statistics. *Transportation Statistics Annual Review*. 2000.
Table 21.10	Florida Department of Transportation. Systems Planning Office. *Generalized Annual Average Daily Volumes for Florida's Urbanized Areas*. 2002.
Table 21.11	U.S. Department of Transportation. Bureau of Transportation Statistics. *Transportation Statistics Annual Review*. 2000.
Table 21.12	California Energy Resources Conservation and Development Commission. *Energy Aware Planning Guide I*. 1993. Available at www.energy.ca.gov/energy_aware_guide/index.html.
Table 21.13	U.S. Department of Transportation. Federal Highway Administration. *Transportation Air Quality: Selected Facts and Figures*. 2002. Available at www.fhwa.dot.gov/environment/aqfactbk/index.htm
Table 21.14	Henkin, Tamar. "Innovative Transportation Revenue Sources." In *3rd National Transportation Finance Conference*. Chicago, 2002.
Table 22.1	U.S. Department of Transportation. Bureau of Transportation Statistics. *National Transportation Statistics*. 2002. Available at www.bts.gov/publications/national_transportation_statistics
Table 22.2	Gesell, Laurence E. *The Administration of Public Airports*. 3rd ed. Chandler, AZ: Coast Aire Publications, 1992.
Table 22.3	Demphsey, Paul Stephen. *Airport Planning and Development Handbook: A Global Survey*. New York: McGraw-Hill, 2000.
Table 22.4	National Association of State Aviation Officials. *Land Use Survey 2004*. Available at meadhunt.com (accessed April 4, 2005).

Table 23.1	Association of American Port Authorities. Available at www.aapa-ports.org
Table 23.2	Association of American Port Authorities. Available at www.aapa-ports.org
Table 23.3	Association of American Port Authorities. Available at www.aapa-ports.org
Table 23.4	Association of American Port Authorities. Available at www.aapa-ports.org
Table 24.1	Altshuler, Alan, and David Luberoff. *Mega-Projects: The Changing Politics of Urban Public Investment.* Washington, DC and Cambridge, MA: Brookings Institution Press and Lincoln Institute of Land Policy, 2003.
Table 25.1	Chung, Connie. *Using Public Schools as Community-Development Tools: Strategies for Community-Based Developers.* Cambridge, MA: Joint Center for Housing Studies of Harvard University and Neighborhood Reinvestment Corporation, 2002.
Table 26.1	Center for Park Excellence, *No Place to Play.* Trust for Public Land, 2004. Available at www.tpl.org/ccpe (accessed July 24, 2008)
Table 26.2	Mertes, James D, and James R Hall. *Park, Recreation, Open Space and Greenway Guidelines* Alexandria, VA: National Recreation and Park Association and the American Academy for Park and Recreation Administration, 1995.
Table 26.3	Mertes, James D, and James R Hall. *Park, Recreation, Open Space and Greenway Guidelines* Alexandria, VA: National Recreation and Park Association and the American Academy for Park and Recreation Administration, 1995.
Table 26.4	City of Santa Monica, CA. *Santa Monica Sustainable City Plan.* 2003.
Table 27.1	American Public Power Association. "Public Power Statistics." 2002.
Table 27.2	U. S. Department of Energy.
Table 27.3	U.S. Department of Energy. Energy Information Administration. *Annual Energy Review.* 2002. Available at www.eia.doe.gov/emeu/aer
Table 27.4	U.S. Department of Energy. *Residential Energy Consumption Survey: Consumption and Expenditure Tables.* 2005. Available at www.eia.doe.gov
Table 27.5	U.S. Department of Energy. *Residential Energy Consumption Survey: Consumption and Expenditure Tables.* 2005. Available at www.eia.doe.gov/
Table 27.6	South Carolina Energy Office. *Energy: Preparing and Energy Element for the Comprehensive Plan.* 2000. Available at www.energy.sc.gov
Table 27.7	New York State Energy Research and Development Authority. Available at www.nyserda.org (accessed August l, 2008).
Table 27.8	Compiled by authors from various sources.
Table 28.1	State of the Internet, 2012 at www.akamai.com
Table 28.2	Chaffee, David and Mitchell Shapiro. *The Municipal and Utility Guidebook to Bringing Broadband Fiber Optics to Your Community.* American Public Power Association, 2008. Available at www.appanet.org
Table 28.3	Asp, Thomas, Harvey L. Reiter, Jerry Schulz, and Ronald L. Vaden. "Broadband Access: Local Government Roles." *IQ Report* 36, no. 2 (2004).
Table 28.4	Chaffee, David and Mitchell Shapiro. *The Municipal and Utility Guidebook to Bringing Broadband Fiber Optics to Your Community.* American Public Power Association, 2008. Available at www.appanet.org
Table 28.5	Meck, Stuart. *Growing Smart Legislative Guidebook 3: Model Statutes for Planning and the Management of Change.* Chicago: APA Planners Press, 2002.
Table 28.6	Gillette, Sharon E., William H. Lehr, and Carlos Osorio. "Local Government Broadband Initiatives." MIT Program on Internet and Telecoms Convergence. 2003.
Table 28.7	West, D.M., *State and Federal Electronic Government in the United States.* Brookings West, 2009.

Index

Please note that page references to Figures or Tables will be in *italics*, whereas those for Notes will have the letter 'n' following the page number. The majority of the content refers to the United States.

Abbott, W. M. 687n
accidents: airport 473; nuclear 582; road 390–1
acquisition and avigation easements 475–6, 656
activated sludge process 302, 655
ADA *see* Americans with Disabilities Act (ADA) 1990
Adams, T. 388
adaptation planning 83–4
Adequate Public Facilities Ordinance (APFO) 108–9, 548, 655
administration costs 242
Advanced Research Projects Agency (ARPA) 618, 619
Adventure Playground *558*
aerodynamic noise 475
Aeronautics Division, Montana 466
Agenda 21, 72
aging population 39–40
agricultural pollution 335
Agriss, T. 686n
Agron, J. 700n
Ai, N. 696n
Air Cargo Deregulation Act 1976 480
Air Commerce Act 1926 479
air pollution 439–41, 470–1
air traffic control and airways 461, 463
aircraft utilization rate 655
Airline Deregulation 1978 480
airmail 452
Airport and Airway Development Act 1970 480
Airport and Airway Development Act Amendment 1976 480
Airport and Airway Trust Fund 454
airport authorities 464–5
Airport Improvement Program (AIP) 458
airport management 451–80; acquisition and avigation easements 475–6, 656; air traffic control and airways 461, 463; airport authorities 464–5; airport sustainability survey, 2008 471; civil aviation categories 458–9; commercial aviation 452, 459; current trends and issues 455–8; demand and supply trends 455; deregulation 454–5; description of airport 459–63; early flights 452; environmentally friendly aircraft 456–7; facilities 459; federal entry into airport business 453; federal legislation/regulation 465, 479–80; financing of airports 463–4; fragmented aviation system planning 457–8; funding shortfall for airports 455–6; general aviation airports 458, 662; general purpose governments 466; and Great Depression 453; history of airports and air travel 451–5; hubbing 460; importance to local practitioners 451; institutions, aviation 463–6; and land use *see* airports and land use; local economic department, marketing and financing 476–8; low-cost and regional carriers 456; marketing an airport 477–8; mechanism of flight 462; from 1960s to the present 454–5; ownership of airports 464–5; personal air travel 668; planning *see* airport planning; postwar period, comprehensive national airport system 453–4; primary airports 669; privatization 465; regional jets and personal travel aircraft 457; regulation 465–6; runways 459–61; security issues 457, 471–2; State Departments of Transportation 465–6; terminals 461;

trust funds 454; users of airports 463; wide-body jets 454; zoning overlay district 476
airport planning: air quality 470; automobiles and parking 469–70; demand management 468; environmental considerations 470–1; expansion, alternatives to 468; fragmented aviation system planning 457–8; intermodal access/transit 470; layout plan 470; location and siting of airports 469–70; master planning 466–72; need identification 467–8; State of Washington selecte model comprehensive plan policies 476; transportation access 469
airports and land use: bird/aircraft strikes 474; building height 473–4; incompatible land use types and issues *472*; noise 474–6; safety zones 473
airside facilities 459
Airways Modernization Act 1957 480
Alameda, California 47, 372, 400, 501, 598; public buildings 511, 512
Alameda Power and Telecom 632
Alamosa Photovoltaic Solar Plant 591
Alaska, climate change 337
Albee, S. 275, 690n
Albuquerque City 290
Alexandria, Virginia 400
Allen, L. 692n
all-terrain bike trail 655
Alonso, W. 679n
alternating current (AC) 580, 586, 590, 655
Altshuler, A. 679n, 687n, 699n
Alvarez-Cohen, L. 691n
America 2050 66
America in Ruins (1981) 35
America Online 622
American Academy for Park and Recreation Administration 566
American Airlines, maintenance operations center 48
American Bell Telephone Company 614
American Institute of Architects 519
American Institute of Certified Planners (AICP) 540
American Planning Association (APA) 17, 76, 244, 540; Adequate Public Facilities Ordinance, model statute 108, 109; *Growing Smart Legislative Guidebook* 35, 78, 549; Sustaining Places Task Force 78
American Public Power Association 597
American Public Transportation Association 418
American Public Works Association (APWA) 331
American Recovery and Reinvestment Act 2009 197
American Recovery and Reinvestment Act 2013 16
American Society for Testing and Materials (ASTM) 570
American Society of Civil Engineers (ASCE) 22, 34; Infrastructure Report Cards 14, *15*, 186
American Telegraph and Telephone Company (AT&T) 614, 615, 621, 639
American Waterworks Association 276, 284
Americans Against Traffic Calming 401
Americans with Disabilities Act (ADA) 1990 392, 409, 418, 442, 570, 576; Architectural Guidelines (ADAAG) 392
Ammons, D. N. 685n
analog telecommunications 615
analytical tools 161–2
Anderson, L. T. 690n, 692n
Anderson, R. 691n
animal feeding operations (AFOs) 335
Anne Arundel County, Maryland *142*, 143
antennas 626, 627
APA *see* American Planning Association (APA)
appropriative water rights 655
aquifers 277, 280, 318, 655
Army Corps of Engineers 22, 53, 326, 330, 493
ARPA (Advanced Research Projects Agency) 618, 619
ARPANET (network) 619, 655
art, public 393, 394, 418
art museums/cultural buildings 509–10
arterial roads 655
artificial recharge, water management 655
Asano, T. 692n
"as-built" project records 159
ASCE *see* American Society of Civil Engineers (ASCE)
Aschauer, D. A. 679n
Ascher, W. 676n
asphalt 326, 342, 387, 389, 578, 656
asset management 139, 156–64, 656; capital budget 138, 158; information management 157–64; landscape asset maintenance 155; linkage with organization's goal-setting process 157; need for focus on infrastructure 154–5
Assured Water Supply program, Arizona 288
AT&T (American Telegraph and Telephone Company) 614, 615, 621, 639
Atlanta, Phoenix 445
Atomic Energy Act 1954 581, 609
Atomic Energy Commission 609
Atoms for Peace Program (Eisenhower) 609
Austin, Texas 48

automobiles: and airport planning 469–70; ownership 417; parking 469–70; pollution measures 342, *343*; recreational automobile era (1920–1945) 414–15; vehicle efficiency improvements 438
avenues 396
average or full cost pricing 215–16, 656, 661
Aviation Safety and Noise Abatement Act 1979 480
avigation easements 476, 656

Baber, C. 692n
Baby Bells 615
Baby Boomers 39, 40, 43, 535, 557
Bailey, S. J. 686n
Baird, J. L. 617
Bakken oil fields, North Dakota *11*
Baldassare, M. 700n
Baltimore, Maryland 155, 485, 491
bandwidth 619–21, 631, 656
banking 656
bankruptcy, municipal 199
Barber, N. L. 690n
Barbour, E. 676n, 681n
Barkley, D. L. 701n
base course 389
baseline conditions, analyzing 92, 94–5
Baton Rouge, Louisiana 277
Bay Area Rapid Transit (BART) District, San Francisco 416, 417, 441
Bay Shore, New York (Hose Company Firehouse) *25*
Becker, B. 321, 682n, 693n, 699n
Beecher, J. A. 686n, 690n, 691n
Belinfante, A. 703n
Bell, A. G. 614, 615
benchmarking 166–7, 656; local benchmarks 186–7; national benchmarks 185–6
Benevento, D. H. 274
Ben-Joseph, E. 696n
Berke, P. K. 682n
Berkeley, California: California Disability Compliance Program 409; César Chávez Park 557; emergency operation center 511; energy costs 601–2; and privatization 257; Public Library *121*
Berkshire Connect project, New England 633
Bhatta, S. D. 680n
bicycles 391, 393, 400, 656, 667
Billington, D. 527, 700n
Binkley, C. 682n
biofuels/biomass 593
bioreactor landfills 378
bio-retention areas 341–2
biosolid (sludge) reuse, centralized 312, 656
bird/aircraft strikes 474
Birmingham, Alabama 253
bits per second (Bps) 656
black water 315, 316, 318, 656
Blaesser, B. W. 687n
Bland, R. L. 685n
Blewitt, R. A. 688n
block rates 220, 659, 663; two-part tariff with 217–18
Bloomington, Illinois 478
Blueprint for Survival report 72
Boeing 787 Dreamliner 456–7
Boise (Idaho), connection fees 323
bond counsel 198
bond issuance process 197–9, 202–5
bonds 21, 189–205; annual municipal sales, dollar volume (1986–2010) *191*; approval of issue by governing body 205; benchmarking 186; Build America Bonds 193, 197; business improvement districts 197, 657; competitive sale 204; consulting engineers 199; cooperative bond pools, use of 202; credit enhancement devices, use of 202; defaults and credit worthiness 199–202; defined 189; demand for/cost of project 203; documentation, preparing 205; double-A-rated corporate 199; financial advisers 198; general facts 189–92; general obligation 189, 192, 193–4, 201, 662; geographically based 189, 196–7, 202; government agencies, issued by 197–8; holders of US municipal securities from 1996 to 2010 *193*; importance to local practitioners 189; industrial restructuring and financing innovations 23; investment grade 200–1; issuance process *see* bond issuance process; issuers 190, 191–2, *194*; junk 201; lease-financing (certificates of participation) 195, 202, 657; liability for payment 203; limited liability *see* revenue bonds; local tax bill, determining impact on 203; maturity structure 666; negotiated sale 204; new security issues *192*; outside experts and professionals, issued by 198–9; parks and recreational facilities management 562; polling expert, issued by 199; private placement 204–5, 669; public schools 538; purchasers of 192; quasi-public buildings 514–15; ratings/rating agencies 199, 200–2; redevelopment 197; revenue 192–3, 194–5, 201–2, 670;

sale of issue 204–5; size of bond market 190, 203; special assessment districts 63, 196–7, 671; stormwater/flood management 345; tax increment 197, 672; taxable 189, 197; tax-exempt 190, 192, 193–4, 196–7; triple-B-rated corporate 199; types 192–7; underwriters 198–9; use of proceeds 191–2
Bonneville Power Administration (BPA) 609
Bonneville Project Act 1937 609
boom and bust cycles 16
BOOT (Build–Own–Operate–Transfer) 254, 657
Border Growth Metros 44, 45
Boston Society of Civil Engineers 22
"bottom line," 508
boulevards 391, 396
Bowyer, R. A. 684n
box, bicycle 391
Bps (bits per second) 656
break bulk cargo 656
bridges 158, 392
British Thermal Units (BTUs) 598
broadband 621, 634
Brookings Institution study 2010 44
Brown, M. 684n
Brown, P. H. 698n
Brundtland Report (*Our Common Future*), 1987 72
Brush, C. 580
Brushy Creek, Austin 305
BTO (Build–Transfer–Operate) 254, 657
"bubble economies," 43
Build America Bonds (BABs) 193, 197
Building America program (U.S. Department of Energy) 524
building and occupancy permit process 118
Building Codes 115, 656, 673; *see also* Green Building Codes
building envelope 520, 656
building permits 132
buildings *see* public buildings
Build–Own–Operate–Transfer (BOOT) 254, 657
Build–Transfer–Operate (BTO) 254, 657
bulk cargo 657
Bulletin or Planning Neighborhoods for Small Houses 1936 409
Bunnell, G. 687n
Burchell, R. W. 174, 675n, 679n, 687n, 688n
Bureau of Economic Statistics 8
Bureau of Labor Statistics 161
Bus Rapid Transit (BRT) 96, 418–19, 657
bus stop/transit shelters 393, *394*
buses 418, 657
business improvement districts (BIDs) 197, 657
Buus, B. 693n, 695n
bypass and blending 657

C40/connecting Delta Cities 83
cable television 618, 623, 657, 663
CAFE (corporate average fuel economy) 438
CAFOs (concentrated animal feeding operations) 335
Cagan, J. 699n
CAL Green 115
California: Aviation System Plan 458; banning of discarded cathode ray tubes from municipal landfills 374; Broadband Task Force 624; carbon emissions reduction 66; charter schools 543; Collaborative for High Performance Schools 546–7; comprehensive plans 76–7; earthquakes 510–11; energy crisis 2001/2001 579; environmental impact assessments 114; gas tax funds 444; impact fees and CIP 238; Integrated Waste Management Board website 696n; local planning 17; Palace of Honor *133*; population growth 45; Roundtable on California's experience with Innovations in Public Financing for Infrastructure 2000 258–9; and water *see* California and water supply; *see also* Alameda, California; Berkeley, California; East Bay Municipal Utility District (EBMUD); East Bay Regional Park District (EBRPD), California; Fresno, California; Long Beach, California; Los Angeles, California; Oakland, California; Orange County, California; Sacramento County, California; San Diego, California; San Francisco, California; Santa Barbara, California; Santa Cruz, California; Santa Monica, California; Sunnyvale, California
California and water supply: aquifers 277; carbon emissions and water 273; conservation 291; growth and water link 273; Merced exchequer dam *270*; Sobrante Ozonation facility, El Sobrante *270*; water supply availability statutes (SB 610 and 221) 288, 289
California Electric Light Company, San Francisco 580
California Environmental Quality Act (CEQA) 114
Caltrans (California Department of Transportation) 428
canals 21
capacity analysis 499–500
capacity building 261–2
Capello, R. 677n

capital budget 137–52; adoption 150–1; citizen and stakeholder involvement process 149; criteria 149; defined 145, 657; developing 148–51; federal government guidelines *139*; forms 149; Government Accounting Standards Board (GASB) Rule 34 and infrastructure budgeting 137, 140, 147–8, 160, 662; local budgeting 137–41; long-term operating budget plan 141–2; and operating budget 145–6; pedestrian needs 400; as practiced by local governments 138; preparing 150; process, developing 149; project-specific capital financing 138; schedule 149; as separate document, recommendations for 146–7; special procedures for capital projects 138; and status of capital improvement plan 137–8; *see also* capital improvements program (CIP)
capital expenditures: capital to operating expenditures for key infrastructure systems 179; community facilities 177; comparison of annual capital needs estimates at local level *187*; by level of government 176; outlays by state and local government in 2009 *178*; state vs. local, on infrastructure 176–7, *178*; transportation 177
capital facilities 322–3, 376–9
capital improvements program (CIP) 137–52, 625, 657; capital needs study 143–5; capital planning 150; committee, designating 148–9; comprehensive plans 140, 476, 525, 658; debt affordability limit 141, *142*, 659; defined 141; designation of percentage of revenues to capital 141; developing 148–51; documentation 141–5; funded and unfunded needs 142; identifying projects and funding options 149; and impact fees, California 238; infrastructure needs, comprehensive assessment 139–40; lead department, designating 148–9; linkage with comprehensive plan 151; local example 184; local government and local infrastructure funds 140–1; Los Angeles Uses Report Card 144; organization 143, 148–9; port planning 497; revenue sources, Fairfax County (2013) *185*; routine, lack of 139; selecting projects 150; status, and budgets 137–8; and strategic planning 80–1; technical tools, improvement requirement 138–9; timeframe 141; *see also* capital budget
capital investment needs 12–15
capital needs study 657
capital projects 138, 143, 657
carbon capture and storage (CCS) 587–8
carbon dioxide (CO_2) emissions 113, 273, 310, 335, 438, 587
carbon emissions 12, 310, 423, 518; low carbon energy and decentralized solutions 591–4; reduction opportunities 66, 606; and water supply 273–4; *see also* greenhouse gas (GHG) emissions
carbon monoxide 440
Carbonell, A. 681n
carbon-neutral cities 16
cargo ports 483–5; capital planning 499–501; finances 492–4
Carrier, B. 699n
cars *see* automobiles
Castells, M. 703n
cathedrals, Middle Ages 6
cathode ray tube (CRT) television 374
Catskill/Delaware Watershed 278, *279*
cell phones 218, 219, 615–16, 657
Center for Neighborhood Technology 433
Center for Performance Measurement, ICMA 167, 526
Center for Responsive Politics 582
Centers for Disease Control and Prevention 399, 549, 550
Central Illinois Regional Airport 478
central office 665
Central Park, New York City 554–5, 571
certificates of participation (lease-financing bonds) 195, 202, 657
Cervero, R. 433, 447, 679n, 697n
César Chávez Park, Berkeley 557
cesspools 24, 296
Chadwick, E. 296
Chandler, A. M. 677n
Chang, S.-L. 682n
Charlotte, North Carolina 81
Charlotte-Mecklenburg Utility Department, North Carolina 256
charter schools 542–3, 657
"chasing smokestacks," 477
Chattanooga *105*
Cheeseman Day, J. 675n
Cherokee County, Georgia 513
Cherry Pointe airport control tower *461*
Chicago, Illinois: bond market 190; commuter rail map *421*; groundwater levels 277; McCormick Place Convention Center 515; parks and recreational facilities management 565; Public Library (*now* Chicago Cultural Center) *25*; Sears Tower 524; stormwater/flood management 155, 342; street management 406; Tunnel and Reservoir Plan 327; Union Station *25*

Chicago School 248
Childers, D. L. 692n
chloramine 270
chlorine 270, 302
Choate, P. and W. S. 678n
cholera 270
Christensen, K. S. 676n
"Cincinnati Plan," 26
CIP *see* capital improvements program (CIP)
Circular Number Five (Subdivision Development), 1935 409
cities: carbon-neutral 16; in Colonial Era United States (1770–1850) 19, 20, 21; "global," 32; and harbors 482; industrial and scientific, rise of (1850s–1930) in United States 22–7; infrastructure challenges, future cities *see* "City of the future", infrastructure challenges; and landowners 688n; and parks 554–6, 566; post-industrial city (1970s–2001), United States 31–5; public school districts, "salo planning" separation 551; and public schools 539–41; and solid waste 352; tourist 558; transportation 423; "walking," 20; worldwide, per capita fees for solid waste management *364*
Cities for Climate Protection campaign (ICLEI) 72
citizens' commission 60, 680n
"City Beautiful" movement 25
City Hall 508, 526, 647
"City of the future," infrastructure challenges: boom and bust cycles 16; capital investment needs 12–15; climate change and environment 11–12; inadequate institutional structure 16–17; local fiscal pressures 16; national framework, lack of 18; water industry 269; *see also* cities
city planning, as profession 22
civic centers 508
civic squares 508, 509
Civil Aeronautics Act 1938 453, 479
Civil Aeronautics Board 453–4
civil engineers 122, 353
Claire, W. H. 677n
Clarke, S. E. 679n
Clarke, W. 687n, 688n
Clarkstown, New York 108
Clean Air Act 1970 32
Clean Water Act 1972 283, 298, 299, 331
Clean Water Act 1977 108, 294, 349
Clean Water State Revolving Fund (CWSRF) programs 347
cleaning and sweeping of streets 405
Cleveland City 48, 597
climate action plans (CAPs) 81, 82
climate change 11–12, 17; climate change planning and sustainability planning 81–3, 518; King County, expected climate change impacts *83*; Kyoto Agreement and Protocol on global warming 73, 439, 521; public buildings and sustainability 518; stormwater 335–7, 347; in United States 583–5; water supply 271–3
cluster development 106
cluster zoning 658
coal 587–8
coaxial cable/coax 658
cogeneration plant 586, 658
cold start 440
Cole, R. J. 675n
Collaborative for High Performance Schools (CHPS) 546–7
collaborative planning 93, 658
Collignon, F. 686n, 688n
Colonial Era (1770–1850), United States 19–22; agriculture-based economy 20; engineers and plans 22; federal role in infrastructure provision 21–2; initial infrastructure 20
Colorado: development density regulations 600; rate shock 285; water conservation 288, 289
Columbia Basin Project 28
combined sewer overflows (CSOs) 303, 304
combined sewer systems (CSSs) 297, 329–30, 331, 347, 658
combined-cycle plant 588, 658
combustion 359
commercial aviation 452, 459
commercial service airports 658
"commons," the 47
Communications Act 1934 622, 639
Communications Satellite Act 1962 639
Community Antenna Television (CATV) 618
community benefits agreements (CBAs) 517
community builders 25
community design and decision making, technology for 99
community facilities 10, 675n
community health centers (CHCs) 514
community parks 658
community process tools 99

community rating system (CRS) 330
community schools 545–6
community water systems 658
compensation, telecommunications systems 629–33
competition: importance to local practitioners 247; internal 256–7; managed 665; *see also* privatization
composting, and food scraps 372–3
Comprehensive Environmental Response, Compensation and Liability Act (CERCLA) 1980 381
Comprehensive Planning Assistance Act 1954 653
comprehensive (general) plans 476, 525, 658; and capital improvement plan 140; current status of local comprehensive land use planning requirements 76; definitions 71, 73–4; environment and infrastructure 75; federal grant requirements 75–6; influences on development 75–6; infrastructure program, importance 88; model ordinances 75; privatization 261; reform efforts 76–9; relationship to development impact fees 240; significant state planning reforms *77*; smart planning and sustainability planning 78–9; state planning requirements for local infrastructure planning 76–7; status of infrastructure and capital planning at state level 78; *see also* plans/planning considerations
computerized maintenance management systems (CMMS) 157, 658
computers, invention 618
concentrated animal feeding operations (CAFOs) 335
concrete 356, 433, 538, 658; airport management 460, 468; stormwater/flood management 326, 342; streets/street management 389, 392
concurrency 108, 234, 658
condemnation 658
condition assessment: defined 658; information management 159–60; infrastructure program 94–5; transportation 430; wastewater planning 321
Conference of Mayors, U.S. 519
congestion, and greenhouse gas emissions 434
congestion pricing 218–19, 434–6, 658
Congressional Budget Office (CBO) 46
conjunctive use 278, 659; water 278
connection fees 659
connector trails 659
conservation: energy 606; land use, and equity 322; of water *see* water conservation
Constitution, U.S., 16th Amendment 23
construction: closeout 133–4; public facilities 117; public infrastructure project, development 123–4, 130–4
construction & demolition (C&D) debris 361
construction contract: and building permits 130–3; entering into 132–3
Construction Engineering Research Lab (CERL), U.S. Army 403
construction management: construction management team 123–4; pollution control 344
consulting engineers, bond issuance 199
consumer price index, trends for public utility services (1970–2007) *211*
Consumer Product Safety Commission (CPSC) 570
container ports 483–4
containers 659
contaminants, water: emergence of 274–5, 303; groundwater 274, 303–4; surface water 303–4
context-sensitive solutions (CSS), street management 398–9
Contract Air Mail Act 1925 479
contracting out, privatization 255–6, 561–2, 659
Contrail, privatization 253
convention centers 515–16
conventional pollutants 299, 659
convergence 659
COOL (Compositing Organics Out of Landfills) 373
cooperative bond pools, use of 202
Cordell, D. 691n
corporate-style benchmarking 166–7
corporation yards 61, 510, *511*
cost information 161
cost-benefit analysis 138, 290
cost-efficiency measures 165
costs: administration 242; allocation of 214; capital, covered by taxpayers 218; delivery 221; energy 601–2, 603–4; full cost of service provision, calculating 221; indirect 222; life-cycle cost analysis 126–30, 162–4, 665; local infrastructure, impact of density on 174; "lumpy," 217; maintenance, covered by users 218; operating, covered by users 218; per dwelling unit 180; per unit of capacity, identifying 242; water, full cost pricing 285–6, 661; *see also* public pricing concepts, and fee types
Council of Educational Facility Planners International (CEFPI) 543, 544
councils of governments (COGs) 65, 66, 659
counties, transportation 423
coupon rate 659
court houses 512

coverage ratio 201
Coy, D. G. 690n
Crandall, R. 703n
Cranz, G. 701n
Crawford, T. 679n
credit enhancement devices, use of 202
credit worthiness, bonds 199–202
Crete, Minoan civilization 296
Creyts, J. 699n
criminal justice facilities 512–14
Crompton, J. L. 702n
cross-functional relationships, promoting 651
crosswalks 390–1, 393
Croton Reservoir, New York 278, *279*
Cruise Line International Association 486
cruise revenues 494–5
cryptosporidiosis 305, 659
cultural and historic districts 517–18
curbs 389–90; curbside source separation 357–8
Curtis, L. W. 693n
cut and cover sanitary landfill method 353
cycle-time measures 165

Daigger, G. 689n, 692n
daily cover, sanitary landfills 358
Daisa, J. M. 696n
Dallas, Texas, infrastructure maintenance needs 155
Damico, J. F. 693n
dams 589
Danielson, M. N. 698n
Darwick, N. 698n
Database of State Incentives for Renewable Energy (DSIRE) 524
DBO (Design–Build–Operate) 254, 660
debt affordability model 141, *142*, 659
debt limit/cap 659
dechlorination 302
dedications 106, 229, 659
deep tunnel concept 327, 329
defaults, bonds 199–202
Definition of Solid Waste (DSW) 356
definitions of infrastructure: current 6–7; from developers 8–9; economic views 7–8
DeKalb County real estate tax statement *183*
Del Porto, D. 692n
Delafons, J. 682n
DeLille, B. 696n
delivery costs 221
demand: future population, translating into 97; and growth *see* demand and growth patterns; induced 431, 659; latent 431, 664; water demand, projecting 284–5
demand and growth patterns, infrastructure requirement: demographic factors 37–49; economy and infrastructure 45–9; human capital, importance 49; importance to local practitioners 37; location decisions 47; national population trends and households 37–40; whether public infrastructure investment causes growth 45–6, 47; rising incomes/rising income inequality 41–3
demand and supply trends, airport management 455
demand forecasting 467
demand management 97, 660; pricing 219–20; transportation *see* demand management transportation
demand management transportation: airports 468; congestion pricing 434–6; GHG emissions and congestion 434; intelligent transportation systems 436–7; parking control 436; system efficiency 439
deMause, N. 699n
demographic factors: aging population 39–40; California, population growth 45; demand, growth and infrastructure requirement 37–49; ethnic diversity, increasing 40–3; future population, projecting 96–7; at metropolitan level 43–5; mobility patterns, changing 39; population trends, households, and infrastructure demand 37–40; post-World War II (1945–1960s), in United States 32; public schools 539–40; and transportation 39
density, impact on local infrastructure costs 174
Denver, Colorado: Art Museum *509*; infrastructure maintenance needs 155; International Airport 459; Lower Downtown Historic District *517*
Denver International Airport 478
Department of Education 536, 537, 543
Department of Energy 583, 600
Department of Energy Organization Act 1977 610
Department of Health and Human Services 550
Department of Transportation (DOT) 53, 157, 422, 449, 454
Depression and World War Two era (1930–1945), United States 22, 180; air travel 453; concepts to be implemented in post-World War II period 29; direct spending 27–8; and park facilities 556–7; regulation of private infrastructure provision 28–9

deregulation: airport management 454–5; defined 660
desalination 278, 660
DESAR (Decentralized Sanitation and Reuse) 314–17
design aspects: community design and decision making, technology for 99; design review and value engineering 127; design team 122; energy/power 601; finalizing of design contract 131; impact fees and exactions 241; incineration 379; landfills 376–9; pollution control 344–5; preliminary design 126; public buildings 527–8; public infrastructure development 122, 126, 127, 131; public schools 544–7; stormwater management and planning 341–2, 344–5; streets 389, 395–6
Design–Build 254, 660; enabling legislation needed 260–1
Design–Build–Operate (DBO) 254, 660
deteriorated infrastructure 14–15, 94, 153–5
Detroit, Michigan 155
developers, definition of infrastructure 8–9
developing countries, privatization in 249
development caps, infrastructure capacity 110, 660
development density regulations 600–1
development impact fees *see* impact fees, development
development impact regulations 107–10
development process 116–18
development regulations/development impact regulations 107–10, 660
development rules: adequacy of local rules, assessment principles 117; Adequate Public Facility Ordinance 108–9; building and occupancy permit process 118; caps, infrastructure capacity 110, 660; cluster development 106; dedications 106, 229, 659; development impact regulations 107–10; development process and infrastructure 116–18; exactions 106; green subdivisions 106; in growing areas 104–7; impact fees 110; level of service standards 109–10; local changes, stormwater planning 342; local responsibilities for development 103–4; planned unit development 106, 668; site improvements 106; smart and sustainable 103–18; subdivision ordinances 104–6, 672; uniform development standards 106–7, 673; for urbanized areas 110–16
development standards 106–7, 660, 673
Dewey, J. 533
digital communications 615
dioxins 300, 359
direct current (DC) 580, 586, 590, 660
disable people 442
disaggregation levels, population projection 96
discretionary permit process 127–8
discrimination, impact fees 234
distillation 660
distributed wastewater management systems 313–14, *315*
diurnal emissions 440
Diverse Giants 44
divestiture, and full privatization 252–3
DNL (day-night average sound level) 475
dockage 493
documentation: bond 205; capital budget as separate document, recommendations for 146–7; capital improvements plan 141–5
Doig, J. W. 698n
Donnelly, S. 701n
Dotterer, S. 683n
double taxation issues, development impact fees 239, 242
double-A-rated corporate bonds 199
Douglas, B. 694n
Dowall, D. E. 676n, 681n, 689n, 700n
Downing, P. B. 687n
drag, air travel 462
Drake, E. 582
drawings, detailed 116–17
Drennan, M. P. 680n
Dresch, M. 687n
drinking water systems 281
du Plessis, C. 675n
Duerksen, C. 683n
Duncan, J. B. 682n
Dzurik, A. 690n

Earth Resource 367
Earth temperature variations *272*
East Bay Municipal Utility District (EBMUD) 218, 313
East Bay Regional Park District (EBRPD), California 572
Economic Development Administration (EDA) 31
economic downturn 16
economies of density 361
economy and infrastructure: importance of human capital for growth 49; whether public infrastructure investment causes growth 45–6, 47; at regional and local level 46–8
"eco-parks"/"eco-tourism," 558

Edison, T. 580, 581
Edmonds, Washington *161*
Edwards, M. 685n
effluent 660
e-government service delivery 623, 635
Eisenhower, D. D. 29, 415, 609
El Sobrante, California *270*
elderly people 39–40, 441, 527
electric street-car era (1890–1920) 413–14
electrical generation facilities 585–6
electricity: currents 580, 586, 590, 655, 660; deregulation 582; early electrical systems in United States 580; Ohm's Law 582, 590; origins 580; public buildings 524; transmission and distribution 589–90; *see also* energy/power
electromagnetic (EM) spectrum 617, 660
electromagnetic radiation 628
electronics recycling 374
Emanuel, K. 694n
emergence of infrastructure, in United States 19–36; Colonial era (1770–1850) 19–22; industrial and scientific city, rise of (1850s–1930) 22–7; Depression and World War Two era (1930–1945) 27–9; post-World War II (1945–1960s) 30–1; post-industrial city (1970s–2001) 31–5; abandonment of infrastructure, in 1950s 33–5; engineers and plans 22, 26–7, 31; extractions (from colonial times until 1960s) 227–8; importance of infrastructure to local practitioners 19; infrastructure systems 24–5, 32; sustainability issues 35–6
emergency management considerations, public buildings 527
emergency operation centers (EOCs) 510, 511
eminent domain 129, 661
Eminent Domain: A Step-by-Step Guide to the Aquisition of Real Property 130
end-of-the-pipe controls, stormwater management 327
Energy Independence and Security Act 2007 611
Energy Policy Act (EPAct) 1992 611
Energy Supply and Environmental Coordination Act (ESECA) 1974 610
Energy Tax Act 1978 610
energy/power 579–611; background, energy element 602–3; building design 601; carbon emissions, and water 310; carbon emissions reduction opportunities 606; coal 587–8; cogeneration plant 586, 658; creating 603–7; decentralized energy sources 594, 601; development density 600–1; electric currents 586; electrical generation facilities 585–6; energy conservation, assessing 606; experimental efforts 593–4; federal regulations 608–11; flows, in United States 583–5; fuel types, in United States 587–9; general purpose governments 596–602; geothermal power 593; history of power 579–83; hydropower 589; importance to local practitioners 579; infrastructure systems 10–11; institutions 57, 594–5; inventory of current usage and projected future needs 604–6; inventory of local energy sources and costs 603–4; land use regulation 598; local renewable resource potential, assessing 606; low carbon or renewable energy sources 591–4; mainstream energy technology and facilities 585–91; municipally-owned systems, control 597; natural gas 582–3, 588, 590–1; nuclear power 588–9; origins of regulation 581–2; peaker plants 586, 668; petroleum 582–3, 588; power plant location 598; producers of power 594–5; public buildings 524; public facilities, decreasing energy costs in 601–2; regulation for local development 599–600; regulators of power 595; rights-of-way and utility corridors 598; solar access ordinances 601; solar power 591–9; specific regulations 600–1; tree ordinances 601; trusts 580–1; uses and sources 583–4; and wastewater 312–13; and water 273–4; wind power 592–3; *see also* electricity
Engineering News Record 161
engineers and plans, in United States 22, 26–7
Enterprise funds 182
enterprise funds 661
entertainment centers 517–18
environment: and climate change challenges 11–12; infrastructure systems 9–10; quality 12; water requirements 274
environmental consultants 123
environmental engineers 122
environmental impact assessments 113–14, 661
Environmental Integrity Project 587
environmental plans 498
Environmental Protection Agency (EPA) 53, 157; funding by 66; noise pollution 474; Office of Wastewater Management 298; solid waste management 354, 356, 365, 374; stormwater/flood management 330–2; wastewater management 298, 299, 305; water supply 281, 283
environmental review *123*, 128

environmentally friendly aircraft 456–7
Envision Utah 66
EPA *see* Environmental Protection Agency (EPA)
Eplan, L. S. 702n
equity concerns, public pricing addressing 220–1
Erie Canal/Erie Canal Commission 20–1
ethanol 438
ethnic diversity, increasing 40–3
Eugene City, Oregon 601
Europe, privatization in 248
eutrophication 308–10; Eutrophication Index 166
Evans, J. 687n, 688n
exactions: defined 229–32, 661; design principles 241; development rules 106; history 227–9; importance to local practitioners 227; recent developments 228–9; *see also* impact fees, development
executive management reporting systems 167–9
expanded capacity, pricing to finance 220
experts, bond issuance 198–9
expressed demand, parks 661
extended product responsibility (EPR) 370–2

facilities: airport *see* airport management; bicycle 391; capital *see* capital facilities; community 10, 675n; construction 117; criminal justice 512–14; electrical generation 585–6; energy 585–91; identifying for projected growth 241; inventory of 94; joint-use, public schools 546; natural gas 590–1; new, timely availability 234; oversizing 97, 98; park *see* parks and recreational facilities management; sizing and locating 97–8; solid waste 358–61; timely availability of new 234; wastewater management/treatment 9, 297–8, 322–3; water, new 287; *see also* public facilities; public schools; wastewater management; water
facility design review 126, 661
failures, infrastructure 35
Fainstein, S. 677n
fair market value 661
Fairfax County, Virginia: budget process 155; CIP revenue sources (2013) 184, *185*
False Creek Energy Center, Washington 312
fare box recovery 661
Farebox Recovery Ratio 444
Federal Aid Highway Act 1944 29
Federal Aid Highway Act 1956 30
Federal Aid Highway Act 1973 449
Federal Aid Road Act 1916 387, 408–9, 448
Federal Airport Act 1946 453, 480
Federal Aviation Act 1958 480
Federal Aviation Administration (FAA) 454, 459–60, 466; noise pollution 474, 475; Reauthorization Act 1996 465
Federal Aviation Agency (FAA) 453, 454
Federal Communication Commission (FCC) 615, 617, 622, 635; broadband definitions 620, 621
Federal Communications Act 1934 28
Federal Emergency Management Agency (FEMA) 330
Federal Energy Regulatory Commission (FERC) 595; Order 636, 1992 611
Federal Highway Administration (FHWA) 397, 422
Federal Housing Administration (FHA) 29, 31, 388, 408, 409
Federal Power Act 1935 609
Federal Power Commission (FPC) 28, 608, 609
Federal Transit Administration (FTA) 174, 422
Federal Water Pollution Act 1972 32
Federal Water Pollution Control Act 1948 293, 349
Federal Water Pollution Control Act Amendments 1972 34, 293, 349
Federal Water Power Act 1920 293, 608
Federal-Aid Highway Act 1938 448
Federal-Aid Highway Program 422
Federal-Aid Highway/Interstate Highway Act 1956 409, 449
fees: connection 659; in lieu 228, 230, 244, 664; linkage 230, 665; mitigation 666; reclaimed water fees 323; regulatory 670; stormwater/flood management 345–7; tipping 363; wastewater management 322, 323; water fee structures 286; *see also* costs; financial context/expenditure; impact fees, development; user fees
Felbinger, C. 7, 675n, 676n
female labor force participation, increasing 39, 41
festival malls 517–18
fiber optic cable 661
fiber to the home (FTTH) 621, 634, 661
Fifth Paradigm: and stormwater 327; water 271
Filardo, M. 700n, 701n
final map 116–17
financial advisers, bond issuance 198
financial context/expenditure 100, 173–88; airport financing 463–4; capital expenditures, historic trends 174; capital spending by level of government 176; capital to operating expenditures for key infrastructure systems 179; comparison of annual capital needs estimates at local level *187*; expanded capacity,

pricing to finance 220; finance team 123; fixed asset expenditures 8; importance to local practitioners 173; infrastructure, amount recommended to be spent on 185–7; investment expenditures on infrastructure (1995–2010) *175*; loans, low cost state 347; local criteria for assessing infrastructure expenditure increases *187*; parks and recreational facilities management 562–5; ports and waterfronts *see* port and waterfront finances; public infrastructure project, development 124; public schools 537–8; public vs. private spending 175–6; revenue sources of local government 179–82; solid waste management 363, *364*, 365; state vs. local capital spending on infrastructure 176–7, *178*; transportation *see* transportation finance; water financing 285–6; *see also* bonds; costs; fees; funding; pricing, public; public pricing concepts, and fee types; user fees
financing innovations, in United States 23–4
Financing the Future (1973) 35
finding of no significant impact (FONSI) 128
fire stations 109–10, 526
First Deficiency Appropriation Act 1949 609
fiscal pressures 16
fiscalization, land use issues 240, 661
Fitch IBCA 199
fixed asset expenditures 8
flat rate fees 217, 661
flight, mechanism of 462
Flint, A. 676n
Flood Control Act 1944 609
flooding, in United States 326, 329–30, 337, 347; *see also* stormwater/flood management
floodways 329
Florida: Adequate Public Facilities Ordinance 108; development caps, infrastructure capacity 110; development impact fees, parks 563–4; local planning 17; privatization 257; public schools 548
Flyuberg, B. 683n
food waste 373
Forester, J. 680n, 683n
Forth Worth 48
fossil fuels 518, 588
Foster, K. A. 680n
four-step method, transportation planning 424–8, 661
Fragile Foundations (1988) 35
fragmented aviation system planning 457–8
France, public–private partnerships 254–5
Frank, J. E. 687n
Frank, S. 682n
Freeman, C. 698n
freeways: defined 661; freeway era (1945–present) 415–16; level of service standards *429*
Fresh Kill, New York 378
Fresno, California 353
Frey, W. H. 678n, 679n
friendly condemnation 130
front loader trucks 357
full cost pricing 215–16, 661
full privatization 252–3, 661
Fulton, W. 696n
fund accounting 662
funding: capital improvements plan 140–1, 142; enterprise funds 661; federal involvement in 32–3; general, special and enterprise funds 182; local governments 56; new schools, bias toward building 544; shortfall for airports 455–6; *see also* costs; fees; financial context/expenditure; impact fees, development
furniture, street 393, 407

Gaile, G. L. 679n
Gainesville-Alachua County (Florida) Airport Authority 478
Gakenheimer, R. 684n
Galardi, D. 693n, 695n
Gallatin, A./Gallatin Plan 18, 21
garbage disposal 352, 357
Garvin, A. 561, 701n
GASB *see* Government Accounting Standards Board (GASB)
gasoline tax 443
Gbps (gigabits per second) 620, 662
General Accounting Office (GAO) 258, 259, 455–6, 539
general aviation airports 662
General Electric (GE) 617
general fund 662
general obligation (GO) bonds 189, 192, 193–4, 201, 662
general plans *see* comprehensive plans
general purpose governments (GPGs): airport management 466; citizens' commission 60, 680n; comprehensive plans 74; defined 662; development responsibilities 103, 104; engineering department 87; inefficient 651–2; information management 157; and local government 140; organizational structure 57–61;

planning department 58–60; port planning 501, 502; ports structured as departments within 489, 491; and power 596–602; public infrastructure project, development 119; public works department 60–1; reporting systems 168; responsibilities *55*; site acquisition and eminent domain 128; stormwater/flood management 331; teambuilding 92
geographic information system (GIS) 99, 400, 566
geographic segmentation of poor people 42
Geological Survey, U.S. 308
geosynchronous orbits 662
geotechnical engineers 122
geothermal power/systems 593, 662
Gerckens, L. C. 677n
Gerhard, C. 169
germ theory of disease 270
Gesell, L. *464*
gigabits per second (Gbps) 620, 662
Gillespie, A. 677n
glazing systems and lighting 520–1, 662
Gleick, P. H. *276*, 690n
global recession 16
global warming 11–12, 440, 584; *see also* climate change challenges; Kyoto Agreement and Protocol on global warming
globalization, impact on infrastructure systems 32
Godschalk, D. R. 682n
Golden Gate Bridge 28
Golden Gate Park, San Francisco 554
Gomez-Ibanez, J. A. 679n, 687n
Gottlieb, R. 695n
Government Accounting Standards Board (GASB), Rule 34, 137, 140, 147–8, 160, 662, 684n
government agencies, bond issuance 197–8
Government Financial Officers Association (GFOA) 166, 167
graffiti removal 405
Graham, S. 7, 675n, 676n, 677n, 681n
Grand Coulee Dam Project 28
Grant Canyon air crash, 1956 453
gray water 315, 317, 318, 662
Great Depression *see* Depression and World War Two era (1930–1945), United States
Great Plains Coal Gasification Plant, privatization 253
green building 662
Green Building Codes 519
Green Building Council 82, 519, 539, 546
Green Building Rating System, LEED 519, 546
green infrastructure 73, 327, 341; San Francisco Green Infrastructure program 522, *523*
Green Revolution 308
green roofs 113, 520, 524
green subdivisions 106
"greenfield" areas 103
greenhouse gas (GHG) emissions 12, 13, 37, 434, 583–4, 607; categories 605–6; environmental and social impacts 437–9; environmental impact assessments 114; goals 585; *see also* carbon emissions
greenways *569*, 647, 662
grids, power 590, 591
Griggs v. Allegheny County 474
Grime Busters (street cleaning program) 405
Gross Domestic Product 515
Gross National Product (GNP) 14, 173, 253
groundside facilities 459
groundwater 275, 277, 280, 662; contaminants 274, 303–4; *see also* surface water
Groundwater Replenishment System (GWRS), Orange County 311–12
growing areas, development rules 104–7
"Growing Greener" initiative, Pennsylvania 562
growth: whether caused by public infrastructure investment 45–6, 47; human capital, importance 49; industrial and scientific city, rise of (1850s–1930) in United States 22–7; new and unevenly distributed 13–14; pent-up demand for, in post-World War II period (United States) 30–1; projected, identification of infrastructure facilities needed 241; and water, on West Coast 273; *see also* demand and growth patterns
growth management tools 107
growth shapers 318

Haarmeyer, D. 690n
Hall, J. R. 702n
Hall, P. 677n
Hall, P. V. 698n
Hall, R. 696n
Hals, A. 692n
Hammer, M. J. 693n
Handbook for Public Playground Safety (CPSC) 570
Hanemann, W. M. 686n
Hannigan, J. 699n
Hansman, R. J. 697n

Harnik, P. 702n
Harrison, B. 679n
Harrison, P. 702n
Hatry, H. 164, 169
hazardous waste 357, 361, 365, 372, 662
headend 663
health centers 513–14
heavy metals, in water supplies 304
Hefetz, A. 689n
Height Limitation Zoning Ordinances (HLZOs) 473–4
Henrico County, Virginia 561
Hermanowicz, S. 692n
hexane 303
High Occupancy Toll Lanes (HOT) 435
Highway Trust Fund (HTF) 416, 443
Hill, J. 580
Hirsch, W. Z. 688n
Hoag, G. 695n
Hoell, K. 682n
Home Box Office (HBO) 618
home-ownership rates 96
Hoover, H. 26
Hoover, J. B. 691n
Hoover Dam hydroelectric power plant 28
Hopper, K. 702n
horizontal consistency *91*, 663
Horrigan, J. B. 703n
horse-drawn carriage 412
hospitals, public 513–14
hot soak 440
hotels 515–16
household hazardous waste (HHW), disposal 357, 361
households: new kinds 38–9; population projection 96; population trends and infrastructure demand 37–40; rates 96
Housing Act 1949, Urban Renewal program 29
housing density, effect on gasoline consumption *439*
housing market collapse 2008 16
housing programs 59
Hoving, T. 558
Hoyle, B. S. 698n
Hoyt, M. *354*
hub airports 460, 663
hub-and-spoke system 663
Huddleston, J. 685n
Hudson river, bird strike 474
human capital, importance for growth 49
Hundley, N. 273
hurricanes 335, 336, 511
Hush, L. W. 683n
HVAC (Heating, Ventilation and Air Conditioning) systems 122, 523, 663
hybrid fiber/coax (HFC) 663
hydraulic efficiency 333
hydrocarbons 272
hydroelectric plants 589, 609,663
hydromodification *333*, 663
hydropower 589
Hyman, L. S. 690n, 691n

immigration 32, 40–3, 45
impact aid for schools 663
impact analysis tools 99
impact fee program: administrative procedures 243–4; development of 234–8
impact fees, development: accounting 243–4; amount 232; anti-growth advocates, placating 234; appropriate services for 236–8; and CIP, California 238; collection 243–4; concurrency 234, 658; defined 229–32, 663; design principles 241; development costs, new revenue source 233–4; development rules 110; discrimination 234; double taxation issues 239, 242; elements in a program 238–43; enforcement 243–4; fiscal impact adjustments 242; history 227–9; importance to local practitioners 227; influence on local economy 240; infrastructure provision, timing 242; land use issues, fiscalization 240, 661; market test 240; parks and recreational facilities management 563–4; payment for 239; policies governing 238–40; population indicators for each land use, developing 241; program *see* impact fee program; projected growth, identification of infrastructure facilities needed 241; quasi-market prices, use as 234; rational nexus 238, 566, 669, 688n; reasons for use 232–4; refund procedure 243–4; relationship to comprehensive plan 240; rough proportionality requirement 239, 670; schedule, preparation 240–3; standards 244; state enabling legislation 231–2; terminology 230–1; timely availability of new facilities 234; timing 232–4, 243; by type and land use *233*; wastewater management 323; *see also* exactions
impact/outcome measures 165–6
in lieu fees 228, 230, 244, 664
incentives, socially best use of infrastructure 214

incineration 94, 359–60, 368, 663; principles 368; and sanitary landfills 353–4; siting and designing incinerators 379
income: inequality, rising 41–3; low-income families 442; rising 41–3; taxation aspects 445
Indian Gaming Regulatory Act 1988 517
Indianapolis, Columbus 48
indicators, performance *see* performance indicators
indirect costs 222
indirect potable reuse (IPR) 311
induced demand 431, 659
industrial and scientific city, rise of (1850s–1930), in United States 22–7; economic and political background 22–3; engineers and plans 26–7; industrial restructuring and financing innovations 23–4; infrastructure systems 24–5; stormwater/flood management 326
Industrial Core Metros 45
Industrial Revolution 12, 22, 27, 72
infiltration 305–6, 663
inflow 305–6, 664
information age telecommunications 618–21
information dissemination, e-government service delivery 635
information highway 33
information management 157–64; condition assessment information 159–60; cost or value information 161; infrastructure inventory 159; infrastructure maintenance, setting investment priorities for 161–4; reporting and evaluation 164
infrastructure: as biophilic 647; deficiencies mitigated during environmental review process 114; definitions *see* definitions of infrastructure; deteriorated 14–15, 94, 153–5; and development process 116–18; financial issues *see under* financial context/expenditure; future city, challenges for 11–18; gray 327; green 341; importance to local practitioners 5; incentives for socially best use 214; lot-specific 117; major systems 9–11; management *see* management of infrastructure; as multi-disciplinary 647; as multi-scalar 647; new paradigm for *see* new paradigm for infrastructure; pent-up demand for, in post-World War II period (United States) 30–1; programs *see* programs, infrastructure; as regional and mega-regional 647; stormwater 326–7; and zoning 112–13; *see also* infrastructure systems
infrastructure function, use of fees by 210–11
Infrastructure Report Cards, American Society of Civil Engineers 14, *15*
infrastructure systems 9–11; energy and power 10–11; impact of globalization on 32; in United States 24–5, 32
Ingram, G. 676n, 685n
injection wells 664
inlets 664
in-lieu recharge, water supply 278
Institute of Museum and Library Services 508
Institute of Public Utilities 211
Institute of Sustainable Infrastructure 115
Institute of Transportation Engineers (ITE) 388, 392, *395*, 396
institutional structure, inadequate 16–17
institutions of infrastructure 51–67; aviation 463–6; energy/power 57, 594–5; evolution 52; federal 53–4; importance to local practitioners 51; intergovernmental context 52–7; local general purpose governments *see* general purpose governments (GPGs); local governments 55–6; parks and recreational facilities 559–61; port and waterfront management 488–92; privatization 258–9, 260–2; providers of infrastructure 52–3; public schools 536–9; regional 57, 65–6; solid waste management 361–6; special districts 61–5; state government 54–5; stormwater/flood management 330–3; streets 396–7; telecommunications systems 57, 621–3; wastewater management 298–9; water 281–4
Insull, S. 581
integrated resource planning (IRP) 284, 664
Integrated Surface Transportation Efficiency Act 1991 498
intelligent transportation systems (ITS) 436–7, 664
interceptors (trunk lines) 300, 301, 318, 323, 664
interjurisdictional coordination, capital facilities planning 138
interLATA 664
Intermodal Surface Transportation Efficiency Act (ISTEA) 1991 403, 416–17, 450
internal plan consistency *91*, 664
Internal Rate of Return (IRR) 138
International City/County Manager Association (ICMA) 630; Center for Performance Measurement 167, 526; *The Practice of Local Government Planning* 34–5; privatization 251, 255, 257, 259
International Civil Aviation Organization 466

International Council for Local Environmental Initiatives (ICLEI) 72, 73, 81, 82, 83, 519; "Cities for Climate Protection" program 602–3
International Green Construction Code 115
International Society of Arboriculture 394
Internet, rise of 618–19
Internet protocol enabled services (IPes) 623
Internet Tax Freedom Act (ITFA) 623
interoperability 664
Inter-Regional Highways Committee 29
intersections 390
Interstate Commerce Commission (ICC) 28
Interstate Highway Act 1956 415
Interstate Highway System 29, 598
IntraLATA 664
investment grade bonds 200–1
irrigation, residential 291
Irvine, California 311

Jackson, W. H. 556
Jacobs, A. 383, 696n
Jacobson, C. D. 676n, 688n
Jacobson, M. T. 688n
jails 512–13, 526
Jargowsky, P. A. 688n
jaywalking 390
Jeer, S. 690n
Jefferson, President T. 7
jobs–housing balance 433, 664
Johnson, F. 700n
Johnson, K. 690n
Jones Act 1920 484
Joseph, J. 686n
Judd, D. R. 677n, 699n, 702n
junk bonds 201

Karif, O. 703n
Kaspersen, J. 693n
Kbps (kilobits per second) 620, 664
Kee, R. 683n
Kelly, E. D. 680n, 682n, 693n, 699n
Kelly, J. M. 168, 685n
Kent, T. J. (Jr.) 34, 73, 677n, 681n
Kentopp, C. M. 687n
Keynesian economic policy 27
kilobits per second (Kkbps) 620, 664
Kindschy, J. W. 695n
King County, Washington: expected climate change impacts *83*; Wastewater Agency 312; Wastewater Plan, 2000–2030 320
Kirk, K. 691n
Kitsap County, Washington 512
Knapp, G. 688n
Koontz, R. A. 698n
Kouri, C. 701n
Krumholz, N. 680n
Krupp, C. 676n
Kyoto Agreement and Protocol on global warming 73, 439, 521

Lackney, J. A. 698n
Lambert, B. 687n
Lancaster, California: Lancaster Turnpike 444; Urban Restructure Program 563
Land and Water Conservation Fund 576
land dedications 106, 229, 659
land use issues: and airports *see* airports and land use; coordination of infrastructure plans with land use policy 100–1; current status of local land use planning requirements 76; energy/power 598; fiscalization 240, 661; inadequate institutional structure 17; incompatible land use types and issues *472*; and local general purpose governments 59; new paradigm for infrastructure 647; port planning 497–8; projections 285; public facilities and land use planning 524–8; and public school planning 547–8; solid waste management 374–5; spatial analysis tools 99; stormwater/flood management 346; transportation 432–4; wastewater planning and land use 318–19; and water supply *see* water supply and land use planning 274; waterfront planning 502–3
landfills/sanitory landfills 94, 358–9, 368, 670; bioreactor landfills 378; gases (LFG) 377; and incineration 353–4; leachate 359, 377–8, 665; principles 368; reuse of site 378; siting and designing 376–9; size, reducing 373
Landis, J. 687n, 697n
Landmark Stables Center Agreement, Los Angeles 517
landscape asset maintenance 155
landscape investment, life-cycle needs 163
large quantity generator (LQG) 664
large urban parks 664
Las Vegas, Nevada 392
latent demand 431, 664

laterals 300–1, 305, 665
Latrobe, Benjamin Henry 22
leachate 359, 377–8, 665
Leadership in Energy/power and Environmental Design (LEED) 82, 115, *520*; Green Building Rating System 519, 546
leak detection and repair system 97
Leaking Underground Storage Tank Fund 443
lease-financing bonds (certificates of participation) 195, 202, 657
Ledford, J. 696n
Lee, D. B. 688n
LeGates, R. T. 681n
Legionnaires' disease 665
Leithe, J. L. 686n
Lenard, T. M. 703n
Lens, P. 692n
Lepatner, B. B. 678n
level of service standards (LOS): defined 665; development rules 109–10; infrastructure programs 95–6; parks and recreational facilities management 566; transportation 400, 428, *429*
Levine, L. 694n
Levy, J. M. 677n, 680n
Lewis, M. 702n
Lewis, P. 681n
libraries, public 109, 508–9, 526
life-cycle analysis (LCA) 154
LIFE-CYCLe budgeting 647, 648
life-cycle cost analysis 71; defined 665; management of infrastructure 162–4; privatization 261; public infrastructure project, development 126–30
lift force, aircraft 665
lift stations 301, 665
light rail transit 665
lighter-aboard-ship (LASH) 486
lighting, street 196, 392
limited liability bonds *see* revenue bonds
Limits to Growth report, 1972 72
Lincoln Institute of Land Policy 185
linkage fees 230, 665
Litvak, D. 686n
load factors 665
local access and transport areas (LATAs) 664
Local Area Networks (LANs) 618
local distribution companies (LCDs) 610
local exchange 665
local general purpose governments (GPGs) *see* general purpose governments (GPGs)
local governments: capital budgeting as practiced by 138; as institutions of infrastructure 55–6; land use planners, considerations for 286–7; and local infrastructure funds 140–1; and privatization 251–2, 257, 261–2; revenue sources *see* revenue sources for local governments; right-of-way management 407; stormwater/flood management 332–3; subdivision ordinances 104; transportation finance 443; wastewater management 299
local governments and telecommunications: broadband supply 634; death of distance 626; e-government service delivery 635; financial subsidies 634; infrastructure, building 634; internal communications 635–6; local government as promotor 633–4; local government as regulator 625–31; local government as telecommunications planner 624–5; local government as telecommunications user 634–6; policy changes 634; zoning ordinances 626–7
location decisions 47
Logan International Airport, Billings 464
London, congestion pricing 435
Long Beach, California, port 484
loop detectors 665
LOS *see* level of service standards (LOS)
Los Angeles, California: condition assessment 159; Department of Water and Power 597; Landmark Stables Center Agreement 517; mass transit system 30; port 484; street management 160, 406; uses report card 144; Vermont Corridor Station 573–4; water supply system 278–9
lot-specific infrastructure and connections 117
Louis XIV, King 387
Loux, J. 690n
low impact development (LID) 327, 338, 340, *341*
low-income families 442
Luberoff, D. 698n, 699n
Lucky, R. L. 703n
Lutsky, N. 697n

MacLean, M. 481
Maco, P. S. 685n
Maddaus, W. 691n
Madden, K. 698n
Magnet Metros, mid-sized 45
magnet schools 543, 665

maintenance backlogs, causes 154, 155–6
maintenance of infrastructure, setting investment priorities for 161–4
maintenance operations center (MOC) 48
Making Peace with the Planet (Commoner) 31
Malmqvist, P. 692n
managed competition 665
management for results (performance management) 79, 156, 164, 168, 668
management of infrastructure 153–70; asset management *see* asset management; benchmarking 166–7; deteriorated infrastructure, problem of 153–5; executive management reporting systems 167–9; implementing of executive management reporting systems 169; importance to local practitioners 153; maintenance backlogs, causes 154, 155–6; performance measurement 164–9; periodic face-to-face meetings 167, 168–9
manholes (utility access ports) 301, 666, 673
Mann, H. 532
Manual of Uniform Traffic Control Devices 392
maps 668; final 116–17; preliminary 116; subdivision 105; tentative 116; "vesting" tentative 117
Marconi, G. 616, 617
marginal cost pricing 216–17, 666
Maricopa County, Arizona 514
maritime businesses, non-traditional 486
market-oriented civic model, parks 562, 666
Marshall, A. 46, 679n
Marshall, M. 169
Marvin, S. 7, 675n, 676n, 677n, 681n
Maryland 17, 66; strategic planning 80–1
Mass Transit Account 443
Massachusetts 115, 190, 374, 547
material bans, solid waste management 372
materials recovery facility (MRF) 358, 360
Mattrick, K. 677n
maturity structure, bonds 666
Mayer, N. S. 687n
Mayors Climate Protection Agreement, 2005 73, 602
Mays, L. W. 694n
Mbps (megabits per second) 620, 666
MBRs (membrane bioreactors) 314, 317
McAdam, J. L. 387
McBurney, W. 163
McCann, B. 696n
McCormick Place Convention Center, Chicago 515
McGregor, B. P. 702n
McKinsey and Company 518
McKoy, D. 700n
McMahon, J. E. 690n
McMillan, T. E. 701n
McNamara, R. 676n
Meck, S. 676n, 681n, 682n, 687n
meetings, executive management reporting systems 168–9
megabits per second (Mbps) 620, 666
mega-projects 61, 133, 514
megaregions 66
Melosi, M. V. 691n, 695n
membrane bioreactors (MBRs) 314, 317
Memphis, Tennessee 297
mercaptan 588, 666
Merced, California *270*
mercury 302
mercury-vapor lamps 666
mergers and acquisitions 23
Mertes, J. D. 702n
metal halide lamps 666
methane 310, 335, 359, 373
metropolitan areas, categorization 44–5
Metropolitan Planning Organization (MPO) 66, 423, 498, 666
Meyers, D. 678n
MGD (gallons per day) 321
Miami International Airport, Florida 464
Miami-Dade County 400
Michigan, Lake 298
micro-turbines 594, 666
middens 351–2, 666
Middle Ages, cathedrals 6
Miles, M. E. 700n
military roads, Roman 386, 389
Millennium Survey, 2000 (APA/AICP) 540
Miller, C. 699n
Milwaukee, Wisconsin: deep tunnel concept 327; failure of sanitary sewer system in 305; groundwater levels 277
mining, stormwater management 335
mini-parks 666
Minneapolis 2011 Sustainability Report 82
Minneapolis Living Well report 82–3
Minnesota: local planning 17; public schools 543, 544
minorities, ethic 40–3

Mishel, L. 679n
Mississippi River aquifers 277
Mississippi River Bridge collapse, Minneapolis (2007) 14
Missouri, Midwestern U.S.A. 563
Mitchell, C. 692n
Mitchell, J. 680n
mitigation fees 666
mobile phones 218, 219, 615–16, 657
mobility patterns, changing 39
model ordinances 75
modem 615, 666
modulator–demodulator 615
Monkkonen, E. H. 676n
monopolies, natural 8, 667
Montana, Aeronautics Division 466
Moody, M. 680n, 688n
Moody's Investors Service (Moody's) 199
Moore, A. T. 697n
"moral entrepreneurs," 555
Morgan, J. P. 580
Morrill Act 1962 27
Morrison, C. 614
Morse, S./Morse Code 614
Moses, R. 27, 556
motorways *see* freeways
Mozdzanowska, A. 697n
MPO (Metropolitan Planning Organization) 66, 423, 498, 666
MTBE (methyl tertiary butyl ether) 274, 303, 666
Muehlmann, C. 677n
Mullen, C. 687n, 688n
Muller, P.O. 696n
Mullins, A. 694n, 695n
multiple chemical sensitivity 666
municipal government buildings: art museums/cultural buildings 509–10; civic centers and City Hall 508, 526, 647; corporation yards 61, 510, *511*; libraries 508–9, 526; performance indicators 526; public safety buildings 510–12; warehouses 510
municipal separate storm sewer systems (MS4s) 328–9, 331, 333, 344
municipal solid waste (MSW) 360, 361
municipal wastewater 666
municipal wastewater, elements of systems 299–303
Muschamp, H. 675n
Musik, N. 676n
Nadal, M. 689n
National Association of Realtors 563
National Association of Telecommunications Officers and Advisors (NATOA) 629
National Association of Water Companies 282
National Cancer Institute 589
National Center for State Courts 511
National Civic Federation 581
National Complete Streets Coalition 400
National Electric Code 115
National Electric Light Association 581
National Endowment for the Arts 517
National Energy Act 1978 610
National Energy Conservation Policy Act 1978 611
National Environmental Policy Act (NEPA) 1969 32, 113, 128
national framework, lack of 18
National Gas Act 1938 609
National Gas Policy Act 1978 611
National Highway Traffic Safety Administration 441
National Housing Act 1934 409
National Industrial Recovery Act (NIRA) 1933 28
National Plan of Integrated Airport Systems (NPIAS) 458
National Pollutant Discharge Elimination Program (NPDES) 89, 298, 331, 349, 667
National Recreation and Park Association (NRPA) 566, 567, 568
National Resources Conservation Service funds, Agriculture Department 330
National Science Foundation 619
National Television System Committee (NTSC) 618
National Transportation Safety Board 454
Nationwide Personal Travel Survey 1995 397
natural gas 582–3, 588, 590–1
Natural Gas Act 1938 582
natural monopoly 8, 667
natural resource areas 667
Natural Resources Defense Council 521
Nebraska, Midwestern United States 526
needs assessment: parks and recreational facilities management 565–72, 566; program, infrastructure 96; solid waste management 368–9
neighborhood parks 667
Neighborhood Unit, The 388
Nelson, A. C. 678n, 680n, 682n, 687n, 688n
neo-bulk cargo 667
Nessie curve 667

Net Present Value (NPV) 138
Netzer, D. 685n
Neuman, W. 697n
Neumark, D. 680n
Nevada 288, 585
New Heartland Metro Areas 44
"New Immigrant" generations 40
New Jersey, local planning 17
New Large Aircraft (NLA) 461
new paradigm for infrastructure 18, 645–54; changes to content and process of what is planned and built 649–50; federal program changes 653; importance to local practitioners 645; improvement for infrastructure decision-making and management tools 650–1; inefficient local general purpose government 651–2; local capacity grants, interdisciplinary efforts 652; long-term structural changes to provision system 651–4; and old paradigm 645–6; smart and sustainable planning 648–51; state-level changes 652–3; time for 645–7; universities and professional organizations 653–4; "virtual jurisdiction," integration of existing infrastructure 648–9
New Urbanism 14, 78, 270, 398, 433, 542
New York City/New York State: Adequate Public Facilities Ordinance 108; bond market 190; Building Codes 115; Central Park 554–5, 571; LaGuardia Airport 453; parks and recreational facilities management 565; PlaNYC 84; port 484; Regional Plan Association 65; solid waste management 352; street management 406; water supply system 278, *279*
New Zealand, privatization in 249, 253
Next Frontier Metro Areas 44
NextGen (satellite-based air traffic control system) 468, 645
niche cargo 484–5, 667
Nicholas, J. C. 243, 688n
9/11 terrorist attacks 457, 511
Nipkow, P. 617
nitrogen/nitrates 300, 309
noise: airports and land use 474–6; DNL (day-night average sound level) 475
Noll, R. G. 699n
nonconventional pollutants 300
non-point source pollution (NPS) 327, 331, 383, 470–1, 667
North American Electrical Reliability Corporation (NERC) regions 595, *596*
North Carolina, broadband feasibility 624
Novotny, V. 689n, 695n
NPDES (National Pollutant Discharge Elimination Program) 89, 298, 331, 349, 667
nuclear accidents 582
nuclear power 588–9
Nunn, S. 679n, 680n

Oakland, California 45, 375, 405, 434, 486
ocean warming 335
off stream storage 667
Office of Road Inquiry, Department of Agriculture 387
off-site infrastructure 9, 105, 112, 117
Ohm's Law 582, 590
oil crisis/spills, 1970s 72–3, 583
Oklahoma City 48
Okun, D. A. 690n
Olmstead, F. L. 554–5, 571
Omnibus Housing Act 1961 416, 449
on-dock warehouses 482
onsite infrastructure 8–9
on-street bikeways 667
on-street parking 401
open space size standards, determining 569, 667
operating budget, and capital budget 145–6
Orange County, California 311–12, 405
Oregon, local planning 17
Orfield, M. 700n
organics 300, 372–3
Osborne, P. 688n
Ottensmann, J. R. 682n
Otterpohl, R. 692n
output measures 165
outsourcing 250
overdrafting 277, 667
oversizing of facilities 97, 98
ozone 270, 302, 584

package units 667
packet switching 619, 667
Palladio, A. 386–7
parallel runways 456
paratransit 442
park trails 667
parking 401–2, 436; automobiles 469–70; bicycles 391
parks and recreational facilities management 10, 553–76; ADA requirements, meeting 570; amount of park and

open spaces in US cities 566; bonds 562; broader role for parks and open space 558–9; budget process 571; and children 561; classifications *568*, *569*; as community facilities 675n; contracting out 561–2, 659; cultural revolution and parks 557–8; development impact fees and mitigations 563–4; economic impact 571–2; expressed demand 661; federal legislation 576; financial considerations 562–5; greenways and parkways classification *569*; guidelines for business plan or economic analysis 571–2; history in United States 554–9; in-house organization 559–60; importance to local practitioners 553; institutional context 559–61; inventory of parks, recreation and open space 569–70; large urban parks 664; latent demand 664; and locality plan 566–7; market-oriented civic model 562, 666; meeting demand for facilities 556–7; mini-parks 666; needs assessment 565–72, 566; neighborhood parks 667; non-tax revenues 564–5; open space assessment districts 563; open space size standards 569, 667; park as nature in city 554–6; park as part of tourist industry 571; park department, relationship with school districts 560–1; per capita expenditures on 562; planning considerations 565–72; private parks 558, 669; property taxes 562; public model 561–2, 669; public–private co-ventures 562, 669; Recreation Activity Menu 567–9, 669; Recreational Fee Demonstration Program 566; reform park (1900–1930) 556; sales tax 562–3; sustainability and green program 572–5; tourist cities 558; user fees 564–5
particulate matter 440
partnerships, public–private 254–5
Passenger Services Act 1886 486
Pasteur, L. 270
pathogens 299, 302, 667
pavement condition index (PCI) 403
pavement management systems (PMS) 158, 403, 668
pavements (sidewalks) 391–2
paving materials, streets 389
pay-as-you go 202, 203
peak load pricing 668
"Peak Phosphorus," 308–10
peaker plants 586, 668
Pearce, F. 702n
Peavey, H. S. 695n
Pecca, S. P. 683n
pedestrian by-ways, users 396–7
Pendall, R. 677n
pension plans 16
perchlorate 668
performance audit 369
performance indicators 165–6, 168, 651, 668; public buildings 525–7
performance management 79, 156, 164, 168, 668
performance measurement 164–9; improvement of process 169; as part of overall management system 168
periodic face-to-face meetings 167, 168–9
permits: building, and construction contract 130–3; discretionary permit process 127–8; specialty 128; use 128
Peroff, K. 683n
Perry, C. 388
Perry, D. 7, 8, 675n, 676n, 680n, 684n, 688n, 699n
personal air travel 668
personal computers (PCs) 618
Peson, H. L. 696n
pestides 300; use reduction 344
Petersen, D. C. 699n
petroleum 588
Pew Center on Global Climate Change 519
Philadelphia, Pennsylvania: oil terminal in 492; Philadelphia Waterworks 22; sale and leaseback bond issuance *196*; stormwater ordinances 521
Phillips' Decision 1954 610
Phoenix, Arizona: five-year facilities management plan 155; privatization 257
phosphorus, wastewater management 300, 308–10
photovoltaic cells 524, 591, 601, 668
Pitkin, J. 678n
Pitt, R. 694n
plain switched telephone network (PSTN) 668
Planned Unit Development (PUD) 106, 668
planning commissions 668
plans/planning considerations: abandonment of infrastructure, in 1950s 33–5; aviation *see* airport planning; capital improvements *see* capital improvements program (CIP); changes to content and process of what is planned and built 649–50; climate change planning 81–3; collaborative planning 93, 658; comprehensive plans *see* comprehensive plans; engineers and plans, in United States 22, 26–7; importance of local plans to local practitioners 71; improvement for infrastructure decision-making and

management tools 650–1; land use *see* land use issues; local planning 71–85; long-term operating budget 141–2; parks and recreational facilities management 565–72; planning process, organizing 92–4; ports and waterfronts *see* port and waterfront planning; privatization 257–9; public infrastructure project, development 131–2; public schools *see* public school planning; redevelopment plans 113; regional level 261; review of plans 131–2; site 113; smart and sustainable planning 78–9, 648–51; special districts 223, 671; specific, in urbanized areas 113; strategic planning and infrastructure 79–84; streets *see* street planning; sustainability 72–3; transportation *see* transportation planning; "virtual jurisdiction," integration of existing infrastructure 648–9; wastewater *see* wastewater planning; water planning 284–6

PlaNYC, New York City 84

Plastrick, P. 688n

plats *see* maps

plumbing systems 668

point source pollution 331, 668

point-to-point communications 614–15

police stations 109, *511*, 526

policy standards 95–6

polling expert, bond issuance 199

pollution: agricultural 335; air 439–41, 470–1; conventional pollutants 299, 659; nonconventional pollutants 300; non-point source 327, 331, 383, 470–1, 667; point source 331, 668; stormwater, pollutants of concern in 346; and stormwater runoff 9, 333; in surface water *333*; thermal 300, 588; toxic pollutants 300, 672; urban runoff 334, 673; *see also* contaminants, water

pollution control/management 668; automobile-related measures 342, *343*; construction management 344; design approaches 344–5; pesticide use reduction 344; planning tools, water quality protection *343*; pollutant accumulation on impervious street and highway surfaces *343*; public education 344; reducing levels of pollutants in stormwater 342–5, *343;* retrofitting existing storm drains 345; solid waste management 365; stream and creak restoration 345; Pollution Prevention Act 1990 354, 381

ponds 329

Poole, R. W. 697n

population: aging 39–40; development impact fees 241; growth, in California 45; projection of future 96–7, 98, 284; trends, households, and infrastructure demand 37–40; in United States 5, 23, 39–40, 45; *see also* demographic factors

port agencies: capital planning of cargo ports 499; purposes, duties and powers 488–9; types 489–92

port and waterfront finances: containerization, impact on revenue structures 493–4; cruise revenues 494–5; ground leases and rents, revenues from 495; subsidizing non-revenue-generating activities 495–6; traditional port revenue sources 492–3

port and waterfront management 481–504; activities on waterfront *488*; containerization 482, 485, 492–3, 493–4; financial aspects *see* port and waterfront finances; history of ports in United States 481–3; importance to local practitioners 481; institutional context 488–92; niche cargo ports 484–5, 667; non-traditional maritime businesses 486–7; planning considerations *see* port and waterfront planning; port district 489; real estate and other new, non-maritime port businesses 487–8; redevelopment of waterfront 483

port and waterfront planning 496–504; capital improvements program 497; cargo ports, capital planning 499–501; competing interests 501–2; diversified waterfront, planning on 501–3; environmental plans 498; land use plans 497–8, 502–3; planning function within ports 499; strategic business plans 497; superior governments, plans required by 498; types of plans 497–8

port tariff 492

Porter, D. R. 683n, 687n

Porter, M. E. 679n

Portland, Oregon 391, 603; stormwater ordinances 521

ports: cargo 483–5, 492–4, 499–501; container 483–4; description 483–8; as destinations 485–8; niche cargo 484–5, 667; planning function within 499; private 492; structured as departments within general purpose governments 489, 491; structured as public authorities 489, 490; structured as special district governments 489, 490–1

Post Office Development Appropriations Bill 1913 448

post-industrial city (1970s–2001), United States: change 31–2; federal involvement in infrastructure funding 32–3; globalization, impact on infrastructure systems 32; infrastructure abandoned, in 1950s 33–5

postwar period, comprehensive national airport system 453–4

post-World War II (1945–1960s), in United States 30–1; airport management 453–4; public schools 534; stormwater/flood management 326
pothole patching 405
power *see* energy/power
Power, T. M. 679n
Power Marketing Administrations (PMA) 609
Practice of Local Government Planning, The (ICMA) 34–5
preliminary map 116
preliminary official statement (POS), bonds 205
Price, S. K. 690n
Price, W. 688n
Price–Anderson Act 1957 610
pricing, public: concepts, and fee types 214–18; congestion pricing 218–19, 434–6, 658; criteria for determining type 213–14; "crowding" issue 218–20; defined 207–9, 669; demand management pricing 219–20; equity concerns, addressing 220–1; expanded capacity, financing 220; fair price for new users 219; functions of prices 212–13; types of infrastructure services appropriate for 213; *see also* public pricing concepts; user fees
primary airports 669
primary treatment 669
prisons 512–13, 526
Private Finance Initiative (PFI), UK 253
private parks 558, 669
private placement 204–5, 669
private ports 492
private pricing 213
private-activity bonds (PABs) 195
privatization: accountability systems 261; airport management 465; capacity building 261–2; contract monitoring function 259; contracting out 255–6, 561–2, 659; costs and benefits 249–50; current status in United States 250–2; evaluation of project 258–9; full 252–3, 661; implementation of project 259–60; importance to local practitioners 247; institutional issues 258–9, 260–2; legislative changes 259, 260–1; local planning 251–2, 257, 261–2; market-based approaches, forms 252–7; parks and recreational facilities management 561–2; in past thirty years 248; planning considerations 257–9; port finances 492; private involvement, involving 257–8; regional level planning 261; resurgence of interest in 248–52; stability of government 260; and water 281, 282; workforce involvement 259; *see also* competition
privy vaults 20, 24, 296
process benchmarking 166
process-efficiency measures 165
product stewardship, and extended product responsibility 370–2
program, infrastructure 87–101; alternatives, identification and evaluation 92–3, 98, 99, 100, 101; baseline conditions, analyzing 92, 94–5; collaborative planning 93, 658; community design and decision making, technology for 99; condition assessment for each system, developing 94–5; control over timing, phasing and cost of development 88; coordination with other infrastructure providers 88; demands and needs, projecting 92, 96–8; federal regulations 94; financing alternatives, identifying 100; future population, projecting 96–7; goals and objectives for system, determining 92, 95–6; identifying alternatives 98; impetus for 89–90; importance to local practitioners 87, 88; intergovernmental considerations 90–1; inventory of existing facilities or system 94; level of service standards, developing 95–6; needs assessment 96; other impacts, identifying 100; planning process, organizing 92–4; preferred alternative, adopting 101; preparation 91, 92; public participation process 93–4; sizing and locating the facilities 97–8; spatial simulation of alternatives 99; status of local programs 89; teambuilding 92–3
Progressive Era 532
Progressive Movement 23
project close-out 133–4
project management systems 168, 669
promissory notes, vs. bonds 189–90
proof-of-payment system 419
property owners, petitioning for special districts 223
property taxes 24, 443, 562, 669; revenue sources for local governments 180–2
proportionality requirement 239, 670
Proposed Recommended Practices (PRP), street planning 398–9
Provus, S. 686n
proximate capitalization 571, 669
PSAP (public safety access point) 635, 636
psychiatric hospitals 513
public art 393, 394, 418
public authorities: defined 669; evolution 64; ports structured as 489, 490; vs. special districts 63–4

public buildings 10; area plans 525; building and occupancy permit process 118; building appearance and design competitions 527–8; building envelope 520, 656; certificates of participation 657; civic squares 509; comprehensive plans 525; convention centers 515–16; criminal justice facilities 512–14; cultural and historic districts 517–18; defined 508; design aspects 527–8; development impact fees 236; elements of a building 520–4; emergency management considerations 527; energy systems and electricity 524; entertainment centers 517–18; festival malls 517–18; glazing systems and lighting 520–1, 662; Green Building Codes 519; heating and cooling 523–4; hotels 515–16; importance to local practitioners 507; land use planning 524–8; Leadership in Energy and Environmental Design (LEED) 82, 115, 519, *520*, 546; local controls *525*; municipal government buildings 508–12; performance indicators 525–7, 668; quasi-public buildings 10, 121, 514–18, 528; retrofilling existing buildings 524; schools *see* schools; size and function 507–8; sports complex or facilities 516–17, 671; and sustainability 518–24; water and wastewater 521–3

public education: pollution management 344; water conservation 290

public entrepreneurs 122

public facilities: construction 117; decreasing energy costs 601–2; and land use planning 524–8

public goods 8

public infrastructure investment: whether causing growth 45–6, 47; development of project *see* public infrastructure project, development; vs. private capital investment 48

public infrastructure project, development 119–34; concept stage 120–1; construction and project closeout 133–4; construction contract and building permits 130–3; construction management team 123–4; design review and value engineering 127; design team 122; discretionary permit process 127–8; entering into construction contract 132–3; environmental consultants 123; environmental review *123*, 128; finalizing of design contract 131; finance team 123; financing 124; importance to local practitioners 119; life-cycle cost analysis 126–30, 162–4, 665; local agency inspections 134; organization chart *121*; organization of project 121–4; overview *120*; preliminary design 126; preparation of detailed plans and specifications 131–2; project close-out 133–4; selection of site 124–5; site acquisition and eminent domain 128–30; site and right-of-way acquisition team 123; specialty permits 128; use permits 128

public model for parks 561–2, 669

public participation: process 94; special districts 223–4

Public Policy Institute of California (PPIC), Statewide Survey on Government 540

public pricing concepts: average or full cost pricing 215–16, 656; less than average cost, including no cost 214–15; marginal cost pricing 216–17, 666; variable rates 217; *see also* costs; user fees

public safety access point (PSAP) 635, 636

public safety buildings 510–12

public school planning: acreage requirements 543–5; community schools 545–6; design practices 544–7; green school design 546–7; green schoolyards 547; high growth areas 548; impediments to small schools 543–4; infill development and adaptive reuse 545; joint-use facilities 546; and land use 547–8; new schools, bias toward building 544; redevelopment plans 550–1; small schools emphasis *see* small schools; Smart Growth 542, 548–50; traffic impacts 549; types of small school being developed 542–3; urban planning and small schools 542; walkable and bikeable schools 549–50

public schools: Adequate Public Facilities Ordinance 655; in cities and metropolitan areas 539–41; current condition of school infrastructure 539; development impact fees 236; establishment of public schools 532–3; facility maintenance and increased capacity, financing 538; finance of 537–8; history, in United States *see* public schools (history of in United States); housing nexus 540–1; importance to local practitioners 531–2; institutional context 536–9; Millennium Survey, 2000 540; and neighborhood demographies 539–40; planning *see* public school planning; as public infrastructure 531–52; revenue mix 538; Smart Growth 542, 544, 545, 548–50; traffic impacts 549; users 537; *see also* school districts

public schools (history of in United States) 532–6; in early nineteenth century 532; progressivism in late nineteenth century 533; in early twentieth century 533–4; post-World War II (1950s) 534; from 1960s to 1970s 534–5; excellence era (1980s–1990s) 535; in twenty-first century 535–6

public transit agencies 414

public transit modes *420*
public utilities commissions (PUCs) 422–3, 595, 609
Public Utilities Holding Company Act (PUHCA) 1935 581, 608–9
Public Utility Regulation Policies Act (PURPA) 1978 610
public utility services, trends in consumer price index (1970–2007) *211*
Public Works Administration (PWA) 27–8
public works, infrastructure as 6–7
public–private co-ventures for parks 562, 669
public–private partnerships 254–5, 516, 669
Puget Sound Regional Council 458
Putman, S. H. 682n

quality of life, infrastructure directly linked to, infrastructure program, importance 88
quantitative research, capital budget 138
quasi-public buildings 10, 121, 514–18, 528
"Quincy Box," 533

racks, bicycle 393
radiation 589, 628
radio 615, 616–17, 622
Radio Act 1912 638
Radio Act 1927 638–9
Radio Corporation of America (RCA) 617
Raffel, J. A. 700n
Rail Passenger Service Act (1970) 449
railroads *29*, 412
railways: commuter rail transit 419–20; light and heavy rail transit 419
rainwater harvesting/water reuse 278, 304
Raleigh-Durham, North Carolina 48
Ramapo phased growth program, New York 108
Randolph, J. 693n, 694n
ratings and rating agencies, bond issuance: cooperative bond pools, use of 202; details of agencies 199; factors considered by rating agencies 201–2, 685n; geographically defined bonds 202; GO (tax-backed) bond rating factors 201; lease-financing bonds (certificates of participation) 202; rating types 200–1; revenue bond rating factors 201–2
rational formulae 669
rational nexus test 238, 566, 669, 688n
Reagan, R. 248
rear loader trucks 357
recession 16
Reclamation Act 1902 293
Recreation Activity Menu (RAM) 567–9, 669
Recreation Enhancement Act 2004 566
recreational automobile era (1920–1945) 414–15
Recreational Fee Demonstration Program 566
recycling 670; distributed recycling system, wastewater management *315*; electronics 374; evaluating 370; mandatory 372; rates *355*
recycling and resource recovery centers 360–1
redevelopment bonds 197
redevelopment plans 113
re-engineering 256–7, 670
reform park (1900–1930) 556
refuse fees, non-collection 215
regional infrastructure institutions 65–6; councils of governments 57, 65; Metropolitan Planning Organization 66; renewed interest in regional efforts in 1990s and 2000s 65–6
regulatory fees 209, 670
Reid, H. E. 684n
Relief and Construction Act 1932 27
reliever airports 670
renewable energy sources 591–4
Repa, E. W. 695n
reporting: and evaluation 164; executive management reporting systems 167–9; types 168
requests for proposals (RFPs) 132
residential irrigation 291
Resource Conservation and Recovery Act (RCRA) 1976 354, 356, 381
resource recover park 670
Resource Recovery Act 1970 381
resource recovery centers 360–1
Retzlaff, R. C. 699n
revenue bonds 192–3, 194–5, 201–2, 670
revenue sources for local governments 179–82; federal or state grants and loans 182, 184; fees and user charges *208*; funds (general, special and enterprise) 182; generation of 214; property taxes 180–2
reverse osmosis 670
Rhode Island, sewer overflow *304*
Richmond, Virginia 413
right-of-way (ROW) 670; local agency inspections 134; management 397, 406–7; preliminary design 126; regulations 114–15; street design 395; telecommunications systems 628–30, 634; and utility corridor 598; wastewater management 305

Right-of-Way Act 1901 608
Riis, J. 556
Rio Earth Summit, 1992 72
riparian rights 670
risk assessment 163–4; infrastructure program 100
Rivenbark, W. C. 168, 685n
Rivers and Harbors Act 1899 293, 348–9, 380
road accidents 390–1
"roads fight blight," 29
roadside 391–2, 395
roadways: and automobiles 417; construction *105*; latent demand 664; level of service 670; military roads, Roman 386, 389; road management policies 342; street management 389
Robinson, S. 684n
Rockefeller, J. 580
"Roll On/Roll Off" (ro/ro), level of service 670
rolling stock 94, 368
Roosevelt, President F. D. 27, 29, 453, 581, 617
Rosenthal, M. 683n
Roslyn Place, Pittsburgh 389
Rostow, W. W. 7–8, 675n
rough proportionality requirement 239, 670
Roundtable on California's experience with Innovations in Public Financing for Infrastructure, 2000 258–9
Royal Road 386
rubberized asphalt concrete 670
Rubenfeld, D. 686n
Rubin, I. S. 685n
running emissions 440
runoff/runoff pollution 9, 304, 323, 673; stormwater/flood management 9, 325, 327, 333–4, 389, 518, 673; surface runoff 327
run-of-river system, hydropower 589
runways 459–61, 468
Rural Electrification Act 1936 609
Rural Electrification Act 1949 639
Rybczynski, W. 527
Rypinski, R. G. 130, 683n

Sabel, C. 681n
Sacramento County, California 563
Safe Drinking Water Act 1974 280, 293
Safe Drinking Water Amendments 1986 294
Safe Drinking Water Amendments 1996 294
safety bicycle 387
safety net hospitals 513
sales tax 444–5, 562–3
Salkin, P. 681n
Salt Lake City, topo map 619, *620*
San Antonio, Texas 346, *527*, 593, 597
San Diego Association of Governments (SANDAG) 8
San Diego, California 243, 291, 434, 487
San Francisco, California: bond market 190; broadband feasibility 624; Department of Parking and Traffic 423; Geneva Yard 418; Golden Gate Park 554; Green Infrastructure program 522, *523*; jail *513*; Ohlone people 352; Parks and Recreation Department 564; port 486, 487, 495; sewer system 297; Telecommunications Plan 625, 636; water reclamation 311; water supply system 278, 290
San Joaquin Valley, California 277
Sanders, H. T. 676n, 699n
Sandlake Galloway Road Project *329*
Sanibel, Florida 110
sanitary sewer overflow (SSO) 303, 304, 670
sanitation, wastewater management: biosolid (sludge) reuse, centralized 312, 656; centralized water reclamation 311–12; decentralized and distributed wastewater management systems 313–14, *315*, 323; energy and wastewater 312–13; financing of sanitary wastewater 322–3; water reuse or reclaimation 310–11; zero emissions wastewater 314–17
Santa Barbara, California 145, 195, 332, *513*
Santa Cruz, California 394
Santa Monica, California 79, 521, 573
Sarte, B. 689n
satellite dishes 626, 627
Savas, E. S. 688n
Savitch, H. V. 680n
school desegration 670
school districts 52, 63, 551; cities, "salo planning" separation 551; consolidation 56, 537; types 536–7
schools, public *see* public schools
Sclar, E. D. 688n
sea level rise 335–6, 337
Sears Tower, Chicago 524
seasonal rates 671
Seattle, Washington: Broadband Task Force 624; General Plan 1996 82; parks and recreational facilities management 565; power and distribution systems 597; Public Library *508*; sewer system 297; solid waste management 369; zoning ordinance 627
secondary treatment 671

Securities and Exchange Commission (SEC) 28, 205
security issues, airport management 457
Sedlak, D. L. 691n
Selzer, E. 681n
Seniors and Americans with Disabilities Act 1990 527
septic tank system 307, 310, 671
service area, wastewater 320–1, 671
service life concept 153–4, *162*
sewers: and alligators, in United States 307; combined sewer systems 297, 329–30, 347, 658; early 296; fees 236, 322; infrastructure systems, history 24–5; line cleaning 305; Milwaukee, failure of sanitary sewer system in 305; municipal separate storm sewer systems (MS4) 328–9, 331, 333, 334; non-collection of fees 215; sewage disposal, history 24; systems overflows 304–5; *see also* solid waste management; waste; wastewater management
Shanaghan, P. 690n
Shapiro, P. S. 697n
Sheffrin, S. M. 687n
Sherman, R. B. 698n
Shiklomanov, I. *276*
Shorr, P. 701n
sick building syndrome 671
sidewalks (pavements) 391–2
Sierra Club (environmental organization) 108, 566
Silent Spring, The (R. Carson) 31
Singer, I. 699n
single-parent households 32
sites: acquisition, and eminent domain 128–30; history of use 125; improvements 106; infrastructure 125; local attitudes 125; organization of project 123; physical features 125; plans 113; selection 124–5; zoning 125
Skilled Anchor Metros 45
slum removal 29, 33, 197
small schools: charter schools 542–3, 657; emphasis on in planning 541–2; impediments to 543–4; magnet schools 543, 665; types being developed 542–3; and urban planning 542; *see also* public schools
Smart Growth (APA) 78, 549
Smart Growth movement 18, 433, 542, 648; public schools 542, 544, 545, 548–50
Snow, J. 270
So, F. S. 678n
Sobrante Ozonation facility, El Sobrante (CA) *270*
social equity 441–2
Society of Civil Engineers, England 22
socioeconomic models 285, 671
sodium lamps 671
Sohagi, M. 686n, 687n
soils engineers 122
solar access ordinances 601
solar collectors 113
solar power 591–9
solar thermal panels 591
Solid Waste Disposal Act 1965 354, 380–1
solid waste management 351–81; ashes 356, 379; business-related studies 369; capital facilities, building or expanding 376–9; case law 361; cities and solid waste 352; collection of solid waste *20*, 357–8, 362–3; collection route analysis 368; construction and demolition wastes 356; curbside source separation 357–8; current management 354, *355*, 356; current program, evaluation 369; definition of solid waste 356, 671; development of local program 369–70; disposal of solid waste 352–3, 362–3; early 351–2; electronics recycling 374; emerging programs/concepts 370–6; facilities 358–61; facility inventory 368; federal regulations 380–1; fee use in 210; financing 363, *364*, 365; flow control decision 362; food scraps and composting 372–3; garbage 352, 356, 357; history 351–4; importance to local practitioners 351; incineration 353–4, 359–60, 368, 379; infrastructure systems 9, 25; institutions 361–6; journey of solid waste *355*; landfills *see* landfills/sanitory landfills; local regulation 366; mandatory recycling and bans 372; materials recovery facility 358, 360; municipal solid waste 360, 361; needs assessment 368–9; organics 372–3; pollution control 365; product stewardship and extended product responsibility 370–2; public sector waste management organization as department of local government 363; recycling and resource recovery centers 360–1; refuse 356; regulation 365–6; Regulations 380–1; regulatory and land use tools 374–5; rubbish 356; special waste 356; state regulation 365–6; systematic disposal methods, need for 352–3; thrift stores 375; tipping fees 363; tonnage targets, translating program into 370; trash 356; treatment-plant waste 356; truck types 357; waste characterization study 368–9; waste diversion and recycling 370, 674; wet/dry collections and processing 358, 373–4; zero waste goals/planning 367–8; *see also* sewers; waste; wastewater management
Sorkin, M. 518, 699n

source control 671
source reduction 671
South Elementary School, Waltham (MA) 547
South Louisiana, Port of 484–5
Southern California Edison Company 597
Southwest Airlines 456
Southworth, M. 696n
special assessment districts 63, 196–7, 671
special assessments 209, 671
special district governments, ports structured as 489, 490–1
special districts 61–5, 671; administrative staff, hiring 224; characteristics 62; evolution 64; functions 61–2; governing board, election 224; issues 64–5; plan for 223; vs. public authorities 63–4; reporting systems 168; school districts 63; setting up 222–4; special assessment districts 63, 196–7, 671; state requirements for establishing, identifying 223
Special funds 182
special obligation (SO) bonds 194; *see also* revenue bonds
special purpose local option taxes 444–5
specialty permits 128
Spector, S. 701n
Spokane, Washington 320–1; Spokane-Rathdrum aquifer 318
sports complex or facilities 516–17, 671
Sprague, F. 413
sprawl 13–14, 174
Sputnik I 618
SSOs (Sanitary Sewer Overflows) 303, 304
stakeholders 93, 149
Standard & Poor's Corporation (S&P) 199
Standard Consumer Safety Performance Specification for Playground Equipment for Public Use (CPSC) 570
Standard State Zoning Enabling Act (SZEA) 74, 111
Stanislaw, J. 688n
Stanley, W. 580
STAR community index 82
state enabling legislation, impact fees 231–2
State of the World report 72
Steffans, L. 556
Stein, C. 387–8
Stein, J. M. 679n
Steiner, R. 683n
Steinfeld, C. 692n
Stephan, J. J. 699n
step-up and step-down transformers 590, 671
Stockholm, congestion pricing 435
stocks, vs. bonds 190
storm drains, retrofitting 345
stormwater management plan (SWMP) 331
stormwater planning: design process 341–2; local development rule changes 342; maintaining pre-development stormwater patterns 338; new development 340–2; "rational formula," 338–40; stormwater management plan (SWMP) 331; street trees 394
stormwater/flood management: and climate change 335–7, 347; combined sewer systems 297, 329–30, 347, 658; current issues 333–7; early responses to rainfall, storms and flooding 326; elements of system 327–30; end-of-the-pipe controls 327; Environmental Protection Agency, role 330–2; faster and more intense runoff 333; federal institutions 330; fees 345–7; Fifth Paradigm, and stormwater 327; financial aspects 345–7; history, stormwater infrastructure 326–7; importance to local practitioners 325; industrial era management 326; institutions, stormwater infrastructure 330–3; level of service standards 109; loans, low cost state 347; local governments, role 332–3; and mining 335; municipal separate storm sewer systems (MS4) 328–9, 331, 333, 344; Phase I and Phase II communities 345; planning considerations *see* stormwater planning; pollutants, reducing of levels 342–5, *343*; pollution and stormwater runoff 333; retrofitting existing storm drains 345; roadway design 389; runoff 9, 323, 327, 333–4, 389, 518, 673; sea level rise 335–6, 337; soft treatment systems, requirement for 347; storage-oriented slow release systems, requirement for 347; stream and creak restoration 345; subdivision ordinances 105–6; urban runoff pollution 334, 673; wetlands 330
strategic benchmarks 167
strategic planning 79–84; adaptation planning 83–4; business plans, ports 497; capital budget 138; and capital improvement plan 80–1; climate change planning and sustainability planning 81–3; communication 80; information and data systems 80; vision 80
stream and creak restoration 345
street art 406
street maintenance: and building 397; cleaning/ sweeping 405; miscellaneous aspects 405–6;

pavement management systems 403, 668; resurfacing and preventive maintenance 402–3
street planning 387–8; bicycle needs 400; context-sensitive solutions 398–9; local 398–402; non-motorized transportation 399–400; pedestrian needs 400–1; Proposed Recommended Practices 398–9
streetcars 413
streets/street management 385–409; alternative level of service criteria for 430; American streets 387; bicycle facilities 391; bus stop/transit shelters 393, *394*; cement trucks for curbs, gutters and sidewalks *106*; classification of streets, newer approach to 396; complete streets 400; components of streetscape 395; context-based framework for design 395–6; context-sensitive solutions 398–9; crosswalks 390–1, 393; curbs 389–90; design aspects 389, 395–6; elements of streetscapes 388–96; federal regulations 408–9; fee use in 210–11; furniture 393, 407; history 386–8; importance to local practitioners 385; infrastructure systems 10; institutions 396–7; intersections 390; lighting 196, 392; maintenance *see* street maintenance; medieval streets 386–7; parking 401–2; paving materials 389; poles and signs 392–3; pollutant accumulation on impervious street and highway surfaces *343*; public art 393, 394; redevelopment of grid 90; Renaissance period 386–7; right-of-way 406–7; roadside 391–2; roadway 389; Roman engineering 386; traditional street hierarchy 395–6; traffic calming 401, 672; traffic control devices 392–3; traffic engineering and street standards 388, 407; traveled way 389–91, 395; trees 394–5; users 396–7
subdivision ordinances 104–6, 672
subgrades 672
sub-mains 300, 301, 328
subnational regions 66
suburbs 65, 413
Sun Oil Company (Sunoco) 492
Sunnyvale, California 626, 633
Supreme Court, United States 21, 23, 361
surface impoundment 672
surface runoff 327
surface water 273, 275, 277, 280, 672; contaminants 303–4; pollution *333*; *see also* groundwater
surveys 159
sustainability: climate change planning and sustainability planning 81–3, 518; local plans 72–3; Minneapolis 2011 Sustainability Report 82; origins of concerns 72; parks and recreational facilities management 572–5; planning considerations 72–3; and public buildings 518–24; smart and sustainable planning 78–9, 648–51; sustainable development 672; in United States 35–6
sustainability indicators 166
"Sustainable Seattle," 167
Sustainable Urban Drainage System (SUDS) 327
Sustaining Places Task Force (APA) 78
Svendsen, K. 683n, 684n
Sweden, ecological villages 317

Tabors, R. D. 682n, 691n, 693n
take-off, air flight 462
Tampa Port Authority 489, 494–5, 698n
Tarr, J. 6, 357, 675n, 676n, 684n, 691n
tax increment financing 197, 672
Tax Reform Act 1986 190
taxation aspects: bonds 189, 190, 192, 193–4, 196–7, 201, 203; capital costs 218; double taxation, development impact fees 239, 242; gasoline tax 443; income, payroll and employer taxes 445; land use issues, fiscalization 240, 661; local fiscal pressures 16; local vehicle and fuel taxes 445; parks, impact on tax base 571; property taxes 24, 180–2, 443, 562, 669; revenue sources for local governments 179; sales tax 444–5, 562–3; special purpose local option taxes 444–5; "tax revolt" (1970s and 1980s) 207; total local revenues and taxes 209–10; user fees 209–10; *see also* costs; fees; financial context/expenditure
taxing districts 63
teams/teambuilding 92–3, 123
technical standards 95
Teitz, M. 676n
Telecommunications Act 1996 622, 625, 627, 639
Telecommunications Plan 624–5
telecommunications systems 10–11, 613–39; analog and digital 615, 655, 660; antennas and satellite dishes, mitigating visual impacts 627; bandwidth and speed 619–21, 631, 656; broadcast TV 617–18; cable telecommunications, franchiser of providers 630–1; cable television 618; and compensation 629–33; content transmission 616–18; electromagnetic (EM) spectrum 617, 660; elements of telecommunications plan 624–5; FCC broadband definitions 621; Federal Communication Commission 621, 622;

federal regulations 638–9; industrial era telecommunications 614–18; information age 618–21; institutions 57, 621–3; Internet, rise of 618–19; Internet protocol enabled services 623; legacy providers 621–2; local government *see* local governments and telecommunications; mast, telecommunications *11*; packet switching 619, 667; planning for telecommunications 624; point-to-point communications 614–15; preparation of telecommunications plan 625; private electrical and telephone utilities, rise of 25; providers of services 621–2; public internet access 636; radio 616–17; regulators of telecommunications 622–3; rights-of-way 628–30; state and local roles 622–3; telegraph 614; telephone 614–15; wireless telephony 615–16; zoning for telecoms 627–8; zoning ordinances 626–8
telegraph communication 614
telephone communication 614–15, 621
television, broadcast 617–18, 621–2
Temel, J. W. 686n
Tennessee Valley Authority Act 1933 28, 608
Tennessee Valley Authority (TVA) 608
tentative map 116
Terabits (trillions of bits per second) 620
terminals, airport management 461
terrorism 457
tertiary treatment 302, 317–18, 672
Texas, local planning 17
Thatcher, M. 248
thermal panels, solar 591
thermal pollution 300, 588
thermal treatment 672
"third pollution," 352
thoroughfares 396
Three Mile Island, nuclear accident 582
"three-legged stool," transportation 438, 584
thrift stores, solid waste management 375
tidal power 606
time switching 672
time-of-day pricing 672
Time-Warner 622
tipping fees 363
Tobin, R. 699n
Todorovich, P. 681n
tolls, transportation 21, 435, 444
Toquop Power Plant, Nevada 585
total maximum daily load (TMDL) 331–2, 672
total water management 283, 672
tourism: cities 558; parks and recreational facilities management 558, 571; post-industrial city (1970s–2001), United States 33; public buildings 514–15
toxic pollutants 300, 672
traffic analysis zones (TAZ) 425
traffic calming 401, 672
traffic control devices 392–3
traffic engineering and street standards 388
transfer stations 94, 368
transit oriented development (TOD) 14, 433–4, 672
transmission 589–90, 620–1, 672
transportation 411–50; access 469; accidents 390–1; agencies, transit 423; and air pollution 439–41; alternate and low-carbon-footprint fuels 438–9; alternative approaches 432–7; alternatives, evaluating 432; buses 418; capacity issues 431, 432; cities and counties 423; CO_2 emissions 438; in Colonial Era (1770–1850) 20–1; condition assessment 430; demand management 434–7; demographic factors 39; development impact fees 236; electric street-car era (1890–1920) 413–14; environmental and social impacts 437–9; facility inventory 430; federal level involvement 416–17, 422; federal regulations 448–50; financial issues *see* transportation finance; freeway era (1945–present) 415–16; importance to local practitioners 411–12; induced demand 431; infrastructure systems 10, 23; jobs-housing balance 433; land use policies 432–4; latent demand 431; level of service standards 428, *429*; local level provision 423; mass transit systems 10, 30; metropolitan, history 412–17; Metropolitan Planning Organization 423; modes 417–20; need identification 428, 430–2; NIMBYism 433, 434; non-motorized 399–400, 420; operating characteristics of public transit modes *420*; planning considerations *see* transportation planning; providers of services 422–3; public schools, traffic impacts 549; public transport *420*; recreational automobile era (1920–1945) 414–15; route selection or trip assignment 427–8; solutions, identifying 432; state level provision 422–3; streets, alternative level of service criteria for 430; system efficiency and demand management tools 439; "three-legged stool," 438, 584; transit 418–20, *421*; transit-oriented development 433–4, 672; usage of public transit modes *420*; vehicle miles traveled, reduction in 439; walking-horse car era (1800–1890) 412–13; wants and needs 430–1

transportation analysis zones (TAZ) 672
transportation engineers 122
Transportation Equity Act 1998 417
transportation finance 442–6; choice of financial mechanisms 445–6; fares 444; gasoline tax 443; income, payroll and employer taxes 445; local vehicle and fuel taxes 445; property taxes 443; sales tax 444–5; special purpose local option taxes 444–5; subsidies and revenue sources 446; tolls 444; user fee model 443
Transportation Infrastructure Finance and Innovation Act (TIFIA) 1998 249
transportation planning 423–8; carbon emissions 423; "conformity," 423; four-step process 424–8, 661; journey to work trips *427*; metropolitan planning organizations 498; mode choice 426–7; purpose of trips *426*, *427*; road capacity evaluation 428; road usage by classification *427*; route selection or trip assignment 427–8; trip distribution 426; trip generation 425, *426*
Transportation Research Board 431, 471
Transportation Security Administration 472
traveled way 389–91, 395
tree ordinances 601
trees, street 394–5
Trees for Green Streets: An Illustrated Guide (City of Portland) 395
trickling filter process 302, 672
trip attractor 673
triple-B-rated corporate bonds 199
trips: assignment 427–8; distribution 426; generation 425; mode choice 426–7
trolley bus 673
trunk fee 323
Trust for Public Lands 564
Tunnel and Reservoir Plan (TARP), Chicago 327
tunnel projects 327
turnpikes 21, 444
"2030 Challenge," 519
two-part tariff: with base rate 217; with block rates 217–18; capital costs covered by taxpayers with maintenance and operation by users 218

ultraviolet light 302
undergrounding 598
underwriters, bond issuance 198–9
unfunded project work plan (UPWP) 142
Uniform Building Code 115, 656, 673; *see also* Green Building Codes
uniform development standards 106–7, 673
Uniform Mechanical Code 115
Uniform Plumbing Code 115
uniform rate structures 673
Union Station, Chicago *25*
United Airlines, maintenance operations center 48
United Kingdom (UK), privatization in 253
United States Army Corps of Engineers (ACE) 673
United States Green Building Council (USGBC) 82, 519, 539, 546
United States (US): electrical systems, early 580; emergence of infrastructure in *see* emergence of infrastructure, in United States; fee use in 209–12; flows of energy and climate change 583–5; fuel types 587–9; infrastructure systems 24–5, 32; petroleum and natural gas in 582–3, 588; population 5; ports in 481–3; privatization, status of 250–2; public schools in, brief history *see* schools, history (in United States); streets in 387; sustainability in 35–6; waste stream composition (1960–2010) *361*; wastewater in 296–7; *see also* American Planning Association; American Society of Civil Engineers; American Waterworks Association; *specific regions/States*
Untermann, R. 696n
uranium fuel 588–9
Urban General Plan, The (Kent) 34
urban growth boundaries 110
Urban Land Institute, Los Angeles 573–4
Urban Mass Transit Administration 416
Urban Mass Transportation Act 1964 416, 449
Urban Park and Recreation Recovery (UPARR) program 1978 576
urban renewal agency 59
Urban Renewal program, Housing Act 1949 29
Urban Restructure Program surcharge, parks 563
urban runoff 334, 673
urbanization 328
urbanized areas, development rules for 110–16; environmental impact assessments 113–14, 661; National Electric Code 115; redevelopment plans 113; right-of-way regulations 114–15; specific plans 113; Uniform Building Code 115; Uniform Mechanical Code 115; Uniform Plumbing Code 115; zoning and infrastructure 112–13; zoning ordinances 111–12
use permits 128

user charges and fees 208
user fees 207–25; base rate with two-part tariff 217; block rates with two-part tariff 217–18; concepts, and fee types 214–18; criteria for determining type 213–14; current charges 209; defined 207–9; developing 221; fee use in United States 209–12; flat rate fees 217, 661; formal policies, adopting 221; full cost of service provision, calculating 221; infrastructure function 210–11; parks and recreational facilities management 564–5; regulatory fees 209, 670; relative levels 211–12; setting 221–2; solid waste management 363; special assessments 209, 671; taxes and total local revenues 209–10; transportation 443; types of fees used at local level 217–18; use in United States 209–12; user charges and fees 208, 673; utility charges 208–9, 673; *see also* prices, public
utilidors 673
utility access ports (UAPs) 301, 666, 673
utility charges 208–9, 666
utility corridors, and rights-of-way 598

vactor trucks 305, *306*
Valderama, A. 694n
value engineering 673
value information 161
Van de Voorde, E. 698n
Vanelis, K. 675n
Varady, D. P. 700n
Vaux, C. 555
Veal, S. 694n, 695n
vehicle efficiency improvements 438
vehicle miles traveled, reduction in 439
Velkoff, V. A. 678n
Vermont Corridor Station, Los Angeles 573
vertical consistency *91*, 673
vertical integration 23
vertical regionalism 76
"vesting" tentative maps 117
Viessman, W. 693n
Vincent, G. K. 678n
Vincent, J. M. 700n
Vincoli, J. W. 695n
"virtual jurisdiction," integration of existing infrastructure 648–9
visualization tools 99
Vogel, R. K. 680n
Vogt, J. A. 684n, 686n
VoIP (Voice over Internet Protocol) 673
volatile organic compounds (VOCs) 440
voltage 590, 615, 673

Wagner, C. 701n
wake vortex 673
Walder, J. 698n
"walkable" neighborhoods 10, 542, 550
walking speeds 400
walking-horse car era (1800–1890) 412–13
Walsh, A. H. 680n
Ward, R. B. 676n
warehouses 510; on-dock 482
Waring, Colonel G. E. (Jr.) 352–3
Warner, M. 689n
Washington, President G. 7
Washington D.C.: climate change 337; fragmented aviation system planning 458; local planning 17; *see also* False Creek Energy Center, Washington; King County, Washington; Seattle, Washington; Spokane, Washington
waste: defined 356; hazardous 357, 361, 365, 372, 663; infrastructure systems 9–10; solid *see* solid waste management; waste stream composition in US (1960–2010) *361*; wastewater *see* wastewater management; *see also* sewers
waste diversion 370, 674
Waste Management, Incorporated (WMI) 361
waste stream 361, *361*, 674
waste-to-energy plants 354, 360, 674
wastewater management 295–324; aging pipes 306–7; capital facilities for wastewater, payment for 322–3; collection systems 300–1; components in system, useful life 307; current issues 303–10; distributed recycling system *315*; energy and wastewater 312–13; eutrophication 308–10; fecal matter 309, 313, 315, 324; fees 322; financing mechanisms 322–3; history of collection and treatment 296–8; importance to local practitioners 295; infill needs 307; infiltration and inflow 305–6; institutions 298–9; municipal wastewater 666; municipal wastewater, elements of systems 299–303; phosphorus 300, 308–10; planning considerations *see* wastewater planning; providers of collection and treatment services 299; public buildings 521–3; regulation of wastewater 298–9; reuse 674; sanitary wastewater, financing 322–3; sanitation, new directions 310–18; septic system, private *308*; system

replacement needs 306–7; treatment *see* wastewater treatment; in United States 296–7; urine 315–16; *see also* sanitation, wastewater management; sewers; solid waste; stormwater/flood management; waste; water; water reclamation
wastewater planning 318–22; content of wastewater plan 319–20; existing facilities, condition assessment 321; future sewer and treatment plant needs 321; key steps 319–22; and land use 318–19; location of wastewater treatment plant 321–2; service area 320–1, 671
wastewater treatment: facilities 9, 297–8; location of wastewater treatment plant 321–2; national systems 317–18; pre-treatment 300, 302–3; primary 302; rural on-site 307–8; secondary 302; tertiary 302, 317–18, 672; treatment plants 307; wastewater treatment plant 301
wastewater treatment plant (WWTP) 301, 310, 313
water 269–94; aging infrastructure and redevelopment/intensification 274; appropriative water rights 655; artificial recharge 655; black 315, 316, 318, 656; and carbon emissions 273–4, 310; climate change 271–3; conjunctive use 278, 659; conservation *see* water conservation; contaminants 274–5; current issues 271–5; cycle 276, 315; demand, projecting 284–5; desalination 278, 660; development impact fees 236; distribution 270, 280; drinking water systems 281; elements of water system 275–80; and energy production 273–4; environmental water need 274; federal regulations 293–4; fee structures 286; Fifth Paradigm 271; financing 285–6; full cost pricing 285–6, 661; gray 315, 317, 318, 662; groundwater *see* groundwater; harvesting 113; history 270–1; importance to local practitioners 269–70; infrastructure systems/trends 9–10, 275; institutions 281–4; key elements of water planning process 287; and land use *see* water supply and land use planning; level of service standards 109; planning issues 284–7; and privatization 281, 282; providers 281–2; public buildings 521–3; publicly owned systems, organizational arrangements 281; regulators 283–4; reuse and rainwater harvesting 278, 304; reviewing applications/proposals for new water facilities 287; rural areas and small community systems 275; sources 275; staffing and governance 281; stormwater *see* stormwater/flood management; supply and distribution of water 270; surface *see* surface water; total water management 284, 672; transportation 278–9; treatment *see* water treatment; users 282–3; wastewater *see* wastewater management; withdrawals 282–3; world supplies *276*; yellow 315, 318
water conservation: in Colorado 289; cost-benefit analysis of options 290; fee structures 286; indoor residential water 290–1; outside water use 291–2; public education programs 290; reduction of losses and leaks 290; residential irrigation facts 291
water cycle 674
water duty factors 674
Water Environment Research Foundation 310, 312
water pipes or lines 674
Water Quality Act 1987 294, 330–1, 349
water reclamation 674; centralized 311–12; or reuse 310–11
Water Resources Development Act 1986 294
Water Reuse Foundation 312
water supply and land use planning: demand management and conservation 289; linkage between land use and water supply, legislation requiring 288; local government land use planners, considerations for 286–7; long-range land use planning 287–9; review of development projects 288–9; reviewing applications/proposals for new water facilities 287; wastewater planning and land use 318–19; water planners influencing planning process 286–7; *see also* land use issues; water
water treatment 270, 279–80, 307
Watkins, A. J. 680n
Watkins, C. 556
websites 619
Weiner, E. 696n
Weingroff, R. F. 697n
Weiss, J. D. 681n
Weiss, M. A. 677n
Weitz, J. 679n, 681n
wellhead prices 609
Wells, A. T. 697n
West Coast (USA), water and growth 273
Western Area Power Administration (WAPA) 610
Westinghouse Company 580
wet well (collection basin) 665
wet/dry collections and processing 373–4
wetlands: natural water treatment systems 317; stormwater/flood management 330, 335
wharfage 493
Wheatstone, C. 614

Wheeler, S. M. 681n
Wheeling, West Virginia 564
Whiteman, K. 700n
Whittington, J. 676n, 681n, 700n
wholesaler 281, 674
wide-body jets 454
Wi-Fi 619, *620*, 633, 636, 674
WIIFM (What's in it for me?) 157
Wilkes-Barre, Pennsylvania 618
Williams, N. 34
WiMax (Worldwide Interoperability for Microwave Access) 674
wind power/turbines 524, 592–3
windows 520–1
wireless e-911, 636
wireless telephony 615–16
Withdrawal Act 1910 608
Wolff, G. 690n
Wood, C. 683n
work order management systems (WMS) 157, 674
workload measures 165
Works Progress Administration (WPA) 27
World Commission on Environment and Development 72
World War Two era (1939–1945), United States: concepts to be implemented in post-World War II period 29; direct spending 27–8; parks, use of 557, 559; regulation of private infrastructure provision 28–9
World Watch Institute 72
Wright brothers 452

Yates, J. 695n
yellow water 315, 318
Yergin, D. A. 688n
Youth Conservation and Service Corps 576
Yudelson, J. 699n

zebra crossings (crosswalks) 390–1
zero emissions wastewater 314–17
Zero Waste movement 366–7
Zimbalist, A. 699n
zoning, and infrastructure 112–13
zoning ordinances 26, 111–12, 626–7, 674; cluster 106; composition 111; conditions to be met 627–8; lot-by-lot approach 106; vs. subdivision ordinances 105; use permits 128
zoning overlay district 476
Zorn, K. C. 686n